TEXTS AND READINGS IN MATHEMATICS **63**

Probability Theory

A Foundational Course

Probability Theory

A Foundational Course

R. P. Pakshirajan

HINDUSTAN
BOOK AGENCY

Published by

Hindustan Book Agency (India)
P 19 Green Park Extension
New Delhi 110 016
India

email: info@hindbook.com
www.hindbook.com

ISBN 978-93-80250-44-1

Dedication

This book is dedicated to the memory of
my wife Kamala Rajan who added an
enormously elegant meaning to my life.

The Divine Injunction

यत्करोषि यदश्नासि यज्जुहोषि ददासि यत् ।
यत्तपस्यसि वैनतेय तत्कुरुष्व मदर्पणम् ॥

Vainatheya,
Make whatever you do an offering unto me

Vainatheya's response submission

यद् यद् कर्म करोमि माधव तत्तदखिलं त्वत्प्रसादात् ।
यद् यद् कार्यं मया कृतं तत्तत्सर्वं त्वदाराधनम् ॥

O, Madhava,
Whatever it is that I am able to do is due to your mercy
Whatever I have done is an offering unto you.

List of books from which theorems have been quoted in the sections labelled **Q**

[D] Doob, J.L. (1952) Stochastic Processes, Wiley

[E] Estermann, T. (1962) Complex Numbers and Functions, Asia Publishing House

[H] Halmos, P.R. (1974) Measure Theory, Springer-Verlag

[K] Kelley, J.L. (1975) General Topology, Springer-Verlag

[P] Parthasarathy, K.R. (2005) Introduction to Probability and Measure, Hindustan Book Agency

[L] Loomis, L.H. (2011) An Introduction to Abstract Harmonic Analysis, Dover

[V] Vaidyanathaswamy, R. (1947) Treatise on Set Topology, Indian Mathematical Society

[T] Titchmarsh, E.C. (1932) The Theory of Functions, Oxford University Press

Preface

The origin of the theory of probability must be traced to the compelling need to develop a mathematical model to analyze chance phenomena. Intuitive approaches had to be the beginning for this. These led to the fashioning of certain crude tools which, over time, were chiseled into the present day accepted model of a basic structure resting on abstraction, definitions and a consistent set of axioms. Measure and integrals play an essential role in the development of this theory. This book is concerned with certain foundational aspects of this development.

The author's intention in writing this book is to help introduce probability theory as a subject of study in the graduate Mathematics and Statistics programs, to help the teacher present the theory with greater ease and to help the student feel comfortable learning the subject. The aim is not to present the frontier results of the day but to gather in one place most of the basic material, a good knowledge of which will translate into the ability to read and digest current literature and provide a strong foundation for pursuing research in the subject.

The main prerequisite for a profitable reading of the book is a focused desire to learn probability theory and to enjoy the glory of it. Additional prerequisites are good grounding in real analysis (including measure and integral) and familiarity with the elements of complex function theory and metric topology. Theorems in these areas, assumed as known and used, are quoted, without proofs, in the sections labelled **Q** (**Q**1 to **Q**15). This is done to make the book largely self contained. Familiarity with the content of the first ten chapters of the book Measure Theory by P.R.Halmos is tacitly assumed and the tools of measure and integration developed there will be used without hesitation.

A brief summary of the chapters would be in order. Chapter 1 is devoted to preliminaries leading to a discussion of the main topic of the chapter which is, product spaces and measures in product spaces, including product measures. The remark under section **(1.11.6)** and the theorem at **(1.11.6)** help us claim (i) in chapter 4 that a sequence of independent random variables with prescibed distributions always exists and (ii) in chapter 7 that a discrete time Markov Chain with arbitrary initial distribution and arbitrary transition probabilities always exists. The concept of the support of a probability measure and of the tightness of a family of probabilty measures are also investigated in this chapter. Section ((**1.8.3**)) may be considered the high spot of this chapter. In Chapter 2, weak convergence of a sequence of distribution functions in the p-dimensional Euclidean space is studied and, via the Helly-Bray theorem, the concept is extended to the weak convergence of probability measures in metric spaces. Sections **(2.7.6)** and **(2.7.12)** may be considered the high spots of this chapter: in **(2.7.6)**, an essential property of the compact sets in the space $C[0,\ 1]$ is established and in section **(2.7.12)** a proof, specific to the $C[0,\ 1]$ space, of Prohorov's theorem is given.

Characteristic functions of distribution functions are studied in detail in Chapter 3. In section **(3.6.5)** a detailed proof is given that a familiar counter example indeed shows that $\varphi'(0)$ may exist finitely but $\mathbb{E}X$ may fail to exist.

Independence of random elements, various modes of convergence, zero-one laws, weak law of large numbers and the strong law of large numbers are treated in Chapter 4. Further, finding explicitly an $a.s.$ convergent subsequence of a sequence converging in probability is illustrated by a non-trivial example, refer section **(4.5)**. In Chapter 5, the classical central limit theorem and its converse, the Erdös-Kac invariance principle and Donsker's functional central limit theorem (FCLT) are presented. A unified proof is given of the FCLT covering all the three cases : the case of variational sequences satisfying Lindeberg condition, the case of iid variables with finite variances and the case of iid variables without variance. Chapter 6 is concerned with the Bulinskii version of the law of the iterated logarithm, including Strassen's functional form. The tool used is Skorohod embedding. Needed information in the topic of conditioning is provided. Chapter 7, the last chapter, is a brief introduction to the study of discrete time Markov chains with a countable state space. The sections on geometric ergodicity and Doeblin's condition deserve enhanced attention.

I am indebted to all the authors of books, notes and research papers which deal with one or more of the topics discussed in the present work, authors on whose shoulders this book rests. I am deeply indebted to my former teachers from whom I imbibed the mathematics culture : My philosopher and guide Professor T. Totadri Iyengar (who taught me Theory of Functions), Professor R. Vaidyanathaswamy (who taught me General topology and Function Spaces) and Professor Herman Rubin (who taught me everything I know in Probability Theory and Stochastic Processes). I am also indebted to Professor A.T. Bharucha-Reid who arranged for me to work under Professor Herman Rubin.

My vague idea of writing a book on probability theory took concrete shape due to the constant and unwaning interest shown by Professor A. Krishnamoorthy. I thank him warmly. In the matter of editing the pages of the book to make it better than the version I presented to him, Professor M. Sreehari was generous with his time, unsparing in his efforts and incisive in his remarks and suggestions. Profound and copious thanks are due to him. He provided the book with a great deal of polish, and my debt to him for his informed participation in my work is enormous. Conventional terms of thankfulness are totally inadequate to express my sense of gratitude to him. With pleasure, I place on record here my appreciation and gratitude to Shri Sanjay Varma for his devoted and competent typesetting work.

R. P. Pakshirajan

Contents

Chapter 1

Probability Measures in Product Spaces

Introduction

All of us have an intuitive notion of chance or randomness when we say, for example, that we are 90% sure that a particular event will occur. Two facts underlie this statement : one, that we are talking about a non-deterministic event and two, that we are quantifying, in a heuristic way, the attitude of our mind. Areas where the notion of chance manifests itself are numerous, some of which are risk assessment, reliability, physics of elementary particles and finance. Probability theory is concerned with developing a mathematical frame work to analyse chance phenomena. Event and probability will be basic notions of that theory. In giving mathematical content and shape to these concepts, we adopt the by now well accepted set theoretic model. In dealing with experiments whose outcomes depend on chance, one considers the set of all outcomes and would like to assign a probability value to *every* possible outcome. In the set theoretical model this is found not possible if the cardinal number of the set of all the outcomes is equal to or exceeds that of the continuum. This in practice turns out to be no infirmity since almost all outcomes of importance to the experiment can be assigned probabilities and the set theoretic model has proved to be highly fruitful. This model rests on the fundamental notions of sets, measures, measurable functions and integrals.

Even though the reader is assumed to be familiar with these concepts, we explain them briefly here to make for smooth reading. In general, we will not repeat theorems available in **[H]**.

We start with some notations and definitions.

Let Ω be an arbitrary abstract set of points, also called elements, denoted by ω. A set of these elements is called a subset of Ω. Subset relation is indicated by the symbol $\subset$. All sets considered are assumed to be subsets of Ω. We write $\omega \in A$ if the element ω is in set A. Set A is called a subset of set B if

($\omega \in A$) implies ($\omega \in B$). In such a case we write $A \subset B$. The empty set, denoted by ϕ, containing no element of Ω would then be a subset of Ω. The set A', called the complement of the set A, consists of all the elements of Ω not in or not belonging to A. Clearly $(A')' = A$. We write $A \cap B$ for the set consisting of all the elements of Ω that lie both in A and B; $A \cup B$ for the set consisting of all the elements of Ω that lie either in A or in B; and $A \sim B$ for the set consisting of all the elements of Ω that lie in A but not in B. We note $A \sim B = A \cap B'$. We denote by R^k the real Euclidean space of k-dimension, $k = 1, 2, 3, \ldots$ The symbol $\Rightarrow$ stands for *implies* or *leads to*. The symbol $\Leftrightarrow$ would then stand for *implies and implied by* or *equivalent*.

1.1 π-system, Dynkin system and σ- fields

Definition 1.1.1. A non-empty collection $\mathfrak{A}$ of subsets Ω is said to be
(i) a π-system or a π-class if it is closed under finite intersections. i.e., if $A \in \mathfrak{A},\ B \in \mathfrak{A} \Rightarrow A \cap B \in \mathfrak{A}$.
(ii) a Dynkin system or a Dynkin class or a λ-system or λ-class if $\Omega \in \mathfrak{A}$, if it is closed under complementation and if it is closed under countable disjoint unions. i.e., if
(α) $\Omega \in \mathfrak{A}$,
(β) $(A \in \mathfrak{A}) \Rightarrow (A' \in \mathfrak{A})$ and
(γ) $(A_n \in \mathfrak{A}, n = 1, 2, \ldots; A_i \cap A_j = \phi, i \neq j) \Rightarrow (\cup_n A_n \in \mathfrak{A})$.
The following three conditions are equivalent to the above three defining conditions :
(α') $\Omega \in \mathfrak{A}$,
(β') $(A,\ B \in \mathfrak{A}$, and $A \subseteq B) \Rightarrow (B \sim A \in \mathfrak{A})$ and
(γ') $(A_n \in \mathfrak{A},\ A_n \subseteq A_{n+1},\ n \geq 1) \Rightarrow (\cup_1^\infty A_n) \in \mathfrak{A})$.
(iii) a field if it is closed under finite unions and complementation.
(iv) a σ-field if it is a field which is closed under countable unions.

Examples and Remarks. Take the real line for Ω. The collection $\mathfrak{A}_1$ of all closed intervals on the line, finite or infinite, is a π-class of sets. It is not a Dynkin class, since the union of two disjoint closed intervals is not a closed interval.

Again let Ω be the real line. Denote by $\mathfrak{A}$ the collection of sets consisting of Ω, $A = [0,\ 3]$, $B = [2,\ 4]$, $A' = (-\infty,\ 0) \cup (3,\ \infty)$ and $B' = (-\infty,\ 2) \cup (4,\ \infty)$. Then $\mathfrak{A}$ is a Dynkin class. It is not a π-class since it is not closed under intersection : $A \cap B = [2,\ 3] \notin \mathfrak{A}$.

Since the set of all the subsets of Ω is a σ-field and since the intersection of an arbitrary family of fields is a field and since the intersection of an arbitrary family of σ-fields is a σ-field , we can talk about the the minimal field $\varphi(\mathbb{A})$[minimal σ-field $\sigma(\mathbb{A})$] containing (or generated by) the collection $\mathbb{A}$ of subsets.

Theorem 1.1.2. *If $\mathfrak{A}$ is both a π-system and a Dynkin system, then it is a σ-field.*

Proof. The hypothesis implies that $\mathfrak{A}$ is closed under finite intersections and complementation. Hence it is closed under relative complementation. i.e., $(A \in \mathfrak{A},\ B \in \mathfrak{A}) \Rightarrow A \sim B = A \cap B' \in \mathfrak{A}$. Also, it follows that $\mathfrak{A}$ is closed under finite unions. For, let $A_n \in \mathfrak{A},\ n = 1,\ 2, \ldots$. Let $B_1 = A_1,\ B_2 = A_2 \cap A_1',\ B_3 = A_3 \cap (A_2 \cup A_1)'$ and so on. Since $\cup_n A_n = \cup_n B_n$, since the B_ns are disjoint and since $\mathfrak{A}$ is Dynkin system, it follows $\cup_n A_n \in \mathfrak{A}$. With this, the proof is complete. □

Let Ω be endowed with a σ-field $\mathscr{A}$. A pair like $(\Omega, \mathscr{A})$ is called a measurable space. Members of $\mathscr{A}$ are called $\mathscr{A}$-measurable sets or simply measurable sets if there is no scope for confusion. They are also called events.

If $E_n \in \mathscr{A}, n \geq 1$, then E^* and E_* belong to $\mathscr{A}$ where $E^* = \bigcap_{n=1}^{\infty} \bigcup_{m=n}^{\infty} E_m$ and $E_* = \bigcup_{n=1}^{\infty} \bigcap_{m=n}^{\infty} E_m$. E^*[respectively, E_*] is called the limit superior [respectively, the limit inferior] set of the sequence (E_n). We write $E^* = \overline{\lim_{n\to\infty}}\ E_n$; $E_* = \underline{\lim_{n\to\infty}}\ E_n$. E^* is the set of all points $\omega \in \Omega$ which belong to infinitely many of the E_ns. We write $E^* = \{\omega : \omega \in E_n$ for infinitely many $n\}$ or $\{\omega : \omega \in E_n$ i.o.$\}$, i.o. standing for *infinitely often* or for infinitely many values of the parameter n. E_* is the set of all the points $\omega \in \Omega$ which belong to all but a finite number of the E_ns. Given a set E, denote by χ_E its indicator function, also called the characterstic function of the set E : $\chi_E(\omega) = 1$ if $\omega \in E$ and $\chi_E(\omega) = 0$ if $\omega \notin E$. For a sequence (E_n) of sets we have:

$$\underline{\lim_{n\to\infty}}\ \chi_{E_n} = \chi_{E_*}; \quad \overline{\lim_{n\to\infty}}\ \chi_{E_n} = \chi_{E^*}.$$

Borel σ-field

When Ω is a topological space, it is natural to take for $\mathscr{A}$ the σ-field generated by the topologically important sets. This is to secure that the measure theoretic and topological structures of the space do not stay unrelated. The topologically important basic sets being the open sets, one should naturally define the Borel σ-field as that generated by all the open sets. This is how it is defined in a metric space. If the metric space Ω is separable, its every open subset is the union of a countable number of open spheres. Hence in this case the Borel σ-field is the σ-field generated by the open spheres, countable in number, with centers at the members of the separability set and rational numbers for radii. We denote by $\mathscr{R}^{(k)}$ the Borel σ-field of R^k, the real Euclidean k- space, endowed with the usual metric.

Amongst the non-metric spaces, it is useful to consider locally compact Hausdorff spaces, which are not metric spaces. For an example of a separable non-metrizable compact Hausdorff space, ref. pp 104; 164 **[K]**.

If Ω is a locally compact Hausdorff space, then there are three σ-fields meriting consideration: the σ-field $\mathscr{B}_1(\Omega)$ generated by the closed sets, the σ-field $\mathscr{B}_2(\Omega)$ generated by the compact sets and the σ-field $\mathscr{B}_3(\Omega)$ generated by the compact g_δ sets (i.e., compact sets which are the limits of decreasing sequences of open sets). $B_3(\Omega)$ is known as the Baire σ-field. Clearly $\mathscr{B}_3(\Omega) \subseteq \mathscr{B}_2(\Omega) \subseteq \mathscr{B}_1(\Omega)$. $\mathscr{B}_1$, $\mathscr{B}_2$ are huge σ-fields. If Ω is separable, then $\mathscr{B}_2(\Omega) = \mathscr{B}_3(\Omega)$ (ref. Theorem E, page 218 **[H]**). If Ω is σ-compact i.e., if Ω is the countable union of compact sets, then $\mathscr{B}_1(\Omega) = \mathscr{B}_2(\Omega)$. A few sections in this book deal with locally compact Hausdorff spaces. These are intended to widen the study field of vision/widen the horizons of the interested reader and are not necessary to understand the rest of this book.

1.2 Product σ-field. Number of components: finite

Let $T = \{1, 2, ..., n\}$. Let $(\Omega_i, \mathscr{A}_i), i \in T$ be measurable spaces. Denote by $\Omega_T =$ the Cartesian product $\times\{\Omega_i : i \in T\}$, consisting of all the finite sequences $(\omega_1, \omega_2, ..., \omega_n)$, $\omega_i \in \Omega_i, 1 \leq i \leq n$. Cartesian products $\times\{A_i : i \in T\}$, where $A_i \in \mathscr{A}_i, 1 \leq i \leq n$, are called the measurable rectangles of Ω_T. The σ-field $\mathscr{A}_T$ generated by the totality of all such rectangles is called the product σ-field. We indicate this by writing $\mathscr{A}_T = \bigotimes_{i=1}^{n} \mathscr{A}_i$. The measurable space $(\Omega_T, \mathscr{A}_T)$ is called the product measurable space. When the $(\Omega_i, \mathscr{A}_i)$s are the same for all the i, say, $(\Omega_i, \mathscr{A}_i) = (U, \mathscr{U})$, then it is usual to write $(\Omega_T, \mathscr{A}_T) = (U^n, \mathscr{U}^n)$. Thus $(R^k, \mathscr{R}^k)$ is the product σ-field of the product of $(R, \mathscr{R})$ with itself k times. Earlier we introduced the notation $\mathscr{R}^{(k)}$ to denote the Borel σ-field of R^k. We will show in **(1.3.3)** that $\mathscr{R}^{(k)} = \mathscr{R}^k$.

Product σ-field. Number of components : infinite

Let T be an arbitrary infinite index set. Let $(\Omega_t, \mathscr{A}_t), t \in T$ be measurable spaces. Denote by Ω_T the Cartesian product $\times\{\Omega_t : t \in T\}$ consisting of all the functions ω defined on T such that $\omega(t) \in \Omega_t$. For $J \subset T$, J finite, define

$\widetilde{\mathbf{a}}_{J;T}$ to be the collection of all subsets of Ω_T, themselves Cartesian products, of the form $\times\{A_t : t \in T\}$, where $A_t \in \mathscr{A}_t$ if $t \in J$ and $A_t = \Omega_t$ if $t \notin J$. Members of $\widetilde{\mathbf{a}}_{J;T}$ are called finite dimensional measurable J-rectangles of Ω_T. Denote by $\mathbf{a}_{J;T}$ the sigma field $\sigma(\widetilde{\mathbf{a}}_{J;T})$ generated by $\widetilde{\mathbf{a}}_{J;T}$. Members of $\mathbf{a}_{J;T}$ are called the finite dimensional measurable J-cylinder subsets of Ω_T. Let $\widetilde{\mathscr{A}_T} = \cup\{\mathbf{a}_{J;T} : J \subset T,\ J \text{ finite }\}$ and note that $\widetilde{\mathscr{A}_T}$ is a field. Denote by $\mathscr{A}_T$ the sigma field $\sigma(\widetilde{\mathscr{A}_T})$ generated by $\widetilde{\mathscr{A}_T}$. $\mathscr{A}_T$ is called the product σ-field of the $\mathscr{A}_t$s and this is indicated by writing $\mathscr{A}_T = \bigotimes_{t\in T} \mathscr{A}_t$.

Let the index set T be uncountable. Let $J \subset T$ be a countable subset. Denote by $\mathscr{A}_{J;T}$ the σ-field $\sigma(\cup\{\mathbf{a}_{F;T}, F \subset J, F \text{ finite}\})$. Then $\cup\{\mathscr{A}_{J;T}, J \subset T,\ J \text{ countably infinite}\}$ is a σ-field and equals $\mathscr{A}_T$. Thus given $A \in \mathscr{A}_T$, we can always find a countable $J,\ J \subseteq T$ such that $A \in \mathscr{A}_{J,T}$.

When $(\Omega_t,\ \mathscr{A}_t) = (\Omega,\ \mathscr{A}),\ t \in T$, then we write $(\Omega^T,\ \mathscr{A}^T)$ for $(\Omega_T,\ \mathscr{A}_T)$.

Let $T = [0,\ 1]$. For $t \in T$, let $\Omega_t = R$, the real line and let $\mathscr{A}_t = \mathscr{R}$, the Borel σ-field of R. Then $\Omega_T = R^T$ consists of all the real valued functions defined on T and $\mathscr{A}_T$ will naturally be denoted by $\mathscr{R}^T$. The next theorem shows that the members of R^T are huge and that certain smaller but desirable sets do not belong to $\mathscr{R}^T$.

Theorem 1.2.1. *$C \notin \mathscr{R}^T$ where $C \subset R^T$ consists of all the real continuous functions defined on R.*

Proof. Let $x \in C$. Suppose $C \in \mathscr{R}^T$. There exists then a countable subset J of T such that $C \in \mathscr{A}_{J,T}$. By the structure of $\mathscr{A}_{J,T}$, it follows that $y \in A,\ A \in \mathscr{R}^T$ arbitrary, if $y(t) = x(t),\ t \in J,\ y(t)$ arbitrary for $t \notin J$. This is to say that if $x \in A$, then necessarily all functions y coinciding with x on J will also be in A. Thus the set C consisting only of continuous functions can not belong to $\mathscr{R}^T$. □

On the same lines it can be shown that $D \notin \mathscr{R}^T$ where D consists of all functions x defined on $[0,\ 1]$ such that they are right continuous at each $t \in [0,\ 1)$, left continuous at 1 and have finite left limits at each $t \in (0,\ 1)$.

Definition 1.2.2. Let $T = [0,\ 1]$ or $[0,\ \infty)$. Let D be a countable dense subset of T. $x \in R^T$ is said to be separable with respect to D or D-separable if to each $t \in T$ there exist points $t_1,\ t_2, \dots$ in D such that $t_n \to t$ and $x(t_n) \to x(t)$. The set $A \subset R^T$ is said to be D- separable if each $x \in A$ is D- separable.

We note that the values taken by a D-separable function x are determined by its values at the points of D. We note further the set of all the D-separable functions can not be a member of $\mathscr{R}^T$.

Theorem 1.2.3. *If x is D-separable, then for every open interval $I \subset T$,*

$$\sup_{t\in I} x(t) = \sup_{t\in I\cap D} x(t); \qquad \inf_{t\in I} x(t) = \inf_{t\in I\cap D} x(t) \tag{1.1}$$

$$\underline{\lim}_{s_j\to t}\, x(s_j) \le x(t) \le \overline{\lim}_{s_j\to t}\, x(s_j),\ t\in T \tag{1.2}$$

Further (1.1) $\Leftrightarrow$ (1.2). *(recall the symbol $\Leftrightarrow$ stands for equivalence)*

Proof. For convenience denote by u and v the left side and the rightside sup values in (1.1). From the definition of u and v, it is immediate that $u \ge v$.

Given $\varepsilon > 0$, a $t_0 \in I$ can be found such that $x(t_0) + \varepsilon > u$. The D-separability of x implies that there exists $s_0 \in D\cap I$ such that $|x(t_0) - x(s_0)| < \varepsilon$, yielding $x(s_0) + 2\varepsilon > u$. Hence $v \ge u$. Thus $u = v$.

On similar lines the second part of (1.1) can be established.

The second part of (1.1) is equivalent to $\inf_{u\in I\cap D} x(u) \le \inf_{u\in I} x(u)$. In this inequality replacing I by $I_n = (t - \frac{1}{n},\ t + \frac{1}{n})$ and letting $n \to \infty$ we arrive at the first part of (1.2). Starting with the first part of (1.1) and reasoning similarly we arrive at the second part of (1.2). Thus (1.1) $\Rightarrow$ (1.2).

Now to prove the reverse implication, let (1.2) hold. Let $t \in I$. Hence $I_n \subset I$ for all n large. The left half of the inequality in (1.2) is equivalent to

$$\lim_{n\to\infty} \inf_{u\in I_n\cap D} x(u) \le x(t).$$

Since $\inf_{u\in I_n\cap D} x(u)$ is a monotonic increasing sequence, it follows that for all n large, $\inf_{u\in I_n\cap D} x(u) \le x(t)$. This implies

$$\inf_{u\in I\cap D} x(u) \le \inf_{u\in I_n\cap D} x(u) \le x(t).$$

Similarly $\sup_{u\in I\cap D} x(u) \ge x(t)$. Thus (1.2) implies the following intermediate result : For every $t \in I$,

$$\inf_{u\in I\cap D} x(u) \le x(t) \le \sup_{u\in I\cap D} x(u). \tag{1.3}$$

It is immediate from this

$$\inf_{u\in I\cap D} x(u) \le \inf_{t\in I} x(t) \le \sup_{t\in I} x(t) \le \sup_{u\in I\cap D} x(u).$$

The reverse inequalities being obvious, this implies (1.1).

With this the proof that (1.1) $\Leftrightarrow$ (1.2) is complete. □

Remarks 1.2.4. (i) Let $x(t) = 0$ if $0 \le t < \frac{1}{\sqrt{2}}$; $x(\frac{1}{\sqrt{2}}) = 1$; $x(t) = 2$ if $\frac{1}{\sqrt{2}} < t \le 1$. Let D be the set of rationals in T $(= [0,\ 1])$. There is no sequence $s_j \in D$ such that $s_j \to \frac{1}{\sqrt{2}}$ and such that $x(s_j) \to x(\frac{1}{\sqrt{2}})$. Hence according to the definition at **(1.2.2)**, x is not separable but x satisfies (1.1).
(ii) Separability of stochastic processes is discussed in section **(5.7.2)**.

1.3 Borel σ-field of the product space

Often the Ω_is are topological spaces and the $\mathscr{A}_i$s are the corresponding Borel σ-fields. The question arises whether the product σ-field $\mathscr{A}_T$ is always equal to the Borel σ-field of Ω_T endowed with the product topology. Below Theorem **(1.3.1)**, Remarks **(1.3.2)**, **(1.3.3)**, Lemma **(1.3.4)** and example **(1.3.5)** elucidate the position.

Theorem 1.3.1. *For $i \in T = \{1,\ 2, \ldots\}$, let $\mathcal{E}_i$ be any collection of subsets of Ω_i with the properties (i) $\sigma(\mathcal{E}_i) = \mathscr{A}_i$ and (ii) $\mathcal{E}_i$ contains a countable subcollection whose union is Ω_i. Denote by $\widehat{\mathbf{a}}_{J;T}$, for $J \subset T$, J finite, the collection of all sets of the type $\times\{A_i : i \in T\}$ where $A_i \in \mathcal{E}_i$ if $i \in J$ and $A_i = \Omega_i$ if $i \notin J$. Let $\mathcal{E} = \cup\{\widehat{\mathbf{a}}_{J;T} : J \subset T, J$ finite $\}$. Then $\mathscr{A}_T = \sigma(\mathcal{E})$*

Proof. Step 1. Fix $p \ge 2$, $J = \{1, 2, \ldots, p\}$. Fix $A_i \in \mathcal{E}_i$, $2 \le i \le p$. Let g_1 be the collection of all sets $A \subset \Omega_1$ such that

$$A \times A_2 \times \cdots \times A_p \in \sigma(\widetilde{\mathcal{E}}_p)$$

where $\widetilde{\mathcal{E}}_p$ is the collection of all sets $B_1 \times B_2 \times \cdots \times B_p$, $B_i \in \mathcal{E}_i$, $1 \le i \le p$. Clearly $g_1 \supset \mathcal{E}_1$. Since

$$\left(\bigcup_1^\infty D_n\right) \times A_2 \times \cdots \times A_p = \bigcup_{n=1}^\infty (D_n \times A_2 \times \cdots \times A_p),$$

it follows that, if $B_n \in g_1$, $n \ge 1$, then $\cup_n B_n \in g_1$. i.e. g_1 is closed under countable union. This implies, in particular, that $\Omega_1 \in g_1$. Since

$$A' \times A_2 \times \cdots \times A_p = (\Omega \times A_2 \times \cdots \times A_p) \sim (A \times A_2 \times \cdots \times A_p),$$

it follows that g_1 is closed under complementation. Thus g_1 is a σ-field. Since $g_1 \supset \mathcal{E}_1$, $g_1 \supset \sigma(\mathcal{E}_1) = \mathscr{A}_1$. Thus $A_1 \times A_2 \times \cdots \times A_p \in \sigma(\widetilde{\mathcal{E}}_p)$ for every $A_1 \in \mathscr{A}_1$ and $A_i \in \mathcal{E}_i$, $2 \le i \le p$. Let g_2 be the collection of all sets of the type

$$A_1 \times A \times A_3 \times \cdots \times A_p \in \sigma(\widetilde{\mathcal{E}}_p)$$

where $A_1 \in \mathscr{A}_1$, $A \subset \Omega_2$, $A_i \in \mathcal{E}_i$, $3 \leq i \leq p$ and argue as before to show that all sets of the type $A_1 \times A_2 \times \cdots \times A_p \in \sigma(\widetilde{\mathcal{E}}_p)$ where $A_1 \in \mathscr{A}_1$, $A_2 \in \mathscr{A}_2$, $A_i \in \mathcal{E}_i$, $3 \leq i \leq p$. This process can be kept up to finally claim $A_1 \times A_2 \times \cdots \times A_p \in \sigma(\widetilde{\mathcal{E}}_p)$ for every $A_i \in \mathscr{A}_i$, $1 \leq i \leq p$. The σ-field generated by these rectangles is precisely the product σ-field $\bigotimes_{i=1}^{p} \mathscr{A}_i$. Hence $\bigotimes_{i=1}^{p} \mathscr{A}_i \subseteq \sigma(\widetilde{\mathcal{E}}_p)$. But since each $\mathcal{E}_i \subseteq \mathscr{A}_i$, $\widetilde{\mathcal{E}}_p \subseteq \bigotimes_{i=1}^{p} \mathscr{A}_i$. Hence

$$\sigma(\widetilde{\mathcal{E}}_p) \subseteq \bigotimes_{i=1}^{p} \mathscr{A}_i. \quad \text{Thus } \bigotimes_{i=1}^{p} \mathscr{A}_i = \sigma(\widetilde{\mathcal{E}}_p).$$

Step 2. $T = \{1, 2, \ldots\}$.

Proceeding as above, we can show that, for $J \subset T$, $\widetilde{\mathbf{a}}_{J;T} \subseteq \sigma(\mathcal{E})$. This implies that

$$\mathscr{A}_T = \sigma(\cup\{\widetilde{\mathbf{a}}_{J;T} : J \subseteq T, J \text{ finite}) \subseteq \sigma(\mathcal{E}).$$

On the other hand, $\mathcal{E}_i \subseteq \mathscr{A}_i$, $i \in T$ and hence $\mathcal{E} \subseteq \bigotimes_{i \in T} \mathscr{A}_i = \mathscr{A}_T$. It follows that $\sigma(\mathcal{E}) \subseteq \mathscr{A}_T$. Thus $\mathscr{A}_T = \sigma(\mathcal{E})$. □

Remark 1.3.2. For each $i \in T$, let $\mathcal{E}_i$ be the collection of all the open sets in Ω_i. The $\mathcal{E}_i$s satisfy the conditions of the theorem. Hence $\mathscr{A}_T = \sigma(\mathcal{E})$. But $\mathcal{E}$ is a base for the product topology of Ω_T. Since an open set in this topology can be an arbitrary union of members of $\mathcal{E}$, it follows that $\mathscr{A}_T$ is a sub σ-field of the Borel σ-field of Ω_T, endowed with the product topology.

Remark 1.3.3. Let each Ω_i be a topological space satisfying the second axiom of countability. (This would be the case if, for example, each Ω_i is a separable metric space.) Denote by $\mathcal{E}_i$ a countable open base for Ω_i. The $\mathcal{E}_i$s satisfy the conditions of the Theorem. Hence $\mathscr{A}_T = \sigma(\mathcal{E})$. We note that $\mathcal{E}$ is a countable open base for the product topology of Ω_T. Thus in this case, the product σ-field is equal to the Borel σ-field of the product space.

If T is a countable infinite set, then R^T is written R^∞ and $\mathscr{R}^T$ is written $\mathscr{R}^\infty$. Since R is a separable metric space, the theorem applies and we have : if T is finite or countably infinite, the product σ-field $\mathscr{R}^T$ coincides with the Borel σ-field $R^{(T)}$ of the product topological space R^T.

Examples of members of $\mathscr{R}^\infty$. If $\mathbf{x} = (x_1, x_2, \ldots)$ represents a generic element of R^∞, then the following sets belong to $\mathscr{R}^\infty$.

(i) $\{\mathbf{x} : \sup_n x_n > a\}$ (ii) $\{\mathbf{x} : \lim_n x_n \text{ exists finitely}\}$

(iii) $\{\mathbf{x} : \sum_1^\infty |x_n| < a\}$ (iv) $\{\mathbf{x} : \sum_1^n x_k = 0 \text{ for at least one } n \geq 1\}$.

Examples 1.3.4. Two examples where the product σ-field is not equal to the Borel σ-field of the product space.

Example 1 Let $T = [0,\ 1]$. Let $I = [0,\ 1]$ and let $\mathscr{I}$ be the Borel σ-field of I. Then the product σ-field $\mathscr{I}^T$ would be a proper subset of the Borel σ-field $\mathscr{I}^{(T)}$ of the product topological space I^T. To see this, let H denote the set of all non-decreasing functions defined on $T = [0,\ 1]$ with values in I. By arguing as in **(1.2.1)**, we see that $H \notin \mathscr{I}^T$. That H is a Borel subset of I^T follows from H being a closed subset of I^T (ref. p 164, **[K]**). Thus in this case we have the strict inclusion $\mathscr{I}^T \subset \mathscr{I}^{(T)}$. Though I, being a separable metric space, satisfies the second axiom of countability, it is not true that $\mathscr{I}^T = \mathscr{I}^{(T)}$. This can only be attributed to the components in the product space being uncountably many.

Example 2. Let **X** be an uncountable set with cardinal number larger than c, the cardinal number of the continuum. Endow **X** with the discrete topology: $d(x,x) = 0,\ d(x,y) = 1$ for all $x, y \in$ **X**, $x \neq y$. Let ρ be a metric on **X** such that (**X**, ρ) satisfies the second axiom of countability. Denote the Borel σ-fields of (**X**,d), (**X**,ρ) and **X** $\times$ **X** (the product metric space) by $\mathcal{B}_1$, $\mathcal{B}_2$ and $\mathcal{B}$. Then $\mathcal{B}_1 \otimes \mathcal{B}_2 \subset \mathcal{B}$, the inclusion being strict.

Proof. By **(1.3.2)**, $\mathcal{B}_1 \otimes \mathcal{B}_2 \subset \mathcal{B}$.

Note that the diagonal set $D = \{(x,x) : x \in \mathbf{X}\}$ is a closed set in **X** $\times$ **X** (ref. p 190, **[V]**).Hence $D \in \mathcal{B}$. We claim that $D \notin \mathcal{B}_1 \otimes \mathcal{B}_2$. Theorem D, p 24 in **[H]** states that if **E** is any class of sets and if $E \in \sigma(\mathbf{E})$, then there exists a countable subclass **D** such that $E \in \sigma(\mathbf{D})$. If $D \in \mathcal{B}_1 \otimes \mathcal{B}_2$ then by this theorem there would exist a countable collection of rectangles

$$\{A_n \times B_n, A_n \in \mathcal{B}_1, B_n \in \mathcal{B}_n,\ n = 1, 2, \dots\}$$

such that D belongs to the σ-field generated by these rectangles. Let $\mathcal{E}_1$ be the σ-field generated by the sequence (A_n) and $\mathcal{E}_2$ the σ-field generated by the sequence (B_n) in **X**. Then it is clear that $D \in \mathcal{E}_1 \otimes \mathcal{E}_2$. We now appeal to Theorem A, page 141 in **[H]** which states that if $E \in \mathscr{A}_1 \otimes \mathscr{A}_2$ then for every $y \in \Omega_2$, $\{x : (x,\ y) \in E\} \in \mathscr{A}_1$ and for every $x \in \Omega_1$, $\{y : (x,\ y) \in E\} \in \mathscr{A}_2$.

Now for $x \in$ **X**, $\{y : (x,\ y) \in D\}$ is the single point set containing x only and it belongs to $\mathcal{E}_1$. If all these single point sets belong to $\mathcal{E}_1$, the cardinal number of $\mathcal{E}_1$ would be larger than c. But $\mathcal{E}_1$ is countably generated. Hence its cardinal number can not be greater than c (ref. **(1.3.5)**). This contradiction establishes the claim. □

It may be noted that the second topology for **X** can be any arbitrary Hausdorff topological space for the validity of the claim. The point of the assumption on ρ is only to stress that even if one of the components satisfies the second axiom of countability, it can happen that $\mathcal{B}$ is strictly larger than $\mathcal{B}_1 \otimes \mathcal{B}_2$.

Lemma 1.3.5. *If $\mathbb{A}$ is countable, then the cardinal number of $\sigma(\mathbb{A})$ is not greater than that of the continuum.*

Proof. Since $\phi, \Omega \in \sigma(\mathbb{A})$, since $\sigma(\mathbb{A}, \phi, \Omega) = \sigma(\mathbb{A})$ and since $\{\mathbb{A}, \phi, \Omega\}$ is a countable collection, we may assume without loss of generality that $\phi, \Omega \in \mathbb{A}$.

Given any collection **C** of sets denote by $\mathbf{C}^*$ the class of all countable union of differences of members of **C** and note that $\mathbf{C} \subset \mathbf{C}^*$. With this notation we define successively $\mathbb{A}_0 = \mathbb{A}$; for any ordinal $\alpha > 0$, finite or non-finite, $\mathbb{A}_\alpha = (\bigcup\{\mathbb{A}_\beta : \beta < \alpha\})^*$ and note that if $\alpha < \beta$, then $\mathbb{A}_\alpha \subseteq \mathbb{A}_\beta \subseteq \sigma(\mathbb{A})$. Write $\mathcal{A} = \bigcup(\mathbb{A}_\alpha : \alpha < \Lambda)$ where Λ is the first uncountable ordinal and note that $\mathcal{A} \subseteq \sigma(\mathbb{A})$. If $A, A_n, n \geq 1$ are sets in $\mathcal{A}$, then there exists a non-finite ordinal α such that these sets belong to $\mathbb{A}_\alpha$. Hence $A' = (\Omega \sim A)$,

$$\bigcup_{n=1}^{\infty} A_n = \bigcup_{n=1}^{\infty} (A_n \sim \phi) \text{ belong to } \mathbb{A}_{\alpha+1}.$$

Thus $\mathcal{A}$ is a σ-field containing $\mathbb{A}$. It follows that $\sigma(\mathbb{A}) \subset \mathcal{A}$. Hence $\sigma(\mathbb{A}) = \mathcal{A}$. Since each $\mathbb{A}_\alpha$ is countable and since each $\alpha < \Lambda$, we conclude that the cardinal number of $\sigma(\mathbb{A})$ does not exceed Λ. □

Theorem 1.3.6. *Non Borel linear sets exist.*

Proof. Let Ω be the interval $(0,\ 1)$. Let $\mathcal{A}$ be the collection of all sub intervals with rational end points. $\mathcal{A}$ is a countable collection. $\sigma(\mathcal{A})$ is the Borel σ-field of Ω endowed with the usual metric. The cardinal number of $\sigma(\mathcal{A})$ will not exceed that of the continuum. The cardinal number of the collection of all the single point subsets of Ω is that of the continuum. Hence the cardinal number of the collection of *all* the subsets of Ω exceeds that of the continuum. This implies that there exist non-Borel sets. i.e. there exist sets $E \subset \Omega$, $E \notin \sigma(\mathcal{A})$. □

1.4 Measurable functions / Random variables / Random elements

Let $(\Omega_i,\ \mathscr{A}_i)$, $i = 1,\ 2,\ 3$, be measurable spaces. A mapping f from Ω_1 into Ω_2 is said to be $\mathscr{A}_1 \backslash \mathscr{A}_2$ measurable (or, simply, measurable if the context is clear) if $f^{-1}(E) \in \mathscr{A}_1$ for every $E \in \mathscr{A}_2$. If $(\Omega_2,\ \mathscr{A}_2) = (R,\ \mathscr{R})$ then we say

f is a random variable (rv). Some situations demand consideration of extended real valued rvs. i.e. we take $(\Omega_2, \mathscr{A}_2) = (\bar{R}, \bar{\mathscr{R}})$ where $\bar{R} = [-\infty, \infty]$ and $\bar{\mathscr{R}} = \sigma(\mathscr{R} \cup \{-\infty\} \cup \{\infty\})$. If $(\Omega_2, \mathscr{A}_2) = (R^k, \mathscr{R}^k)$, $k \geq 2$, then we say f is a random vector (rV). In all other cases f will be called a random element (*rE*). The expression 'random element' will also be used as a generic term referring to all the three cases. The collection $f^{-1}(\mathscr{A}_2) = \{f^{-1}(E) : E \in \mathscr{A}_2\}$ of subsets of $\mathscr{A}_1$ is easily seen to be a σ-field. If $(f_\alpha, \alpha \in \Lambda)$ is a family of $\mathscr{A}_1 \backslash \mathscr{A}_2$ measurable functions, then $\sigma(\cup_{\alpha \in \Lambda} f_\alpha^{-1}(\mathscr{A}_2))$ is called the σ-field generated by the family (f_α). It is the smallest sub σ-field in $\mathscr{A}_1$ with respect to which each f_α is measurable. If f_ν is $\mathscr{A}_\nu \backslash \mathscr{A}_{\nu+1}$ measurable, $\nu = 1, 2$, then the mapping $f_2(f_1)$, taking Ω_1 into Ω_3 is $\mathscr{A}_1 \backslash \mathscr{A}_3$ measurable. If $\mathbb{D} \subset \mathscr{A}_2$, if $\sigma(\mathbb{D}) = \mathscr{A}_2$ and if $f^{-1}(\mathbb{D}) \subset \mathscr{A}_1$ then f is $\mathscr{A}_1 \backslash \mathscr{A}_2$ measurable. This assertion follows from the observations that if $f^{-1}(\mathbb{D}) \subset \mathscr{A}_1$, then $\sigma(f^{-1}(\mathbb{D})) \subset \mathscr{A}_1$ and that $f^{-1}(\mathscr{A}_2) = f^{-1}(\sigma(\mathbb{D})) = \sigma(f^{-1}(\mathbb{D}))$.

Thus, for example, if Ω_1, Ω_2 are metric spaces and $\mathscr{A}_1$, $\mathscr{A}_2$ their Borel σ-fields and if f is a continuous mapping of Ω_1 into Ω_2, then f is $\mathscr{A}_1 \backslash \mathscr{A}_2$ measurable. For, since f is a continuous function, $f^{-1}(G)$ is an open set in Ω_1 if G is an open set in Ω_2 and hence $f^{-1}(G) \in \mathscr{A}_1$. Further $\sigma(\mathbb{D}) = \mathscr{A}_2$ where $\mathbb{D}$ is the collection of all the open subsets of Ω_2.

We note that if $(\Omega_2, \mathscr{A}_2) = (R, \mathscr{R})$ then f is a rv if $\{\omega : f(\omega) \leq x\} \in \mathscr{A}_1$ for every x, $-\infty < x < \infty$, since $\sigma((-\infty, x], x \in R) = \mathscr{R}$. If Ω_2 is a separable metric space and if $\mathcal{A}$ is the collection of all the spheres in Ω_2 then $\sigma(\mathcal{A}) = \mathscr{A}_2$. Hence, in this case, $f^{-1}(\mathcal{A}) \subset \mathscr{A}_1$ implies f is $\mathscr{A}_1 \backslash \mathscr{A}_2$ measurable.

To see that this result is not true if the metric space Ω_2 is not separable, consider the following example. For $0 < a < 1$ define function x_a on $I = (0, 1)$ thus: $x_a(t) = 0$ if $0 < t < a$ and $x_a(t) = 1$ if $a \leq t < 1$. Denote by Ω_2 the collection of all such functions x_a. Define $d(x, y) = \sup\limits_{0<t<1} |x(t) - y(t)|$, $x, y \in \Omega_2$. We see that (Ω_2, d) is a metric space and that $d(x, y) = 1$, $x \neq y$, $x, y \in \Omega_2$. Hence under this metric Ω_2 is not separable. Further *all* subsets of Ω_2 are open sets. Hence the Borel σ-field of (Ω_2, d) is the totality of all the subsets of Ω_2 Take $\Omega_1 = I$ with the usual metric. Let $\mathscr{A}_1$ be its Borel σ-field. Define function f mapping Ω_1 onto Ω_2: $f(a) = x_a$. Let S be an open sphere of radius r and center x_a in Ω_2. If $r < 1$ then $f^{-1}(S) = \{a\} \in \mathscr{A}_1$. If $r \geq 1$, then $f^{-1}(S) = \Omega_1$. Thus $f^{-1}(S) \in \mathscr{A}_1$ for all spheres S. There is an obvious one-to-one correspondence between the collection of all the subsets of Ω_1 and the collection of all the subsets of (Ω_2, d). Let E be a non-Borel subset of Ω_1 (ref. **(1.3.6)**). Let $F = \{x_a : a \in E\}$. Then $F \in \mathscr{A}_2$ but $f^{-1}(F) = E \notin \mathscr{A}_1$. i.e. f is not measurable.

Henceforth, $(M,\ d)$ will denote a metric space and $\mathfrak{m}$ its Borel σ-field. That all real continuous functions defined on M are $\mathfrak{m}$-measurable was noted in the last section. The following theorem presents a converse to this property.

Theorem 1.4.1. *Denote by $\mathcal{B}$ the σ-field generated by the collection C_0 of all the real continuous functions defined on M. i.e., it is the minimal σ-field with respect to which every $f \in C_0$ is measurable. Then $\mathcal{B} = \mathfrak{m}$.*

Proof. The definition of $\mathfrak{m}$ and of a continuous function imply $\mathcal{B} \subseteq \mathfrak{m}$. Given $A \subseteq M$ a closed set, define $f_A(x) = f(x) = 1 - d(x,\ A),\ x \in M$ if $d(x,\ A) \leq 1$ and $f(x) = 0$ otherwise and note that f is a continuous function. Hence it is $\mathcal{B}$ measurable. This implies $A = f^{-1}(\{1\}) \in \mathcal{B}$. Thus all closed sets belong to $\mathcal{B}$. This, in turn, implies $\mathfrak{m} \subseteq \mathcal{B}$. □

To prove the next theorem we need the following result from Topology, quoted without proof.

Q1 (Theorem B, p216, **[H]**) *Let C be a compact subset and F a closed subset of a locally compact Hausdorff space Ω with $C \cap F = \phi$. Then there exists a real continuous function f on Ω such that $0 \leq f(\omega) \leq 1$ and such that $f(\omega) = 0$ if $\omega \in C$ and $f(\omega) = 1$ if $\omega \in F$.(It is not claimed that $\{\omega : f(\omega) = 0\} = C$).*

Theorem 1.4.2. *Let Ω be a locally compact Hausdorff space, $\mathcal{B}_3$ its Baire σ-field and $\mathfrak{m}$ the minimal σ-field with respect to which all the real continuous functions defined on Ω are measurable. Then*
(i) $\mathcal{B}_3 \subseteq \mathfrak{m}$
(ii) If Ω, in addition, is separable and σ-compact then $\mathfrak{m} \subseteq \mathcal{B}_3$
(iii) If Ω is a compact Hausdorff space (not necessarily separable) then $\mathfrak{m} \subseteq \mathcal{B}_3$

Proof. (i) Let $C \subseteq \Omega$ be a compact g_δ set. There exists then a strictly decreasing sequence (G_n) of open sets such that $C = \cap_n G_n$. For each n, the compact set C and the closed set $\Omega \sim G_n$ are disjoint sets. Hence there exists a continuous function f_n, $0 \leq f_n(\omega) \leq 1$ such that $f_n(\omega) = 0$ if $\omega \in C$ and $f_n(\omega) = 1$ if $\omega \in \Omega \sim G_n$. Write $f(\omega) = \sum\limits_{n=1}^{\infty} \frac{1}{2^n} f_n(\omega)$ and note that $0 \leq f(\omega) \leq 1$ and that f is a continuous function. Further $f(\omega) = 0$ if $\omega \in C$. If $\omega \in \Omega \sim C$, then there exists $N = N(\omega)$ such that $\omega \in \Omega \sim G_n$ for all $n \geq N$. This implies that if $\omega \in \Omega \sim C$, then for some $N \geq 1,\ f(\omega) \geq \sum\limits_{n=N}^{\infty} \frac{1}{2^n} > 0$. It follows that $C = \{\omega : f(\omega) = 0\}$. This proves that all compact g_δ sets belong to $\mathfrak{m}$ and therefore $\mathcal{B}_3 \subseteq \mathfrak{m}$.

(ii) As noted earlier (ref. page 4) the hypothesis implies $\mathcal{B}_1 = \mathcal{B}_3$. Since all continuous functions are $\mathcal{B}_1$ measurable, it follows that $\mathfrak{m} \subseteq \mathcal{B}_1 = \mathcal{B}_3$.
(iii) Let f be a real continuous function on Ω. For c an arbitrary real number and $n \geq 1$, define $A = \{\omega : f(\omega) \leq c\}$ and $A_n = \{\omega : f(\omega) < c + \frac{1}{n}\}$. We note that A is a compact set, that each A_n is an open set and that $A_n \downarrow A$. Thus A is a g_δ set. This implies that $\mathfrak{m} \subseteq \mathcal{B}_3$. □

Remark 1.4.3. We record here for ready reference the consequence that if Ω is a compact Hausdorff space then $\mathcal{B} = \mathfrak{m}$.

Theorem 1.4.4. *Let $\Omega_1,\ \Omega_2$ be compact Hausdorff spaces and Ω their topological product. (Ω will necessarily be a compact Hausdorff space). Let $\mathcal{B}_1,\ \mathcal{B}_2,\ \mathcal{B}$ be the Baire σ-fields respectively of $\Omega_1,\ \Omega_2$ and Ω. Let $\widehat{\mathcal{B}}$ be the product σ-field $\mathcal{B}_1 \bigotimes \mathcal{B}_2$. Then $\widehat{\mathcal{B}} = \mathcal{B}$.*

Proof. Let $\mathcal{E}_i,\ i = 1,\ 2$ denote the collection of all the compact g_δ subsets of Ω_i and note that $\Omega_i \in \mathcal{E}_i$. We follow the notation and the result of **(1.2)**. We note that $\sigma(\mathcal{E}_i) = \mathcal{B}_i$ and that every member of $\mathcal{E}$ is a compact g_δ subset of Ω. Hence $\widehat{\mathcal{B}} = \sigma(\mathcal{E}) \subseteq \mathcal{B}$.

Let $H_i,\ i = 1,\ 2$ and H denote the collection of all real continuous functions on Ω_i and Ω. We note that for each $i,\ \Omega_i$, being a compact Hausdorff space, is completely regular (ref. p 146, **[K]** or p 216, **[H]**). Hence members of H_i distinguish the points of Ω_i. We will treat members of H_i as members also of H as follows. $f \in H_1$ will be identified with the function $f^* \in H$ where for $x = (x_1,\ x_2) \in \Omega,\ f^*(x) = f^*(x_1,\ x_2) = f(x_1)$. Similarly functions in H_2. Let $\widetilde{H} = H_1 \cup H_2$ and note $\widetilde{H} \subseteq H$. Further (i) constant functions belong to $\widetilde{H}$ and (ii) members of $\widetilde{H}$ distinguish the points of Ω. Hence by the Stone-Weierstrass theorem (ref. Problem R, p 244, **[K]**) $\widetilde{H}$ is a dense sub collection of H endowed with the sup norm. Now by **(1.4.3)** every $f \in H_1$ is $\mathcal{B}_1$-measurable. Hence every $f \in \widetilde{H}$ is $\widehat{\mathcal{B}}$ measurable. Since members of H are pointwise limits of sequences in $\widetilde{H}$, it follows that members of H are $\widehat{\mathcal{B}}$ measurable. Since $\mathcal{B}$ is the smallest σ-field with respect to which all the members of H are measurable, it follows $\mathcal{B} \subseteq \widehat{\mathcal{B}}$. □

Remark 1.4.5. The extension of Theorem **(1.4.4)** to countably infinitely many component compact Hausdorff spaces is straight forward.

Remark 1.4.6. All rvs, rVs, and rEs are supposed defined on a basic measurable space $(\Omega,\ \mathscr{A})$.

Let $(\Omega,\ \mathscr{A}), (\Omega_i,\ \mathscr{A}_i),\ 1 \leq i < \infty$ be measurable spaces. Let $(\Omega_T,\ \mathscr{A}_T)$ be the product measurable space $(\times \Omega_i,\ \bigotimes \mathscr{A}_i)$.

Theorem 1.4.7. *A mapping $\mathfrak{X} = (X_1, X_2, \ldots)$ from Ω to Ω_T is measurable if and only if (iff) each X_i is measurable.*

Proof. Let the notation be as in section **(1.2)**. Since $\sigma(\cup\{\mathbf{a}_{J;T} : J \subset T, J$ finite $\}) = \mathscr{A}_T$, it follows that $\mathfrak{X}$ is measurable *iff* if $\mathfrak{X}^{-1}(\mathbf{a}_{J;T}) \subseteq \mathscr{A}$ for every $J \subset T$, J finite. Without loss of generality, after relabelling if necessary, take $J = \{1, 2, \ldots, m\}$. Let $Y_n = (X_1, X_2, \ldots, X_m)$. Thus $\mathfrak{X}$ is measurable *iff* Y_m is $\mathscr{A} \setminus \bigotimes_1^m \mathscr{A}_i$ measurable. Since $\bigotimes_1^m \mathscr{A}_i$ is the σ-field generated by rectangles of the type $\times_1^m A_i$, $A_i \in \mathscr{A}_i$, $1 \leq i \leq m$, the measurability of $\mathfrak{X}$ happens *iff* each X_i is measurable. □

Definition 1.4.8. A simple rE is one which takes a finite number of values. An elementary rE is one which takes a countably infinite number of values.

Thus X is a simple rv *iff* there exists an integer $k \geq 1$, mutually exclusive sets $E_i \in \mathscr{A}$, $1 \leq i \leq k$, and real numbers $a_i \neq 0$ such that $X(\omega) = \sum_1^k a_i \chi_i(\omega)$ where χ_i is the characteristic function of the set E_i. Similarly X would be an elementary rv *iff* there exist mutually exclusive sets $E_i \in \mathscr{A}$, $i \geq 1$ and real numbers $a_i \neq 0$ such that $X(\omega) = \sum_1^\infty a_i \chi_i(\omega)$. In both the cases rv $X \geq 0$ *iff* all the a_is are non-negative.

Let $E_i \in \mathscr{A}$, $1 \leq i \leq p$, the E_is not necessarily mutually exclusive, $\cup_1^k E_i = \Omega$ and let $X(\omega) = \sum_1^p a_i \chi_{E_i}(\omega)$, $a_i \neq 0$. Now we show that X can be written $X(\omega) = \sum_1^q b_i \chi_i(\omega)$ where the χ_is are the indicator functions of mutually exclusive sets $F_j \in \mathscr{A}$, $1 \leq j \leq q$ and no b_i is zero. The method is by induction and here is how it can be done. Let $p = 2$. Let $X = a_1\chi_{E_1} + a_2\chi_{E_2}$. Then clearly $X = a_1\chi_{E_1 \sim E_2} + (a_1 + a_2)\chi_{E_1 \cap E_2} + a_2\chi_{E_2 \sim E_1}$. Omitting the middle term in case $a_1 + a_2 = 0$, we see X can be written in the desired form, with $q = 2$ or 3. Suppose now the claim has been established for $p = r$. Let then $X = \sum_1^r a_i\chi_{E_i} + a_{r+1}\chi_{E_{r+1}}$ where $E_1, E_2, \ldots, E_r$ are mutually exclusive and $a_i \neq 0$, $1 \leq i \leq r+1$. On simplification $X = \sum_1^r a_i\chi_{E_i \sim E_{r+1}} + \sum_1^r (a_i + a_{r+1})\chi_{E_i \cap E_{r+1}} + a_{r+1}\chi_{E_{r+1} \sim \{\cup_1^r E_i\}}$. As before, omitting the terms in which $a_i + a_{r+1} = 0$, we see X assumes the desired form.

In view of the above result, our definition of a simple rv, from now on, would be as follows : A simple rv or a simple real measurable function is *any* finite linear combination of indicator functions.

It is immediate from this that the sum of two simple rvs is a simple rv.

Theorem 1.4.9. (i) *Every rv X is the limit of a sequence of simple rvs.*
(ii) *Every rE taking values in a compact metric space is the uniform limit of a sequence of simple rEs.*
(iii) *every rv is the uniform limit of a sequence of elementary rvs.*
(iv) *If $X,\ Y,\ X_1,\ X_2,\ldots$ are rvs then the following are rvs.*
(p) $\max(X,\ Y)$, $\min(X,\ Y)$, X^+, X^- *where* $X^+ = \max(X,\ 0)$ *and* $X^- = -\min(X,\ 0)$
(q) $U = \sup\limits_{1\leq k<\infty} X_k$, $V = \inf\limits_{1\leq k<\infty} X_k$,
(r) $\xi = \overline{\lim\limits_{n\to\infty}} X_n$,
(s) $\eta = \underline{\lim\limits_{n\to\infty}} X_n$ *and* χ_A *where* $A = \{\omega : \lim\limits_{n\to\infty} X_n(\omega)$ *exists*$\} \in \mathscr{A}$
(t) $X \pm Y$.
(v) *It is possible X_t is a rv for every $t \in T$ but* $\sup\limits_{t\in T} X_t$ *is not a rv if T is an uncountable set.*

Proof. (i) Given n, define sets $E_{-n} = \{\omega : X(\omega) \leq -n\}$; $E_n = \{\omega : X(\omega) \geq n\}$; $E_{-i} = \{\omega : -\frac{i}{2^n} < X(\omega) \leq -\frac{i-1}{2^n}\}, E_i = \{\omega : \frac{i-1}{2^n} < X(\omega) \leq \frac{i}{2^n}\}$, $i = 1,\ 2,\ \ldots,\ n2^n$. All these $n2^{n+1}+2$ sets are measurable sets, since X is a measurable function. Hence if

$$x_n(\omega) = \begin{cases} -n & \text{if } \omega \in E_{-n} \\ -\frac{i-1}{2^n} & \text{if } \omega \in E_i,\ i = 1,\ 2, \ldots,\ n2^n \\ \frac{i-1}{2^n} & \text{if } \omega \in E_i,\ i = 1,\ 2, \ldots,\ n2^n \\ n & \text{if } X(\omega \in E_n \end{cases}$$

then clearly x_n is a simple rv. Fix ω. Suppose $X(\omega) \geq 0$. In that case $0 \leq x_n(\omega) \leq X(\omega)$ for all n. Further it is easily verified that $x_n(\omega)$ is a monotonic non-decreasing sequence. Also for all n sufficiently large, largeness depending on ω, $0 \leq X(\omega) - x_n(\omega) \leq \frac{1}{2^n}$. Similar arguments hold when $X(\omega) < 0$. This completes the proof of (i).

(ii) Let (M, d) be compact metric space, let $\mathfrak{m}$ be its Borel σ-field and let f be a measurable mapping of Ω into M. i.e. f is a rE in M. Given a positive integer N, we use the compactness of M to find a finite number of open spheres of radius $\frac{1}{N}$ which will cover M. Out of these, a finite number of disjoint members of $\mathfrak{m}$, say, $A_1, A_2, \ldots, A_q$, $q = q(N)$ can be constructed, whose union would be M. Choose $\omega_i \in f^{-1}(A_i)$ arbitrarily. Let $f(\omega_i) = t_i$, $t_i \in A_i$, $1 \leq i \leq q$. Define $g_N(\omega) = t_i$ if $\omega \in f^{-1}(A_i)$, $1 \leq i \leq q$

and note that g_N is defined for all $\omega \in \Omega$ and that it is a simple function. Now $d(f(\omega), g_N(\omega)) \leq \max_{1\leq i\leq q} \text{diam} A_i \leq \frac{2}{N}$ for all ω. It follows now that $g_N(\omega) \to f(\omega)$ as $N \to \infty$ uniformly in ω.

(iii) Define x_n: for $n = 1, 2, \ldots$.

$$x_n(\omega) = \begin{cases} \frac{i-1}{2^n} & \text{if } \frac{i-1}{2^n} \leq X(\omega) < \frac{i}{2^n},\ i = 1, 2, \ldots, \\ -\frac{i-1}{2^n} & \text{if } -\frac{i}{2^n} < X(\omega) \leq -\frac{i-1}{2^n},\ i = 1, 2, \ldots, \end{cases}$$

As argued in part (i), the sets involved are measurable and x_n is an elementary set. The proof is completed by noting that $|x(\omega) - x_n(\omega)| \leq \frac{1}{2^n}$ for every $\omega \in \Omega$.

(iv) Apply definition. $\{\omega : \max(X, Y)(\omega) \leq x\} = \{\omega : X(\omega) \leq x\} \cap \{\omega : Y(\omega) \leq x\} \in \mathscr{A}$. Proof for $\min(X, Y)$ is similar. Claims regarding X^+, X^- follow by taking $Y \equiv 0$ in the above result. This completes the proof of (p).

(q) Let $U_n = \max_{1\leq k\leq n} X_k$. Then $U_n(\omega) \uparrow U(\omega)$. By (p), each U_n is a *rv*. U may be extended valued.

$$\{\omega : U(\omega) = \infty\} = \cap_{k=1}^{\infty} \{\omega : \lim_{n\to\infty} U_n(\omega) \geq k\}$$
$$= \bigcap_{k=1}^{\infty} \bigcup_{r=1}^{\infty} \bigcap_{n=r}^{\infty} \{\omega : U_n(\omega) \geq k\} \in \mathscr{A}.$$

Hence in order to show that U is an extended real valued *rv* it remains to show that $\{\omega : c < U(\omega) < \infty\}$ lies in $\mathscr{A}$. But this is obvious since $\{\omega : c < U(\omega) < \infty\} = \bigcup_{n=1}^{\infty} \{\omega : X_n(\omega) > c\}$. Proof for V is similar.

(r and s) That ξ, η are *rv*s, possibly extended valued, follows from the relations $\xi = \inf_{n\geq 1} \sup_{m\geq n} X_m$ and $\eta = \sup_{n\geq 1} \inf_{m\geq n} X_n$. Now, $A \in \mathscr{A}$ since $A = \{\omega : \xi(\omega) = \eta(\omega)\}$.

(t) Let X_n, Y_n be simple *rv*s such that $\lim_{n\to\infty} X_n(\omega) = X(\omega)$, and $\lim_{n\to\infty} Y_n(\omega) = Y(\omega)$, $\omega \in \Omega$. Since X, Y are finite valued it follows that $X \pm Y = \lim_{n\to\infty} (X_n \pm Y_n)$. By **(1.4.8)** $X_n \pm Y_n$, for each n, is a simple *rv*. Hence $X \pm Y$, being limits of sequences of *rv*s, are *rv*s.

(v) Let $\Omega = [0, 1]$ and $\mathscr{A}$ its Borel σ-field. Let $\Lambda = [0, 1]$ be the index set. Let $A \subset \Lambda$ be a non-Borel set. Refer to **(1.3.4)** for assurance that such a set A exists. For $t \in A'$ define $X_t(\omega) = 0$ for all $\omega \in \Omega$. For $t \in A$, $t \neq \omega$, define $X_t(\omega) = 0$. For $t \in A$, and $t = \omega$ define $X_t(\omega) = 1$. For t fixed and $t \notin A$, $X_t \equiv 0$ and hence is a *rv*. For t fixed, $t \in A$, and X_t takes only two values, 0 and 1. Further $\{\omega : X_t(\omega) = 1\} = \{t\}$, a Borel set. Thus for all t, X_t is a rv. Let $Y(\omega) = \sup_{t \in A} X_t(\omega)$. We note that Y takes only two values, 0 and 1 and that $\{\omega : Y(\omega) = 1\} = A$. Hence Y is not a rv. □

Remarks 1.4.10. Results (i) and (iii) of **(1.4.9)** have obvious extensions to rVs.

The result **(1.4.9)**(ii) implies that every bounded rV is the uniform limit of a sequence of simple rVs.

The *proof* in **(1.4.9)**(iii) depends on the property of the real line that it is well ordered. To stress that the claim depends only on the fact that R is a separable metric space (and not on the cited special property of the real line), we argue as follows.

Let $(M,\ d)$ be a separable metric space, with its Borel σ-field denoted, as usual, by $\mathcal{m}$. Let $\{r_n, n = 1, 2, \ldots\}$ be a separability set for M. Denote by $S_{n,k}$ the open sphere with centre r_n, radius $\frac{1}{k}$ and note that $\bigcup_{n=1}^{\infty} S_{n,k} = M$, $k = 1, 2, \ldots$. Write $A_{n,1} = S_{n,1}$, $A_{n,2} = S'_{n,1} \cap S_{n,2}$, $A_{n,3} = S'_{n,1} \cap S'_{n,2} \cap S_{n,3}, \ldots$ and note that the $A_{n,k}$s are Borel subsets, that for k fixed they are disjoint, that none of them are null and that $\bigcup_{n=1}^{\infty} A_{n,k} = M$. Given X, a M-valued *rE*, define $E_{n,k} = X^{-1} A_{n,k}$ and note that for k fixed the $E_{n,k}$s are disjoint non-null members of $\mathscr{A}$ and that $\bigcup_{n=1}^{\infty} E_{n,k} = \Omega$. Choose elements $a_{n,k} \in A_{n,k}$ arbitrarily and define $X_k(\omega) = a_{n,k}$ if $\omega \in E_{n,k}$, $n \geq 1$. Each X_k is clearly an elementary rE. Further $d(X(\omega),\ X_k(\omega)) \leq \frac{1}{k}$ for all $\omega \in \Omega$.

Define the range of a rE as $\{X(\omega) : \omega \in \Omega\}$.

The above result shows that the closure of the range of a rE in a separable metric space is separable.

Theorem 1.4.11. *Let* $(\Omega,\ \mathscr{A})$, $(U,\ \mathscr{U})$ *be arbitrary measurable spaces. Let* X *be a measurable mapping of* Ω *into* U. *Denote by* $\mathscr{B} \subseteq \mathscr{A}$ *the* σ*-field generated by* X. *i.e.,* $\mathscr{B} = X^{-1}(\mathscr{U})$. *Then* Y *is a* rv *on* $(\Omega,\ \mathscr{B})$ *iff there exists a* rv g *on* $(U,\ \mathscr{U})$ *such that* $Y = g(X)$.

Proof. The forward result is trivially true. For, if E is a linear Borel set and if $Y = g(X)$, then $Y^{-1}(E) = X^{-1}(g^{-1}(E)) \in \mathscr{B}$.

To prove the converse, assume that Y is a rv on $(\Omega,\ \mathscr{B})$. For $n = 1, 2, \ldots$. and $i = 0, \pm 1, \pm 2, \ldots$, define $A_{i,n} = \{\omega : \frac{i}{2^n} < Y(\omega) \leq \frac{i+1}{2^n}\}$. For n

fixed, the $A_{i,n}$s are disjoint with $\cup_i A_{i,n} = \Omega$. By hypothesis all the $A_{i,n}$s belong to $\mathscr{B}$. Since $\mathbf{X}$ generates $\mathscr{B}$, corresponding to each $A_{i,n}$, there exists $C_{i,n} \in \mathscr{U}$ such that $\mathbf{X}^{-1}(C_{i,n}) = A_{i,n}$. For n fixed, the $C_{i,n}$s are disjoint since the $A_{i,n}$s are so. Further if $C_n = \cup_i C_{i,n}$, then $\mathbf{X}^{-1}(C_n) = \cup_i \mathbf{X}^{-1}(C_{i,n}) = \cup_i A_{i,n} = \Omega$. This shows that C_n is independent of n; equal to, say, C. Also $\{\mathbf{X}(\omega) : \omega \in \Omega\} = C$. Define g_n on U : for $u \in C$, $g_n(u) = \frac{i}{2^n}$ if $u \in C_{i,n}, i = 0, \pm 1, \pm 2, \ldots$ and $= 0$ if $u \notin C$. We note each g_n is a rv. Hence, by the forward part, $y_n = g_n(X)$ is a rv. Now, $y_n(\omega) = \frac{i}{2^n}$ if $\omega \in A_{i,n}$. For each ω, $y_n(\omega) \to Y(\omega)$, since $|y_n(\omega) - Y(\omega)| \leq \frac{1}{2^n}$.

If $u \in C'$, then $g_n(u) = 0$. If $u \in C$, then $g_n(u) = g_n(X(\omega))$ for some $\omega) = y_n(\omega)$. Since $y_n(\omega) \to Y(\omega)$ for all $\omega \in \Omega$, it follows that for every u, $\lim_{n\to\infty} g_n(u)$ exists finitely. Denote the limit function by g. Since the g_ns are rvs, so is g. Now,

$$\begin{aligned} g(X(\omega)) = g(u) &= \lim_{n\to\infty} g_n(u) = \lim_{n\to\infty} g_n(\mathbf{X}(\omega)) \\ &= \lim_{n\to\infty} y_n(\omega) = Y(\omega) \end{aligned}$$

and the proof is complete. □

1.5 Probability measure spaces and distributions

Let $\mathbf{E}$ be a collection of subsets of a set Ω. A non-negative function μ defined on $\mathbf{E}$ is said to be finitely additive if, for every $m \geq 1$ and every collection of disjoint sets E_n, $1 \leq n \leq m$ in $\mathbf{E}$, it is true that $\mu(\bigcup_{n=1}^{m} E_n) = \sum_{n=1}^{m} \mu(E_n)$ whenever $\bigcup_{n=1}^{m} E_n$ is in $\mathbf{E}$; it is said to be countably additive if for every disjoint sequence (E_n) of sets in $\mathbf{E}$, it is true that

$$\mu(\bigcup_{n=1}^{\infty} E_n) = \sum_{n=1}^{\infty} \mu(E_n)$$

whenever $(\bigcup_{n=1}^{\infty} E_n)$ belongs to $\mathbf{E}$.

If $\mathbf{E}$ is a field, then finite additivity of μ implies that μ is monotonic. For, let $A,\ B \in \mathbf{E}$ and $A \subset B$. Hence $B \sim A \in \mathbf{E}$. And then, $\mu(B) = \mu(A \cup (B \sim A)) = \mu(A) + \mu(B \sim A) \geq \mu(A)$.

If $\mathbf{E}$ is a field and if $\mu(\Omega) < \infty$, then (i) finite additivity of μ implies $\mu(\phi) = 0$ and (ii) countable additivity of μ implies its finite additivity. For, since Ω and ϕ are two mutually exclusive members of **E** with Ω for their union, we have $\mu(\Omega) = \mu(\Omega \cup \phi) = \mu(\Omega) + \mu(\phi)$. Since $\mu(\Omega) < \infty$, claim (i) follows. If μ is countably additive, $\mu(\Omega) = \mu(\Omega \cup \phi \cup \phi \cup \ldots) = \mu(\Omega) + \mu(\phi) + \mu(\phi) + \ldots$. Since $\mu(\Omega) < \infty$, this implies $\mu(\phi) + \mu(\phi) + \cdots = 0$. Since μ is non-negative set function, it follows $\mu(\phi) = 0$. Let, for $n \geq 1$ arbitrary, A_i, $1 \leq i \leq n$ be n mutually exclusive members of **E**. Then $\mu(\cup_1^n A_i) = \cup_1^\infty A_i$ where $A_r = \phi$ for all $r \geq n+1$, leading to

$$\mu(\cup_1^n A_i) = \sum_1^\infty \mu(A_i) = \sum_1^n \mu(A_i).$$

Thus claim (ii) is true.

If **E** is a field and if $\mu(\Omega) < \infty$, then countable additivity of μ is equivalent to continuity of μ at the null set. i.e. $E_n \downarrow \phi$ implies $\mu(E_n) \to 0$ (necessarily monotonically). To see this, assume first that μ is countably additive. Let $E_n \downarrow \phi$, $E_n \in$ **E**. Hence $E_n' \uparrow \Omega$ or $\Omega = \cup_1^\infty (E_n' \sim E_{n-1}')$, $E_0 = \phi$. Since the sets $E_n' \sim E_{n-1}'$, $n \geq 1$ are mutually exclusive and belong to **E**, we get, by the hypothesis,

$$\mu(\Omega) = \sum_1^\infty \mu(E_n' \sim E_{n-1}\prime) = \sum_1^\infty \{\mu(E_n') - \mu(E_{n-1}')\}$$

$$= \lim_{n\to\infty} \sum_1^n \{\mu(E_n') - \mu(E_{n-1}')\} = \lim_{n\to\infty} \mu(E_n'),$$

or, what is same, $\lim_{n\to\infty} \mu(E_n) = \mu(\Omega') = \mu(\phi) = 0$, as was to be proved. The steps are reversible and hence the converse part follows.

The measure μ is said to be σ-finite if there exist $E_n \in$ **E**, $n = 1, 2, \ldots$ such that $\mu(E_n) < \infty, n \geq 1$ and $\bigcup_{n=1}^\infty E_n = \Omega$; it is said to be a probability measure (*pm*) if $\mu(\Omega) = 1$. $(\Omega, \mathscr{A}, \mu)$ is called a measure space if $(\Omega, \mathscr{A})$ is a measurable space and if μ is a measure on $\mathscr{A}$. A measure space is called a probability measure space or, simply, a probability space if μ is a probability measure (pm). The Lebesgue measure $\mathcal{L}$ on the Borel σ-field $\mathscr{R}^p$ of Euclidean p-space R^p is an example of a non-finite σ-finite measure. $\mathcal{L}$ is that measure such that $\mathcal{L}(I) = \Pi_{i=1}^p (b_i - a_i)$ where I is the 'interval' $\times_{i=1}^p (a_i, b_i]$ and such that for arbitrary $\mathbf{a} \in R^p$ and arbitrary Borel set E, $\mathcal{L}(E + \mathbf{a}) = \mathcal{L}(E)$. For the construction of the Lebesgue measure the reader is referred to pp 32–36, **[H]**. Integration with respect to the Lebesgue measure is denoted also by $d\mathbf{x}$.

The following two results are simple consequences of the definition of a π-system, a Dynkin system and a measure. The method of proof of the first result is a useful technique. For other applications of this method, refer **(1.5.5)**(Part 2) and **(1.11.8)**.

(i) Let $\mathcal{C}$ be a π-system and let $\mathfrak{A}$ be the σ-field generated by $\mathcal{C}$. Let $\mu,\ \nu$ be two *pms* on $\mathfrak{A}$ such that $\mu(A) = \nu(A),\ A \in \mathcal{C}$. Then $\mu(A) = \nu(A),\ A \in \mathfrak{A}$. In words : if μ and ν agree on $\mathcal{C}$, then they agree on $\mathfrak{A}$.

Proof. Let $\mathcal{B}$ be the collection of all the sets in $\mathfrak{A}$ such that μ and ν agree on $\mathcal{B}$ and such that if $A,\ B \in \mathcal{B}$, then $\mu(A \cap B) = \nu(A \cap B)$. Clearly, $\mathcal{C} \subset \mathcal{B}$.

Let $A \in \mathcal{B}$. Hence $\mu(A) = \nu(A)$. This implies, since $\mu,\ \nu$ are finite measures, $\mu(A') = 1 - \mu(A) = 1 - \nu(A) = \nu(A')$. Hence $\mathcal{B}$ is closed under complementation. If $A,\ B\ \in \mathcal{B}$, then $A \sim B \in \mathcal{B}$, since $\mathcal{B}$ is closed under complementation and intersection. This implies that a countable union of members of $\mathcal{B}$ can be expressed as the countable union of disjoint members of $\mathcal{B}$. Since $\mu,\ \nu$ are measures, it follows now that $\mathcal{B}$ is closed under countable unions. Thus $\mathcal{B} \supset \mathcal{C}$ is a σ-field on which μ agrees with ν. Since $\mathfrak{A}$ is the σ-field generated by $\mathcal{C}$, it follows that $\mathcal{B} = \mathfrak{A}$ and, hence, that μ agrees with ν on $\mathfrak{A}$.

(ii) Let Ω be the unit interval, , $\mathscr{A}$ its Borel σ-field and , $\mathcal{L}$ the Lebesgue measure. Let $\mathbf{I} = \{(a,b), [a,b), (a,b], [a,b]\ :\ 0 < a \leq b < 1\}$ and observe that $\mathbf{I}$ is closed under finite intersections and that $\mathscr{A}$ is the σ-field generated by $\mathbf{I}$. Let μ be another measure on $\mathscr{A}$ such that the μ-measure of any interval is its length. If μ agrees with $\mathcal{L}$ on $\mathbf{I}$, then $\mu \equiv \mathcal{L}$. i.e., μ agrees with $\mathcal{L}$ on $\mathscr{A}$. In other words the Lebesgue measure is unique.

Proof. The claim is immediate from (i), applying it to the situation on hand. □

Definition 1.5.1. (i) Suppose that a proposition is described by the points of an event E in a measure space $(\Omega, \mathscr{A}, \mu)$ and that $\mu(E^{'}) = 0$. If μ is an arbitrary measure, then we say that the proposition is true almost everywhere $(a.e.)\ \mu$; if μ is a *pm*, then we say that the proposition is true with probability 1 $(wp1)$ or almost surely $(a.s.)$.

For example, suppose $\{X_n, n \geq 0\}$ is a sequence of *rvs* defined on a probability space $(\Omega, \mathscr{A}, P)$. Let $E = \{\omega : X_n(\omega) \to X_0(\omega)\}$. By **(1.4.3)**(iv)(s), $E \in \mathscr{A}$. If $P(E) = 1$, we say that X_n converges to X_0, $wp1$ and write $X_n \stackrel{wp1}{\longrightarrow} X_0$. If the underlying measure μ is not a probability measure, if $F = \{\omega\ :\ X_n(\omega) \nrightarrow X_0(\omega)\}$ and if $\mu(F) = 0$ then we say that X_n converges to X_0 almost everywhere and write $X_n \stackrel{a.e.}{\longrightarrow} X_0$.

(ii) **Completion of a measure space**.

A measure space $(\Omega,\ \mathscr{A},\ \mu)$ is said to be complete if every subset of a set of measure zero is measurable (and consequently has measure zero). Define $\mathcal{B}$ to be the class of all sets of the form $E\Delta N$, where $E \in \mathscr{A}$ and N is a subset of a set of measure zero in $\mathscr{A}$. Since $E\Delta\phi = E$, it follows that $\mathscr{A} \subset \mathcal{B}$. It can be shown that $\mathcal{B}$ is a σ-field and that ν defined on it through $\nu(E\Delta N) = \mu(E)$ is a complete measure. For proof refer Theorem B, p 55, **[H]**. $(\Omega,\ \mathcal{B},\ \nu)$ is called the completion of $(\Omega,\ \mathscr{A},\ \mu)$.

Absolute continuity of a pm in R^p

Theorem 1.5.2. *Let $A \in \mathscr{R}^p$. Then for every $\boldsymbol{u} \in R^p$, $A + \boldsymbol{u} \in \mathscr{R}^p$ where $A + \boldsymbol{u} = \{\boldsymbol{x} + \boldsymbol{u}\ :\ \boldsymbol{x} \in A\}$.*

Proof. Define $f_{\mathbf{u}}$ mapping R^p onto R^p : $f_{\mathbf{u}}(\mathbf{x}) = \mathbf{x} - \mathbf{u},\ \mathbf{x} \in R^p$. Being a continuous function $f_{\mathbf{u}}$ is measurable. Hence $f_{\mathbf{u}}^{-1}(A) \in \mathscr{R}^p$. But $f_{\mathbf{u}}^{-1}(A) = A + \mathbf{u}$. □

To prove **(1.5.3)** we need the following result (viz. **Q**2(i)) from integration theory (ref. Theorem B, p 112, **[H]**). Along with it we quote two other related theorems for later use.

Q2 Let $(\Omega,\ \mathscr{A},\ \mu)$ be an arbitrary measure space.
(i) (Monotone convergence Theorem) *If (f_n) is an increasing sequence of extended real valued non-negative measurable functions, if $\lim_n f_n(\omega) = f(\omega)$ a.e. μ then $\lim_n \int_\Omega f_n(\omega)\ d\mu(\omega) = \int_\Omega f(\omega)\ d\mu(\omega)$.*
(ii) (Bounded Convergence Theorem) (Theorem D, p 110 **[H]**) *If (f_n) is a sequence of integrable functions converging to f a.e. μ and if g is an integrable function such that $|f_n(\omega)| \le g(\omega)$ a.e., then f is integrable and the sequence (f_n) converges to f in mean of order 1 i.e.* $\int_\Omega |f_n(\omega) - f(\omega)|\ d\mu(\omega) \to 0$.

Note. We have

$$\lim_{n\to\infty} \left| \int_\Omega f_n(\omega) d\mu(\omega) - \int_\Omega f(\omega)\ d\mu(\omega) \right| \le \lim_{n\to\infty} \int_\Omega \left| f_n(\omega) - f(\omega) \right| d\mu(\omega) \to 0.$$

(iii) (Fatou's lemma)(Theorem F, p. 113 **[H]**) *If (f_n) is a sequence of non-negative integrable functions for which $\underline{\lim}_{n\to\infty} \int_\Omega f_n(\omega)\ d\mu(\omega) < \infty$, then the* *function* f *defined by* $f(\omega) = \underline{\lim}_{n\to\infty} f_n(\omega)$ *is integrable and* $\int_\Omega f(\omega)\, d\mu(\omega) \leq \underline{\lim}_{n\to\infty} \int_\Omega f_n(\omega)\, d\mu(\omega)$.

Definitions 1.5.3. By $\mathbf{a} < \mathbf{b}$ for $\mathbf{a}, \mathbf{b} \in R^p$, we understand $a_i < b_i$, $1 < i < p$ where $a_i,\ b_i$ are the ith components of $\mathbf{a}$ and of $\mathbf{b}$. When $p \geq 2$, we write $\mathbf{a} \leq \mathbf{b}$ if $a_i \leq b_i$ for all i and $a_i < b_i$ for at least one i. If $\mathbf{a} < \mathbf{b}$ then by the p-dimensional 'interval' $(\mathbf{a}, \mathbf{b})$ we understand the set $\{\mathbf{x} : a_i < x_i < b_i,\ 1 \leq i \leq p\}$.

Given a real valued function g defined on R^p and $\mathbf{a},\ \mathbf{b}\ \in R^p,\ \mathbf{a} \leq \mathbf{b}$ write

$$\begin{aligned}\Delta_{\mathbf{a}}^{\mathbf{b}}(g) &= g(b_1, b_2, \ldots, b_p) - \sum_{j=1}^{p} g(b_1, b_2, \ldots, b_{j-1}, a_j, b_{j+1}, \ldots, b_p) \\ &\quad + \sum_{1\leq j<k\leq p} g(b_1, b_2, \ldots, b_{j-1}, a_j, b_{j+1}, \ldots, b_{k-1}, a_k, b_{k+1}, \ldots, b_p) \\ &\quad - \cdots + (-1)^p g(a_1, a_2, \ldots, a_p).\end{aligned}$$

g is said to be absolutely continuous if for every $\varepsilon > 0$ there corresponds a $\delta > 0$ such that

$$\sum_{r=1}^{n} |\Delta_{a_r}^{b_r}(g)| < \varepsilon$$

for every finite disjoint class $\{(\mathbf{a}_r,\ \mathbf{b}_r)\ :\ 1 \leq r \leq n\}$ of bounded open 'intervals' for which $\sum_{r=1}^{n} \mathcal{L}(\mathbf{a}_r\ , \mathbf{b}_r) < \delta$.

The distribution function (df) $\mathbb{F}$ associated with or generated by a pm μ on $\mathscr{R}^p$ is defined thus:

$$\mathbb{F}(\mathbf{x}) = \mathbb{F}(x_1,\ x_2, \ldots\ ,\ x_p) = \mu(I_{\mathbf{x}})$$

where $I_{\mathbf{x}} = \times_1^p(-\infty,\ x_i]$. We note $\Delta_{\mathbf{a}}^{\mathbf{b}}(\mathbb{F}) = \mu(\times_1^p(a_i,\ b_i]) \geq 0$.

Lemma 1.5.4 (Part 1). *Let* μ *be a finite measure on* $\mathscr{R}^p$ *and let* $\mathbb{F}$ *be the associated* df : $\mathbb{F}(\boldsymbol{x}) = \mu(I_{\boldsymbol{x}})$. *Then* $\mathbb{F}$ *is a measurable function.*

Proof. Let $\sigma > 0$,

$$\varphi_\sigma(y) = \frac{1}{\sigma\sqrt{2\pi}} e^{-\frac{1}{2\sigma^2}y^2}, \quad -\infty < y < \infty,$$

$$\Phi_\sigma^{(p)}(\mathbf{x}) = \int_{I_{\mathbf{x}}} \prod_{i=1}^{p} \varphi_\sigma(x_i)\mathbf{x}$$

and

$$\mathbb{G}_\sigma^{(p)}(\mathbf{x}) = \int_{R^p} \Phi_\sigma^{(p)}(\mathbf{x}-\mathbf{y})\, d\mathbb{F}(\mathbf{y}).$$

Since $\Phi_\sigma^{(p)}$ is a continuous function, we have, by an application of the bounded convergence theorem, $\mathbb{G}_\sigma^{(p)}$ is a continuous, and hence a measurable, function. It follows that $\lim_{\sigma\to 0} \mathbb{G}_\sigma^{(p)}$ is a measurable function. Here and below, limit as $\sigma \to 0$ is assumed to be taken along a sequence. Again by the bounded convergence theorem,

$$\begin{aligned}\lim_{\sigma\to 0} \mathbb{G}_\sigma^{(p)}(\mathbf{x}) &= \int_{R^p} \lim_{\sigma\to 0} \Phi_\sigma^{(p)}(\mathbf{x}-\mathbf{y})\, d\mathbb{F}(\mathbf{y}) \\ &= \int_{\cap_1^p (y_i\le x_i)} \lim_{\sigma\to 0} \Phi_\sigma^{(p)}(\mathbf{x}-\mathbf{y})\, d\mathbb{F}(\mathbf{y}).\end{aligned}$$

For $p = 2$, the right side takes the form

$$\begin{aligned}&\iint_{(-\infty,\, x)\times(-\infty,\, y)} dF(u,\, v) + \iint_{(-\infty,\, x)\times\{y\}} \frac{1}{2} dF(u,\, v) \\ &\quad + \iint_{\{x\}\times(-\infty,\, y)} \frac{1}{2}\, dF(u,\, v) + \iint_{\{(x,\, y)\}} \frac{1}{4}\, dF(u,\, v) \\ &= F(x-,\, y-) + \frac{1}{2}\{F(x-,\, y) - F(x-,\, y-)\} \\ &\quad + \frac{1}{2}\{F(x,\, y-) - F(x-,\, y-)\} + \frac{1}{4}\mu(\{(x,\, y)\}) \\ &= \frac{1}{2}\{F(x,\, y-) + F(x-,\, y)\} + \frac{1}{4}\mu(\{(x,\, y)\}) \\ &= -\frac{1}{2}\mu(\{(x,\, y)\}) + \frac{1}{2}\{F(x,\, y) + F(x-,\, y-)\} \\ &\quad + \frac{1}{4}\mu(\{(x,\, y)\}) \\ &= \frac{1}{2}\{F(x,\, y) + F(x-,\, y-)\} - \frac{1}{4}\mu(\{(x,\, y)\}).\end{aligned}$$

Define $\hat{F} \;:\; \hat{F}(x,\, y) = F(x,\, y) + F(x-,\, y-)$. Since the integral on the left side is a measurable function, we can conclude that $\hat{F}$ is a measurable function if we show that $\tilde{F}$ is measurable where $\tilde{F}(x,\, y) = \mu(\{(x,\, y)\})$. Now if $\{(x, y) : \tilde{F}(x, y) > t\}$ contains n or more elements, then necessarily $nt \leq 1$. Hence $n < \frac{1}{t}$, showing that the set is finite and hence is measurable.

The measurability of $\hat{F}$ follows now and it implies that of F since for each (x, y),

$$\begin{aligned} 2F(x, y) &= \lim_{k\to\infty} \{F(x + \frac{1}{k}, \overline{y + \frac{1}{k}}) + F(\overline{x + \frac{1}{k}-}, \overline{y + \frac{1}{k}-})\} \\ &= \lim_{k\to\infty} \hat{F}(x + \frac{1}{k}, y + \frac{1}{k}). \end{aligned}$$

For all positive integers p, parallel steps will lead us to establish $\mathbb{F}(\mathbf{x}) + \mathbb{F}(\mathbf{x}-)$ is measurable. The proof that this implies $\mathbb{F}$ is measurable follows the same lines as in the case $p = 2$. □

Lemma 1.5.5 (Part 2). *Let μ be a pm on $\mathscr{R}^p$. Then for any fixed $E \in \mathscr{R}^p$, $\mu(\mathbf{x} + E)$ is a real measurable function of $\mathbf{x}$.* (A reference to **(1.5.8)**(a) would be in order)

Proof. Let $\mathbb{F}$ be the generated *df*.

Let $\mathcal{E}$ denote the collection of all members of $\mathscr{R}^p$ for which the Lemma is true.

(i) Clearly $R^p \in \mathcal{E}$ since $\mu(\mathbf{x} + R^p) = \mu(R^p) = 1$.

(ii) Let E be a rectangle of the type $E = \times_1^p(u_i, v_i]$. Then $\mu(\mathbf{x}+E) = \Delta_{\mathbf{x}+\mathbf{u}}^{\mathbf{x}+\mathbf{v}}\mathbb{F}$. Each term of the defining sum being measurable, (ref. **(1.5.4)** Part 1) it follows that $\mu(. + E)$ is measurable. Thus intervals of the above type belong to $\mathcal{E}$.

(iii) Let (E_n), $E_n \in \mathcal{E}$ be a monotonic sequence in $\mathscr{R}^p$ with E for its limit. Hence $\mu(\mathbf{x}+E) = \mu(\lim_{n\to\infty}(\mathbf{x}+E_n)) = \lim_{n\to\infty}\mu(\mathbf{x}+E_n)$. Thus in this case $\mu(.+E)$ is the limit of a sequence of measurable functions and hence is measurable. This implies, in particular that all intervals belong to $\mathcal{E}$.

(iv) Since $\mu(\mathbf{x}+E') = \mu(\{\mathbf{x}+E\}') = 1-\mu(\mathbf{x}+E)$, it follows that $\mathcal{E}$ is closed under complementation.

(v) Let (E_n) be a disjoint sequence in $\mathcal{E}$ and E their union. Since for each n, $\mu(\mathbf{x}+E_n)$ is a measurable function of $\mathbf{x}$ and since $\mu(\mathbf{x}+E) = \sum_n \mu(\mathbf{x}+E_n)$, it follows that $\mathcal{E}$ is closed under countable union of disjoint members. This together with (iii) implies that all open subsets in R^p lie in $\mathcal{E}$. Then (iv) would imply that all closed sets would lie in $\mathcal{E}$. Later in **(1.9.2)** we will be showing that given $A \in \mathscr{R}^p$ we can find an increasing sequence (C_n) of closed sets such that $\mu(C_n) \uparrow \mu(A)$. This implies $\mu(\mathbf{x}+C_n) \uparrow \mu(\mathbf{x}+A)$. Hence $A \in \mathcal{E}$. Thus $\mathcal{E} = \mathscr{R}^p$. □

We write $\mathcal{L}$ for $\mathcal{L}^p$ the Lebesgue measure on $\mathscr{R}^p$.

Theorem 1.5.6. *Let μ be a pm on $\mathscr{R}^p$ with df $\mathbb{F}$. Define set function φ on*

$$\mathscr{R}^p : \varphi(E) = \int_{R^p} \mu(\boldsymbol{x} + E)\, d\mathcal{L}(\boldsymbol{x}),$$

where $\mathcal{L}$ is the Lebesgue measure in R^p. Then φ is a translation invariant non-atomic measure with finite values for compact sets.

Proof. We note that φ is well defined on $\mathscr{R}^p$ and is non-negative, with $\varphi(E)$ finite or infinite.

$$\varphi(\{\mathbf{a}\}) = \int_{R^p} \mu(\{\mathbf{x} + \mathbf{a}\})\, d\mathcal{L}(\mathbf{x}) = \int_{A-\mathbf{a}} \mu(\{\mathbf{x} + \mathbf{a}\})\, d\mathcal{L}(\mathbf{x})$$

where A is the set of all the atoms of μ. A is a countable set. (For the definition of an atom and a proof of this assertion, ref. **(1.6.2)** and **(1.6.3)**. $A - \mathbf{a}$ being a countable set, its Lebesgue measure is zero. Hence $\varphi(\{\mathbf{a}\}) = 0,\ \mathbf{a} \in R^p$. Thus φ is atom free.

Write $\varphi(\mathbf{E}) = \int_{R^p} g(\mathbf{x})\, d\mathcal{L}(\mathbf{x})$. Then $\varphi(\mathbf{a} + E) = \int_{R^p} g(\mathbf{a} + \mathbf{x})\, d\mathcal{L}(\mathbf{x})$

$= \int_{R^p} g(f(\mathbf{x}))\, d\mathcal{L}(\mathbf{x})$, where $f(\mathbf{x}) = \mathbf{a} + \mathbf{x}$

$= \lim_{\lambda\to\infty} \int_{[-\lambda,\ \lambda]^p} g(f(\mathbf{x}))\, d\mathcal{L}(\mathbf{x})$ (valid since the integrand is non-negative)

$= \lim_{\lambda\to\infty} \int_{\mathbf{a}+[-\lambda,\ \lambda]^p} g(\mathbf{x})\, d\mathcal{L}f^{-1}(\mathbf{x})$ by **(2.3.6)**

$= \lim_{\lambda\to\infty} \int_{\mathbf{a}+[-\lambda,\ \lambda]^p} g(\mathbf{x})\, d\mathcal{L}(\mathbf{x})$ since $f^{-1}(x) = \mathbf{x} - \mathbf{a}$

and since $\mathcal{L}$ is translation invariant

$= \int_{R^p} g(\mathbf{x})\, d\mathcal{L}(\mathbf{x}) = \varphi(E)$. Thus φ is translation invariant.

Let (E_n), $n \geq 1$ be a sequence of disjoint members of $\mathscr{R}^p$ and let $E = \cup_n E_n$. We have

$$\varphi(E) = \int_{R^p} \mu(\mathbf{x} + E)\, d\mathcal{L}(\mathbf{x}) = \int_{R^p} \mu(\mathbf{x} + \lim_{n\to\infty} \cup_{k=1}^n E_k)\, d\mathcal{L}(\mathbf{x})$$

$$= \int_{R^p} \lim_{n\to\infty} \mu(\mathbf{x} + \cup_{k=1}^n E_k)\, d\mathcal{L}(\mathbf{x})$$

$$= \lim_{n\to\infty} \int_{R^p} \mu(\mathbf{x} + \cup_{k=1}^n E_k)\, d\mathcal{L}(\mathbf{x}) \text{ (ref, } \mathbf{Q2}\text{)}$$

$$= \lim_{n\to\infty} \int_{R^p} \{\sum_{k=1}^n \mu(\mathbf{x} + E_k)\}\, d\mathcal{L}(\mathbf{x}) \text{ (since the sets are disjoint)}$$

$$= \lim_{n\to\infty} \sum_{k=1}^{n} \int_{R^p} \mu(\mathbf{x} + E_k)\, d\mathcal{L}(\mathbf{x})$$

$$= \sum_{k=1}^{\infty} \int_{R^p} \mu(\mathbf{x} + E_k)\, d\mathcal{L}(\mathbf{x}) = \sum_{k=1}^{\infty} \varphi(E_k).$$

Thus φ is countably additive.

For $a > 0$, let E_n stand for the rectangle $[na,\ (n+1)a]^p$. Now

$$\begin{aligned}
\varphi([0,\ a]^p) = \varphi(E_0) &= \int_{R^p} \mu(\mathbf{x} + E_0)\, d\mathcal{L}(\mathbf{x}) \\
&= \sum_{\substack{-\infty < n_i = n < \infty \\ 1 \le i \le p}} \int_{\substack{na \le x_i \le (n+1)a \\ 1 \le i \le p}} \mu(\mathbf{x} + E_0)\, d\mathcal{L}(\mathbf{x}) \\
&\le \sum_{\substack{-\infty < n_i = n < \infty \\ 1 \le i \le p}} \int_{\substack{na \le x_i \le (n+1)a \\ 1 \le i \le p}} \mu(E_{n_i})\, d\mathcal{L}(\mathbf{x}) \\
&= \sum_{\substack{-\infty < n_i = n < \infty \\ 1 \le i \le p}} \int_{\substack{na \le x_i \le (n+1)a \\ 1 \le i \le p}} \mu(E_{n_i}) a^p \\
&\le a^p \mu(R^p) = a^p.
\end{aligned}$$

This result, together with the translation invariant property of φ yields $\varphi([-a,\ a]^p) < \infty$, $a > 0$. Hence φ is finite for all compact subsets of R^p. □

The reader can pursue this line of argument and show that $\varphi(E) = c\mathcal{L}(E)$ where $c = \varphi([0,\ 1]^p)$.

For a simple and an alternative proof and for the evaluation of c, refer to item (iii) under *Some applications* under **Q**8.

Let us evaluate the integral in the following numerical case. Let $p = 1$ and let $\mu(\{n\}) = a_n$, $n \ge 1$, $\sum_n a_n = 1$. For this μ,

$$\begin{aligned}
\varphi([0,1]) &= \int_{-\infty}^{\infty} \mu([x,\ x+1])\, dx = \int_{0}^{\infty} \mu([x,\ x+1])\, dx \\
&= \int_{A} \mu([x,\ x+1])\, dx
\end{aligned}$$

where A' is the set of the integers $0,\ 1,\ 2, \ldots$. This step is valid since A', being a countable set, has Lebesgue measure 0. Thus

$$\varphi([0,\ 1]) = \sum_{n=0}^{\infty} \int_{(n,\ n+1)} \mu([x,\ x+1]) = \sum_{n=1}^{\infty} a_n = 1.$$

Corollary 1.5.7. *If $\mathcal{L}(E) = 0$ then by the theorem $\varphi(E) = 0$. This implies $\mathcal{L}(A) = 0$ where $A = \{x : \mu(x+E) > 0\}$.*

Theorem **(1.5.8)** discusses conditions for a *pm* on $\mathscr{R}^p$ to be absolutely continuous with respect to the Lebesgue measure. To prove **(1.5.8)** we need the following results.

Q3 (i) Radon-Nikodym derivative. (Theorem B, pp 128–9 **[H]**)
Let $\mu,\ \nu$ be σ-finite measures on a measurable space $(\Omega,\ \mathscr{A})$. We say ν is absolutely continuous with respect to μ ($\nu << \mu$) if $\nu(A) = 0$ whenever $\mu(A) = 0$; we say $\mu,\ \nu$ are mutually singular ($\mu \perp \nu$) or, each of the two measures is singular with respect to the other, if there exists $A \in \mathscr{A}$ such that $\mu(A') = 0 = \nu(A)$.

If ν is a finite measure and if $\nu << \mu$ then, corresponding to every $\varepsilon > 0$, there exists $\delta > 0$ such that $\nu(E) < \varepsilon$ for every $E \in \mathscr{A}$ for which $\mu(E) < \delta$. A measure μ defined on $\mathscr{R}^p$ is said to be absolutely continuous if it is so with respect to the p-dimensional Lebesgue measure.

If $\nu << \mu$ then there exists a finite valued non-negative measurable function f on Ω such that for every $E \in \mathscr{A}$, $\nu(E) = \int_E f(\omega)\ d\mu(\omega)$. f will be μ-integrable iff $\nu(\Omega) < \infty$. f is unique in the sense that if also $\nu(E) = \int_E g(\omega)\ d\mu(\omega)$ then $\mu\big(\omega; f(\omega) \neq g(\omega)\big) = 0$.

(ii) *A property of the space $L^p = L^p(R^k;\ \mathcal{L}^k)$ where $\mathcal{L}^k$ is the Lebesgue measure on $\mathscr{R}^k$, $1 \leq p \leq \infty$.* (Theorem under **30C**, p 118 **[L]**).
Given a real measurable function g on R^k, define $g_{\mathbf{s}}$, $\mathbf{s} \in R^k$ by setting $g_{\mathbf{s}}(\boldsymbol{t}) = g(\mathbf{s} + \boldsymbol{t})$ and note that $g_{\mathbf{s}} \in L^p$ iff $g \in L^p$. Then $g_{\mathbf{s}}$ as an element of L^p is a continuous function of $\mathbf{s}$.

(iii) *Lebesgue Decomposition* (Theorem C, p 134 **[H]**).
Let $\mu,\ \nu$ be two σ-finite measures on a measurable space $(\Omega,\ \mathscr{A})$. Then there exist two uniquely determined σ-finite measures $\nu_0,\ \nu_1$ such that $\nu = \nu_0 + \nu_1$ and such that $\nu_0 \perp \mu$ and $\nu_1 << \mu$.

Theorem 1.5.8. *Let μ, $\mathcal{L}^p$ denote respectively a pm and the Lebesgue measure on $\mathscr{R}^p$. Then*

(a) If $\mu << \mathcal{L}^p$, then $\mu(\mathbf{x} + E)$ is a continuous function of $\mathbf{x} \in R^p$ for every $E \in \mathscr{R}^p$ (compare **(1.5.5)**) *and*
(b) The two conditions that μ has compact support and $\mu(\mathbf{x} + E)$ is a continuous function of $\mathbf{x}$ for every $E \in \mathscr{R}^p$ together imply $\mu << \mathcal{L}^p$.

Proof. (a) Let $\mu << \mathcal{L}^p$. Then by **Q3**(i) above there exists a real finite valued $\mathscr{R}^p$ measurable non-negative function f such that for every $E \in \mathscr{R}^p$, $\mu(E) = \int_E f(\mathbf{t})d\mathcal{L}^p(\mathbf{t})$. Since μ is a pm, f is integrable over R^p. Hence by **Q3**(ii), $f_{\mathbf{x}}$, $\mathbf{x} \in R^p$ as an element of L^1 is a continuous function of $\mathbf{x}$ where $f_{\mathbf{x}}(\mathbf{t}) = f(\mathbf{t} + \mathbf{x})$.

$$\text{i.e.,} \quad \|f_{\mathbf{x}} - f_{\mathbf{y}}\| \to 0 \text{ as } \|\mathbf{x} - \mathbf{y}\| \to 0.$$

i.e., given $\varepsilon > 0$, a δ can be found such that

$$\int_{R^p} |f(\mathbf{x} + \mathbf{t}) - f(\mathbf{y} + \mathbf{t})| \, d\mathcal{L}(\mathbf{t}) < \varepsilon \text{ whenever } \|\mathbf{x} - \mathbf{y}\| < \delta.$$

The continuity of $\mu(\mathbf{x} + E)$ follows now since

$$\mu(\mathbf{x} + E) = \int_{\mathbf{x}+E} f(\mathbf{t})d\mathcal{L}(\mathbf{t}) = \int_E f(\mathbf{x} + \mathbf{t}) \, d\mathcal{L}^p(\mathbf{t})$$
$$= \int_{R^p} \chi_E(\mathbf{t}) f_{\mathbf{x}}(\mathbf{t}) \, d\mathcal{L}^p(\mathbf{t})$$

and hence

$$|\mu(\mathbf{x} + E) - \mu(\mathbf{y} + E)| \leq \int_{R^p} |f(\mathbf{x} + \mathbf{t}) - f(\mathbf{y} + \mathbf{t})| \, d\mathcal{L}^p(\mathbf{t}) < \varepsilon$$

whenever $\|\mathbf{x} - \mathbf{y}\| < \delta$.
(b) Let the conditions hold. Let $\mu = \mu_1 + \mu_2$ be the Lebesgue decomposition of μ with $\mu_1 \perp \mathcal{L}^p$ and $\mu_2 << \mathcal{L}^p$. We note μ_1, μ_2 are finite measures with compact supports. We claim $\mu_1 \equiv 0$. If not, let $\mu_1(E) > 0$ for some $E \in \mathscr{R}^p$ with $\mathcal{L}^p(E) = 0$. (Such an E exists since $\mu_1 \perp \mathcal{L}^p$). By hypothesis $\mu_1(x + E) + \mu_2(x + E)$ is a continuous function of x. Since $\mu_2 << \mathcal{L}^p$, it follows by the forward result that $\mu_2(x + E)$ is continuous. Hence $\mu_1(x + E)$ is continuous, so $G = \{x : \mu_1(x + E) > 0\}$ is an open set. G is non-null since $0 \in G$. Hence $\mathcal{L}^p(G) > 0$. But by **(1.5.7)** applied to the measure μ_1, $\mathcal{L}^p(G) = 0$. This contradiction proves that $\mu_1 \equiv 0$. That $\mu = \mu_2 << \mathcal{L}^p$ follows now. □

Let μ be a *pm* on $\mathscr{R}^p$ and let $\mathbb{F}$ be its associated *df*. As earlier, $\mathcal{L}^p$ denotes the Lebesgue measure. The following theorem ensures that the absolute continuity of $\mathbb{F}$ (according to the definition at **(1.5.3)**) and the absolute continuity of μ (as defined in **Q**3) imply each other.

Theorem 1.5.9. *A necessary and sufficient condition that* $\mathbb{F}$ *be absolutely continuous is that* $\mu << \mathcal{L}$.

Proof. Suppose $\mu << \mathcal{L}^p$. Hence (ref. **Q**3) to every $\varepsilon > 0$, there corresponds a $\delta > 0$ such that $E \in \mathscr{R}^p$, $\mathcal{L}^p(E) < \delta$ imply $\mu(E) < \varepsilon$. If now $\{(\mathbf{a}_r,\ \mathbf{b}_r],\ 1 \leq r \leq n\}$ is a disjoint class of bounded intervals for which

$$\mathcal{L}^p\Big(\bigcup_1^n(\mathbf{a}_r,\ \mathbf{b}_r]\Big) = \sum_{r=1}^n \mathcal{L}^p(\mathbf{a}_r,\ \mathbf{b}_r] < \delta,$$

then

$$\sum_{r=1}^n \Delta_{\mathbf{a}_r}^{\mathbf{b}_r}\mathbb{F} = \sum_{r=1}^n \mu(\mathbf{a}_r,\ \mathbf{b}_r] = \mu\Big(\bigcup_{r=1}^n(\mathbf{a}_r,\ \mathbf{b}_r]\Big) < \varepsilon$$

implying the absolute continuity of $\mathbb{F}$.

Conversely suppose $\mathbb{F}$ is absolutely continuous. Continue the notation above. Given $\varepsilon > 0$, we can find a $\delta > 0$ such that

$$\sum_{r=1}^n \mathcal{L}^p(\mathbf{a}_r,\ \mathbf{b}_r] < \delta \text{ implies } \sum_{r=1}^n \Delta_{\mathbf{a}_r}^{\mathbf{b}_r}\mathbb{F} < \varepsilon.$$

If $E \in \mathscr{R}^p$ with $\mathcal{L}^p(E) = 0$, then there exists a disjoint sequence $\{(\mathbf{a}_r,\ \mathbf{b}_r]\}$ of semiclosed intervals such that

$$E \subset \bigcup_{r=1}^n(\mathbf{a}_r,\ \mathbf{b}_r] \text{ and } \sum_{r=1}^n \mathcal{L}(\mathbf{a}_r,\ \mathbf{b}_r] < \varepsilon.$$

Hence

$$\sum_{r=1}^n \Delta_{\mathbf{a}_r}^{\mathbf{b}_r}\mathbb{F} < \varepsilon.$$

It follows from this that

$$\mu(E) \leq \sum_{r=1}^n \mu(\mathbf{a}_r,\ \mathbf{b}_r] = \sum_{r=1}^n \Delta_{\mathbf{a}_r}^{\mathbf{b}_r}\mathbb{F} < \varepsilon.$$

Since $\varepsilon > 0$ is arbitrary, we conclude $\mu(E) = 0$. □

Remark 1.5.10. If *df* $\mathbb{F}$ in R^p is absolutely continuous, then so will be each of the marginals but the converse is not true.

Proof. Take $p = 2$ to avoid notation clutter. The absolute continuity of $\mathbb{F}$ implies that there exists an integrable function f such that

$$\mathbb{F}(x, y) = \int_{-\infty}^{x} \int_{-\infty}^{y} f(t, u)\, dt\, du.$$

Hence

$$F_1(x) = \int_{-\infty}^{x} g(t)\, dt \text{ where } g(t) = \int_{-\infty}^{\infty} f(t, u)\, du$$

which is finite valued. This shows that F_1, and similarly F_2, is absolutely continuous.

The following simple example shows that the converse is not true. Let X be any rv whose *df* G is absolutely continuous. Then the bivariate variable $(X, Y) = (X, X)$ has *df* $\mathbb{F}(x, y) = G(\min(x, y))$. Any linear combination $aX + bY = (a+b)X$ of the component variables has an absolutely continuous *df*. But $\mathbb{F}$ is not absolutely continuous. For, if μ is the pm generated by $\mathbb{F}$ and if $D = \{(x, x) : x \in R\}$, then trivially $\mu(D) = 1$. Since $\mathcal{L}^{(2)}(D) = 0$, where $\mathcal{L}^{(2)}$ is the Lebesgue measure in R^2, it follows that μ is singular with respect to $\mathcal{L}^{(2)}$ $(\mu \perp \mathcal{L}^{(2)})$. □

1.6 Atoms of a probability measure (pm)

Definition 1.6.1. $A \in \mathscr{A}$ is said to be an atom of a *pm* μ or of the probability space $(\Omega, \mathscr{A}, \mu)$ if $\mu(A) > 0$ and for every $B \in \mathscr{A}, B \subset A$, either $\mu(B) = \mu(A)$ or $\mu(B) = 0$.

The definition implies that if $\mu(A) > 0$ and if A contains no measurable non-empty subset, then A is an atom. If A is an atom and if $B \in \mathscr{A}$ with $\mu(B) = 0$, then $A \cup B$ is an atom. Measurable single point sets with positive μ-measure will be atoms. If A is an atom, then $\mu(A)$ is called the *saltus* of μ at A. Two atoms A, B will be considered the same if $\mu(A \cap B) > 0$ and will be considered distinct if $\mu(A \cap B) = 0$. Clearly the union of two distinct atoms can not be an atom. If A, B are distinct atoms, then $\mu(A \cup B) = \mu(A) + \mu(B)$. If a pm μ has no atoms, it will be called atom-free.

Theorem 1.6.2. *The distinct atoms of μ are at most countable but can form a dense subset.*

Proof. Let $\mathcal{D}_k$ be the collection of all the distinct atoms each with saltus $\geq \frac{1}{k}$. Then the set of all the atoms of μ is $\mathcal{D} = \cup_{k=1}^{\infty} \mathcal{D}_k$. Let $A_1, A_2, \ldots, A_n$ be any n members of $\mathcal{D}_k$. We have

$$1 \geq \mu(\cup_1^n A_j) = \sum_1^n \mu(A_j) \geq \frac{n}{k}.$$

Hence $n \leq k$. i.e. each $\mathcal{D}_k$ can have utmost k members. Hence $\mathcal{D}$ is countable.

Let μ be the *pm* on $\mathscr{R}$ defined through setting $\mu(\{r_j\}) = \frac{1}{2^j}, j = 1, 2, \ldots$ where (r_j) is a listing of the rationals in R. This example shows that the set of all the atoms can be dense. □

Theorem 1.6.3. *Let (M, d) be a separable metric space, and μ a pm on $\mathfrak{m}$. Then to each atom A, there corresponds a unique element $a \in A$ such that $\mu(\{a\}) = \mu(A)$. Conversely, if every single point set in M has zero μ-measure then the space is atom free.*

[In that case μ is said to be non-atomic.]

Proof. Let A_0 be an atom. Let $\Lambda = \{a_1, a_2, \ldots\}$ be a separability set for M. For $j = 1, 2, \ldots$, denote by $S_{j,n}$ the closed sphere with centre a_j and radius $\frac{1}{2n}$. Since $\bar{\Lambda} = M$, we have

$$M = \cup_{j=1}^{\infty} S_{j,n}, \quad n = 1, 2, \ldots.$$

$A_0 = \cup_{j=1}^{\infty}(A_0 \cap S_{j,1})$ implies that $0 < \mu(A_0) \leq \sum_{j=1}^{\infty} \mu(A_0 \cap S_{j,1})$. There necessarily therefore exists at least one j such that $\mu(A_1) > 0$ where $A_1 = A_0 \cap S_{j,1}$. Since $A_1 \subseteq A_0$, since $\mu(A_1) > 0$ and since A_0 is an atom, it follows that $\mu(A_1) = \mu(A_0)$. We note that $\text{diam}(A_1) \leq 1$. Now start with $A_1 = \cup_{j=1}^{\infty}(A_1 \cap S_{j,2})$ and repeat the argument to arrive at $A_2 \subseteq A_1$ with $\mu(A_2) = \mu(A_1) = \mu(A_0)$ and with $\text{diam}(A_2) \leq \frac{1}{2}$. This process can be kept up indefinitely to arrive at a decreasing sequence $(A_n, n \geq 1)$ of subsets of A_0 such that $\mu(A_n) = \mu(A_0)$ and $\text{diam}(A_n) \leq \frac{1}{n}, n \geq 1$. If $A = \cap_{n=1}^{\infty} A_n$, then it follows that $\mu(A) = \mu(A_0) > 0$ (so A is non-null) and $\text{diam}(A) = 0$. Hence A consists of a single point. We have found a point $x_0 \in A_0$ such that $\mu(\{x_0\}) = \mu(A_0)$. The uniqueness of the point follows immediately from the fact that A_0 is an atom.

To establish the converse, assume that for every $x \in M$, $\mu(\{x\}) = 0$. If possible let A be an atom. Hence by the forward part, there exists an $x_0 \in A$ such that $\mu(\{x_0\}) = \mu(A) > 0$, a contradiction. □

Remark 1.6.4. In a non-metric space, it is possible single point sets are not measurable. Consider $\Omega = \{1, 2, 3\}$. Endow it with the σ-field $\{A, A^{'}, \Omega, \phi\}$ where $A = \{1, 2\}$, and let $P(A) = P(A') = \frac{1}{2}$.

1.7 Discontinuities of distribution functions

1.7.1 Marginals of dfs / pms.

We recall that the distribution function (*df*) $\mathbb{F}$ associated with a pm μ in R^p was defined thus:

$$\mathbb{F}(x_1,\ x_2, \ldots,\ x_p) = \mu\left((-\infty,\ x_1] \times (-\infty,\ x_2] \times \cdots \times (-\infty,\ x_p]\right).$$

Let π_j denote the projection mapping: $\pi_j(x_1,\ x_2, \ldots,\ x_p) = x_j$. The pm $\mu_{p,j} = \mu\pi_j^{-1}$ is called j^{th} component or the jth marginal of μ. The *df* associated with $\mu_{p,j}$ is $F_{p,j}$:

$$F_{p,j}(x) = \lim_{\substack{x_r \to \infty \\ 1 \le r \le p,\ r \ne j}} \mathbb{F}(x_1, \ldots,\ x_{j-1},\ x,\ x_{j+1}, \ldots,\ x_p).$$

1.7.2 Properties of dfs

(i) Every *df* $\mathbb{F}$ is monotonic non-decreasing.
To see this let $\mathbf{a} = (a_1,\ a_2, \ldots,\ a_p)$, $\mathbf{b} = (b_1,\ b_2, \ldots,\ b_p)$ and $\mathbf{a} \le \mathbf{b}$. The claim follows since

$$\mathbb{F}(\mathbf{b}) - \mathbb{F}(\mathbf{a}) = \mu\left(\times_1^p(-\infty,\ b_i] \sim \times_1^p(-\infty,\ a_i]\right) \ge 0.$$

(ii) Every *df* $\mathbb{F}$ is right continuous and has left limits.
Because of the monotonicity of $\mathbb{F}$ it is enough to show that $\alpha_n \to \alpha$ where $\alpha_n = \mathbb{F}(a_1 + \frac{1}{n}, \ldots,\ a_p + \frac{1}{n})$ and $\alpha = \mathbb{F}(a_1, \ldots,\ a_p)$ for every choice of $(a_1,\ a_2, \ldots,\ a_p) \in R^p$. This is true since $\alpha_n = \mu(A_n)$, $\alpha = \mu(A)$ and $A_n \downarrow A$ where $A_n = \times_1^p(-\infty,\ a_i + \frac{1}{n}]$ and $A = \times_1^p(-\infty,\ a_i]$.

The left limits exist at all points because of monotonicity of $\mathbb{F}$ and the left limit of $\mathbb{F}$ at $\mathbf{a}$ will be denoted by

$$\mathbb{F}(\mathbf{a}-) \quad \text{or} \quad \mathbb{F}(a_1-,\ a_2-, \ldots,\ a_p-).$$

(iii) Since $\mathbb{F}$ is monotonic non-decreasing and since

$$(-\infty,\ x_1] \times (-\infty,\ x_2] \times \cdots \times (-\infty,\ x_p] \to R^p$$

as $\min\limits_{1\le i\le p} x_i \to \infty$, it follows that

$$\lim_{\substack{\min x_i \to \infty \\ 1\le i\le p}} \mathbb{F}(x_1, \ldots, x_p) = 1.$$

Similarly since $\mathbb{F}$ is monotonic non-decreasing and since

$$(-\infty, x_1] \times (-\infty, x_2] \times \cdots \times (-\infty, x_p] \to \phi \text{ as } \min_{1\le i\le p} x_i \to -\infty,$$

it follows that

$$\lim_{\substack{\min x_i \to -\infty \\ 1\le i\le p}} \mathbb{F}(x_1, \ldots, x_p) = 0.$$

Let $\mathbf{a} = (a_1, a_2, \ldots, a_p)$, $\mathbf{b} = (b_1, b_2, \ldots, b_p)$ and $\mathbf{a} \le \mathbf{b}$. We note that $\Delta_{\mathbf{a}}^{\mathbf{b}}\mathbb{F} = \mu(\times_1^p(a_i < x_i \le b_i])$. Thus

$$\mu(\{\mathbf{b}\}) = \lim_{\substack{\mathbf{a}\to\mathbf{b} \\ \mathbf{a}<\mathbf{b}}} \Delta_{\mathbf{a}}^{\mathbf{b}}\mathbb{F}$$

which we denote by $\Delta_{\mathbf{b}\text{-}}^{\mathbf{b}}\mathbb{F}$. Written explicitly when $p = 2$, we have

$$\mu(\{\mathbf{b}\}) = \mathbb{F}(b_1, b_2) - \mathbb{F}(b_1-, b_2) - \mathbb{F}(b_1, b_2-) + \mathbb{F}(b_1-, b_2-).$$

(iv) $\mathbf{b}$ is a point of continuity of $\mathbb{F}$ *iff* if $\mathbb{F}(\mathbf{b}) - \mathbb{F}(\mathbf{b}-) = 0$. Since $\Delta_{\mathbf{b}-}^{\mathbf{b}} \le \mathbb{F}(\mathbf{b}) - \mathbb{F}(\mathbf{b}-)$, it follows that an atom is necessarily a point of discontinuity. But the converse is not true, *except* when $p = 1$. That a point of discontinuity need not be an atom when $p \ge 2$ is made clear by the following example. Let $p = 2$. Let $E = \{(x, 0) : 0 \le x \le 1\}$ and $F = \{(x, 1) : 0 \le x \le 1\}$. Let $\mathcal{L}$ denote linear Lebesgue measure. Define μ on $\mathscr{R}^2$ as follows. For $A \in \mathscr{R}^2$, $\mu(A) = \frac{1}{2}\{\mathcal{L}(A \cap E) + \mathcal{L}(A \cap F)\}$. So defined, μ is easily verified to be a *pm*. Consider the corresponding *df* $\mathbb{F}$:

$$\mathbb{F}(x, y) = \mu\{(-\infty, x] \times (-\infty, y]\} = \begin{cases} 0 & \text{if } x < 0 \text{ or } y < 0 \\ \frac{x}{2} & \text{if } 0 \le x < 1 \text{ and } 0 \le y < 1 \\ x & \text{if } 0 \le x < 1 \text{ and } y \ge 1 \\ \frac{1}{2} & \text{if } x \ge 1 \text{ and } 0 \le y < 1 \\ 1 & \text{if } x \ge 1 \text{ and } y \ge 1. \end{cases}$$

Every point in E and every point in F and no other point is a discontinuity point of $\mathbb{F}$. Since for each $(x_0, y_0) \in E \cup F$, $\mu(\{(x_0, y_0)\}) = 0$, it follows that $\mathbb{F}$ does not have an atom. In the case $p = 1$, the converse is easily seen to be true.

(v) If $\mathbf{b} = (b_1, b_2, \ldots, b_p)$ is an atom of $\mathbb{F}$, then b_k is an atom of the marginal F_k, $1 \le k \le p$. The converse is false.
Proof. $\mu_{p,j}(\{b_j\}) = \mu\pi_j^{-1}(\{b_j\}) \ge \mu(\{\mathbf{b}\}) > 0$. Hence b_j is an atom of $F_{p,j}$.

The following example shows that the converse to the above is not true. Let $p = 2$, $\mu(\{(0, 0)\}) = \mu(\{(0, 1)\}) = \mu(\{(1, 0)\}) = \frac{1}{3}$. This implies $\mu_{2,1}(\{1\}) = \frac{1}{3}$; $\mu_{2,2}(\{1\}) = \frac{1}{3}$. Hence 1 is an atom of $\mu_{2,1}$ and of $\mu_{2,2}$. But $(1, 1)$ is not an atom of $\mathbb{F}$ since $\mu(\{(1, 1)\}) = 0$.

Note. It is easily verified $(1, 1)$ is a discontinuity point of $\mathbb{F}$.

Let us examine further the example in (iv). The marginal distributions are:

$$F_1(x) = \begin{cases} 0 & \text{if } x < 0; \\ x & \text{if } 0 \le x < 1; \\ 1 & \text{if } x \ge 1. \end{cases}$$

$$F_2(y) = \begin{cases} 0 & \text{if } y < 0; \\ \frac{1}{2} & \text{if } 0 \le y < 1; \\ 1 & \text{if } y \ge 1. \end{cases}$$

Thus F_1 is a continuous *df* while F_2 has atoms at 0 and 1 with saltus $\frac{1}{2}$ each.

This example raises the following questions for *df*s in R^p, $p \ge 2$. Is it correct to say (i) that discontinuities can be uncountable? (ii) that the p-dimensional Lebesgue measure of the set of discontinuities is 0? (iii) that if $(x_1, x_2, \ldots, x_p)$ is a discontinuity point of $\mathbb{F}$, then for some i, $1 \le i \le p$, the one dimensional marginal F_i will have a discontinuity at x_i? We will establish that the answer is *yes* to each of these questions.

(vi) For a *df* $\mathbb{F}$, let $C_{\mathbb{F}}$ denote the set of all of its continuity points. Then $\widehat{C_{\mathbb{F}}} = \times_{i=1}^{p} C_{F_{p,i}} \subseteq C_{\mathbb{F}}$. The inclusion can be strict.
Proof. Let b_i be a point of continuity of the marginal $F_{p,i}$, $1 \le i \le p$. i.e., $\mathbf{b} = (b_1, b_2, \ldots, b_p) \in \widehat{C_{\mathbb{F}}}$. Let $a_i \le b_i$. Define for $i = 1, 2, \ldots, p$, $A_i = \pi_i^{-1}(-\infty, a_i]$ and $B_i = \pi_i^{-1}(-\infty, b_i]$. Then

$$\begin{aligned} 0 &\le \mathbb{F}(b_1, b_2, \ldots, b_p) - \mathbb{F}(a_1, a_2, \ldots, a_p) \\ &= \mu(\bigcap_{i=1}^{p} B_i) - \mu(\bigcap_{i=1}^{p} A_i) = \mu\{(\bigcap_{i=1}^{p} B_i) \sim (\bigcap_{i=1}^{p} A_i)\} \\ &= \mu(\bigcup_{r=1}^{p}(\bigcap_{i=1}^{p} B_i \cap A_r')) \le \sum_{r=1}^{p} \mu(\bigcap_{i=1}^{p} B_i \cap A_r') \\ &\le \sum_{r=1}^{p} \mu(B_r \cap A_r') = \sum_{r=1}^{p} \{F_{p,r}(b_r) - F_{p,r}(a_r)\}. \end{aligned}$$

Taking limit as $a_i \to b_i$, we see that $\mathbf{b} \in C_{\mathbb{F}}$.

To appreciate that strict inclusion is possible, consider the following example. Let $E_1 = \{0\} \times [0,1]$ and $E_2 = \{1\} \times [0,1]$. Let $A \in \mathcal{R}^2$. Note that $A \cap E_1$ and $A \cap E_2$ can be written uniquely as $\{0\} \times B_1$and $\{1\} \times B_2$. Denote by L the linear Lebesgue measure and define $\mu(A) = \frac{1}{2}\{L(B_1) + L(B_2)\}$. Note that, so defined, μ is a *pm*. We have,

$$\mathbb{F}(x,y) = \mu\{(-\infty, x] \times (-\infty, y]\}$$
$$= \begin{cases} 0 & \text{if } x < 0 \text{ or } y < 0; \\ \frac{1}{2}\min(y,1) & \text{if } 0 \leq x < 1 \text{ and } y \geq 0; \\ \min(y,1) & \text{if } x \geq 1 \text{ and } y \geq 0. \end{cases}$$

$$F_{2,1}(x) = \begin{cases} 0 & \text{if } x < 0; \\ \frac{1}{2} & \text{if } 0 \leq x < 1; \\ 1 & \text{if } x \geq 1. \end{cases}$$

$$F_{2,2}(y) = \begin{cases} 0 & \text{if } y \leq 0; \\ y & \text{if } 0 \leq y \leq 1; \\ 1 & \text{if } y \geq 1. \end{cases}$$

$C_{F_{2,1}} = (-\infty, 0) \cup (0,1) \cup (1,\infty)$; $C_{F_{2,2}} = (-\infty, \infty)$. $(0,0)$ is a continuity point for $\mathbb{F}$. But this point does not belong to $C_{F_{2,1}} \times C_{F_{2,2}}$. Thus in this case $C_{F_{2,1}} \times C_{F_{2,2}} \subsetneqq C_{\mathbb{F}}$. □

(vii) Each $C_{F_{p,i}}$, being the complement of a countable set, is a dense set in R. Being the Cartesian product of the $C_{F_{p,i}}$s, $\widehat{C_{\mathbb{F}}}$ is dense in R^p.

(viii) If all the $F_{p,i}$s are continuous everywhere, then $\mathbb{F}$ is continuous everywhere.

(ix) If $(u_1, u_2, \ldots, u_p)$ is a discontinuity point of $\mathbb{F}$, then u_i is a discontinuity point of $F_{p,i}$ for at least one i, $1 \leq i \leq p$.

(x) Suppose x is a discontinuity point of F_i. Then there exist $y_1, y_2, \ldots$, $y_{i-1}, y_{i+1}, \ldots, y_p$ such that $(y_1, y_2, \ldots, y_{i-1}, x, y_{i+1}, \ldots, y_p)$ is a discontinuity point of $\mathbb{F}$.

Proof of (x). There is no loss in generality if we take $i = 1$. If possible let $(x, u_2, u_3, \ldots, u_p)$ be a point of continuity of $\mathbb{F}$ for every choice of $(u_2, u_3, \ldots, u_p)$. Let $\varepsilon > 0$ be given. Since

$$F_{p,1}(x) = \lim_{\min\{x_r : 2 \leq r \leq p\} \to \infty} \mathbb{F}(x, x_2, \ldots, x_j, \ldots, x_p),$$

we can find $(u_2, u_3, \ldots, u_p)$ such that

$$0 \leq F_{p,1}(x) - \mathbb{F}(x, u_2, u_3, \ldots, u_p) < \varepsilon.$$

By reason of the assumed continuity of $\mathbb{F}$ at $(x, u_2, u_3, \ldots, u_p)$, a δ can be found such that $\mathbb{F}(x, u_2, u_3, \ldots, u_p) - \mathbb{F}(x - \delta, u_2, u_3, \ldots, u_p) < \varepsilon$. This leads to:

$$\begin{aligned} 0 < \alpha =& F_{p,1}(x) - F_{p,1}(x-) \leq F_{p,1}(x) - F_{p,1}(x - \delta) \\ &\leq \varepsilon + \mathbb{F}(x, u_2, u_3, \ldots, u_p) - \mathbb{F}(x - \delta, u_2, u_3, \ldots, u_p) \\ &< \varepsilon + \varepsilon, \end{aligned}$$

a contradiction since ε is arbitrary. □

(xi) A consequence of the last result is : If $\mathbb{F}$ is continuous everywhere then each F_i is continuous everywhere. Compare with (viii) and conclude that in this case $\widehat{C_{\mathbb{F}}} = C_{\mathbb{F}}$.

(xii) Let μ be a pm in R^p, $\mathbb{F}$ the df associated with it and $\mu_{p,j}$, $F_{p,j}$, $1 \leq j \leq p$ the marginals. For $\lambda > 0$, denote by I_λ the interval$[-\lambda, \ \lambda]$. Let $A_\lambda = I_\lambda^p$. Given $\varepsilon > 0$, there exists a $\lambda > 0$ such that $\mu(A_\lambda) > 1 - \varepsilon$.

Proof. Since $I_\lambda \uparrow R$ as $\lambda \uparrow \infty$, we can, given $\varepsilon > 0$, find a λ such that $\mu_{p,j}(I_\lambda) > 1 - \frac{\varepsilon}{p}$. As noted in (vii), the set of continuity points of $F_{p,j}$ is the complement of a countable set. Hence $-\lambda, \ \lambda$ can be chosen as continuity points of $F_{p,j}$. By (vi), it follows that the vertices of A_λ are continuity points of $\mathbb{F}$. Further since I_λ is a rectangle,

$$A_\lambda = \bigcap_1^p \pi_j^{-1}(I_\lambda).$$

We have

$$\mu(A_\lambda) = \mu\Big(\bigcap_1^p \pi_j^{-1}(I_\lambda)\Big).$$

This leads to

$$\mu(A_\lambda') \leq \sum_1^p \mu\pi_j^{-1}(I_\lambda') = \sum_1^p \mu_{p,j}(I_\lambda') \leq \sum_1^p \frac{\varepsilon}{p},$$

This gives

$$\Delta_{-\Lambda}^{\Lambda}(\mathbb{F}) = \mu(A_\lambda) > 1 - \varepsilon$$

where $\Lambda = (1, 1, \ldots, 1)\lambda$. □

(xiii) Let $\mathfrak{L}^p$ denote the Lebesgue measure in R^p. Then the set of all the discontinuities of $\mathbb{F}$ has $\mathfrak{L}^p$ measure 0.

Proof. If $(u_1, u_2, \ldots, u_p)$ is a discontinuity point of $\mathbb{F}$, then it is true by (ix) that u_i is a discontinuity point of F_i for some i. This means that the set D of all the discontinuity points of $\mathbb{F}$ is contained in the set $\cup_{i=1}^{p} A_i$, where A_i is the union of the countable number of the hyperplanes $x_i = r_{i,j}$. Here $r_{i,j},\ j = 1,\ 2, \ldots$ are the discontinuity points of F_i. Since the $\mathfrak{L}^p$ measure of each of these hyperplanes is zero, the claim follows. □

1.7.3 Remarks

(i) If a real function $\mathbb{F}$ defined on R^p satisfies the condition $\Delta_{\mathbf{a}}^{\mathbf{b}}\mathbb{F} \geq 0$ for every pair of vectors $\mathbf{a} \leq \mathbf{b}$, if it is right continuous and if $\mathbb{F}$ possesses the two properties detailed in **(1.7.2)**(iii)), then it can be shown that $\mathbb{F}$ determines a unique pm μ on $\mathscr{R}^p$ such that the df of μ is $\mathbb{F}$. Hence the function $\mathbb{F}$ is justifiably called a df.

Consequently, if $\mathbb{F}_1,\ \mathbb{F}_2$ are dfs in R^p, then for every $\alpha,\ \beta$ with $0 \leq \alpha,\ \beta \leq 1,\ \alpha + \beta = 1,\ \alpha\mathbb{F}_1 + \beta\mathbb{F}_2$ is a df.

Let $S \subset R$ be a countable dense subset. Suppose a function $\mathbb{F}$ defined on S^p satisfies the following conditions:

$\mathbb{F}$ is monotonic non-decreasing

$\mathbb{F}$ is right continuous

$$\lim_{\min\{s_1, s_2, \ldots, s_p\} \to \infty} \mathbb{F}(s_1, s_2, \ldots,\ s_p) = 1$$

$$\lim_{\min\{s_1, s_2, \ldots, s_p\} \to -\infty} \mathbb{F}(s_1, s_2, \ldots, s_p) = 0.$$

$\Delta_{\mathbf{a}}^{\mathbf{b}}\mathbb{F} \geq 0,\ \mathbf{a} \leq \mathbf{b},\ \mathbf{a}, \mathbf{b} \in S^p$.

Then $\mathbb{F}$ can be extended to all of R^p so as to be a df.

The proof is straightforward. The details are omitted.

With every pm μ in a measurable space $(\mathbf{Y},\ \mathfrak{Y})$ a rE $\mathbf{X}$ can always be associated such that μ is the measure induced on $\mathfrak{Y}$ by $\mathbf{X}$. This is done by taking the basic probability space $(\Omega,\ \mathscr{A},\ P) = (\mathbf{Y},\ \mathfrak{Y},\ \mu)$ and $\mathbf{X}(y) = y,\ y \in \mathbf{Y}$.

(ii) Let $F_1,\ F_2, \ldots,\ F_p$ be dfs in R. Define

$$\mathbb{F}(x_1,\ x_2, \ldots,\ x_p) = \times_1^p F_i(x_i).$$

Then $\mathbb{F}$ is a df in R^p.

Proof. Satisfaction of the other conditions being easily verified, we note

$$\Delta_{\mathbf{a}}^{\mathbf{b}}\mathbb{F} = \times_1^p\{F_i(b_i) - F_i(a_i)\} \geq 0.$$

1.8 Carrier of a pm

By the carrier or the support or the spectrum of a measure we understand the minimal closed set which carries the entire mass of the measure. The next three theorems deal with the existence of the carrier of a measure in three different settings.

Definition 1.8.1. A measure μ on the Borel σ-field $\mathscr{A}$ of a Hausdorff topological space Ω is said to possess property $\mathcal{R}$ if $\mu(K) < \infty$ for every compact set K and if for every $A \in \mathscr{A}$, $\mu(A) = \sup\{\mu(K) : K \subseteq A,\ K \text{ compact}\}$.

Theorem 1.8.2. *Let Ω be a compact Hausdorff space and $\mathcal{A}$ be its Borel σ-field. Let μ be an arbitrary measure on $\mathcal{A}$ possessing property $\mathcal{R}$. Then there exists a unique closed set A_μ (the carrier or the support of μ) such that $\mu(\Omega \sim A_\mu) = 0$ and $\mu(U) > 0$ for every non-null relatively open subset U of A_μ.*

Proof. We note that $\mu(\Omega) < \infty$ since μ has property $\mathcal{R}$ and Ω is a compact set.

Let $\mathcal{F}$ be the family of closed sets F with $\mu(F) = \mu(\Omega)$ or, equivalently, $\mu(\Omega \sim F) = 0$. $\mathcal{F}$ is not empty since $\Omega \in \mathcal{F}$. It is obvious that $\mathcal{F}$ has the finite intersection property, since

$$\mu(\Omega \sim (F_1 \cap F_2)) \leq \mu(\Omega \sim F_1) + \mu(\Omega \sim F_2) = 0.$$

Since Ω is a compact set, the finite intersection property implies that $A = \cap\{F : F \in \mathcal{F}\}$ is not empty (ref. Theorem 1, p 136 **[K]**). We will show that this A is the required A_μ. We claim $\mu(A) = \mu(\Omega)$. If $\mu(A) \neq \mu(\Omega)$, then $\mu(A') > 0$. Since μ has property $\mathcal{R}$, there exists a compact set $K \subset A'$ such that $\mu(K) > 0$. Consider the collection $\mathcal{F}^*$ of closed sets : $\mathcal{F}^* = \{K \cup F : F \in \mathcal{F}\}$. For every $F \in \mathcal{F}$, $\mu(K \cup F) = \mu(\Omega)$ since for every $B \in \mathcal{F}$, $\mu(B) = \mu(\Omega)$. This implies that $K \cup F \in \mathcal{F}$. The inclusion $\mathcal{F}^* = \{K \cup F : F \in \mathcal{F}\} \subset \mathcal{F}$, implies $A \cup K \subseteq A$. Hence $A \cup K = A$. This is not possible since $K \neq \phi$ and $A \cap K = \phi$. Hence $\mu(A) = \mu(\Omega)$.

The existence of a non-null relatively open subset G of A with $\mu(G) = 0$ would imply that the closed proper subset $A \sim G$ of A is such that $\mu(A \sim G) = \mu(A) = \mu(\Omega)$. i.e. $A \sim G \in \mathcal{F}$, $A \sim G \subsetneqq A$, a contradiction of the fact that A is the intersection of all the members of $\mathcal{F}$.

Finally, A is unique. For if possible let B be a second closed set with the specified properties. Since $A \neq B$, at least one of the two sets $A \sim B$ and $B \sim A$ is non-null. Suppose $A \sim B \neq \phi$. As a non-null relatively open subset of A it has positive μ-measure. As a set disjoint from B. it has μ- measure zero. This contradiction arises because of the assumption $A \neq B$. Hence $A = B$ and the claim of uniqueness of the support follows. □

Theorem 1.8.3. *Let Ω be a locally compact σ-compact Hausdorff space. Let $\mathcal{A}$ be its Borel σ-field. Let μ be an arbitrary measure on $\mathcal{A}$ possessing property*

$\mathcal{R}$. Then there exists a unique closed set A_μ (the carrier of μ) such that $\mu(\Omega \sim A_\mu) = 0$ and $\mu(U) > 0$ for every non-null relatively open subset U of A_μ.

Proof. As $\mu \not\equiv 0$ and as μ has property $\mathcal{R}$, we can find a compact set C with $\mu(C) > 0$. As the space is locally compact we can, by considering the covering of C by compact neighbourhoods of the points of C, find an open set $D \supseteq C$ with $\bar{D}$ compact. It is clear $\mu(D) > 0$ since $D \supseteq C$. By Theorem **(1.8.2)** there exists a compact set $D_1 \subseteq \bar{D}$ such that every non-null relatively open subset of D_1 has positive μ-measure and $\mu(\bar{D} \sim D_1) = 0$. Write $E = D \cap D_1$. The sets D, D_1 belong to $\mathcal{A}$. Hence $E \in \mathcal{A}$. Further

$$\mu(D \sim E) = \mu(D \sim D_1) \leq \mu(\bar{D} \sim D_1) = 0.$$

Also every non-null relatively open subset of E is a non-null relatively open subset of D_1 and hence has positive μ-measure.

Let $\mathcal{A}_\mu$ be the completion of the σ-field $\mathcal{A}$ under μ and let $\bar{\mu}$ be the complete measure on $\mathcal{A}_\mu$ obtained from μ. Consider now the collection of all open subsets A which admit of a compact subset B of $\bar{A} \in \mathcal{A}_\mu$ whose non-null relatively open subsets have positive measure and $\bar{\mu}(\bar{A} \sim B) = 0$. Let $\mathcal{E}$ denote the family of all such pairs (A, B). In view of earlier remarks this class is non-empty. Partially order this class by:

$$(A_1, B_1) \leq (A_2, B_2) \textit{ iff } A_1 \subseteq A_2 \text{ and } B_1 = A_1 \cap B_2.$$

We will show that every chain in $\mathcal{E}$ has an upper bound. It will then follow that $\mathcal{E}$ has a maximal element. Let $\{(A_\alpha, B_\alpha), \alpha \in \Sigma)\}$ be a chain in $\mathcal{E}$. Write

$$A = \bigcup_{\alpha \in \Sigma} A_\alpha \text{ and } B = \bigcup_{\alpha \in \Sigma} B_\alpha.$$

Being an open set, $A \in \mathcal{A}$.

To claim that $B \in \mathcal{A}_\mu$, we argue as follows. Let μ^*, μ_* be respectively the outer and inner measures induced by μ. Since $B_\alpha \subseteq A_\alpha$ for each α, $B \subseteq A$. Hence $\mu^*(B) \leq \mu(A)$. Now $\mu_*(B) = \sup\{\mu(F) : F \subseteq B, F \in \mathcal{A}\}$. But since μ possesses property $\mathcal{R}$, this supremum is equal to $\sup\{\mu(C) : C \subseteq B, C \text{ compact}\}$. Hence $C \subseteq A$ and $A_\alpha, \alpha \in \Sigma$ is an open covering for the compact set C. A finite subcovering can be extracted out of this. As the A_αs are linearly ordered, it follows that $C \subseteq A_\theta$ for some $\theta \in \Sigma$. Now

$$\mu(C) = \mu(C \cap A_\theta) = \mu(C \cap B_\theta)$$

since $\mu(A_\theta \sim B_\theta) = 0$. Since $C \cap B_\theta$ is a Borel set and since $C \cap B_\theta \subseteq B$, it follows from the definition of μ_* that

$$\mu_*(B) \geq \mu(C \cap B_\theta) = \mu(C).$$

Since this is true for every compact set $C \subseteq A$ and since μ possesses property R, we have $\mu_*(B) \geq \mu(A)$. Thus $\mu_*(B) = \mu^*(B) = \mu(A)$. This implies that $B \in \mathcal{A}_\mu$ and $\bar{\mu}(A \sim B) = 0$. Suppose V is an open set with $B \cap V \neq \phi$. Hence there is a $B_\alpha, \alpha \in \Sigma$ such that $B_\alpha \cap V \neq \phi$. Now, $\mu(B_\alpha \cap V) > 0$ by the construction of the set B_α. Therefore $\bar{\mu}(B \cap V) \geq \mu(B_\alpha \cap V) > 0$. Thus $(A, B) \in \mathcal{E}$. To show that $(A_\alpha, B_\alpha) \leq (A, B)$ for all $\alpha \in \Sigma$ we argue as follows. Fix $\alpha \in \Sigma$. We must show that $A_\alpha \subseteq A$ (which is obvious) and that $B_\alpha = A_\alpha \cap B$. For every $\theta \in \Sigma$, $(A_\alpha, B_\alpha) \leq (A_\theta, B_\theta)$ or $(A_\theta, B_\theta) \leq (A_\alpha, B_\alpha)$, since $\{(A_t, B_t),\ t \in \Sigma\}$ is a chain. Let Λ consist of those $\theta \in \Sigma$ for which the partial order $(A_\alpha, B_\alpha) \leq (A_\theta, B_\theta)$ holds and $W = \Sigma \sim \Lambda$. Trivially, $A_\alpha \cap B \supseteq A_\alpha \cap B_\alpha = B_\alpha$. Now, $A_\alpha \cap B = \cup_{\theta \in \Sigma}(A_\alpha \cap B_\theta)$. If $\theta \in \Lambda$, then $A_\alpha \cap B_\theta = B_\alpha$. If $\theta \in W$, i.e. if $(A_\theta, B_\theta) \leq (A_\alpha, B_\alpha)$, then $B_\theta = A_\theta \cap B_\alpha$. Hence $A_\alpha \cap B_\theta = A_\alpha \cap A_\theta \cap B_\alpha = A_\theta \cap B_\alpha \subseteq B_\alpha$. These results imply that $A_\alpha \cap B \subseteq B_\alpha$. Consequently, $A_\alpha \cap B = B_\alpha$ and the proof is complete that every chain in $\mathcal{E}$ has an upper bound. This implies (ref. Zorn Lemma, p 33 **[K]**) that $\mathcal{E}$ has a maximal element. Denote it by (M, N).

We claim $M = \Omega$. For, if not, two cases can arise.

Case (i) $\mu(\Omega \sim M) = 0$.

In this case we see immediately that $(\Omega, N) \in \mathcal{E}$. Further trivially $(M, N) \leq (\Omega, N)$ and the two elements are not the same, thus contradicting the maximality of (M, N).

Case (ii) $\mu(\Omega \sim M) > 0$.

There would exist then a compact set $K \subseteq \Omega \sim M$ with $\mu(K) > 0$. There exists then an open set $U \supseteq K$ such that $\bar{U}$ is compact and a set $F \in \mathcal{A}_\mu$ such that $(U, F) \in \mathcal{E}$. Let $M_1 = M \cup U$ and $N_1 = N \cup F_1$ where $F_1 = F \sim M$. Consider (M_1, N_1). M_1, being the union of the open sets M and U, is an open set. Every non-null relatively open subset of $N \cup F$ contains a non-null relatively open subset of N or of F and hence has positive μ-measure. Further,

$$\begin{aligned}
\bar{\mu}(M_1 \sim N_1) &= \bar{\mu}\left((M \cup U) \cap (N' \cap F_1')\right) \\
&\leq \bar{\mu}(M \cap N' \cap F_1') + \bar{\mu}(U \cap N' \cap F_1') \\
&\leq \bar{\mu}(M \cap N') + \bar{\mu}\left(U \cap N' \cap (F' \cup M)\right) \\
&\leq 0 + \bar{\mu}(M \cap U \cap N') + \bar{\mu}(U \cap N' \cap F') \\
&\leq \bar{\mu}(M \cap N') + \bar{\mu}(U \cap F') = 0.
\end{aligned}$$

Hence $(M_1, N_1) \in \mathcal{E}$. Since

$$M \cap N_1 = M \cap (N \cup F_1) = (M \cap N) \cup (M \cap F_1) = (M \cap N) \cup \phi = N.$$

Hence $(M, N) \leq (M_1, N_1)$. We note M_1 is strictly larger than M since U contains a compact set disjoint from M. Consequently, (M, N) and (M_1, N_1)

are distinct members of $\mathcal{E}$. This violates the maximality of (M, N).Therefore $M = \Omega$ as claimed.

Let $A_\mu = \bar{N}$. Then $\mu(\Omega \sim A_\mu) = 0$. Every open subset having a non-null intersection with $\bar{N}$ has a non-null intersection with N. Hence every non-null relatively open subset of A_μ has positive measure.

Proof of the uniqueness of A_μ follows the same lines as in **(1.8.2)**. □

Theorem 1.8.4. *Let (M, d) be a separable metric space. Let $\mathcal{A}$ be its Borel σ-field. Let μ be an arbitrary measure on $\mathcal{A}$. Then there exists a unique closed set A_μ (called the carrier of μ) such that $\mu(\Omega \sim A_\mu) = 0$ and $\mu(G) > 0$ for every non-null relatively open subset G of A_μ.*

Proof. M has a countable base. Every open subset U of M is the union of a countable collection of the base open sets. Let $\mathcal{U}$ consist of those members of the base open sets with measure 0. Let $V = \cup\{U : U \in \mathcal{U}\}$ then note that V is an open set. Further $\mu(V) = 0$ since $\mathcal{U}$ is a countable collection. Denote by A_μ the closed set $M \sim V$. If G is an open subset and if $G \cap A_\mu \neq \phi$ (so $G \cap A_\mu$ is a non-null relatively open subset of A_μ), then we claim $\mu(G \cap A_\mu) > 0$. For, if $\mu(G \cap A_\mu) = 0$, then $\mu(G) = \mu(G \cap A_\mu) + \mu(G \cap A_\mu') \leq 0 + \mu(V) = 0$. This would imply that $G \in \mathcal{U}$ and then the intersection of G with A_μ would be null, a contradiction.

The proof of the uniqueness of A_μ is on the same lines as in **(1.8.2)**. □

Remark 1.8.5. In **(1.8.4)**, an isolated point $a \in A_\mu$ is necsaarily an atom of μ. For, (a is an isolated point of A_μ) $\Rightarrow$ (there exists an open set J such that $J \cap A_\mu = \{a\}$). Since $J \cap A_\mu$ is a non-null relatively open subset of A_μ and since A_μ is the support of μ, it follows that $\mu(\{a\}) = \mu(J \cap A_\mu) > 0$, proving that a is an atom of the measure μ.

1.9 Tightness of a family of *pms*.

The concept of tightness is a fundamental notion in the study of weak convergence of sequences of *pms* in metric spaces. The importance of the concept stems from the fact that weakly convergent sequences of *pms* in complete separable metric spaces are necessarily tight, (ref. **(1.9.4)**) and **(2.4.5)**. Tight sequences of *pms* arise naturally in the study of functional central limit theorem, ref. **(5.6.4)**. We now define 'tightness' and discuss some of its properties.

Definitions 1.9.1. (i) A family (μ_α) of *pms* on the Borel σ-field m of a metric space $(M,\ d)$ is said to be tight if, given $\varepsilon > 0$, a compact set K can be found such that $\mu_\alpha(K) > 1 - \varepsilon$ for every α in the index set.

(ii) A family (μ_α) of pms on $\mathscr{R}^p$ is said to be shift tight with shift constant vectors $a_\alpha \in R^p$ if the family (ν_α) of pms is tight where $\nu_\alpha(A) = \mu_\alpha(A + a_\alpha)$, $A \in \mathscr{R}^p$.
(iii) The distribution or the distribution measure (dm) of a rE $\mathbf{X}$ in $(M,\ d)$ is the pm μ_X generated or induced on $\mathcal{M}$ by $\mathbf{X}: \mu_X(A) = P(\mathbf{X}^{-1}(A))$.

The distribution function $\mathbb{F}$ of a rV X in R^p is the *df* associated with μ_X. The *df* $\mathbb{F}$ determines a pm $\mu_{\mathbb{F}}$ on $\mathscr{R}^p$. It may be noted that $\mu_{\mathbf{X}} \equiv \mu_{\mathbb{F}}$.

A useful property of probability measures in arbitrary metric spaces is established in the following theorem.

Theorem 1.9.2. *Let μ be an arbitrary pm defined on the Borel σ-field $\mathcal{M}$ of a metric space $(M,\ d)$. Then*
(α) μ is both inner regular and outer regular. i.e.

$$(i)\ \mu(A) = \sup_{\substack{C \subseteq A \\ C\ closed}} \mu(C) \quad and \quad (ii)\ \mu(A) = \inf_{\substack{G \supseteq A \\ G\ open}} \mu(G)$$

for every $A \in \mathcal{M}$.
(β)

$$\mu(A) = \sup_{\substack{K \subseteq A \\ K\ compact}} \mu(K) \qquad if\ \mu\ is\ tight$$

In other words, a tight measure μ on $\mathcal{M}$ has property $\mathcal{R}$.
(γ) If $\mu,\ \nu$ are tight pms on $\mathcal{M}$ and if $\mu(K) = \nu(K)$ for every compact set K, then $\mu \equiv \nu$.

Proof. (α) Let $\mathcal{V} \subset \mathcal{M}$ be the collection of all sets $A \in \mathcal{M}$ for which (i) and (ii) hold. If A is an arbitrary closed set, define $G_r = \{x : d(x, A) < \frac{1}{r}\}$ for $r = 1, 2 \ldots$. Note that each G_r is an open set and $G_r \downarrow A$ as $r \uparrow \infty$. Continuity of μ entails $\mu(G_r) \downarrow \mu(A)$. Hence $A \in \mathcal{V}$ i.e. all closed sets belong to $\mathcal{V}$. That if $E \in \mathcal{V}$ then $E' \in \mathcal{V}$ is obvious. i.e. $\mathcal{V}$ is closed under the complementation operation. Let $A, B \in \mathcal{V}$ and let $E = A \cap B$. Given $\epsilon > 0$ we can find closed sets C_1, C_2 and open sets G_1, G_2 such that $C_1 \subseteq A \subseteq G_1$, $C_2 \subseteq B \subseteq G_2$, $\mu(G_i \sim C_i) < \epsilon$, $i = 1, 2$. Let $C = C_1 \cap C_2$ and $G = G_1 \cap G_2$ and note that C is a closed set, G an open set and $C \subseteq E \subseteq G$. $\mu(G \sim C) \leq \mu(G_1 \sim C_1) + \mu(G_2 \sim C_2) < 2\epsilon$. This shows that $E \in \mathcal{V}$. Thus $\mathcal{V}$ is closed under finite intersections. Thus $\mathcal{V}$ is a field. Let now $E_n \in \mathcal{V}$ and let $E_n \downarrow E$. We will show that $E \in \mathcal{V}$. Given $\epsilon > 0$ we can find, for $n \geq 1$, a closed set C_n and an open set H_n such that $C_n \subseteq E_n \subseteq H_n$ and $\mu(H_n \sim C_n) < \frac{\epsilon}{2^n}$. If $C = \cap_n C_n$ then $C \subseteq E$ and C is a closed set. Further $\mu(E \sim C) \leq \sum_n \mu(E_n \sim C_n) < \epsilon$. Since $E_n \downarrow E$, we can

find $N = N(\epsilon)$ such that $\mu(E_N) < \mu(E) + \frac{\epsilon}{2}$. Since $E_N \supseteq E$, we have $\mu(E_N \sim E) < \frac{\epsilon}{2}$. Now $H_N \supseteq E_N \supseteq E$, and $\mu(H_N \sim E_N) < \frac{\epsilon}{2^N} < \frac{\epsilon}{2}$. $\mu(H_N \sim E) = \mu(H_N \sim E_N) + \mu(E_N \sim E) < \epsilon$. It follows $\mathcal{V}$ is a σ-field, containing the closed sets and contained in $\mathcal{m}$. Hence $\mathcal{V} = \mathcal{m}$.
(β) By part (α), $\mu(A) = \sup\{\mu(C) : C \text{ closed}, \ C \subseteq A\}$. Hence given $\varepsilon > 0$ there exists a closed set $C \subseteq A$ such that $\mu(A) \leq \mu(C) + \varepsilon$. Further the tightness of μ implies that there exists a compact set K with $\mu(K) > 1 - \varepsilon$. Now $\mu(C) = \mu(C \cap K) + \mu(C \cap K') \leq \mu(C \cap K) + \varepsilon$. Hence, noting that $C \cap K(\subseteq C \subseteq A)$ is a compact set and that $\mu(A) \leq \mu(C \cap K) + 2\varepsilon$, we complete the proof of (β).
(γ) Immediate from (β). □

Remarks 1.9.3. (i) Consider the relation $(\alpha)(ii)$. For each $n \geq 1$, we can find an open set $U_n \supset A$ such that

$$\mu(U_n) < \mu(A) + \frac{1}{n}.$$

Let $G_n = \cap_1^n U_k$. Then

$$G_n \supseteq A, \ \mu(G_n) < \mu(A) + \frac{1}{n} \text{ and } G_n \downarrow$$

with $\lim\limits_n G_n = G^*$, say. Then clearly

$$\mu(A) = \mu(G^*), \ \mu(G^* \sim A) = 0.$$

Similar remarks can be made about the other two relations.
(ii) Every *pm* μ on $\mathscr{R}$ is tight. For, if F is the *df* corresponding to μ, then

$$\mu([-a, \ a]) \geq F(a) - F(-a) \to 1 \text{ as } a \to \infty.$$

[This property holds also in arbitrary complete separable metric spaces, refer **(1.9.4)**]

To prove **(1.9.4)**, where we present an important property of *pms* in complete separable metric spaces, the following result from topology is needed.

Q4 *Let (M, d) be a metric space.*
(i) If $A \subset M$ is a compact set, then A has the form

$$A = \bigcap_{n=1}^{\infty} \bigcup_{j=1}^{k_n} \overline{S}_{n,j}$$

where $\overline{S}_{n,j}$ *are closed spheres of radius* $\frac{1}{n}$ *for some choice of* $k_n \to \infty$.
(ii) If (M, d) *is a complete metric space and if* $A \subset M$ *has the above form, then* A *is compact.*

Theorem 1.9.4. *Every* pm μ *on the Borel* σ*-field* $\mathcal{M}$ *of a complete separable metric space* (M, d) *is tight.*

Proof. Let $(a_1, a_2, \ldots,)$ be a separability set for M. Let $\overline{S}_{n,j}$ be the closed sphere of radius $\frac{1}{n}$ with centre a_j. Then

$$M = \bigcup_{j=1}^{\infty} \overline{S}_{n,j}, \quad \text{for each } n.$$

Given $\epsilon > 0$ we can find k_n such that

$$\mu(B_n) > 1 - \frac{\epsilon}{2^n} \text{ where } B_n = \bigcup_{j=1}^{k_n} \overline{S}_{n,j}.$$

(The sequence (k_n) will of course depend on the pm μ.) Define $K = \bigcap_n B_n$. Since M is a complete metric space it follows that K is a compact set. Now

$$\mu(K') \leq \sum_n \mu(B'_n) \leq \sum_{n=1}^{\infty} \frac{\epsilon}{2^n} < \epsilon,$$

showing that μ is tight. □

Remarks 1.9.5. (i) That the separability-completeness condition in **(1.9.4)** is only a sufficient condition is brought out by the following example. Let $M = R \sim \{0\}$ with d denoting the usual distance. (M, d) is not a complete metric space (but is separable). However, it being a σ-compact space under the metric d, all *pms* on $\mathcal{M}$ are tight.
(ii) Let $(M_i,\ d_i,\ \mathcal{M}_i,\ i = 1,\ 2)$ be complete and separable metic spaces and let $(M = M_1 \times M_2, \mathcal{M} = \mathcal{M}_1 \otimes \mathcal{M}_2)$ be the product measurable space. Endowed with the product topology, M will be a complete and separable metric space. If μ is a pm on $\mathcal{M}$, then it will be a tight measure. Hence by **(1.9.2)**, for any $\mathbf{A} \in \mathcal{M}$,

$$\mu(\mathbf{A}) = \sup_{\substack{\mathbf{K} \subset \mathbf{A} \\ \mathbf{K} \text{ compact}}} \mu(\mathbf{K}).$$

In the case $\mathbf{A}$ is of the type $A_1 \times A_2,\ A_i \in \mathcal{M}_i$, that the supremum can be taken over a smaller class of compact sets is argued now. Let $\pi_i,\ i = 1,\ 2$ be the projection mapping of M onto M_i. Let K be a compact subset of M and $\pi_i(K) = B_i$. We note that B_i is a compact set of M_i, that $B_1 \times B_2 \subset M$ is a compact subset and that $\mathbf{K} \subset B_1 \times B_2 \subset A_1 \times A_2$. Hence

$$\mu(A_1 \times A_2) = \sup_{\substack{B_1 \times B_2 \subset A_1 \times A_2 \\ B_1,\ B_2 \text{ compact}}} \mu(B_1 \times B_2).$$

The arguments go over for any finite number of factors.

1.10 Structure of separable metric spaces with non-atomic pms

In this section, we show that, in a sense to be made precise shortly, a separable and complete metric space endowed with a non-atomic pm is no different from the unit interval endowed with the Lebesgue measure.

Definition 1.10.1. Two probability spaces $(\Omega_i,\ \mathscr{A}_i,\ \mu_i)$, $i = 1,\ 2$ are said to be isomorphic if
(i) there exist $N_i \in \mathscr{A}_i$ with $\mu_i(N_i) = 0$
(ii) there exists a one-to-one onto mapping f from $\Omega_1 \sim N_1$ to $\Omega_2 \sim N_2$ such that f is

$$(\Omega_1 \sim N_1) \cap \mathscr{A}_1 \backslash (\Omega_2 \sim N_2) \cap \mathscr{A}_2$$

measurable and f^{-1} is

$$(\Omega_2 \sim N_2) \cap \mathscr{A}_2 \backslash (\Omega_1 \sim N_1) \cap \mathscr{A}_1$$

measurable.
(iii) $\mu_1 f^{-1} = \mu_2$.
(It would, of course, follow that $\mu_2 f = \mu_1$).

To prove theorem **(1.10.2)** we need the following result from topology.

Q5 (Ex. 28, p. 202 **[V]**) *Let $I = [0,\ 1]$. Let M be a separable metric space. Then (i) there corresponds a Borel (in fact, a g_δ) subset $A \subseteq I^\infty$ and a homeomorphic mapping f of M onto A and (ii) there exists a Borel set $B \subseteq$ and a one-to-one mapping g of I^∞ on to B such that both g and g^{-1} are measurable.*

The following theorem uses the above result to connect measurable functions to continuous functions.

Theorem 1.10.2 (Lusin). *Let $(M_i,\ d_i)$, $i = 1, 2$ be complete separable metric spaces. Denote their Borel σ-fields by $\mathcal{m}_i,\ i = 1,\ 2$. Let μ be a pm on $\mathcal{m}_1$*

and let f be a measurable mapping of M_1 into M_2. Then for each $\varepsilon > 0$, there corresponds a compact set K such that $\mu(K) > 1 - \varepsilon$ and f restricted to K is continuous.

Proof. By **Q5**, there exists a homeomorphic mapping g taking M_2 into a subset of the compact set I^∞. Write $h = g \circ f$ and note that h is a measurable mapping from M_1 into I^∞ endowed with the product topology.

Let us first prove result when f is a simple function. In that case note that h is a simple function. Let h take the values $a_1, a_2, \ldots, a_r$. Let

$$A_j = h^{-1}(\{a_j\}), \quad 1 \leq j \leq r.$$

These are disjoint sets.

Let $\varepsilon > 0$ be given. Since, by **(1.9.4)** μ is tight, we can find, corresponding to each j, a compact set $C_j \subset A_j$ such that $\mu(A_j \sim C_j) < \frac{\varepsilon}{r}$. The C_js are disjoint closed sets. On each C_j, h, being a 'constant', is continuous. Hence h is continuous on the compact set

$$C = \cup_{j=1}^r C_j.$$

$\mu(C) > 1 - \varepsilon$ since

$$\mu(C') = \mu(C' \cap \cup_1^r A_j) = \sum_1^r \mu(A_j \sim C_j) < \sum_1^r \frac{\varepsilon}{r} = \varepsilon.$$

By **(1.4.3)**(ii), h is the uniform limit of a sequence, say, (h_n) of simple functions. By what has just been established, corresponding to each h_n we can find a compact set K_n such that $\mu(K_n) > 1 - \frac{\varepsilon}{2^n}$ and such that h_n, restricted to K_n, is a continuous function. Let $K = \cap_n K_n$. Then K is a compact set and

$$\mu(K') < \sum_{n=1}^\infty \mu(K_n') < \sum_{n=1}^\infty \frac{\varepsilon}{2^n} = \varepsilon.$$

Since all the h_ns are continuous when restricted to K and since for $\omega \in M_1$, $\lim_{n\to\infty} h_n(\omega) = h(\omega)$ uniformly in ω, it follows that h is continuous on K. This together with the fact that g is a homeomorphism implies that f restricted to K is a continuous function. □

Sections **(1.10.3)** and **(1.10.4)** are devoted to the study, in greater detail, of the support of a pm on $\mathscr{R}$.

Definition 1.10.3. For a *df* F on the real line, *rext*F and *lext*F are defined as follows. If a finite number β exists such that $F(\beta) = 1$ and if $F(\beta) - F(\beta - \varepsilon) > 0$ for every $\varepsilon > 0$, then β is defined to be the right extremity of F or *rext*F. If no such finite number β exists then it is clear that $F(x) < 1$ for every finite x and in that case we define *rext*F to be ∞.

It may be noted that, for every *df* F, *rext*F is uniquely defined. The left extremity of F or *lext*F is defined as $-$*rext*G where G is the *df* given at all its continuity points x by $G(x) = 1 - F(-x)$.

Theorem 1.10.4. Let μ be a *pm* on the Borel σ-field $\mathcal{R}$ of the real line R. Let A_μ be the support of μ. Let F be the *df* corresponding to μ. Then (i) $\sup\{x : x \in A_\mu\} =$ *rext*F and $\inf\{x : x \in A_\mu\} =$ *lext*F (ii) At no three points $x < y < z$; $x,\ y,\ z \in A_\mu$, F will have equal value.

Proof. (i) Let F be the *df* corresponding to $\mu : F(x) = \mu((-\infty, x])$. Let A_μ be the support of μ. Let *rext*$F = \beta$ and let $\sup\{x; x \in A_\mu\} = \gamma$. We claim $\beta = \gamma$. Suppose $\beta < \infty$. Then necessarily $\gamma < \infty$. For, $\gamma = \infty$ would imply that the open infinite interval (β, ∞) will have non-null intersection with A_μ and hence will have positive measure. But this is not possible by the definition of *rext*F. Thus $\gamma < \infty$. Again arguing as before, we conclude that $\gamma \leq \beta$. Suppose $\gamma < \beta$. By the definition of *rext*F, the interval $(\gamma, \beta]$ will have positive measure. But this interval is a subset of A'_μ which has measure 0. Consequently, $\gamma = \beta$. If $\beta = \infty$, then $\gamma = \infty$ too, as otherwise the interval (γ, ∞) will have measure 0 and that would force β to be finite. In the same way, we can show that *lext*$F = \inf\{x : x \in A_\mu\}$.
(ii) If possible let $x < y < z$ be three points in A_μ such that $a = F(x) = F(y) = F(z)$. This implies that the μ-measure of the interval $(x, z]$ is equal to $F(z) - F(x) = 0$. But the open interval (x, z) has non-null intersection with A_μ, since the intersection contains the point y and has therefore a positive measure. This contradiction establishes the claim. □

Theorems **(1.10.5)** and **(1.10.6)** lead to **(1.10.7)** where we show that $(M, m,\ \mu)$ has the same structure as $(I,\ \mathcal{I},\ \mathcal{L})$ if μ is a non-atomic *pm* and if M is a complete separable metric space.

Theorem 1.10.5. *Let μ be a non-atomic pm on the Borel σ-field $\mathcal{R}$ of the real line R. Let I be the unit interval, $\mathcal{I}$ its Borel σ-field and let $\mathcal{L}$ denote the Lebesgue measure. Then $(R, \mathcal{R}, \mu)$ and $(I, \mathcal{I}, \mathcal{L})$ are isomorphic.*

Proof. Let A_μ denote the support of μ. A'_μ, being an open set, is the union of a countable number of disjoint open intervals, say, $J_n,\ n = 1,\ 2,\ \ldots\ldots\ldots,\ J_m \cap$

$J_n = \phi$, $m \neq n$. No two of these intervals can have a common end point, say, a. (i.e., the distance between the closures of any two of these intervals is positive). For, such a point a will be in A_μ and will be an isolated point of A_μ. By **(1.8.5)**, a will be an atom of μ. But μ is given to be non-atomic. Hence A_μ is the countable union of disjoint non-degenerate closed intervals. Remove from A_μ the end points of each of the closed interval and denote by B_μ what is left of A_μ. Since μ is non-atomic, what is removed is a set of μ-measure 0. Let G be the restriction of F to B_μ where F is the *df* associated with μ. It is easily seen that $\{G(x) : x \in A_\mu\} = (0,1)$ and that G maps B_μ onto $(0,1)$ in a one-to-one manner. Further G and G^{-1} map intervals onto similar intervals. Hence they are measurable mappings. Thus $(B_\mu, B_\mu \cap \mathcal{R})$ and $((0,1), (0,1) \cap \mathcal{I})$ are isomorphic measurable spaces. Further $\mu(B_\mu^{'}) = 0$ and $\mathcal{L}(\{0,1\}) = 0$. Also if $0 < a < b < 1$, then there exist unique points $u < v$ in B_μ such that $G(u) = a$ and $G(v) = b$;

$$\begin{aligned}\mu(G^{-1}[a,b]) &= \mu(G^{-1}[G(u), G(v)]) = \mu([u,v]) \\ &= F(v) - F(u) = b - a = \mathcal{L}([a,b]).\end{aligned}$$

With this the proof is complete that $(R, \mathcal{R}, \mu)$ and $(I, \mathcal{I}, L)$ are isomorphic. □

Theorem 1.10.6. *Let (M,d) be a complete and separable metric space and let μ be a pm on its Borel σ-field $\mathfrak{m}$. Then there exists a measurable and one-to-one mapping f of M into I such that $(M, \mathfrak{m}, \mu)$ and $(I, \mathcal{I}, \mu f^{-1})$ are isomorphic.*

Proof. By **(Q5)**, there exists a one-to-one measurable mapping f of M into I. Tightness of μ and Lusin's theorem at **(1.10.2)** imply that, given $\varepsilon > 0$, a compact set $K_\varepsilon \subseteq M$ can be found such that $\mu(K_\varepsilon) > 1 - \varepsilon$ and such that f restricted to K_ε is continuous. It follows that if $M_1 = \cup_n K_{\frac{1}{n}}$ then $\mu(M_1) = 1$, f restricted to M_1 is continuous, $f(M_1)$ is σ-compact and since f is one-to-one, f and f^{-1} are measurable. These properties imply that $(M, \mathfrak{m}, \mu)$ and $(I, \mathcal{I}, \mu f^{-1})$ are isomorphic *pm* spaces. □

Corollary 1.10.7. *If μ in the theorem is, additionally, non-atomic then the pm spaces $(M, \mathfrak{m}, \mu)$ and $(I, \mathcal{I}, \mathcal{L})$ are isomorphic.*

Proof. Let $a \in I$. If $a \notin f(M)$, then $f^{-1}(\{a\}) = \phi$ and $\mu f^{-1}(\{a\}) = 0$. If $a \in f(M)$, then there exists a unique point, say, $\alpha \in M$ such that $f^{-1}(\{a\}) = \alpha$. Since μ is non-atomic, we have $\mu f^{-1}(\{a\}) = \mu(\{\alpha\}) = 0$. i.e. μf^{-1} is a non-atomic measure on $\mathcal{I}$. Hence by **(1.10.6)**, $(I, \mathcal{I}, \mu f^{-1})$ and $(I, \mathcal{I}, \mathcal{L})$ are isomorphic. The claim follows now that $(M, \mathfrak{m}, \mu)$ and $(I, \mathcal{I}, \mathcal{L})$ are isomorphic since isomorphism is a transitive relation. □

Given a linear interval I of finite length α and λ, $0 < \lambda < 1$, it is trivial to see that we can always mark off a sub interval of length (or, Lebesgue measure) $\lambda\alpha$. That this result is true also when I is replaced by a Borel set A is the content of the next theorem.

Theorem 1.10.8. *Let the Borel set A of the unit interval have positive Lebesgue measure: $\alpha = \mathcal{L}(A) > 0$. Let $0 < t < \alpha$. Then there exists $B \subset A, B$ Borel, such that $\mathcal{L}(B) = t$.*

Proof. We note first that theorem is trivially true if A is an interval. Since I is a complete separable metric space, we conclude, by **(1.9.4)**, that $\mathcal{L}$ is a tight pm. This implies that there exists a compact set $K \subseteq A$ such that $t \leq \mathcal{L}(K) \leq \mathcal{A} = \alpha$. Since K', being an open set, is the union of a countable collection of disjoint open intervals, it follows that K is the union of a countable collection (J_n) of disjoint closed intervals. We note that $\sum_n \mathcal{L}(J_n) = \mathcal{L}(K) = \beta$, say. From each non-degenerate interval J_n mark a sub interval B_n of length $\frac{t}{\beta}\mathcal{L}(J_n)$ and let $B = \cup_n B_n$. Since the B_ns are disjoint, we have

$$\mathcal{L}(B) = \sum_n \frac{t}{\beta}\mathcal{L}(J_n) = \frac{t}{\beta}\mathcal{L}(K) = t. \qquad \square$$

Theorem 1.10.9. *Let μ be a non-atomic pm on the Borel σ-field $\mathfrak{m}$ of a complete separable metric space $(M,\ d)$. Let $A \in \mathfrak{m}$ and let $0 < t < \mu(A)$. Then there exists $C \in \mathfrak{m}$, $C \subset A$ such that $\mu(C) = t$.*

Proof. By (**Q5**), there exists a mapping f taking M into I such that $(M,\ \mathfrak{m},\ \mu)$ and $(I,\ \mathcal{I}, \mathcal{L})$ are isomorphic. This implies that if $A \in \mathfrak{m}$, then $\mu(A) = \mathcal{L}(f(A))$. Thus if $0 < t < \mu(A)$, we have $0 < t < \mathcal{L}(f(A))$. By (**(1.10.8)**), there exists $B \in \mathcal{I}$, $B \subset f(A)$ such that $\mathcal{L}(B) = t$. Since f is an isomorphic mapping, $C \in \mathfrak{m}$, $C \subset A$ exists such that $f(C) = B$. Hence $\mu(C) = t$. $\square$

Theorem 1.10.10. *Let f be a real valued $\mathfrak{m}$-measurable function defined on the probability space$(M,\ \mathfrak{m},\ \mu)$ where M is a complete separable metric space. Let $A \subset R$, $A \notin \mathscr{R}$ be such that $E = f^{-1}(A) \in \mathfrak{m}$. Then there exist linear Borel sets $A_1,\ A_2$ such that $A_1 \subset A \subset A_2$ and $\mu f^{-1}(A_2 \sim A_1) = 0$.*

Proof. μ is necessarily a tight measure (ref. **(1.9.4)**. By **(1.9.2)**(β) there exist compact sets $C_n \subset E,\ C_n \uparrow$ such that $\mu(E \sim C_n) \to 0$. By Lusin theorem (ref. **(1.10.2)**) there exist compact sets $D_n,\ D_n \uparrow$ such that $\mu(M \sim D_n) \to 0$ and f restricted to D_n is continuous. Let $K_n = C_n \cap D_n$. We note K_ns are compact subsets of $E,\ K_n \uparrow,\ \mu(E \sim K_n) \to 0$ and f restricted to K_n is continuous. Since $K_n \subset D_n$ and since f restricted to D_n is continuous, it follows that $B_n = f(K_n)$ is a (linear) compact set. Define $A_1 = \cup_n B_n$. We note that A_1 is a Borel set and that

$$A_1 = \cup_n f(K_n) = f(\cup_n K_n) \subset f(E) = A.$$

Hence $f^{-1}(A_1) \supset \cup_n K_n$. We note

$$f^{-1}(A_1) \subset f^{-1}(A) = E$$

$$\begin{aligned}
\text{and}\quad \mu(E \sim f^{-1}(A_1)) &\le \mu(E \sim \cup_n K_n) \\
&= \mu(\cap_n (E \sim K_n)) \\
&= \lim_{n\to\infty} \mu(E \sim K_n)(\text{since } (E \cap K_n') \downarrow) \\
&= 0. \\
\text{Thus}\quad \mu(f^{-1}(A) \sim f^{-1}(A_1)) &= 0.
\end{aligned}$$

Now consider the set A' and note that $f^{-1}(A') \in \mathcal{m}$ since $f^{-1}(A') = (f^{-1}(A))' = E'$. Hence, arguing as above, we can find a Borel set $A_2' \subset A'$ such that

$$\mu\left(f^{-1}(A') \sim f^{-1}(A_2')\right)) = 0. \text{ i.e. } \mu\left(f^{-1}(A_2) \sim f^{-1}(A)\right) = 0.$$

These two results imply

$$\mu\left(f^{-1}(A_2) \sim f^{-1}(A_1)\right) = 0. \text{ i.e } \mu(f^{-1}(A_2 \sim A_1) = 0. \qquad \square$$

1.11 Probability measures in product spaces

Let T be an arbitrary index set. Let $(\Omega_t,\ \mathscr{A}_t)$, $\tilde{\mathbf{a}}_{J;T}$, $\mathbf{a}_{J;T}$, $\widetilde{\mathscr{A}_T}$, $\mathscr{A}_T$ be as per notation introduced in section **(1.2)**. Suppose μ_T is a *pm* on $\mathscr{A}_T$. Trivially $\mu_{J:T}$, the restriction of μ_T to the sub σ-field $\mathbf{a}_{J;T}$, is a *pm*. We note that the collection $\{\mu_{J;T},\ J \subset T,\ J \text{ finite}\}$ of *pms* is a consistent family. i.e. it possesses the following two properties. (i) The correspondence between two finite subsets, $\tilde{J}$ and J, 1 of T, each being a permutation of the other, determines a unique one-to-one onto mapping τ on R^p where p is the number of members in J. The property we draw attention to is that if $E \in \mathscr{R}^p$, then $\mu_{\tilde{J}:T}(E) = \mu_{\tau(J):T}(E) = \mu_{J:T}(\tau^{-1}(E))$. (ii) if $J_1 \subset J_2 \subset T,\ J_1,\ J_2$ finite sets, then $\mu_{J_2;T}$ restricted to $\mathbf{a}_{J_1;T}$ is $\mu_{J_1;T}$.

The *pms* $\mu_{J;T}$ are called the finite dimensional distributions (*fdd*) of μ. We ask if the reverse is true. i.e. suppose we start with *pms* $\mu_{J;T}$, defined on $\mathbf{a}_{J:T}$,

$J \subset T,\ J$ a finite set, for all possible choices of J. We ask : Does there exist a pm μ_T on $\mathscr{A}_T$ such that μ_T restricted to $\mathbf{a}_{J;T}$ is $\mu_{J;T}$ for each J? Obviously these pms can not be totally arbitrary. Clearly they must form a consistent family. What other condition should be satisfied? We develop below the needed infrastructure to answer this question.

Theorem **(1.11.1)** is the basis for claiming the existence of real valued stochastic processes with prescribed finite dimensional distributions. i.e. a family $(x_t,\ t \in T)$ of real valued rvs such that for every choice of positive integer k and every choice of parameter values $\{t_1,\ t_2, \ldots,\ t_k\}$ the probability distribution of $(x_{t_1},\ x_{t_2}, \ldots,\ x_{t_k})$ is preassigned in a consistent manner.

To prove **(1.11.1)**, we need the following result.

Q6 (ref. 9F and 13A **[H]**) *Let $\mathcal{A}$ be the σ-field generated by a field $\widehat{\mathcal{A}}$ of subsets of an abstract set Ω. Let μ be a non-negative finitely additive set function defined on $\widehat{\mathcal{A}}$ with $\mu(\Omega) = 1$. Then μ can be extended uniquely to a pm on $\mathcal{A}$ iff μ is continuous from above at the null set ϕ. i.e. whenever $A_n \in \widehat{\mathcal{A}},\ A_n \downarrow \phi$ implies* $\lim\limits_n \mu(A_n) = 0$.

Consequently, two pms $\mu_1,\ \mu_2$ defined on $\mathcal{A}$ are identical if they coincide on $\widehat{\mathcal{A}}$. [*Compare with the theorem in section* **(1.5)**]

Theorem 1.11.1. *For each $t \in T$ let Ω_t be a complete separable metric space and let $\mathscr{A}_t$ denote its Borel σ-field. Let the family of pms $\{\mu_{J;T},\ J \subset T,\ J$ finite$\}$, defined on $\mathbf{a}_{J;T}$, be a consistent family of pms. Then there exists a unique pm μ on $\mathscr{A}_T$ such that its restriction to $\mathbf{a}_{J;T}$ is $\mu_{J;T}$.*

Proof. Let $\widetilde{\mathscr{A}_T} = \cup\{\mathbf{a}_{J;T} :\ J$ finite $J \subset T\}$. Note that $\widetilde{\mathscr{A}_T}$ is a field and that the $\mu_{J;T}$s determine on $\widetilde{\mathscr{A}_T}$ a non-negative set function $\widetilde{\mu}$ such that it is finitely additive and $\widetilde{\mu}(\Omega_T) = 1$.

Let $A_n \in \widetilde{\mathscr{A}_T},\ n \geq 1,\ A_n \downarrow \phi$. Since $\widetilde{\mu}$ is additive, it follows that $(\widetilde{\mu}(A_n))$ is a monotonic decreasing sequence. If $\lim\limits_n \widetilde{\mu}(A_n) = 0$ then we are done, by **Q6**.

If the limit is not zero, then $\widetilde{\mu}(A_n) \geq \varepsilon$ for some $\varepsilon > 0$ for all $n \geq 1$. Let $A_n \in \mathbf{a}_{J_n;T}$. With no loss of generality we may assume $J_n \subset J_{n+1},\ n \geq 1$. Let $S = \cup_n J_n$. Define $\mathbf{b}_{R;S},\ R \subset S,\ R$ finite on the lines of $\mathbf{a}_{J;T}$. Let $\widetilde{\mathscr{B}_S} = \cup\{\mathbf{b}_{R;S} :\ R$ finite, $R \subset S\}$. We note that $\widetilde{\mathscr{B}_S}$ is a field and the σ-field it generates is

$$\mathscr{B}_S = \bigotimes_{t \in S} \mathscr{A}_t.$$

It is clear that to each $A \in \widetilde{\mathscr{A}_T}$ there corresponds a unique $B \in \widetilde{\mathscr{B}_S}$ such that

$A = B \times \Omega_{T\sim S}$. Let B_n correspond to A_n, $n \geq 1$. We note $B_n \in \mathbf{b}_{J_n;S}$. The *pm* $\mu_{J_n;T}$ determines uniquely a *pm* $\nu_{J_n;S}$ on $\mathbf{b}_{J_n;S}$ such that $\nu_{J_n;S}(B_n) = \mu_{J_n;S}(A_n)$. Let $\widetilde{\nu}$ denote the finitely additive set function on $\widetilde{\mathscr{B}}_S$ whose restriction to $\mathbf{b}_{J_n;S}$ is the *pm* $\nu_{J_n;S}$.

We thus have:

$$B_n \in \widetilde{\mathscr{B}}_S,\ B_n \downarrow \phi, \widetilde{\nu}(B_n) \geq \varepsilon,\ n \geq 1.$$

By an application of **(1.9.4)**, we can find

$$C_n \subset B_n,\ C_n \in \mathbf{b}_{J_n;S} \text{ with } \widetilde{\nu}_{J_n;S}(B_n \sim C_n) \leq \frac{\varepsilon}{2^{n+1}}$$

where for a compact subset

$$K_n \subset \Omega_{J_n},\ C_n = K_n \times \Omega_{S\sim J_n}.$$

Let

$$C_n^* = \bigcap_1^n C_r.$$

We note that $C_n,\ C_n^*$ are in $\mathbf{b}_{J_n;S}$. Now

$$B_n \sim \cap_1^n C_i = B_n \cap \cup_1^n C_i' = \cup_1^n (B_n \sim C_i) \subset \cup_1^n (B_i \sim C_i).$$

Hence

$$\widetilde{\nu}(B_n \sim C_n^*) \leq \sum_1^n \nu_{J_i;S}(B_i \sim C_i) \leq \sum_1^n \frac{\varepsilon}{2^{i+1}} \leq \frac{\varepsilon}{2}.$$

This implies

$$\varepsilon \leq \widetilde{\nu}(B_n) = \widetilde{\nu}(B_n \cap C_n^*) + \widetilde{\nu}(B_n \sim C_n^*) \leq \widetilde{\nu}(C_n^*) + \frac{\varepsilon}{2}.$$

i.e.,

$$\widetilde{\nu}(C_n^*) \geq \frac{\varepsilon}{2},\ n \geq 1.$$

For $R \subset S$, let $\pi_R(E)$ be the projection image of the set E into Ω_R. Since $\nu_{J_n;S}(C_n^*) > 0$, it follows that $\pi_{J_n}(C_n^*) \neq \phi$. This implies $\pi_{J_n\sim J_{n-1}}(C_n^*) \neq \phi$. This, in turn, implies that the compact set

$$\pi_{J_n\sim J_{n-1}}(C_n) \neq \phi.$$

Let

$$K = \times_1^\infty \pi_{J_n\sim J_{n-1}}(K_n),\ J_0 = \phi$$

and note that K is a non-null compact subset of Ω_S. To C_n^* there correspond $D_n \in \mathscr{B}_{J_n}$ and $E_n \in \Omega_{S \sim J_n}$ such that (i) $C_n^* = D_n \times E_n$ and (ii) $D_n \subset \times_1^n \pi_{J_r \sim J_{r-1}}(K_r)$. We have from this

$$C_n^* \cap K = D_n \times_{n+1}^{\infty} \pi_{J_r \sim J_{r-1}}(K_r).$$

Thus $C_n^* \cap K$ is a monotonic decreasing sequence of non-null compact sets. Hence $\cap_n C_n^* \neq \phi$ (finite intersection property). This is not possible since $C_n \subset B_n$ and $B_n \downarrow \phi$. Thus $\widetilde{\mu}(A_n) \to 0$ and the proof is complete by **Q6**. $\square$

Note. The following theorem can be proved using similar arguments (ref. proposition 27.4, p. 119, **[P]**).

Let $(X,\ d)$ be a complete separable metric space. Let $\mathcal{X}$ denote its Borel σ-field. For $n = 0,\ 1, \ldots$, let $\mathcal{X}_n$ be a sub σ-field of $\mathcal{X}$ such that $\mathcal{X}_n \subset \mathcal{X}_{n+1}$ and such that the field $\cup_n \mathcal{X}_n$ generates $\mathcal{X}$. For each n let μ_n be a pm on $\mathcal{X}_n$ such that (μ_n) is consistent. i.e., μ_n restricted to $\mathcal{X}_{n-1}$ is μ_{n-1}. Then there exists a pm μ on $\mathcal{X}$ such that μ restricted to $\mathcal{X}_n$ is μ_n.

In **(1.11.6)** we show that the theorem is true with the Ω_ts being abstract spaces and not necessarily complete and separable metric spaces provided T is a countable index set and provided the pms μ_n are constructed in a certain way. **(1.11.6)** is the basis for claiming, in later chapters, the existence of (i) a sequence of independent rvs with preassigned distributions and (ii) a Markov chain with a countable state space.

Transition probability measures and the case of abstract component spaces

Definition 1.11.2. Let $(\Omega_i,\ \mathscr{A}_i),\ i = 1,\ 2$ be measurable spaces.
(i) A function q_1 defined on $\Omega_1 \times \mathscr{A}_2$ is called a transition probability measure if for $A \in \mathscr{A}_2$ fixed, $q(.,\ A)$ is real valued and $\mathscr{A}_1$-measurable and if for $\omega \in \Omega_1$ fixed, $q(\omega,\ .)$ is a pm on $\mathscr{A}_2$.
(ii) For $E \subset \Omega_1 \times \Omega_2$ and $x \in \Omega_1$, denote by E_x or by $E(x)$ the set $\{y : y \in \Omega_2,\ (x,\ y) \in E\}$. E_y is analogously defined. $E_x,\ E_y$ are called respectively the x-section and the y-section of the set E.

Let f be a function defined on a subset $E \subseteq \Omega_1 \times \Omega_2$ and taking values in a measurable space. The function f_x defined on E_x is called the x-section of f where $f_x(y) = f(x,\ y)$. The y-section of f is similarly defined.

We note that if $E \in \mathscr{A}_1 \otimes \mathscr{A}_2$ then $E_x \in \mathscr{A}_2$ and $E_y \in \mathscr{A}_1$. Also f_x is $\mathscr{A}_2$-measurable and f_y is $\mathscr{A}_1$-measurable.

In **(1.11.3)** below, we construct a probability measure in the product space, using a transition probability measure. To do this, the following result is needed.

Q7 *(i) Let $\tilde{\mathcal{E}} \subset \mathscr{A}_1 \otimes \mathscr{A}_2$ denote the class of all finite unions of disjoint measurable rectangles $A \times B$. It is known to be a ring (ref. Theorem 33E* **[H]***). (It is in fact a field since $\Omega_1 \times \Omega_2$, being a measurable rectangle, belongs to it). Further the σ-field generated by $\tilde{\mathcal{E}}$ is $\mathscr{A}_1 \otimes \mathscr{A}_2$.*

These results have natural extensions to n-dimensional product spaces, $n > 2$.

(ii) A monotone class of sets is one which is closed under monotone limits. A monotone class containing a field $\tilde{\mathcal{E}}$ of sets contains the σ-field $\sigma(\tilde{\mathcal{E}})$ generated by $\tilde{\mathcal{E}}$ (ref. pp. 27–28 **[H]**).

Theorem 1.11.3. *Let $(\Omega, \mathscr{A})$, $(U, \mathscr{U})$ be measurable spaces and q_1 be a transition pm on $\Omega \times \mathscr{U}$. If the real valued function f defined on $\Omega \times U$ is bounded (i.e. $\|f\| = \sup_{\omega,u} |f(\omega, u)| < \infty$) and is $\mathcal{B} = \mathscr{A} \otimes \mathscr{U}$-measurable, then $\hat{f}(\omega) = \int_U f(\omega, u) q_1(\omega, du)$ is a bounded and $\mathscr{A}$-measurable function of ω. Further, if q is a pm on $\mathscr{A}$ and if μ is defined on $\mathcal{B}$ through:*

$$\mu(E) = \int_\Omega q(d\omega) \int_U \chi_E(\omega, u) q_1(\omega, du) \tag{1.4}$$

then μ is a pm on $\mathcal{B}$.

Proof. Step 1. Let $\mathcal{E}$ be the collection of all sets $E \in \mathcal{B}$ such that $\hat{f}$ is measurable $\mathscr{A}$ where $f = \chi_E$. If $E = \Omega \times U$ then $f \equiv 1$, also $\hat{f} \equiv 1$. Hence $\Omega \times U \in \mathcal{E}$. If $E \in \mathcal{E}$, then $E' \in \mathcal{E}$, since $\int_U \chi_{E'}\ q(\omega, \mathrm{d}u) = 1 - \hat{f}$ is $\mathscr{A}$-measurable. If $E, F \in \mathcal{E}$ and if $E \cap F = \phi$, then $f = \chi_{E \cup F} = \chi_E + \chi_F$ $(= f_1 + f_2$, say) and then it follows that $\hat{f}$ is $\mathscr{A}$-measurable, since each of $\hat{f}_1$ and $\hat{f}_2$ has this property and $\hat{f} = \hat{f}_1 + \hat{f}_2$. This trivially extends to finite number of disjoint sets. If $E = A \times B$, $A \in \mathscr{A}$ and $B \in \mathscr{U}$ then $f = \chi_E(\omega, u) = \chi_A(\omega)\, \chi_B(u)$ and $\hat{f}(\omega) = \chi_A(\omega) q_1(\omega, B)$, a $\mathscr{A}$-measurable function. Thus all measurable rectangles in $\mathcal{B}$ lie in $\mathcal{E}$.

To summarize, $\mathcal{E}$ contains all the measurable rectangles, is closed under complementation and is closed also under finite disjoint unions. Hence the field $\tilde{\mathcal{E}} \subset \mathcal{E}$ where $\tilde{\mathcal{E}}$ is as in **Q7**.

Let (E_r) be a monotonic sequence in $\mathcal{E}$ with $\lim_r E_r = E_0$. Write $f_r = \chi_{E_r}$. We have

$$\begin{aligned}\hat{f}_0(\omega) &= \int_U \chi_{E_0}(\omega,\ u)q_1(\omega,\ du)\\ &= \int_U \lim_{r\to\infty} \chi_{E_r}(\omega,\ u)q_1(\omega,\ du)\\ &\qquad\qquad \text{(by the bounded convergence theorem for integrals)}\\ &= \lim_{r\to\infty} \int_U \chi_{E_r}(\omega,\ u)q_1(\omega,\ du) = \lim_{r\to\infty} \hat{f}_r(\omega).\end{aligned}$$

$\hat{f}_0$, being thus the limit of a sequence of measurable functions, is measurable. This result shows that $\mathcal{E}$ is a monotone class. It follows now by **Q**7 that $\mathcal{E}$ contains (and is therefore equal to) $\mathcal{B}$. The boundedness of $\hat{f}$ is obvious since

$$|\hat{f}(\omega)| \leq \int_U |f(\omega, u)| q_1(\omega, \mathrm{d}u) \leq \|f\| \int_U q_1(\omega, \mathrm{d}u) = \|f\|.$$

Step 2. An immediate and obvious consequence of the above result is that the the first part of the theorem is true for simple functions. If f is an arbitrary bounded $\mathcal{B}$-measurable function, then a sequence (f_n) of $\mathcal{B}$-measurable simple functions dominated by f can be found such that f_n converges to f pointwise. Hence $\hat{f}(\omega) = \lim_{n\to\infty} \hat{f}_n(\omega)$ is as claimed in the first part of the theorem.

Step 3. It is now sufficient to show that μ is countably additive to qualify to be a pm.

Let $E_n,\ n \geq 1$ be a sequence of mutually exclusive events in $\mathcal{B}$. Let $F_n = \cup_1^n E_r$. Let $f_n = \chi_{F_n}$ Then

$$\begin{aligned}\mu(\cup_n E_n) &= \int_\Omega [\int_U \chi_{\{\cup_n E_n\}}(\omega, u) q_1(\omega, \mathrm{d}u)] q(\mathrm{d}\omega)\\ &= \int_\Omega [\int_U \lim_{n\to\infty} \chi_{F_n}(\omega, u) q_1(\omega, \mathrm{d}u)] q(\mathrm{d}\omega)\\ &= \text{(by the bounded convergence theorem)}\\ &\quad \int_\Omega \lim_{n\to\infty} [\int_U \chi_{F_n}(\omega, u) q_1(\omega, \mathrm{d}u)] q(\mathrm{d}\omega)\end{aligned}$$

$$= \text{(again by the bounded convergence theorem,}$$

$$\text{which applies since the } \hat{f}_n\text{s are bounded by 1)}$$

$$\lim_{n\to\infty} \int_{\Omega} [\int_{U} \chi_{F_n}(\omega, u) q_1(\omega, du)] q(d\omega)$$

$$= \lim_{n\to\infty} \int_{\Omega} \{\int_{U} \sum_{r=1}^{n} \chi_{E_r} q_1(\omega, \ du)\} q(\ d\omega)$$

$$= \lim_{n\to\infty} \sum_{r=1}^{n} \int_{\Omega} \{\int_{U} \chi_{E_r} q_1(\omega, \ du)\} q(\ d\omega)$$

$$= \lim_{n\to\infty} \sum_{r=1}^{n} \mu(E_r) = \sum_{1}^{\infty} \mu(E_r).$$

That $\mu(\Omega \times U) = 1$ follows immediately from (1.4). The proof that μ is a *pm* on $\mathcal{B}$ is now complete. □

Remark 1.11.4. Let $(\Omega_n, \ \mathscr{A}_n), \ n \geq 1$ be abstract measurable spaces. Let q_n be a transition pm on $\Omega_n \times \mathscr{A}_{n+1}$. Let q be a pm on $\mathscr{A}_1$. For $A \in \bigotimes_{1}^{n} \mathscr{A}_i$, define

$$\mu_n(A) = \int_{\Omega_1} q(d\omega_1) \Big\{ \int_{\Omega_2} q_1(\omega_1, d\omega_2) \Big\{ \int_{\Omega_3} q_2(\omega_2, d\omega_3)$$

$$\cdots \int_{\Omega_n} q_{n-1}(\omega_{n-1}, d\omega_n) \chi_A(\omega_1, \omega_2, \omega_3, \ldots, \omega_{n-1}, \omega_n) \Big\} \Big\}.$$

That the integral exists finitely is seen by a repeated application of **(1.11.3)**. Proof on similar lines can be constructed to show that μ_n is a pm.

Remark 1.11.5. Let $\mu(E) \geq 2\varepsilon$ and let $F = \{\omega : q_1(\omega, \ E_\omega) \geq \varepsilon\}$. Then $q(F) \geq \varepsilon$.

Proof. We have

$$\begin{aligned}2\varepsilon \le \mu(E) &= \int_{\Omega} q_1(\omega,\, E_\omega)q(\,d\omega)\\ &= \int_F q_1(\omega,\, E_\omega)q(\,d\omega) + \int_{F'} q_1(\omega,\, E_\omega)q(\,d\omega)\\ &\le \int_F q_1(\omega,\, E_\omega)q(\,d\omega) + \varepsilon\\ &\le q(F) + \varepsilon.\end{aligned}$$

Hence $q(F) \ge \varepsilon$. □

These *pms* μ_n can be considered as defined in a natural way on the σ-fields $\mathbf{a}_{J_n;T}$ where $J_n = \{1,\, 2, \ldots,\, n\}$ and $T = \{1,\, 2, \ldots\}$. We note

$$\mu_{n+1}(A \times \Omega_{n+1}) = \mu_n(A) \text{ for every } A \in \bigotimes_{r=1}^{n} \mathscr{A}_r.$$

This observation helps us see that the *pms* μ_n constitute a consistent family.

Theorem 1.11.6. *The μ_ns determine a unique pm μ on $\mathscr{A}_T$ such that its restriction to $\mathbf{a}_{J_n;T}$ is μ_n.*

Proof. Since it is a consistent family, the μ_ns determine a unique non-negative and finitely additive set function μ on $\widetilde{\mathscr{A}_T}$ such that $\mu(\Omega_T) = 1$ and such that μ restricted to $\mathbf{a}_{J_n;T}$ is μ_n. Let $(E_j,\, j \ge 1)$ be an arbitrary sequence in $\widetilde{A}_T$ monotonically decreasing to the null set. By **Q6**, the claim would follow if we show that $\mu(E_j) \downarrow 0$. Since μ is additive, $\mu(E_j) \downarrow$. If $\lim_{j\to\infty} \mu(E_j) \ne 0$, there exists $\varepsilon > 0$ such that $\mu(E_j) \ge \varepsilon$, $j \ge 1$.

For $n \ge 1$, let $J_n = \{1,\, 2, \ldots,\, n\}$ and then $(\Omega_{T\sim J_n},\, \widetilde{\mathscr{A}_{T\sim J_n}})$ be the analogue of $(\Omega_T,\, \widetilde{\mathscr{A}_T})$ constructed from $(\Omega_t,\, \mathscr{A}_t)$, $t \ge n+1$. Given $x \in \Omega_n$, we define the non-negative finitely additive set function $\mu^{(n)(x)}$ on $\widetilde{\mathscr{A}_{T\sim J_n}}$ analogous to μ on $\widetilde{\mathscr{A}_T}$ starting with the *pm* $q_n(x,\, .)$ on $\mathscr{A}_{n+1}$ and using the transition *pms* q_m, $m \ge n+1$. We note that if $E \in \widetilde{\mathscr{A}_T}$ and given $(x_1,\, x_2, \ldots,\, x_n) \in \times_1^n \Omega_r$, then the section $E(x_1,\, x_2, \ldots,\, x_n) \in \widetilde{\mathscr{A}_{T\sim J_n}}$.

Define

$$F_j = \{x : x \in \Omega_1,\, \mu^{(1)(x)}(E_j(x)) \ge \frac{\varepsilon}{2}\}.$$

Then

$$\begin{aligned}
\mu(E_j) &= \int_{\Omega_1} \mu^{(1)(x)}(E_j(x))q(dx) \\
&= \int_{F_j} \mu^{(1)(x)}(E_j(x))q(dx) + \int_{F'_j} \mu^{(1)(x)}(E_j(x))q(dx) \\
&\leq q(F_j) + \frac{\varepsilon}{2}.
\end{aligned}$$

Hence $q(F_j) \geq \frac{\varepsilon}{2}$. We use the fact $E_j \downarrow$ and the definition of F_j to conclude that $F_j \downarrow$. The F_js are members of $\mathscr{A}_1$ and q is a pm. These facts imply that $\cap_j F_j \neq \phi$. Hence there exists $x_1 \in F_j,\ j \geq 1$ such that, for all $j \geq 1,\ \mu^{(1)(x_1)}(E_j(x_1)) \geq \frac{\varepsilon}{2}$.

Since $(E_j(x_1))$ is a decreasing sequence of sets in $\widetilde{\mathscr{A}_{T\sim J_1}}$, the arguments applied to $\Omega_T,\ (E_j,\ j \geq 1),\ q,\ (q_n,\ n \geq 1), \varepsilon$ may be repeated for $\Omega_{T\sim J_1}$, $(E_j(x_1), j \geq 1), q_1(x_1,\ .), (q_n,\ n \geq 2),\ \frac{\varepsilon}{2}$. We obtain $x_2 \in \Omega_2$ such that

$$\mu^{(2)(x_1,\ x_2)}(E_j(x_1,\ x_2)) \geq \frac{\varepsilon}{4},\quad j \geq 1.$$

This process can be continued indefinitely and we arrive at a point $(x_1,\ x_2, \dots) \in \Omega_T$ and

$$\mu^{(n)(x_1,\ x_2,\dots,\ x_n)}(E_j(x_1,\ x_2, \dots,\ x_n)) \geq \frac{\varepsilon}{2^n},\quad j \geq 1.$$

Given $j \geq 1$, we can find $n \geq 1$ such that $E_j \in \mathbf{a}_{J_n,T}$. The fact that

$$\mu^{(n)(x_1,\dots,\ x_n)}(E_j(x_1,\ x_2, \dots,\ x_n)) > 0$$

implies that E_j contains at least one point whose first n co-ordinates are x_1, $x_2, \dots, x_n$. But $E_j \in \mathbf{a}_{J_n,T}$. i.e. it is a cylinder set. Hence $(x_1,\ x_2, \dots) \in E_j,\ j \geq 1$, showing $\cap_j E_j \neq \phi$, contradicting the assumption that $E_j \downarrow \phi$.

The unique pm of the theorem will also be denoted by μ. □

Remark 1.11.7. Suppose that for each $n,\ q_n$ is a pm on $\mathscr{A}_{n+1}$ independent of ω_n. Then it is easy to see that for measurable rectangles $E = \times_1^n A_i$, where $A_i \in \mathscr{A}_i,\ \mu_n(E) = q(A_1)q_1(A_2)\dots q_{n-1}(A_n)$. In this case μ_n is called the product measure of the components $q,\ q_i,\ 1 \leq i \leq n-1$ and μ the product measure of the components $q,\ q_i,\ 1 \leq i < \infty$. In symbols we write

$$\mu_n = q \times q_1 \times q_2 \times \cdots \times q_{n-1}$$

and

$$\mu = q \times q_1 \times q_2 \times \dots$$

$(\times_1^n \Omega_i, \bigotimes_1^n \mathscr{A}_i, \mu_n)$, $(\Omega_T, \mathscr{A}_T, \mu)$ are called the corresponding product measure spaces.

Remark 1.11.8. Let $(\Omega, \mathscr{A}, \mu_1)$, $(U, \mathscr{U}, \mu_2)$ be probability measure spaces and let $(\Omega \times U, \mathcal{B}, \mu_1 \times \mu_2)$, $(U \times \Omega, \mathcal{C}, \mu_2 \times \mu_1)$ be the product measure space where $\mathcal{B} = \mathscr{A} \bigotimes \mathscr{U}$ and $\mathcal{C} = \mathscr{U} \bigotimes \mathscr{A}$. We note the obvious one-to-one correspondence between $\mathcal{B}$ and $\mathcal{C}$ through the mapping $(\omega, u) \leftrightarrow (u, \omega)$. Under this mapping let the image of E be denoted by $\hat{E}$. Then $(\mu_1 \times \mu_2)(E) = (\mu_2 \times \mu_1)(\hat{E})$.

Proof. Trivially, if $E = A \times B$, $A \in \mathscr{A}$ and $B \in \mathscr{U}$, then

$$\begin{aligned}(\mu_1 \times \mu_2)(E) &= (\mu_1 \times \mu_2)(A \times B) = \mu_1(A) \times \mu_2(B) \\ &= (\mu_2 \times \mu_1)(B \times A) = (\mu_2 \times \mu_1)(\hat{E}).\end{aligned}$$

Thus, the measurable rectangles belong to $\mathcal{E}$ the collection of all sets $E \in \mathcal{B}$ with $(\mu_1 \times \mu_2)(E) = (\mu_2 \times \mu_1)(\hat{E})$. The fact that $\mathcal{E}$ is a monotone class is a consequence of the monotone convergence theorem in integration theory (ref. **Q**2). Using the fact that $(\mu_1 \times \mu_2)$ is a pm, it is easy to see that $\mathcal{E}$ is a field. Since it is, additionally, a monotone class it is a σ-field. With this the proof is complete. □

Remark 1.11.9. It is immediate from **(1.11.8)** that

$$(\mu_1 \times \mu_2)(E) = \int_{\Omega} \mu_2(E_x)\, d\mu_1(x) = \int_{U} \mu_1(E_u)\, d\mu_2(u).$$

Let $(\Omega, \mathscr{A}) = (U, \mathscr{U}) = (R^p, \mathscr{R}^p)$, let $E \in \mathscr{R}^p$ and let $A = \{(\mathbf{x}, \mathbf{y}) : \mathbf{x} + \mathbf{y} \in E\}$. (It may be noted that $\chi_A(\mathbf{x}, \mathbf{y}) = \chi_E(\mathbf{x} + \mathbf{y})$). Then the above conclusion would read

$$(\mu_1 \times \mu_2)(A) = \int_{R^p} \mu_2(E - \mathbf{x})\, d\mu_1(\mathbf{x}) = \int_{R^p} \mu_1(E - \mathbf{x})\, d\mu_2(\mathbf{x}).$$

Q8 For proof of the following well known theorem, refer to p148 **[H]**.

Theorem 1.11.10 (Fubini). *With notation as in **(1.11.8)**, let h be a $\mu_1 \times \mu_2$ integrable function on $\Omega \times U$. Then section h_ω is μ_2 integrable and section h_u*

is μ_1 integrable. If $f(\omega) = \int_U h(\omega, u)\, d\mu_2(u)$ and $g(u) = \int_\Omega h(\omega, u)\, d\mu_1(\omega)$, then $f,\ g$ are measurable and are respectively μ_1 and μ_2 integrable. Further,

$$\int_{\Omega\times U} h(\omega, u)\, d(\mu_1 \times \mu_2)(\omega, u) = \int_\Omega f(\omega)\, d\mu_1(\omega) = \int_U g(u)\, d\mu_2(u). \quad (1.5)$$

The relation (1.5) holds also if h is a non-negative measurable function on $\Omega \times U$. (We note such a function h is always integrable with possibly ∞ for the value of the integral)

Some applications

(i) Let F be an arbitrary *df* and $c > 0$, arbitrary. Then

$$\int_{-\infty}^{\infty} \{F(x+c) - F(x)\}\, dx$$

$$= \int_{-\infty}^{\infty} \{ \int_{(x,\, x+c]} dF(y)\}\, dx = \int_{-\infty}^{\infty} \{\int_{y-c}^{y} dx\}\, dF(y)$$

$$= c \int_{-\infty}^{\infty} dF(y) = c.$$

[As a numerical illustration, take

$$F(x) = 1 - e^{-x}, \quad x \geq 0.$$

Then $\displaystyle\int_{-\infty}^{\infty} \{F(x+c) - F(x)\}\, dx$

$$= \int_{-c}^{0} F(x+c)\, dx + \int_{0}^{\infty} \{F(x+c) - F(x)\}\, dx$$

$$= \int_{-c}^{0} \{1 - e^{-(x+c)}\}\, dx + \int_{0}^{\infty} \{e^{-x} - e^{-(x+c)}\}\, dx$$

$$= c - (1 - e^{-c}) + (1 - e^{-c}) = c]$$

(ii) If F is a continuous df then

$$G_n(x) = \int_{-\infty}^{x} F^n(y)\, dF(y) = \frac{F^{n+1}(x)}{n+1}, \quad n \geq 0. \tag{*}$$

To see this we first note that since F is a bounded measurable function and since the integration is with respect to a df, $G_n(x)$ exists finitely for all x and for all $n \geq 0$.

We note that for any two continuous dfs F_1, F_2,

$$\int_{-\infty}^{x} F_1(y)\, dF_2(y) = \int_{-\infty}^{x} \{\int_{-\infty}^{y} dF_1(t)\}\, dF_2(y)$$

$$= \int_{-\infty}^{x} \{\int_{t}^{x} dF_2(y)\}\, dF_1(t) = \int_{-\infty}^{x} \{F_2(x) - F_2(t)\}\, dF_1(t).$$

We establish (*) by induction. For $n = 0$ the claim is obvious. Suppose it has been shown that

$$G_k(x) = \frac{F^{k+1}(x)}{k+1}, \; k = 0,\, 1, \ldots,\, n-1;\; n \geq 1.$$

The induction hypothesis gives

$$F^n(t) = n \int_{-\infty}^{t} F^{n-1}(u)\, dF(u).$$

Hence by Radon-Nikodym Theorem

$$d\mu^{(n)}(x) = nF^{n-1}(x)\, d\mu(x)$$

where $\mu^{(n)}$, μ are respectively the measures generated by F^n and F. We have

$$G_n(x) = \int_{-\infty}^{x} F^n(y)\, dF(y) = \int_{-\infty}^{x} \{F(x) - F(t)\}\, dF^n(t)$$

$$= F^{n+1}(x) - \int_{-\infty}^{x} F(t)\, dF^n(t)$$

$$= F^{n+1}(x) - \int_{-\infty}^{x} nF(t)F^{n-1}(t)\, dF(t)$$

$$= F^{n+1}(x) - nG_n(x).$$

Thus

$$G_n(x) = \frac{1}{n+1} F^{n+1}(x).$$

(iii) *An alternate proof of* **(1.5.6)**.
In the notation of **(1.5.6)**,

$$\varphi(E) = \int_{R^p} \{ \int_{y \in \mathbf{x}+E} d\mu(\mathbf{y}) \} \, d\mathbf{x}$$

$$= \int_{R^p} \{ \int_{\mathbf{x} \in \mathbf{y}-E} d\mathbf{x} \} \, d\mu(\mathbf{y}) \text{ (by the last part of } \mathbf{Q}8)$$

$$= \int_{R^p} L(\mathbf{y} - E) \, d\mu(\mathbf{y}) = \int_{R^p} L(E) \, d\mu(\mathbf{y})) = L(E).$$

Remark 1.11.11. (i) For two other methods of proving the result, refer to **(2.3.7)** and to **(3.1.3)**(vi)
(ii) The condition of continuity imposed on F is essential, as the following example makes it clear. Consider

$$F(x) = \begin{cases} 0 & \text{if } x < 0; \\ p & \text{if } 0 \leq x < 1; \text{ and} \\ 1 & \text{if } x \geq 1; \ 0 < p < 1. \end{cases}$$

Then

$$I = \int_{-\infty}^{\infty} F(x) \, dF(x) = F(0)p + F(1)(1-p) = p^2 + 1 - p.$$

For $p = \frac{1}{2}$, $I = \frac{3}{4}$.

1.12 Convolution and shift tightness

Let μ_i, $i = 1, 2$ be *pms* on $\mathscr{R}^p$. Consider the integral

$$I = \int_{R^p} \mu_1(E - \mathbf{y}) \, d\mu_2(\mathbf{y}), \quad E \in \mathscr{R}^p.$$

By **(1.5.3)**, the integrand is a measurable function; further it is bounded-by-one, since μ_1 is a *pm*. Hence the integral exists. By **(1.11.9)**,

$$I = \int_{R^p} \mu_2(E - \mathbf{y}) \, d\mu_1(\mathbf{y}).$$

Define $\mu(E)$ to be this common value : for $E \in \mathscr{R}^p$,

$$\mu(E) = \int\limits_{R^p} \mu_1(E - \mathbf{y})\, d\mu_2(\mathbf{y}) = \int\limits_{R^p} \mu_2(E - \mathbf{x})\, d\mu_1(\mathbf{x}).$$

Using the facts that $E' - \mathbf{x} = (E - \mathbf{x})'$, that $\cup_n E_n - \mathbf{x} = \cup_n (E_n - \mathbf{x})$ and that μ_1 is a pm and using the monotone convergence theorem (ref. **Q2**), it is straightforward to show that μ is a pm.

Definition 1.12.1. The pm μ is called the convolution or the composition of the pms μ_1 and μ_2.

We indicate this by writing $\mu = \mu_1 * \mu_2$. We note that μ is also equal to $\mu_2 * \mu_1$ and therefore that the convolution operation $*$ is commutative.

If $\mu_1,\ \mu_2,\ \mu_3$ are pms on $\mathscr{R}^p$ and if $E \in \mathscr{R}^p$, then

$$\begin{aligned}
((\mu_1 * \mu_2) * \mu_3)(E) &= \int\limits_{R^p} (\mu_1 * \mu_2)(E - \mathbf{x})\, d\mu_3(\mathbf{x}) \\
&= \int\limits_{R^p} \{ \int\limits_{R^p} \mu_2(E - \mathbf{x} - \mathbf{y})\, d\mu_1(\mathbf{y})\}\, d\mu_3(\mathbf{x}) \\
&= \int\limits_{R^p} \{ \int\limits_{R^p} \mu_2(E - \mathbf{x} - \mathbf{y})\, d\mu_3(\mathbf{x})\}\, d\mu_1(\mathbf{y})
\end{aligned}$$

(by Fubini's theorem. ref. **Q8**)

$$\begin{aligned}
&= \int\limits_{R^p} (\mu_2 * \mu_3)(E - \mathbf{x})\, d\mu_1(\mathbf{x}) \\
&= (\mu_2 * \mu_3) * \mu_1(E) = \mu_1 * (\mu_2 * \mu_3)(E).
\end{aligned}$$

This shows that the operation $*$ is associative.

Denote by $F_1,\ F_2,\ F$ the dfs associated with the pms $\mu_1,\ \mu_2$ and μ respectively. Taking for E the interval $(-\infty,\ \mathbf{x}]$, we have

$$F(\mathbf{x}) = \int\limits_{R^p} F_1(\mathbf{x} - \mathbf{y})\, d\mu_2(\mathbf{y});$$

this is also equal to

$$\int\limits_{R^p} F_2(\mathbf{x} - \mathbf{y})\, d\mu_1(\mathbf{y}).$$

This is written in symbols as $F = F_1 * F_2 = F_2 * F_1$. If this relation is written as $F(x) = \int_{R^p} F_1(\mathbf{x}-\mathbf{y})\, dF_2(\mathbf{y})$ the integral should be understood as the same as $\int_{R^p} F_1(\mathbf{x}-\mathbf{y})\, d\mu_2(\mathbf{y})$ and not as a Riemann-Stieltjes integral. Except in special situations like when F_1 or F_2 is a continuous function, $F(\mathbf{x})$ may fail to exist as a Riemann-Stieltjes integral. For example, let F be the distribution function on the real line defined thus:

$$F(x) = \begin{cases} 0 & \text{if } x < 0 \\ 1 & \text{if } x \geq 1. \end{cases}$$

Then $\int\limits_R F(x-y)\, dF(y)$ does not exist as a Riemann-Stieltjes integral for $x = 0$.

Theorem 1.12.2. *Let (μ_n), (ν_n) be two tight sequences of pms in $\mathscr{R}^p$. Then (λ_n) is a tight sequence too, where $\lambda_n = \mu_n * \nu_n$.*

Proof. Fix $\varepsilon > 0$. The hypothesis ensures the existence of a compact set K such that, for all n, $\mu_n(K) > 1-\varepsilon$ and $\nu_n(K) > 1-\varepsilon$. There is no loss in generality in taking K to be a closed symmetric interval : $K = \times_1^p[-a \leq x_i \leq a]$. Define compact set $C = K - K = \times_1^p[-2a \leq x_i \leq 2a]$ and note that for every x $\in K$, we have $C' -$ x $\subset K'$. Now

$$\begin{aligned}
\lambda_n(C') &= \int\limits_{R^p} \mu_n(C'-\mathrm{x})\, d\nu_n(\mathrm{x}) \\
&= \int\limits_K \mu_n(C'-\mathrm{x})\, d\nu_n(\mathrm{x}) + \int\limits_{K'} \mu_n(C'-\mathrm{x})\, d\nu_n(\mathrm{x}) \\
&\leq \int\limits_K \mu_n(C'-\mathrm{x})\, d\nu_n(\mathrm{x}) + \int\limits_{K'} d\nu_n(\mathrm{x}) \\
&\leq \int\limits_K \mu_n(C'-\mathrm{x})\, d\nu_n(\mathrm{x}) + \varepsilon \\
&\leq \int\limits_K \mu_n(K')\, d\nu_n(\mathrm{x}) + \varepsilon \leq 2\varepsilon.
\end{aligned}$$

□

Theorem 1.12.3. *Let (μ_n), (ν_n) be two sequences of pms in $\mathscr{R}^p$. Let $\lambda_n = \mu_n * \nu_n$. If (λ_n) is a tight sequence then the sequences (μ_n), (ν_n) are shift tight.*

Proof. Let (λ_n) be a tight sequence. Hence we can find a compact set K_r such that

$$\lambda_n(K_r) > 1 - \frac{1}{3^{r+1}}.$$

Define

$$A_{n;r} = \{\mathbf{x} : \mu_n(K_r - \mathbf{x}) > 1 - \frac{1}{2^r}\}.$$

Thus

$$\begin{aligned} 1 - \frac{1}{3^{r+1}} < \lambda_n(K_r) &= \int_{R^p} \mu_n(K_r - \mathbf{x})\, d\nu_n(\mathbf{x}) \\ &= \int_{A_{n;r}} \mu_n(K_r - \mathbf{x})\, d\nu_n(\mathbf{x}) + \int_{A'_{n;r}} \mu_n(K_r - \mathbf{x})\, d\nu_n(\mathbf{x}) \\ &\leq \nu_n(A_{n;r}) + (1 - \frac{1}{2^r})\nu_n(A'_{n;r}) \\ &= 1 - \frac{1}{2^r}\nu_n(A'_{n;r}), \end{aligned}$$

leading to

$$\nu_n(A'_{n;r}) \leq \frac{1}{3}(\frac{2}{3})^r.$$

Define

$$A_n = \bigcap_{r=1}^{\infty} A_{n;r}$$

$$\text{and note that} \quad \nu_n(A'_n) \leq \sum_{r=1}^{\infty} \nu_n(A'_{n;r}) \leq \frac{1}{3}\sum_{r=1}^{\infty}(\frac{2}{3})^r = \frac{2}{3}.$$

This implies that $A_n \neq \phi$. Choose and fix $\mathrm{x}_n \in A_n$. Since $\mathrm{x}_n \in A_{n;r}$ for every $r \geq 1$, we have

$$\mu_n(K_r - \mathbf{x}_n) > 1 - \frac{1}{2^r}.$$

This is equivalent to claiming that the sequence (μ_n) is a shift tight. Similarly we can find y_n such that for each $r \geq 1$ fixed,

$$\nu_n(K_r - \mathbf{y}_n) > 1 - \frac{1}{2^r} \quad \text{for all } n,$$

proving the shift tightness of the sequence (ν_n). □

Chapter 2

Weak Convergence of Probability Measures

2.1 Weak convergence of distribution functions

Let $\mathbf{X} = (X_1, X_2, \ldots, X_p)$ be a p-vector variable with *df* $\mathbb{F}$ and *dm* denoted by $\mu_{\mathbf{X}}$ or $\mu_{\mathbb{F}}$. The *df* F_j of X_j is called the jth marginal of **X** or of $\mathbb{F}$ or of $\mu_{\mathbb{F}}$, $1 \leq j \leq p$.

$$F_j(x) = \lim_{\min\{x_r:1\leq r\leq p,\ r\neq j\}\to\infty} \mathbb{F}(x_1, \ldots, x_{j-1}, x, x_{j+1}, \ldots, x_p).$$

Let $(\mu_n, n \geq 0)$ be a sequence of *pms* in R^p with associated *df*s $(\mathbb{F}_n)$. Since we will be defining the weak convergence of μ_n to μ_0 as pointwise convergence of $\mathbb{F}_n$ to $\mathbb{F}_0$ at the continuity points of $\mathbb{F}_0$, it is desirable to study first the discontinuity points of a general *df* in R^p, relating them to those of its one-dimensional marginals.

Let $(\mathbb{X}_n, n \geq 0)$ be a sequence of random vectors in R^p with corresponding *df*s $(\mathbb{F}_n)$ and induced probability measures (μ_n).

Definition 2.1.1. Sequence $(\mathbb{X}_n)$ is said to converge in distribution (or, weakly) to $\mathbb{X}_0$ ($\mathbb{X}_n \overset{d}{\to} \mathbb{X}_0$), also $\mathbb{F}_n$ is said to converge weakly to $\mathbb{F}_0$ ($\mathbb{F}_n \overset{d}{\to} \mathbb{F}_0$), also μ_n is said to converge weakly to μ_0 ($\mu_n \overset{d}{\to} \mu_0$) if $\mathbb{F}_n(\mathbf{x}) \to \mathbb{F}_0(\mathbf{x})$ for every $\mathbf{x} \in C_{\mathbb{F}_0}$, the set of all the continuity points of $\mathbb{F}_0$.

Remarks 2.1.2.
(i) Let

$$\Phi(x) = \int_{-\infty}^{x} \frac{1}{\sqrt{2\pi}} e^{-\frac{1}{2}y^2} dy, -\infty < x < \infty.$$

(Φ is called the (one dimensional) standard normal *df* and will be denoted by $\mathcal{N}(0,\ 1)$). Let $\sigma > 0$. Then the *df* $\psi(x) = \Phi(\frac{x-\theta}{\sigma})$ will be denoted by

$\mathcal{N}(\theta, \sigma^2)$. It is easy to verify that $\int\limits_R x\, d\psi(x)$ (called the mean or the first moment of the *df* ψ) is θ and that $\int\limits_R (x-\theta)^2\, d\psi(x)$ (called the variance or the second moment about the mean) is σ^2. Define $F_n(x) = \frac{1}{2}\{\Phi(n\overline{x-1}) + \Phi(nx)\}$. Let $G(x) = 0$ for $x < 0$; $= \frac{1}{4}$ for $x = 0$; $= \frac{1}{2}$ for $0 < x < 1$; $\frac{3}{4}$ for $x = 1$; and $= 1$ for $x > 1$. It is easy to verify that $\lim\limits_{n\to\infty} F_n(x) = G(x)$ for all x. But we do not say that $F_n \overset{d}{\to} G$ since G is not right continuous. If $F_0(x) = 0$ for $x < 0$; $= \frac{1}{2}$ for $0 \le x < 1$; and $= 1$ for $x \ge 1$, then clearly F_0 is right continuous and $\lim\limits_{n\to\infty} F_n(x) = F_0(x)$ for all x except 0 and 1. i.e. at all the continuity points of F_0. We say $F_n \overset{d}{\to} F_0$.

Since a *df* is determined completely by its values at its continuity points, it follows that if $F_n \overset{d}{\to} F$ and $F_n \overset{d}{\to} G$, then $F \equiv G$. The following example presents another possibility. Let

$$F_n(x) = \begin{cases} 0 & \text{for } x < -n; \\ \frac{1}{4} & \text{for } -n \le x < 0; \\ \frac{3}{4} & \text{for } 0 \le x < n \\ 1 & \text{for } x \ge n \end{cases}$$

$$F(x) = \begin{cases} \frac{1}{4} & \text{for } x < 0, \\ \frac{3}{4} & \text{for } x \ge 0. \end{cases}$$

It is clear that for every x, $F_n(x) \to F(x)$ but we can not say that $F_n \overset{d}{\to} F$ since F is not a *df*.

(ii) Suppose $F_n \overset{d}{\to} F$ and suppose $u \notin C_F$, where C_F is the set of all the continuity points of the *df* F. Then what can we say about the sequences $(F_n(u))$ and $(F_n(u-))$? It is possible that $F_n(u) \to F(u)$ and $F_n(u-) \to F(u-)$ as in the case $F_n = F$. It is possible that either or both the sequences fail to converge. And even if convergent, the limit may not be equal to $F(u)$ or $F(u-)$ as the case may be. Witness the following example. Define *df*s G_1, G_2 as follows.

$$G_1(x) = \begin{cases} \frac{1}{6} \int\limits_{-\infty}^{x} e^{-|x|} dx, & \text{if } x < 0 \\ \frac{1}{6}, & \text{if } 0 \le x < 1 \\ 1, & \text{if } x \ge 1 \end{cases}$$

$$G_2(x) = \begin{cases} \frac{1}{3\sqrt{2\pi}} \int\limits_{-\infty}^{x} e^{-\frac{y^2}{2}} dy & \text{if } x < 0 \\ \frac{5}{6} & \text{if } x = 0 \\ \frac{5}{6} + \frac{1}{3\sqrt{2\pi}} \int\limits_{0}^{x} e^{-\frac{y^2}{2}} & \text{if } x > 0. \end{cases}$$

Define $F_{2n+1}(x) = G_1(nx)$, $n \geq 0$ and $F_{2n}(x) = G_2(nx)$, $n \geq 1$. If

$$F_0(x) = \begin{cases} 0 & \text{for } x < 0 \\ 1 & \text{for } x \geq 1, \end{cases}$$

then it is easily verified that $F_n \overset{d}{\to} F_0$. 0 is a discontinuity point of F_0. $F_n(0-) = \frac{1}{6}$ for all n; $F_{2n+1}(0) = \frac{1}{6}$; $F_{2n}(0) = \frac{5}{6}$; $F(0) = 1$.

(iii) Let F be a non-degenerate df. Define dfs $F^{\times a}$, $a > 0$: $F^{\times a}(x) = F(ax)$; $F^{+a} : F^{+a}(x) = F(x + a)$, a real.

Let $F_n \overset{d}{\to} F$. Then

(α) $F_n^{\times a_n}$ converges weakly *iff* $\lim\limits_{n\to\infty} a_n = a$ exists, $0 < a < \infty$ and then $F_n^{\times a_n} \overset{d}{\to} F^{\times a}$.

(β) $F_n^{+\theta_n}$ converges weakly *iff* $\lim\limits_{n\to\infty} \theta_n = \theta$ exists finitely and then $F_n^{+\theta_n} \overset{d}{\to} F^{+\theta}$.

Proof. (α) Let $F_n^{\times a_n} \overset{d}{\to} G$. Let a subsequence (a_m) have limit zero. Given $\varepsilon > 0$ and $x > 0$ a continuity point of G, choose $\eta \leq \varepsilon$ such that $x\eta$ is a continuity point of F. We note for all m large $a_m \leq \eta$. We have

$$G(x) = \lim_{m\to\infty} F_n(xa_n) \leq F_n(x\eta) \to F(x\eta) \leq F(x\varepsilon).$$

Since $\varepsilon > 0$ is arbitrary, this leads to $G(x) \leq F(0)$, $x > 0$. Letting $x \to \infty$, we conclude $F(0) = 1$. Parallel arguments for $x < 0$ lead to $F(0-) = 0$. This is not possible since F is a non-degenerate df. Thus $\varliminf\limits_{n\to\infty} a_n > 0$.

Similarly it can be shown that $\varlimsup\limits_{n\to\infty} a_n < \infty$.

Let $\lim\limits_{m\to\infty} a_m = a$, $0 < a < \infty$. Given $\varepsilon > 0$ let $0 < \eta < \varepsilon$ arbitrary. We will have, for all n large, $a - \eta < a_n < a + \eta$. Let $x > 0$ be a continuity point of G and of $F^{\times a}$. Choose η such that $a - \eta$, $a + \eta$ are continuity points of F.

Now,

$$G(x) = \lim_{m\to\infty} F_m(xa_m) \leq \lim_{m\to} F_m(x(a + \eta) = F(x\overline{a + \eta}).$$

Let $\eta \to 0$ and conclude $G(x) \le F(ax)$; starting from $G(x) \ge F_m(x\overline{a - \eta})$, we conclude $G(x) = F^{\times a}(x)$. Thus $G(x) = F(ax)$. Similar arguments hold for $x < 0$. Thus if $a_m \to a$ then $G(x) = F(ax)$ at all the common continuity points of G and $F^{\times a}$. Since these points form a dense subset of R, it follows $G(x) = F(ax)$, $x \in R$. This result holds for every limit point a of the sequence (a_n). This is possible only if $\lim_{n\to\infty} a_n = a$ exists and is positive and finite. And then $G(x) = F(ax)$. i.e. the claim is true.

(β) Arguments are similar. Details are omitted. □

Theorem 2.1.3. *If* $\mathbb{F}_n \overset{d}{\to} \mathbb{F}_0$ *then* $F_{n,j} \overset{d}{\to} F_{0,j}$, $1 \le j \le p$, *the* $F_{n,j}$*s being the one-dimensional marginals of* $\mathbb{F}_n$, $n \ge 0$.

[*Note.* ***(2.4.2)****(vii) read with* ***(2.1.12)*** *provides another proof of this theorem*].

Proof. There is no loss in generality in proving result for $j = 1$. Let $\mathbb{H}_n$ denote the $(p-1)$ dimensional marginal of $\mathbb{F}_n$ corresponding to the second, the third,. . ., the pth components. We note

$$F_{n,1}(x) \ge \mathbb{F}_n(x, \lambda_2, \lambda_3, \ldots, \lambda_p)$$

and also,

$$1 - F_{n,1}(x) \ge \mathbb{H}_n(\lambda_2, \lambda_3, \ldots, \lambda_p) - \mathbb{F}_n(x, \lambda_2, \lambda_3, \ldots, \lambda_p)$$

for all values of the arguments. Let $x, \lambda_2, \lambda_3, \ldots, \lambda_p$ be continuity points respectively of $F_{0,1}, F_{0,2}, \ldots, F_{0,p}$. Hence, by **(2.1.1)**,

$$(x, \lambda_2, \lambda_3, \ldots, \lambda_p) \in C_{\mathbb{F}_0}$$

and

$$(\lambda_2, \lambda_3, \ldots, \lambda_p) \in C_{\mathbb{H}_0}.$$

From these we get:
(i)

$$\underline{\lim}_{n\to\infty} F_{n,1}(x) \ge \lim_{n\to\infty} \mathbb{F}_n(x, \lambda_2, \lambda_3, \ldots, \lambda_p) = \mathbb{F}_0(x, \lambda_2, \lambda_3, \ldots, \lambda_p).$$

Take $\lambda_2, \lambda_3, \ldots, \lambda_p \to \infty$, to get $\underline{\lim}_{n\to\infty} F_{n,1}(x) \ge F_{0,1}(x)$.

Similarly,

$$\begin{aligned}\underline{\lim}_{n\to\infty} \mathbb{H}_n(\lambda_2, \lambda_3, \ldots, \lambda_p) &\ge \underline{\lim}_{n\to\infty} \mathbb{F}_n(\lambda, \lambda_2, \lambda_3, \ldots, \lambda_p) \text{ (for all } \lambda) \\ &= \mathbb{F}_0(\lambda, \lambda_2, \lambda_3, \ldots, \lambda_p) \text{ (choosing } \lambda \in C_{F_{0,1}}) \\ &\to \mathbb{H}_0(\lambda_2, \lambda_3, \ldots, \lambda_p), \text{ as } \lambda \to \infty.\end{aligned}$$

(ii)

$$\begin{aligned} 1-\overline{\lim_{n\to\infty}}F_{n,1}(x) &= \underline{\lim_{n\to\infty}}\{1-F_{n,1}(x)\} \\ &\geq \underline{\lim_{n\to\infty}}\,\mathbb{H}_n(\lambda_2,\lambda_3,\ldots,\lambda_p)-\lim_{n\to\infty}\mathbb{F}_n(x,\lambda_2,\lambda_3,\ldots,\lambda_p) \\ &\geq \mathbb{H}_0(\lambda_2,\lambda_3,\ldots,\lambda_p)-\mathbb{F}_0(x,\lambda_2,\lambda_3,\ldots,\lambda_p). \end{aligned}$$

Take $\lambda_2,\lambda_3,\ldots,\lambda_p\to\infty$, to get $1-\overline{\lim_{n\to\infty}}F_{n,1}(x)\geq 1-F_{0,1}(x)$. Combining the two inequalities we get

$$F_{n,1}(x)\to F_{0,1}(x) \text{ for all } x\in C_{F_{0,1}}. \qquad \square$$

Remarks 2.1.4.
(i) Let $(x,\ \lambda_2,\ \lambda_3,\ \ldots,\ \lambda_p)\in C_{\mathbb{F}_0}$. Taking $n\to\infty$ and then $x\to\infty$ in the inequalities

$$\begin{aligned} \mathbb{F}_n(x,\ \lambda_2,\ \lambda_3,\ \ldots,\ \lambda_p) &\leq \mathbb{H}_n(\lambda_2,\ \lambda_3,\ \ldots,\ \lambda_p) \\ &\leq 1-F_{n,1}(x)+\mathbb{F}(x,\ \lambda_2,\ \lambda_3,\ \ldots,\ \lambda_p), \end{aligned}$$

we get

$$\overline{\lim_{n\to\infty}}\,\mathbb{H}_n(\lambda_2,\lambda_3,\ldots,\lambda_p)\leq \mathbb{H}_0(\lambda_2,\lambda_3,\ldots,\lambda_p).$$

Thus $\mathbb{H}_n \overset{d}{\to} \mathbb{H}_0$.
(ii) Arguing similarly it can be shown that every q-dimensional marginal of $\mathbb{F}_n$, $q\leq p$, converges weakly to the corresponding q-dimensional marginal of $\mathbb{F}_0$.
(iii) All these results can be obtained as corollaries of a more general theorem to be proved later (ref. **(2.4.2)**).
(iv) Given $\mathbf{x}\in R^p$, denote by $V_{p,\,\mathbf{x}}$ the collection of all the 2^p symbols $\mathbf{x}^l=(x_1^l,x_2^l,\ldots,x_p^l)$ where $x_i^l=x_i$ or x_i-. Denote $F(x_1^l,x_2^l,\ldots,x_p^l)$ by $\mathbb{F}(\mathbf{x}^l)$. Thus in the case $p=2$, the 2^2 $(=4)$ possible values of $F(x^l,\ y^l)$ are $F(x,\ y)$, $F(x,\ y-)$, $F(x-,\ y)$, and $F(x-,\ y-)$.

(v) Again arguing on similar lines, the following proposition can be proved : If $\mathbb{F}_n(\mathbf{x}^l)\to\mathbb{F}_0(\mathbf{x}^l)$ for all $\mathbf{x}\in R^p$ and a particular x^l, then for $1\leq j\leq p$ and any u^l, $F_{n,j}(u^l)\to F_{0,j}(u^l)$ for all $u\in R$.

(vi) Next we investigate conditions to ensure uniform convergence. i.e. to ensure that $\sup_{\mathbf{x}\in R^p}|\mathbb{F}_n(\mathbf{x})-\mathbb{F}_0(\mathbf{x})|\to 0$. A necessary condition obviously is : $|\mathbb{F}_n(\mathbf{x})-\mathbb{F}_0(\mathbf{x})|\to 0$ for every $\mathbf{x}\in R^p$. That this condition is not sufficient is brought out by the following example. Let $p = 1$,

$$F_n(x) = \begin{cases} 0, & \text{if } x < -\frac{1}{n}; \\ \frac{1}{2}, & \text{if } -\frac{1}{n} \leq x < 1; \\ 1 & \text{if } x \geq 1 \end{cases}$$

and

$$F(x) = \begin{cases} 0 & \text{if } x < 0; \\ \frac{1}{2} & \text{if } 0 \leq x < 1; \\ 1 & \text{if } x \geq 1. \end{cases}$$

Then $F_n(x) \to F(x)$ for all x but $\sup_{x \in R} |F_n(x) - F(x)| = \frac{1}{2}$.

Theorem 2.1.5. *If $\mathbb{F}_n \xrightarrow{d} \mathbb{F}$ and if $\boldsymbol{x} \in C_{\mathbb{F}}$, then $\mathbb{F}_n(\boldsymbol{x}^l) \to \mathbb{F}(\boldsymbol{x})$, $\boldsymbol{x}^l \in V_{\boldsymbol{x}}$.*

Proof. Since $\mathbb{F}_n(\mathbf{x}^l) \leq \mathbb{F}_n(\mathbf{x})$, we have

$$\overline{\lim_{n\to\infty}} \mathbb{F}_n(\mathbf{x}^l) \leq \lim_{n\to\infty} \mathbb{F}_n(\mathbf{x}) = \mathbb{F}(\mathbf{x}).$$

Choose $\mathbf{x}_k \in C_F$, $\mathbf{x}_k \ll \mathbf{x}$, $\mathbf{x}_k \to \mathbf{x}$ as $k \to \infty$. This is always possible since $C_{\mathbb{F}}$ is dense in R^p (ref. **(1.7.2)**(vii)). Since $\mathbb{F}_n(\mathbf{x}^l) \geq \mathbb{F}_n(\mathbf{x}_k)$, we have

$$\underline{\lim_{n\to\infty}} \mathbb{F}_n(\mathbf{x}^l) \geq \lim_{n\to\infty} \mathbb{F}_n(\mathbf{x}_k) = \mathbb{F}(\mathbf{x}_k) \to \mathbb{F}(\mathbf{x}). \qquad \square$$

We now present a necessary and sufficient condition for the uniform convergence of a weakly convergent sequence of *df*s. Clearly there must be convergence at every point and not only at the continuity points of $\mathbb{F}_0$. But that this condition is not enough has been already commented on (ref. **(2.1.4)**(vi)).

The condition stated in the theorem **(2.1.5)** is automatically satisfied in the case the limit *df* is continuous everywhere. It is pointed out that even if the limit *df* has discontinuities, uniform convergence is possible, as in the trivial case where $F_n = F$, $n \geq 1$. The case of distributions in R is presented in the next theorem and the case of distributions in R^p, $p \geq 2$ in theorem **(2.1.7)**.

Theorem 2.1.6. *Let $p = 1$. Let, for all x, $F_n(x) \to F(x)$ and $F_n(x-) \to F(x-)$. Then $\sup_{x \in R} |F_n(x) - F(x)| \to 0$ and $\sup_{x \in R} |F_n(x-) - F(x-)| \to 0$.*

Proof. Given $\varepsilon > 0$, find $a < b$, continuity points of F, such that $F(a) < \varepsilon$ and $1 - F(b) < \varepsilon$. Find N_1 such that $|F_n(a) - F(a)| < \varepsilon$ and $|F_n(b) - F(b)| < \varepsilon$ for all $n \geq N_1$. Then for $x \leq a$,

$$\begin{aligned} |F_n(x) - F(x)| &\leq F_n(x) + F(x) \leq F_n(a) + F(a) \\ &\leq F(a) + \varepsilon + F(a) < 3\varepsilon, \end{aligned}$$

for all $n \geq N_1$. Similarly, for $x \geq b$,

$$|F_n(x) - F(x)| \leq 1 - F_n(x) + 1 - F(x) \\ \leq 1 - F_n(b) + 1 - F(b) < 3\varepsilon,$$

for all $n \geq N_1$. Next let $a \leq x \leq b$. Let $c = F(b) - F(a)$. Choose integer k such that $\frac{c}{k} < \varepsilon$.

Define $a_0 < a_1 < a_2 < \cdots < a_k$ uniquely through the following inequalities. $F(a_j-) \leq F(a) + \frac{jc}{k} \leq F(a_j)$. Note that $a_0 = a$ and $a_k = b$. Consequent to the definition of the a_js, we have $F(a_{j+1}-) - F(a_j) \leq \frac{c}{k} < \varepsilon$, $j = 0, 1, 2, \ldots, k-1$. Choose integer N_2 such that $|F_n(a_j) - F(a_j)| < \varepsilon$ and $|F_n(a_j-) - F(a_j-)| < \varepsilon$ for all $n \geq N_2$ and for all $j, 0 \leq j \leq k$. Consider region $a_j < x < a_{j+1}, j = 0, 1, 2, \ldots, k-1$. For such an x, we have $F_n(a_j) - F(a_{j+1}-) \leq F_n(x) - F(x) \leq F_n(a_{j+1}-) - F(a_j)$. This implies

$$F_n(a_j) - F(a_j) - \{F(a_{j+1}-) - F(a_j)\} \\ \leq F_n(x) - F(x) \leq F_n(a_{j+1}-) - F(a_{j+1}-) + F(a_{j+1}-) - F(a_j).$$

Thus $-2\varepsilon \leq F_n(x) - F(x) \leq 2\varepsilon$ for all $n \geq N_2$. Collecting the results,

$$\sup_{x \in R} |F_n(x) - F(x)| < 3\varepsilon \text{ for all } n \geq N = \max(N_1, N_2),$$

which is equivalent to the first claim.

The same steps go through with x replaced by $x-$. Hence the second claim too is true. □

We next study the conditions to ensure uniform convergence of $\mathbb{F}_n$ to $\mathbb{F}_0$ in R^p.

Theorem 2.1.7. *If for every $\boldsymbol{x} \in R^p$, and every $\boldsymbol{x}^l \in V_{\boldsymbol{x}}$, $\mathbb{F}_n(\boldsymbol{x}^l) \to \mathbb{F}_0(\boldsymbol{x}^l)$ then $\sup_{\boldsymbol{x} \in R^p} |\mathbb{F}_n(\boldsymbol{x}^l) - \mathbb{F}_0(\boldsymbol{x}^l)| \to 0$. [Consequently, in view of* **(2.1.5)**, *this uniform convergence would hold when $\mathbb{F}_0$ is continuous everywhere]*

Proof. We establish the claim by induction on p. For $p = 1$, the theorem was proved in **(2.1.6)**. Let now $p \geq 2$. Our induction hypothesis is that the claim is true in space R^{p-1} when the corresponding conditions are satisfied.

Step 1. First we show that, for t fixed,

$$\sup_{x_1, x_2, \ldots, x_{p-1}} |\mathbb{F}_n(x_1^l, x_2^l, \ldots, x_{p-1}^l, t) - \mathbb{F}_0(x_1^l, x_2^l, \ldots, x_{p-1}^l, t)| \to 0.$$

Let $F_{n,j}$, $1 \leq j \leq p$ denote the one dimensional marginals of $\mathbb{F}_n$.
Case 1. $F_{0,p}(t) = 0$. This implies that

$$\mathbb{F}_0(x_1^l, x_2^l, \ldots, x_{p-1}^l, t) \leq F_{0,p}(t) = 0$$

for all $x_1, x_2, .., x_{p-1}$. The claim follows because

$$\mathbb{F}_n(x_1^l, x_2^l, \ldots, x_{p-1}^l, t) \leq F_{n,p}(t) \to F_{0,p}(t) = 0$$

(trivially uniformly in $x_1, x_2, .., x_{p-1}$)
Case 2. $F_{0,p}(t) > 0$. We can find N_1 such that, for all $n \geq N_1$, $F_{n,p}(t) > 0$. For $n = 0$ and $n \geq N_1$, $\mathbb{G}_n$ defined through

$$\mathbb{G}_n(x_1^l, x_2^l, \ldots, x_{p-1}^l) = \frac{F_n(x_1^l, x_2^l, \ldots, x_{p-1}^l, t)}{F_{n,p}(t)},$$

is a proper *df* in R^{p-1}. Note that $\mathbb{G}_n(\mathbf{u}^l) \to \mathbb{G}_0(\mathbf{u}^l)$ for all $\mathbf{u} \in R^{p-1}$ and all $\mathbf{u}^l \in V_{p-1,\,\mathbf{u}}$. Hence by the induction hypothesis

$$\sup_{x_1, x_2, \ldots, x_{p-1}} |\mathbb{G}_n(x_1^l, x_2^l, \ldots, x_{p-1}^l) - \mathbb{G}_0(x_1^l, x_2^l, \ldots, x_{p-1}^l)| \to 0.$$

Since

$$\begin{aligned}
&|\mathbb{F}_n(x_1^l, x_2^l, \ldots, x_{p-1}^l, t) - \mathbb{F}_0(x_1^l, x_2^l, \ldots, x_{p-1}^l, t)| \\
&\quad = |F_{n,p}(t)\mathbb{G}_n(x_1^l, x_2^l, \ldots, x_{p-1}^l) - F_{0,p}(t)\mathbb{G}_0(x_1^l, x_2^l, \ldots, x_{p-1}^l)| \\
&\quad \leq F_{n,p}(t)|\mathbb{G}_n(x_1^l, x_2^l, \ldots, x_{p-1}^l) - \mathbb{G}_0(x_1^l, x_2^l, \ldots, x_{p-1}^l)| \\
&\qquad + |F_{n,p}(t) - F_{0,p}(t)|,
\end{aligned}$$

the claim follows.

Step 2. The above result and the arguments leading to it are valid also when t is replaced by $t-$.

Step 3. Given $\varepsilon > 0$ find $a < b$, continuity points of $F_{0,p}$ such that $F_{0,p}(a) < \varepsilon$ and $1 - F_{0,p}(b) < \varepsilon$. Divide R^p into three regions: $x_p < a$; $x_p > b$ and $a \leq x_p \leq b$. Choose N_1 such that, for all $n \geq N_1$,

$$|F_{n,p}(a) - F_{0,p}(a)| < \varepsilon, \quad |F_{n,p}(b) - F_{0,p}(b)| < \varepsilon,$$
$$\sup_{x_1, x_2, \ldots, x_{p-1}} |\mathbb{F}_n(x_1^l, x_2^l, \ldots, x_{p-1}^l, a) - \mathbb{F}_0(x_1^l, x_2^l, \ldots, x_{p-1}^l, a)| < \varepsilon$$

and

$$|\mathbb{F}_n(x_1^l, x_2^l, \ldots, x_{p-1}^l, b) - |\mathbb{F}_0(x_1^l, x_2^l, \ldots, x_{p-1}^l, b)| < \varepsilon.$$

Step 4. In the region $x_p < a$,

$$\begin{aligned}|\mathbb{F}_n(x_1, x_2, \ldots, x_p) - \mathbb{F}_0(x_1, x_2, \ldots, x_p)|& \\ &\leq \mathbb{F}_n(x_1^l, x_2^l, \ldots, x_p^l) + \mathbb{F}_0(x_1^l, x_2^l, \ldots, x_p^l) \\ &\leq F_{n,p}(x_p) + F_{0,p}(x_p) \\ &\leq \varepsilon + 2F_{0,p}(x_p) \leq \varepsilon + 2F_0(a) < 3\varepsilon\end{aligned}$$

for all $n \geq N_1$.

Step 5. In the region $x_p > b$, for all $n \geq N_1$,

$$\begin{aligned}&|\mathbb{F}_n(x_1^l, x_2^l, \ldots, x_p^l) - \mathbb{F}_0(x_1^l, x_2^l, \ldots, x_p^l)| \\ &\quad\leq |\mathbb{F}_n(x_1^l, x_2^l, \ldots, x_p^l) - \mathbb{F}_n(x_1^l, x_2^l, \ldots, x_{p-1}^l, b)| \\ &\qquad + |\mathbb{F}_n(x_1^l, x_2^l, \ldots, x_{p-1}^l, b) - \mathbb{F}_0(x_1^l, x_2^l, \ldots, x_{p-1}^l, b)| \\ &\qquad + |\mathbb{F}_0(x_1^l, x_2^l, \ldots, x_{p-1}^l, b) - \mathbb{F}_0(x_1^l, x_2^l, \ldots, x_p^l)| \\ &\quad\leq 1 - F_{n,p}(b) + \sup_{x_1, x_2, \ldots, x_{p-1}} |\mathbb{F}_n(x_1^l, x_2^l, \ldots, x_{p-1}^l, b) \\ &\qquad - \mathbb{F}_0(x_1^l, x_2^l, \ldots, x_{p-1}^l, b)| + 1 - F_{0,p}(b) \\ &\quad< \varepsilon + 1 - F_{0,p}(b) + \varepsilon + \varepsilon < 4\varepsilon.\end{aligned}$$

Step 6. To deal with the region $a \leq x_p \leq b$, the method adopted in **(2.1.6)** will be used. Let $c = F_{0,p}(b) - F_{0,p}(a)$. Choose integer $k > \frac{c}{\varepsilon}$. Define $a = a_0 < a_1 < a_2 < \cdots < a_k = b$ uniquely in the same way as in **(2.1.6)** using *df* $F_{0,p}$ in place of F. We will have

$$F_{0,p}(a_i-) - F_{0,p}(a_{i-1}) \leq \frac{c}{k} < \varepsilon \text{ for all } i.$$

A consequence is that, for all $(x_1, x_2, \ldots, x_{p-1}) \in R^{p-1}$,

$$\begin{aligned}\mathbb{F}_0(x_1^l, x_2^l, \ldots, x_{p-1}^l, a_i-) - \mathbb{F}_0(x_1^l, x_2^l, \ldots, x_{p-1}^l, a_{i-1})& \\ \leq F_{0,p}(a_i-) - F_{0,p}(a_{i-1}) < \varepsilon.&\end{aligned}$$

Choose N_2 such that, for all $n \geq N_2$ and for all $i, 0 \leq i \leq k$,

$$\sup_{x_1, x_2, \ldots, x_{p-1}} |\mathbb{F}_n(x_1^l, x_2^l, \ldots, x_{p-1}^l, a_i) - \mathbb{F}_0(x_1^l, x_2^l, \ldots, x_{p-1}^l, a_i)| < \varepsilon$$

and

$$\sup_{x_1,x_2,\ldots,x_{p-1}} |\mathbb{F}_n(x_1^l, x_2^l, \ldots, x_{p-1}^l, a_i-) - \mathbb{F}_0(x_1^l, x_2^l, \ldots, x_{p-1}^l, a_i-)| < \varepsilon.$$

Then for $a_{i-1} < x_p < a_i$,

$$|\mathbb{F}_n(x_1^l, x_2^l, \ldots, x_p^l) - \mathbb{F}_0(x_1^l, x_2^l, \ldots, x_p^l)| \leq \max(\alpha, \beta)$$

where

$$\alpha = |\mathbb{F}_n(x_1^l, x_2^l, \ldots, x_{p-1}^l, a_i-) - \mathbb{F}_0(x_1^l, x_2^l, \ldots, x_p^l)|$$

and

$$\beta = |\mathbb{F}_n(x_1^l, x_2^l, \ldots, x_{p-1}^l, a_{i-1}) - \mathbb{F}_0(x_1^l, x_2^l, \ldots, x_p^l)|.$$

Now

$$\begin{aligned}\alpha &\leq |\mathbb{F}_n(x_1^l, x_2^l, \ldots, x_{p-1}^l, a_i-) - \mathbb{F}_0(x_1^l, x_2^l, \ldots, x_{p-1}^l, a_i-)| \\ &\quad + |\mathbb{F}_0(x_1^l, x_2^l, \ldots, x_{p-1}^l, a_i-) - \mathbb{F}_0(x_1^l, x_2^l, \ldots, x_p^l)| \\ &< \varepsilon + |\mathbb{F}_0(x_1^l, x_2^l, \ldots, x_{p-1}^l, a_i-) - \mathbb{F}_0(x_1^l, x_2^l, \ldots, x_p^l)| \\ &\leq \varepsilon + |\mathbb{F}_0(x_1^l, x_2^l, \ldots, x_{p-1}^l, a_i-) - \mathbb{F}_0(x_1^l, x_2^l, \ldots, x_{p-1}^l, a_{i-1})| \\ &< 2\varepsilon.\end{aligned}$$

Similarly, $\beta < 2\varepsilon$.

Step 7. If $n \geq \max(N_1, N_2)$, then $\mathbb{F}_n(\mathbf{x}^l) - \mathbb{F}_0(\mathbf{x}^l)| < 4\varepsilon$ for all $\mathbf{x} \in R^p$. □

Theorem 2.1.8. *Let assertion (i) be* $\sup_{\boldsymbol{x} \in R^p} |\mathbb{F}_n(\boldsymbol{x}) - \mathbb{F}_0(\boldsymbol{x})| \to 0$ *and let assertion (ii) be* $\sup_{\boldsymbol{x} \in R^p} |\mathbb{F}_n(\boldsymbol{x}^l) - \mathbb{F}_0(\boldsymbol{x}^l)| \to 0$. *Then (i) implies (ii) for all* x^l *while (i) will follow if (ii) is true for any one* x^l.

Proof. Let (i) hold and let x^l be fixed. For $\mathbf{x}^l$ to be different from $\mathbf{x}$, there has to be at least one i such that $x_i^l = x_i-$. Relabelling, if necessary, assume $x_i^l = x_i-, 1 \leq i \leq r$ and $x_i^l = x_i, r+1 \leq i \leq p$ if $r < p$. In view of **(2.1.7)**, to establish (ii), it is enough to show that for each $\mathbf{x} \in R^p$ and for every x^l, $\mathbb{F}_n(\mathbf{x}^l) \to \mathbb{F}_0(\mathbf{x}^l)$. Fix $\mathbf{u} \in R^p$. Given $\varepsilon > 0$, find $\eta > 0$ such that $\mathbb{F}_0(\mathbf{u}^l) - \mathbb{F}_0(\mathbf{u} - \mathbf{a}(\eta)) < \varepsilon$ where $\mathbf{a}(\eta) = (a_1(\eta), a_2(\eta), \ldots, a_p(\eta)), a_i(\eta) = \eta, 1 \leq i \leq r$ and $a_i(\eta) = 0, r+1 \leq i \leq p$ if $r < p$. We have

$$\sup_{0 \leq \delta \leq \eta} |\mathbb{F}_n(\mathbf{u} - \mathbf{a}(\delta)) - \mathbb{F}_0(\mathbf{u} - \mathbf{a}(\delta))| \leq \sup_{\mathbf{x} \in R^p} |\mathbb{F}_n(\mathbf{x}) - \mathbb{F}_0(\mathbf{x})| < \varepsilon,$$

for all n large, say, $n \geq N$. Hence, for $n \geq N$,

$$\begin{aligned}\sup_{0\leq\delta\leq\eta} &|\mathbb{F}_n(\mathbf{u}-\mathbf{a}(\delta)) - \mathbb{F}_0(\mathbf{u}^l)| \\ &\leq \sup_{0\leq\delta\leq\eta} |\mathbb{F}_n(\mathbf{u}-\mathbf{a}(\delta)) - \mathbb{F}_0(\mathbf{u}-\mathbf{a}(\delta))| \\ &\quad + \sup_{0\leq\delta\leq\eta} |\mathbb{F}_0(\mathbf{u}-\mathbf{a}(\delta)) - \mathbb{F}_0(\mathbf{u}^l)| < 2\varepsilon.\end{aligned}$$

From the definition of $\mathbb{F}_n(\mathbf{u}^l)$, it is immediate that for each n, a $\delta_n < \eta$ can be found such that $\mathbb{F}_n(\mathbf{u}^l) - \mathbb{F}_n(\mathbf{u}-\mathbf{a}(\delta_n)) < \varepsilon$. Hence for all $n \geq N$,

$$\begin{aligned}|\mathbb{F}_n(\mathbf{u}^l) &- \mathbb{F}_0(\mathbf{u}^l)| \\ &\leq \mathbb{F}_n(\mathbf{u}^l) - \mathbb{F}_n(\mathbf{u}-\mathbf{a}(\delta_n)) + |\mathbb{F}_n(\mathbf{u}-\mathbf{a}(\delta_n)) - \mathbb{F}_0(\mathbf{u}^l)| \\ &\leq \mathbb{F}_n(\mathbf{u}^l) - \mathbb{F}_n(\mathbf{u}-\mathbf{a}(\delta_n)) + \sup_{0\leq\delta\leq\eta} |\mathbb{F}_n(\mathbf{u}-\mathbf{a}(\delta)) - \mathbb{F}_0(\mathbf{u}^l)| \\ &< 3\varepsilon.\end{aligned}$$

Since $\varepsilon > 0$ is arbitrary, it follows that $\lim_{n\to\infty} \mathbb{F}_n(\mathbf{u}^l) = \mathbb{F}_0(\mathbf{u}^l)$. If $r = p$, then $a_i(\eta) = \eta$ for all i and the result follows.

The second part can be established arguing on similar lines, using the fact that if $\mathbf{x} \ll \mathbf{y}$, then $\lim_{\mathbf{y}\to\mathbf{x}} \mathbb{F}(\mathbf{y}^l) = \mathbb{F}(\mathbf{x})$ for every y^l. The details are omitted. □

Corollary 2.1.9. *(i) If* $\sup_{x\in R^p} |\mathbb{F}_n(\mathbf{x}^l) - \mathbb{F}_0(\mathbf{x}^l)| \to 0$ *for one* $\mathbf{x}^l$ *then it is true for all* $\mathbf{x}^l$.
(ii) By **(2.1.7)***,* $\sup_{x\in R^p} |\mathbb{F}_n(\mathbf{x}) - \mathbb{F}_0(\mathbf{x})| \to 0$ *if* $\mathbb{F}_0$ *is continuous everywhere and* $\mathbb{F}_n \xrightarrow{d} \mathbb{F}_0$.

Remark 2.1.10. Let $\mathbb{F}_0$ be a continuous function and let $\mathcal{I}$ be the collection of all semi infinite rectangles $\times_1^p(-\infty, x_i]$. Then the above result can be written

$$\sup_{A\in\mathcal{I}} |\mu_n(A) - \mu_0(A)| \to 0$$

where μ_n is the pm generated by $\mathbb{F}_n$. We will show in Theorem (**2.7.15**)

$$\sup_{A\in\mathcal{J}} |\mu_n(A) - \mu_0(A)| \to 0$$

for a class $\mathcal{J}$ $(\supset \mathcal{I})$ of Borel sets.

Let $p = 1$, and let F be an arbitrary *df* and u an arbitrary point on the line. Let $\widehat{C}_F(u) = \{x : x \in C_F, 2u - x \in C_F\}$ and note that $R \sim \widehat{C_F(u)}$ is atmost countable. As such $\widehat{C}_F(u)$ is dense in R. In p-dimension, define $\widehat{C}_{\mathbb{F}}(\mathbf{u})$ $= \times_1^p \widehat{C}_{F_i}(u_i)$ and note that $\widehat{C}_{\mathbb{F}}(\mathbf{u})$ is dense in R^p and that $\widehat{C}_{\mathbb{F}}(\mathbf{u}) \subseteq C_{\mathbb{F}}$.

Theorem 2.1.11. $\mathbb{F}_n \overset{d}{\to} \mathbb{F}_0$ *if and only if* $\mathbb{F}_n(\boldsymbol{x}) \to \mathbb{F}_0(\boldsymbol{x})$ *for every* $\boldsymbol{x} \in \widehat{C_{\mathbb{F}_0}} = \times_{j=1}^p C_{F_{0,j}}$.

Proof. Since $\widehat{C_{\mathbb{F}_0}} \subseteq C_{\mathbb{F}_0}$, implication one way is trivial. Suppose now $\mathbb{F}_n(\mathbf{x}) \to \mathbb{F}_0(\mathbf{x})$ for every $\mathbf{x} \in \widehat{C_{\mathbb{F}_0}}$. Let $\mathbf{u} \in C_{\mathbb{F}_0}$. Given $\varepsilon > 0$, we can then find a δ such that $|\mathbb{F}_0(\mathbf{u}) - \mathbb{F}_0(\mathbf{v})| < \varepsilon$ whenever $\|\mathbf{u} - \mathbf{v}\| < \delta$. Let $D = \{\mathbf{x} : \mathbf{x} \in \widehat{C_{\mathbb{F}_0}}(\mathbf{u}), \|\mathbf{u} - \mathbf{x}\| < \delta\}$ and note that D is non-null because $\widehat{C_{\mathbb{F}_0}}(\mathbf{u})$ is dense in R^p. Let $\mathbf{a} \in D$. Hence $\mathbf{a} \in \widehat{C_{\mathbb{F}_0}}(\mathbf{u})$ and $\|\mathbf{u} - \mathbf{a}\| < \delta$. Define **y**, **z** through prescribing $y_i = \min\{a_i, 2u_i - a_i\}$ and $z_i = \max\{a_i, 2u_i - a_i\}$, $i = 1, 2, \ldots, p$. Then clearly

(i) $\mathbf{y}, \mathbf{z} \in C_{\mathbb{F}_0}(\mathbf{u})$

(ii) $\|\mathbf{u} - \mathbf{y}\| = \|\mathbf{u} - \mathbf{z}\| = \|\mathbf{u} - \mathbf{a}\| < \delta$

(iii) $\mathbf{y} \ll \mathbf{u} \ll \mathbf{z}$.

These imply

(α) $\lim_{n\to\infty} \mathbb{F}_n(\mathbf{y}) = \mathbb{F}_0(\mathbf{y})$ and $\lim_{n\to\infty} \mathbb{F}_n(\mathbf{z}) = \mathbb{F}_0(\mathbf{z})$;

(β) $|\mathbb{F}_0(\mathbf{u}) - \mathbb{F}_0(\mathbf{y})| < \varepsilon$ and $|\mathbb{F}_0(\mathbf{u}) - \mathbb{F}_0(\mathbf{z})| < \varepsilon$;

(γ) $\mathbb{F}_n(\mathbf{y}) \le \mathbb{F}_n(\mathbf{u}) \le \mathbb{F}_n(\mathbf{z}), n \ge 0$.

These, in turn, lead to, for all n large, say, $n \ge N = N(\mathbf{y}, \mathbf{z}, \varepsilon)$,

$$\begin{aligned} -\varepsilon + \mathbb{F}_0(\mathbf{u}) &\le \mathbb{F}_0(\mathbf{y}) \le \mathbb{F}_n(\mathbf{y}) + \varepsilon \\ &\le \mathbb{F}_n(\mathbf{u}) + \varepsilon \le \mathbb{F}_n(\mathbf{z}) + \varepsilon \\ &\le \mathbb{F}_0(\mathbf{z}) + 2\varepsilon \le \mathbb{F}_0(\mathbf{u}) + 3\varepsilon. \end{aligned}$$

Taking $n \to \infty$ in the relevant inequalities, we get

$$-2\varepsilon + \mathbb{F}_0(\mathbf{u}) \le \underline{\lim}_{n\to\infty} \mathbb{F}_n(\mathbf{u}) \le \overline{\lim}_{n\to\infty} \mathbb{F}_n(\mathbf{u}) \le \mathbb{F}_0(\mathbf{u}) + 2\varepsilon.$$

ε being arbitrary, we conclude that $\lim_{n\to\infty} \mathbb{F}_n(\mathbf{u})$ exists and is equal to $\mathbb{F}_0(\mathbf{u})$. Since $\mathbf{u} \in C_{\mathbb{F}_0}$ is arbitrary, it follows that $\mathbb{F}_n \overset{d}{\to} \mathbb{F}_0$. □

The next theorem provides a necessary and sufficient condition for the weak convergence of *df*s in R^k in terms of certain integrals. The importance of this theorem stems from the fact that it forms the basis for the definition of weak convergence of *pm*s in metric spaces (ref. **(2.4)**). Further, this theorem brings out better the topological basis of weak convergence than is seen in the definition **(2.1.1)**.

Theorem 2.1.12 (Helly-Bray).

$$\mathbb{F}_n \overset{d}{\to} \mathbb{F}_0 \text{ iff } \int_{R^p} g(\boldsymbol{x})d\mathbb{F}_n(\boldsymbol{x}) \to \int_{R^P} g(\boldsymbol{x})d\mathbb{F}_0(\boldsymbol{x})$$

for every $g \in C_1$, the collection of all real bounded continuous functions defined on R^p.

Proof. We note first that the integrals involved exist even as Riemann-stieltjes integrals, since g is a continuous function.

Let $\mathbb{F}_n \overset{d}{\to} \mathbb{F}_0$. Let $g \in C_1$. Let c be an upper bound for $|g|$. Given $\varepsilon > 0$, we can find $-a, a$ continuity points of the marginal $F_{0,i}$ for each $i, 1 \leq i \leq p$ such that $\Delta^{\mathbf{a}}_{-\mathbf{a}}(\mathbb{F}_0) = \mu_0(A) > 1 - \varepsilon$, where $\mathbf{a}$ is the point in R^p with all its coordinates equal to a and $A = \times_1^p(-a,\ a]$. We note that every $\mathbf{a}^l$ is a continuity point of $\mathbb{F}_0$. (ref **(2.1.4)**(iv) for notation). Since $\mu_n(A) = \Delta^{\mathbf{a}}_{-\mathbf{a}}(\mathbb{F}_n) \to \Delta^{\mathbf{a}}_{-\mathbf{a}}(\mathbb{F}_0)$, it follows that there exists N such that for all $n \geq N$, $\mu_n(\bar{A}) > 1 - 2\varepsilon$. Then for all $n \geq N$,

$$\begin{aligned}
&|\int_{R^p} g(\mathbf{x})\mathrm{d}\mathbb{F}_n(\mathbf{x}) - \int_{R^p} g(\mathbf{x})\mathrm{d}\mathbb{F}_0(\mathbf{x})| \\
&\leq \int_{A'} |g(\mathbf{x})|\mathrm{d}\mathbb{F}_n(\mathbf{x}) + \int_{A'} |g(\mathbf{x})|\mathrm{d}\mathbb{F}_0(\mathbf{x}) + |\int_A g(\mathbf{x})\mathrm{d}\mathbb{F}_n(\mathbf{x}) \\
&\qquad - \int_A g(\mathbf{x})\mathrm{d}\mathbb{F}_0(\mathbf{x})| \\
&\leq c\{\mu_n(A') + \mu_0(A')\} + |\int_A g(\mathbf{x})\mathrm{d}\mathbb{F}_n(\mathbf{x}) - \int_A g(\mathbf{x})\mathrm{d}\mathbb{F}_0(\mathbf{x})| \\
&\leq 2c\varepsilon + I
\end{aligned}$$

where $I = |\int_A g(\mathbf{x})\mathrm{d}\mathbb{F}_n(\mathbf{x}) - \int_A g(\mathbf{x})\mathrm{d}\mathbb{F}_0(\mathbf{x})|$. The integral I can be evaluated as a Riemann-Stieltjes integral since g is a continuous function. g, being continuous on $\bar{A}$, is uniformly continuous on $\bar{A}$. Hence a δ can be found such that $|g(\mathbf{x}) - g(\mathbf{y})| < \varepsilon$ whenever $\|\mathbf{x} - \mathbf{y}\| < \delta$. Divide the interval $[-a, a]$ on the ith co-ordinate axis through the points $-a = a_{i,0} < a_{i,1} < \cdots < a_{i,k_i} = a$ such that (i) the division points are continuity points of $F_{0,i}$ and (ii) $\max_{1 \leq j \leq k_i} (a_{i,j} - a_{i,j-1}) < \frac{\delta}{\sqrt{p}}$. These $k_1 \times k_2 \times \ \ldots \ \times k_p$ points on the p co-ordinate axes determine a grid of 'left open and right closed' rectangles in A with their union equal to A. List them as $A_1, A_2, \ldots, A_M$ and note that

$\mu_n(A_j) \to \mu_0(A_j), 1 \le j \le M$ where μ_n and μ_0 are the *pms* corresponding to the *df*s $\mathbb{F}_n$ and $\mathbb{F}_0$ respectively. Hence for all n not less than a certain $N_1 = N_1(\varepsilon, M)$ and for all $j, 1 \le j \le M$, $|\mu_n(A_j) - \mu_0(A_j)| < \frac{\varepsilon}{M}$. Let $\mathbf{u}_j$ be an arbitrary point in $A_j, 1 \le j \le M$. Define $g_{_M}(\mathbf{x}) = g(\mathbf{u}_j)$ if $\mathbf{x} \in A_j$. Then it is clear that $\sup\limits_{\mathbf{x}\in A} |g(\mathbf{x}) - g_{_M}(\mathbf{x})| < \varepsilon$. Now

$$
\begin{aligned}
I \le & \int\limits_A |g(\mathbf{x}) - g_{_M}(\mathbf{x})| d\mathbb{F}_n(\mathbf{x}) + |\int\limits_A g_{_M}(\mathbf{x}) d\mathbb{F}_n(\mathbf{x}) \\
& - \int\limits_A g_{_M}(\mathbf{x}) d\mathbb{F}_0(\mathbf{x})| + \int\limits_A |g(\mathbf{x}) - g_{_M}(\mathbf{x})| d\mathbb{F}_0(\mathbf{x}) \le 2\varepsilon + |I_1|
\end{aligned}
$$

where

$$
\begin{aligned}
I_1 &= \int\limits_A g_{_M}(\mathbf{x}) d\mathbb{F}_n(\mathbf{x}) - \int\limits_A g_{_M}(\mathbf{x}) d\mathbb{F}_0(\mathbf{x}) \\
&= \sum_{j=1}^{M} \{ \int\limits_{A_j} g_{_M}(\mathbf{x}) d\mathbb{F}_n(\mathbf{x}) - \int\limits_{A_j} g_{_M}(\mathbf{x}) d\mathbb{F}_0(\mathbf{x}) \} \\
&= \sum_{j=1}^{M} \{ g(\mathbf{u}_j)\mu_n(A_j) - g(\mathbf{u}_j)\mu_0(A_j) \}.
\end{aligned}
$$

Hence, for all $n \ge N_1$,

$$|I_1| \le \sum_{j=1}^{M} c \frac{\varepsilon}{M} = c\varepsilon.$$

This proves that

$$\int\limits_{R^p} g(\mathbf{x}) d\mathbb{F}_n(\mathbf{x}) \to \int\limits_{R^P} g(\mathbf{x}) d\mathbb{F}_0(\mathbf{x}).$$

To prove the sufficiency part, define in R^p the distance function $d : d(\mathbf{x}, \mathbf{y}) = \max\limits_{1 \le i \le p} |x_i - y_i|$ and note that the topology induced by this metric is the same as that induced by the norm metric. Let $\mathbf{u}$ be a continuity point of $\mathbb{F}_0$. Define $E = \times_1^p (-\infty, u_i]$ and $D = \times_1^p (-\infty, u_i + \varepsilon]$. We note that E, D are closed sets. Given $\varepsilon > 0$, arbitrary, define function g thus : $g(\mathbf{x}) = 1 - \frac{1}{\varepsilon} d(\mathbf{x}, E)$ if $d(\mathbf{x}, E) \le \varepsilon$ and $g(\mathbf{x}) = 0$ otherwise and note that, so defined, g is a bounded continuous function. Now

$$\mathbb{F}_n(\mathbf{u}) = \int\limits_E d\mathbb{F}_n(\mathbf{x}) = \int\limits_E g(\mathbf{x}) d\mathbb{F}_n(\mathbf{x}).$$

Hence,

$$\overline{\lim_{n\to\infty}}\,\mathbb{F}_n(\mathbf{u}) = \overline{\lim_{n\to\infty}} \int_E g(\mathbf{x})\mathrm{d}\mathbb{F}_n(\mathbf{x}) \le \overline{\lim_{n\to\infty}} \int_{R^p} g(\mathbf{x})\mathrm{d}\mathbb{F}_n(\mathbf{x})$$
$$= \lim_{n\to\infty} \int_{R^p} g(\mathbf{x})\mathrm{d}\mathbb{F}_n(\mathbf{x}) = \int_{R^p} g(\mathbf{x})\mathrm{d}\mathbb{F}_0(\mathbf{x}) = \int_D g(\mathbf{x})\mathrm{d}\mathbb{F}_0(\mathbf{x})$$
$$\le \int_D \mathrm{d}\mathbb{F}_0(\mathbf{x}) = \mathbb{F}_0(\mathbf{u} + \varepsilon\mathbf{1})$$

(where $\mathbf{1} = (1, 1, \ldots, 1)) \to \mathbb{F}_0(\mathbf{u})$ as $\varepsilon \to 0$ (since $\mathbb{F}_0$ is right continuous).

Now take $E = \times_1^p(-\infty, u_i - \varepsilon]$ and $D = \times_1^p(-\infty, u_i]$. Define g as before in terms of the new set E. Then

$$\mathbb{F}_n(\mathbf{u}) = \int_D \mathrm{d}\mathbb{F}_n(\mathbf{x}) \ge \int_D g(\mathbf{x})\mathrm{d}\mathbb{F}_n(\mathbf{x}) = \int_{R^p} g(\mathbf{x})\mathrm{d}\mathbb{F}_n(\mathbf{x}),$$

leading to,

$$\underline{\lim_{n\to\infty}}\,\mathbb{F}_n(\mathbf{u}) \ge \int_{R^p} g(\mathbf{x})\mathrm{d}\mathbb{F}_0(\mathbf{x}) \ge \int_E \mathrm{d}\mathbb{F}_0(\mathbf{x})$$
$$= \mathbb{F}_0(\mathbf{u} - \varepsilon\mathbf{1}) \to \mathbb{F}_0(\mathbf{u})$$

as $\varepsilon \to 0$ (because $\mathbf{u}$ is a continuity point of $\mathbb{F}_0$).

With this, the proof of the sufficiency part and also of the theorem is complete. □

Remark 1. Let $F_n \xrightarrow{d} F$. To stress that $\int_R g(x)\,\mathrm{d}F_n(x)$ may fail to converge to $\int_R g(x)\,\mathrm{d}F(x)$ when g, though bounded, is not a continuous function, we present the following example. Let

$$F_n(x) = \begin{cases} 0, & \text{if } x < \frac{1}{n} \\ 1, & \text{if } x \ge \frac{1}{n} \end{cases}; \quad F(x) = \begin{cases} 0 & \text{if } x < 0 \\ 1 & \text{if } x \ge 0 \end{cases};$$

$$g(x) = \begin{cases} 0 & \text{if } x \le 0 \\ 1 & \text{if } x > 0; \end{cases}.$$

μ_n, μ denote the *pms* corresponding to F_n and F. We note $F_n \xrightarrow{d} F$. We note too that $\int_R g(x)\,\mathrm{d}F_n(x)$, $\int_R g(x)\,\mathrm{d}F(x)$ do not exist as Riemann-Stieltjes integrals. Now, $\int_R g(x)dF_n(x) = 1$ and $\int_R g(x)dF(x) = 0$.

Remark 2. $\mathbb{F}_n \xrightarrow{d} \mathbb{F}_0$ implies also

$$\int_B g(\mathbf{x})\,\mathrm{d}\mathbb{F}_n(\mathbf{x}) \to \int_B g(\mathbf{x})\,\mathrm{d}\mathbb{F}_0(\mathbf{x}),\ g \in C_1$$

for certain types of subsets $B \subseteq R^p$. For example, $B = (-\infty,\ \mathbf{a}]$ where $\mathbf{a}$ is a continuity point of $\mathbb{F}_0$. This topic will be discussed in a more general setting in **(2.4.3)**.

In the next theorem, we show that when the measures μ_n on $\mathscr{R}$ are concentrated in a closed interval, then the class C_1 in Helly-Bray theorem can be replaced by the smaller class of powers of x.

Theorem 2.1.13. *Let F_n, $n \geq 0$ be dfs such that for some $a < b$ and for all n, $F_n(a) = 0$ and $F_n(b) = 1$. Then $\int_{[a,\,b]} x^k dF_n(x) \to \int_{[a,\,b]} x^k dF_0(x)$ for every integer $k \geq 1$ iff $F_n \xrightarrow{d} F_0$.*

Proof. Let

$$\int_a^b x^k \mathrm{d}F_n(x) \to \int_a^b x^k \mathrm{d}F_0(x)$$

for every $k \geq 1$. This implies that for every polynomial Q_ν of degree ν, $\int_a^b Q_\nu(x)\mathrm{d}F_n(x) \to \int_a^b Q_\nu(x)\mathrm{d}F_0(x)$. It is immediate from this $\int_a^b g(x)\mathrm{d}F_n(x) \to \int_a^b g(x)\mathrm{d}F_0(x)$ for every continuous function g defined on $[a,\ b]$ since such a g is the uniform limit of a sequence of polynomials. Thus $\int_a^b x^k\mathrm{d}F_n(x) \to \int_a^b x^k\mathrm{d}F_0(x)$ for every $k \geq 1$ *iff* $\int_a^b g(x)\mathrm{d}F_n(x) \to \int_a^b g(x)\mathrm{d}F_0(x)$ for every continuous function g on $[a,\ b]$.

Define, for all n, $F_n^*(x) = 0$ if $x < 0$; $F_n^*(x) = F_n(x)$ if $a \leq x \leq b$; $F_n^*(x) = 1$ if $x > b$. So defined the F_n^*s are proper dfs on R. We note that $F_n^* \xrightarrow{d} F_0^*$ *iff* $F_n \xrightarrow{d} F_0$. For h an arbitrary bounded continuous function defined on R,

$$\int_R h(x)\,\mathrm{d}F_n^*(x) = \int_{[a,\,b]} h(x)\,\mathrm{d}F_n^*(x) = \int_{[a,\,b]} h(x)\,\mathrm{d}F_n(x)$$

$$\to \int_{[a,\,b]} h(x)\,\mathrm{d}F_0(x) = \int_{[a,\,b]} h(x)\,\mathrm{d}F_n^*(x) = \int_R h(x)\,\mathrm{d}F_0^*(x).$$

The Helly-Bray theorem applies to the sequence (F_n^*) and the proof is complete. □

Theorem **(2.1.14)** stated in terms of rvs would read: $\frac{Y_n - a_n}{b_n} \xrightarrow{d} Y_0$ and $\frac{Y_n - \alpha_n}{\beta_n} \xrightarrow{d} Y_0$, where Y_0 is a non-degenerate rv, is possible only if $b_n \sim \beta_n$ and $\frac{\alpha_n - a_n}{b_n} = o(1)$. In other words, the 'norming constants' $a_n,\ b_n$ are 'unique'.

Theorem 2.1.14. *Let $F_n,\ n \geq 0$ be dfs. Let $b_n > 0,\ \beta_n > 0,\ a_n,\ \alpha_n$ be constants. Define dfs $G_n : G_n(x) = F_n(b_n x + a_n)$ and $H_n : H_n(x) = F_n(\beta_n x + \alpha_n),\ n \geq 1$.*
(i) Let

$$G_n \xrightarrow{d} F \tag{2.1}$$

$$H_n \xrightarrow{d} F \tag{2.2}$$

where F is a proper or non-degenerate distribution. If both (2.1) *and* (2.2) *hold, then*

$$\beta_n \sim b_n \quad \text{and} \quad \frac{a_n - \alpha_n}{b_n} = o(1) \tag{2.3}$$

(ii) Under (2.3), (2.1) ⇔ (2.2) .

Proof. (i) Taking a subsequence, if necessary, assume, with no loss of generality, $\lim_{n\to\infty} \frac{b_n}{\beta_n}$ and $\lim_{n\to\infty} \frac{a_n - \alpha_n}{\beta_n}$ exist, finite or infinite. Define $B_n = \frac{b_n}{\beta_n}$ and $A_n = \frac{a_n - \alpha_n}{\beta_n}$. Write $A = \lim A_n$ and $B = \lim B_n$.

Step 1. We show first $B \neq \infty$. If, possible, let $B = \infty$. Let $\lambda = \sup\{x : \overline{\lim_{n\to\infty}} (B_n x + A_n) < \infty\}$.

Suppose $\lambda = -\infty$. Let $u_r \downarrow -\infty,\ x,\ u_r \in C_F$. We have

$$F(x) = \lim_{n\to\infty} H_n(B_n x + A_n) \leq \lim_{n\to\infty} H_n(u_r) = F(u_r).$$

This being true for all $r \geq 1$, we conclude $F(x) = 0,\ x \in C_F$, which contradicts the fact that F is a proper df. Hence $-\infty < \lambda \leq \infty$. Now, suppose $\lambda = \infty$. Hence, $-\infty < v < x < \infty$ leads to $\overline{\lim_{n\to\infty}} (B_n v + A_n) < \infty$ and at the same time

$$\overline{\lim_{n\to\infty}} (B_n v + A_n) = \overline{\lim_{n\to\infty}} \{B_n(v - x) + (B_n x + A_n)\} = -\infty.$$

Thus $-\infty < \lambda < \infty$. This is inconsistent with $B_n \to \infty$. Hence $B < \infty$.

Step 2. ($B < \infty$ and $A = \infty$) $\Rightarrow F(x) = 1$ for all x while ($B < \infty$ and $A = -\infty$) $\Rightarrow F(x) = 0$ for all x. Hence $|A| < \infty$.

Step 3. If $B = 0$, then given $x > 0$, $\varepsilon > 0$, we can find $N = N(x, \varepsilon)$ such that $A - \varepsilon \leq B_n x + A_n \leq A + \varepsilon$, leading to $F_n(A - \varepsilon) \leq F_n(B_n x + A_n) \leq F_n(A + \varepsilon)$. Choose x, $A \pm \varepsilon \in C_F$ and take limit as $n \to \infty$ to get $F(A - \varepsilon) \leq F(x) \leq F(A + \varepsilon)$. Successively taking x to ∞ and to $-\infty$ through members of C_F, we conclude $F(A-\varepsilon) = 0$ and $F(A+\varepsilon) = 1$. Since $\varepsilon > 0$ is arbitrary, this implies $F(A-) = 0$ and $F(A) = 1$. This contradicts the assumption that F is a non-degenerate distribution. Hence $B > 0$.

Collecting the results, we have : A is finite and B positive and finite.

Step 4. The hypothesis can now be written $H_n \xrightarrow{d} F$ and $H_n(B_n x + A_n) \to F(x)$, $x \in C_F$.

Since $B_n x + A \to Bx + A$, it follows $Bx + A - \varepsilon \leq B_n x + A_n \leq Bx + A + \varepsilon$. Hence $H_n(Bx + A - \varepsilon) \leq H_n(B_n x + A_n) \leq H_n(Bx + A + \varepsilon)$. Let $\varepsilon_r \downarrow 0$, x, $Bx + A$, $Bx + A \pm \varepsilon_r \in C_F$. Take limit as $n \to \infty$ to get $F(Bx + A - \varepsilon_r) \leq F(x) \leq F(Bx + A + \varepsilon_r)$. Now letting $r \to \infty$ gives $F(x) = F(Bx + A)$.

Step 5. In this step we show $B = 1$ and $A = 0$.

Repeated application of the relation $F(x) = F(Bx + A)$ gives

$$F(x) = F(B^2 x + A(1 + B)) = \dots$$
$$= F(B^n x + A(1 + B + \dots + B^{n-1})) = F(B^n x + A\frac{1 - B^n}{1 - B}).$$

If $0 < B < 1$, then this leads to $F(x) = F(\frac{A}{1-B})$. This is not possible since F is a proper df. Hence $B \geq 1$.

We also have $F(Bx+A) = F(x)$ or $F(x) = F(\frac{1}{B}x - \frac{A}{B})$. Since $0 < \frac{1}{B}, 1$, we arrive at a contradiction, following the case for $B < 1$. Hence $B = 1$.

If $A \neq 0$, then $F(x) = F(x + nA)$. Let $nA \to \infty$ and then let $nA \to -\infty$ and obtain $F(x) = F(\infty) = F(-\infty)$. As before, this is not possible. Hence $A = 0$.

Step 6. We have shown that for every subsequence (B_{n_k}, A_{n_k}) for which the limit exists, the limit is $(1, 0)$. Hence $\lim_{n\to\infty} (B_n, A_n)$ is $(1, 0)$.

(ii) Let (2.3) and (2.1) hold. Given any three points $x < y < z$, (2.3) implies the following for all n sufficiently large: $x < \frac{\beta_n}{b_n} y + \frac{\alpha_n - a_n}{b_n} < z$ or, what is same, $b_n x + a_n < \beta_n y + \alpha_n < b_n z + a_n$. This leads to $F_n(b_n x + a_n) \leq$

$F_n(\beta_n y + \alpha_n) \leq F_n(b_n z + a_n)$. Now choose $x,\ y,\ z$ to be in C_F. Take limit as $n \to \infty$ to get, using (2.1),

$$F(x) \leq \varliminf_{n\to\infty} F_n(\beta_n y + \alpha_n) \leq \varlimsup_{n\to\infty} F_n(\beta_n y + \alpha_n) \leq F(z).$$

Now let $x \to y$ and $z \to y$. Since the two extreme terms become $F(y)$, we conclude (2.2) follows. □

2.2 Tightness and weak convergence in R^p

The next theorem studies the relation between the tightness of (μ_n) (defined on $\mathscr{R}^p$) and that of its marginals.

Theorem 2.2.1. *(i) Every finite collection of pms μ_k, $k = 1, 2, \ldots, n$ on $\mathscr{R}^p$ is tight.*
(ii) Let $(\mathbb{F}_n)$ be a sequence of dfs in R^p and let the μ_ns denote the associated induced pms. Let $(F_{n,j}), (\mu_{n,j}), 1 \leq j \leq p$ correspond to the marginals. Then (μ_n) is a tight sequence iff each sequence $(\mu_{n,j})$ is tight.
(iii) If $\mathbb{F}_n \overset{d}{\to} \mathbb{F}_0$ then (μ_n) is a tight sequence. $\left[\right.$ This result is true in all complete separable metric spaces, ref. **(2.4.5)** $\left.\right]$

Proof. (*i*) The tightness of a single pm on $\mathscr{R}^p$ is an immediate consequence of property **(1.7.2)**(iii) of dfs. The tightness of a finite family of pms follows from the fact that the union of a finite number of compact sets is compact.

(*ii*) Suppose (μ_n) is a tight sequence. Given $\varepsilon > 0$, a compact set K_ε can be found such that $\mu_n(K_\varepsilon) > 1 - \varepsilon$. Let $K_{\varepsilon,j} = \pi_j(K_\varepsilon)$ and note that it is a compact set, since π_j is a continuous function. Note also that $\pi_j^{-1}(K_{\varepsilon,j}) \supseteq K_\varepsilon$. Now $\mu_{n,j}(K_{\varepsilon,j}) = \mu_n \pi_j^{-1}(K_{\varepsilon,j}) \geq \mu_n(K_\varepsilon) > 1 - \varepsilon$, proving that the marginal *pms* are tight.

Conversely, let each sequence $(\mu_{n,j})$ be tight. Then given $\varepsilon > 0$, we can find $a > 0$ such that $\mu_{n,j}([-a, a]) > 1 - \frac{\varepsilon}{p}, 1 \leq j \leq p$. Let K_ε be the cube in R^p with the intervals $[-a, a]$ for its sides and note that K_ε is a compact set. Tightness of the sequence (μ_n) follows now because

$$\mu_n(K_\varepsilon') \leq \sum_{j=1}^{p} \mu_{n,j}(K_{\varepsilon,j}') \leq \sum_{j=1}^{p} \frac{\varepsilon}{p} = \varepsilon.$$

(*iii*) Given $\varepsilon > 0$, there exists (ref. proof of **(2.1.12)**) a number N and a compact set $\bar{A}$ such that $\mu_n(\bar{A}) > 1 - 2\varepsilon$ for all $n \geq N + 1$. The finite family

μ_ν, $1 \le \nu \le N$ is tight by (i). Hence there is a compact set B such that $\mu_\nu(B) > 1 - \varepsilon$, $1 \le \nu \le N$. We then have ; $C = \bar{A} \cup B$ is a compact set and $\mu_\nu(C) > 1 - 2\varepsilon$. This is equivalent to the claim. □

The following theorem is true also in arbitrary complete separable metric spaces. But we will not prove it in that generality. We will prove it (ref. **(2.7.12)**) when the *pms* μ_n are defined on $(C,\ \rho;\ \mathscr{C})$.

Theorem 2.2.2. *A tight sequence of pms* $(\mu_n, n \ge 1)$ *in* R^p *always admits of a weakly convergent subsequence.*

Proof. Let $\mathbb{F}_n$ be the *df* associated with μ_n. i.e. $\mathbb{F}_n(\mathbf{x}) = \mu_n\{\mathbf{y} : \mathbf{y} \le \mathbf{x}\}$. Let $D = (\mathbf{a}_i, i \ge 1)$ be a countable dense subset of R^p. Since $(\mathbb{F}_n(\mathbf{a}_1))$ is a sequence in the compact interval $[0, 1]$, this sequence contains a convergent subsequence, say, $(\mathbb{F}_{1,m}(\mathbf{a}_1), m \ge 1)$. Similarly the sequence $(\mathbb{F}_{1,m}(\mathbf{a}_2))$ of real numbers in $[0, 1]$ contains a convergent subsequence, say, $\mathbb{F}_{2,m}(\mathbf{a}_2)$. And then both the sequences $(\mathbb{F}_{2,m}(\mathbf{a}_i), i = 1, 2)$ would be convergent. This procedure can be kept up indefinitely. It follows that the diagonal sequence $(\mathbb{F}_{n,n})$ is a subsequence of the given sequence and converges on D.

Define function $\mathbb{F}_D$ on D by setting $\mathbb{F}_D(\mathbf{a}_i) = \lim_{n\to\infty} \mathbb{F}_{n,n}(\mathbf{a}_i)$. Since (μ_n) is a tight sequence we can, given $\varepsilon > 0$, find $\mathbf{a}, \mathbf{b} \in D, \mathbf{a} \ll \mathbf{b}$ such that $\Delta_{\mathbf{a}}^{\mathbf{b}}\mathbb{F}_{n,n} > 1 - \varepsilon$. Hence, taking limit as $n \to \infty$, $\Delta_{\mathbf{a}}^{\mathbf{b}}\mathbb{F}_D \ge 1 - \varepsilon$. This implies that $\lim \mathbb{F}_D(\mathbf{a})$ is 1 or 0 according as $\min(a_1, a_2, \ldots, a_p) \to \infty$ or $-\infty$. Define, for all $\mathbf{x} \in R^p$, including those in D, the function $\mathbb{F}$ thus : $\mathbb{F}(\mathbf{x}) = \inf_{\mathbf{x} \le \mathbf{a}_i} \mathbb{F}_D(\mathbf{a}_i)$. So defined, $\mathbb{F}$ is easily verified to possess the properties needed to qualify to be a *df* (ref. **(1.7.2)**). We will now prove that $\mathbb{F}_n \xrightarrow{d} \mathbb{F}$. We note that $\mathbb{F}(\mathbf{a}) = \mathbb{F}_D(\mathbf{a})$, for $\mathbf{a} \in D$ and that if $\mathbf{x}$ is a continuity point of $\mathbb{F}$ then $\mathbf{a}_n \ll \mathbf{x}$, $\mathbf{a}_n \uparrow \mathbf{x}$ would imply $\lim_{n\to\infty} \mathbb{F}(\mathbf{a}_n) = \mathbb{F}(\mathbf{x})$. Let $\mathbf{x} \in C_{\mathbb{F}}$. Let $\mathbf{a}_k$, $\mathbf{b}_k$ $\in D$ and with $\mathbf{a}_k \ll \mathbf{x} \ll \mathbf{b}_k$, $k \ge 1$, $\mathbf{a}_k \uparrow \mathbf{x}$ and $\mathbf{b}_k \downarrow \mathbf{x}$. Since $\mathbb{F}_{n,n}(\mathbf{a}_k) \le \mathbb{F}_{n,n}(\mathbf{x}) \le \mathbb{F}_{n,n}(\mathbf{b}_k)$, it follows that

$$\mathbb{F}_D(\mathbf{a}_k) \le \varliminf_{n\to\infty} \mathbb{F}_{n,n}(\mathbf{x}) \le \varlimsup_{n\to\infty} \mathbb{F}_{n,n}(\mathbf{x}) \le \mathbb{F}_D(\mathbf{b}_k).$$

Now take $k \to \infty$ and obtain

$$\mathbb{F}(\mathbf{x}) = \lim_{n\to\infty} \mathbb{F}_{n,n}(\mathbf{x}), \mathbf{x} \in C_{\mathbb{F}}.$$ □

Theorem 2.2.3. *Let* $(\mathbb{F}_n)$ *be a sequence of dfs on* $\mathscr{R}^p$ *such that every one of its subsequence admits of a further subsequence which is weakly convergent and such that all such weak limits are the same. Then the sequence* $(\mathbb{F}_n)$ *is weakly convergent.*

Proof. Denote by $\mathbb{F}$ the common limit of all the weakly convergent subsequences. Let $\mathbf{x} \in C_{\mathbb{F}}$. Let $\alpha = \overline{\lim_{n\to\infty}} \mathbb{F}_n(\mathbf{x})$. We can find $n_k \uparrow \infty$ such that $\mathbb{F}_{n_k}(\mathbf{x}) \to \alpha$. By hypothesis $\mathbb{F}_{n_k}$ admits of a subsequence, say, $\mathbb{F}_r$ converging weakly to $\mathbb{F}$. Hence $\lim_{r\to\infty} \mathbb{F}_r(\mathbf{x}) = \mathbb{F}(\mathbf{x})$. This implies that $\alpha = \mathbb{F}(\mathbf{x})$. Similarly $\underline{\lim}_{n\to\infty} \mathbb{F}_n(\mathbf{x}) = \mathbb{F}(\mathbf{x})$. □

2.3 Characteristic functions, tightness and weak convergence in R^p

Later (ref. **(2.4.2)**(ii)) we will see that in the theorem **(2.1.12)** the collection C_1 can be replaced by the smaller collection C_2 of bounded uniformly continuous functions. In R^p, it is possible to replace C_2 by a still smaller class C_4 of functions given by $C_4 = \{\cos(\mathbf{t}'\mathbf{x}),\ \sin(\mathbf{t}'\mathbf{x}),\ \mathbf{t} \in R^p\}$. This section is devoted to establishing this claim.

Definition 2.3.1. The characteristic function (*cf*) **f** of a *df* $\mathbb{F}$ is defined thus: for $\mathbf{t} \in R^p$, $\mathbf{f}(\mathbf{t}) = \int_{R^p} e^{i\mathbf{t}'\mathbf{x}}\, d\mathbb{F}(\mathbf{x})$.

It is immediate from the definition that the *cf* f_j of the jth marginal of $\mathbb{F}$ is given by $f_j(t) = \int_{-\infty}^{\infty} e^{itx} dF_j(x) = \boldsymbol{f}(0,0,\ldots,t,0,\ldots,0)$, t being in the jth place. Also immediate from the definition is that all cfs are uniformly continuous over R^p. For,

$$|f(\mathbf{t}+\mathbf{u}) - f(\mathbf{t})| \le \int_{R^p} |2\sin(\frac{1}{2}\mathbf{u}'\mathbf{x})| d\mathbb{F}(\mathbf{x})$$

Hence

$$\begin{aligned}\limsup_{\mathbf{u}\to\mathbf{0}}\sup_{\mathbf{t}\in R^p} |f(\mathbf{t}+\mathbf{u}) - f(\mathbf{t})| &\le \lim_{\mathbf{u}\to\mathbf{0}} \int_{R^p} |2\sin(\tfrac{1}{2}\mathbf{u}'\mathbf{x})| d\mathbb{F}(\mathbf{x}) \\ &= \int_{R^p} \lim_{\mathbf{u}\to\mathbf{0}} |2\sin(\tfrac{1}{2}\mathbf{u}'\mathbf{x})|\, d\mathbb{F}(\mathbf{x}) \text{ (by the bounded convergence theorem)} \\ &= 0\end{aligned}$$

Theorem 2.3.2. *Let $\Re(z)$, $\Im(z)$ denote respectively the real and imaginary part of the complex number z. Let μ, μ_n be pms on $\mathscr{R}$; F, F_n and f, f_n the corresponding dfs and cfs. Let u be a positive number. Then*

(i) $\int_{|x|\ge\frac{1}{u}} dF(x) \le \frac{2}{u}\int_0^u \{1 - \Re(f(t))\}dt.$

*[(ref. **(4.4.1)**) for another inequality for $P(|X| \geq \lambda)$]*
(ii) For $|t|$ small,

$$|1 - f(t)| \leq 1 - \cos\sqrt{|t|} + \sin\sqrt{|t|} + 3 \int\limits_{|x| \geq \frac{1}{\sqrt{|t|}}} dF(x).$$

(iii) The family (F_α) of dfs is a tight family iff the corresponding family (f_α) of cfs has the property: $\lim_{t\to 0} \sup_\alpha |1 - f_\alpha(t)| = 0$. This property is called the equicontinuity of the family at 0. (ref. **Q**9).
(iv) If there exists an interval $[-a,\ a]$ and a function h defined and continuous on that interval and if $f_n(t) \to h(t)$ for all $t \in [-a,\ a]$, then (μ_n) is a tight sequence.

Proof (*i*)

$$I = \frac{1}{u}\int_0^u \{1 - \Re(f(t))\}\mathrm{d}t = \frac{1}{u}\int_0^u \{\int_{-\infty}^{\infty} (1 - \cos tx)\mathrm{d}F(x)\}\mathrm{d}t$$

$$= \int_{-\infty}^{\infty} \{1 - \frac{\sin ux}{ux}\}\mathrm{d}F(x) \quad \text{(by Fubini's theorem)}$$

$$\geq \int\limits_{|x| \geq \frac{2}{u}} \{1 - \frac{\sin ux}{ux}\}\mathrm{d}F(x) \quad \text{(since the integrand is non-negative)}$$

$$= \int\limits_{|x| \geq \frac{2}{u}} \{1 - \frac{\sin|ux|}{|ux|}\}\mathrm{d}F(x).$$

Now the the integrand is $\geq 1 - \frac{1}{|ux|} \geq \frac{1}{2}$. Hence for all values of x in the range, the integrand is $\geq \frac{1}{2}$ and then $I \geq \frac{1}{2} \int\limits_{|x| \geq \frac{2}{u}} \mathrm{d}F(x)$. i.e. the claim is true.

(*ii*) Let t be close to 0. Then, $1 - \Re(f(t)) = \int\limits_R (1 - \cos tx)\,\mathrm{d}F(x)$

$$= \int\limits_{|x| \leq \frac{1}{\sqrt{|t|}}} (1 - \cos tx)\,\mathrm{d}F(x) + \int\limits_{|x| \geq \frac{1}{\sqrt{|t|}}} (1 - \cos tx)\,\mathrm{d}F(x)$$

$$\leq 1 - \cos\sqrt{|t|} + 2 \int\limits_{|x| \geq \frac{1}{\sqrt{|t|}}} \mathrm{d}F(x);$$

$$|\Im(f(t))| \leq \int\limits_R |\sin tx|\,\mathrm{d}F(x)$$

$$= \int\limits_{|x| \le \frac{1}{\sqrt{|t|}}} |\sin tx| \, \mathrm{d}F(x) + \int\limits_{|x| \ge \frac{1}{\sqrt{|t|}}} |\sin tx| \, \mathrm{d}F(x)$$

$\le \sin \sqrt{|t|} + \int\limits_{|x| \ge \frac{1}{\sqrt{|t|}}} \mathrm{d}F(x)$. Hence the stated inequality.

(*iii*) Suppose (F_α) is a tight family. Given $\varepsilon > 0$, we can then find $t_0 > 0$ such that for all t with $|t| \le t_0$ and allα,

$$\int\limits_{|x| \ge \frac{1}{\sqrt{|t|}}} dF_\alpha(x) \le \varepsilon.$$

Now use (*ii*) to get

$$|1 - f_\alpha(t)| \le 1 - \cos \sqrt{|t|} + \sin \sqrt{|t|} + 3\varepsilon$$

for all t with $|t| \le t_0$ and all n. That the sequence (f_α) is equicontinuous at 0 follows immediately from this. Conversely, suppose that the family (f_α) is equicontinuous at 0. Hence given $\varepsilon > 0$, we can find a $\tau > 0$ such that $\sup\limits_{|t| \le \tau} |1 - f_\alpha(t)| \le \varepsilon$ for all α. Now by (*i*),

$$\int\limits_{|x| \ge \frac{1}{\tau}} \mathrm{d}F_\alpha(x) \le \frac{2}{\tau} \int_0^\tau \{1 - \Re(f_\alpha(t))\} dt$$

$$\le \frac{2}{\tau} \int_0^\tau |1 - \Re(f_\alpha(t))| dt \le \frac{2}{\tau} \int_0^\tau |1 - f_\alpha(t)| dt$$

$$\le \frac{2}{\tau} \int_0^\tau \varepsilon \, \mathrm{d}t = 2\varepsilon,$$

proving that (F_α) is a tight family.

(*iv*) With no loss of generality, we may assume $a < 1$. Given $\varepsilon > 0$, find $\tau,\ 0 < \tau < a$ such that

$$\sup_{t \in [-\tau,\, \tau]} \{1 - \Re(h(t))\} < \varepsilon.$$

This is possible because $\Re(h)$ is continuous on $[-a,\ a]$ and because $\Re(h(0)) = \lim\limits_{n \to \infty} \Re(f_n(0)) = 1$. By the bounded convergence theorem,

$$\int_0^\tau \{1 - \Re(f_n(t))\} dt \to \int_0^\tau \{1 - \Re(h(t))\} dt.$$

Find $N = N(\tau, \varepsilon)$ such that for all $n \geq N$,

$$\left| \int_0^\tau \{1 - \Re(f_n(t))\}\mathrm{d}t - \int_0^\tau \{1 - \Re(h(t))\}\mathrm{d}t \right| < \varepsilon.$$

We then have, for $n \geq N$,

$$\int_{|x| \geq \frac{1}{\tau}} \mathrm{d}F_n(x) \leq \frac{2}{\tau} \int_0^\tau \{1 - \Re(f_n(t))\}\mathrm{d}t \qquad \text{(by(i))}$$

$$\leq 2\varepsilon + \frac{2}{\tau} \int_0^\tau \{1 - \Re(h(t))\}\mathrm{d}t \leq 3\varepsilon.$$

This proves that (μ_n) is a tight sequence. □

Result 2.3.3. *We recall some known results from complex function theory. For $\sigma > 0$, define*

$$g_\sigma(x) = \begin{cases} \frac{x}{\sigma^2} e^{-\frac{x}{\sigma}} & \text{if } x \geq 0 \\ 0 & \text{if } x < 0 \end{cases} \qquad \hat{g}_\sigma(x) = \begin{cases} \frac{|x|}{\sigma^2} e^{-\frac{|x|}{\sigma}} & \text{if } x \leq 0 \\ 0 & \text{if } x > 0 \end{cases}$$

and

$$u_\sigma(t) = \int_{-\infty}^{\infty} e^{itx} g_\sigma(x)dx \qquad \hat{u}_\sigma(t) = \int_{-\infty}^{\infty} e^{itx} \hat{g}_\sigma(x)dx$$

Then the following results can be established by contour integration.

$$(i)\ u_\sigma(t) = (1 - \sigma it)^{-2}; \qquad (ii)\ \hat{u}_\sigma(t) = (1 + i\sigma t)^{-2}$$

For all x,

$$(iii)\ \frac{1}{2\pi} \int_{-\infty}^{\infty} e^{-itx} u_\sigma(t)dt = g_\sigma(x); \ (iv) \frac{1}{2\pi} \int_{-\infty}^{\infty} e^{-itx} \hat{u}_\sigma(t)dt = \hat{g}_\sigma(x).$$

Characteristic functions play a crucial role in probability theory, a major role in central limit theorems (ref. Chapter 5). Study of convolutions of distribution functions is reduced to the study of the products of their cfs (ref. ***(2.3.10)****). This reduction, a helpful facilitator, is a great advantage, since characteristic functions, as Fourier transforms, is a well researched area in mathematics. That a df uniquely determines its cf is obvious, being a simple consequence of the definition of an integral. The validity of the converse proposition needs to be established before the theory can proceed further. The next theorem,* ***(2.3.4)****, establishes that two distinct dfs can not have the same cf.*

More definitive results will follow later.

Theorem 2.3.4 (Uniqueness theorem for cf). *Let f_1, f_2 be the cfs of the dfs $\mathbb{F}_1$, $\mathbb{F}_2$. If $f_1(\boldsymbol{t}) = f_2(\boldsymbol{t})$ for all $\boldsymbol{t} \in R^p$ then $\mathbb{F}_1 \equiv \mathbb{F}_2$.*

Proof. Suppose $f_1(\mathbf{t}) = f_2(\mathbf{t})$ for all $\mathbf{t} \in R^p$. We must show that $\mathbb{F}_1(\mathbf{x}) = \mathbb{F}_2(\mathbf{x})$ for every $\mathbf{x} \in R^p$. Fix $\mathbf{x}$. Let $a_\sigma(\mathbf{y}) = \times_{i=1}^{p}\gamma_\sigma(y_i)$ where $\gamma_\sigma(y_i) = g_\sigma(y_i)$ or $\hat{g_\sigma}(y_i)$ according as $y_i \geq 0$ or $y_i < 0$. Let

$$b_\sigma(\mathbf{t}) = \times_{i=1}^{j} u_\sigma(t_i) \times_{i=j+1}^{p} \hat{u}_\sigma(t_i);\, |b_\sigma(t)| = \times_{j=1}^{p} \frac{1}{1+\sigma^2 t_j^2}.$$

Let $A_\sigma(\mathbf{v}) = \int\limits_{\mathbf{y}\leq\mathbf{v}} a_\sigma(\mathbf{y})\, d\mathbf{y}$ and verify that $\lim\limits_{\sigma\to 0} A_\sigma(v_1, v_2, \ldots, v_p) = 1$ if $\min\limits_{1\leq i\leq p} v_i > 0$ and is 0 if $\min\limits_{1\leq i\leq p} v_i < 0$. We note that $|b_\sigma(\mathbf{t}) f_i(\mathbf{t})|, i = 1, 2$ is Lebesgue integrable over R^p, since it is dominated by the integrable function $|b_\sigma(\mathbf{t})|$. Now for $r = 1, 2$, let

$$\Lambda_r(\mathbf{v}, \sigma) = \int\limits_{R^p} e^{-i\mathbf{t}'\mathbf{v}} b_\sigma(\mathbf{t}) f_r(\mathbf{t}) d\mathbf{t} = \int\limits_{R^p} e^{-i\mathbf{t}'\mathbf{v}} b_\sigma(\mathbf{t}) \{ \int\limits_{R^p} e^{i\mathbf{t}'\mathbf{z}} d\mathbb{F}_r(\mathbf{z})\} d\mathbf{t}$$

$$= \int\limits_{R^p} J(\mathbf{v},\mathbf{z}) d\mathbb{F}_r(\mathbf{z}) \text{ (by Fubini's Theorem)}$$

$$\text{(where } J(\mathbf{v},\mathbf{z}) = \int\limits_{R^p} e^{-i\mathbf{t}'(\mathbf{v}-\mathbf{z})} b_\sigma(\mathbf{t}) d\mathbf{t})$$

$$= a_\sigma(\mathbf{v} - \mathbf{z}).$$

Thus

$$\lim_{\sigma\to 0} \int\limits_{\mathbf{v}\leq\mathbf{x}} \Lambda_r(\mathbf{v}, \sigma)\, d\mathbf{v} = \lim_{\sigma\to 0} \int\limits_{\mathbf{v}\leq\mathbf{x}} \Big\{ \int\limits_{R^p} a_\sigma(\mathbf{v} - \mathbf{z})\, d\mathbb{F}_r(\mathbf{z}) \Big\}\, d\mathbf{v}$$

$$= \lim_{\sigma\to 0} \int\limits_{R^p} A_\sigma(\mathbf{x} - \mathbf{z})\, d\mathbb{F}_r(\mathbf{z})$$

by Fubini theorem, which is applicable, the integrand being non-negative

$$= \int\limits_{R^p} \lim_{\sigma\to 0} A_\sigma(\mathbf{x} - \mathbf{z})\, d\mathbb{F}_r(\mathbf{z}) = \mathbb{F}_r(\mathbf{x})$$

(ref.: bounded convergence theorem). Now, $(f_1 \equiv f_2) \Rightarrow \{\Lambda_1(\mathbf{v}, \sigma) = \Lambda_2(\mathbf{v}, \sigma)\}$ for all $\mathbf{v}$ and σ. Hence $\mathbb{F}_1(\mathbf{x}) = \mathbb{F}_2(\mathbf{x})$. □

Theorem 2.3.5 (Continuity theorem). *Let f, f_n be the cfs respectively of the dfs $\mathbb{F}$ and $\mathbb{F}_n$.*

$$f_n(\boldsymbol{t}) \to f(\boldsymbol{t}),\ \boldsymbol{t} \in R^p \text{ iff } \mathbb{F}_n \stackrel{d}{\to} \mathbb{F}.$$

Proof. Let $f_n(\mathbf{t}) \to f(\mathbf{t}), \mathbf{t} \in R^p$. Denote the marginals by f_j, $f_{n;j}$, F_j, $F_{n;j}$. The hypothesis implies that $f_{n;j}(t) \to f_j(t)$, $1 \le j \le p$. By **(2.3.2)**(iv) this implies that $(F_{n;j},\ n = 1,\ 2,\ \ldots)$ is a tight sequence for each j. Hence, by **(2.2.1)**(ii), $(\mathbb{F}_n)$ is a tight sequence. If $(\mathbb{F}_{n_k})$ is a weakly convergent subsequence (such subsequences exist by **(2.2.2)**) converging to, say, *df* $\mathbb{G}$ with *cf* g, then (by Helly-Bray theorem) $f_{n_k}(\mathbf{t}) \to g(\mathbf{t})$, $\mathbf{t} \in R^p$. But $f_{n_k}(\mathbf{t}) \to f(\mathbf{t})$ by hypothesis, leading to $f \equiv g$. It follows now by (2.3.4) that $\mathbb{F} \equiv \mathbb{G}$. Since every weakly convergent subsequence of the sequence $(\mathbb{F}_n)$ converges to the same *df*, namely, $\mathbb{F}$, it follows by **(2.2.3)** that $\mathbb{F}_n \stackrel{d}{\to} \mathbb{F}$.

The converse part trivially follows from the Helly-Bray theorem. □

Notation. We write $\varphi_n \stackrel{d}{\to} \varphi_0$ when φ_n, $n \ge 0$ are cfs and $\varphi_n(\mathbf{t}) \to \varphi_0(\mathbf{t})$, $\mathbf{t} \in R^p$.

The next result enables one to evaluate the integrals of rvs over abstract probability spaces as integrals of corresponding real functions over the real line with respect to the induced pm. (ref. **(2.3.7)** Remark (ii))

Theorem 2.3.6 (Formula for change of variables). *Let $(\Omega,\ \mathscr{A},\ \mu)$ be a probability measure space, $(\boldsymbol{X},\ \mathcal{X})$ a measurable space, f a measurable mapping from Ω into $\boldsymbol{X}$, $\hat{B} \in \mathcal{X}$ and g a Borel measurable mapping of $\boldsymbol{X}$ into the real line. Let $B = f^{-1}(\hat{B})$. Then $\int_{\hat{B}} g(x)d(\mu f^{-1})(x) = \int_B g(f(\omega))d\mu(\omega)$ in the sense that if either side exists then the other side will also exist and the two sides will be equal.*

Proof. If the claim is true for two measurable functions g_1, g_2, then it is trivial to verify that it is true for $ag_1 + bg_2$; a, b arbitrary real numbers. Hence it is enough to establish theorem assuming g is non-negative. Let $E \in \mathcal{X}$. Then

$$\begin{aligned}
&\int_{\hat{B}} \chi_E(x)\mathrm{d}(\mu f^{-1})(x) \\
&\qquad = (\mu f^{-1})(E \cap \hat{B}) = \mu(f^{-1}(E \cap \hat{B})) = \mu(f^{-1}(E) \cap B) \\
&\qquad = \int_B \chi_{f^{-1}(E)} \mathrm{d}\mu(\omega) = \int_B \chi_E(f(\omega))\mathrm{d}\mu(\omega).
\end{aligned}$$

This shows that the claim is true if g is the indicator function of a member of $\mathcal{X}$ and hence true if g is a simple function. If g is a non-negative measurable function, then it is the limit everywhere of a monotonic increasing sequence of (non-negative) simple functions. Hence, by the monotone convergence theorem, the claim is true for non-negative measurable functions. □

Remarks 2.3.7.
(i) If g is a complex valued function, the theorem can be applied separately to the real part and to the imaginary part.
(ii) In the theorem take $(\mathbf{X},\ \mathscr{X}) = (R,\ \mathscr{R})$; $f = X$ a rv and $g(x) = x,\ x \in R$. Let F denote the *df* of X : $F(x) = \mu(X^{-1}(-\infty,\ x])$ and let ν denote the pm induced by X on $\mathscr{R}$: $\nu = \mu X^{-1}$. Let $\hat{B}$ be a linear Borel set and let $B = f^{-1}(\hat{B})$. Then, in the sense that if either side exists so does the other, we have $\int_B X(\omega)\,\mathrm{d}\mu(\omega) = \int_{\hat{B}} x\,\mathrm{d}\nu(x)$ which is often written, in the case $\hat{B} = R$, as $\int_R x\,\mathrm{d}F(x)$ (ref. final lines of **(1.12.1)**).
(iii) Let $\mathbf{X}$ be a rV in R^p with dm μ and with *cf* φ. Then $(\mathfrak{a})$ ψ is a *cf* where $\psi(\mathbf{t}) = \varphi(-\mathbf{t})$ and $(\mathfrak{b})$ φ is real valued *iff* $\mathbf{X}$ is symmetrically distributed. i.e., *iff* $\mu(E) = \mu(-E),\ E \in \mathscr{R}^p$.
[Symmetric measures are discussed again in **(3.4.3)** and **(3.9)**]

Proof.
$(\mathfrak{a})$ Define $\nu(E) = \mu(-E),\ E \in \mathscr{R}^p$ and note that ν is a pm. If ψ is its *cf* then

$$\psi(\mathbf{t}) = \int_{R^p} e^{i\mathbf{t}'\mathbf{x}} d\nu(\mathbf{x}) = \int_{R^p} g(f(\mathbf{x})) d\nu(\mathbf{x})$$

where $g(\mathbf{x}) = e^{-i\mathbf{t}'\mathbf{x}}$ and $f(\mathbf{x}) = -\mathbf{x}$. We note $f^{-1}(\mathbf{x}) = -\mathbf{x}$ and $\nu f^{-1} = \mu$. Then, by **(2.3.6)**,

$$\psi(\mathbf{t}) = \int_{R^p} e^{-i\mathbf{t}'\mathbf{x}} d\nu f^{-1}(\mathbf{x}) = \int_{R^p} e^{-i\mathbf{t}'\mathbf{x}} d\mu(\mathbf{x}) = \varphi(-\mathbf{t}).$$

$(\mathfrak{b})$ If φ is real valued then $\varphi(\mathbf{t}) = \varphi(-\mathbf{t}) = \psi(\mathbf{t})$, Since $\varphi \equiv \psi$, it follows by **(2.3.4)** that $\mu \equiv \nu$. i.e., $\mathbf{X}$ is symmetrically distributed.

Conversely, if $\mathbf{X}$ is symmetrically distributed, i.e., if $\mu \equiv \nu$, then

$$\begin{aligned}\varphi(\mathbf{t}) &= \int_{R^p} e^{i\mathbf{t}'\mathbf{x}} d\mu(\mathbf{x}) = \int_{R^p} e^{i\mathbf{t}'\mathbf{x}} d\nu(\mathbf{x}) \\ &= \int_{R^p} e^{i\mathbf{t}'\mathbf{x}} d\mu f^{-1}(\mathbf{x}) = \int_{R^p} e^{-i\mathbf{t}'\mathbf{x}} d\mu(\mathbf{x}) = \varphi(-\mathbf{t}).\end{aligned}$$ □

(iv) Refer to (1.11.11 (ii)). Here is a second proof of the relation

$$\int_{-\infty}^{x} F^n(y)\,\mathrm{d}F(y) = \tfrac{1}{n+1}F^{n+1}(x)$$

if F is a continuous *df*. For yet another proof, refer to **(3.1.3)**(vi).

Let μ denote the pm generated by F. Since F is a continuous *df* its range is the interval $I = (0,\ 1)$ or its semi-closed or closed version. μF^{-1} is a pm on the Borel σ-field of I. For $0 < a < 1$, define $x_a^- = \inf\{x : F(x) = a\}$ and $x_a^+ = \sup\{x : F(x) = a\}$. We note that if $0 < a < b < 1$, then $F^{-1}(a,\ b) = (x_a^-,\ x_b^+)$ and further that $\mu(x_a^-,\ x_b^+) = F(x_b^+) - F(x_a^-) = b - a$. This shows that μF^{-1} is the same as the Lebesgue measure on I. Hence

$$\begin{aligned}\frac{1}{n+1}F^{n+1}(x) &= \int_0^{F(x)} y^n\,\mathrm{d}y = \int_0^1 y^n \chi_{(0,\ F(x))}(y)\,\mathrm{d}y\\ &= \int_0^1 y^n \chi_{(0,\ F(x))}(y)\,\mathrm{d}\mu F^{-1}(y)\\ &= \int_{-\infty}^{\infty} F^n(t)\chi_{(-\infty,\ x)}(y) d\mu(t)\end{aligned}$$

which is the desired result.

Theorems ***(2.3.8)*** *and* ***(2.3.9)*** *are important in themselves and are also further illustrations of applications of* ***(2.3.6)***.

Theorem 2.3.8. *Treat* $\mathbf{t} \in R^p$ *as a function defined on* $R^p : \mathbf{t}(\mathbf{x}) = \mathbf{t}'\mathbf{x}$. *Let* (μ_n) *be a sequence of pms on* $\mathscr{R}^p$. *Then sequence* (μ_n) *has a weak limit iff the sequence* $(\mu_n \mathbf{t}^{-1})$ *has a weak limit.*

Proof. Let $\mu_n \overset{d}{\to} \mu_0$. Hence, by **(2.3.5)**,

$$\int_{R^p} e^{it\mathbf{t}(\mathbf{x})}\,\mathrm{d}\mu_n(\mathbf{x}) \to \int_{R^p} e^{it\mathbf{t}(\mathbf{x})}\,\mathrm{d}\mu_0(\mathbf{x}).$$

In **(2.3.6)** if we take $(\Omega,\ \mathscr{A},\ P) = (R^p,\ \mathscr{R}^p,\ \mu_n)$; g defined on R, $g(y) = e^{ity}$, we get

$$\int_R g(y)\,\mathrm{d}\mu_n \mathbf{t}^{-1}(y) = \int_{R^p} g(\mathbf{t}(\mathbf{x}))\,\mathrm{d}\mu_n(\mathbf{x}).$$

$$\text{i.e. } \int_R e^{ity}\,\mathrm{d}\mu_n \mathbf{t}^{-1}(y) = \int_{R^p} e^{it\mathbf{t}(\mathbf{x})}\,\mathrm{d}\mu_n(\mathbf{x}).$$

This implies

$$\int_R e^{ity}\, d\mu_n \mathbf{t}^{-1}(y) \to \int_R e^{ity}\, d\mu_0 \mathbf{t}^{-1}(y).$$

Appeal to **(2.3.5)** to conclude from this that $\mu_n \mathbf{t}^{-1} \overset{d}{\to} \mu_0 \mathbf{t}^{-1}$.

Conversely, suppose that for every $\mathbf{t} \in R^p$, sequence $(\mu_n \mathbf{t}^{-1})$ has a weak limit. Hence, by **(2.3.5)**, $\lim_n \int_R e^{ity}\, d\mu_n \mathbf{t}^{-1}(y)$ exists for each $t \in R$, and each $\mathbf{t} \in R^p$. This is equivalent (ref. **(2.3.6)**) to saying that $\lim_n f_n(t\mathbf{t})$ exists for all $t \in R$, and each $\mathbf{t} \in R^p$ where $f_n(\mathbf{t}) = \int_{R^p} e^{i\mathbf{t}(\mathbf{x})}\, d\mu_n(\mathbf{x})$, the *cf* of μ_n. Let $\lim_n f_n(\mathbf{t}) = a(\mathbf{t})$.

Let $(\mu_{n;j},\ 1 \le j \le p)$ denote the marginals. If $\mathbf{t}_1 = (1, 0, 0, \ldots, 0)'$, then $\mu_n \mathbf{t}_1^{-1} = \mu_{n;1}$. The existence of the weak limit of $\mu_{n;1}$ implies that $(\mu_{n;1})$ is a tight sequence. Similarly all the sequences $(\mu_{n;j},\ 2 \le j \le p)$ are tight. Hence, by **(2.2.1)**(ii), (μ_n) is a tight sequence. Let μ_{n_k} be a weakly convergent subsequence converging to, say, ν. This implies, by **(2.3.5)**, that $\mathbf{f}_{n_k}(\mathbf{t}) \to b(\mathbf{t})$ where $b(.)$ is the *cf* of ν. This shows that $a(.) \equiv b(.)$ and $a(.)$ is the *cf* of ν. Hence $\mu_n \overset{d}{\to} \nu$. Since this is true for every weakly convergent subsequence, it follows (by **(2.2.3)**) that the weak limit of μ_n exists. □

(For a different proof of this result ref. **(2.4.2)**(vii))

Theorem 2.3.9. *Let* $\mathbb{F} = \mathbb{F}_1 * \mathbb{F}_2$. *Then*
(i) $F_i = F_{1,i} * F_{2,i},\ 1 \le i \le p$.
(ii) $\mathbb{F}$ *is a continuous df if* $\mathbb{F}_1$ *or* $\mathbb{F}_2$ *is a continuous df.*
(iii) The converse of the proposition (ii) is true if $p = 1$ *and is false if* $p \ge 2$.

Proof. (*i*) With no loss of generality, take $i = 1$. Then $F_1(x) = \lim_{x_2, x_3, \ldots, x_p \to \infty} \mathbb{F}(x,\ x_2,\ \ldots,\ x_p)$. By the bounded convergence theorem, we have

$$\begin{aligned} F_1(x) &= \int_{R^p} \lim_{x_2, x_3, \ldots, x_p \to \infty} \mathbb{F}_1(x - y_1,\ x_2 - y_2,\ \ldots,\ x_p - y_p) \\ &\qquad d\mathbb{F}_2(y_1,\ y_2,\ \ldots,\ y_p) \\ &= \int_{R^p} F_{1,1}(x - y_1)\, d\mathbb{F}_2(y_1,\ y_2,\ \ldots,\ y_p) \\ &= \int_R F_{1,1}(x - y)\, d\mu_{2,1}(y) = F_{1,1} * F_{2,1}(x). \end{aligned}$$

(*ii*) Let $\mathbf{x}_n \to \mathbf{x}$. Let $\mathbb{F}_2$ be a continuous *df*. Appealing to the Bounded Convergence Theorem and using the fact that $\mathbb{F}_2$ is a continuous function, we have

$$\lim_{n\to\infty} \mathbb{F}(\mathbf{x}_n) = \int_{R^p} \lim_{n\to\infty} \mathbb{F}_2(\mathbf{x}_n - \mathbf{y})\, \mathrm{d}\mathbb{F}_1(\mathbf{y}) = \int_{R^p} \mathbb{F}_2(\mathbf{x} - \mathbf{y})\, \mathrm{d}\mathbb{F}_1(\mathbf{y}) = \mathbb{F}(\mathbf{x}).$$

Since $\mathbb{F} = \mathbb{F}_1 * \mathbb{F}_2 = \mathbb{F}_2 * \mathbb{F}_1$, the proof is complete.
(*iii*) Suppose $\mathbb{F}$ is a continuous *df*. Hence (ref. **(1.7.2)**(xi)) each marginal F_i is a continuous *df*. Let $\mu,\ \mu_1,\ \mu_2$ denote the pms generated by $\mathbb{F},\ \mathbb{F}_1$ and $\mathbb{F}_2$.

Case $p = 1$. If possible let F_1 and F_2 be discontinuous dfs. Let then $\mu_1(\{a\}) > 0$ and $\mu_2(\{b\}) > 0$. Since

$$\begin{aligned}\mu(\{a+b\}) &= \int_{R} \mu_1(\{a+b\} - y)\, \mathrm{d}F_2(y)\\ &\geq \int_{\{b\}} \mu_1(\{a+b\} - y)\, \mathrm{d}F_2(y) = \mu_1(\{a\})\mu_2(\{b\}) > 0.\end{aligned}$$

This shows that if F is a continuous *df* then either F_1 or F_2 must be a continuous *df*.

Let $p = 2$. Let $F_1,\ F_2$ be continuous dfs in R. Let G be the *df* given by $G(x) = 0$ if $x < 0$ and $G(x) = 1$ if $x \geq 0$. Define dfs in

$$R^2 : \mathbb{F}_1(x,\ y) = F_1(x)G(y);\ \mathbb{F}_2(x,\ y) = G(x)F_2(y).$$

That these are proper dfs follows from **(1.7.3)**(ii). Let $\alpha,\ \beta,\ \gamma$ be the pms induced in R respectively by $F_1,\ F_2$ and G. Then the pms in R^2 determined by $\mathbb{F}_1,\ \mathbb{F}_2$ are $\alpha \times \gamma$ and $\gamma \times \beta$.

Clearly $\mathbb{F}_1,\ \mathbb{F}_2$ have discontinuities . For example every point of the set $\{(x,\ 0) : F_1(x) > 0\}$ is a discontinuity point of $\mathbb{F}_1$. Similarly every point of the set $\{(0,\ y) : F_2(y) > 0\}$ is a discontinuity point of $\mathbb{F}_2$. Let us find now $\mathbb{F} = \mathbb{F}_1 * \mathbb{F}_2$. We have

$$\begin{aligned}\mathbb{F}(x_1,\ x_2) &= \int_{R^2} \mathbb{F}_1(x_1 - y_1,\ x_2 - y_2) d(\gamma \times \beta)(y_1,\ y_2)\\ &= \int_{R^2} F_1(x_1 - y_1)G(x_2 - y_2) d(\gamma \times \beta)(y_1,\ y_2)\\ &= \int_{R} F_1(x_1 - y_1) d\gamma(y_1) \int_{R} G(x_2 - y_2) d\beta(y_2)\\ &\qquad\qquad \text{(by Fubini's theorem)}\\ &= F_1(x_1)F_2(x_2).\end{aligned}$$

It is easily seen now from this that $\mathbb{F}$ is a continuous *df*.
If $p \geq 3$, a counter example can be constructed as follows. Let $\mathbb{F}_1, \mathbb{F}_2$ be as above. Let $\mathbb{H}$ be a continuous *df* in R^{p-2}. Define *df*s Λ_1, Λ_2 in R^p as follows. $\Lambda_i(x_1, x_2, ..., x_p) = \mathbb{F}_i(x_1, x_2)\mathbb{H}(x_3, ..., x_p), i = 1, 2$. By **(1.7.2)**(ii), each Λ_i is a *df* in R^p. Each of them is discontinuous due to the factor $\mathbb{F}_i$. But $\Lambda = \Lambda_1 * \Lambda_2 = (\mathbb{F}_1 * \mathbb{F}_2)(\mathbb{H} * \mathbb{H})$. The first factor is continuous by the case $p = 2$ and the second factor is continuous by part (ii). Hence Λ is a continuous *df*. □

Theorem 2.3.10. *Let μ_1, μ_2 be pms on $\mathscr{R}^p$ with cfs $\boldsymbol{f}_1$ and $\boldsymbol{f}_2$. Let $\mu = \mu_1 * \mu_2$ and let $\boldsymbol{f}$ be the cf of μ. Then $\boldsymbol{f} \equiv \boldsymbol{f}_1\boldsymbol{f}_2$.*

Proof. By direct computation (ref. **(1.11.9)**) we see that for $E \in \mathscr{R}^p$,

$$\mu(E) = \int_{R^p} \chi_E(\mathbf{u})\, d\mu(\mathbf{u}) = \int_{R^p}\int_{R^p} \chi_E(\mathbf{x} + \mathbf{y})\, d\mu_1(\mathbf{x})\, d\mu_2(\mathbf{y}).$$

This relation obviously extends to simple functions. Then by the customary process of monotonic limits it extends to bounded non-negative measurable functions and finally to all bounded measurable functions g. Thus

$$\mathbf{f}(\mathbf{t}) = \int_{R^p} e^{i\mathbf{t}'\mathbf{u}}\, d\mu(\mathbf{u}) = \int_{R^p}\int_{R^p} e^{i\mathbf{t}'(\mathbf{x}+\mathbf{y})}\, d\mu_1(\mathbf{x})\, d\mu_2(\mathbf{y})$$

$$= \int_{R^p} e^{i\mathbf{t}'\mathbf{x}}\, d\mu_1(\mathbf{x}) \int_{R^p} e^{i\mathbf{t}'\mathbf{y}}\, d\mu_2(\mathbf{y}) = \mathbf{f}_1(\mathbf{t})\mathbf{f}_2(\mathbf{t}).$$ □

As applications of the above result we prove

Theorem 2.3.11. *(i) If φ is a cf then so is $e^{\lambda(\varphi-1)}$ for every $\lambda > 0$ and (ii) If $\mathbb{F}_n \overset{d}{\to} \mathbb{F}_0$ and if $\mathbb{G}_n \overset{d}{\to} \mathbb{G}_0$, then $\mathbb{H}_n \overset{d}{\to} \mathbb{H}_0$ where $\mathbb{H}_n = \mathbb{F}_n * \mathbb{G}_n$, $n \geq 0$. The converse of (ii) is not true.*

Proof. (i) Let μ on $\mathscr{R}^p$ be the pm with $\mu(\{\mathbf{0}\}) = 1$. Let *df* $\mathbb{F}_1$ correspond to μ and note that its *cf* is identically 1. Let φ be the *cf* of the *df* $\mathbb{F}_2$. Define $\mathbb{F}_n = (1 - \frac{\lambda}{n})\mathbb{F}_1 + \frac{\lambda}{n}\mathbb{F}_2$ and note that it is a *df* for all $n \geq \lambda$ (ref. **(1.7.3)**). The *cf* φ_n of $\mathbb{F}_n$ is $1 + \frac{\lambda(\varphi-1)}{n}$. Let

$$G_n = \underbrace{\mathbb{F}_n * \mathbb{F}_n * \cdots * \mathbb{F}_n}_{n \text{ factors}}.$$

By **(2.3.8)**, the *cf* g_n of G_n is $(1 + \frac{\lambda(\varphi-1)}{n})^n$ which converges pointwise to the continuous function $e^{\lambda(\varphi-1)}$. Hence (ref. **(2.3.2)**(iv) read with **(2.2.1)**(ii)) (G_n)

is a tight sequence. Let (G_{n_k}) be a weakly convergent subsequence. Since the corresponding *cf* sequence converges pointwise to $e^{\lambda(\varphi-1)}$, the claim that this is a *cf* follows.
(*ii*) Let $\mathfrak{f}_n$, $\mathfrak{g}_n$ and $\mathfrak{h}_n$ denote the *cf*s respectively of $\mathbb{F}_n$, $\mathbb{G}_n$ and $\mathbb{H}_n$. The hypothesis implies

$$\mathfrak{f}_n(\mathbf{t}) \to \mathfrak{f}_0 \quad \text{and} \quad \mathfrak{g}_n(\mathbf{t}) \to \mathfrak{g}_0(\mathbf{t}) \tag{2.4}$$

The forward part is trivial since all we need show is that $\mathfrak{h}_n(\mathbf{t}) \to \mathfrak{h}_0(\mathbf{t})$, $\mathbf{t} \in R^p$ (ref. **(2.3.5)**). But this is obvious, since $\mathfrak{h}_n(\mathbf{t}) = \mathfrak{f}_n(\mathbf{t})\mathfrak{g}_n(\mathbf{t})$, $n \geq 0$ (ref. **(2.3.10)**) and since the information (2.4) is available.

To show that the converse need not hold, let $p = 1$, $F_n(x) = F(x - n)$ and $G_n(x) = F(x + n)$. If φ is the *cf* of F, then the *cf* of F_n is $e^{itn}\varphi(t)$ and that of G_n is $e^{-itn}\varphi(t)$. Hence the *cf* of $F_n * G_n$ is φ^2. But neither of the component sequences (F_n) and (G_n) is weakly convergent. □

2.4 Weak convergence of *pms* in metric spaces

In section **(2.1)** weak convergence of a sequence (μ_n) of *pms* in R^p was defined using the associated sequence of *df*s. In the present section we investigate an appropriate definition of weak convergence of *pms* in abstract metric spaces which will coincide with our earlier definition when the metric space is R^p. The Helly-Bray theorem at **(2.1.12)** provides the lead. Let (M, d) be an arbitrary metric space. Denote its Borel σ-field by $\mathcal{M}$. Let μ with or without suffix denote a *pm* on $\mathcal{M}$. Let C denote the family of all the real continuous functions defined on M. Let $C_1 \subset C$ denote the collection of those members of C which are bounded. Let $C_2 \subset C_1$ consist of those members of C_1 which are uniformly continuous. Let C_3 denote the family of all functions $f_{a,F}$ defined on M where $0 < a < \infty$, F is a closed subset of M and, for $x \in M$, $f_{a,F}(x) = 1 - \frac{1}{a}d(x, F)$ if $d(x, F) \leq a$ and equals 0 if $d(x, F) \geq a$. Since $|f_{a,F}(x) - f_{a,F}(y)| \leq \frac{1}{a}|d(x, F) - d(y, F)| \leq \frac{1}{a}d(x, y)$, it follows that $C_3 \subset C_2$.

Definition 2.4.1. μ_n converges weakly to μ_0 ($\mu_n \overset{d}{\to} \mu_0$) if

$$\lim_{n\to\infty} \int_M g(x)\mathrm{d}\mu_n(x) = \int_M g(x)\mathrm{d}\mu_0(x) \tag{2.5}$$

for every $g \in C_1$.

Theorem 2.4.2. *The following statements are equivalent.*
(i) $\lim_{n\to\infty} \int_M g(x)d\mu_n(x) = \int_M g(x)d\mu_0(x)$ *for every* $g \in C_1$.
(ii) $\lim_{n\to\infty} \int_M g(x)d\mu_n(x) = \int_M g(x)d\mu_0(x)$ *for every* $g \in C_2$.
(iii) $\lim_{n\to\infty} \int_M g(x)d\mu_n(x) = \int_M g(x)d\mu_0(x)$ *for every* $g \in C_3$.
(iv) $\underline{\lim}_{n\to\infty} \mu_n(U) \geq \mu_0(U)$ *for every open set* U.
(v) $\overline{\lim}_{n\to\infty} \mu_n(F) \leq \mu_0(F)$ *for every closed set* F.
(vi) $\lim_{n\to\infty} \mu_n(A) = \mu_0(A)$ *for every* μ_0*-continuous sets* A. *i.e. sets* A *with* $\mu_0(Bd(A)) = 0$ *where* $Bd(A)$ *is the boundary of the set* A.
(vii) For every $f \in C$, $F_n \overset{d}{\to} F_0$ *where* F_n, $n \geq 0$ *is the df corresponding to the pm* $\mu_n f^{-1}$.

Proof. Since $C_3 \subset C_2 \subset C_1$, it follows that *(i)* $\Rightarrow$ *(ii)* $\Rightarrow$ *(iii)*. Since open sets and closed sets are complements of each other, *(iv)* $\Leftrightarrow$ *(v)*.
Suppose (iv) holds. Hence (*v*) holds. Let $A \in \mathcal{m}$ and let $\mu_0(Bd(A)) = 0$. We recall that $Int(A) = A \sim Bd(A)$ and $\bar{A} = A \cup Bd(A)$ (ref. page 46 [K]). Hence $\mu_0(Int(A)) = \mu_0(A) = \mu_0(\bar{A})$. An appeal to (*iv*) yields $\underline{\lim}_{n\to\infty} \mu_n(A) \geq \underline{\lim}_{n\to\infty} \mu_n(Int(A)) \geq \mu_0(Int(A)) = \mu_0(A)$. Now an appeal to (*v*) yields $\overline{\lim}_{n\to\infty} \mu_n(A) \leq \overline{\lim}_{n\to\infty} \mu_n(\bar{A}) \leq \mu_0(\bar{A}) = \mu_0(A)$. These two inequalities show that (*iv*) $\Rightarrow$ (*vi*). Suppose now (*vi*) holds. Let F be a closed subset of M. For $\delta > 0$, define $A(\delta) = \{x : d(x, F) = \delta\}$. $\{A(\delta), \delta > 0\}$ is an uncountable collection of disjoint sets in $\mathcal{m}$. Atmost countably many of them can have μ_0 measure positive. Hence we can find $\delta_k, k = 1, 2, \ldots, \delta_k \downarrow 0$ such that $\mu_0(A(\delta_k)) = 0$. Define $F_k = \{x : d(x, F) \leq \delta_k\}$ and note that $Bd(F_k) = A(\delta_k)$, and $F \subset F_k$. Now $\overline{\lim}_{n\to\infty} \mu_n(F) \leq \overline{\lim}_{n\to\infty} \mu_n(F_k) = \mu_0(F_k)$ (applying (*vi*), which is valid since $\mu_0(Bd(F_k)) = 0$). Take $k \to \infty$, observe that $F_k \downarrow F$ and conclude that $\overline{\lim}_{n\to\infty} \mu_n(F) \leq \mu_0(F)$. i.e. (*vi*) $\Rightarrow$ (*v*).

Thus (*iv*) $\Leftrightarrow$ (*v*) $\Leftrightarrow$ (*vi*).

Suppose now (*v*) holds. We will show that this implies (*i*). Let $g_1 \in C_1$. Since g_1 is a bounded function we can find α, β such that $0 < g < 1$ where $g = \alpha g_1 + \beta$. Showing $\int g_1(x)d\mu_n(x) \to \int g_1(x)d\mu_0(x)$ is equivalent to showing $\int g(x)d\mu_n(x) \to \int g(x)d\mu_0(x)$. Choose and fix a positive integer N. Define, for $k = 0, 1, 2, \ldots, N-1, E_k = \{x : \frac{k}{N} \leq g(x) < \frac{k+1}{N}\}$; $F_k = \{x : g(x) \geq \frac{k}{N}\}$. Note that $F_N = \phi$, that $F_0 = M$, that the F_ks are closed sets and that $E_k = F_k \sim F_{k+1}$. Now,

$$\int_M g(x)\,\mathrm{d}\mu_n(x) = \sum_{k=0}^{N-1} \int_{E_k} g(x)\,\mathrm{d}\mu_n(x)$$

$$
\begin{aligned}
&\le \textstyle\sum_{k=0}^{N-1} \frac{k+1}{N}\mu_n(E_k)\\
&= \tfrac{1}{N}\textstyle\sum_{k=0}^{N-1}(k+1)\{\mu_n(F_k)-\mu_n(F_{k+1})\}\\
&= \tfrac{1}{N}\textstyle\sum_{k=0}^{N-1} k\{\mu_n(F_k)-\mu_n(F_{k+1})\}\\
&\qquad\qquad\qquad + \tfrac{1}{N}\{\mu_n(F_0)-\mu_n(F_N)\}\\
&= \tfrac{1}{N}\textstyle\sum_{0}^{N-1} \mu_n(F_k).
\end{aligned}
$$

Hence

$$
\begin{aligned}
\varlimsup_{n\to\infty} \int g(x)\,\mathrm{d}\mu_n(x) &\le \tfrac{1}{N}\textstyle\sum_{0}^{N-1} \varlimsup_{n\to\infty}\mu_n(F_k)\\
&\le \tfrac{1}{N}\textstyle\sum_{k=0}^{N-1} \mu_0(F_k)\\
&= \tfrac{1}{N}\textstyle\sum_{k=0}^{N-1}(k+1)\mu_0(E_k)\\
&= \textstyle\sum_{k=0}^{N-1} \frac{k}{N}\mu_0(E_k) + \frac{1}{N}\\
&\le \textstyle\sum_{k=0}^{N-1} \int_{E_k} g(x)d\mu_0(x) + \frac{1}{N}\\
&= \textstyle\int_M g(x)d\mu_0(x) + \frac{1}{N}.
\end{aligned}
$$

Now define

$$G_k = \{x : g(x) > \tfrac{k}{N}\}, \quad Q_k = \{x : \tfrac{k}{N} < g(x) \le \tfrac{k+1}{N}\}.$$

Note $G_0 = M$; $G_N = \phi$.

$$\textstyle\int_M g(x)d\mu_n(x) = \sum_{k=0}^{N-1}\int_{Q_k} g(x)d\mu_n(x) \ge \sum_{k=0}^{N-1}\frac{k}{N}\mu_n(Q_k).$$

Use (*iv*) to get

$$
\begin{aligned}
\varliminf_{n\to\infty}\int_M g(x)d\mu_n(x) &\ge \varliminf_{n\to\infty}\frac{1}{N}\sum_{k=0}^{N-1} k\{\mu_n(G_k)-\mu_n(G_{k+1})\}\\
&= \frac{1}{N}\varliminf_{n\to\infty}\sum_{0}^{N-1}\mu_n(G_k) - \frac{1}{N}\\
&\ge \text{(by Fatou's Lemma)}\ \frac{1}{N}\sum_{0}^{N-1}\varliminf_{n\to\infty}\mu_n(G_k) - \frac{1}{N}\\
&\ge \frac{1}{N}\sum_{0}^{N-1}\mu_0(G_k) - \frac{1}{N} = \sum_{0}^{N-1}\frac{k+1}{N}\mu_0(Q_k) - \frac{1}{N}\\
&\ge \sum_{0}^{N-1}\int_{Q_k} g(x)d\mu_0(x) - \frac{1}{N} = \int_M g(x)d\mu_0(x) - \frac{1}{N}.
\end{aligned}
$$

N being arbitrary, it follows that (*v*) $\Rightarrow$ (*i*).
Let now (*iii*) hold. Given F, a closed set, define $V_k = \{x : d(x, F) \leq \frac{1}{k}\}$ and note that $V_k \downarrow F$. Now,

$$\mu_n(F) = \int\limits_F \mathrm{d}\mu_n(x) = \int\limits_F f_{\frac{1}{k},F}(x)\mathrm{d}\mu_n(x) \leq \int\limits_M f_{\frac{1}{k},F}(x)\mathrm{d}\mu_n(x).$$

Hence

$$\begin{aligned}\overline{\lim_{n\to\infty}}\mu_n(F) &\leq \lim_{n\to\infty}\int\limits_M f_{\frac{1}{k},F}(x)\mathrm{d}\mu_n(x) = \int\limits_M f_{\frac{1}{k},F}(x)\mathrm{d}\mu_0(x)\\ &= \text{(applying (}iii\text{))} \int\limits_{V_k} f_{\frac{1}{k},F}(x)\mathrm{d}\mu_0(x) \leq \int\limits_{V_k} \mathrm{d}\mu_0(x)\\ &= \mu_0(V_k) \to \mu_0(F) \text{ as } k \to \infty.\end{aligned}$$

Thus (*iii*) $\Rightarrow$ (*v*). Consequently (*i*) $\Leftrightarrow$ (*ii*) $\Leftrightarrow$ (*iii*) $\Leftrightarrow$ (iv) $\Leftrightarrow$ (v) $\Leftrightarrow$ (vi). Suppose now that (*vii*) holds. Hence $\int\limits_R g(x)\,\mathrm{d}F_n(x) \to \int\limits_R g(x)\,\mathrm{d}F_0(x)$ (ref. **(2.1.12)**) for every real bounded continuous function g defined on R. Or, what is same $\lim\limits_{n\to\infty}\int\limits_R g(x)\,\mathrm{d}\mu_n f^{-1}(x) = \int\limits_R g(x)\,\mathrm{d}\mu_0 f^{-1}(x)$. This means that with $M = R$, (*i*) is true for the sequence $(\mu_n f^{-1})$. Hence for this sequence (*v*) is true. This implies that for the single point closed set $\{1\}$, we have $\overline{\lim\limits_{n\to\infty}}\mu_n f^{-1}(\{1\}) \leq \mu_0 f^{-1}(\{1\})$. Now, given a closed subset $F \subset M$, take $f = f_{1,F}$ and note that $f^{-1}\{1\} = F$. We thus get $\overline{\lim\limits_{n\to\infty}}\mu_n(F) \leq \mu_0(F)$, proving that (*vii*) $\Rightarrow$(*v*).

Conversely suppose now(*v*) holds. Hence (vi) also holds. Let $f \in C$. Let the F_ns be as described in (*vii*). We must show that $F_n \xrightarrow{d} F_0$. Let x be a continuity point of F_0. i.e., $\mu_0 f^{-1}(\{x\}) = 0$. This implies that $f^{-1}((-\infty, x])$ is a μ_0-continuous subset of M, since
$Bd\{f^{-1}(-\infty, x]\} = f^{-1}(\{x\})$. Hence, by (vi),
$\lim\limits_{n\to\infty}\mu_n f^{-1}((-\infty, x]) = \mu_0 f^{-1}((-\infty, x])$, which is the desired result. □

Note. The result (*v*) $\Rightarrow$ (*vii*) *is true also when* f *is a continuous map of* M *into any metric space.*

Remarks 2.4.3. (i) Suppose that $(\mu_n, n \geq 1)$ is a tight sequence and that $\mu_n \xrightarrow{d} \mu_0$. Then μ_0 is a tight measure.
Proof. By hypothesis, given $\varepsilon > 0$, there can be found a compact set K such that $\mu_n(K) > 1-\varepsilon$ for all $n \geq 1$. Since K is a closed set and since $\mu_n \xrightarrow{d} \mu_0$,

$$1-\varepsilon \leq \underline{\lim_{n\to\infty}}\,\mu_n(K) \leq \overline{\lim_{n\to\infty}}\,\mu_n(K) \leq \mu_0(K),$$

by (*v*) of **(2.4.2)**.
(ii) Let $x_n \in M, n \geq 0$. Let *pm* μ_n be degenerate at x_n. i.e. $\mu_n(\{x_n\}) = 1$. Then $\mu_n \overset{d}{\to} \mu_0$ *iff* $d(x_n, x_0) \to 0$.
Proof. Let U be an open neighborhood of x_0. Suppose $\mu_n \overset{d}{\to} \mu_0$. This implies that $\underline{\lim}_{n\to\infty} \mu_n(U) \geq \mu_0(U) = 1$. Hence $\mu_n(U) \to 1$. Consequently, there exists $N = N(U)$ such that $\mu_n(U) > 0$ for all $n \geq N$. But $\mu_n(U) > 0$ *iff* $x_n \in U$. Thus $x_n \in U$ for all $n \geq N$. Since neighbourhood U is arbitrary, it follows that $x_n \to x_0$.

Conversely, let $d(x_n, x_0) \to 0$. Hence for every real continuous function f, $f(x_n) \to f(x_0)$. This translates to :
$\lim_{n\to\infty} \int_M f(x) \mathrm{d}\mu_n(x) = \int_M f(x) \mathrm{d}\mu_0(x)$. This is equivalent to $\mu_n \overset{d}{\to} \mu_0$.
(iii) $\{$*This subsection may be omitted in the first reading*$\}$
Part of the result at **(2.3.8)** was that in R^p, the weak limit of μ_n exists if the weak limit of $\mu_n \mathbf{t}^{-1}$ exists for every $\mathbf{t} \in R^p$. Considering R^p as a (finite dimensional) Hilbert space, we ask if the above result holds in the case of an infinite dimensional Hilbert space. To see that the answer is 'No', consider the separable Hilbert space l_2, the space of all real sequences $(a_1, a_2, \dots)$ with $\sum_1^\infty a_k^2 < \infty$. Let α_n be the point in l_2 with its nth co-ordinate equal to 1 and all other co-ordinates zeroes. Let $\mu_n(\{\alpha_n\}) = 1$. For $\mathbf{t}, \mathbf{x} \in l_2$, $\mathbf{t}(\mathbf{x}) = \sum_{i=1}^\infty t_i x_i$. Thus considered as a function defined on l_2, every $\mathbf{t}$ is a continuous function.

For any $\mathbf{t} \in l_2$, $\mu_n \mathbf{t}^{-1}$ is the pm with $\mu_n \mathbf{t}^{-1}(\{t_n\}) = 1$. Now, $t_n \to 0$, since $\sum_i t_n^2 < \infty$. Hence if $\nu_{\mathbf{t}}$ is the pm in R with $\nu_{\mathbf{t}}(\{0\}) = 1$, then $\mu_n \mathbf{t}^{-1} \overset{d}{\to} \nu_{\mathbf{t}}$. But μ_n does not converge weakly to any pm μ. For if $\mu_n \overset{d}{\to} \mu$, then for every $\mathbf{t}$, $\mu \mathbf{t}^{-1}(\{0\}) = 1$. Hence $\mu(\{\mathbf{0}\}) = 1$. For the closed set $S = \{\mathbf{x} : \|\mathbf{x}\| = 1\}$ the relation $\overline{\lim}_{n\to\infty} \mu_n(S) \leq \mu(S)$ does not hold since $\mu_n(S) = 1$ for all n since $\alpha_n \in S$ but $\mu(S) = 0$ since $\mathbf{0} \notin S$.

Let us explore the reason for the failure. Analogous to R^p, define the *cf* f of a pm μ on the Borel σ-field $\mathscr{H}$ of a real separable Hilbert space H as the function on H given by $\mathbf{f}(\mathbf{t}) = \int_H e^{i\langle \mathbf{t}, \mathbf{x}\rangle} \mathrm{d}\mu(\mathbf{x})$. Let μ_n, $n \geq 0$ be pms on $\mathscr{H}$ with *cf*s $\mathbf{f}_n$. If $\mu_n \overset{d}{\to} \mu_o$ then, since $\cos\langle \mathbf{t}, \mathbf{x}\rangle$, $\sin\langle \mathbf{t}, \mathbf{x}\rangle$ are bounded continuous functions of $\mathbf{x}$, it follows that $\mathbf{f}_n(\mathbf{t}) \to \mathbf{f}_0(\mathbf{t})$. In R^p we established the result that $f_n(\mathbf{t}) \to f(\mathbf{t})$ for every $\mathbf{t} \in R^p$ implies that $\mu_n \overset{d}{\to} \mu_0$. That this converse result is not true in H without additional conditions is brought out by the example in the last paragraph. There we find $\mathbf{f}_n(\mathbf{t}) = e^{it_n} \to 1 = \mathbf{f}_0(\mathbf{t})$ for every $\mathbf{t}$ but $\mu_n \overset{d}{\to} \mu_0$ does not hold.

(iv) If every subsequence of the sequence (μ_n) admits of a further subsequence converging weakly to the same *pm* μ_0 then $\mu_n \overset{d}{\to} \mu_0$.
Proof. Let C be an arbitrary closed set. Let $c = \overline{\lim_{n\to\infty}} \mu_n(C)$. Let (n') be a subsequence such that $\lim_{n'\to\infty} \mu_{n'}(C) = c$. By hypothesis, $\mu_{n''} \overset{d}{\to} \mu_0$ where (n'') is a subsequence of (n'). We then have

$$\overline{\lim_{n\to\infty}} \mu_n(C) = c = \lim_{n''\to\infty} \mu_{n''}(C) = \overline{\lim_{n''\to\infty}} \mu_{n''}(C) \le \mu_0(C).$$

This being true for every closed set C the proof is complete (ref. **(2.4.2)**).
(v) (*Continuation of Remark 2 under* **(2.1.12)**).

Let $B \subseteq M$ be a μ_0-continuity set.Then for every g bounded and continuous on B,

$$\int_B g(x)\,\mathrm{d}\mu_n(x) \to \int_B g(x)\,\mathrm{d}\mu_0(x) \tag{2.6}$$

Proof. If $\mu_0(B) = 0$, the claim is obvious. Assume $\mu_0(B) > 0$. Since B is μ_0-continuous, so is $\bar{B}$, the closure of B. Hence $\lim_n \mu_n(\bar{B}) = \mu_0(\bar{B})$. This implies $\mu_n(\bar{B}) > 0$ for all n large. Since $\mu_n(\bar{B} \sim B) = \mu_n(\bar{B}) - \mu_n(B) \to 0$ and since g is bounded, it follows that (2.6) will be true *iff*

$$\int_{\bar{B}} g(x)\,\mathrm{d}\mu_n(x) \to \int_{\bar{B}} g(x) d\mu_0(x) \tag{2.7}$$

is true. Since every closed subset of the metric space $(\bar{B},\ d)$ is a closed subset of $(M,\ d)$, every Borel subset of $(\bar{B},\ d)$ is a Borel subset of $(M,\ d)$. On the Borel σ-field $\mathcal{B}$ of $\bar{B}$, define *pm*

$$\nu_n : \nu_n(A) = \frac{\mu_n(A)}{\mu_n(\bar{B})},\ A \in \mathcal{B},\ n \ge 0.$$

Since $\mu_n \overset{d}{\to} \mu_0$ and since $\mu_n(\bar{B}) \to \mu_0(\bar{B})$, it follows that $\overline{\lim_n} \nu_n(C) \le \nu_0(C)$ for every closed set $C \subseteq \bar{B}$. i.e. $\nu_n \overset{d}{\to} \nu_0$. Hence by **(2.4.2)**,

$$\int_{\bar{B}} g(x)\,\mathrm{d}\nu_n(x) \to \int_{\bar{B}} g(x) d\nu_0(x).$$

This implies (2.7), since $\mu_n(\bar{B}) \to \mu_0(\bar{B})$.

(vi) Let $(M_i,\ d_i)$ be metric spaces and $\mathfrak{m}_i,\ i=1,\ 2$ their Borel σ-fields. Let μ be a pm on $\mathfrak{m}_1$. Let $f_n,\ n \geq 0$ be continuous mappings of M_1 into M_2 such that $f_n(x) \to f_0(x),\ x \in M_1$. Then $\mu f_n^{-1} \overset{d}{\to} \mu f_0^{-1}$.

If we show that $\int\limits_{M_2} g(y)\,\mathrm{d}\mu f_n^{-1}(y) \to \int\limits_{M_2} g(y)\,\mathrm{d}\mu f_0^{-1}(y)$, then the claim follows. i.e. (by **(2.3.6)**) if we show

$$\int\limits_{M_1} g(f_n(x))\,\mathrm{d}\mu(x) \to \int\limits_{M_1} g(f_0(x))\,\mathrm{d}\mu(x)$$

for every bounded continuous function g. But this is true by the bounded convergence theorem. □

The following Lemma is needed for the proof of the next theorem which will have the consequence that a weakly convergent sequence of pms in a complete separable metric space will be automatically tight.

Lemma 2.4.4. *Suppose $(M,\ d)$ is a metric space. Let $K \subset M$ be a compact set. Let $x_n,\ n \geq 1$ be elements such that $d(x_n, K) \to 0$. Then the sequence (x_n) admits of a subsequence converging to an element in K.*

Proof. We need consider only the case where all but a finite number of the x_ns lie outside K. By the definition of $d(x, K)$, there corresponds to each x_n an element $y_n \in K$ such that $d(x_n, K) = d(x_n, y_n)$. This is so because K is a closed set. Since K is a compact set, (y_n) admits of a subsequence $(y_{n'})$ converging to an element $y_0 \in K$. Now

$$\lim_{n' \to \infty} d(x_{n'}, y_0) \leq \lim_{n' \to \infty} \{d(x_{n'}, y_{n'}) + d(y_{n'}, y_0)\} = 0. \qquad \square$$

Theorem 2.4.5. *Let each pm $\mu_n,\ n \geq 0$ on $\mathfrak{m}$ be tight and let $\mu_n \overset{d}{\to} \mu_0$. Then the family $(\mu_n, n \geq 0)$ is tight.*
[*By **(1.9.4)**, it would then follow that in a complete separable metric space, $\mu_n \overset{d}{\to} \mu_0$ implies the family $(\mu_n,\ n \geq 0)$ is tight*]

Proof. Tightness of μ_0 implies that, given $\epsilon > 0$, a compact set K_0 can be found such that $\mu_0(K_0) > 1 - \epsilon$. Define, for $r = 1, 2, \ldots, G_r = \{x : d(x, K_0) < \frac{1}{r}\}$. G_r is an open set, $G_r \supset K_0$. $\mu_0(G_r) > 1 - \epsilon$. Also $\mu_0(G_r \sim K_0) < \epsilon$ for all r. Since $\mu_n \overset{d}{\to} \mu_0$, $\underline{\lim}_{n \to \infty} \mu_n(G_r) \geq \mu_0(G_r) > 1 - \epsilon$. This implies that, for r fixed, there exists n_r such that $\mu_n(G_r) > 1 - 2\epsilon$ for all $n \geq n_r$. There is no loss in generality in assuming that (n_r) is a strictly increasing sequence. Consider n with $n_r \leq n < n_{r+1}$. μ_n is a tight *pm* and

$G_r \in \mathcal{M}$. Hence by **(1.9.2)** (β), a compact set $K_{r,n}$ can be found such that $\mu_n(K_{r,n}) > 1 - 3\epsilon$ with $K_{r,n} \subset G_r$, for $n_r \leq n < n_{r+1}$. Write

$$K_r = K_0 \cup \bigcup_{n=n_r}^{n_{r+1}-1} K_{r,n}.$$

K_r is a compact subset of G_r and

$$\mu_n(K_r) \geq \mu_n(K_{r,n}) > 1 - 3\epsilon.$$

$\{\mu_n, 1 \leq n < n_{r_0}\}$ is a finite collection of tight measures. A compact set C can therefore be found to satisfy $\mu_n(C) > 1 - \epsilon$, $1 \leq n < n_{r_0}$. Define $K = C \cup \bigcup_{r=r_0}^{\infty} K_r$. Clearly $\mu_n(K) > 1 - 3\epsilon$ for all n. The proof will be complete if we show that K is a compact set. Let $(x_\nu, \nu \geq 1)$ be a sequence in K. If infinitely many of the x_νs lie in C or in some K_m, the compactness of these sets ensures the existence of a subsequence converging to a point in the relevant set. In the contrary case no K_r will contain infinitely many x_νs. To avoid notation clutter, assume $x_r \in K_r, r = 1, 2, \ldots$. Hence $x_r \in G_r$ and then $d(x_r, K_0) < \frac{1}{r}$. It now follows from Lemma **(2.4.4)** that a subsequence of (x_ν) converges to some point in K_0. Since, for all r, $K_0 \subseteq K_r$, the limit point lies in K. This completes the proof that K is compact. □

Theorem 2.4.6. *Let* $(\mu_n,\ n \geq 0)$ *be a tight family of pms on* $\mathcal{M}$ *and let* $\mu_n \overset{d}{\to} \mu_0$. *Let f be a real function defined on M such that it is continuous on a set E of μ_0-measure* 1. *i.e.* $x_n \in E,\ n \geq 0$ *and* $d(x_n,\ x_0) \to 0$ *imply* $f(x_n) \to f(x_0)$. *Then* $\mu_n f^{-1} \overset{d}{\to} \mu_0 f^{-1}$.

Proof. The tightness of the family, together with the fact that μ is regular on E (ref. **(1.9.2)**) implies that, given $\varepsilon > 0$, we can find a compact set $K \subset E$ such that $\mu_n(K) > 1 - \varepsilon$, $n \geq 0$. We note f is continuous on K.

Now, let A be a closed linear set. We claim $f^{-1}(A) \cap K$ is a closed set. To see this let $x_n \in f^{-1}(A) \cap K$, $n \geq 1$ and let $d(x_n,\ x) \to 0$. We must show that $x \in f^{-1}(A) \cap K$. Since K is a closed set, $x \in K$. Hence $f(x_n) \to f(x)$. Now $f(x_n) \in A$ since $x_n \in f^{-1}(A)$. Since A is a closed set and since $f(x_n) \to f(x)$, it follows that $f(x) \in A$. This implies $x \in f^{-1}(A)$. Thus $x \in f^{-1}(A) \cap K$. We then have

$$\begin{aligned}\mu_n f^{-1}(A) &= \mu_n\big(f^{-1}(A) \cap K\big) + \mu_n\big(f^{-1}(A) \cap K'\big)\\ &\leq \mu_n\big(f^{-1}(A) \cap K\big) + \varepsilon.\end{aligned}$$

Hence

$$\overline{\lim_{n\to\infty}} \mu_n f^{-1}(A) \leq \overline{\lim_{n\to\infty}} \mu_n\big(f^{-1}(A) \cap K\big) + \varepsilon$$
$$\leq \mu_0\big(f^{-1}(A) \cap K\big) + \varepsilon$$

which, since $\mu_n \stackrel{d}{\to} \mu_0$ and since $f^{-1}(A) \cap K$ is a closed set, is $\leq \mu_0 f^{-1}(A) + \varepsilon$. The claim follows since $\varepsilon > 0$ is arbitrary. □

Theorem 2.4.7. (i) *Let the metric space M be separable. Then given μ, there can be found a sequence (A_n) of disjoint μ-continuous sets such that $\cup_n A_n = M$.*
(ii) *Define a family $\mathfrak{A}$ of continuous mappings of M into R^p to be equicontinuous at each point $x \in M$ if $\sup_{f\in\mathfrak{A}} \|f(x)\| < \infty$ and if, given $\varepsilon > 0$, there exists a neighbourhood $N_x = \{y : d(x, y) < r\}$ of x such that $\|f(x) - f(y)\| < \varepsilon$ for all $y \in N_x$ and for all $f \in \mathfrak{A}$. If $\mathfrak{A}$ is a family of equicontinuous mappings as defined above then, given $\varepsilon > 0$, the sequence (A_n) in (ii) can be chosen such that that if $x, y \in A_n$, then $\|f(x) - f(y)\| < \varepsilon$ for every $f \in \mathfrak{A}$.*

Proof. (i) We recall some properties of the boundary of a set.
$BdA = BdA'; Bd(A \cup B) \subset BdA \cup BdB.$
It follows from these two properties that $Bd(A \cap B) = Bd(A' \cup B')' = Bd(A' \cup B') \subset BdA' \cup BdB' = BdA \cup BdB$. The second property readily extends to a finite number n of sets : $Bd(\cup_{k=1}^n A_k) \subset \cup_{k=1}^n BdA_k$. [That this inclusion property does not hold for infinitely many sets is seen by the following example. Let $(r_n, n = 1, 2, ...)$ be a listing of the rationals in R. Let $A_n = \{r_n\}$, so $BdA_n = A_n$ and $\cup_1^\infty BdA_n = \{r_n, n = 1, 2, ...\}$ while $Bd(\cup_1^\infty A_n) = R$]

Thus if A, B, A_n are μ-continuous sets, then

$$A', B', A \cap B, \cup_1^n A_k \textit{ etc.} \text{ are too.} \tag{2.8}$$

Denote by $S(x, \eta)$ the open sphere with centre at $x \in M$ and η as radius. Fix $\delta > 0$. Given $x \in M$, consider the collection of the boundary sets : $\{BdS(x, \eta), 0 < \eta < \delta\}$. This is an uncountable collection. Hence the μ-measure of every member of this collection can not be positive. We can therefore find $\eta < \delta$, $\eta = \eta(x, \delta)$ such that $\mu(BdS(x, \eta)) = 0$. These open spheres $S(x, \eta)$, $x \in M$, is clearly an open cover of M. Since M is separable, it admits of a countable subcover (by Lindelöf theorem, page 49, [K]). Denote the subcover by $\{A(x_i), i = 1, 2, ...\}$. Define $B_1 = A(x_1)$; $B_2 = A(x_2) \cap (A(x_1))'$; $B_3 = A(x_3) \cap (\cup_{i=1}^2 A(x_i))'$; $B_n = A(x_n) \cap (\cup_{i=1}^{n-1} A(x_i))'$, $n \geq 2$,

note that $\mu(BdA(x_i)) = 0$ for all i, that the B_is are mutually exclusive and that $\cup_i B_i = M$. Further each B_i is a μ-continuity set, by the properties noted in (2.8) With this the proof of part (i) is complete.
(ii) $\mathfrak{A}$ is given to be an equicontinuous family. Hence given $\varepsilon > 0$ and a_r, we can find ε_{r,ν^*} such that $\sup_{y \in B_{r,\nu^*},\, f \in \mathfrak{A}} \|f(a_r) - f(y)\| < \frac{\varepsilon}{2}$. This implies $\sup_{x,\, y \in B_{r,\nu^*},\, f \in \mathfrak{A}} \|f(a_r) - f(y)\| < \varepsilon$. Form the C_ns, as before, but now from the countable collection $B_{r,\nu}$, $\nu \geq \nu^*(r)$; $r = 1, 2, \ldots$ and from the C_ns form the A_ns. The A_ns thus obtained form the required sequence. $\square$

A justification for the nomenclature 'weak convergence'

Let (M, d) be a metric space, $\mathcal{m}$ its Borel σ-field and let C_1 be the space of all the real bounded continuous functions defined on M. We note that, under the sup norm, C_1 is a Banach space. Let C_1^* denote its conjugate space. Let $\mathcal{M}$ denote the collection of all *pms* on $\mathcal{m}$. Corresponding to $\mu \in \mathcal{M}$ and $g \in C_1$, define: $\mu(g) = \int_M g(x) d\mu(x)$, $g \in C_1$. This is a linear functional on C_1. Further $|\mu(g)| \leq ||g||$. Thus $\mathcal{M} \subset C_1^*$. Endow C_1^* with the weak* topology. i.e. it is the topology for which a subbase for the neighbourhood system at a point $x_0^* \in C_1^*$ is the following family of sets. $\{x^* : x^* \in C_1^*, |x^*(x) - x_0^*(x)| < \varepsilon\}$, $x \in C_1$, $\varepsilon > 0$. In this topology a sequence $(x_n^*) \subset C_1^*$ converges to $x_0^* \in C_1^*$ *iff* $x_n^*(x) \to x_0^*(x)$ for each $x \in C_1$. In particular, μ_n converges to μ_0 in this topology *iff* $\mu_n(g) \to \mu_0(g)$ for every $g \in C_1$.

$$\text{i.e. } \textit{iff} \int_M g(u) \mathrm{d}\mu_n(u) \to \int_M g(u) \mathrm{d}\mu_0(u).$$

Convergence in this weak* topology of C_1^*, of which $\mathcal{M}$ is a subset, is known as weak convergence in probability theory.

2.5 Convergence of medians under weak convergence

Definition 2.5.1. m is said to be a median of a *rv* X or its *df* F if $P(X \geq m) \geq \frac{1}{2}$ and $P(X \leq m) \geq \frac{1}{2}$. Equivalently, m is a median if $F(m) \geq \frac{1}{2}$ and $F(m-) \leq \frac{1}{2}$.

Remarks. (i) Every *df* has at least one median. *Proof.* Since F is a non-decreasing function with $F(-\infty) = 0$ and $F(\infty) = 1$, there necessarily exists an x such that $F(x) \geq \frac{1}{2}$. Then $m = \inf\{x : F(x) \geq \frac{1}{2}\}$ is clearly a median.
(ii) A *df* F can have many medians as is clear from the following example.

$F(x) = 0$ for $x < 0$; $= x$ for $0 \leq x \leq \frac{1}{2}$; $= \frac{1}{2}$ for $\frac{1}{2} \leq x \leq \frac{3}{4}$; $= 2x - 1$ for $\frac{3}{4} \leq x \leq 1$ and $= 1$ for $x > 1$. For this *df* every point m, $\frac{1}{2} \leq m \leq \frac{3}{4}$ is a median.

When the median is not unique, let A be the collection of all the medians of F. Directly from the definition, we see that if $m_1 < m_2$ are medians of F then every $m, m_1 \leq m \leq m_2$ is a median too. Let $b = \sup_{m \in A} m$; $a = \inf_{m \in A} \{m\}$. Then b has to be a finite number, as otherwise arbitrarily large numbers λ will be medians. But that is not possible, since that would imply $F(\lambda-) \geq \frac{1}{2}$. Similarly a is a finite. Thus A is a finite closed interval.

Theorem 2.5.2. *Let $F_n \xrightarrow{d} F$. Let m_n be a median of F_n. Then (i) (m_n) is a bounded sequence and (ii) $m_n \to m$ implies that m is a median of F.*

Proof (i) If possible, let $\lim_{n\to\infty} m_{k(n)} = \infty$, for a subsequence $(k(n))$. Let u be an arbitrary continuity point of F with $F(u) > \frac{1}{2}$. For all n large, $m_{k(n)} > u$. Hence
$1 - F(u) = \lim_{n\to\infty} \{1 - F_{k(n)}(u)\} \geq \underline{\lim}_{n\to\infty}\{1 - F_{k(n)}(m_{k(n)}-)\} \geq \frac{1}{2}$, leading to $1 - F(u) \geq \frac{1}{2}$, a contradiction. Hence $\overline{\lim_{n\to\infty}}\, m_n < \infty$. Similarly, $\underline{\lim_{n\to\infty}}\, m_n < \infty$ □

Remark. This result implies that $\cup_n A_n$ is a bounded set where A_n is the set of all the medians of F_n.
(ii) Let $\varepsilon_r \downarrow 0$ be numbers such that for each r, $m - \varepsilon_r, m + \varepsilon_r$ are continuity points of F. Given ε_r, we can find $N = N(r)$ such that for all $n \geq N$, we will have $m - \varepsilon_r < m_n < m + \varepsilon_r$. Hence for $n \geq N$, $\frac{1}{2} \leq \{1 - F_n(m_n-)\} \leq 1 - F_n(m - \varepsilon_r)$, leading to $\frac{1}{2} \leq \lim_{n\to\infty} \{1 - F_n(m - \varepsilon_r)\} = 1 - F(m - \varepsilon_r)$. Take $r \to \infty$ to get $1 - F(m-) \geq \frac{1}{2}$ or $F(m-) \leq \frac{1}{2}$. Similar steps lead to $\frac{1}{2} \leq F(m)$. Thus m is a median of F.

Remark. An example to show that even though $F_n \xrightarrow{d} F_0$ and even though the F_ns possess unique medians m_n, sequence (m_n) does not converge to m_0. Define, for $n \geq 2$,

$$G_n(x) = \begin{cases} 0 \text{ if } x < 0 \\ \frac{1}{2} + (-1)^{n-1}\frac{1}{n} \text{ if } 0 \leq x < 1 \\ 1 \text{ if } x \geq 1 \end{cases} \qquad G_0(x) = \begin{cases} 0 \text{ if } x < 0 \\ \frac{1}{2} \text{ if } 0 \leq x < 1 \\ 1 \text{ if } x \geq 1 \end{cases}$$

We observe that $G_n \xrightarrow{d} G_0$, that the distributions have unique medians $m_{2n} = 1$; $m_{2n+1} = 0; n \geq 2$ and $m_0 = 0$. (m_n) is not a convergent sequence.

2.6 Moments

Definitions 2.6.1. Let X be a rv defined on the basic probability space $(\Omega,\ \mathscr{A},\ P)$. The moment of order 1 (or the expected value) of X or of its *df* F orof its $dm\ \mu$ is denoted by $\mathscr{E}X$ and is defined as the integral $\int\limits_\Omega X\,\mathrm{d}P$, provided the integral exists. i.e., provided $\int\limits_\Omega |X|\,\mathrm{d}P < \infty$. Sometimes the definition of existence is extended to include the following cases : (i) $\int\limits_\Omega X^+\,\mathrm{d}P = \infty$ and $\int\limits_\Omega X^-\,\mathrm{d}P < \infty$ and then we set $\mathbb{E}X = \infty$. (ii) $\int\limits_\Omega X^+\,\mathrm{d}P < \infty$ and $\int\limits_\Omega X^-\,\mathrm{d}P = \infty$ and then we set $\mathbb{E}X = -\infty$. For n an integer, $\mathbb{E}X^n$ is called the moment of order n of X and for $\lambda \geq 0$, $\mathbb{E}|X|^\lambda$ is called the absolute moment of X of order λ.

Apply **(2.3.6)** and note that $\int\limits_\Omega X\,\mathrm{d}P = \int\limits_R x\,\mathrm{d}\mu$, $\int\limits_\Omega |X|^\lambda\,\mathrm{d}P = \int\limits_R |x|^\lambda\,\mathrm{d}\mu$, $\int\limits_\Omega X^+\,\mathrm{d}P = \int\limits_{[0,\ \infty)} x\,\mathrm{d}\mu$. Since the integrands are continuous functions of x, $\int\limits_R x\,\mathrm{d}\mu$ and $\int\limits_R |x|^\lambda\,\mathrm{d}\mu$ can be evaluated as the Riemann-Stieltj's integrals $\int\limits_R x\,\mathrm{d}F(x)$ and $\int\limits_R |x|^\lambda\,\mathrm{d}F(x)$ respectively. Again, since the integrand is zero at $x = 0$, $\int\limits_{[0,\ \infty} x\,\mathrm{d}\mu = \int\limits_{(0,\ \infty)} x\,\mathrm{d}\mu = \int\limits_{(0,\ \infty)} x\,\mathrm{d}F(x)$.

An example of a rv X for which $\mathbb{E}X$ does not exist is as follows. Let F have frequncy function $f\ :\ f(x) = \frac{1}{2|x|^2}$ for $|x| \geq 1$ and $= 0$, otherwise, since $\mathbb{E}X^+ = \infty = \mathbb{E}X^-$.

When $\mathbb{E}X$ exists finitely, $\mathbb{E}(X - \mathbb{E}X)^2$ is called the variance of X or of F. When finite, it is denoted by σ_X^2 or σ_F^2 or var$X$; σ_X or σ_F, the positive square root of the variance is called the standard deviation of X or F. If $(X,\ Y)$ has *df* $\mathbb{F}$ and if $\mathbb{E}X = \mathbf{0} = \mathbb{E}Y$, then the covariance cov$(X,\ Y)$ or $\sigma_{X,Y}$ between X and Y is defined as $\mathbb{E}XY$. i.e. by $\iint\limits_{R^2} xy\,\mathrm{d}\mathbb{F}(x,\ y)$. This integral will exist finitely if $\sigma_X < \infty$ and $\sigma_Y < \infty$. (ref. **(2.6.5)**).

Let $\mathbf{X} = (X_1,\ X_2,\ \ldots,\ X_p)'$ be a rV with $\mathbb{E}X_i = 0$ and $\mathbb{E}X_i^2 = \sigma_i^2 < \infty$, $1 \leq i \leq p$. Let $\sigma_{i,j} = \mathbb{E}X_iX_j$. The matrix $\Sigma = (\sigma_{i,j})$ is obviously symmetric and is called the variance-covariance matrix of $\mathbf{X}$ or of its *df* $\mathbb{F}$. An essential property of variance-covariance matrices is that they are non-negative definite. The truth of this is seen by observing that for $\mathbf{a} \in R^p$ arbitrary, we have $0 \leq \text{var}(\mathbf{a}'\mathbf{X}) = \mathbf{a}'\Sigma\mathbf{a}$.

An example to show that the absolute moment of order α of a *df* can be infinite for every $\alpha \neq 0$ is as follows.

Let μ be the *pm* defined by setting

$$\mu(\{n\}) = \mu(\{\frac{1}{n}\}) = \frac{c}{n(\log n + 1)^2}, \quad n \geq 2$$

where $\sum_{2}^{\infty} \frac{c}{n(\log n+1)^2} = \frac{1}{2}$. Here $\int_{0}^{\infty} x^\alpha \, \mathrm{d}\mu(x) = \infty$ for all $\alpha \neq 0$. In view of this example the following theorem assumes importance.

Theorem 2.6.2. *Given a df F, there can always be found a positive function ψ defined on $(-\infty, \infty)$ such that $\lim_{|t| \to \infty} \psi(t) = \infty$ and such that*

$$\int_{-\infty}^{\infty} \psi(x) dF(x) < \infty.$$

Proof. Consider first the range $(0, \infty)$. Clearly, we can assume that $F(x) < 1$ for all x as otherwise the result is trivial, one choice for ψ being $\psi(t) = t$. Choose $c_k \downarrow 0$ and $\alpha_k \uparrow \infty$ satisfying $\sum c_k \alpha_k < \infty$. Take $x_0 = 0$ and define $x_k = \inf\{x : 1 - F(x) \leq c_k\}, k \geq 1$. We note $x_k \uparrow \infty$. Define $\psi(t) = \alpha_k$ if $x_k < t \leq x_{k+1}$. Then we note that $\psi(t) \to \infty$ as $t \to \infty$. Further

$$\int_{(x_k, x_{k+1}]} \psi(t) \mathrm{d}F(t) = \alpha_k\{F(x_{k+1}) - F(x_k)\} \leq \alpha_k\{1 - F(x_k)\} \leq \alpha_k c_k.$$

Hence

$$\int_{0}^{\infty} \psi(t) \mathrm{d}F(t) < \infty.$$

Similarly, one can construct a positive ψ defined on $(-\infty, 0)$ such that $\psi(t) \uparrow \infty$ as $t \downarrow -\infty$ and $\int_{-\infty}^{0} \psi(t) \mathrm{d}F(t) < \infty$.

Thus given any *df F*, we can construct a positive ψ such that $\psi(t) \uparrow \infty$ as $|t| \uparrow \infty$ and $\int_{-\infty}^{\infty} \psi(t) \mathrm{d}F(t) < \infty$. □

Remark Suppose g is non-negative and $m = \int_{-\infty}^{\infty} g(t) \mathrm{d}F(t) < \infty$. Then starting with the *df* $\frac{1}{m} \int_{(-\infty, x]} g(t) \mathrm{d}F(t)$, one can construct a ψ with the above properties such that $\int_{-\infty}^{\infty} \psi(t) g(t) \mathrm{d}F(t) < \infty$.

Integration by parts; evaluating expected values by Riemann integration

Lemma 2.6.3. *Let* $\alpha > 0$. *Then for all values of* $t > 0$, $I(t) = 0$ *where*

$$I(t) = \int_{(0,t]} x^\alpha dF(x) - t^\alpha F(t) + \int_0^t \alpha x^{\alpha-1} F(x) dx. \tag{2.9}$$

Proof. Let $h > 0$ be small.

$$\begin{aligned}
I(t+h) - I(t) &= \int_{(t,t+h]} x^\alpha \mathrm{d}F(x) - (t+h)^\alpha F(t+h) + t^\alpha F(t) \\
&\quad + \int_t^{t+h} \alpha x^{\alpha-1} F(x) \mathrm{d}x \\
&= \int_{(t,t+h]} x^\alpha \mathrm{d}F(x) - (t+h)^\alpha \int_{(t,t+h]} \mathrm{d}F(x) - (t+h)^\alpha F(t) \\
&\quad + t^\alpha F(t) + \int_t^{t+h} \alpha x^{\alpha-1} F(x) \mathrm{d}x \\
&= \int_{(t,t+h]} \{x^\alpha - (t+h)^\alpha\} \mathrm{d}F(x) - F(t) \int_t^{t+h} \alpha x^{\alpha-1} \mathrm{d}x \\
&\quad + \int_t^{t+h} \alpha x^{\alpha-1} F(x) \mathrm{d}x \\
&= \int_{(t,t+h]} \{x^\alpha - (t+h)^\alpha\} \mathrm{d}F(x) - \int_t^{t+h} \alpha x^{\alpha-1} \{F(t) - F(x)\} \mathrm{d}x.
\end{aligned}$$

The first integral in absolute value does not exceed

$$\begin{aligned}
&\{(t+h)^\alpha - t^\alpha\}\{F(t+h) - F(t)\} \\
&\qquad = O(h)\{F(t+h) - F(t)\} = o(h).
\end{aligned}$$

The second integral in absolute value does not exceed

$$\{F(t+h) - F(t)\} \int_t^{t+h} \alpha x^{\alpha-1} \, \mathrm{d}x = o(h).$$

Thus $|I(t+h) - I(t)| = o(h)$. Through a parallel reasoning we show that $|I(t-h) - I(t)| = o(h)$. Hence $I'(t) = 0$ for all t, leading to $I(t) \equiv \theta$, a constant. Since $\lim_{t \downarrow 0} I(t) = 0$, it follows that $\theta = 0$. □

Note. We could have proved the Lemma with an integrand g more general than x^α [the following result is parallel to (2.9)

$$\int_{(a,\,b]} g(x)\,dF(x) - \{g(b)F(b) - g(a)F(a)\} + \int_a^b g'(x)F(x)\,dx \equiv 0 \textit{ for all } a < b,$$

provided g is a bounded function and possesses a continuous derivative.] but it was felt that with this particular integrand the principle underlying integration by parts would become easy to understand. In this book the other integrands g used are $e^{-\lambda x}$ (ref. **(4.5)**, **(5.5.4)** and **(6.8.4)**), $\frac{1}{x^2}$ (ref. **(5.4.3)**) and $\frac{x^2}{1+x^2}$ (ref. **(5.1.3)**) *Proof of the claimed relation in each of those cases can be constructed exactly on the same lines.*

Theorem 2.6.4. *For any $\alpha > 0$, let $I_\alpha = \int_0^\infty x^\alpha dF(x)$ and $J_\alpha = \alpha \int_0^\infty x^{\alpha-1}\{1 - F(x)\}dx$. Then $I_\alpha = J_\alpha$ in the sense that if either integral converges so does the other. (It may be noted that for every $\varepsilon > 0$, $\int_0^\varepsilon x^{\alpha-1}\{1 - F(x) + F(-x)\}dx$ exists finitely.)*

Proof. By the Lemma,

$$\begin{aligned} \int_{(0,t]} x^\alpha \mathrm{d}F(x) &= t^\alpha F(t) - \alpha \int_0^t x^{\alpha-1} F(x)\mathrm{d}x \\ &= -t^\alpha\{1 - F(t)\} + \alpha \int_0^t x^{\alpha-1}\{1 - F(x)\}\mathrm{d}x. \end{aligned}$$

Trivially,

$$\int_{(0,t]} x^\alpha \mathrm{d}F(x) \le \alpha \int_0^t x^{\alpha-1}\{1 - F(x)\}\mathrm{d}x.$$

Hence I_α converges if J_α does. Further, convergence of I_α implies $\lim_{b\to\infty} \int_b^\infty x^\alpha \mathrm{d}F(x) = 0$. It follows from this that $\lim_{b\to\infty} b^\alpha\{1 - F(b)\} \to 0$. Hence $I_\alpha = J_\alpha$. Arguments can be appropriately reversed to show that if I_α converges, then so does J_α and that the two integrals are equal. □

Remarks. On similar lines, it can be shown that

$$\int_{-\infty}^{0} (-x)^{\alpha} \mathrm{d}F(x) = \alpha \int_{-\infty}^{0} (-x)^{\alpha-1} F(x) \mathrm{d}x = \alpha \int_{0}^{\infty} x^{\alpha-1} F(-x) \mathrm{d}x.$$

Combining the two results we have, in the same sense as in the theorem:

$$\int_{-\infty}^{\infty} |x|^{\alpha} \mathrm{d}F(x) = \alpha \int_{0}^{\infty} x^{\alpha-1} \{1 - F(x) + F(-x)\} \mathrm{d}x,$$

The right side integral can be written

$$\alpha \int_{0}^{\infty} x^{\alpha-1} \{P(X > x) + P(X \leq -x)\}\, \mathrm{d}x \text{ or } \alpha \int_{0}^{\infty} x^{\alpha-1} \{P(|X| \geq x)\, \mathrm{d}x,$$

since $\{x \ : \ P(|X| = x) > 0\}$ is atmost countable and hence has Lebesgue measure zero.

From

$$\int_{0}^{\infty} x^{\alpha-1} P(|X| \geq x)\, \mathrm{d}x = \sum_{n=0}^{\infty} \int_{n}^{n+1} x^{\alpha-1} P(|X| \geq x)\, \mathrm{d}x$$

we get

$$\frac{1}{2^{\alpha-1}} \sum_{n=2}^{\infty} n^{\alpha-1} P(|X| \geq n) \leq \int_{0}^{\infty} x^{\alpha-1} P(|X| \geq x) dx$$

$$\leq 1 + 2^{\alpha-1} \sum_{n=1}^{\infty} n^{\alpha-1} P(|X| \geq n),$$

leading to

$$\mathbb{E}|X|^{\alpha} < \infty \Leftrightarrow \int_{0}^{\infty} x^{\alpha-1} P(|X| \geq x)\, \mathrm{d}x < \infty$$

$$\Leftrightarrow \sum_{n=1}^{\infty} n^{\alpha-1} P(|X| \geq n) < \infty.$$

Theorem 2.6.5 (Inequalities. Set I). *Let $p > 1,\ q > 1$ be such that $\frac{1}{p} + \frac{1}{q} = 1$. Let $a,\ b,\ a_i,\ b_i,\ i = 1, 2, \ldots, n$ be any non-negative numbers. Then*
(i) $ab \leq \frac{a^p}{p} + \frac{b^q}{q}$.
(ii) $\sum_{i=1}^{n} a_i b_i \leq (\sum_{i=1}^{n} a_i^p)^{\frac{1}{p}} (\sum_{i=1}^{n} b_i^q)^{\frac{1}{q}}$.
(iii) *(Hölder inequality) Let $X,\ Y$ be rvs with $\mathbb{E}|X|^p < \infty$, and $\mathbb{E}|Y|^q < \infty$. Then $\mathbb{E}|XY| < \infty$ and $\mathbb{E}|XY| \leq (\mathbb{E}|X|^p)^{\frac{1}{p}} (\mathbb{E}|Y|^q)^{\frac{1}{q}}$.*
(iv) *Let $u,\ v$ be any two real numbers. then $|u + v|^p \leq 2^{p-1}(|u|^p + |v|^p)$.*
(v) *(Minkowski inequality) If $\mathbb{E}|X|^p < \infty$, and if $\mathbb{E}|Y|^p < \infty$, then $\mathbb{E}|X + Y|^p < \infty$. Further $(\mathbb{E}|X + Y|^p)^{\frac{1}{p}} \leq (\mathbb{E}|X|^p)^{\frac{1}{p}} + (\mathbb{E}|Y|^p)^{\frac{1}{p}}$.*
(vi) *If $a \geq 0,\ b \geq 0$ and $0 \leq \alpha \leq 1$, then $(a + b)^\alpha \leq (a^\alpha + b^\alpha)$.*
(iv) *and* (vi) *are called the c_r inequalities.*

Proof. (i) Consider the function φ defined on $(0, \infty)$: $\varphi(t) = \frac{t^p}{p} + \frac{t^{-q}}{q}$ and note that $\varphi'(t) = 0$ only for $t = 1$ and $\varphi''(1) = p + q > 0$. Hence φ has a minimum at $t = 1$. It follows that

$$\frac{t^p}{p} + \frac{t^{-q}}{q} = \varphi(t) \geq \varphi(1) = \frac{1}{p} + \frac{1}{q} = 1.$$

For any two positive numbers a, b we then have, writing

$$t = \frac{a^{\frac{1}{q}}}{b^{\frac{1}{p}}}, \quad 1 \leq \frac{a^{p-1}}{bp} + \frac{b^{q-1}}{aq}.$$

This reduces to $ab \leq \frac{a^p}{p} + \frac{b^q}{q}$. If a or b is zero, the inequality is obvious.
(ii) Suppose for some $i,\ j\ \leq n,\ a_i \neq 0,\ b_j \neq 0$. Write

$$u_i = \frac{a_i}{(\sum_{j=1}^{n} a_j^p)^{\frac{1}{p}}} \text{ and } v_i = \frac{b_i}{(\sum_{j=1}^{n} b_j^q)^{\frac{1}{q}}}.$$

Substituting u_i for a and v_i for b in the above inequality and summing over i, we get

$$\sum_{i=1}^{n} a_i b_i \leq (\sum_{i=1}^{n} a_i^p)^{\frac{1}{p}} (\sum_{i=1}^{n} b_i^q)^{\frac{1}{q}}.$$

If each $a_i = 0$ or each $b_j = 0$, the inequality is obvious.
(iii) If X or Y is degenerate at 0, the inequality is obvious. Hence let $\mathbb{E}|X|^p > 0$ and $\mathbb{E}|Y|^q > 0$. In (i) take

$$a = \frac{|X|}{(\mathbb{E}|X|^p)^{\frac{1}{p}}} \text{ and } b = \frac{|Y|}{(\mathbb{E}|Y|^q)^{\frac{1}{p}}}$$

to get

$$\frac{|XY|}{(\mathbb{E}|X|^p)^{\frac{1}{p}}(\mathbb{E}|Y|^q)^{\frac{1}{q}}} \le \frac{1}{p}\frac{|X|^p}{\mathbb{E}|X|^p} + \frac{1}{q}\frac{|Y|^q}{\mathbb{E}|Y|^q}.$$

This inequality shows that $\mathbb{E}|XY| < \infty$ and gives,

$$\frac{\mathbb{E}|XY|}{(\mathbb{E}|X|^p)^{\frac{1}{p}}(\mathbb{E}|Y|^q)^{\frac{1}{q}}} \le \frac{1}{p} + \frac{1}{q} = 1,$$

(obtained on taking expected values of both sides) the desired result.
(iv) If $u + v = 0$, the claim is obvious. Let $|u + v| > 0$.

$$\begin{aligned}|u+v|^p &\le |u+v|^{p-1}(|u| + |v|) = |u||u+v|^{p-1} + |v||u+v|^{p-1} \\ &\le (|u|^p + |v|^p)^{\frac{1}{p}}(|u+v|^{q(p-1)} + |u+v|^{q(p-1)})^{\frac{1}{q}} \\ &= 2^{\frac{1}{q}}|u+v|^{p-1}(|u|^p + |v|^p)^{\frac{1}{p}}\end{aligned}$$

Hence $|u + v| \le 2^{\frac{1}{q}}(|u|^p + |v|^p)^{\frac{1}{p}}$ which is the desired result.
(v) If $\mathbb{E}|X+Y| = 0$, the result is obvious. Let $\mathbb{E}|X+Y| > 0$. The above result with $u,\ v$ replaced by the *rvs* $X,\ Y$ would read $|X+Y|^p \le 2^{p-1}(|X|^p+|Y|^p)$. This shows that $\mathbb{E}|X|^p < \infty,\ \mathbb{E}|Y|^p < \infty$ imply $\mathbb{E}|X+Y|^p < \infty$. From the first step in (iv), we have

$$\begin{aligned}\mathbb{E}|X+Y|^p &\le \mathbb{E}|X||X+Y|^{p-1} + \mathbb{E}|Y||X+Y|^{p-1} \\ &\le \text{(applying (iii) to each term)} \\ &\quad (\mathbb{E}|X|^p)^{\frac{1}{p}}(\mathbb{E}|X+Y|^{q(p-1)})^{\frac{1}{q}} + (\mathbb{E}|Y|^p)^{\frac{1}{p}}(\mathbb{E}|X+Y|^{q(p-1)})^{\frac{1}{q}} \\ &= (\mathbb{E}|X|^p)^{\frac{1}{p}}(\mathbb{E}|X+Y|^p)^{\frac{1}{q}} + (\mathbb{E}|Y|^p)^{\frac{1}{p}}(\mathbb{E}|X+Y|^p)^{\frac{1}{q}},\end{aligned}$$

leading to

$$(\mathbb{E}|X+Y|^p)^{1-\frac{1}{q}} \le (\mathbb{E}|X|^p)^{\frac{1}{p}} + (\mathbb{E}|Y|^p)^{\frac{1}{p}},$$

the desired result.
(vi) If $a = 0$ or $b = 0$, the claim is obvious. Assume therefore $a > 0$ and $b > 0$. With no loss of generality assume $a > b$. If the claim is not true, we will have $(a + b)^\alpha > a^\alpha + b^\alpha$. This implies $(1 + c)^\alpha > 1 + c^\alpha$ where $0 < c = \frac{b}{a} < 1$. Hence $1 + c > (1 + c^\alpha)^{\frac{1}{\alpha}} \ge (1 + c^\alpha)^{[\frac{1}{\alpha}]}$. The integer part $[\frac{1}{\alpha}]$ of $\frac{1}{\alpha}$ is ≥ 1. Expanding by the binomial theorem and, omitting all the terms except the first and the last, we get $1+c > 1+c^{\alpha[\frac{1}{\alpha}]} \ge 1+c^{\alpha\frac{1}{\alpha}}$ (since $c < 1$). Thus we have arrived at the contradiction $1 + c > 1 + c$. □

Theorem 2.6.6. *Let* $0 < \alpha < \beta$. *Then (i)* $\mathbb{E}|X|^\beta < \infty$ *implies* $\mathbb{E}|X|^\alpha < \infty$, *and (ii)* $(\mathbb{E}|X|^\alpha)^{\frac{1}{\alpha}} \le (\mathbb{E}|X|^\beta)^{\frac{1}{\beta}}$ *and (iii)* $\mathbb{E}|X|^\alpha < \infty$ *for all* $\alpha > 0$ *and* $\lim\limits_{\alpha\to\infty}(\mathbb{E}|X|^\alpha)^{\frac{1}{\alpha}} < \infty$ *imply that* $lextX$, $rextX$ *are finite.*

Proof. (i) This is an immediate consequence of the fact that over the region $\{|x| > 1\}$, $|x|^\alpha < |x|^\beta$.
(ii) Let $p = \frac{\beta}{\alpha}$ and $q = \frac{\beta}{\beta-\alpha}$ We note that $p > 1$; $q > 1$; and $\frac{1}{p} + \frac{1}{q} = 1$. We note too that $\mathbb{E}(|X|^\alpha)^p = \mathbb{E}|X|^\beta < \infty$. Hence taking $|X|^\alpha$ in place of X and 1 in place of Y in Hölder inequality (ref: **(2.6.5)**), we get $\mathrm{E}|X|^\alpha \le (\mathrm{E}|X|^\beta)^{\frac{\alpha}{\beta}}$.
(iii) Suppose the conditions hold. X has finite *lext and rext iff rext* of $|X|$ is finite. If not true for $|X|$, then $1 - G(x) > 0$ for all $x > 0$ where G is the *df* of $|X|$. Now, for $\lambda > 1$, arbitrary,

$$(\mathbb{E}|X|^\alpha)^{\frac{1}{\alpha}} \ge (\int\limits_{x\ge\lambda} x^\alpha \mathrm{d}G(x))^{\frac{1}{\alpha}} \ge \{\lambda^\alpha[1 - G(\lambda)]\}^{\frac{1}{\alpha}} \ge \lambda\{1 - G(\lambda)\}^{\frac{1}{\alpha}}.$$

Letting $\alpha \to \infty$, we get $\underline{\lim}_{\alpha\to\infty}(\mathbb{E}|X|^\alpha)^{\frac{1}{\alpha}} \ge \lambda$. Since $\lambda > 1$ is arbitrary, this implies $\lim\limits_{\alpha\to\infty}(\mathbb{E}|X|^\alpha)^{\frac{1}{\alpha}} = \infty$. This contradiction implies that G has the claimed property. □

Example 2.6.7. Two distinct distributions can have the same moment sequence. Let

$$g(x) = \frac{1}{\sqrt{2\pi}}\frac{1}{x}e^{-\frac{1}{2}(\log x)^2},\ 0 < x < \infty.$$

For $-1 < \alpha < 1$, let

$$f_\alpha(x) = g(x)\{1 + \alpha\sin(2\pi\log x)\},\ 0 < x < \infty.$$

Note that $f_0(x) = g(x)$, that $f_\alpha(x) \ge 0$ for all x, that

$$\begin{aligned}\int\limits_0^\infty x^k g(x)\sin(2\pi\log x)\mathrm{d}x &= \int\limits_{-\infty}^\infty \frac{1}{\sqrt{2\pi}}e^{ky}e^{-\frac{1}{2}y^2}\sin(2\pi y)\mathrm{d}y \\ &= \frac{1}{\sqrt{2\pi}}e^{\frac{1}{2}k^2}\int\limits_{-\infty}^\infty e^{-\frac{1}{2}(y-k)^2}\sin(2\pi y)\mathrm{d}y \\ &= \frac{1}{\sqrt{2\pi}}e^{\frac{1}{2}k^2}\int\limits_{-\infty}^\infty e^{-\frac{1}{2}y^2}\sin(2\pi(y+k))\mathrm{d}y\end{aligned}$$

$$= \frac{1}{\sqrt{2\pi}} e^{\frac{1}{2}k^2} \int_{-\infty}^{\infty} e^{-\frac{1}{2}y^2} \sin(2\pi y) \mathrm{d}y$$
$$= 0$$

and that

$$\int_{-\infty}^{\infty} f_\alpha(x) \mathrm{d}x = 1.$$

Further

$$\int_0^{\infty} x^k g(x)\, \mathrm{d}x = e^{\frac{1}{2}k^2}.$$

Thus for the family of *df*s $\{F_\alpha, -1 < \alpha < 1\}$, $F_\alpha(x) = \int_0^x f_\alpha(y)\, \mathrm{d}y$ the moment sequence is the same, namely, $(e^{\frac{1}{2}k^2},\ k \geq 0)$.

The question naturally arises: Suppose the *df* F has finite moments of all orders. Under what condition will the moments determine F? This is addressed in section **(3.8)**.

Convergence of moments under weak convergence of dfs.

That weak convergence of *df*s does not automatically imply the convergence of the moments is brought out by the following example. Consider *df*s F_n, $n \geq 1$:

$$F_n(x) = \begin{cases} 0 & \text{if } x < 0; \\ 1 - \frac{1}{n} & \text{if } 0 \leq x < n; \\ 1 & \text{if } x \geq n, . \end{cases}$$

$$F_0(x) = \begin{cases} 0 & \text{if } x < 0 \\ 1 & \text{if } x \geq 0. \end{cases}$$

Clearly $F_n \overset{d}{\to} F_0$. But $1 = \int_R x\, \mathrm{d}F_n(x) \nrightarrow \int_R x\, \mathrm{d}F_0(x) = 0$. Other possibilities are as follows. Let

$$f(x) = \begin{cases} \frac{1}{2|x|^2} & \text{for } |x| \geq 1 \\ 0 & \text{otherwise.} \end{cases}$$

Define $F(x) = \int\limits_{-\infty}^{x} f(y)\,\mathrm{d}y$ and note that $\int\limits_R x\,\mathrm{d}F(x)$ does not exist. Define

$$F_n(x) = \begin{cases} 0 & \text{for } x < -n; \\ \frac{F(x)-F(-n)}{F(n)-F(-n)} & \text{for } -n \le x < n; \\ 1 & \text{for } x \ge n. \end{cases}$$

We note that $F_n \xrightarrow{d} F$ and that $\int\limits_R x\,\mathrm{d}F_n(x)$ exists finitely, equal to 0. Again, with F same as above, define $F_n(x) = F(nx),\ x \in R$. Then $F_n \xrightarrow{d} G$ where $G(x) = 0$ if $x < 0$ and $G(x) = 1$ if $x \ge 0$. Here $\int\limits_R x\,\mathrm{d}F_n(x)$ does not exist while $\int\limits_R x\,\mathrm{d}G(x) = 0$. Another possibility is brought out by the following example. Let $P(X_n = k) = \frac{6}{\pi^2}\frac{C(n)}{k^{2+\frac{1}{n}}}$, $k \ge 1$ and $n \ge 1$. Of course, $\frac{1}{C(n)} = \frac{6}{\pi^2}\sum\limits_{k=1}^{\infty}\frac{1}{k^{2+\frac{1}{n}}}$. Since $\frac{1}{k^{2+\frac{1}{n}}} \le \frac{1}{k^2}$, limit can be taken under the summation sign and we get $\lim\limits_{n\to\infty} C(n) = 1$. If $P(X_0 = k) = \frac{6}{\pi^2}\frac{1}{k^2}, k \ge 1$, then $X_n \xrightarrow{d} X_0$. In this case $EX_n < \infty, n \ge 1$ but $EX_0 = \infty$.

We now investigate conditions that will ensure the convergence of the moments.

Let T be an index set with a generic point in it denoted by a.

Definition 2.6.8. A real measurable function g is said to be uniformly integrable with respect to a family $(\mu_a,\ a \in T)$ of *pms* on $\mathscr{R}$ if $\int_R |g(x)|\,\mathrm{d}\mu_a(x) < \infty$ for each a and

$$\lim_{r\to\infty}\sup_a \int\limits_{|g(x)|\ge r} |g(x)|\mathrm{d}\mu_a(x) = 0.$$

Taking $g(x) \equiv 1$, we see that the uniform integrability of 1 is equivalent to the family (μ_a) being tight. If $0 < q < p$, then, using **(2.6.6)**, it is not hard to see that the

$$\text{uniform integrability of} |x|^p \text{ implies that of} |x|^q \tag{2.10}$$

If X_a is a *rv* with *dm* μ_a and if g is a real Borel measurable function, then we say that the family $(g(X_a))$ is uniformly integrable if $g(x)$ is uniformly integrable with respect to the family (μ_a) of dms of the X_as. This condition can equivalently be written

$$\int_\Omega |g(X_a(\omega))|\,\mathrm{d}P(\omega) < \infty,\ a \in T \tag{2.11}$$

and

$$\sup_a \int_{|g(X_a)|>r} |g(X_a(\omega))|\mathrm{d}P(\omega) \to 0 \text{ as } r \to \infty \tag{2.12}$$

In this terminology, (2.10) would read

$$\text{uniform integrability of} |X_a|^p \text{ implies that of} |X_a|^q$$

Let $(|X_a|^\alpha)$ be uniformly integrable. Then, given $\varepsilon > 0$, we can find r such that $\sup_a \int_{|X_a|\geq r} |X_a|^\alpha \mathrm{d}P < \varepsilon$. It follows from this that

$$\sup_a \int_\Omega |X_a|^\alpha \mathrm{d}P \leq r^\alpha + \varepsilon.$$

Thus

$$c_\alpha = \sup_a \mathbb{E}|X_a|^\alpha < \infty \tag{2.13}$$

An immediate consequence is that if $(|X_n|^\alpha)$ is a uniformly integrable sequence for some $\alpha > 0$, then (X_n) is a tight sequence, since $P(|X_n| \geq \lambda) \leq \frac{c_\alpha}{\lambda^\alpha}$.

Another equivalent definition of uniform integrability is as follows. The family $(|X_a|^\alpha)$ is uniformly integrable (i.e., (2.11) and (2.12) hold, with $g(X_a) = |X_a|^\alpha$) *iff* (2.13) holds and given $\varepsilon > 0$, a $\delta > 0$ can be found such that

$$\sup_a \int_A |X_a|^\alpha\,\mathrm{d}P < \varepsilon \tag{2.14}$$

whenever $P(A) < \delta$.

Let us prove this assertion.

Suppose (2.11) and (2.12) hold. Hence (2.13) holds. Given $\varepsilon > 0$, find $\lambda > 0$ such that $\sup_a \int_{|X_a|>\lambda} |X_a|^\alpha\,\mathrm{d}P < \frac{\varepsilon}{2}$, possible by (2.12). Now choose δ such that $\lambda^\alpha \delta < \frac{\varepsilon}{2}$. We have,

$$\begin{aligned}\sup_a \int_A |X_a|^\alpha\,\mathrm{d}P &= \sup_a\{\int_{A\cap(|X_a|\leq\lambda)} |X_a|^\alpha dP + \int_{A\cap(|X_a|>\lambda)} |X_a|^\alpha\,\mathrm{d}P\} \\ &\leq \sup_a \lambda^\alpha P(A) + \frac{\varepsilon}{2} < \varepsilon \text{ whenever } P(A) < \delta.\end{aligned}$$

Conversely let (2.13) and (2.14) hold. Trivially (2.13) implies (2.11). Given $\varepsilon > 0$, let δ be as guaranteed in (2.14). Choose λ large such that $\frac{c_\alpha}{\lambda^\alpha} < \delta$. We note

$$\int_\Omega |X_a|^\alpha \, \mathrm{d}P \geq \int_{|X_a|>\lambda} |X_a|^\alpha \, \mathrm{d}P \geq \lambda^\alpha P(|X_a| > \lambda).$$

(This inequality is called the Tchebichev inequality. ref. **(4.4.1)**). For $a \in T$ fixed write $A_a = \{\omega : |X_a(\omega)| > \lambda\}$ and note $P(A_a) \leq \delta$. Hence by (2.14), $\sup_a \int_{A_a} |X_a|^\alpha \, \mathrm{d}P < \varepsilon$. This is equivalent to (2.12). With this the proof is complete that $\{(2.11) \text{ and } (2.12)\} \Leftrightarrow \{(2.13) \text{ and } (2.14)\}$.

Remark Let $\alpha > 0$. Let $\int_{-\infty}^{\infty} |x|^\alpha \mathrm{d}F_n(x) < \infty$, $n \geq 1$. Then $|x|^\alpha$ is uniformly integrable with respect to the sequence (F_n) of *df*s *iff*

$$\sup_n \int_r^\infty x^{\alpha-1}\{F_n(-x) + 1 - F_n(x)\}\mathrm{d}x \to 0 \text{ as } r \to \infty.$$

Proof. By Lemma **(2.6.3)**, for $t > r$,

$$\int_{r<|x|\leq t} |x|^\alpha \mathrm{d}F_n(x) = \int_{0<|x|\leq t} |x|^\alpha \mathrm{d}F_n(x) - \int_{0<|x|\leq r} |x|^\alpha \mathrm{d}F_n(x)$$

$$= -t^\alpha\{1 - F_n(t) + F_n(-t)\} + \alpha \int_0^t x^{\alpha-1}\{1 - F_n(x) + F_n(-x)\}\mathrm{d}x$$

$$+ r^\alpha\{1 - F_n(r) + F_n(-r)\} - \alpha \int_0^r x^{\alpha-1}\{1 - F_n(x) + F_n(-x)\}\mathrm{d}x.$$

Hence

$$\int_{|x|>r} |x|^\alpha \mathrm{d}F_n(x) = r^\alpha\{1 - F_n(r) + F_n(-r)\}$$

$$+ \alpha \int_r^\infty x^{\alpha-1}\{1 - F_n(x) + F_n(-x)\}dx \quad (2.15)$$

Since

$$\int_{|x|>r} |x|^\alpha \mathrm{d}F_n(x) \geq \alpha \int_r^\infty x^{\alpha-1}\{1 - F_n(x) + F_n(-x)\}\mathrm{d}x,$$

the necessity of the condition follows.

Suppose now

$$\sup_n \int_r^\infty x^{\alpha-1}\{1 - F_n(x) + F_n(-x)\}\mathrm{d}x \to 0 \text{ as } r \to \infty.$$

Since $\{1 - F_n(x) + F_n(-x)\}$ is a monotonic decreasing function of x, we have

$$\int_{\frac{t}{2}}^{t} \alpha x^{\alpha-1}\{1 - F_n(x) + F_n(-x)\}\mathrm{d}x$$

$$\geq \{1-F_n(t)+F_n(-t)\}\int_{\frac{t}{2}}^{t} \alpha x^{\alpha-1}\mathrm{d}x = \{1-F_n(t)+F_n(-t)\}t^\alpha(1-\frac{1}{2^\alpha}).$$

This, together with the hypothesis, implies

$$\lim_{r\to\infty} \sup_n \int_{|x|>r} |x|^\alpha \mathrm{d}F_n(x) = 0,$$

in view of (2.14). □

Theorem 2.6.9. *Let $F_n \xrightarrow{d} F_0$ and let $|x|^\alpha$, $\alpha > 0$ be uniformly integrable with respect to the sequence $(F_n, n \geq 1)$. Then*
(i) *$\int_R |x|^\alpha \, dF_0(x) < \infty$ and*
(ii) *$\int_R |x|^\alpha \, dF_n(x) \to \int_R |x|^\alpha \, dF_0(x)$. Also, if α is an integer then*
(iii) *$\int_R x^\alpha \, dF_n(x) \to \int_R x^\alpha \, dF_0(x)$.*
If $\int_R |x|^\alpha \, dF_n(x) < \infty$, $n \geq 0$ and if $\int_R |x|^\alpha \, dF_n(x) \to \int_R |x|^\alpha \, dF_0(x)$, then
(iv) *$|x|^\alpha$ is uniformly integrable with respect to the sequence (F_n).*

Proof. (i) Let $\pm\lambda$ be arbitrary continuity points of F_0. Define

$$G_n(x) = \begin{cases} 0, & \text{if } x \leq -\lambda \\ \frac{F_n(x) - F_n(-\lambda)}{F_n(\lambda) - F_n(-\lambda)}, & \text{if } -\lambda \leq x \leq \lambda \\ 1, & \text{if } x \geq \lambda \end{cases}$$

and note that G_n is a *df* and that $G_n \xrightarrow{d} G_0$. Hence by Helly-Bray theorem,

$$\int_{-\lambda}^{\lambda} |x|^\alpha \, dG_n(x) \to \int_{-\lambda}^{\lambda} |x|^\alpha \, dG_0(x).$$

Equivalently

$$\int_{|x| \leq \lambda} |x|^\alpha \, dF_n(x) \to \int_{|x| \leq \lambda} |x|^\alpha \, dF_0(x).$$

Hence given $\varepsilon > 0$ we can find $N = N(\lambda, \varepsilon)$ such that for all $n \geq N$,

$$\int_{|x| \leq \lambda} |x|^\alpha \, dF_n(x) \geq \int_{|x| \leq \lambda} |x|^\alpha \, dF_0(x) - \varepsilon.$$

The uniform integrability condition in the hypothesis implies that

$$c_\alpha = \sup_{n \geq 1} \int_R |x|^\alpha \, dF_n(x) < \infty.$$

Hence

$$c_\alpha \geq \int_{|x| \leq \lambda} |x|^\alpha \, dF_N(x) \geq \int_{|x| \leq \lambda} |x|^\alpha \, dF_0(x) - \varepsilon.$$

Let $\lambda \to \infty$ to conclude $\int_R |x|^\alpha \, dF_0(x) < \infty$.

(ii) Fix a target error $\varepsilon > 0$. Then choose $\lambda = \lambda(\varepsilon)$ such that $\pm\lambda$ are continuity points of F_0 and such that

$$\sup_{n \geq 0} \int_{|x| > \lambda} |x|^\alpha \, dF_n(x) < \varepsilon.$$

This is possible because of the assumption that $|x|^\alpha$ is uniformly integrable with respect to $\{F_n, \; n \geq 1\}$ and by part (i). Choose $N = N(\lambda, \varepsilon)$ as in (i).

Let $n \geq N$. Then

$$\left|\int_R |x|^\alpha \,\mathrm{d}F_n(x) - \int_R |x|^\alpha \,\mathrm{d}F_0(x)\right|$$
$$\leq \left|\int_{|x|\leq\lambda} |x|^\alpha \,\mathrm{d}F_n(x) - \int_{|x|\leq\lambda} |x|^\alpha \,\mathrm{d}F_0(x)\right|$$
$$+ \int_{|x|>\lambda} |x|^\alpha \,\mathrm{d}F_n(x) + \int_{|x|>\lambda} |x|^\alpha \,\mathrm{d}F_0(x) = a_n + b_n + c_n, \text{ say.}$$

By Helly-Bray theorem, $a_n < \varepsilon$; by the choice of λ, $b_n + c_n < 2\varepsilon$. Since $\varepsilon > 0$ is arbitrary, the proof is complete.

(iii) The proof of this assertion follows on similar lines:

$$\left|\int_R x^\alpha \,\mathrm{d}F_n(x) - \int_R x^\alpha \,\mathrm{d}F_0(x)\right| \leq \left|\int_{|x|\leq\lambda} x^\alpha \,\mathrm{d}F_n(x) - \int_{|x|\leq\lambda} x^\alpha \,\mathrm{d}F_0(x)\right|$$
$$+ \int_{|x|>\lambda} |x|^\alpha \,\mathrm{d}F_n(x) + \int_{|x|>\lambda} |x|^\alpha \,\mathrm{d}F_0(x).$$

(iv) Write for $n \geq 0$, $\beta_n(\alpha) = \int_R |x|^\alpha \,\mathrm{d}F_n(x)$. Given $\varepsilon > 0$, choose N_1 such that $|\beta_n(\alpha) - \beta_0(\alpha)| < \varepsilon$ for all $n \geq N_1$. Choose λ, $-\lambda$ continuity points of F_0 such that $\int_{|x|>\lambda} |x|^\alpha \mathrm{d}F_0(x) < \varepsilon$. By Helly-Bray theorem,

$$\int_{-\lambda}^{\lambda} |x|^\alpha \mathrm{d}F_n(x) \to \int_{-\lambda}^{\lambda} |x|^\alpha \mathrm{d}F_0(x).$$

Choose N_2 such that

$$\left|\int_{-\lambda}^{\lambda} |x|^\alpha \mathrm{d}F_n(x) - \int_{-\lambda}^{\lambda} |x|^\alpha \mathrm{d}F_0(x)\right| < \varepsilon \text{ for all } n \geq N_2.$$

Now,

$$\begin{aligned}\int\limits_{|x|>\lambda} |x|^\alpha \mathrm{d}F_n(x) &\leq |\int\limits_{|x|>\lambda} |x|^\alpha \mathrm{d}F_n(x) - \int\limits_{|x|>\lambda} |x|^\alpha \mathrm{d}F_0(x)| \\ &\quad + \int\limits_{|x|>\lambda} |x|^\alpha \mathrm{d}F_0(x) \\ &\leq |\beta_n(\alpha) - \beta_0(\alpha)| + |\int\limits_{|x|\leq\lambda} |x|^\alpha \mathrm{d}F_n(x) \\ &\quad - \int\limits_{|x|\leq\lambda} |x|^\alpha \mathrm{d}F_0(x)| + \int\limits_{|x|>\lambda} |x|^\alpha \mathrm{d}F_0(x) \\ &< 3\varepsilon, \quad n \geq N(= \max(N_1,\ N_2)).\end{aligned}$$

This is equivalent to the claim $|x|^\alpha$ is uniformly integrable with respect to the sequence $\{F_n,\ n \geq N\}$ and hence with respect to the sequence $\{F_n,\ n \geq 1\}$ since integrability with respect to each of a finite number of *df*s implies its uniform integrability with respect to them. □

2.7 Weak convergence in the metric space $C[0, 1]$

Let $C = C[0, 1]$ denote the collection of all the real valued continuous functions defined on the interval [0, 1] and vanishing at zero.

We note that the only constant function in C is the one which is identically zero.

Endow C with the uniform metric ρ : for $x,\ y \in$ C, $\rho(x,\ y) = \|x - y\| = \sup\limits_{0\leq t\leq 1} |x(t) - y(t)|$. Then (C, ρ) would be a separable and complete metric space. Consequently every single pm on its Borel σ-field $\mathscr{C}$ would be tight (ref. **(1.9.4)**). The separability of (C, ρ) implies that $\mathscr{C}$ is generated by the spheres with rational numbers for their radii and centers at the points of a separability set.

Let for $0 \leq t_i \leq 1,\ 1 \leq i \leq p,\ \pi_{t_1,t_2,\ldots,t_p}$ denote the projection mapping of C into R^p and note that it is a continuous mapping (and hence is $\mathscr{C}$-measurable).

Let $T = \{t_1, t_2, \ldots\}$ be a countable dense subset of [0, 1]. Let $\mathscr{C}_1$ be the minimal σ-field with respect to which the $\pi_t,\ t \in T$ are measurable. Hence $\mathscr{C}_1 \subset \mathscr{C}$. Since $t_n \to t$ implies $\pi_{t_n} \to \pi_t$ pointwise, it follows that all the $\pi_t,\ t \in [0,\ 1]$ are $\mathscr{C}_1$-measurable.

Consider a sphere $S = \{x : \|x - a\| \le r\}$, $a \in$C, fixed. Now

$$S = \{x : \sup_{t\in[0,\,1]} |x(t) - a(t)| \le r\} = \{x : \sup_{t\in T} |x(t) - a(t)| \le r\}$$
$$= \lim_{n\to\infty} \bigcap_{j=1}^{n} A_j$$

where $A_j = \{x : |x(t_j) - a(t_j)| \le r\}$. Since each $A_j \in \mathscr{C}_1$, it follows that $S \in \mathscr{C}_1$ and hence $\mathscr{C} \subset \mathscr{C}_1$. Thus we see that $\mathscr{C}$ is generated by the functions $\{\pi_t,\ t \in T\}$ (Compare: **(1.4.1)**).

Let $\mu_n,\ n \ge 0$ be *pms* on $\mathscr{C}$. In this section we find tractable necessary and sufficient conditions for $\mu_n \xrightarrow{d} \mu_0$, after noting the difficulty in checking the satisfaction of any of the seven equivalent conditions in **(2.4.2)** because of the vast number of relations to be verified.

The modulus of continuity function δ_c and characterization of compact sets in C.

Given $c > 0$ and $x \in C$ define

$$\delta_c(x) = \sup_{\substack{0\le t,\, u\le 1\\ |t-u|\le c}} |x(t) - x(u)|$$

and note

$$\text{(i) that } \lim_{c\to 0} \delta_c(x) = 0 \tag{2.16}$$

which follows from the uniform continuity of x over [0, 1]

$$\text{(ii) that } \delta_c \downarrow \text{as } c \downarrow \tag{2.17}$$

and

$$\text{(iii) that, for } c \text{ fixed, } \delta_c \text{ is a continuous mapping of } C \text{ into } R. \tag{2.18}$$

For, let $\|x_n - x\| \to 0$. Now

$$\delta(c;\ x_n) = \delta_c(x_n) = \sup_{\substack{0\le t,\, u\le 1\\ |t-u|\le c}} |x_n(t) - x_n(u)|$$
$$\le \sup_{\substack{0\le t,\, u\le 1\\ |t-u|\le c}} \{|x_n(t) - x(t)| + |x(t) - x(u)| + |x_n(u) - x(u)|\}$$
$$\le 2\|x_n - x\| + \delta_c(x).$$

This leads to $\overline{\lim_{n\to\infty}}\ \delta_c(x_n) \le \delta_c(x)$. On similar lines we have $\delta_c(x) \le 2\|x_n - x\| + \delta_c(x_n)$, leading to $\delta_c(x) \le \underline{\lim_{n\to\infty}}\ \delta_c(x_n)$. These two inequalities imply $\lim_{n\to\infty} \delta_c(x_n) = \delta_c(x)$.

Arzela-Ascoli characterization of compact sets in $C[0,\ 1]$ is quoted below for future reference.

Q9. Definitions and a **Theorem.**

(i) **Definition.** $A \subset C$ is said to be bounded if $\sup_{x \in A} \|x\| < \infty$

(ii) **Definition.** ($\mathfrak{a}$)$A \subset C$ is said to be equiconinuous at $t_0 \in [0,\ 1]$ if, given $\varepsilon > 0$, a $\delta > 0,\ \delta = \delta(t_0,\ \varepsilon)$, can be found such that $\sup_{x \in A} |x(t) - x(t_0)| < \varepsilon$ whenever $|t - t_0| < \delta$.

($\mathfrak{b}$)$A \subset C$ is said to be equicontinuous in $[0,\ 1]$ if it is equicontinuous at each point of $[0,\ 1]$.

($\mathfrak{c}$)$A \subset C$ is said to be uniformly equicontinuous in $[0,\ 1]$ if : for every $\varepsilon > 0$, there exists $c > 0$ (which depends only on ε) such that

$$\sup_{x \in A} \delta_c(x) < \varepsilon \text{ where } \delta_c(x) = \sup_{\substack{0 \le t,\ u \le 1 \\ |t-u| \le c}} |x(t) - x(u)|.$$

In this book, we write 'equicontinuous' for 'uniformly equicontinuous'.

Note For any $c > 0$ and any $\varepsilon > 0$, $A = \{x\ :\ \delta_c(x) \le \varepsilon\}$ is a closed and bounded set.

In particular, an equicontinuous set A is a bounded set.

Proof. Since δ_c is a continuous function (ref. (2.18)), it follows that A is a closed set.

Recall $x(0) = 0$ for all $x \in C$. Let $0 < t \le 1$ and let $x \in A$. Divide the interval $[0,\ t]$ through the points $0,\ c,\ 2c,\ \ldots,\ Nc,\ t$ where $N = [\frac{t}{c}]$. Now

$$|x(t)| = |x(t) - x(Nc) + \sum_{j=1}^{N} \{x(jc) - x(\overline{j-1}c)\}| \le \varepsilon + N\varepsilon \le \frac{c+1}{c}\varepsilon.$$

Since this is true for all $t \in [0,\ 1]$ and for all $x \in A$, the boundedness of the set A follows.

(iii) **Theorem** (Arzela-Ascoli) A subset $A \subset$ C is compact *iff* it is closed and equicontinuous.

Since $A \subset C$ is equicontinuous *iff* $\bar{A}$ is equicontinuous, (iii) can be stated equivalently thus: $A \subset C$ has compact closure *iff* it is equicontinuous.

Some applications of the Arzela-Ascoli Theorem

Theorem 2.7.1. *Let (ε_j), (η_j) be real sequences strictly monotonically tending to 0. Define, for $j = 1,\ 2, \ldots,\ A_j = \{x\ :\ x \in C,\ \delta(\eta_j;\ x) \le \varepsilon_j\}$ and $B_j = \cap_{k=j}^{\infty} A_k$. Then each B_k is a compact set.*

Proof. We first note that, since δ_c is a continuous mapping, each A_j (and hence each B_j) is a closed set. Further we note that $B_j \uparrow$. Hence it is enough to show that B_r is a compact set for all r large. Given $\varepsilon > 0$, find $j = j(\varepsilon)$ such that $\varepsilon_j < \varepsilon$. Take $c = \eta_j$ and note that c depends only on ε. Let $r > j$. Let $x \in B_r$ be arbitrary. Since $B_r \subset A_r$, $x \in A_r$. Hence $\delta(c;\ x) < \varepsilon_r < \varepsilon_j < \varepsilon$. i.e. given $\varepsilon > 0$, we have found a $c > 0$ such that $\sup\limits_{x\in B_r} \delta(c;\ x) < \varepsilon$. By Arzela-Ascoli, this, together with the fact that B_r is a closed set, is equivalent to claiming that B_r is a compact set. □

Theorem 2.7.2. *For every $\alpha > 1$ and for every positive number τ, the set $A_{\alpha,\tau} = \{x : x \in C, x$ absolutely continuous, $\int\limits_0^1 |\frac{dx}{dt}|^\alpha\, dt \le \tau\}$ is a closed set.*

Proof. Let $x_n \in C, n \ge 1, x_n$ absolutely continuous with

$$\int_0^1 |\frac{dx_n}{dt}|^\alpha\, dt \le \tau, \quad n \ge 1$$

and let $\|x_n - x_0\| \to 0$. We will show first that x_0 is absolutely continuous. To do this we must, given $\varepsilon > 0$, find a $\delta > 0$ such that

$$\sum_{i=1}^m |x_0(b_i) - x_0(a_i)| < \varepsilon$$

whenever $0 \le a_1 < b_1 \le a_2 < \cdots \le a_m < b_m \le 1$ and whatever the integer m may be, provided

$$\sum_{i=1}^m (b_i - a_i) < \delta \qquad (*).$$

Write $p = \alpha$ and $q = \frac{\alpha}{\alpha-1}$ and note that $q > 1$ and $\frac{1}{p} + \frac{1}{q} = 1$. Let $\varepsilon > 0$ be given. For δ arbitrary and for a partition $(a_1,\ b_1,\ a_2,\ b_2,\ ...,\ a_m,\ b_m)$ for which $(*)$ holds, we have

$$\begin{aligned}\sum_{i=1}^m |x_0(b_i) - x_0(a_i)| &= \sum_{i=1}^m \frac{|x_0(b_i) - x_0(a_i)|}{(b_i - a_i)^{\frac{1}{q}}} (b_i - a_i)^{\frac{1}{q}} \\ &\le \Big\{\sum_{i=1}^m \frac{|x_0(b_i) - x_0(a_i)|^p}{(b_i - a_i)^{\frac{p}{q}}}\Big\}^{\frac{1}{p}} \{\sum_{i=1}^m (b_i - a_i)\}^{\frac{1}{q}} \\ &\qquad \text{(by the Hölder inequality)} \\ &\le \Big\{\sum_{i=1}^m \frac{|x_0(b_i) - x_0(a_i)|^p}{(b_i - a_i)^{\frac{p}{q}}}\Big\}^{\frac{1}{p}} \delta^{\frac{1}{q}}\end{aligned}$$

$$= \lim_{k\to\infty} \Big\{\sum_{i=1}^{m}(b_i - a_i)^{-\frac{p}{q}}|x_k(b_i) - x_k(a_i)|^p\Big\}^{\frac{1}{p}} \delta^{\frac{1}{q}}$$

$$= \lim_{k\to\infty} \Big\{\sum_{i=1}^{m}(b_i - a_i)^{-\frac{p}{q}}\Big|\int_{a_i}^{b_i} \frac{dx_k}{dt}dt\Big|^p\Big\}^{\frac{1}{p}} \delta^{\frac{1}{q}}$$

$$\leq \lim_{k\to\infty} \Big\{\sum_{i=1}^{m}(b_i - a_i)^{-\frac{p}{q}}\Big[\int_{a_i}^{b_i} \Big|\frac{dx_k}{dt}\Big|dt\Big]^p\Big\}^{\frac{1}{p}} \delta^{\frac{1}{q}}$$

$$\leq \lim_{k\to\infty} \Big\{\sum_{i=1}^{m}(b_i - a_i)^{-\frac{p}{q}}\Big[\Big(\int_{a_i}^{b_i} \Big|\frac{dx_k}{dt}\Big|^p dt\Big)^{\frac{1}{p}}\Big(\int_{a_i}^{b_i} 1^q dt\Big)^{\frac{1}{q}}\Big]^p\Big\}^{\frac{1}{p}} \times \delta^{\frac{1}{q}}$$

(again by the Hölder inequality)

$$= \lim_{k\to\infty} \Big\{\sum_{i=1}^{m}\int_{a_i}^{b_i} \Big|\frac{dx_k}{dt}\Big|^p dt\Big\}^{\frac{1}{p}} \delta^{\frac{1}{q}}$$

$$\leq \lim_{k\to\infty} \Big\{\int_{0}^{1} \Big|\frac{dx_k}{dt}\Big|^p dt\Big\}^{\frac{1}{p}} \delta^{\frac{1}{q}} \leq \tau^{\frac{1}{\alpha}} \delta^{\frac{1}{q}},$$

Now choose $\delta < (\frac{\varepsilon}{\tau^{\frac{1}{\alpha}}})^q$ and conclude that x_0 is absolutely continuous.

Also from the above steps we have

$$\sum_{i=1}^{m}(b_i - a_i)^{-\frac{p}{q}}|x_0(b_i) - x_0(a_i)|^p \leq \tau.$$

Define

$$y_n(t) = \frac{x_0(\frac{i}{2^n}) - x_0(\frac{i-1}{2^n})}{\frac{1}{2^n}} \text{ for } \frac{i-1}{2^n} < t \leq \frac{i}{2^n},\ 1 \leq i \leq 2^n.$$

Note that the y_ns are step functions and that

$$y_n(t) \to \frac{dx_0(t)}{dt}.$$

Now,

$$\int_0^1 |y_n(t)|^p \mathrm{d}t = \sum_{i=1}^{2^n} \frac{|x_0(\frac{i}{2^n}) - x_0(\frac{i-1}{2^n})|^p}{\frac{1}{2^{np}}} \frac{1}{2^n}$$

$$= \sum_{i=1}^{2^n} \frac{|x_0(\frac{i}{2^n}) - x_0(\frac{i-1}{2^n})|^p}{(\frac{1}{2^n})^{\frac{p}{q}}} \leq \tau.$$

It follows now by Fatou's Lemma that

$$\int_0^1 \left|\frac{\mathrm{d}x_0(t)}{\mathrm{d}t}\right|^p \mathrm{d}t = \int_0^1 \lim_{n\to\infty} |y_n(t)|^p \mathrm{d}t \leq \underline{\lim}_{n\to\infty} \int_0^1 |y_n(t)|^p \mathrm{d}t \leq \tau.$$

Thus $x_0 \in A_{\alpha,\tau}$ and the proof that $A_{\alpha,\tau}$ is a closed set is complete. □

Remark

Separability of $A_{\alpha,\,\tau}$ follows now since it is a subset of a separable metric space, namely C (ref. An Introduction To The Theory Of Integration by A.C. Zaanen, Theorem 1, p40).

Theorem 2.7.3. *The sets $A_{\alpha,\tau}$ in* **(2.7.2)** *are compact subsets of (C, ρ).*

Proof. Let $\alpha > 1$, $\tau > 0$ be fixed. Define p, q as in **(2.7.2)**. For $x \in A_{\alpha,\tau}$ and $0 \leq a < b \leq 1$,

$$|x(a) - x(b)| = \left|\int_a^b \frac{\mathrm{d}x(t)}{\mathrm{d}t}\mathrm{d}t\right| \leq \int_a^b \left|\frac{\mathrm{d}x(t)}{\mathrm{d}t}\right|\mathrm{d}t$$

$$\leq \left\{\int_a^b |\frac{\mathrm{d}x(t)}{\mathrm{d}t}|^p\right\}^{\frac{1}{p}} \left\{\int_a^b \mathrm{d}t\right\}^{\frac{1}{q}}$$

(appealing to the Hölder inequality)

$$\leq (b-a)^{\frac{1}{q}} \tau^{\frac{1}{p}}.$$

This being true for every $x \in A_{\alpha,\tau}$, it follows by Arzela-Ascoli theorem that $A_{\alpha,\tau}$ has compact closure. Since the set has already been shown to be a closed set (ref. **(2.7.2)**), it is a compact set. □

The separability of $K_{\alpha,a}$ can be concluded from the fact that it is a compact metric space. We present now explicitly a separability set for $K_{\alpha,a}$. We claim that the collection of all functions $x_m \in C$, $m = 1, 2, \ldots$ satisfying the following conditions is dense in $K_{\alpha,a}$: $x_m(\frac{i}{m})$ is rational, x_m is linear on $[\frac{i-1}{m}, \frac{i}{m}]$, $i = 1, 2, \ldots, m$ and $\int_0^1 |\frac{\mathrm{d}x_m(t)}{\mathrm{d}t}|^\alpha \,\mathrm{d}t \le a^2$.

The proof is as follows.

The mentioned collection is clearly non-null. Let $y \in K_{\alpha,a}$. Define y_m : $y_m(\frac{i}{m}) = y(\frac{i}{m})$ and y_m is linear elsewhere, $m = 1, 2, 3, \ldots,$. We invoke the standard result for uniformly continuous functions:

$$\lim_{m\to\infty} \|y - y_m\| = 0. \tag{2.19}$$

Recall notation $p = \alpha$ and $q = \frac{\alpha}{\alpha-1}$.

$$\int_0^1 |y_m'(t)|^p \,\mathrm{d}t = \sum_{i=1}^m \int_{\frac{i-1}{m}}^{\frac{i}{m}} |y_m'(t)|^p \mathrm{d}t = \sum_{i=1}^m \Big|\frac{y_m(\frac{i}{m}) - y_m(\frac{i-1}{m})}{\frac{1}{m}}\Big|^p \frac{1}{m}$$

$$= \sum_{i=1}^m \Big|\frac{y(\frac{i}{m}) - y(\frac{i-1}{m})}{\frac{1}{m}}\Big|^p \frac{1}{m} = m^{p-1} \sum_{i=1}^m \Big|\int_{\frac{i-1}{m}}^{\frac{i}{m}} y'(t)\,\mathrm{d}t\Big|^p$$

$$\le m^{p-1} \sum_{i=1}^m \Big(\int_{\frac{i-1}{m}}^{\frac{i}{m}} |y'(t)|\,\mathrm{d}t\Big)^p$$

$$\le m^{p-1} \sum_{i=1}^m \Big\{\Big(\int_{\frac{i-1}{m}}^{\frac{i}{m}} |y'(t)|^p \,\mathrm{d}t\Big)^{\frac{1}{p}} \Big(\int_{\frac{i-1}{m}}^{\frac{i}{m}} 1^q \,\mathrm{d}t\Big)^{\frac{1}{q}}\Big\}^p$$

by **(2.6.5)**(iii)

$$= \sum_{i=1}^m m^{p-1} \sum_{i=1}^m \int_{\frac{i-1}{m}}^{\frac{i}{m}} |y'(t)|^p \,\mathrm{d}t \times \frac{1}{m^{\frac{p}{q}}}$$

$$= \sum_{i=1}^m \int_{\frac{i-1}{m}}^{\frac{i}{m}} |y'(t)|^p \,\mathrm{d}t = \int_0^1 |y'(t)|^p \,\mathrm{d}t \le a^2.$$

This shows $y_m \in K_{\alpha,a}$. Choose $0 \le u_{i,m} \le \frac{1}{m}$ such that $x_m(\frac{i}{m}) = y_m(\frac{i}{m}) + u_{i,m}$ is a rational number. Define polygonal function x_n with values at $\frac{i}{m}$, $1 \leq i \leq m$ as above and linear elsewhere on $[0,\ 1]$. Since, trivially, $\frac{dx_m}{dt} = \frac{dy_m}{dt}$, it follows $x_m \in K_{\alpha,a}$. Now $\|x_m - y_m\| \leq \frac{1}{m} \to 0$ as $m \to \infty$. This together with (2.16) implies $\|x_m - y\| \to 0$ □

Theorem 2.7.4. *Every pm μ on $\mathscr{C}$ is tight.*

Proof. [Note. Since $(C,\ \rho)$ is a complete separable metric space, the theorem follows immediately from **(1.9.4)**. Here we give an alternate proof which is $(C,\ \rho)$-specific.]

Let $\varepsilon_j \downarrow 0$. Let β be an arbitrary small number. For j fixed, let $E_k = \{x : \delta(\frac{1}{k};\ x) \leq \varepsilon_j\}$. Since $\delta(c;\ x) \downarrow 0$ as $c \downarrow 0$, it follows that $E_k \uparrow C$. Hence given $\beta > 0$ we can find a k_j such that $\mu(E_{k_j}) > 1 - \frac{\beta}{2^{j+1}}$. Let $\eta_j = \frac{1}{k_j}$. Write $E_{k_j} = A_j$ and $B = \cap_j A_j$. Then by **(2.7.1)**, B is a compact set. Tightness of μ follows now because

$$\mu(B') \leq \sum_{j=1}^{\infty} \mu(A_j') \leq \sum_{j=1}^{\infty} \frac{\beta}{2^{j+1}} < \beta.$$ □

The above steps also show that given $\varepsilon > 0,\ \eta > 0$, a number $c > 0$ can be found such that $\mu\big(x\ :\ \delta_c(x) < \varepsilon\big) > 1 - \eta$.

Remark 2.7.5. The above steps show that the theorem is equivalent to the following statement.

Given $\varepsilon > 0,\ \eta > 0$, there exists a $c > 0$ such that $\mu(\{x : \delta_c(x) < \varepsilon\}) > 1 - \eta$

It is then trivial to see that a family (μ_α) of pms on $\mathscr{C}$ would be tight if, given $\varepsilon > 0,\ \eta > 0$, we can find a c independent of α such that $\mu_\alpha(\{x : \delta_c(x) < \varepsilon\}) > 1 - \eta$ for every α.

A basic property of compact sets in C is presented next.

In the rest of this section, let $T = \{t_1,\ t_2,\ \ldots..\}$ be an arbitrary but a fixed countable dense subset of [0, 1]. Also μ with or without suffix will denote a pm on $\mathscr{C}$.

Theorem 2.7.6. *Let $K \subset C$ be a compact set. Let $\pi_{t_1,\ t_2,\ \ldots,\ t_n}$ denote the projection maps : $\pi_{t_1,\ t_2,\ \ldots,\ t_n}(x) = (x(t_1),\ x(t_2),\ \ldots,\ x(t_n))$ taking C into R^n, $n \geq 1$. Then*

$$K = \bigcap_{n=1}^{\infty} \pi^{-1}_{t_1,\ t_2,\ \ldots,\ t_n} \pi_{t_1,\ t_2,\ \ldots,\ t_n} K. \tag{2.20}$$

Proof. Since $\pi^{-1}_{t_1, t_2, \ldots, t_n}\pi_{t_1, t_2, \ldots, t_n}K \supset K$, it follows that K is a subset of the set on the right side of (2.20).

Let now x be an arbitrary member of the right side of (2.20). Hence, for every n, $x \in \pi^{-1}_{t_1, t_2, \ldots, t_n}\pi_{t_1, t_2, \ldots, t_n}K$. There exists therefore $y_n \in K$ such that $\pi_{t_1, t_2, \ldots, t_n}x = \pi_{t_1, t_2, \ldots, t_n}y_n$. Since K is a compact set, (y_n) admits of a convergent subsequence, say, (y_m) converging to, say, $y_0 \in K$ in the metric ρ. This implies $y_m(t) \to y_0(t)$ for all $t \in [0, 1]$. Fix r. Then for all $m > r$, $x(t_r) = y_m(t_r)$. Let $m \to \infty$ and conclude $x(t_r) = y_0(t_r)$. Since this is true for every $t_r \in T$, since T is dense in [0, 1] and since the functions are continuous, we see that $x \equiv y_0$. Thus $x \in K$ and the proof is complete. □

Definition 2.7.7. The members of the family $(\mu\pi^{-1}_{r_1, r_2, \ldots, r_n}, n \geq 1$, all r_js in [0, 1]), of *pms* in R^n are called the *fdds* of μ.

Theorem 2.7.8. *(i) Two pms μ, ν with the same fdd are identical.*
(ii) If, for every choice of $r \geq 1$, $\mu_n\pi^{-1}_{t_1,t_2,\ldots,t_r} \overset{d}{\to} \mu$, then for every compact set K, $\overline{\lim_{n\to\infty}}\, \mu_n(K) \leq \mu(K)$.

Proof. (i) We recall that μ, ν are tight measures. Hence claim follows if we show that $\mu(K) = \nu(K)$ for every compact set K. Let $E_n = \pi_{t_1,t_2,\ldots,t_n}K$ and $Q_n = \pi^{-1}_{t_1,t_2,\ldots,t_n}E_n$ and note that $Q_n \downarrow K$. Then $\mu(K) = \mu(\lim Q_n)$ $= \lim\mu(Q_n) = \lim\nu(Q_n)$ (since the *fdd* of ν are the same as those of μ) $=$ $\nu(\lim Q_n) = \nu(K)$.
(ii) Let E_r, Q_r be as defined above. We note E_r is a compact subset, and hence a closed subset, of R^r. Now for every $r \geq 1$, $\pi^{-1}_{t_1,t_2,\ldots,t_r}(E_r) \supset K$. Hence $\mu_n(K) \leq \mu_n\pi^{-1}_{t_1,t_2,\ldots,t_r}(E_r)$. Hence, by hypothesis read with **(2.4.2)**,

$$\overline{\lim_{n\to\infty}}\, \mu_n(K) \leq \mu_n\pi^{-1}_{t_1,t_2,\ldots,t_r}(E_r) \leq \mu\pi^{-1}_{t_1,t_2,\ldots,t_r}(E_r) = \mu(Q_r).$$

Allow $r \to \infty$ and conclude $\overline{\lim_{n\to\infty}}\, \mu_n(K) \leq \mu(K)$. □

Theorem 2.7.9. *For each finite set $A \subset T = [0, 1]$ let μ_A be a pm on $\mathscr{R}^p$ where p is the cardinal number of A. Assume that this family of pms is consistent (ref.* **(1.11)***). Further assume that for some positive constants c, α, β,*

$$\iint_{R^2} |x - y|^\beta \, d\mu_{s,t}(x, y) \leq c|t - s|^{1+\alpha} \tag{2.21}$$

for every choice of s, $t \in T$. Then there is a unique probability measure μ on $(C, \mathscr{C})$ with the μ_As for its finite dimensional distributions.

Proof. Let $T_n = \{\frac{j}{2^n},\ 0 \le j \le 2^n\}$ and note that $T_n \uparrow \tilde{T} = \cup_n T_n$ is a countable dense subset of T. Form the product measurable space $(U,\ \mathcal{A})$ where $U = \times_{t\in\tilde{T}} R$ and $\mathcal{A} = \bigotimes_{t\in\tilde{T}} \mathscr{R}$. By **(1.11.1)**, the family $(\mu_B,\ B \subset \tilde{T},\ B$ finite) of *pms* determines a unique *pm* Q on $\mathcal{A}$ such that the *fdd* of Q are the μ_Bs. We note that, for each $u \in U$, $(u(t),\ t \in \tilde{T})$ is a countable set of real numbers. For $t_j \in \tilde{T},\ 1 \le j \le p$, denote by $\mathbb{P}_{t_1,\ \dots,\ t_p}$ the projection operator defined on $U : \mathbb{P}_{t_1,\ \dots,\ t_p} u = (u(t_1),\ \dots,\ u(t_p))$; note that each $\mathbb{P}_{t_1,\ t_2,\ \dots,\ t_p}$ is a rV on U. Given $u \in U$, define u_n on T_n as follows. $u_n(t) = u(t),\ t \in T_n$ and linear elsewhere. If $t \in T \sim T_n$, then $u_n(t)$ is a linear combination of two rvs $u(t_1),\ u(t_2)$ for some $t_1,\ t_2 \in T_n$ and hence is a rv. We note that for $m > n,\ u_m = u_n$ on T_n. Since $u_n,\ u_{n+1}$ are continuous functions

$$\sup_{t\in T} |u_n(t) - u_{n+1}(t)| = \max_{t\in\tilde{T}} |u_n(t) - u_{n+1}(t)|$$

and hence is a rv on U. Since $u_n,\ u_{n+1}$ are polygonal functions coinciding on T_n, it is easy to see that

$$\begin{aligned}\sup_{t\in T} |u_n(t) - u_{n+1}(t)| &\le \max_{1\le j\le 2^n} |u_n(\frac{2j-1}{2^{n+1}}) - u_{n+1}(\frac{2j-1}{2^{n+1}})| \\ &\le \max_{1\le j\le 2^n} \max\{|u(\frac{2j-1}{2^{n+1}}) - u(\frac{2j}{2^{n+1}})|;\ |u(\frac{2j-1}{2^{n+1}}) - u(\frac{2j-2}{2^{n+1}})|\}.\end{aligned}$$

Let γ be an arbitrary positive number less than $\frac{\alpha}{\beta}$. We have

$$\begin{aligned}&Q\{u : \sup_{t\in T} |u_n(t) - u_{n+1}(t)| \ge 2^{-\gamma n}\} \\ &\le Q\{u : \max_{1\le j\le 2^{n+1}} |u(\frac{j}{2^{n+1}}) - u(\frac{j-1}{2^{n+1}})| \ge 2^{-\gamma n}\} \\ &\le \sum_{j=1}^{2^{n+1}} Q\{u : |u(\frac{j}{2^{n+1}}) - u(\frac{j-1}{2^{n+1}})| \ge 2^{-\gamma n}\} \\ &\le \sum_{j=1}^{2^{n+1}} 2^{\gamma n\beta} \mathbb{E}|u(\frac{j}{2^{n+1}}) - u(\frac{j-1}{2^{n+1}})|^{\beta}\end{aligned}$$

(by Tchebichev inequality **(4.4.1)**; expected value is with respect to the measure Q or, what is same, with respect to the measure $\mu_{s,t}$ for appropriate values of s and t)

$$
\begin{aligned}
&\leq 2^{\gamma n\beta}\sum_{j=1}^{2^{n+1}} c(\frac{1}{2^{n+1}})^{1+\alpha} \text{ (by (2.21))}\\
&= c2^{\gamma n\beta}2^{n+1}2^{-(1+\alpha)(n+1)}\\
&= c_1 2^{-n(\alpha-\beta\gamma)} = c_1 2^{-n\delta},\ \delta > 0 \quad \text{by choice of } \gamma.
\end{aligned}
$$

Hence

$$\sum_n Q\{u : \sup_{t\in T}|u_n(t) - u_{n+1}(t)| \geq 2^{-\gamma n}\} < \infty.$$

This implies, by the Borel-Cantelli lemma, that there exists $E \in \mathcal{A}$ such that $Q(E) = 1$ and such that for each $u \in E$,

$$\sup_{t\in T}|u_n(t) - u_{n+1}(t)| \leq 2^{-\gamma n}$$

for all large n. For $m < n$, we have, for each $u \in E$,

$$\sup_{t\in T}|u_n(t) - u_m(t)| \leq \sum_{r=m}^{n-1}|u_r(t) - u_{r+1}(t)| \leq \sum_{m}^{n-1} 2^{-\gamma r} \to 0$$

as $m \to \infty$. i.e. for each $u \in E$, (u_n) is a Cauchy sequence in C. Since C, under the uniform metric, is complete there exists $u^* \in C$ such that $\sup_{t\in T}|u_n - u^*| \to 0$. Thus to every $u \in E$, there corresponds a unique $u^* \in C$.
We note that $u(t) = u^*(t)$, $t \in \tilde{T}$, that $(U \cap E,\ \mathcal{A} \cap E,\ Q)$ is a probability space and that every $u \in U \cap E$ can be extended uniquely to a continuous function u^*. Define mapping f on this space : $U \cap E \to C$; $f(u) = u^*$. Let us show that this mapping is measurable. Let $u_0^* \in f(E)$ be fixed. Let $S = \{x : x \in C, \|x - u_0^*\| \leq r\}$, the closed sphere.

$$
\begin{aligned}
f^{-1}(S) = f^{-1}(S \cap f(U \cap E)) &= f^{-1}\{u^* : \|u^* - u_0^*\| \leq r\}\\
&= f^{-1}\{u^* : \sup_{t\in\tilde{T}}|u^*(t) - u_0^*(t)| \leq r\}\\
&= \{u : \sup_{t\in\tilde{T}}|u(t) - u_0(t)| \leq r\} \in \mathcal{A} \cap E,
\end{aligned}
$$

where $f(u_0) = u_0^*$, proving that f is measurable.

Let $\pi_{t_1,\, t_2,\, \ldots,\, t_p}$ denote the projection operator taking C into R^p. Denote by ν the pm induced by f on $\mathscr{C}$: $\nu(D) = Qf^{-1}(D)$. We note that if $t_j \in \tilde{T}$, $1 \leq j \leq p$, then $\pi_{t_1,\, t_2,\ldots,t_p} f = \mathbb{P}_{t_1,\, t_2,\ldots,t_p}$. It follows from this that, for $B \subset \tilde{T}$, B finite, the *fdd* ν_B of ν is the same as μ_B. If $t_j \in T \sim \tilde{T}$, $1 \leq j \leq p$, then we can find for each t_j a sequence $(t_{j,n})$ in $\tilde{T}$ converging to t_j. By **(2.4.3)**(vi), $\mu\mathbb{P}^{-1}_{t_{1,n},\, t_{2,n},\ldots,\, t_{p,n}} \xrightarrow{d} \mu\mathbb{P}^{-1}_{t_1,\, t_2,\ldots,\, t_p}$ and a similar result holds for ν. It then follows by the earlier remark that $\nu\pi^{-1}_{t_1,\, t_2,\ldots,\, t_p} = \mu\pi^{-1}_{t_1,\, t_2,\ldots,\, t_p}$. The *pm* ν has the desired properties. **(2.7.8)** guarantees that ν is unique. □

Note. (i) For another version of this theorem ref. **(5.7.4)**.
(ii) An example of a consistent family of finite dimensional distributions μ_A satisfying (2.21): Let $A = \{t_1,\, t_2,\, \ldots,\, t_p\}$ and for $E \in \mathscr{R}^p$, let

$$\mu_A(E) = \int_E \frac{1}{(2\pi)^{\frac{p}{2}}} \frac{1}{|\Sigma|^{\frac{1}{2}}} e^{-\frac{1}{2}\mathbf{U}'\Sigma^{-1}\mathbf{U}}\, du_1\, du_2 \ldots du_p$$

where $\mathbf{U} = (u_1,\, u_2,\, \ldots,\, u_p)'$ and matrix $\Sigma = (\min(t_i,\, t_j))$. In this case, for $s < t$, and $A = \{s,\, t\}$

$$\Sigma = \begin{pmatrix} s & s \\ s & t \end{pmatrix} \qquad \text{and } \Sigma^{-1} = \frac{1}{s(t-s)} \begin{pmatrix} t & -s \\ -s & s \end{pmatrix}$$

We then have

$$\mu_A(E) = \iint_E \frac{1}{2\pi} \frac{1}{\sqrt{st - s^2}} e^{-\frac{1}{2}\frac{1}{st-s^2}[(t-s)x^2 + s(x-y)^2]} dxdy.$$

From this

$$\iint_{R^2} |x-y|^4 \frac{1}{2\pi} \frac{1}{\sqrt{st-s^2}} e^{-\frac{1}{2}\frac{1}{st-s^2}[(t-s)x^2+s(x-y)^2]}\, dx\, dy = q(t-s)^2.$$

Thus (2.21) is satisfied with $\beta = 4$ and $\alpha = 1$. The resulting unique *pm* μ on $\mathscr{C}$ guaranteed by the theorem in this case is called the Wiener measure.

Theorem 2.7.10. *A sequence (μ_n) of pms on $\mathscr{C}$ has a weak limit iff the following three conditions are satisfied:*
(i) (μ_n) is a tight sequence.
(ii) For each $r \geq 1$, sequence $(\mu_n \pi^{-1}_{t_1,\, t_2,\, \ldots,\, t_r})$ has a weak limit.
(iii) There exists a pm, say, $\hat{\mu}$ (necessarily unique by **(2.7.8)***) having the weak limits in (ii) for its fdd.*

Proof. Let $\mu_n \xrightarrow{d} \mu_0$. Since every single *pm* on $\mathscr{C}$ is tight (ref. **(2.7.4)**), the hypothesis implies (ref. **(2.4.5)**) (μ_n) is a tight sequence. That (ii) is implied by the hypothesis follows from **(2.4.2)**(vii). (iii) is trivially true.

Let now the sufficient conditions (i), (ii) and (iii) hold. (ii) and (iii) imply that for every compact set K we have $\overline{\lim\limits_{n\to\infty}}\ \mu_n(K) \leq \hat{\mu}(K)$. Now (i) implies that, given $\varepsilon > 0$, a compact set K can be found such that $\mu_n(K) > 1 - \varepsilon$ for all n. Let C be an arbitrary closed set. Then

$$\mu_n(C) = \mu_n(C\cap K)+\mu_n(C\cap K') \leq \mu_n(C\cap K)+\mu_n(K') \leq \mu_n(C\cap K)+\varepsilon.$$

Observe that $C \cap K$ is compact and let $n \to \infty$ to get

$$\overline{\lim_{n\to\infty}}\ \mu_n(C) \leq \hat{\mu}(C \cap K) + \varepsilon \leq \hat{\mu}(C) + \varepsilon.$$

Since $\varepsilon > 0$ is aritrary, we have

$$\overline{\lim_{n\to\infty}}\ \mu_n(C) \leq \hat{\mu}(C)$$

and this implies (ref. **(2.4.2)**(v)) $\mu_n \overset{d}{\to} \hat{\mu}$. □

Remark 2.7.11. Conditions (ii) and (iii) alone are not sufficient to ensure $\mu_n \overset{d}{\to} \hat{\mu}$, as is clear from the following example. Let $x_n(t) = \frac{nt}{1+n^2t^2}$. Let $x_0 \equiv 0$. Let $\mu_n(\{x_n\}) = 1,\ n \geq 0$. Since $x_n(t) \to 0$ for all t, $\big(x_n(t_1),\ x_n(t_2),\ \ldots,\ x_n(t_k)\big) \to (0,\ 0,\ \ldots,\ 0)$ for every $k \geq 1$ and every choice of $t_1,\ t_2,\ \ldots,\ t_k$ in [0, 1]. Since $\mu_n\pi^{-1}_{t_1,t_2,\ldots,t_k}(0,\ 0,\ \ldots,\ 0) = 1,\ n \geq 0$. Hence (ref. **(2.4.3)**(ii)) $\mu_n\pi^{-1}_{t_1,t_2,\ldots,t_k} \overset{d}{\to} \hat{\mu}\pi^{-1}_{t_1,t_2,\ldots,t_k}$. But $\mu_n \overset{d}{\to} \hat{\mu}$ fails to hold. For $\mu_n \overset{d}{\to} \hat{\mu}$ would imply $\rho(x_n,\ x_0) \to 0$ (again ref. **(2.4.3)**(ii)). But this is not possible since $\rho(x_n,\ x_0) = \frac{1}{2}$.

Refer **(2.2.2)**. The promised extension to the particular complete separable metric space $(C,\ \mathscr{C})$ is presented now.

Theorem 2.7.12 (Prohorov). *Every tight sequence (μ_n) of pms on $\mathscr{C}$ admits of a (a necessarily tight) weakly convergent subsequence.*

Proof. Recall $T = \{t_1,\ t_2,\ t_3,\ \ldots\}$ is a dense subset of $[0,\ 1]$.

Let $\varepsilon > 0$ be given. Since the sequence (μ_n) is tight, a compact set $K \subset C$ can be found such that $\mu_n(K) > 1-\varepsilon$ for all $n \geq 1$. Define $C_k = \pi_{t_1,t_2,\ldots,t_k}K$ and note that it is a compact subset of R^k. Trivially $\pi^{-1}_{t_1,t_2,\ldots,t_k}C_k \supset K$. Hence

$$\mu_n\pi^{-1}_{t_1,t_2,\ldots,t_k}C_k \geq \mu_n(K) > 1 - \varepsilon.$$

This shows that $(\mu_n\pi^{-1}_{t_1,t_2,\ldots,t_k},\ n \geq 1)$ is a tight sequence of *pms* in R^k and admits, by **(2.2.2)**, of a weakly convergent subsequence.
Let $(\mu_{1,n}\pi^{-1}_{t_1})$ be then a weakly convergent subsequence of $(\mu_n\pi^{-1}_{t_1})$. The

tightness of the sequence $(\mu_{1,n})$ implies the tightness of the sequence $(\mu_{1,n}\pi^{-1}_{t_1,t_2})$, which would then admit of a weakly convergent subsequence, say, $(\mu_{2,n}\pi^{-1}_{t_1,t_2})$. Now we start with the sequence $(\mu_{2,n})$ and arguing similarly arrive at $(\mu_{3,n})$ which is such that $(\mu_{3,n}\pi^{-1}_{t_1})$, $(\mu_{3,n}\pi^{-1}_{t_1,t_2})$ and $(\mu_{3,n}\pi^{-1}_{t_1,t_2,t_3})$ are weakly convergent sequences in R^1, R^2, R^3 respectively. In this way we determine a family of sequences $(\mu_{j,n},\ n = 1, 2, \dots)$, $j = 1, 2, \dots$, where $(\mu_{j+1,n})$ is a subsequence of $(\mu_{j,n})$. The diagonal sequence $\nu_n (= \mu_{n,n})$ will have the property that $(\nu_n \pi^{-1}_{t_1,t_2,\dots,t_k})$ is a weakly convergent sequence for every $k \geq 1$. Further (ν_n), being a subsequence of the sequence (μ_n), is a tight sequence.

Let $\nu_n \pi^{-1}_{t_1,t_2,\dots,t_k} \overset{d}{\to} \alpha_k$. Let $\mathcal{C}_k = \nu_n \pi^{-1}_{t_1,t_2,\dots,t_k}(\mathscr{R}^k)$. We note that $\mathcal{C}_k$ is a sub σ-field of $\mathscr{R}$ and that $\mathcal{C}_k \subset \mathcal{C}_{k+1}$. Let $\tilde{\mathcal{C}} = \cup_k \mathcal{C}_k$. Since T is a dense subset, it follows $\sigma(\tilde{\mathcal{C}})$, the σ-field generated by $\tilde{\mathcal{C}}$ is $\mathscr{C}$. Now, α_k can be thought of as a pm defined on $\mathcal{C}_k$. Let λ_{k+1} denote the projection operator mapping R^{k+1} onto R^k. Then $\lambda_{k+1}\pi_{t_1,\ t_2,\ \dots,\ t_k,\ t_{k+1}} = \pi_{t_1,\ t_2,\ \dots,\ t_k}$. Hence we have : α_k, the weak limit of $\mu_n \pi^{-1}_{t_1,\ t_2,\ \dots,\ t_k}$, is the same as the weak limit of $\mu_n \pi^{-1}_{t_1,\ t_2,\ \dots,\ t_k,\ t_{k+1}} \lambda^{-1}_{k+1}$.
i.e., of $(\mu_n \pi^{-1}_{t_1,\ t_2,\ \dots,\ t_k,\ t_{k+1}})\lambda^{-1}_{k+1}$, which (ref. **(2.4.2)**) is $\alpha_{k+1}\lambda^{-1}_{k+1}$. This shows that the family (α_k) defined on the $\mathcal{C}_k$s is a consistent family of measures. The *Note* under **(1.11.1)** applies and we conclude that there exists a μ on $\mathscr{C}$ such that $\mu\pi^{-1}_{t_1,\ t_2,\ \dots,\ t_k} = \alpha_k$ for every choice of positive integer k and every choice of $t_1,\ t_2,\ \dots,\ t_k$ in T.

That $\nu_n \overset{d}{\to} \mu$ follows now by **(2.7.10)**. □

2.8 Uniform convergence of weakly convergent measures

Let $(M,\ d)$ be a separable metric space and let $\mathfrak{m}$ be its Borel σ-field. μ_n, μ are pms defined on $\mathfrak{m}$.

Theorem 2.8.1. $\mu_n \overset{d}{\to} \mu_0$ *implies*

$$\lim_{n\to\infty} \sup_{f\in\mathfrak{A}} \left| \int_M f\, d\mu_n(x) - \int_M f\, d\mu_0(x) \right| = 0 \tag{2.22}$$

for every uniformly bounded equicontinuous family $\mathfrak{A}$ *of real functions.*
[By the uniform boundedness of a family $\mathfrak{A}$ of functions, we understand that there exists a constant h such that $\sup_{x\in M,\ f\in\mathfrak{A}} |f(x)| \leq h$]

Proof. Given $\mathfrak{A}$ and given $\varepsilon > 0$, let (A_j) be the sequence of disjoint μ_0-continuous sets constructed in **(2.4.7)**(iii). For each j, choose $x_j \in A_j$ arbitrarily. For each $n \geq 0$, construct discrete *pms* μ_n^* thus : $\mu_n^*(\{x_j\}) = \mu_n(A_j)$. We have

$$\begin{aligned}
&\Big|\int_M f(x)\,\mathrm{d}\mu_n(x) - \int_M f(x)\,\mathrm{d}\mu_n^*(x)\Big| \\
&\leq \sum_{j=1}^{\infty}\Big|\int_{A_j} f(x)\,\mathrm{d}\mu_n(x) - f(x_j)\mu_n(A_j)\Big| \\
&\leq \sum_{j=1}^{\infty}\int_{A_j} |f(x) - f(x_j)|\,\mathrm{d}\mu_n(x) \leq \sum_{j=1}^{\infty} \varepsilon\mu_n(A_j) \leq \varepsilon.
\end{aligned}$$

Again,

$$\begin{aligned}
&\Big|\int_M f(x)\,\mathrm{d}\mu_n^*(x) - \int_M f(x)\,\mathrm{d}\mu_0(x)\Big| \\
&\leq \sum_{j=1}^{\infty}\Big|\int_{A_j} f(x)\,\mathrm{d}\mu_n^*(x) - \int_{A_j} f(x)\,\mathrm{d}\mu_0^*(x)\Big| \\
&\leq \sum_{j=1}^{\infty} |f(x_j)||\mu_n(A_j) - \mu_0(A_j)| \leq h\sum_{j=1}^{\infty} |\mu_n(A_j) - \mu_0(A_j)|.
\end{aligned}$$

The proof will be completed if we show that the sum of the infinite series on the right side tends to 0 as $n \to \infty$. We now proceed to do this.

Endow $\mathbf{A} = \{x_1, x_2, \ldots\}$ with the σ-field $\mathcal{A}$ consisting of all the subsets of $\mathbf{A}$. Define *pms* ν_n on $\mathcal{A}$ thus : $\nu_n(\{x_j\}) = \mu_n(A_j), n = 0, 1, 2, \ldots; j = 1, 2, \ldots$. and note that $\nu_n(\{x_j\}) \to \nu_0(\{x_j\})$. The Radon-Nikodym derivative g_n of ν_n with respect to ν_0 is : $g_n(x_j) = \frac{\mu_n(A_j)}{\mu_0(A_j)}$, $j = 1, 2, \ldots$. We note $g_0 \equiv 1$ and

$$\begin{aligned}
\int_{\mathbf{A}} |g_n(x) - g_0(x)| d\nu_0(x) &= \sum_{j=1}^{\infty} \int_{\{x_j\}} |g_n(x) - g_0(x)| d\nu_0(x) \\
&= \sum_{j=1}^{\infty} |\mu_n(A_j) - \mu_0(A_j)|.
\end{aligned}$$

It remains to show that

$$\lim_{n\to\infty} \int_{\mathbf{A}} |g_n(x) - g_0(x)| d\nu_0(x) = 0.$$

We note $u_n(x) = g_n(x) - g_0(x) \to 0$ as $n \to \infty$ for all $x \in \mathbf{A}$. Now, $u_n(x) \geq -g_0(x)$. Hence $u_n^-(x) = \max(-u_n(x),\ 0) \leq g_0(x) = 1$. Thus we have : $u_n^-(x) \to 0$ and since $u_n^-(x) \leq 1$ for all $x \in \mathbf{M}$, it follows by the bounded convergence theorem that $\int_{\mathbf{A}} u_n^-(x)\, \mathrm{d}\nu_0(x) \to 0$. Since $\int_{\mathbf{A}} g_n(x)\, \mathrm{d}\nu_0(x) = 1,\ n \geq 0$, it follows that $\int_{\mathbf{A}} u_n(x)\, \mathrm{d}\nu_0(x) = 0$. This, together with the earlier result on u_n^-, implies $\int_{\mathbf{A}} u_n^+(x)\, \mathrm{d}\nu_0(x) \to 0$ and then we will have $\int_{\mathbf{A}} (u_n^+(x) + u_n^-(x))\, \mathrm{d}\nu_0(x) \to 0$. i.e. $\int_{\mathbf{A}} |g_n(x) - g_0(x)|\, \mathrm{d}\nu_0(x) \to 0$, as was to be proved. □

Let the sequence of continuous functions $(f_n),\ n \geq 0$ mapping M into a metric space $(\tilde{M},\ \tilde{d})$ converge to f_0 uniformly on compacta. i.e., $\lim_{n\to\infty} \sup_{x\in K} \tilde{d}(f_n(x),\ f_0(x)) = 0$ for each compact subset K of M and let g be a real bounded uniformly continuous function defined on $\tilde{M}$. Clearly, $g(f_n)$ converges to $g(f_0)$ uniformly over compacta. Then the family $\mathfrak{A} = \{g(f_n),\ n \geq 0\}$ is easily seen to be uniformly bounded and equicontinuous.

Theorem 2.8.2. *($\mu_n \overset{d}{\to} \mu_0$) implies ($\mu_n f_n^{-1} \overset{d}{\to} \mu_0 f_0^{-1}$).*

Proof. Let g be an arbitrary bounded uniformly continuous mapping of $\tilde{M}$ into the real line. What is required is to show

$$\lim_{n\to\infty} \int_{\tilde{M}} g(y)\, \mathrm{d}\mu_n f_n^{-1}(y) = \int_{\tilde{M}} g(y) d\mu_0 f_0^{-1}(y)$$

(ref. **(2.4.2)**(ii)). Equivalently (ref. **(2.3.6)**) we must show

$$\lim_{n\to\infty} \int_{M} g(f_n(x))\, \mathrm{d}\mu_n(x) = \int_{M} g(f_0(x))\, \mathrm{d}\mu_0(x).$$

The hypothesis implies (ref. (2.22))

$$\lim_{n\to\infty} \sup_{u\in\mathfrak{A}} \Big| \int_{M} u(x)\, \mathrm{d}\mu_n(x) - \int_{M} u(x)\, \mathrm{d}\mu_0(x) \Big| = 0.$$

Choosing the function $g(f_n)$ in place of u, we get

$$\lim_{n\to\infty} \Big| \int_M g(f_n(x))\, \mathrm{d}\mu_n(x) - \int_M g(f_n(x))\, \mathrm{d}\mu_0(x) \Big| = 0.$$

Since, by the bounded convergence theorem,

$$\lim_{n\to\infty} \int_M g(f_n(x))\, \mathrm{d}\mu_0(x) = \int_M g(f_0(x))\, \mathrm{d}\mu_0(x),$$

the proof is complete. □

Let $\mathfrak{A}$ be the collection of all the continuous mappings of M into R^p. Define a sequence (f_n) in $\mathfrak{A}$, $n \geq 0$ to be convergent to f_0 if $f_n(x) \to f_0(x)$ uniformly over compact subsets of M. This convergence scheme determines a topology on $\mathfrak{A}$ (ref. chapter XI, [V]). Let $K \subset \mathfrak{A}$ be an arbitrary compact subset. Let $\mathcal{E}$ be the collection of all the R^p-rectangles $\times_1^p(-\infty, a_j]$ for all possible choices of the a_js. Define $\mathbf{S}_K = \cup\{f^{-1}(\mathcal{E}) : f \in K\}$. For every $f \in K$, let the *df* $\mathbb{F}_0^{(f)}$ in R^p corresponding to $\mu_0 f^{-1}$ be continuous everywhere. Then

Theorem 2.8.3. $(\mu_n \overset{d}{\to} \mu_0)$ *implies* $\lim_{n\to\infty} \sup_{A\in S_K} |\mu_n(A) - \mu_0(A)| = 0.$

Proof. We note

$$\sup_{A\in S_K} |\mu_n(A) - \mu_0(A)| = \sup_{f\in K} \sup_{E\in\mathcal{E}} |\mu_n f^{-1}(E) - \mu_0 f^{-1}(E)|.$$

The hypothesis $(\mu_n \overset{d}{\to} \mu_0)$ implies (ref. **(2.4.2)**) $(\mu_n f^{-1} \overset{d}{\to} \mu_0 f^{-1})$ for every $f \in K$. Since the *df* $\mathbb{F}_0^{(f)}$ corresponding to $\mu_0 f^{-1}$ is assumed to be continuous everywhere,

$$\lim_{n\to\infty} \sup_{E\in\mathcal{E}} |\mu_n f^{-1}(E) - \mu_0 f^{-1}(E)| = 0.$$

We claim

$$\lim_{n\to\infty} \sup_{f\in K} \sup_{E\in\mathcal{E}} |\mu_n f^{-1}(E) - \mu_0 f^{-1}(E)| = 0.$$

If the claim is not admitted, then there exists a $\delta > 0$ and a subsequence (n') such that

$$\sup_{f\in K} \sup_{E\in\mathcal{E}} |\mu_{n'} f^{-1}(E) - \mu_0 f^{-1}(E)| > \delta \quad \text{for all } n'.$$

For n' fixed, this implies that there exists $f \in K$, labelled $f_{n'}$ such that, for all n'

$$\sup_{E\in\mathcal{E}} |\mu_{n'} f_{n'}^{-1}(E) - \mu_0 f_{n'}^{-1}(E)| > \delta. \tag{2.23}$$

Since M is a separable metric space, compact subsets of $\mathfrak{A}$ are sequentially compact. Since $(f_{n'})$ is a sequence in the compact set K it will necessarily admit of a convergent subsequence, say, (f_m) converging to, say, $f_0 \in K$. i.e., $f_n(x) \to f_0(x)$ uniformly over compact subsets of M. Inequality (2.23) applied to the sequence (m) leads to

$$\begin{aligned}\delta &< \sup_{E\in\mathcal{E}} |\mu_m f_m^{-1}(E) - \mu_0 f_m^{-1}(E)| \\ &\le \sup_{E\in\mathcal{E}} |\mu_m f_m^{-1}(E) - \mu_0 f_0^{-1}(E)| + \sup_{E\in\mathcal{E}} |\mu_0 f_m^{-1}(E) - \mu_0 f_0^{-1}(E)|\end{aligned}$$

for all m. Now let $m \to \infty$. Since f_m, f_0 are continuous functions, since $f_m(x) \to f_0(x)$, $x \in M$ and since the *df* $\mathbb{F}_0^{(f_0)}$ is continuous everywhere, it follows (ref. **(2.4)**(vi))

$$\lim_{m\to\infty} \sup_{E\in\mathcal{E}} |\mu_0 f_m^{-1}(E) - \mu_0 f_0^{-1}(E)| = 0.$$

The f_ms satisfy the conditions of the theorem **(2.8.2)**. Hence

$$\lim_{m\to\infty} \sup_{E\in\mathcal{E}} |\mu_m f_m^{-1}(E) - \mu_0 f_0^{-1}(E)| = 0.$$

These two limit results lead to the contradiction $\delta < 0$. The desired result follows. □

Let $\mathbf{A} = \{\mathbf{a} : \mathbf{a} \in R^p, \|\mathbf{a}\| = 1\}$ and $\mathbb{A} = \times_1^m \mathbf{A}$. Let $\mathcal{E}$ denote the collection of all the rectangles in R^m of the type $\times_{j=1}^m (-\infty, t_j]$. Treat each $\lambda \in \mathbb{A}$, $\lambda = (\mathbf{a}_1, \mathbf{a}_2, \ldots, \mathbf{a}_m)$ (a vector in $(R^p)^m$), as a map of R^p into R^m : $\lambda(\mathbf{x}) = (\mathbf{a}_1'\mathbf{x}, \mathbf{a}_2'\mathbf{x}, \ldots, \mathbf{a}_m'\mathbf{x})$ and note that λ is a continuous map. Given a linear functional $\mathbf{a}$ on R^p, $\mathbf{a} = (a_1, a_2, \ldots, a_p)'$, define $E_{\mathbf{a},\, t} = \{\mathbf{x} : \mathbf{x} \in R^p,\ \mathbf{a}'\mathbf{x} \le t$, i.e., $a_1x_1 + a_2x_2 + \ldots + a_px_p \le t\}$, $-\infty < t < \infty$. Sets $E_{\mathbf{a},\, t}$ are called half spaces of R^p. Treating $\mathbf{a}$ as a functional on R^p, the above set can be written $\mathbf{a}^{-1}((-\infty, t])$. Clearly the half spaces induced by $\mathbf{a}$ are the same as the half spaces induced by $\tilde{\mathbf{a}} = \frac{\mathbf{a}}{\|\mathbf{a}\|}$. Let $\mathfrak{A}_m$ be the collection of all possible finite intersections (not exceeding m) of half spaces. Let μ_n, $n \ge 0$ be *pms* on $\mathscr{R}^p$. We assume that for every $\mathbf{a}$, the *df* corresponding to $\mu_0\mathbf{a}^{-1}$ is continuous everywhere. Then

Theorem 2.8.4. *($\mu_n \xrightarrow{d} \mu_0$) implies* $\lim_{n\to\infty} \sup_{A\in\mathfrak{A}_m} |\mu_n(A) - \mu_0(A)| = 0.$

Proof. We note

$$\mathfrak{A}_m = \bigcup_{\lambda \in (R^p)^m} \lambda^{-1}(\mathcal{E}).$$

Since $\mu_0 \mathbf{a}^{-1}$ is a continuous distribution, it follows that $\mu_0 \lambda^{-1}$ is a (m-variate) continuous distribution. Hence by **(2.1.9)**

$$\lim_{n \to \infty} \sup_{A \in \mathbb{A}} |\mu_n \lambda^{-1}(A) - \mu_0 \lambda^{-1}(A)| = 0.$$

What needs to be proved is equivalent to proving

$$\lim_{n \to \infty} \sup_{\lambda \in (R^p)^m} \sup_{A \in \mathbb{A}} |\mu_n \lambda^{-1}(A) - \mu_0 \lambda^{-1}(A)| = 0.$$

If this is not true, then there exist $\delta > 0$ and a subsequence (n') such that for all n',

$$\sup_{\lambda \in (R^p)^m} \sup_{A \in \mathbb{A}} |\mu_{n'} \lambda^{-1}(A) - \mu_0 \lambda^{-1}(A)| > \delta.$$

This implies that for each n' there can be found $\lambda_{n'}$ such that,

$$\sup_{A \in \mathbb{A}} |\mu_{n'} \lambda_{n'}{}^{-1}(A) - \mu_0 \lambda_{n'}^{-1}(A)| > \delta.$$

Let $\lambda_{n'} = (\mathbf{a}_{1,n'}, \mathbf{a}_{2,n'}, \ldots, \mathbf{a}_{m,n'})$. Since each of the m sequences $(\mathbf{a}_{\nu,n'})$ is contained in a compact subset of R^p, we can find a further subsequence, say, (k) such that (λ_k) converges pointwise to a limit, say, $\lambda^* = (\mathbf{a}_1, \mathbf{a}_2, \ldots, \mathbf{a}_m)$. Also we will have, for all k,

$$\sup_{A \in \mathbb{A}} |\mu_k \lambda_k^{-1}(A) - \mu_0 \lambda_k^{-1}(A)| > \delta.$$

Since the pointwise convergence of λ_k to λ^* is uniform over compact subsets of R^p, **(2.8.2)** applies. And then arguing as in **(2.8.3)** we arrive at a contradiction and there by complete the proof. □

Let $\mu,\ \mu_n,\ n = 1,\ 2,\ ...$ be *pms* on the Borel σ-field $\mathcal{m}$ of a metric space $(M,\ \rho)$. We recall $A \in \mathcal{m}$ is said to be a continuity set for μ if $\mu(BdA) = 0$. Denote by $\mathfrak{A}_\mu$ the collection of all μ-continuous sets. We note that if $A \in \mathfrak{A}_\mu$ and if $\mu_n \overset{d}{\to} \mu$, then

$$\lim_{n \to \infty} |\mu_n(A) - \mu(A)| \to 0 \tag{2.24}$$

This convergence may not be uniform in A over $\mathfrak{A}_\mu$.

Definition 2.8.5. *We say $\mathfrak{B} \subset \mathfrak{A}_\mu$ is a uniformity class for μ or a μ-uniformity class if*

$$\lim_{n\to\infty} \sup_{A\in\mathfrak{B}} |\mu_n(A) - \mu(A)| = 0 \tag{2.25}$$

whenever

$$\mu_n \xrightarrow{d} \mu \tag{2.26}$$

In order to develop a necessary and sufficient condition for a given $\mathfrak{B} \subset \mathfrak{A}_\mu$ to be a μ-uniformity class, we first note that the condition $\sup_{A\in\mathfrak{B}} \mu(BdA) = 0$ is no different from from the condition which is already possessed when (2.26) holds, namely, $\mu(BdA) = 0$ for every $A \in \mathfrak{B}$. So we proceed as follows.

For $\delta > 0$, define $Bd_\delta A = \{x : x \in M, \rho(x, BdA) < \delta\}$ and note that, since $Bd_\delta A \downarrow BdA$,

$$\mu(BdA) = 0 \textit{ iff } \lim_{\delta\downarrow 0} \mu(Bd_\delta A) = 0.$$

This gives us the idea to propose and prove the following

Theorem 2.8.6. *$\mathfrak{B}$ is a μ-uniformity class iff*

$$\lim_{\delta\downarrow 0} \sup_{A\in\mathfrak{B}} \mu(Bd_\delta A) = 0. \tag{2.27}$$

We will prove only the sufficiency part of the theorem and do it after establishing some needed auxiliary results.

The proof of the necessity of the condition (2.27) is avoided since it needs a vast build up in the form of theorems in the topology of separable metric spaces. For details, see Billingsley, P and Topsøe, F : *Uniformity in Weak Convergence*, Zeit. Wahr. verw. Geb 7, 1 - 16 (1967)

Lemma 2.8.7. *Let $0 \le X_n$, $\mathbb{E}X_n < \infty$, $n \ge 1$. Then $X_n \xrightarrow{L_1} X$ iff $X_n \xrightarrow{pr} X$ and $\mathbb{E}X_n \to \mathbb{E}X$.*

Proof. Necessity of the condition follows from the inequalities

$\mathbb{E}|X| \le \mathbb{E}|X - X_n| + \mathbb{E}|X_n|$.

and $\mathbb{E}|X_n| \le \mathbb{E}|X_n - X| + \mathbb{E}|X|$.

leading to $|\mathbb{E}X_n - \mathbb{E}X| \le \mathbb{E}|X_n - X| \to 0$.

For this result, the condition $X_n \ge 0$ is not necessary,

Let us prove the converse part now. Now the part of the hypothesis $X_n \xrightarrow{pr} X$ implies (hint : consider subsequence $X_{n'} \xrightarrow{wp1} X$) $X \ge 0$ $wp1$. Appeal to Fatou's Lemma and conclude $\mathbb{E}X < \infty$. Further this hypothesis implies $X - X_n \overset{pr}{\to} 0$. Hence $(X - X_n)^+ \overset{pr}{\to} 0$. If $X - X_n > 0$, then $(X - X_n)^+ = X - X_n \leq X$. If $X - X_n < 0$, then $(X - X_n)^+ = 0 \leq X$. Hence by the bounded convergence theorem, $\int\limits_{\Omega} (X - X_n)^+ \, dP \to 0$. The second part of the hypothesis is $\mathbb{E}(X - X_n) \to 0$. Since $X - X_n = (X - X_n)^+ - (X - X_n)^-$, it follows $\mathbb{E}(X - X_n)^- \to 0$. From these two results, we get $\mathbb{E}|X - X_n| = \mathbb{E}(X - X_n)^+ + \mathbb{E}(X - X_n)^- \to 0$. With this the proof of the converse is complete. □

Lemma 2.8.8. *(Scheffe's Theorem).*

Let $(\Omega, \mathscr{A}, \mu)$ be an arbitrary measure space. Let ν, ν_n be finite measures on $(\Omega, \mathscr{A})$ such that $\nu_n(\Omega) = \nu(\Omega)$ and such that they are absolutely continuous with respect to μ with corresponding nonnegative densities $g, g_n, n \geq 1$. If $g_n \to g$ a.s.$[\mu]$, then

$$\sup_{A \in \mathscr{A}} |\nu_n(A) - \nu(A)| \to 0.$$

Proof. We note $\int\limits_{\Omega} g_n \, d\mu = \int\limits_{\Omega} g \, d\mu$. Also we are given $g_n \to g$ a.s.$[\mu]$. It follows that the sufficient conditions of **(2.8.6)** are satisfied. Consequently, $\int\limits_{\Omega} |g_n - g| \, d\mu \to 0$.

Now,
$$\begin{aligned}\sup_{A \in \mathscr{A}} |\nu_n(A) - \nu(A)| &= \sup_{A \in \mathscr{A}} \Big|\int\limits_{A} g_n \, d\mu - \int\limits_{A} g \, d\mu\Big| \\ &= \sup_{A \in \mathscr{A}} \Big|\int\limits_{A} (g_n - g) \, d\mu\Big| \\ &\leq \int\limits_{\Omega} |g_n - g| \, d\mu \to 0.\end{aligned}$$
□

Let (M, d) be a separable metric space and μ a *pm* on m. In **(2.4.7)** we established the existence of a sequence (B_n) of disjoint μ-continuous sets such that $\cup_n B_n = M$. The sequence depends on a preassigned small number δ. Let $\mathcal{B}_\delta$ be the σ-field generated by the B_ns. We note that $\mathcal{B}_\delta$ consists of unions, finite or countably infinite, of the B_ns.

Theorem 2.8.9. *$\mathcal{B}_\delta$ is a μ- uniformity class. i.e.,* (2.25) *holds with $\mathfrak{B}$ replaced by $\mathcal{B}_\delta$, whenever* (2.26) *is true.*

Proof. . Let *pms* μ_n on $\mathcal{B}_\delta$ satisfying (3) be given. Define f_n on M : for $x \in B_i$, $i = 1, 2, \ldots$, $f_n(x) = \frac{\mu_n(B_i)}{\mu(B_i)}$ or 0 according as $\mu(B_i) >$ or 0. It is immediate that $f_n(x) \xrightarrow{\text{a.s. } [\mu]} f(x) (\equiv 1)$. Also $\int\limits_{M} f_n(x) \, d\mu(x) = \sum\limits_{i} \int\limits_{B_i} \frac{\mu_n(B_i)}{\mu(B_i)} \, d\mu(x) = \sum\limits_{i} \mu_n(B_i) = 1$. It now follows by Scheffe's theorem that

$$\lim_{n\to\infty} \sup_{A\in\mathcal{B}_\delta} \left| \int_A f_n(x)\,\mathrm{d}\mu(x) - \int_A f(x)\,\mathrm{d}\mu(x) \right| = 0.$$

i.e.,

$$\lim_{n\to\infty} \sup_{A\in\mathcal{B}_\delta} |\mu_n(A) - \mu(A)| = 0 \tag{2.28}$$

The proof is now complete that $\mathcal{B}_\delta$ is a μ-uniformity class. □

Proof of the sufficiency part of Theorem **(2.8.6)** .

Given $A \in \mathfrak{B}$, let $V = \cup\{B_i \,:\, A \cap B_i \neq \phi\}$ and note that $A \subset V$ and that $V \in \mathcal{B}_\delta$.

Suppose $x \in V$. Hence $x \in A$ or $x \in V \sim A$. If $x \in A$, then $x \in A^{(\delta)}$ $(= \{x \,:\, \rho(x,\, A) < \delta\})$. Suppose $x \in V \sim A$. There exists then B_i such that $x \in (B_i \sim A)$. This implies, since the diameter of B_i is $< 2\delta$, that $x \in A^{2\delta}$. Thus $A \subset V \subset A^{(2\delta)}$.

Next, let $x \in V \sim A$. This is equivalent to $x \in (B_i \sim A)$. If A is not a closed set and $x \in BdA$, then $x \in (BdA)^{(\delta)}$. If $x \in (B_i \sim \bar{A})$, then, since the diameter of B_i is $< 2\delta$ and since $\rho(x,\, \bar{A}) = \rho(x,\, BdA)$, it follows that $x \in (BdA)^{(2\delta)}$.

Thus $A \subset V \subset A^{(2\delta)}$ and $(V \sim A) \subset (BdA)^{(2\delta)}\big(= Bd_{2\delta}A\big)$.

On the same lines, we can find $W' \in \mathcal{B}_\delta$ such that $A' \subset W'$ and $W' \sim A' \subset (BdA')^{(2\delta)}$. Taking complements, $W \in \mathcal{B}_\delta$, $W \subset A$ and $(A \sim W) \subset (BdA)^{2\delta}\big(= Bd_{2\delta}A\big)$.

Let $\mu_n \overset{d}{\to} \mu$ and let (2.27) hold. Given $\varepsilon > 0$, find $\delta > 0$ such that

$$\sup_{A\in\mathfrak{B}} \mu(Bd_{2\delta}A) < \varepsilon \tag{2.29}$$

Let the μ-uniformity class $\mathcal{B}_\delta$ be as arrived at in **(2.8.9)**. There exists then $N = N(\varepsilon)$ such that, for all $n \geq N$,

$$\sup_{F\in\mathcal{B}_\delta} |\mu_n(F) - \mu(F)| < \varepsilon. \tag{2.30}$$

As earlier explained, given $A \in \mathfrak{B}$, we can find $E \in \mathcal{B}_\delta$ such that $A \subset E$ and $E \sim A \subset Bd_\delta A$. We have, $|\mu_n(A) - \mu(A)| = |\mu_n(E) - \mu_n(E \sim A) - \mu(E) + \mu(E \sim A)| \leq |\mu_n(E) - \mu(E)| + \mu_n(E \sim A) + \mu(E \sim A) < \sup_{F\in\mathcal{B}_\delta} |\mu_n(F) - \mu(F)| + \mu_n(B_\delta A) + \mu(B_\delta A) \leq \varepsilon + \mu_n(Bd_\delta A) + \varepsilon$ for $n \geq N$.

Since $\mu_n \overset{d}{\to} \mu$, we have $\overline{\lim_{n\to\infty}}\, \mu_n(\overline{B_\delta A}) \leq \mu(\overline{B_\delta A}) \leq \mu(Bd_{2\delta}A) < \varepsilon$. Thus for all n sufficiently large (largeness independent of A), $|\mu_n(A) - \mu(A)| < 3\varepsilon$. Since $\varepsilon > 0$, the (2.27) follows. □

Chapter 3

Characteristic Functions

Characteristic functions were defined in **(2.3.1)** and a limited study of them was made there, confined to finding equivalent conditions for the weak convergence of the corresponding *df*s. Here we make a systematic study of their properties.

3.1 Remarks on the uniqueness theorem and the continuity theorem

In **(2.3.4)** we proved that if the cfs $f,\ g$ of two *df*s $F,\ G$ are identical, i.e., if $f(t) = g(t)$ for *all* $t \in R$, then $F \equiv G$. We ask now if this result holds when $f(t) = g(t)$ only for t in a neighborhood $[-a,\ a]$ of the origin. It is natural to raise this question since many properties of the *df* F of a rv X are determined by the behavior of φ, its cf, near the origin. Examples : (i) If the moments of X exist (moments of any positive integer order, absolute moments of any positive order), they can be obtained if φ is known in a neighborhood of the origin (ref. **(3.6.1)** and **(3.6.9)**). To determine if a family (F_α) of *df*s is tight it is enough if we know the behavior of the corresponding family (φ_α) of *cf*s in a constant interval around the origin (ref. **(2.3.2)**(iii)). The question can be rephrased as follows. Is a *cf* completely determined by its values in a bounded interval around the origin? That the answer is NO and that it is possible for two distinct cfs to coincide on an interval around the origin is brought out by the following example.

Examples 3.1.1.
(i) Let $X,\ Y$ be rvs with cfs ψ and φ where

$$\psi(t) = \frac{1}{2} + \sum_{k=0}^{\infty} \frac{4\cos(2k+1)\pi t}{(2k+1)^2\pi^2}, \qquad -\infty < t < \infty$$

and

$$\varphi(t) = \begin{cases} 1 - \frac{|t|}{\pi} & \text{for } |t| \le \pi \\ 0 & \text{for } |t| \ge \pi. \end{cases}$$

Clearly X is the variable which takes the values $0, \pm(2k+1)\pi, k = 0, 1, 2, \ldots$ with $P(X = 0) = \frac{1}{2}$, $P(X = \pm(2k+1)\pi) = \frac{2}{(2k+1)^2\pi^2}$. It can be verified that φ is the cf corresponding to the df F where $F(x) = \int\limits_{-\infty}^{x} \frac{1-\cos\pi y}{\pi^2 y^2}\, dy$. φ has the Fourier series $\frac{1}{2}a_0 + \sum\limits_{k=1}^{\infty} a_k \cos kt$ where $a_0 = 2\int\limits_0^1 (1-t)dt = 1$; $a_k = \frac{1}{\pi}\int\limits_{-\pi}^{\pi}\{1 - \frac{|t|}{\pi}\}\cos kt dt$. a_k equals 0 if k is even and equals $\frac{4}{k^2\pi^2}$ if k is odd. Thus the Fourier series of φ is

$$\frac{1}{2} + \sum_{0}^{\infty} \frac{4\cos(2k+1)t}{\pi^2(2k+1)^2} \text{ for } |t| \le \pi.$$

Let us now examine if this series converges to φ. Clearly there is convergence at $t = 0, t = \pm\pi$, since $\sum\limits_{k=0}^{\infty} \frac{1}{(2k+1)^2} = \frac{\pi^2}{8}$. In the regions $(-\pi,\ 0)$ and $(0,\ \pi)$, φ is monotonic and continuous. Hence from general results in the theory of Fourier series , there is convergence (in fact, uniform convergence) of the series to φ over these intervals. Thus, if $\tilde{\varphi}(t) = \varphi(\pi t)$, then $\tilde{\varphi}$ is a cf and

$$\tilde{\varphi}(t) = \frac{1}{2} + \sum_{0}^{\infty} \frac{4\cos(2k+1)\pi t}{\pi^2(2k+1)^2}.$$

$\tilde{\varphi}$, ψ are two distinct cfs but coinciding over the interval $[-1,\ 1]$.

(ii) It is worth recording the obvious corollary from **(2.3.2)** (iii)) that a family $\{\mu_n\}$ of pms is tight if the sequence of its cfs of $\{\mu_n\}$ converges to a function continuous on an interval around 0.

(iii) In **(2.3.5)** we showed that if $\mathbb{F}_n \xrightarrow{d} \mathbb{F}_0$, then $f_n(\mathbf{t}) \to f_0(\mathbf{t})$, $\mathbf{t} \in R^p$. Now we show the convergence of f_n to f_0 is uniform over closed spheres in R^p. i.e., given $\lambda > 0$, we show that

$$\lim_{n\to\infty} \sup_{||\mathbf{t}|| \le \lambda} |f_n(\mathbf{t}) - f_0(\mathbf{t})| = 0.$$

Equivalently we show that $||\mathbf{t}_n|| \le \lambda$, $\mathbf{t}_n \to \mathbf{t}_0$ implies $f_n(\mathbf{t}_n) \to f_0(\mathbf{t}_0)$.

Since $\mathbb{F}_n \overset{d}{\to} \mathbb{F}_0$, it follows that $(\mathbb{F}_n)$ is a tight sequence (ref. **(2.7.10)**). Hence given $\varepsilon > 0$, we can find a $\alpha > 0$ such that

$$\int_{\|\mathbf{x}\|\geq\alpha} d\mathbb{F}_n < \varepsilon \text{ for all } n.$$

Choose α such that the set $A_\alpha = \{\mathbf{x} : \|\mathbf{x}\| \leq \alpha\}$ is an $\mathbb{F}_0$-continuity set. i.e. $\mu_0(Bd(A_\alpha)) = 0$. This can always be arranged, since all these boundary sets $\{\mathbf{x} : \|\mathbf{x}\| = \alpha, \alpha > 0\}$ are uncountably many and not more than countably many of them can have positive μ_0- measure. Now,

$$a_n = |f_n(\mathbf{t}_n) - f_0(\mathbf{t}_0)| \leq I_n + \int_{\|\mathbf{x}\|\geq\alpha} d\mathbb{F}_n(\mathbf{x}) + \int_{\|\mathbf{x}\|\geq\alpha} d\mathbb{F}_0(\mathbf{x}) \leq I_n + 2\varepsilon$$

where

$$\begin{aligned}
I_n &= \Big| \int_{\|\mathbf{x}\|\leq\alpha} e^{i\mathbf{t}_n'\mathbf{x}}\, d\mathbb{F}_n(\mathbf{x}) - \int_{\|\mathbf{x}\|\leq\alpha} e^{i\mathbf{t}_0'\mathbf{x}}\, d\mathbb{F}_0(\mathbf{x}) \Big| \\
&\leq \Big| \int_{\|\mathbf{x}\|\leq\alpha} e^{i\mathbf{t}_n'\mathbf{x}}\, d\mathbb{F}_n(\mathbf{x}) - \int_{\|\mathbf{x}\|\leq\alpha} e^{i\mathbf{t}_0'\mathbf{x}}\, d\mathbb{F}_n(\mathbf{x}) \Big| + \Big| \int_{\|\mathbf{x}\|\leq\alpha} e^{i\mathbf{t}_0'\mathbf{x}}\, d\mathbb{F}_n(\mathbf{x}) \\
&\qquad - \int_{\|\mathbf{x}\|\leq\alpha} e^{i\mathbf{t}_0'\mathbf{x}} d\mathbb{F}_0(\mathbf{x}) \Big| \\
&\leq \Big| \int_{\|\mathbf{x}\|\leq\alpha} e^{i\mathbf{t}_n'\mathbf{x}}\, d\mathbb{F}_n(\mathbf{x}) - \int_{\|\mathbf{x}\|\leq\alpha} e^{i\mathbf{t}_0'\mathbf{x}}\, d\mathbb{F}_n(\mathbf{x}) \Big| + \varepsilon
\end{aligned}$$

for all n large, say, $n \geq N_1$ (For this claim, use is made of the special choice of α, of the convergence $\mathbb{F}_n \overset{d}{\to} \mathbb{F}_0$ and **(2.4.3)**(v)

$$\begin{aligned}
&\leq \Big| \int_{\|\mathbf{x}\|\leq\alpha} \{e^{i(\mathbf{t}_n-\mathbf{t}_0)'\mathbf{x}} - 1\}\, d\mathbb{F}_n(\mathbf{x}) \Big| + \varepsilon \\
&\leq \int_{\|\mathbf{x}\|\leq\alpha} 2\Big|\sin\Big(\frac{(\mathbf{t}_n - \mathbf{t}_0)'\mathbf{x}}{2}\Big)\Big|\, d\mathbb{F}_n(\mathbf{x}) + \varepsilon.
\end{aligned}$$

Find $N > N_1$ such that for all $n \geq N$, $\|\mathbf{t}_n - \mathbf{t}_0\| < \frac{\varepsilon}{\alpha}$. Hence over $\|\mathbf{x}\| \leq \alpha$ and $n \geq N$, $|\sin(\frac{(\mathbf{t}_n-\mathbf{t}_0)'\mathbf{x}}{2})| \leq \frac{\varepsilon}{2}$. Hence $I_n \leq 2\varepsilon$, leading to $a_n \leq 4\varepsilon$, $n \geq N$. □

(iv) To see that a sequence of *cf*s may converge pointwise to a function which is not continuous, let $\varphi_n(t) = 1 - n|t|$ for $|t| \leq \frac{1}{n}$ and 0 elsewhere (ref. **(3.2)**(v)). It is easy to see that $\lim_{n\to\infty} \varphi_n(0) = 1$ and for $t \neq 0$, $\lim_{n\to\infty} \varphi_n(t) = 0$.

We now redefine and strengthen the uniqueness theorem proved in **(2.3.4)** by establishing a procedure for reconstructing the *df* F from its *cf* φ.

Theorem 3.1.2 (Inversion theorem). *Let $a < b$ be continuity points of F. Then*

$$F(b) - F(a) = \lim_{u\to\infty} I_u$$

where

$$I_u = \frac{1}{2\pi}\int_{-u}^{u} \frac{e^{-ita} - e^{-itb}}{it}\varphi(t)dt.$$

Proof. Writing $b = \alpha + c$ and $a = \alpha - c$,

$$I_u = \frac{1}{\pi}\int_{-u}^{u} e^{-it\alpha}\frac{\sin ct}{t}\varphi(t)\mathrm{d}t = \frac{1}{\pi}\int_{-u}^{u} e^{-it\alpha}\frac{\sin ct}{t}\{\int_{-\infty}^{\infty} e^{itx}\mathrm{d}F(x)\}\mathrm{d}t$$

$$= \int_{-\infty}^{\infty}\{\frac{1}{\pi}\int_{-u}^{u} e^{it(x-\alpha)}\frac{\sin ct}{t}\mathrm{d}t\}\mathrm{d}F(x).$$

Noting that $|e^{it(x-\alpha)}\frac{\sin ct}{t}| \leq c$ and hence noting that it is integrable with respect to the product measure $\mathrm{d}F \times \mathrm{d}L$, where L is the Lebesgue measure, over the region $(-\infty, \infty) \times [-u, u]$, we see that the change of order of integration is justified.

$$I_u = \int_{-\infty}^{\infty}\{\frac{2}{\pi}\int_{0}^{u} \frac{\sin ct}{t}\cos t(x-\alpha)\mathrm{d}t\}\mathrm{d}F(x) = \int_{-\infty}^{\infty} J(u,x)\mathrm{d}F(x)$$

where

$$J(u,x) = \frac{1}{\pi}\int_{0}^{u} \frac{\sin(c+x-\alpha)t + \sin(c-x+\alpha)t}{t}\mathrm{d}t$$

$$= \frac{1}{\pi}\int_{0}^{u(x-a)} \frac{\sin t}{t}\mathrm{d}t + \frac{1}{\pi}\int_{0}^{u(b-x)} \frac{\sin t}{t}\mathrm{d}t.$$

Let us recall now some known facts. The Riemann integral $\int_0^\lambda \frac{\sin t}{t} dt$ is positive for all $\lambda > 0$ and negative for all $\lambda < 0$. In both the cases, the integrand, in absolute value, is bounded by 1. Further, the improper Riemann integral

$$\int_0^\infty \frac{\sin \theta t}{t} dt = \begin{cases} \frac{\pi}{2} & \text{if } \theta > 0; \\ 0 & \text{if } \theta = 0; \\ -\frac{\pi}{2} & \text{if } \theta < 0. \end{cases}$$

Thus $J(u, x)$ is bounded uniformly in u. Therfore the bounded convergence theorem applies and limit $u \to \infty$ can be taken under the integral sign of I_u. Consequently,

$$\lim_{u \to \infty} I_u = \int_{-\infty}^{\infty} \{\lim_{u \to \infty} J(u, x)\} \, dF(x).$$

Now,

$$\lim_{u \to \infty} \int_0^{u(x-a)} \frac{\sin t}{t} dt = \begin{cases} \frac{\pi}{2} & \text{if } x > a; \\ 0 & \text{if } x = a; \\ -\frac{\pi}{2} & \text{if } x < a. \end{cases}$$

Similar results hold for

$$\lim_{u \to \infty} \int_0^{u(b-x)} \frac{\sin t}{t} dt.$$

Collecting the results we have

$$\lim_{u \to \infty} J(u, x) = \begin{cases} 0 & \text{if } x < a \text{ or } x > b \\ 1 & \text{if } a < x < b. \end{cases}$$

Remembering that a, b are continuity points of F, we conclude that $\lim_{u \to \infty} I_u = F(b) - F(a)$. □

Suppose the *cf* φ of a *df* F is given. By reason of the inversion theorem, it should technically be possible to answer all questions on F by inverting the *cf*. But, except in simple cases, it is a daunting exercise, if not an impossible one, to invert a *cf* and obtain F explicitly and establish properties of F, like deciding if F is continuous everywhere. It is therefore desirable to relate as many properties of F as possible to those of φ without obtaining F explicitly.

3.1.3. Connecting properties of φ to those of F.
(i) The saltus at a given point u:

$$F(u) - F(u-) = \lim_{T\to\infty} \frac{1}{2T} \int_{-T}^{T} e^{-itu}\varphi(t)\mathrm{d}t.$$

Proof. We have

$$\lim_{T\to\infty} \frac{1}{2T} \int_{-T}^{T} e^{-itu}\varphi(t)\mathrm{d}t = \lim_{T\to\infty} \frac{1}{2T} \int_{-T}^{T} e^{-itu}\{\int_{-\infty}^{\infty} e^{itx}\mathrm{d}F(x)\}\mathrm{d}t$$

$$= \lim_{T\to\infty} \int_{-\infty}^{\infty} \{\frac{1}{2T} \int_{-T}^{T} e^{it(x-u)}\mathrm{d}t\}\mathrm{d}F(x)$$

(by Fubini's theorem, since the function $e^{-itu}e^{itx}$ is integrable with respect to the product measure $\mathrm{d}F(x)\,\mathrm{d}\mathcal{L}(t)$)

$$= \int_{-\infty}^{\infty} \{\lim_{T\to\infty} \frac{1}{2T} \int_{-T}^{T} e^{it(x-u)}\mathrm{d}t\}\mathrm{d}F(x)$$

$$= \int_{-\infty}^{\infty} \{\lim_{T\to\infty} \rho(T,x)\}\mathrm{d}F(x)$$

(where $\rho(T,x) = 1$ if $x = u$ and $\rho(T,x) = \frac{\sin T(x-u)}{T(x-u)}$ if $x \neq u$)

$$= \int_{-\infty}^{\infty} \rho_u(x)\mathrm{d}F(x)$$

(where $\rho_u(x) = 0$ if $x \neq u$ and $\rho_u(u) = 1$)

$$= F(u) - F(u-).$$ □

Corollary.
F is continuous at u *iff*

$$\lim_{T\to\infty} \frac{1}{2T} \int_{-T}^{T} e^{-itu}\varphi(t)\mathrm{d}t = 0.$$

(ii) Let a_n, $n = 1, 2, 3, \ldots$ be the set of all the atoms of F. Then

$$\lim_{T\to\infty} \frac{1}{2T} \int_{-T}^{T} |\varphi(t)|^2 \mathrm{d}t = \sum_n \{F(a_n) - F(a_n-)\}^2.$$

Proof.

$$\frac{1}{2T} \int_{-T}^{T} |\varphi(t)|^2 \mathrm{d}t = \frac{1}{2T} \int_{-T}^{T} \{ \int_{-\infty}^{\infty} e^{itx} \mathrm{d}F(x) \int_{-\infty}^{\infty} e^{-ity} \mathrm{d}F(y)\} \mathrm{d}t$$

$$= \int_{-\infty}^{\infty} \int_{-\infty}^{\infty} \{\frac{1}{2T} \int_{-T}^{T} e^{it(x-y)} \mathrm{d}t\} \mathrm{d}F(x) \mathrm{d}F(y)$$

$$= \int_{-\infty}^{\infty} \int_{-\infty}^{\infty} \{\frac{\sin T(x-y)}{T(x-y)}\} \mathrm{d}F(x) \mathrm{d}F(y)$$

(the integrand to be taken as 1 for $x = y$)

$$\to \iint_{R^2} g(x, y) \mathrm{d}F(x) \mathrm{d}F(y)$$

as $T \to \infty$ where $g(u, u) = 1$, $-\infty < u < \infty$, and $g(u, v) = 0$ for $u \neq v$, $-\infty < u, v < \infty$. Denote by E the diagonal set in R^2: $E = \{(x, x), -\infty < x < \infty\}$. Denote by μ the *pm* generated by F. We have, with μ denoting the *pm* generated by F,

$$\lim_{T\to\infty} \frac{1}{2T} \int_{-T}^{T} |\varphi(t)|^2 \mathrm{d}t = \iint_E g(x,\ y)\, \mathrm{d}\mu(x)\, \mathrm{d}\mu(y)$$

$$= \iint_{R^2} \chi_E(x,\ y)\, \mathrm{d}\mu(x)\, \mathrm{d}\mu(y)$$

$$= \int_{-\infty}^{\infty} \mu(E_x) \mathrm{d}\mu(x) \quad \text{(by Fubini's Theorem)}$$

Now $E_x = \{x\}$ and $\mu(\{x\}) = 0$ if x is a continuity point of F and

$\mu(\{x\}) = F(x) - F(x-)$ if x is an atom of F. Thus,

$$\lim_{T\to\infty} \frac{1}{2T} \int_{-T}^{T} |\varphi(t)|^2 \mathrm{d}t = \int_{-\infty}^{\infty} \{F(x) - F(x-)\} \mathrm{d}F(x)$$

$$= \sum_n \{F(a_n) - F(a_n-)\}^2. \quad \square$$

Corollary
F is continuous everywhere *iff*

$$\lim_{T\to\infty} \frac{1}{2T} \int_{-T}^{T} |\varphi(t)|^2 \mathrm{d}t = 0.$$

Remark.
A sufficient condition for F to be continuous everywhere is

$$\lim_{|t|\to\infty} |\varphi(t)| = 0.$$

Proof. Note $|\varphi(t)| = |\varphi(-t)|$. Given $\varepsilon > 0$, find $t_0 > 0$ such that $|\varphi(t)| < \varepsilon$ for all t with $|t| > t_0$. For $T > \frac{t_0}{\varepsilon}$, we have

$$\frac{1}{T} \int_{-T}^{T} |\varphi(t)|^2 \mathrm{d}t = \frac{2}{T} \int_{0}^{T} |\varphi(t)|^2 \mathrm{d}t$$

$$= \frac{2}{T} \int_{0}^{t_0} |\varphi(t)|^2 \mathrm{d}t + \frac{2}{T} \int_{t_0}^{T} |\varphi(t)|^2 \mathrm{d}t \le \frac{2t_0}{T} + 2\varepsilon^2 < 4\varepsilon.$$

ε being arbitrary, it follows that

$$\lim_{T\to\infty} \frac{1}{T} \int_{-T}^{T} |\varphi(t)|^2 \mathrm{d}t = 0.$$

By the corollary, the claim follows now. $\square$

(iii) The following example enhances our understanding. Let $\varphi(t) = \frac{1}{1+t^2}$. By direct computation we see that this is the *cf* of the *df* F:

$$F(x) = \int_{-\infty}^{x} \frac{1}{2} e^{-|u|} \mathrm{d}u.$$

By **(2.3.11)**, $g = e^{\varphi - 1}$ is a *cf*. We ask : is the *df* G corresponding to g continuous? We note that $\lim_{|t| \to \infty} g(t) = e^{-1} \neq 0$. From this we are unable to conclude that G is *not* continuous everywhere. Since $g(t) \to e^{-1}$ as $|t| \to \infty$, we can, given $\varepsilon > 0$, find a t_0 such that $(1-\varepsilon)e^{-1} \leq g(t) \leq (1+\varepsilon)e^{-1}$ for all $|t| \geq t_0$. Now taking $T > t_0$,

$$\frac{1}{T}\int_{-T}^{T} (g(t))^2 \, dt \geq \frac{2}{T}\int_{t_0}^{T} (1-\varepsilon)^2 e^{-2} \, dt = \frac{2(T-t_0)(1-\varepsilon)^2}{Te^2}.$$

Hence G is not continuous everywhere. Let us locate a discontinuity point of G. i.e. a point u at which the saltus is positive. We note g is an even function. The saltus at 0 is

$$\alpha = \lim_{T\to\infty} \frac{1}{2T}\int_{-T}^{T} g(t) \, dt = \lim_{T\to\infty} \frac{1}{T}\int_{0}^{t_0} g(t) \, dt + \lim_{T\to\infty} \frac{1}{T}\int_{t_0}^{T} g(t) \, dt$$

$$\geq \lim_{T\to\infty} \frac{1}{T}\int_{t_0}^{T} (1-\varepsilon)e^{-1} \, dt = (1-\varepsilon)e^{-1}.$$

Again arguing similarly $\alpha \leq (1+\varepsilon)e^{-1}$. Thus the saltus at 0 is e^{-1}. □

(iv) $F'(a)$ exists *iff*

$$\lim_{h\to 0} \lim_{T\to\infty} \int_{-T}^{T} \frac{1 - e^{-iuh}}{iuh} e^{-iua} \varphi(u) du$$

exists.

Proof. This is obtained from the inversion formula by taking $b = a + h$ and using the definition of a differential coefficient. □

(v) If $\int_{-\infty}^{\infty} |\varphi(t)| dt < \infty$, then $f(x) = F'(x)$ exists for all x and is bounded and continuous on R. Further

$$f(x) = F'(x) = \frac{1}{2\pi}\int_{-\infty}^{\infty} e^{-iux}\varphi(u) du.$$

If $\int_{-\infty}^{\infty} |t^n \varphi(t)| \, dt < \infty$, $n \geq 1$, then f is continuously differentiable n times and each differential coefficient $f^{(\nu)}$, $1 \leq \nu \leq n$ is a bounded function and has the representation $f^{(\nu)}(x) = \frac{(-i)^\nu}{2\pi} \int_{-\infty}^{\infty} u^\nu \varphi(u) \, du$.

Proof.

$$\lim_{h\to 0}\lim_{T\to\infty}\frac{1}{2\pi}\int_{-T}^{T}\frac{1-e^{-iuh}}{iuh}e^{-iua}\varphi(u)du$$

$$=\lim_{h\to 0}\frac{1}{2\pi}\int_{-\infty}^{\infty}\frac{1-e^{-iuh}}{iuh}e^{-iua}\varphi(u)du$$

$$=\frac{1}{2\pi}\int_{-\infty}^{\infty}e^{-iua}\varphi(u)du,$$

the first equality being valid since the integrand is bounded by the integrable function $|\varphi|$ and the second equality being valid by the bounded convergence theorem. That $f(x) = F^{'}(x)$ exists finitely for all x and is given by the stated form follows from (iv). The boundedness of f follows from the fact that φ is integrable. The continuity everywhere of f follows by the bounded convergence theorem. This completes the proof of the claim when $n = 0$. For $n \geq 1$, proof is by induction.

$$\lim_{h\to 0}\frac{f^{(n)}(x+h)-f^{(n)}(x)}{h}$$

$$=\lim_{h\to 0}\frac{(-i)^n}{2\pi}\int_{-\infty}^{\infty}\frac{e^{-iuh}-1}{h}e^{-iux}u^n\varphi(u)\,du$$

$$=\frac{(-i)^{n+1}}{2\pi}\int_{-\infty}^{\infty}u^{n+1}\varphi(u)\,du,$$

by the bounded convergence theorem, which is applicable since the integrand is bounded by the function $u^n\varphi(u)$, which is integrable by the induction hypothesis. Thus

$$f^{n+1}(x)=\frac{(-1)^{n+1}}{2\pi}\int_{-\infty}^{\infty}e^{-iux}u^n\varphi(u)\,du,$$

as was to be proved. The boundedness and continuity of $f^{(n+1)}$ derive from the integrability of $u^{n+1}\varphi(u)$ and the bounded convergence theorem. □

Remark

We stress that $\int_{-\infty}^{\infty} |\varphi(u)|\mathrm{d}u < \infty$ is only a sufficient condition, and not a necessary condition, for the existence, continuity and boundedness of f. This truth is brought home by the following examples.

Let $f(x) = \frac{1}{2}$ if $-1 \le x \le 1$ and 0 otherwise. $F(x) = \int_{-\infty}^{x} f(y)\mathrm{d}y$. Then $\varphi(t) = \frac{\sin t}{t}$ and $\int_{-\infty}^{\infty} |\varphi(t)|\mathrm{d}t = \infty$.

Let $f(x) = e^{-x}$ if $x \ge 0$ and 0 otherwise. $F(x) = \int_{-\infty}^{x} f(y)\mathrm{d}y$. In this case $\varphi(t) = \frac{1}{1-it}$. $|\varphi(t)| = \frac{1}{\sqrt{1+t^2}}$ and $\int_{-\infty}^{\infty} |\varphi(t)|\mathrm{d}t = \infty$.

(vi) If F is a continuous *df* then

$$I_n(x) = \int_{-\infty}^{x} F^n(y)\,\mathrm{d}F(y) = \frac{F^{n+1}(x)}{n+1}.$$

Proof. Let X be a rv with *df* F. Let rv Y be independent of X and distributed as $\frac{1}{k}\xi$. Then G_k the *df* of $X+Y$ is absolutely continuous. This claim follows from (v) above since the cf of $X+Y$ in absolute value is dominated by the integrable cf of Y. We note $G_k \overset{d}{\to} F$. Since F is a continuous *df* this implies (ref. **(2.1.9)**)

$$a_k = \sup_{x\in R} |G_k(x) - F(x)| \to 0 \text{ as } k \to \infty.$$

Given $\varepsilon > 0$, find K such that $a_k < \varepsilon$ for all $k \ge K$.

Since $G_k \overset{d}{\to} F$ and since F^n is a bounded continuous function, we have

$$\begin{aligned} I_n(x) &= \lim_{k\to\infty} \int_{-\infty}^{x} F^n(y)\,\mathrm{d}G_k(y) \\ &= \lim_{k\to\infty} \{ \int_{-\infty}^{x} (F^n(y) - G_k^n(y))\,\mathrm{d}G_k(y) + \int_{-\infty}^{x} G_k^n(y)\,\mathrm{d}G_k(y)\}. \end{aligned}$$

The first integral under the double bracket in absolute value is $\le na_k = o(1)$ as $k \to \infty$. Hence

$$\begin{aligned} I_n(x) &= \lim_{k\to\infty} \int_{-\infty}^{x} G_k^n(y)\,\mathrm{d}G_k(y) = \lim_{k\to\infty} \int_{-\infty}^{x} G_k^n(y)\frac{\mathrm{d}}{\mathrm{d}y}G_k(y)\,\mathrm{d}y \\ &= \lim_{k\to\infty} \int_{-\infty}^{x} \frac{1}{n+1}\frac{\mathrm{d}}{\mathrm{d}y}G_k^{n+1}(y)\,\mathrm{d}y = \lim_{k\to\infty} \frac{1}{n+1}G_k^{n+1}(x) \\ &= \frac{1}{n+1}F^{n+1}(x). \end{aligned}$$

□

For two other solutions to this problem, refer to **(2.3.7)** *and to 'some applications' under* ***Q8***.

Definition 3.1.4. When F is absolutely continuous, $f(x) = \frac{\mathrm{d}F(x)}{\mathrm{d}x}$ is called the frequency function or the density function of F or of the associated *rv*. If μ is the pm generated by F, then f will be the Radon-Nikodym derivative of μ with respect to the Lebesgue measure.

Definition 3.1.5. A *df* F is said to be bounded or finite if there exist finite numbers $a < b$ such that $F(b) = 1$ and $F(a) = 0$.

If no such b $[a]$ exists, then $1 - F(x) > 0$ $[F(-x) > 0]$ for all $x > 0$. F is a bounded distribution *iff* $lextF$ and $rextF$ are finite.

When *lext*F and *rext*F are finite, the next theorem provides a formula for determining them from the *cf* φ of F.

Theorem 3.1.6. *If F is a bounded df, then the definition of its cf φ can be extended to all complex values of t and the resulting function, $\tilde{\varphi}$, would be an integral function (i.e., a function analytic at all points of the finite part of the z-plane) with $|\tilde{\varphi}(z)| \leq e^{\beta|z|}$ for some $\beta > 0$. Further lext$F = -\overline{\lim_{r\to\infty}} \frac{1}{r} log\varphi(ir)$ and rext$F = \overline{\lim_{r\to\infty}} \frac{1}{r} log\varphi(-ir)$.*

The following limited converse will be proved. *If*

$$\tilde{\varphi}(z) = \int_{-\infty}^{\infty} e^{izx}\, \mathrm{d}F(x)$$

exists for all z as an integral function and if $|\tilde{\varphi}(z)| \leq Ae^{B|z|}$, for some $A > 0$, $B > 0$, then F is a bounded distribution.

Proof. Avoiding a triviality, we assume $lextF = a < b = rextF$ finite numbers. The integral defining the *cf* φ takes the form $\varphi(t) = \int_a^b e^{itx}\mathrm{d}F(x)$. The function $\tilde{\varphi}(z) = \int_a^b e^{izx} dF(x)$ is defined for all complex numbers z and $\tilde{\varphi}(t) = \varphi(t)$ for all real t. $\tilde{\varphi}$ is analytic over the entire complex plane. Also $|\tilde{\varphi}(z)| \leq e^{\beta|z|}$ where $\beta = \max(|a|, |b|)$.

If the cf φ (of a bounded distribution F) is known i.e. if $\varphi(t)$ is known for all real values of t then $\tilde{\varphi}(z)$ can be obtained from φ by replacing t by z in $\varphi(t)$. This is because the coefficient of t^n in the infinite series expansion for φ around the origin is the same as the coefficient of z^n in the infinite series expansion of $\tilde{\varphi}$ around the origin, $n \geq 0$. Thus the first part of the forward claim is proved.

Let $a = lextF$ and $b = rextF$. Let $r > 0$. From $e^{ra} \leq \int_a^b e^{rx} dF(x) \leq e^{rb}$, we get $a \leq \frac{1}{r}\log\varphi(-ir) \leq b$, leading to $\overline{\lim_{r\to\infty}} \frac{1}{r}\log\varphi(-ir) = \xi \leq b$. We claim $\xi = b$. If $\xi < b$, we can find $u \in C_F$ such that $\xi < u < b$. We note that, since $b = rextF$, $F(b) - F(u) = 1 - F(u) > 0$. Now,

$$\varphi(-ir) = \int_a^b e^{rx}\mathrm{d}F(x) \geq \int_u^b e^{rx}\mathrm{d}F(x) \geq e^{ru}\{1 - F(u)\}.$$

Hence

$$\frac{1}{r}\log\varphi(-ir) \geq u + \frac{1}{r}\log\{1 - F(u)\},$$

yielding

$$\xi = \overline{\lim_{r\to\infty}} \frac{1}{r}\varphi(-ir) \geq u,$$

a contradiction. Hence $\overline{\lim_{r\to\infty}} \frac{1}{r}\varphi(-ir) = rextF.$

Similarly, $\overline{\lim_{r\to\infty}} \frac{1}{r}\varphi(ir) = -lextF.$

Now the converse part. Let $r > 0,\ \pm u, \pm v \in C_F$ with $B < u < v$. We have

$$Ae^{Br} \geq \varphi(-ir) = \int_{-\infty}^{\infty} e^{rx} dF(x) \geq \int_u^v e^{rx} dF(x)$$
$$\geq e^{ru}\{F(v) - F(u)\}.$$

Hence $F(v) - F(u) \leq Ae^{r(B-u)}$. Taking $r \to \infty$ we see that $F(v) - F(u) = 0$. This implies that $rext\ F$ is finite.

Again,

$$Ae^{rB} \geq \varphi(ir) = \int_{-\infty}^{\infty} e^{-rx}\mathrm{d}F(x) \geq \int_{-v}^{-u} e^{-rx}\mathrm{d}F(x)$$
$$\geq e^{ru}\{F(-u) - F(-v)\}.$$

As before we take $r \to \infty$ and conclude that $lextF$ is finite.

It follows now that F is a bounded *df.* □

Theorem 3.1.7. *The product $\varphi_0 = \varphi_1\varphi_2$ of two cfs φ_1, φ_2 is the cf of a bounded df F_0 if and only if φ_i corresponds to a bounded df F_i, $i = 1, 2$. When the boundedness condition holds, $a_0 = a_1 + a_2$ and $b_0 = b_1 + b_2$ where $a_j = lextF_j$ and $b_j = rextF_j, j = 0, 1, 2$.*

Proof. Section **(2.3.10)** assures us that the product of two *cf*s is a *cf.* For $i = 1, 2$ let F_i be a bounded *df* with $rextF_i = b_i$ and $lextF_i = a_i$. Let $\varepsilon > 0$ be arbitrary and small. Then

$$\begin{aligned} F_0(a_1 + a_2 - \varepsilon) &= \int_{-\infty}^{\infty} F_1(a_1 + a_2 - \varepsilon - y)\mathrm{d}F_2(y) \\ &= \int_{(-\infty, a_2-\varepsilon]} F_1(a_1 + a_2 - \varepsilon - y)\mathrm{d}F_2(y) \\ &\quad + \int_{(a_2-\varepsilon, \infty)} F_1(a_1 + a_2 - \varepsilon - y)\mathrm{d}F_2(y) = 0. \end{aligned}$$

(The first integral is 0 because $F_2(y) = 0$ for $y \le a_2 - \varepsilon$; the second integral is 0 because for $y > a_2 - \varepsilon$, $F_1(a_1 + a_2 - \varepsilon - y) < F_1(a_1 - \delta) = 0$ for some $\delta > 0$). Hence $a_0 = lextF_0$ is finite and $a_0 \ge a_1 + a_2$. It can be shown on similar lines that $b_0 = rextF_0$ is finite and that $b_0 \le b_1 + b_2$. It follows that F_0 is a bounded *df.*

Suppose now that F_0 is a bounded *df.* If $lextF_0 = a_0$ then $0 = F_0(a_0 - \varepsilon) = \int_{-\infty}^{\infty} F_1(a_0 - \varepsilon - y)\mathrm{d}F_2(y)$. This implies that $F_1(a_0 - \varepsilon - y) = 0$ almost surely (*a.s.*) F_2. Since there exists at least one y with $F_1(a_0 - \varepsilon - y) = 0$, it follows that $lextF_1$ is finite. Similarly, if $rextF_0 = b_0$, then from $0 = 1 - F_0(b_0) = \int_{-\infty}^{\infty} \{1 - F_1(b_0 - y)\}\mathrm{d}F_2(y)$, we conclude that $rextF_1$ is finite. Thus F_1 is a bounded *df.* On similar lines F_2 can be shown to be a bounded *df.*

Let now the boundedness condition hold. By the first part $a_0 \ge a_1 + a_2$. We claim equality. If the claim is not admitted, let $a_0 > a_1 + a_2$. We can then find $\delta > 0$ such that $2\delta < b_2 - a_2$ and $a_1 + a_2 < a_1 + a_2 + 2\delta < a_0$. Since a_1 is $lextF_1$, we have $F_1(a_1 + \delta) > 0$. Since a_0 is $lextF_0$, we have

$$\begin{aligned} 0 = F_0(a_1 + a_2 + 2\delta) &= \int_{-\infty}^{\infty} F_1(a_1 + a_2 + 2\delta - y)\,\mathrm{d}F_2(y) \\ &= \int_{[a_2,\, b_2]} F_1(a_1 + a_2 + 2\delta - y)\,\mathrm{d}F_2(y) \\ &\ge \int_{[a_2,\, a_2+\delta]} F_1(a_1 + a_2 + 2\delta - y)\,\mathrm{d}F_2(y) \\ &\ge F_1(a_1 + \delta) \int_{[a_2,\, a_2+\delta]} \mathrm{d}F_2(y). \end{aligned}$$

Since a_2 is the $lext$ of the *df* F_2, it follows that

$$\int_{[a_2,\ a_2+\delta]} \mathrm{d}F_2(y) > 0.$$

Hence $F_1(a_1 + \delta) = 0$, a contradiction. Hence necessarily $a_0 = a_1 + a_2$.
Similarly $b_0 = b_1 + b_2$. □

Remarks 3.1.8. (i) Recall from **(2.3.7)** (iii) that $\bar{\varphi}$ ($\bar{\varphi}(t) = \varphi(-t)$) is a *cf* if φ is one. Hence (ref. **(2.3.10)**) $\varphi_0 = \varphi_1\bar{\varphi}_1$ ($= |\varphi_1|^2$) is a *cf*. Then $b_0 = b_1 - a_1$ and $a_0 = a_1 - b_1$ where $a_i = lext\varphi_i$, and $b_i = rext\varphi_i$, $i = 0,\ 1$.

(ii) The theorem is easily extended to the product of a finite number *cf*s.

(iii) If a bounded *df* F with *cf* φ has $rextF = b$ and $lextF = a$, then the *df* G with *cf* $g(t) = e^{-it\frac{a}{b-a}}\varphi(\frac{t}{b-a})$ has $rextG = 1$ and $lextG = 0$. And then the *df* H with *cf* $h = g\bar{g}$ has $rextH = 1$ and $lextH = -1$.

3.2 Some important characteristic functions

(i) Let μ be the pm in R with

$$\mu(\{j\}) = \binom{n}{j} p^j q^{n-j},\ j = 0,\ 1, \ldots, n, p, q \geq 0,\ p + q = 1.$$

This is called the binomial distribution with parameters n and p. The *cf* of μ is $(q + pe^{it})^n$.

(ii) Let $\mu(\{j\}) = e^{-\lambda}\frac{\lambda^j}{j!},\ j = 0,\ 1, \ldots, \lambda > 0$. The corresponding *cf* is $e^{\lambda(e^{it}-1)}$. This is called the Poisson distribution with parameter λ.

(iii) Let $\mu(A) = \int_A \mathrm{d}x, A \subseteq [0,\ 1], A$ Borel. This is called the uniform distribution on the interval $[0,\ 1]$. Its *cf* is $\frac{e^{it}-1}{it}$.

(iv) Let $\mu(A) = \int_A (1 - |x|)\,\mathrm{d}x,\ A \subseteq [-1,\ 1]$, A Borel. This is called the triangular distribution over $[-1,\ 1]$. Its *cf* is $\frac{2(1-\cos t)}{t^2}$.

(v) Associated with the triangular distribution is the following: $\mu(A) = \int_A \frac{1}{\pi}\frac{1-\cos x}{x^2}\,\mathrm{d}x$, A Borel, $-\infty < x < \infty$. Its *cf* is $1 - |t|$ for $|t| \leq 1$ and is 0 for $|t| \geq 1$.

(vi) For the gamma distribution

$$\mu(A) = \int_A \frac{1}{\Gamma(\gamma)} x^{\gamma-1} e^{-x}\, dx,\ x > 0,\ \gamma > 0,$$

A Borel. Its *cf* is $\frac{1}{(1-it)^\gamma}$. Here that branch of the complex valued function $\frac{1}{(1-it)^\gamma}$ is understood which is continuous and which is 1 at $t = 0$.

A related distribution is the χ^2 distribution given by

$$\mu(A) = \int_A \frac{1}{2^{\frac{n}{2}}\Gamma(\frac{n}{2})} x^{\frac{n}{2}-1} e^{-\frac{x}{2}}\, dx$$

where n, called the degrees of freedom of the distribution, is a positive integer. Its *cf* is $\frac{1}{(1-2it)^{\frac{n}{2}}}$.

(vii) The bilateral exponential distribution μ is given by

$$\mu(A) = \int_A \frac{1}{2} e^{-|x|}\, dx, \quad -\infty < x < \infty,$$

A Borel. Its *cf* is $\frac{1}{1+t^2}$.

(viii) Associated with the last one is the Cauchy distribution:

$$\text{For } A \text{ Borel, } \mu(A) = \int_A \frac{1}{\pi}\frac{1}{1+x^2}\, dx, \quad -\infty < x < \infty,$$

Its *cf* is $e^{-|t|}$.

(ix) It is hyperbolic cosine distribution when

$$\mu(A) = \int_A \frac{1}{\pi}\frac{1}{\cosh x}\, dx,\ -\infty < x < \infty,$$

A Borel. Its *cf* is $\frac{1}{\cosh(\frac{\pi t}{2})}$.

(x) The distribution with

$$\mu(A) = \int_A \frac{1}{\sqrt{2\pi}} e^{-\frac{x^2}{2}}\, dx,\ -\infty < x < \infty, \quad A \text{ Borel},$$

is called the standard Gaussian or the standard normal distribution. It has mean 0 and variance 1. Its *cf* is $e^{-\frac{t^2}{2}}$. If X is a rv with a standard normal distribution, then X^2 is χ^2 distributed with 1 degree of freedom.

(xi) The p-variate normal distribution with mean vector $\mathbf{0}$ and variance covariance matrix $\Sigma = (\sigma_{i,j})$ is the pm μ in R^p such that for every choice of $\mathbf{a} \in R^p$, $\mu(\mathbf{a}'\mathbf{x})^{-1}$ is a normal distribution in R with mean 0 and variance $\mathbf{a}'\Sigma\mathbf{a}$. We need to show that such a distribution exists and that it exists uniquely. To do this we consider two cases.

Case 1. Σ positive definite.
In this case $|\Sigma| = \det(\Sigma) \neq 0$, and Σ^{-1} exists. For $A \in \mathscr{R}^p$, define

$$\mu(A) = \int_A \frac{1}{(2\pi)^{\frac{p}{2}}} \frac{1}{|\Sigma|} e^{-\frac{1}{2}\mathbf{x}'\Sigma^{-1}\mathbf{x}}\, \mathbf{dx}.$$

It is easily verified that μ so defined is indeed a pm and that its *cf* is $e^{-\frac{1}{2}\mathbf{t}'\Sigma\,\mathbf{t}}$. The *cf* of $\mu(\mathbf{a}'\mathbf{x})^{-1}$ is $\int_R e^{ity}\, \mathrm{d}\mu(\mathbf{a}'\mathbf{x})^{-1}(y) =$ (by the change of variables theorem **(2.3.6)**) $\int_{R^p} e^{it\mathbf{a}'\mathbf{x}}\, \mathrm{d}\mu(\mathbf{x}) = e^{-\frac{t^2}{2}\mathbf{a}'\Sigma\mathbf{a}}$, which is the *cf* of a univariate normal with mean 0 and variance $\mathbf{a}'\Sigma\mathbf{a}$. Thus in this case there exists a μ satisfying the requirements.
Case 2. $|\Sigma| = 0$.
Let the rank of Σ be $m < p$. Given the symmetric matrix Σ, we can find a non-singular matrix B such that $B'\Sigma B = \Lambda$, a diagonal $p \times p$ matrix of rank m. There is no loss of generality in assuming that $\alpha_1,\ \alpha_2,\ \ldots,\ \alpha_m$, the first m diagonal elements of Λ, are positive and that the remaining elements 0. Denote by Λ^* the $m \times m$ diagonal matrix with its diagonal elements $\alpha_1^{-1},\ \alpha_2^{-1}, \ldots, \alpha_m^{-1}$.

Let $\mathcal{A}$ be the subspace of R^p consisting of all the points with their last $p - m$ co-ordinates zeroes. Clearly R^m and $\mathcal{A}$ are in one-to-one correspondence which is a measurable mapping. Let $\mathcal{S} = \{B\mathbf{y} : \mathbf{y} \in \mathcal{A}\}$ i.e. $\mathcal{S} = B\mathcal{A}$. Since matrix B is non-singular there is a measurable one-to-one correspondence between $\mathcal{A}$ and $\mathcal{S}$. In $\mathcal{A}$ define pm μ_1 as follows. Given a Borel set $A \subseteq \mathcal{A}$, denote by $\hat{A}$ the corresponding set in R^m

Define

$$\mu_1(A) = \int_{\hat{A}} \frac{1}{(2\pi)^{\frac{m}{2}}} \frac{1}{\sqrt{\alpha_1\alpha_2\ldots\alpha_m}} e^{-\frac{1}{2}\mathbf{y}'\Lambda^*\mathbf{y}}\, \mathbf{dy}.$$

The *cf* ψ of μ_1 then is:

$$\psi(\mathbf{u}) = e^{-\frac{1}{2}\sum_1^m \alpha_k u_k^2} = e^{-\frac{1}{2}\mathbf{u}'\Lambda\mathbf{u}}.$$

Hence the *cf* of μ is

$$e^{-\frac{1}{2}(B^{-1}\mathbf{t})'\Lambda B^{-1}\mathbf{t}} = e^{-\frac{1}{2}\mathbf{t}'(B')^{-1}\Lambda B^{-1}\mathbf{t}} = e^{-\frac{1}{2}\mathbf{t}'\Sigma\mathbf{t}}.$$

Since the *cf* is determined and since the *cf* determinines the *df* uniquely, it follows that a unique μ satisfying the requirements exists in this case also. To summarise: The *cf* of a multivariate normal distribution as described above with prescribed variance covariance matrix Σ is $e^{-\frac{1}{2}\mathbf{t}'\Sigma\mathbf{t}}$, whether Σ is singular or non-singular.

For future ready reference we record the following obvious result.

Theorem 3.2.1. *If φ is a cf then so is $\Re(\varphi)$, the real part of φ.*

Proof. Let φ be the *cf* of the pm μ on R^p. Define $pm\ \nu : \nu(E) = \mu(-E)$, $E \in \mathscr{R}^p$. We note that for the $pm\ \alpha = \frac{1}{2}\{\mu + \nu\}$, the *cf* is $\frac{1}{2}\{\varphi + \bar{\varphi}\} = \Re(\varphi)$. □

Polya's sufficient condition for characteristic functions

Given a complex valued continuous function φ, defined for all $t \in R$, the question arises: how to decide if φ is a *cf*? We quote a necessary and sufficient condition.

Q10.(i) **Definition** A possibly complex valued function φ defined on the real line is said to be non-negative-definite if for every choice of integers $n \geq 1$ and every choice of real numbers t_j, $1 \leq j \leq n$, and every choice of real or complex numbers a_j, $1 \leq j \leq n$,

$$\sum_{1 \leq j,k \leq n} \varphi(t_j - t_k) a_j \bar{a}_k \geq 0.$$

(ii) **Theorem** (Bochner) A necessary and sufficient condition for φ with $\varphi(0) = 1$ to be a *cf* is that it is nonnegative-definite.

But this condition proves to be difficult to verify. The following sufficient condition may prove easier of verification. It is applicable to real functions φ with $\varphi(0) = 1$ which are convex.

Q11. (i) **Definition**. A convex function ξ defined on R is a real continuous function such that if $a < b$, then

$$\xi\left(\frac{a+b}{2}\right) \leq \frac{\xi(a) + \xi(b)}{2}.$$

(ii) **Theorem**. A convex function ξ has a finite right derivative ξ^+, which is non-decreasing and continuous from the right; also, it has a finite left derivative ξ^-, which is non-decreasing and continuous from the left. Further

$$\xi(y) - \xi(x) = \int_x^y \xi^+(t)\,\mathrm{d}t = \int_x^y \xi^-(t)\,\mathrm{d}t.$$

Theorem 3.2.2. *A continuous even function g is a cf if it satisfies the following conditions:* (α) $g(0) = 1$; (β) $\lim_{t\to\infty} g(t) = 0$; (γ) *g is convex for $t > 0$.*

Proof. Let g possess the properties stated. Let g' denote its right derivative and note that g' is non-decreasing, since g is a convex function. Hence $\lim_{t\to\infty} g'(t)$ exists. $g'(t)$ can not be positive for any t since $g'(t_0) > 0$ would imply g is monotonic increasing over $[t_0,\ \infty)$, violating condition (β). Hence $\lim_{t\to\infty} g'(t) = -a \leq 0$. Hence g monotonically decreases to 0. Further, condition (β) implies that $a = 0$. For, if $a > 0$, then for all $t > 0$,

$$g(t) - g(0) \leq \int_0^t (-a)\,\mathrm{d}y = -at$$

or, $g(t) \leq 1 - at$. Letting $t \to \infty$ in this inequality, we arrive at a contradiction.

Let $f(t) = -g'(t)$ and note that $f(t) \downarrow 0$ as $t \uparrow \infty$. For $x > 0$ and $t > 0$, define $\lambda(x;t) = f(t) - f(t + \frac{\pi}{x}) + f(t + \frac{2\pi}{x}) - f(t + \frac{3\pi}{x}) + \ldots$. This is an alternating series of positive terms and the nth term tends to 0 as $n \to \infty$. Hence $\lambda(x;t)$ is well defined. Let $q_n(x;t)$ denote the partial sum to n-terms of this series. We note that $q_{2n}(x;t) \geq 0$, that $q_{2n}(x;t) \uparrow \lambda(x;t)$ and that $q_{2n}(x;t) \leq f(t)$. Hence

$$\begin{aligned}
0 \leq \int_0^{\frac{\pi}{x}} \lambda(x;t)\sin xt\,\mathrm{d}t &= \int_0^{\frac{\pi}{x}} \lim_{n\to\infty} q_{2n}(x;t)\sin xt\,\mathrm{d}t \\
&= \lim_{n\to\infty} \sum_{k=0}^{2n-1} (-1)^k \int_0^{\frac{\pi}{x}} f(t + \frac{k\pi}{x})\sin xt\,\mathrm{d}t \\
&= \lim_{n\to\infty} \sum_{k=0}^{2n-1} (-1)^k \int_{\frac{k\pi}{x}}^{\frac{(k+1)\pi}{x}} f(u)(-1)^k \sin xu\,\mathrm{d}u
\end{aligned}$$

$$= \lim_{n\to\infty} \int_0^{\frac{2n\pi}{x}} f(u) \sin xu \, du = \int_0^{\infty} f(u) \sin xu \, du$$

$$= - \int_0^{\infty} g'(u) \sin xu \, du$$

$$= -g(u) \sin xu \Big|_0^{\infty} + x \int_0^{\infty} g(t) \cos xt \, dt.$$

Thus we have shown that

$$g_1(x) = \frac{1}{2\pi} \int_{-\infty}^{\infty} e^{-itx} g(t) \, dt = \frac{1}{\pi} \int_0^{\infty} g(t) \cos xt \, dt$$

exists finitely for every $x \neq 0$, that $g_1(x) = g_1(-x)$ and that $g_1(x) \geq 0$. It now follows by Fourier's theorem that $g(t) = \int_{-\infty}^{\infty} e^{itx} g_1(x) \, dx$. □

Some examples of functions g satisfying the conditions of the theorem are:
(i) $\varphi(t) = 1 - |t|$ if $|t| \leq 1$ and $\varphi(t) = 0$ if $|t| \geq 1$.
(ii) $\varphi(t) = \frac{1}{1+|t|}$, $-\infty < t < \infty$. (iii) $\varphi(t) = e^{-|t|}$, $-\infty < t < \infty$.

3.3 Infinite divisibility

Definition 3.3.1. A *rv* X or its *df* F or its *cf* φ in R^p is said to be infinitely divisible (*id*) if φ has the following property. For each integer $n \geq 1$, there exists a *cf* φ_n such that $\varphi = \varphi_n^n$.

For the role of infinitely divisible *df*s in the study of weak limits of sums of infinitesimal *rv*s, refer to section **(5.2)**.

It is easy to verify that if φ is *id*, (*i*) then so is $\overline{\varphi}$; (*ii*) that ψ is *id* where $\psi(\mathbf{t}) = e^{i\mathbf{t}'\mathbf{b}}\varphi(a\mathbf{t})$, $a, \mathbf{b}$ being, respectively, an arbitrary real number and an arbitrary real vector (*iii*) that ψ is *id* where $\psi(\mathbf{t}) = \varphi(\mathbf{t}'\mathbf{b})$, $\mathbf{b}$ an arbitrary real vector and (*iv*) that the product of a finite number of *id cf*s is *id* too. (*iii*) can be stated in another way : if $(X_1, X_2, ..., X_p)$ is an *id* vector, then $a_1X_1 + a_2X_2 + ... + a_pX_p$ is an *id rv* for every $(a_1, a_2, ..., a_p) \in R^p$.

Let the *cf* φ be *id*. Hence for each $k \geq 1$, there exists a *cf* φ_k such that $\varphi(t) = \{\varphi_k(t)\}^k$ for all $t \in R$. This implies a branch of $\varphi^{\frac{1}{k}}$ is a *cf*. By $\varphi^{\frac{1}{k}}$, we always mean this branch.

Theorem 3.3.2. (i) *If φ is a cf, then for each $\lambda > 0$, $g = e^{\lambda(\varphi-1)}$ is an id cf.* (ii) *Let φ_n, $n \geq 0$ be cfs, for $n \geq 1$, let φ_n be id and let $\lim\limits_{n\to\infty} \varphi_n(t) = \varphi_0(t)$ for all t. Then φ_0 is i.d. (This property is studied further in* **(3.4.9)**)

Proof. (i) That g is a *cf* was proved in **(2.3.11)**. Since for each $n \geq 1$, $g = g_n^n$ where $g_n = e^{\frac{\lambda}{n}(\varphi-1)}$ is a *cf*, the claim follows directly from the definition of infinite divisibility.

(ii) Let μ_n be the pm whose cf is φ_n. The hypothesis is equivalent to $\mu_n \xrightarrow{d} \mu_0$. This implies (μ_n) is a tight sequence, ref. **(2.4.5)** read with**(1.9.4)**. Let integer $k \geq 2$ be fixed. Since φ_n is id there exists a cf $\varphi_{n,k}$ such that $\varphi_n = \{\varphi_{n,k}\}^k$. Or, $\mu_n = \underbrace{\mu_{n,k} * \mu_{n,k} * \ldots * \mu_{n,k}}$ (k factors), where $\mu_{n,k}$ is the measure with cf $\varphi_{n,k}$. Since (μ_n) is tight, we conclude from **(1.12.3)** that $(\mu_{n,k})$ is a shift tight sequence. Hence there exist vectors $\mathbf{a}_r$ and a subsequence (n_r) of integers, $n_r \uparrow \infty$ as $r \uparrow \infty$ such that the sequence $(\nu_{r,k})$ of pms is weakly convergent, where $\nu_{r,k}(A) = \mu_{n_r,k}(A - \mathbf{a}_r)$, $A \in \mathscr{R}^p$, ref. **(2.2.2)**. Let the limit cf be ψ_k. Define transformation T_r : $T_r\mathbf{x} = \mathbf{x} + \mathbf{a}_r$, so $T_r^{-1}\mathbf{x} = \mathbf{x} - \mathbf{a}_r$. Make use of the formula at **(2.3.6)** to conclude that $\nu_{r,k} = \mu_{n_r,k}T_r^{-1}$ and that the cf of $\nu_{r,k}$ is $e^{i\mathbf{t}'\mathbf{a}_r}\varphi_{n_r,k}$. Thus $\lim\limits_{r\to\infty} e^{i\mathbf{t}'\mathbf{a}_r}\varphi_{n_r,k}(\mathbf{t}) = \psi_k(t)$. Hence $\{\psi_k(\mathbf{t})\}^k = \lim\limits_{r\to\infty} e^{ik\mathbf{t}'\mathbf{a}_r}\{\varphi_{n_r,k}(\mathbf{t})\}^k = \lim\limits_{r\to\infty} e^{ik\mathbf{t}'a_r}\varphi_{n_r}(\mathbf{t}) = \lim\limits_{r\to\infty} e^{ik\mathbf{t}'\mathbf{a}_r}\varphi_0(\mathbf{t})$. This shows that $\lim\limits_{r\to\infty} e^{ik\mathbf{t}'\mathbf{a}_r}$ exists, equal to, say, $h(\mathbf{t})$. Thus for all $\mathbf{t} \in R^p$, $\{\psi_k(\mathbf{t})\}^k = h(\mathbf{t})\varphi_0(t)$. Hence h is a continuous function and $h(0) = 1$. Since $(e^{ik\mathbf{t}'\mathbf{a}_r})$ is a sequence of cfs converging to a continuous function, it follows, ref.**(2.3.2)**(iv), that the sequence (α_r) of measures degenerate at $k\mathbf{a}_r$ is a tight sequence. This implies that $(\mathbf{a}_r)$ is a bounded sequence. Taking a subsequence, if necessary, we conclude that $\varphi_0(\mathbf{t})$ is of the form $\{e^{i\mathbf{t}'\mathbf{a}}\psi_k(\mathbf{t})\}^k$ for some $\mathbf{a} \in R^p$. That φ_0 is id is immediate now. □

Remark 3.3.3. In the theorem the assumption that the limit function φ_0 is a *cf* can not be dispensed with. i.e., it is not claimed that if a sequence of φ_n of *id cf*s converges pointwise to a function φ_0, then it is a *cf* and hence is *id*. To see this clearly, let φ_n, $n \geq 0$ be as in the example at **(3.1.1)**(iv). By **(2.3.11)** $\psi_n = e^{\varphi_n - 1}$, $n \geq 1$ are *id* cfs. $\lim\limits_{n\to\infty} \psi_0(t) = e^{\varphi_0(t)-1}$ which, being equal to 1 for $t = 0$ and equal to e^{-1} for $t \neq 0$, is not a *cf*.

The next theorem proves that if a non-degenerate *df* F is id, then it is not possible for both *lext*F and *rext*F to be finite.

Theorem 3.3.4. *A bounded non-degenerate df F can not be infinitely divisible.*

Proof. It is obvious that it is enough to consider the one-dimensional case. Let $p = 1$. If possible, let F be an infinitely divisible, non-degenerate and bounded *df* with *lext*$F = a$ and *rext*$F = b$ and with *cf* φ. We note (i) that

$$\psi(t) = e^{-i\frac{b+a}{2(b-a)}t}\varphi\Big(\frac{t}{b-a}\Big)$$

is the *cf* of the bounded *df* F_1 where

$$F_1(x) = F\Big(\frac{2\overline{b-a}x + b + a}{2}\Big)$$

with *lext*$F_1 = -\frac{1}{2}$ and *rext*$F_1 = \frac{1}{2}$ and (ii) that if $g(t) = |\psi(t)|^2$ then g is the *cf* of the bounded *df* G with *lext*$G = -1$ and *rext*$G = 1$ where $G(x) = \int_R F_1(x-y)\,\mathrm{d}F_1(y)$, by **(3.1.12)** and **(2.3.10)**. We note further that g is *id* and non-degenerate. From $g(t) = \int_{-1}^{1} \cos tx \mathrm{d}G(x)$, $g'(0) = 0$ and $g''(0) = -\int_{-1}^{1} x^2 \mathrm{d}G(x)$. Because G is *id*, there exists, corresponding to each $n \geq 1$, a *df* G_n with cf g_n such that $g = g_n^n$. Since g is real valued, g_n will be so too and hence an even function. Further, by **(3.1.11)** G_n is a bounded distribution and $rextG_n = \frac{1}{n}, lextG_n = -\frac{1}{n}$. We then have $g_n'(0) = 0$ and $g_n''(0) = -\int_{-\frac{1}{n}}^{\frac{1}{n}} x^2 \mathrm{d}G_n(x)$. Since $g''(0) = ng_n''(0)$, we have

$$\int_{-1}^{1} x^2 \mathrm{d}G(x) = n \int_{-\frac{1}{n}}^{\frac{1}{n}} x^2 \mathrm{d}G_n(x) \leq \frac{1}{n} \to 0 \text{ as } n \to \infty.$$

Thus $\int_{-1}^{1} x^2 \mathrm{d}G(x) = 0$. This is possible if and only if $G(0) - G(0-) = 1$. i.e. G is degenerate at 0, a contradiction. □

Lemma 3.3.5. *Let* φ *be a* cf *in* R^p. *Then for every* $\boldsymbol{t} \in R^p$ *and for every positive integer* m,
$1 - \Re(\varphi(2^m \boldsymbol{t})) \leq 4^m\{1 - \Re(\varphi(\boldsymbol{t}))\}$.

Proof.

$$
\begin{aligned}
1-\Re(\varphi(2\mathbf{t})) &= \int_{-\infty}^{\infty}\{1-\cos 2\mathbf{t}'\mathbf{x}\}d\mathbb{F}(\mathbf{x})\\
&= 2\int_{-\infty}^{\infty}\sin^2\mathbf{t}'\mathbf{x}\, d\mathbb{F}(\mathbf{x})\\
&= 2\int_{-\infty}^{\infty}(1-\cos\mathbf{t}'\mathbf{x})(1+\cos\mathbf{t}'\mathbf{x})d\mathbb{F}(\mathbf{x})\\
&\le 4\int_{-\infty}^{\infty}(1-\cos\mathbf{t}'\mathbf{x})d\mathbb{F}(\mathbf{x})\\
&= 4\{1-\Re(\varphi(\mathbf{t}))\}.
\end{aligned}
$$

Repeatedly using this inequality m times, the claim is established. □

Theorem 3.3.6. (i) *An id cf φ never vanishes.*
(ii) *The real part of an id cf may fail to be id.*

Proof. (i) Equivalently, we will show that the non-negative *id cf* $\psi = |\varphi|^4$ never vanishes.

Step 1.
Since $|\varphi|^2$ is id, there exists, corresponding to any given integer $k \ge 1$, a cf φ_k such that $|\varphi|^2 = \{\varphi_k\}^k$. Hence $\psi = |\varphi|^2 = |\varphi_k(\mathbf{t})|^k$, leading to $|\psi| = |\varphi|^4 = \{|\varphi_k|^2\}^k = (\psi_k)^k$, say. We note $\psi_k = |\varphi_k|^2$ is a non-negative cf.

Step 2.
Since ψ is a cf, we can, given $\varepsilon > 0$, find a positive number a such that $\psi(\mathbf{t}) \ge \varepsilon$ for all $\mathbf{t}$ with $||\mathbf{t}|| \le a$. This implies that, for $||\mathbf{t}|| \le a$, $\psi_k(\mathbf{t}) \ge \varepsilon^{\frac{1}{k}}$. Or, $1-\psi_k(\mathbf{t}) \le 1-\varepsilon^{\frac{1}{k}}$. Since ψ_k is a real cf, we have by Lemma **(3.3.5)**, $1-\psi_k(2^m\mathbf{t}) \le 4^m\{1-\psi_k(\mathbf{t})\} \le 4^m(1-\varepsilon^{\frac{1}{k}})$, where m is an arbitrary positive integer and $||\mathbf{t}|| \le a$. This can be written $1-\psi_k(\mathbf{t}) \le 4^m(1-\varepsilon^{\frac{1}{k}})$ for $||\mathbf{t}|| \le 2^m a$. Thus $\lim\limits_{k\to\infty}\psi_k(\mathbf{t}) = 0$ for all $\mathbf{t}$.

Step 3.
If $\psi(\mathbf{t}_0) = 0$ for some $\mathbf{t}_0$, we end up with a contradiction since in that case $\psi_k(\mathbf{t}_0) = 0$ for all k and then $\lim\limits_{k\to\infty}\psi_k(\mathbf{t}_0) = 0$.

(ii) In **(3.2.1)** we proved that the real part of a *cf* is a *cf*. Consider *id cf* $\varphi(t) = e^{\frac{\pi}{\sqrt{2}}(e^{it}-1)}$. Its real part is $e^{\frac{\pi}{\sqrt{2}}(\cos t-1)}\cos(\frac{\pi}{\sqrt{2}}\sin t)$. This vanishes at $t=\frac{\pi}{4}$. Hence the real part of this $\varphi(t)$ can not be id. □

3.4 Canonical form of infinitely divisible characteristic functions

In the last section we saw that the property of infinite divisibility of a *cf* φ ensures that it does not vanish anywhere in R^p. In the present section we show that there exists a unique continuous function ψ such that, for all **t**, $\varphi(t) = e^{\psi(\mathbf{t})}$. We call ψ the distinguished logarithm of φ. Let

$$J(\mathbf{t},\mathbf{u}) = \{e^{i\mathbf{t}'\mathbf{u}} - 1 - \frac{i\mathbf{t}'\mathbf{u}}{1+|\mathbf{u}|^2}\}\frac{1+|\mathbf{u}|^2}{|\mathbf{u}|^2}.$$

Theorem 3.4.1 (The Levy-Khintchine canonical representation). *A cf φ in R^p is id iff it has the form e^{ψ} where*

$$\psi(\boldsymbol{t}) = i\gamma'\boldsymbol{t} - \frac{1}{2}\boldsymbol{t}'Q\boldsymbol{t} + \int_{R^p\sim\{\boldsymbol{0}\}} J(\boldsymbol{t},\boldsymbol{u})d\mathbb{G}(\boldsymbol{u})$$

for some vector $\gamma \in R^p$, some non-negative definite $p\times p$ matrix Q, and a finite measure $\mathbb{G}$ on $\mathscr{R}^p$ with $\mathbb{G}(\{\boldsymbol{0}\}) = 0$.

Proof. Step 0. For all real x, $|\frac{e^{ix}-1-ix}{x^2}| \le 3$.

This claim can be established as follows. $|e^{ix}-1-ix| \le |1-\cos x| + |\sin x - x|$. Now, for all x, $\frac{1-\cos x}{x^2} \le \frac{1}{2}$. If $|x|<1$, then $|\sin x - x| < \frac{1}{6}x^3$. If $|x|\ge 1$, then $|\frac{\sin x - x}{x^2}| \le |\frac{\sin x}{x}| + 1 \le 2$. The desired result follows now.

Step 1. If $0<|\mathbf{u}|<1$, then

$$\begin{aligned}
|J(\mathbf{t},\mathbf{u})| &\le |e^{i\mathbf{t}'\mathbf{u}} - 1 - i\mathbf{t}'\mathbf{u} + i\mathbf{t}'\mathbf{u}\frac{|\mathbf{u}|^2}{1+||\mathbf{u}|^2}|\frac{1+|\mathbf{u}|^2}{|\mathbf{u}|^2}\\
&\le |e^{i\mathbf{t}'\mathbf{u}} - 1 - |i\mathbf{t}'\mathbf{u}|\frac{2}{|\mathbf{u}|^2} + ||\mathbf{t}'\mathbf{u}|\\
&= |\frac{e^{i\mathbf{t}'\mathbf{u}} - 1 - i\mathbf{t}'\mathbf{u}}{(\mathbf{t}'\mathbf{u})^2}|\frac{2(\mathbf{t}'\mathbf{u})^2}{|\mathbf{u}|^2} + |\mathbf{t}'\mathbf{u}|\\
&\le 3\frac{2}{|\mathbf{u}|^2}(\mathbf{t}'\mathbf{u})^2 + ||\mathbf{t}||\mathbf{u}|\\
&\le 6|\mathbf{t}|^2 + |\mathbf{t}|.
\end{aligned}$$

If $|\mathbf{u}| > 1$,

$$|J(\mathbf{t}, \mathbf{u})| \leq 2\left(2 + \frac{|\mathbf{t}'\mathbf{u}|}{1 + |\mathbf{u}|^2}\right) \leq 2(2 + |\mathbf{t}|).$$

Thus for all $\mathbf{u}$, $|J(\mathbf{t}, \mathbf{u})| \leq 6|\mathbf{t}|^2 + 2|\mathbf{t}| + 4$.

Step 2. For each $n \geq 1$, there exists a cf φ_n such that $\varphi(\mathbf{t}) = \varphi_n^n(\mathbf{t}) = \lim_{n\to\infty} \{1 + \frac{\varphi_n(\mathbf{t})-1}{n}\}^n$. Since this limit exists, it follows that $\psi(\mathbf{t}) = \lim_{n\to\infty} n\{\varphi_n(t) - 1\}$ exists and $\varphi(\mathbf{t}) = e^{\psi(\mathbf{t})}$. Hence

$$\psi(\mathbf{t}) = \lim_{n\to\infty} n\{\varphi_n(\mathbf{t}) - 1\} = \lim_{n\to\infty} n \int_{R^p} \{e^{i\mathbf{t}'\mathbf{x}} - 1\} d\mathbb{F}_n(\mathbf{x})$$

(where $\mathbb{F}_n$ denotes the *df* corresponding to φ_n)

$$= \lim_{n\to\infty} n \int_{R^p \sim \{\mathbf{0}\}} \{e^{i\mathbf{t}'\mathbf{x}} - 1\} d\mathbb{F}_n(\mathbf{x})$$

$$= \lim_{n\to\infty} \int_{R^p \sim \{\mathbf{0}\}} \{e^{i\mathbf{t}'\mathbf{x}} - 1\} \frac{1 + |\mathbf{x}|^2}{|\mathbf{x}|^2} d\mathbb{G}_n(\mathbf{x})$$

where $\mathbb{G}_n$ is the measure on $\mathscr{R}^p$ given by

$$\mathbb{G}_n(A) = n \int_A \frac{|\mathbf{u}|^2}{1 + |\mathbf{u}|^2} d\mathbb{F}_n(\mathbf{u}), \quad A \in \mathscr{R}^p.$$

By **(2.3.11)** , $e^{n(\varphi_n - 1)}$ is a cf and by **(3.1.1)**(iii) the convergence is uniform over bounded 'intervals'/spheres of R^p. For $\tau > 0$, define $A(\tau) = \{\mathbf{x} : \mathbf{x} \in R^p,\ 0 < |\mathbf{x}| < \tau\}$. Let i_j be the p-vector with 1 for its jth component and zeros for all other components. Let $A_j(1) = \{\mathbf{x} : \mathbf{x} \in A(1), |x_j| \geq \frac{|\mathbf{x}|}{\sqrt{p}}\}$, $1 \leq j \leq p$ and notice that $A(1) \subset \cup_{1 \leq j \leq p} A_j(1)$. Hence

$$\mathbb{G}_n(A(1)) \leq \sum_{j=1}^{p} \mathbb{G}_n(A_j(1)).$$

From

$$-\Re(\psi(\mathbf{t})) = \lim_{n\to\infty} n\{1 - \Re(\varphi_n(\mathbf{t}))\}$$

$$= \lim_{n\to\infty} \int_{R^p} \{1 - \cos \mathbf{t}'\mathbf{x}\} \frac{1 + |\mathbf{x}|^2}{|\mathbf{x}|^2} d\mathbb{G}_n(\mathbf{x})$$

we get, for $1 \leq j \leq p$, for $\eta > 0$ and for all n large,

$$
\begin{aligned}
-\Re(\psi(i_j)) + \eta &\geq \int_{R^p \sim \{\mathbf{0}\}} \{1 - \cos x_j\} \frac{1 + |\mathbf{x}|^2}{|\mathbf{x}|^2} \mathrm{d}\mathbb{G}_n(\mathbf{x}) \\
&\geq \int_{A_j(1)} \{1 - \cos x_j\} \frac{1 + |\mathbf{x}|^2}{|\mathbf{x}|^2} \mathrm{d}\mathbb{G}_n(x) \\
&\geq \int_{A_j(1)} \frac{1 - \cos \frac{|\mathbf{x}|}{\sqrt{p}}}{|\mathbf{x}|^2} \mathrm{d}\mathbb{G}_n(\mathbf{x}) = \frac{1}{2p} \int_{A_j(1)} \frac{\sin^2 \frac{|\mathbf{x}|}{2\sqrt{p}}}{\frac{|\mathbf{x}|^2}{4p}} \mathrm{d}\mathbb{G}_n(x) \\
&\geq \frac{1}{2p} \int_{A_j(1)} \frac{4}{\pi^2} \mathrm{d}\mathbb{G}_n(\mathbf{x}) \geq \frac{1}{8p} \mathbb{G}_n(A_j(1)).
\end{aligned}
$$

Hence

$$
\mathbb{G}_n(A(1)) \leq \sum_{j=1}^{p} \mathbb{G}_n(A_j(1)) \leq 8p^2\eta + 8p \sum_{j=1}^{p} \{-\log|\varphi(i_j)|\}.
$$

Consequently $\sup_n \mathbb{G}_n(A(1)) < \infty$.

Step 3. Define, for $r > 0$, $B(r) = \{\mathbf{x} : \mathbf{x} \in R^p,\ |\mathbf{x}| \geq r\}$. For $j = 1, 2, ..., p$, let

$$
b_j(s) = -\Re(\psi(si_j)) = \lim_{n \to \infty} \int_{R^p \sim \{\mathbf{0}\}} \{1 - \cos sx_j\} \frac{1 + |\mathbf{x}|^2}{|\mathbf{x}|^2} \mathrm{d}\mathbb{G}_n(\mathbf{x})
$$

and recall that the convergence is uniform over $[0, 2\sqrt{p}]$. Define $C_j(r) = \{\mathbf{x} : \mathbf{x} \in B(r),\ |x_j| \geq \frac{|\mathbf{x}|}{\sqrt{p}}\}$ and observe that $B(r) \subset \cup_j C_j(r)$. Choose and fix $\eta > 0$. For n sufficiently large and for all $s \in [0, 2\sqrt{p}]$,

$$
\eta + b_j(s) \geq \int_{C_j(r)} (1 - \cos(sx_j)) \frac{1 + |\mathbf{x}|^2}{|\mathbf{x}|^2} \mathrm{d}\mathbb{G}_n(\mathbf{x}).
$$

Noting that b_j is a continuous function we get, on integration,

$$
\begin{aligned}
\eta + \frac{r}{2\sqrt{p}} \int_0^{\frac{2\sqrt{p}}{r}} b_j(s)\mathrm{d}s &\geq \int_{C_j(r)} \{1 - \frac{\sin(\frac{2x_j\sqrt{p}}{r})}{\frac{2x_j\sqrt{p}}{r}}\} \frac{1 + |\mathbf{x}|^2}{|\mathbf{x}|^2} \mathrm{d}\mathbb{G}_n(\mathbf{x}) \\
&\geq \frac{1}{2} \mathbb{G}_n(C_j(r)),
\end{aligned}
$$

since

$$\left| \frac{\sin(\frac{2\sqrt{p}x_j}{r})}{\frac{2\sqrt{p}x_j}{r}} \right| \leq \frac{1}{\frac{2\sqrt{p}|x_j|}{r}} \leq \frac{r}{2|\mathbf{x}|} \leq \frac{1}{2}.$$

It follows that

$$\mathbb{G}_n(B(r)) \leq 2p\eta + \frac{r}{\sqrt{p}} \sum_1^p \int_0^{\frac{2\sqrt{p}}{r}} b_j(s)\mathrm{d}s.$$

Hence $\sup_n \mathbb{G}_n(B(r)) < \infty$. This result for $r = 1$ combined with the previous result yields: If $a_n = \mathbb{G}_n(R^p)$, then $\sup_n a_n < \infty$.

Step 4. Since $\lim_{y\to 0} \frac{1}{y} \int_0^y b_j(s)\mathrm{d}s = b_j(0) = 0$ and since η is arbitrary, $\lim_{r\to\infty} \sup_n \mathbb{G}_n(B(r)) = 0$. It now follows that $\frac{1}{a_n}\mathbb{G}_n$ is a tight sequence of pms in R^p. For a subsequence (m) of (n), we have : (a_m) is a convergent sequence converging to, say, a and $(\frac{1}{a_m}\mathbb{G}_m)$ is a weakly convergent sequence of pms, converging to the pm, say, $\frac{1}{a}\mathbb{G}_0$.

Step 5. Recall

$$\psi(\mathbf{t}) = \lim_{n\to\infty} \int_{R^p} \left(e^{i\mathbf{t}'\mathbf{u}} - 1\right) n\mathrm{d}\mathbb{F}_n(\mathbf{u}).$$

Helly-Bray theorem is not applicable to this expression or to any of its subsequences. For, even though the integrand is a bounded continuous function, (nF_n) is not a tight sequence of pms. So we write

$$\begin{aligned}\psi(\mathbf{t}) &= \lim_{n\to\infty} \int_{R^p\sim\{\mathbf{0}\}} \left(e^{i\mathbf{t}'\mathbf{u}} - 1\right) n\,\mathrm{d}\mathbb{F}_n(\mathbf{u}) \\ &= \lim_{n\to\infty} \int_{R^p\sim\{\mathbf{0}\}} \left(e^{i\mathbf{t}'\mathbf{u}} - 1\right) \frac{1+|\mathbf{u}|^2}{|\mathbf{u}|^2} \mathrm{d}\mathbb{G}_n(\mathbf{u})\end{aligned}$$

[Now, atleast the subsequence $(\mathbb{G}_m)$ is weakly convergent. But the integrand is not a bounded function. Hence, again Helly -Bray theorem is not applicble. We therefore write this as]

$$= \lim_{n\to\infty} \left\{ \int\limits_{R^p\sim\{\mathbf{0}\}} \left(e^{i\mathbf{t}'\mathbf{u}} - 1 - \frac{i\mathbf{t}'\mathbf{u}}{1+|\mathbf{u}|^2} \right) \frac{1+|\mathbf{u}|^2}{|\mathbf{u}|^2} d\mathbb{G}_n(\mathbf{u}) \right.$$

$$\left. + \int\limits_{R^p\sim\{\mathbf{0}\}} \frac{i\mathbf{t}'\mathbf{u}}{|\mathbf{u}|^2} d\mathbb{G}_n(\mathbf{u}) \right\}$$

$$= \lim_{n\to\infty} (I_{n,1} + iI_{n,2}),$$

say. This step is valid since one of the intgrals, namely

$$I_{n,2} = \int\limits_{R^p\sim\{\mathbf{0}\}} \frac{\mathbf{t}'\mathbf{u}}{1+|\mathbf{u}|^2} n d\mathbb{F}_n(\mathbf{u}),$$

exists finitely. $I_{n,2}$ is a linear functional of $\mathbf{t}$. Hence there exists $\alpha_n \in R^p$ such that $I_{n,2} = \alpha_n'\mathbf{t}$. Now,

$$I_{n,1} = I_{n,3} - \frac{1}{2} I_{n,4}$$

$$\text{where } I_{n,3} = \int\limits_{R^p\sim\{\mathbf{0}\}} \left\{ J(\mathbf{t},\mathbf{u}) + \frac{1}{2}\frac{(\mathbf{t}'\mathbf{u})^2}{|\mathbf{u}|^2} \right\} d\mathbb{G}_n(\mathbf{u});$$

$$\text{and } I_{n,4} = \int\limits_{R^p\sim\{\mathbf{0}\}} \frac{(\mathbf{t}'\mathbf{u})^2}{|\mathbf{u}|^2} d\mathbb{G}_n(\mathbf{u}).$$

This step is valid because one of the integrals, namely

$$I_{n,4} = \int\limits_{R^p\sim\{\mathbf{0}\}} \frac{(\mathbf{t}'\mathbf{u})^2}{1+|\mathbf{u}|^2} n d\mathbb{F}_n(\mathbf{u}),$$

exists finitely. $I_{n,3}$ can be written

$$\int\limits_{R^p} \left\{ J(\mathbf{t},\mathbf{u}) + \frac{1}{2}\frac{(\mathbf{t}'\mathbf{u})^2}{|\mathbf{u}|^2} \right\} d\mathbb{G}_n(\mathbf{u}),$$

defining the integrand to be 0 at $\mathbf{u} = \mathbf{0}$. So defined the integrand is a continuous function of $\mathbf{u}$, bounded by $7|\mathbf{t}|^2 + 2|\mathbf{t}| + 4$. Hence by Helly-Bray theorem,

$\lim_{m\to\infty} I_{m,3}$ exists and is equal to

$$I_3 = \int_{R^p} \left\{ J(\mathbf{t},\mathbf{u}) + \frac{1}{2}\frac{(\mathbf{t}'\mathbf{u})^2}{|\mathbf{u}|^2} \right\} d\mathbb{G}_0(\mathbf{u}).$$

Now,

$$I_{n,4} = \int_{R^p \sim \{\mathbf{0}\}} \frac{(\mathbf{t}'\mathbf{u})^2}{1+|\mathbf{u}|^2} n d\mathbb{F}_n(\mathbf{u})$$

is a non-negative definite quadratic form and hence can be written $\mathbf{t}'B_n\mathbf{t}$ for some non-negative definite matrix B_n. If $b(j,k;n)$ is the jth-row, k^{th}-column element of B_n, then, for $1 \le j, k \le p$,

$$\begin{aligned}
|b(j,k;n)| &\le |\text{coefficient of } t_j t_k| \le n \int_{R^p} \frac{|u_j u_k|}{1+|\mathbf{u}|^2} dF_n(\mathbf{u}) \\
&\le n \int_{R^p} \frac{|\mathbf{u}|^2}{1+|\mathbf{u}|^2} d\mathbb{F}_n(\mathbf{u}) = \mathbb{G}_n(R^p) \\
&\le \sup_n \mathbb{G}_n(R^p) < \infty.
\end{aligned}$$

There exists therefore a subsequence (m') of (m) such that $B_{m'}$ converges pointwise to a non-negative definite matrix B.

It now follows that $\lim_{m'\to\infty} I_{m',1}$ exists and is equal to $I_3 - \frac{1}{2}\mathbf{t}'B\mathbf{t}$. This implies that $\lim_{m'\to\infty} I_{m',2}$ exists finitely, equal to, say, $\gamma'\mathbf{t}$ for some $\gamma \in R^p$.

Further,

$$\begin{aligned}
I_3 &= \int_{R^p \sim \{\mathbf{0}\}} \left\{ J(\mathbf{t},\mathbf{u}) + \frac{1}{2}\frac{(\mathbf{t}'\mathbf{u})^2}{|\mathbf{u}|^2} \right\} d\mathbb{G}_0(\mathbf{u}) \\
&= \int_{R^p \sim \{\mathbf{0}\}} J(\mathbf{t},\mathbf{u}) d\mathbb{G}_0(\mathbf{u}) + \int_{R^p \sim \{\mathbf{0}\}} \frac{1}{2}\frac{(\mathbf{t}'\mathbf{u})^2}{|\mathbf{u}|^2} d\mathbb{G}_0(\mathbf{u}) \\
&= \int_{R^p \sim \{\mathbf{0}\}} J(\mathbf{t},\mathbf{u}) d\mathbb{G}_0(\mathbf{u}) + \frac{1}{2}\mathbf{t}'B_1\mathbf{t}
\end{aligned}$$

where B_1 is a non-negative definite $p \times p$ matrix. Now,

$$\mathbf{t}'B\mathbf{t} = \lim_{m'\to\infty} \int\limits_{R^p} \frac{(\mathbf{t}'\mathbf{u})^2}{1+|\mathbf{u}|^2} m' \mathrm{d}F_{m'}(\mathbf{u})$$

$$= \lim_{m'\to\infty} \left\{ m' \int\limits_{\{\mathbf{0}\}\cup A(\tau)} \frac{(\mathbf{t}'\mathbf{u})^2}{1+|\mathbf{u}|^2} \mathrm{d}F_{m'}(\mathbf{u}) + \int\limits_{B(\tau)} \frac{(\mathbf{t}'\mathbf{u})^2}{1+|\mathbf{u}|^2} m' \mathrm{d}F_{m'}(\mathbf{u}) \right\}$$

$$= \lim_{m'\to\infty} \left\{ m' \int\limits_{\{\mathbf{0}\}\cup A(\tau)} \frac{(\mathbf{t}'\mathbf{u})^2}{1+|\mathbf{u}|^2} \mathrm{d}F_{m'}(\mathbf{u}) + \int\limits_{B(\tau)} \frac{(\mathbf{t}'\mathbf{u})^2}{|\mathbf{u}|^2} \mathrm{d}G_{m'}(\mathbf{u}) \right\}.$$

for all $\tau > 0$. Choose τ such that the G_0 measure of the boundary of $B(\tau)$ is 0. The boundary of $B(\tau)$ is $\{\mathbf{u}; |\mathbf{u}| = \tau\}$. The limit of the second term in the braces would then be $\int\limits_{B(\tau)} \frac{(\mathbf{t}'\mathbf{u})^2}{|\mathbf{u}|^2} \mathrm{d}G_0(\mathbf{u})$. Necessarily the first term in the braces has a limit for every τ chosen as above. But the first term is a positive definite quadratic form in $\mathbf{t}$ for each m'. Take τ to 0 along a monotonically decreasing sequence (τ_n) such that each $B(\tau_n)$ is a continuity set of G_0. The first term, on $m' \to \infty$ and then on $\tau_n \downarrow 0$, converges to $\mathbf{t}'Q_1\mathbf{t}$ for some non-negative definite matrix Q_1. Thus

$$\mathbf{t}'B\mathbf{t} = \mathbf{t}'Q_1\mathbf{t} + \lim_{\nu\to\infty} \int\limits_{B(\tau_\nu)} \frac{(\mathbf{t}'\mathbf{u})^2}{|\mathbf{u}|^2} \mathrm{d}G_0(\mathbf{u}) = \mathbf{t}'Q_1\mathbf{t} + \int\limits_{R^p\sim\{\mathbf{0}\}} \frac{(\mathbf{t}'\mathbf{u})^2}{|\mathbf{u}|^2} \mathrm{d}G_0(\mathbf{u})$$

(invoking the continuity property of measures)

$$= \mathbf{t}'Q_1\mathbf{t} + \mathbf{t}'B_1\mathbf{t}.$$

We thus arrive at

$$\psi(\mathbf{t}) = i\gamma'\mathbf{t} - \frac{1}{2}\mathbf{t}'\Sigma\mathbf{t} + \int\limits_{R^p\sim\{\mathbf{0}\}} J(\mathbf{t}, \mathbf{u})\mathrm{d}G(\mathbf{u}).$$

Σ is the Q of the statement. □

To prove the converse, let

$$\Theta(t) = \int\limits_{R^p\sim\{\mathbf{0}\}} \{e^{i\mathbf{t}'\mathbf{u}} - 1 - \frac{i\mathbf{t}'\mathbf{u}}{1+|\mathbf{u}|^2}\} \frac{1+|\mathbf{u}|^2}{|\mathbf{u}|^2} \mathbb{G}(\mathbf{u})$$

where $\mathbb{G}$ is a finite measure on $\mathscr{R}^p$ with $\mathbb{G}(\{\mathbf{0}\}) = 0$. The integrand may be taken to be 0 at $\mathbf{0}$ with no loss of generality. The integrand is bounded and measurable and hence is $\mathbb{G}$-integrable. The integral can therefore be evaluated as

$$\lim_{n\to\infty} \Theta_n(\mathbf{t}) = \lim_{n\to\infty} \int_{A_n} \{e^{i\mathbf{t}'\mathbf{u}} - 1 - \frac{i\mathbf{t}'\mathbf{u}}{1+|\mathbf{u}|^2}\}\frac{1+|\mathbf{u}|^2}{|\mathbf{u}|^2}\, d\mathbb{G}(\mathbf{u}).$$

Here n is a positive integer and $A_n = \{\mathbf{x} : \frac{1}{n} \leq |x_i| \leq n,\ 1 \leq i \leq p\}$. We note that A_n is the closed region between two rectangles around the origin. In this region the integrand is a continuous function. Hence $\Theta_n(t)$ can be evaluated as a Riemann-Stieltjes integral. Divide A_n into small rectangles of at least one side of length $\frac{1}{k}$ and others of length not exceeding $\frac{1}{k}$. List these rectangles as $B_{j,k},\ 1 \leq j \leq j_k$. Let $\mathbf{u}_{j,k}$ be an arbitrary interior point of $B_{j,k}$. Then

$$\Theta_n(\mathbf{t}) = \lim_{k\to\infty} \sum_{j=1}^{j_k} \{e^{i\mathbf{t}'\mathbf{u}_{j,k}} - 1 - \frac{i\mathbf{t}'\mathbf{u}_{j,k}}{1+|\mathbf{u}_{j,k}|^2}\}\frac{1+|\mathbf{u}_{j,k}|^2}{|\mathbf{u}_{j,k}|^2}\mathbb{G}(B_{j,k}).$$

The jth term $\Theta_{n;j}(\mathbf{t})$ of this sum is of the form $i\mathbf{t}'\mathbf{a}_{j,k} + \lambda_{j,k}(e^{i\mathbf{t}'\mathbf{u}_{j,k}} - 1)$ where $\lambda_{j,k}$ is a positive number. Hence, by **(3.3.2)**(i), $e^{\Theta_{n;j}(\mathbf{t})}$ is an *id cf*. It follows, by **(3.3.2)**(ii), that $e^{\Theta_n((\mathbf{t}))}$ is an *id cf*. Again appealing to the same section we conclude that $e^{\Theta(\mathbf{t})}$ is an *id cf*. Since $e^{-\frac{1}{2}\mathbf{t}'\Sigma\mathbf{t}}$ and $e^{i\gamma'\mathbf{t}}$ are *id cf*s, it follows that $e^{\psi(\mathbf{t})}$ is an *id cf*. □

Remark 3.4.2. The term $-\frac{1}{2}\mathbf{t}'Q\mathbf{t}$, which may be absent, is called the normal component of the cf (ref. **(3.2)**(xi)).

Theorem 3.4.3. *The representation of an $id\ cf$ in Theorem **(3.4.1)** is unique.*

Proof. Equivalently, we show that ψ is unique. Let

$$\psi(\mathbf{t}) = i\gamma'\mathbf{t} - \frac{1}{2}\mathbf{t}'Q\mathbf{t} + \int_{R^p \sim \{\mathbf{0}\}} J(\mathbf{t},\mathbf{u}) d\mathbb{G}(\mathbf{u})$$

for some $\gamma,\ Q$, and $\mathbb{G}$. Hence

$$\Re(\psi(\mathbf{t})) = -\frac{1}{2}\mathbf{t}'Q\mathbf{t} + \int_{R^p \sim \{\mathbf{0}\}} \{\cos \mathbf{t}'\mathbf{x} - 1\}\frac{1+|\mathbf{x}|^2}{|\mathbf{x}|^2} d\mathbb{G}(\mathbf{x}).$$

Since $\mathbb{G}$ is a finite measure we can, given $\varepsilon > 0$, find a δ such that

$$\int_{0<|\mathbf{x}|\leq\delta} d\mathbb{G}(\mathbf{x}) < \varepsilon.$$

Now

$$\lim_{\lambda\to\infty} -\frac{1}{\lambda^2}\Re(\psi(\lambda\mathbf{t})) = \frac{1}{2}\mathbf{t}'Q\mathbf{t} + I_1 + I_2 \text{ where}$$

$$I_1 = \lim_{\lambda\to\infty}\frac{1}{\lambda^2}\int\limits_{0<|\mathbf{x}|\le\delta}\{1-\cos\lambda\mathbf{t}'\mathbf{x}\}\frac{1+|\mathbf{x}|^2}{|\mathbf{x}|^2}\,\mathrm{d}\mathbb{G}(\mathbf{x})$$

$$\le 4\lim_{\lambda\to\infty}\frac{1}{\lambda^2}\int\limits_{0<|\mathbf{x}|\le\delta}\{\sin^2\frac{1}{2}\lambda\mathbf{t}'\mathbf{x}\}\frac{1}{|\mathbf{x}|^2}\,\mathrm{d}\mathbb{G}(\mathbf{x}) \le |\mathbf{t}|^2\varepsilon;$$

$$I_2 = \lim_{\lambda\to\infty}\frac{1}{\lambda^2}\int\limits_{|\mathbf{x}|>\delta}\{1-\cos\lambda\mathbf{t}'\mathbf{x}\}\frac{1+|\mathbf{x}|^2}{|\mathbf{x}|^2}\,\mathrm{d}\mathbb{G}(\mathbf{x})$$

$$\le \lim_{\lambda\to\infty} 2\frac{1+\delta^2}{\delta^2}\frac{1}{\lambda^2} = 0.$$

Thus

$$\lim_{\lambda\to\infty} -\frac{1}{\lambda^2}\Re(\psi(\lambda\mathbf{t})) = \frac{1}{2}\mathbf{t}'Q\mathbf{t},$$

showing that φ determines Q.

Let $V(\mathbf{t}) = -\Re(\psi(\mathbf{t})) - \frac{1}{2}\mathbf{t}'Q\mathbf{t}$. We note that V is determined by φ and that V is a continuous function. Then $V^*(\mathbf{t})$ is determined by V and hence by φ where

$$V^*(\mathbf{t}) = V^*(t_1, t_2, ..., t_p)$$

$$= \int\limits_{t_1-1}^{t_1+1}\int\limits_{t_2-1}^{t_2+1}\cdots\int\limits_{t_p-1}^{t_p+1} V(x_1, x_2, \ldots, x_p)\mathrm{d}x_1\mathrm{d}x_2\cdots\mathrm{d}x_p$$

$$= 2^p\int\limits_{R^p\sim\{\mathbf{0}\}}\left\{\cos(\mathbf{t}'\mathbf{u})\times_{j=1}^{p}\frac{\sin u_j}{u_j} - 1\right\}\frac{1+|\mathbf{u}|^2}{|\mathbf{u}|^2}\mathrm{d}\mathbb{G}(\mathbf{u})$$

$$= 2^p\int\limits_{R^p\sim\{\mathbf{0}\}}\{\cos(\mathbf{t}'\mathbf{u}) - 1\}\frac{1+|\mathbf{u}|^2}{|\mathbf{u}|^2}\mathrm{d}\mathbb{G}(u)$$

$$- 2^p\int\limits_{R^p\sim\{\mathbf{0}\}}\cos\mathbf{t}'\mathbf{u}\,(1-\times_{j=1}^{p}\frac{\sin u_j}{u_j})\frac{1+|\mathbf{u}|^2}{|\mathbf{u}|^2}\mathrm{d}\mathbb{G}(u)\}$$

$$= 2^pV(\mathbf{t}) - 2^p\int\limits_{R^p\sim\{\mathbf{0}\}}\cos(\mathbf{t}'\mathbf{u})\mathrm{d}\mathbb{G}_1(\mathbf{u}).$$

Note that G_1 is a finite measure on the Borel σ-field of $R^p \sim \{\mathbf{0}\}$. Define measure $\mathbb{G}_2$ thus : $\mathbb{G}_2(\{\mathbf{0}\}) = 0$; for every $A \in \mathscr{R}^p$, take $\mathbb{G}_2(A) = \mathbb{G}_1(A \sim \{\mathbf{0}\})$. Obviously, $\mathbb{G}_1$ and $\mathbb{G}_2$ determine each other. $\int_{R^p} \cos(\mathbf{t}'\mathbf{u})d\mathbb{G}_3(\mathbf{u})$ is the Fourier transform of a symmetric measure $\mathbb{G}_3$ which determines and is determined by $\mathbb{G}_2$. Thus the cf of $\mathbb{G}_3$, and hence $\mathbb{G}_3$, is determined by φ. Since $\mathbb{G}$, $\mathbb{G}_1$, $\mathbb{G}_2$, $\mathbb{G}_3$ determine one another, the proof that φ determines $\mathbb{G}$ is complete. Since φ determines Q and $\mathbb{G}$, it follows that φ determines γ too, proving the claim. □

Note
Henceforth, ψ *and* $\log \varphi$ *will be used interchangeably*

Remarks 3.4.4. (i) The canonical representation of a real $id\ cf\ g$ takes the form

$$\log g(\mathbf{t}) = -\frac{1}{2}\mathbf{t}'Q\mathbf{t} + \int_{R^p \sim \{\mathbf{0}\}} \{\cos \mathbf{t}'\mathbf{u} - 1\}\frac{1+|\mathbf{u}|^2}{|\mathbf{u}|^2} d\mathbb{G}(\mathbf{u}) \tag{3.1}$$

(ii) This representation for real *id* laws is unique.

Proof of this assertion can be given on the same lines as the proof of the theorem.
(iii) Let the *id cf g* have the canonical representation

$$\log g(\mathbf{t}) = i\gamma'\mathbf{t} - \frac{1}{2}\mathbf{t}'Q\mathbf{t} + \int_{R^p \sim \{\mathbf{0}\}} \{e^{i\mathbf{t}'\mathbf{u}} - 1 - \frac{i\mathbf{t}'\mathbf{u}}{1+|\mathbf{u}|^2}\}\frac{1+|\mathbf{u}|^2}{|\mathbf{u}|^2} d\mathbb{G}(\mathbf{u}) \tag{3.2}$$

Then $g(\mathbf{t})e^{-i\gamma'\mathbf{t}}$ is real valued *iff* the measure G is symmetric (ref. **(2.3.7)**(iii)).

Proof. Stated in terms of a rV **X** with *cf g*, what is claimed is that the distribution of **X** is symmetric about the vector γ *iff* the measure G is symmetric about the origin.

Suppose G is a symmetric measure. If f is the mapping $R^p \xrightarrow{f} R^p$: $f(\mathbf{x}) = -\mathbf{x}$, then $f^{-1}(\mathbf{x}) = -\mathbf{x}$ and $Gf^{-1} = G$. We have from (3.2),

$$\begin{aligned}
&\log g(\mathbf{t}) \\
&= i\gamma'\mathbf{t} - \frac{1}{2}\mathbf{t}'Q\mathbf{t} + \int_{R^p \sim \{\mathbf{0}\}} \{e^{i\mathbf{t}'\mathbf{u}} - 1 - \frac{i\mathbf{t}'\mathbf{u}}{1+|\mathbf{u}|^2}\}\frac{1+|\mathbf{u}|^2}{|\mathbf{u}|^2} d\mathbb{G}f^{-1}(\mathbf{u}) \\
&= i\gamma'\mathbf{t} - \frac{1}{2}\mathbf{t}'Q\mathbf{t} + \int_{R^p \sim \{\mathbf{0}\}} \{e^{-i\mathbf{t}'\mathbf{u}} - 1 - \frac{-i\mathbf{t}'\mathbf{u}}{1+|\mathbf{u}|^2}\}\frac{1+|\mathbf{u}|^2}{|\mathbf{u}|^2} d\mathbb{G}(\mathbf{u})
\end{aligned}$$

(applying **(2.3.6)**)

$$= i\gamma'\mathbf{t} - \frac{1}{2}\mathbf{t}'Q\mathbf{t} + \{\log g(-\mathbf{t}) + i\gamma'\mathbf{t} + \frac{1}{2}\mathbf{t}'Q\mathbf{t}\},$$

leading to $\log g(t) - i\gamma'\mathbf{t} = \log g(-\mathbf{t}) + i\gamma'\mathbf{t}$, which is equivalent to the claim.
Conversely, if $g(\mathbf{t})e^{-i\gamma'\mathbf{t}}$ is real number for every $\mathbf{t} \in R^p$, then

$$\int\limits_{R^p \sim \{\mathbf{0}\}} \{e^{i\mathbf{t}'\mathbf{u}} - 1 - \frac{i\mathbf{t}'\mathbf{u}}{1+|\mathbf{u}|^2}\}\frac{1+|\mathbf{u}|^2}{|\mathbf{u}|^2}\mathrm{d}\mathbb{G}(\mathbf{u})$$

$$= \int\limits_{R^p \sim \{\mathbf{0}\}} \{e^{-i\mathbf{t}'\mathbf{u}} - 1 - \frac{-i\mathbf{t}'\mathbf{u}}{1+|\mathbf{u}|^2}\}\frac{1+|\mathbf{u}|^2}{|\mathbf{u}|^2}\mathrm{d}\mathbb{G}(\mathbf{u})$$

$$= \int\limits_{R^p \sim \{\mathbf{0}\}} \{e^{i\mathbf{t}'\mathbf{u}} - 1 - \frac{i\mathbf{t}'\mathbf{u}}{1+|\mathbf{u}|^2}\}\frac{1+|\mathbf{u}|^2}{|\mathbf{u}|^2}\mathrm{d}\mathbb{G}f^{-1}(\mathbf{u}).$$

By the uniqueness theorem proved at **(3.4.3)**, it follows that $G \equiv Gf^{-1}$, which means that G is a symmetric measure. □

The Kolmogorov representation

Theorem 3.4.5. *If* $\mathbb{E}|\boldsymbol{X}|^2 < \infty$ *for an id rV with cf* φ*, then log* φ *can be put in the form*

$$\log\varphi(\boldsymbol{t}) = i\gamma'\boldsymbol{t} - \frac{1}{2}\boldsymbol{t}'Q\boldsymbol{t} + \int\limits_{R^p \sim \{\boldsymbol{0}\}} (e^{i\boldsymbol{t}'\boldsymbol{u}} - 1 - i\boldsymbol{t}'\boldsymbol{u})\frac{1}{|\boldsymbol{u}|^2}d\mathbb{G}(\boldsymbol{u})$$

for some $\gamma \in R^p$*, some non-negative definite* $p \times p$ *matrix* Q *and some finite measure* G *on* $\mathscr{R}^p$*. This representation is unique.*

Proof. The claim, including the uniqueness part, can be established following the same lines as for Levy-Khintchine representation except that we define $\mathbf{G}_n(A) = n\int\limits_A |\mathbf{u}|^2\mathrm{d}\mathbb{F}_n(u)$. □

Remarks 3.4.6. (i) Upon differentiation of $\log\varphi$ and evaluation at $\mathbf{t} = \mathbf{0}$, we find $\mathbb{E}X = \gamma$ and $\mathbb{E}XX' = Q + \int\limits_{R^p \sim \{\mathbf{0}\}} \frac{\mathbf{u}\mathbf{u}'}{|\mathbf{u}|^2}\mathrm{d}\mathbb{G}(\mathbf{u})$.
(ii) There can exist *id rv* X with $\mathbb{E}|X|^2 = \infty$ and with a representation as above but with G not a finite measure. To see this consider *rv* X having frequency function $f(x) = \frac{1}{|x|^3}$, $|x| \geq 1$ and $f(x) = 0$ elsewhere. Denote its

df by F and its cf by φ_1. Then we know by **(2.3.11)** and by the definition of an *id cf* that ξ is an $id\ cf$ where $\xi(t) = e^{\varphi_1(t)-1}$. We have :

$$\log \xi(t) = \int_{-\infty}^{\infty} (e^{itx} - 1)\mathrm{d}F(x) = \int_{-\infty}^{\infty} (e^{itx} - 1 - itx)\mathrm{d}F(x)$$

(because

$$\int_{-\infty}^{\infty} x\mathrm{d}F(x) = 0) = \int_{-\infty}^{\infty} (e^{itx} - 1 - itx)\frac{1}{x^2}\mathrm{d}G(x)$$

where

$$G(x) = \int_{-\infty}^{x} y^2\mathrm{d}F(y)).$$

G is not a finite measure.
(iii) We saw in **(3.2)**(xi) that a multivariate normal distribution with mean vector γ and variance-covariance matrix Σ is one with *cf* $\varphi = e^{i\gamma'\mathbf{t} - \frac{1}{2}t'\Sigma\mathbf{t}}$. That this *cf* is *id* follows immediately from the definition of infinite divisibility. From the uniqueness of Kolmogorov canonical representation, it is clear that the finite measure $\mathbb{G}$ occurring in the representation is identically zero.

Cramér's theorem for id laws

Suppose φ_1, φ_2 are *id cf*s and that $\varphi = \varphi_1\varphi_2$ is multivariate normal then each of φ_1 and φ_2 is multivariate normal.

Proof. Let the parameters in the Levy-Khintchin Canonical representation of φ_i be γ_i, Q_i and $\mathbb{G}_i$, $i = 1,\ 2$. By the uniqueness of the representation, the finite measure $\mathbb{G}$ corresponding to φ is $\mathbb{G}_1 + \mathbb{G}_2$, which is identically zero since φ is multivariate normal. Hence each of the measures $\mathbb{G}_1$ and $\mathbb{G}_2$ is identically zero. The stated assertion follows now. □

(iv) Let $p = 1$. The Poisson *cf* $\varphi(t) = e^{i\gamma t + \lambda(e^{it}-1)}$ is obviously *id* with canonical representation parameters $\gamma = \frac{\lambda}{2}$, $\sigma = 0$ and $G(\{1\}) = \frac{\lambda}{2}$ and $G(\{1\}') = 0$.

Raikov's theorem for id laws

Suppose the Poisson *cf* φ above is the product $\varphi_1\varphi_2$ of two *id cf*s φ_1 and φ_2. Then each is a Poisson *cf*.

Proof. Argue as in (iii) above. $G = G_1 + G_2$ and G is degenerate at 1 imply G_1 and G_2 are degenerate at 1. Let $G_i(\{1\}) = \alpha_i,\ \alpha_i \geq 0,\ i = 1,\ 2,\ \alpha_1 + \alpha_2 = \frac{\lambda}{2}$. Further $\sigma_1^2 + \sigma_2^2 = \sigma^2 = 0$. Thus $\varphi_1,\ \varphi_2$ are Poisson *cf*s. □

Linnik's theorem for id laws

(v) On similar lines it can be proved that if φ is the product of a normal *cf* and a Poisson *cf*, $\varphi = \varphi_1\varphi_2$, where $\varphi_1,\ \varphi_2$ are *id cf*s, then each of φ_1 and φ_2 is the product of a normal *cf* and a Poisson *cf*.

The transform φ of a *df* $\mathbb{F}$ in R^p with respect to a kernel function $K(\mathbf{t},\ \mathbf{x})$, $\mathbf{t} \in R^p$, $\mathbf{x} \in R^p$ is defined thus : $\varphi(\mathbf{t}) = \int_{R^p} K(\mathbf{t},\ \mathbf{x})\, \mathrm{d}\mathbb{F}(\mathbf{x})$. The uniqueness theorem and the continuity theorem for the kernel function $K(\mathbf{t},\ \mathbf{x}) = e^{i\mathbf{t}'\mathbf{x}}$ were proved respectively in **(2.3.4)** and **(2.3.5)**. In **(3.4.3)** the uniqueness theorem for the kernel function $K(\mathbf{t},\ \mathbf{x}) = J(\mathbf{t},\ \mathbf{x})$ (ref. **(3.4.1)**) was proved. We now prove the continuity theorem for this kernel function.

Let $\mu_n,\ n \geq 0$ be measures in $\mathscr{R}^p$ with $0 < \mu_n(R^p) = a_n < \infty$ and with $\mu_n(\{\mathbf{0}\}) = 0$. Let $\mathbb{G}_n(\mathbf{x}) = \mu_n(I_\mathbf{x})$ where $I_\mathbf{x} = \times_1^p(-\infty,\ x_j]$. We write simply $\mu_n \overset{d}{\to} \mu_0$ or $\mathbb{G}_n \overset{d}{\to} \mathbb{G}_0$ if $a_n \to a_0$ and if $a_n^{-1}\mu_n \overset{d}{\to} a_0^{-1}\mu_0$. We recall (ref. **(3.4.1)**) that for each $\mathbf{t} \in R^p$, $J(\mathbf{t},\ \mathbf{x})$ is a bounded function of $\mathbf{x}$. Let $A(\varepsilon) = \{\mathbf{x} : 0 < |\mathbf{x}| \leq \varepsilon\}$.

Lemma 3.4.7. *Recall* $A(1) = \{\boldsymbol{x}\ :\ 0 < |\boldsymbol{x}| < 1\}$ *Suppose that* $\lambda = \sup_{n\geq 1} \mu_n(A(1)) < \infty$. *Then*

$$g(\boldsymbol{t}) = \lim_{\varepsilon\to 0}\lim_{n\to\infty} \int_{A(\varepsilon)} J(\boldsymbol{t},\ \boldsymbol{x})\, d\mu_n(\boldsymbol{x}) \text{ exists finitely iff}$$

$$h(\boldsymbol{t}) = \lim_{\varepsilon\to 0}\lim_{n\to\infty} \int_{A(\varepsilon)} \frac{(\boldsymbol{t}'\boldsymbol{x})^2}{|\boldsymbol{x}|^2}\, d\mu_n(\boldsymbol{x}) \text{ exists finitely.}$$

And when the limits exist, $g(\boldsymbol{t}) = -\frac{1}{2}h(\boldsymbol{t})$.

Proof. $g(\mathbf{t})$ exists finitely *iff*

$$\lim_{\varepsilon\to 0}\lim_{n\to\infty} \int_{A(\varepsilon)} \{i\mathbf{t}'\mathbf{x} - \frac{1}{2}(\mathbf{t}'\mathbf{x})^2\frac{1+|\mathbf{x}|^2}{|\mathbf{x}|^2} + \frac{1+|\mathbf{x}|^2}{|\mathbf{x}|^2}\sum_{k=3}^{\infty}\frac{(i\mathbf{t}'\mathbf{x})^k}{k!}\}\, \mathrm{d}\mu_n(\mathbf{x})$$

exists finitely. Now

$$\lim_{\varepsilon\to 0}\lim_{n\to\infty} \int_{A(\varepsilon)} |i\mathbf{t}'\mathbf{x}|\, \mathrm{d}\mu_n(\mathbf{x}) \leq \lim_{\varepsilon\to 0}\lim_{n\to\infty} |\mathbf{t}|\varepsilon\lambda = 0.$$

Also

$$\lim_{\varepsilon\to 0}\lim_{n\to\infty}\int\limits_{A(\varepsilon)}\{\frac{1+|\mathbf{x}|^2}{|\mathbf{x}|^2}\sum_{k=3}^{\infty}\frac{(|\mathbf{t}'\mathbf{x}|)^k}{k!}\}\,\mathrm{d}\mu_n(\mathbf{x})\le\lim_{\varepsilon\to 0}\lim_{n\to\infty}\varepsilon\sum_{k=3}^{\infty}\frac{|\mathbf{t}|^k}{k!}\lambda=0.$$

Again,

$$\lim_{\varepsilon\to 0}\lim_{n\to\infty}\int\limits_{A(\varepsilon)}|\mathbf{t}'\mathbf{x}|^2\,\mathrm{d}\mu_n(\mathbf{x}\le\lim_{\varepsilon\to 0}\lim_{n\to\infty}|\mathbf{t}|^2\varepsilon^2\lambda=0.$$

Hence $g(\mathbf{t})$ exists finitely *iff*

$$\lim_{\varepsilon\to 0}\lim_{n\to\infty}\int\limits_{A(\varepsilon)}\frac{(\mathbf{t}'\mathbf{x})^2}{|\mathbf{x}|^2}\,\mathrm{d}\mu_n(\mathbf{x})$$

exists finitely. □

An example to show that $h(\boldsymbol{t})$ may not exist when $p>1$.

Let $p=2$, $\mu_{2k}(\{(\frac{1}{2k},\ \frac{1}{2k})\})=1$, $\mu_{2k-1}(\{(\frac{1}{2k},\ \frac{1}{k})\})=1$, $k\ge 1$, $\mu_0(\{\mathbf{0}\})=1$. That $\mu_n\overset{d}{\to}\mu_0$ is obvious (ref. **(2.4.3)**(ii)). But

$$\int\limits_{R^2\sim\{\mathbf{0}\}}\frac{(\mathbf{t}'\mathbf{x})^2}{|\mathbf{x}|^2}\,\mathrm{d}\mu_n(\mathbf{x})=\frac{(t_1+t_2)^2}{2}\ \text{or}\ \frac{(t_1+2t_2)^2}{5}$$

according as n is even or odd.

Theorem 3.4.8. *Define $B(r)=\{\boldsymbol{x}:|\boldsymbol{x}|\ge r\}$. If $\Re(\psi_n(\boldsymbol{t}))\to\Re(\psi_0(\boldsymbol{t}))$, $\boldsymbol{t}\in R^p$ then (i) $\sup\limits_{n\ge 1}\mu_n(A(1))<\infty$ (ii) $\sup\limits_{n\ge 1}\mu_n(B(1))<\infty$ and (iii)* $\lim\limits_{r\to\infty}\sup\limits_{n\ge 1}\mu_n(B(r))=0.$

Proof. (i) We note that at every $\mathbf{x}=(x_1,\ x_2,\ldots,\ x_p)$, it is not possible to have $|x_j|<\frac{|\mathbf{x}|}{\sqrt{p}}$ for all j, $1\le j\le p$. Hence $R^p\sim\{\mathbf{0}\}=\cup_{j=1}^{p}\Lambda_j$ where $\Lambda_j=\{\mathbf{x}\ :\ \mathbf{x}\ne\mathbf{0}$ and $|x_j|\ge\frac{|\mathbf{x}|}{\sqrt{p}}\}$. Define, for $1\le j\le p$, $S_j=\{\mathbf{x}:0<|\mathbf{x}|<1,\ |x_j|\ge\frac{|\mathbf{x}|}{\sqrt{p}}\}$ and note that $A(1)=\cup_1^p S_j$. Denote by $\mathbf{v}_j$, $1\le j\le p$, the jth column vector of the $p\times p$ identity matrix. Hence, given $\eta>0$, we have for all n large,

$$\begin{aligned}
\eta - \Re(\psi_0(\mathbf{v}_j)) &\geq -\Re(\psi_n(\mathbf{v}_j)) \\
&= \frac{1}{2}\sigma_{j,j}(n) + \int_{R^p \sim \{\mathbf{0}\}} (1 - \cos x_j)\frac{1 + |\mathbf{x}|^2}{|\mathbf{x}|^2}\, \mathrm{d}\mu_n(\mathbf{x}) \\
&\geq \int_{S_j} (1 - \cos x_j)\frac{1 + |\mathbf{x}|^2}{|\mathbf{x}|^2}\, \mathrm{d}\mu_n(\mathbf{x}) \geq \int_{S_j} \frac{1 - \cos \frac{|\mathbf{x}|}{\sqrt{p}}}{|\mathbf{x}|^2}\, \mathrm{d}\mu_n(\mathbf{x}) \\
&\geq \frac{1}{2p}\int_{S_j} \frac{1 - \cos \frac{|\mathbf{x}|}{\sqrt{p}}}{\frac{|\mathbf{x}|^2}{2p}}\, \mathrm{d}\mu_n(\mathbf{x}) \geq \frac{1}{2p}\int_{S_j} \frac{4}{\pi^2}\, \mathrm{d}\mu_n(\mathbf{x}) \\
&\geq \frac{1}{8p}\mu_n(S_j).
\end{aligned}$$

This implies

$$\mu_n(A(1)) \leq \sum_{j=1}^{p} \mu_n(S_j) \leq 8p^2\eta + 8p\sum_{1}^{p}\{-\Re(\psi_0(\mathbf{v}_j))\} < \infty.$$

(ii) For $r > 0$ and $1 \leq j \leq p$, define $T_j(r) = \{\mathbf{x} : |\mathbf{x}| \geq r,\ |x_j| \geq \frac{|\mathbf{x}|}{\sqrt{p}}\}$ and note that $B(r) = \cup_1^p T_j(r)$. Fix target error $\eta > 0$. Remembering that convergence of ψ_n to ψ_0 is necessarily uniform over bounded **t**-'intervals', we have : For n large and for all $s \in [0,\ \sqrt{2p}]$,

$$\eta - \Re(\psi_0(sv_j)) \geq \int_{T_j(r)} (1 - \cos sx_j)\, \mathrm{d}\mu_n(\mathbf{x}).$$

We recall that ψ_0 is a continuous function. Hence integrating over $[0,\ \frac{2\sqrt{p}}{r}]$, we get

$$\begin{aligned}
\eta + \frac{r}{2\sqrt{p}} &\int_0^{\frac{2\sqrt{p}}{r}} \{-\Re(\psi_0(s\mathbf{v}_j))\}\, \mathrm{d}s \\
&\geq \int_{T_j(r)} \left\{1 - \frac{\sin(\frac{2x_j\sqrt{p}}{r})}{\frac{2x_j\sqrt{p}}{r}}\right\} \mathrm{d}\mu_n(\mathbf{x}) \\
&\geq \frac{1}{2}\mu_n(T_j(r)).
\end{aligned}$$

(by Fubini's theorem). Hence

$$\mu_n(B(r)) \leq 2p\eta + \frac{r}{\sqrt{p}} \sum_{j=1}^{p} \int_0^{2\frac{\sqrt{p}}{r}} \{-\Re(\psi_0(s\mathbf{v}_j))\}\, \mathrm{d}s$$

for all n large. (ii) follows now by putting $r = 1$. (iii) follows by noting that, for each j,

$$\lim_{r\to\infty} \frac{r}{\sqrt{p}} \int_0^{2\frac{\sqrt{p}}{r}} \{-\Re(\psi_0(s\mathbf{v}_j))\}\, \mathrm{d}s = 0$$

since $\lim_{s\to 0} \psi_0(s\mathbf{v}_j) = 0$. □

In **(3.3.2)** it was shown that a cf which is the limit of a sequence of id cfs is id. The theorem below in **(3.4.9)** shows that, in such a situation more can be said, that the corresponding parameters converge.

Theorem 3.4.9. *Let*

$$\psi_n(\mathbf{t}) = i\boldsymbol{a}_n'\boldsymbol{t} - \frac{1}{2}\boldsymbol{t}'Q_n\boldsymbol{t} + \int_{R^p\sim\{\boldsymbol{0}\}} J(\boldsymbol{t},\, \boldsymbol{x})\, d\mu_n(\boldsymbol{x}),\ n \geq 0.$$

Then (i) ⇔ ((ii), (iii) and (iv)) where

(i) $\psi_n(\boldsymbol{t}) \to \psi_0(\boldsymbol{t})$, $\boldsymbol{t} \in R^p$

(ii) *:* $\mu_n \overset{d}{\to} \mu_0$.

(iii) *:*

$$\boldsymbol{t}'Q_n\boldsymbol{t} + \int_{R^p\sim\{\boldsymbol{0}\}} \frac{(\boldsymbol{t}'\boldsymbol{x})^2}{|\boldsymbol{x}|^2}\, d\mu_n(\boldsymbol{x}) \to \boldsymbol{t}'Q_0\boldsymbol{t} + \int_{R^p\sim\{\boldsymbol{0}\}} \frac{(\boldsymbol{t}'\boldsymbol{x})^2}{|\boldsymbol{x}|^2}\, d\mu_0(\boldsymbol{x})$$

and

(iv) *:* $\boldsymbol{a}_n \to \boldsymbol{a}_0$.

Proof. Let (i) hold. The claim that $\sup_{n\geq 1} \mu_n(R^p) < \infty$ follows from **(3.4.8)** since $A(1) \cup B(1) = R^p$. Again from the same theorem we know that, given $\varepsilon > 0$ we can find $r > 0$ such that $\sup_{n\geq 1} \mu_n(B(r)) < \varepsilon$. Hence, **(2.2.2)** with an obvious modification applies and there exists a subsequence $(\mu_{n_k},\ k \geq 1)$ and finite measure μ^* such that $\mu_{n_k} \overset{d}{\to} \mu^*$. We note that $\mu^*(\{\boldsymbol{0}\})$ may not be 0. Write

$$\Re(\psi_{n_k}(t)) = -\tfrac{1}{2}\mathbf{t}'Q_{n_k}\mathbf{t} + \int\limits_{R^p} \left(\cos\mathbf{t}'\mathbf{x} - 1 + \tfrac{1}{2}\tfrac{(\mathbf{t}'\mathbf{x})^2}{1+|\mathbf{x}|^2}\right)\tfrac{1+|\mathbf{x}|^2}{|\mathbf{x}|^2}\, \mathrm{d}\mu_{n_k}(\mathbf{x})$$
$$- \tfrac{1}{2} \int\limits_{R^p\sim\{\mathbf{0}\}} \tfrac{(\mathbf{t}'\mathbf{x})^2}{|\mathbf{x}|^2}\, \mathrm{d}\mu_{n_k}(\mathbf{x})$$

where the integrand in the first integral is defined to be zero at $\mathbf{0}$. So defined, it is a bounded continuous function. The splitting of the integral is justified : each integral exists finitely since each integrand is a bounded measurable function. The integrand being a bounded continuous function, Helly-Bray theorem applies to the first integral. Its limit exists and is equal to

$$\int\limits_{R^p} \left(\cos\mathbf{t}'\mathbf{x} - 1 + \frac{1}{2}\frac{(\mathbf{t}'\mathbf{x})^2}{1+|\mathbf{x}|^2}\right)\frac{1+|\mathbf{x}|^2}{|\mathbf{x}|^2}\, \mathrm{d}\mu^*(\mathbf{x}).$$

This implies that if

$$\mathbf{t}'J_n\mathbf{t} = \int\limits_{R^p\sim\{\mathbf{0}\}} \frac{(\mathbf{t}'\mathbf{x})^2}{|\mathbf{x}|^2}\, \mathrm{d}\mu_n(\mathbf{x}),$$

then $\lim\limits_k\{\mathbf{t}'(Q_{n_k} + J_{n_k})\mathbf{t}\}$ exists, equal to, say, $\mathbf{t}'\tilde{J}\mathbf{t}$. Letting $k \to \infty$, we get

$$\Re(\psi_0(\mathbf{t})) = -\frac{1}{2}\mathbf{t}'\tilde{J}\mathbf{t} + \int\limits_{R^p} \left(\cos\mathbf{t}'\mathbf{x} - 1 + \frac{1}{2}\frac{(\mathbf{t}'\mathbf{x})^2}{1+|\mathbf{x}|^2}\right)\frac{1+|\mathbf{x}|^2}{|\mathbf{x}|^2}\, \mathrm{d}\mu^*(\mathbf{x}).$$

Writing

$$\mathbf{t}'J_0\mathbf{t} = \int\limits_{R^p\sim\{\mathbf{0}\}} \frac{(\mathbf{t}'\mathbf{x})^2}{|\mathbf{x}|^2}\, \mathrm{d}\mu^*(\mathbf{x}),$$

the above relation takes the form :

$$\Re(\psi_0(\mathbf{t})) = -\frac{1}{2}\mathbf{t}'(\tilde{J} - J_0)\mathbf{t} + \int\limits_{R^p\sim\{\mathbf{0}\}} (\cos\mathbf{t}'\mathbf{x} - 1)\frac{1+|\mathbf{x}|^2}{|\mathbf{x}|^2}\, \mathrm{d}\mu^*(\mathbf{x}).$$

Appealing to **(3.4.3)** or arguing on the lines of that section, we can prove $\tilde{J} - J_0 = Q_0$ and $\mu^* = \mu_0$.

It follows now that every weakly convergent subsequence of the tight sequence (μ_n) converges to the same measure μ_0. This is equivalent to : $\mu_n \overset{d}{\to} \mu_0$ (ref. **(2.4.3)**(iv)). Thus (i) $\Rightarrow$ (ii). We have:

$$- \Re(\psi_n(t)) = -\frac{1}{2}\mathbf{t}'(Q_n + J_n)\mathbf{t}$$
$$+ \int\limits_{R^p} \left(1 - \cos\mathbf{t}'\mathbf{x} - \frac{1}{2}\frac{(\mathbf{t}'\mathbf{x})^2}{1+|\mathbf{x}|^2}\right)\frac{1+|\mathbf{x}|^2}{|\mathbf{x}|^2}\, \mathrm{d}\mu_n(\mathbf{x}).$$

Since it has been shown that $\mu_n \xrightarrow{d} \mu_0$ and since the integrand is a bouned continuous function Helly-Bray theorem applies and we conclude that the integral converges to

$$\int_{R^p} (1 - \cos \mathbf{t}'\mathbf{x} - \frac{1}{2}\frac{(\mathbf{t}'\mathbf{x})^2}{1+|\mathbf{x}|^2})\frac{1+|\mathbf{x}|^2}{|\mathbf{x}|^2}\, \mathrm{d}\mu_0(\mathbf{x}) = -\Re\psi_0(\mathbf{t}) - \frac{1}{2}\mathbf{t}'(Q_0 + J_0)\mathbf{t}.$$

Since $\Re(\psi_n(\mathbf{t})) \to \Re(\psi_0(\mathbf{t}))$, this implies that $\lim_{n\to\infty} (\mathbf{t}'(Q_n + J_n)\mathbf{t})$ exists and that this limit is equal to $\mathbf{t}'(Q_0 + J_0)\mathbf{t}$. Thus (i) $\Rightarrow$ (iii).

We have

$$\begin{aligned}\psi_n(\mathbf{t}) = i\mathbf{a}_n'\mathbf{t} - \frac{1}{2}\mathbf{t}'(Q_n + J_n)\mathbf{t}\\ + \int_{R^p} \{e^{i\mathbf{t}'\mathbf{x}} - 1 - \frac{i\mathbf{t}'\mathbf{x}}{1+|\mathbf{x}|^2} + \frac{1}{2}\frac{(\mathbf{t}'\mathbf{x})^2}{1+|\mathbf{x}|^2}\}\frac{1+|\mathbf{x}|^2}{|\mathbf{x}|^2}\, \mathrm{d}\mu_n(\mathbf{x}),\end{aligned}$$

the integrand being taken as zero at $\mathbf{x} = \mathbf{0}$. By Helly-Bray theorem, the limit of the integral exists. $\lim_n \psi_n(\mathbf{t})$ exists by hypothesis. That $\lim_n(Q_n + J_n)$ exists has been shown. Hence $\lim_n \mathbf{a}_n$ exists, equal to $\mathbf{b}$, say. This implies that $\mathbf{t}'(\mathbf{a}_0 - \mathbf{b}) = 0$ for all $\mathbf{t}$. Hence $\mathbf{b} = \mathbf{a}_0$. Thus (i) $\Rightarrow$ (iv)

To prove the converse, assume (ii), (iii) and (iv) hold. It is noted that (ii) implies $\sup_n \mu_n(R^p) = \beta < \infty$.

Choose a sequence of τ-values, monotonically decreasing to 0 such that for each τ in the sequence, $B(\tau)$ is a continuity set for μ_0. Now

$$\psi_n(\mathbf{t}) - i\mathbf{a}_n'\mathbf{t} = -\frac{1}{2}\mathbf{t}'Q_n\mathbf{t} + \int_{A(\tau)} J(\mathbf{t},\, \mathbf{x})\, \mathrm{d}\mu_n(\mathbf{x}) + \int_{B(\tau)} J(\mathbf{t},\, \mathbf{x})\, \mathrm{d}\mu_n(\mathbf{x}).$$

The last integral converges, as $n \to \infty$, to $\int_{B(\tau)} J(\mathbf{t}, \mathbf{x})\, \mathrm{d}\mu_0(\mathbf{x})$ (ref. **(2.4.3)**(v)). As $\tau \downarrow 0$ along the chosen sequence, this tends to

$\int_{R^p \sim \{\mathbf{0}\}} J(\mathbf{t}, \mathbf{x})\, d\mu_0(\mathbf{x})$. Now,

$$-\frac{1}{2}\mathbf{t}'Q_n\mathbf{t} + \int\limits_{A(\tau)} J(\mathbf{t}, \mathbf{x})\, d\mu_n(\mathbf{x})$$

$$= -\frac{1}{2}\mathbf{t}'(Q_n + J_n)\mathbf{t} + \frac{1}{2}\int\limits_{B(\tau)} \frac{(\mathbf{t}'\mathbf{x})^2}{|\mathbf{x}|^2}\, d\mu_n(\mathbf{x})$$

$$+ \int\limits_{A(\tau)} \{e^{i\mathbf{t}'\mathbf{x}} - 1 - i\mathbf{t}'\mathbf{x} + \frac{1}{2}(\mathbf{t}'\mathbf{x})^2\}\frac{1 + |\mathbf{x}|^2}{|\mathbf{x}|^2}\, d\mu_n(\mathbf{x})$$

$$+ \int\limits_{A(\tau)} i\mathbf{t}'\mathbf{x}\, d\mu_n(\mathbf{x}) - \frac{1}{2}\int\limits_{A(\tau)} (\mathbf{t}'\mathbf{x})^2\, d\mu_n(\mathbf{x}).$$

By (iii), limit of the first term on the right side is $-\frac{1}{2}\mathbf{t}'(Q_0 + J_0)\mathbf{t}$; the second term tends to $\frac{1}{2}\int_{B(\tau)} \frac{(\mathbf{t}'\mathbf{x})^2}{|\mathbf{x}|^2}\, d\mu_0(\mathbf{x})$ by (ii) and **(2.4.3)**(v); as $\tau \downarrow 0$ along the chosen sequence, this expression tends to $\frac{1}{2}\mathbf{t}'J_0\mathbf{t}$; the integrals in the third, fourth and the last terms are bounded respectively by $\frac{1}{6}q|\mathbf{t}|(1+\tau^2)\tau$, $q|\mathbf{t}|\tau$ and $\frac{1}{2}q|\mathbf{t}|^2\tau^2$. Hence each of the last three terms has limit zero as $n \to \infty$ followed by $\tau \to 0$ along the chosen sequence. Thus

$$\lim_n\{\psi_n(\mathbf{t}) - i\mathbf{a}_n'\mathbf{t}\} = -\frac{1}{2}\mathbf{t}'Q_0\mathbf{t} + \int\limits_{R^p \sim \{\mathbf{0}\}} J(\mathbf{t}, \mathbf{x})\, d\mu_0(\mathbf{x}).$$

Since, by (iv), $\mathbf{a}_n \to \mathbf{a}_0$, the proof is complete that (ii), (iii) and (iv) imply (i). □

Remark 3.4.10. In the case $p = 1$, condition (iii) reduces to $\sigma_n^2 \to \sigma_0^2$.

3.5 Discrete distributions and lattice distributions

Definition 3.5.1. A rV **X** is said to be discrete valued or to have a discrete *df* if the range of **X** is countable. i.e. all the mass of the distribution is concentrated on its atoms.

The range of a discrete valued X can be a dense subset of R^p. To see this let $r_1, r_2, \ldots$ be a listing of the rationals in R and consider rv X where $P(X = r_n) = \frac{6}{\pi^2}\frac{1}{n^2}, n = 1, 2, \ldots$.

Definition 3.5.2. A discrete univariate rv X is said to have a lattice distribution or said to be lattice if all its atoms are situated at the points $a+nb$ for some constants $b \neq 0$ and a where n runs through integers, not necessarily through all the integers. A discrete rV is said to have a lattice distribution or said to be lattice *iff* each of its components is lattice.

Theorem 3.5.3. *Let X be a rv with df F and cf φ. Then X is lattice iff $|\varphi(u)| = 1$ for some $u \neq 0$.*

Proof. Let X be lattice with range $a + nb$ where $b \neq 0$ and $n \in \mathcal{N}$, $\mathcal{N}$ being a set of integers. Then for

$$\varphi(\frac{2\pi}{b}) = \sum_{n\in\mathcal{N}} P(X = a + nb)e^{i\frac{2\pi}{b}a+i2n\pi}$$

$$= e^{i\frac{2\pi}{b}a} \sum_{n\in\mathcal{N}} P(X = a + nb) = e^{i\frac{2\pi}{b}a}.$$

Hence $|\varphi(\frac{2\pi}{b})| = 1$.

Conversely, suppose $|\varphi(u)| = 1$, $u \neq 0$. Hence $\varphi(u) = e^{iv}$ for some v real. With $v = au$, we have $1-e^{-iua}\varphi(u) = 0$. i.e. $\int_R\{1-e^{iu(x-a)}\}dF(x) = 0$. This implies that

$$\int_R \{1 - \cos u(x - a)\}dF(x) = 0.$$

Since the integrand is non-negative, it follows that μ_F is concentrated on a set contained in the set of points x satisfying $\cos u(x - a) = 1$.
i.e., in the set $\{a + \frac{2\pi}{u}n\}$, n an integer. Thus X is lattice. □

Remarks 3.5.4. (1) Let rV $\mathbf{X} \in R^p$, $p \geq 2$ have *df* $\mathbb{F}$ and *cf* φ. If $\mathbf{X}$ is lattice, then there exists $\mathbf{u} \neq \mathbf{0}$ such that $|\varphi(\mathbf{u})| = 1$.

Proof. Directly from definition, it follows component X_1 is lattice. By the result for univariate lattice variables, there exists $u \neq 0$ such that the absolute value of its *cf* at u is 1. i.e., $|\varphi(u,\ 0,\ 0, \ldots,\ 0)| = 1$.

(2) The converse of the result in (1) is generally not true. To see this, let X be a standard normal variable. Then the *cf* $\varphi(t,\ u)$ of the bivariate variable $(X,\ -X)$ is $e^{-\frac{1}{2}(t-u)^2}$, which is equal to 1 for all $t = u$ but $(X,\ -X)$ is not lattice.

However, it is not difficult to prove that if $\mathbf{X}$ has independent components, then the converse is true.

(3) In R, we defined X to be lattice if the measure μ_F is concentrated in a set of the type $\{a + nb\}$ as n runs over some integers. In R^p, $p \geq 2$, let us examine a parallel definition of lattice variables : Say a discrete rV **X** with *df* $\mathbb{F}$ is lattice if there exist vectors $\mathbf{b} \neq \mathbf{0}$ and **a** such that the measure $\mu_{\mathbb{F}}$ is concentrated in the set $\{\mathbf{a} + n\mathbf{b}\}$ as n runs over a set of integers. If any component b_i of **b** is zero, then it implies that the component X_i of **X** is degenerate at a_i where $\mathbf{a} = (a_1, a_2, \ldots, a_p)'$ and then the problem reduces to one in R^{p-1}. Hence we assume $\min\limits_{1 \leq i \leq p} |b_i| > 0$. We note

$$\varphi(\mathbf{t}) = \sum_n P(\mathbf{X} = \mathbf{a} + n\mathbf{b})e^{i\mathbf{t}'(\mathbf{a}+n\mathbf{b})}.$$

Setting $\mathbf{u} = (\frac{2\pi}{b_1}, \frac{2\pi}{b_2}, \ldots, \frac{2\pi}{b_p})'$, we get

$$\begin{aligned}\varphi(\mathbf{u}) &= e^{i\mathbf{u}'\mathbf{a}} \sum_{n=1}^{\infty} P(\mathbf{X} = \mathbf{a} + n\mathbf{b})e^{in\mathbf{u}'\mathbf{b}} \\ &= e^{i\mathbf{u}'\mathbf{a}} \sum_{n=1}^{\infty} P(\mathbf{X} = \mathbf{a} + n\mathbf{b})e^{2i\pi n} = e^{i\mathbf{u}'\mathbf{a}}.\end{aligned}$$

Thus $|\varphi(\mathbf{u})| = 1$.

Using the same counter example detailed above, we see the converse is not true for this definition of lattice variables too.

One other point deserves notice. If a p-vector variable is lattice according to the new definition, then clearly, each of its p component variables is lattice. But the converse is not true. For, let X take the values $-1,\ 1$ and let Y also take the values $-1,\ 1$, then both X and Y are lattice. Now, $(X,\ Y)$ takes the 4 values $(-1,\ -1)$, $(-1,\ 1)$, $(1,\ -1)$, and $(1,\ 1)$. We claim that $(X,\ Y)$ is not lattice. For, if possible, let there exist vectors $(a,\ b)$; $(c,\ d)$ and integers $n_1,\ n_2,\ n_3,\ n_4$ such that the 4 values of $(X,\ Y)$ can be put in the form $(a,\ b) + n_i(c,\ d)$. This would imply the relations $-1 = b + n_1 d$; $1 = b + n_2 d$; $-1 = b + n_3 d$; $1 = b + n_4 d$. Hence $(n_2 - n_4)d = 0$; and $(n_1 - n_3)d = 0$.

Case 1. $d = 0$. This implies $b = 1$ and $b = -1$ simultaneously.

Case 2. $d \neq 0$. This implies $n_1 = n_3$ and $n_2 = n_4$. This is not possible since $n_1 = n_3$ leads to $a + n_1 c = 1$ and $a + n_1 c = -1$ simultaneously. Thus $(X,\ Y)$ is not lattice.

Again, the above example shows that the new definition fails to meet one's natural expectation for $(X,\ Y)$ to be considered lattice distributed.

3.6 Moments and characteristic functions

Theorem 3.6.1. *If, for $n \geq 1$, $\mathbb{E}|X|^n < \infty$, then the cf φ of X is continuously differentiable everywhere n times. Further*

$$\varphi^{(n)}(t) = i^n \int_{-\infty}^{\infty} e^{itx} x^n dF(x).$$

In particular $\varphi^{(n)}(0) = i^n EX^n$.

Proof. We note

$$\left|\frac{\varphi(t+h) - \varphi(t)}{h}\right| \leq \int_{-\infty}^{\infty} \left|\frac{e^{ixh} - 1}{ihx}\right| |x| \mathrm{d}F(x) \leq \int_{-\infty}^{\infty} \left|\frac{\sin(\frac{hx}{2})}{\frac{hx}{2}}\right| |x| \mathrm{d}F(x).$$

Since $|\frac{\sin\theta}{\theta}| \leq 1$ for all θ and since by hypothesis $\int_{-\infty}^{\infty} |x| \mathrm{d}F(x) < \infty$, the bounded convergence theorem applies and we have

$$\lim_{h \to 0} \frac{\varphi(t+h) - \varphi(t)}{h} = \int_{-\infty}^{\infty} \lim_{h \to 0} e^{itx} \frac{e^{ihx} - 1}{h} \mathrm{d}F(x).$$

Thus $\varphi'(t)$ exists finitely for all t and equals $\int_{-\infty}^{\infty} e^{itx} x \mathrm{d}F(x)$. The boundedness and continuity of φ' and the relation $\varphi'(0) = i\mathrm{E}X$ follow from this.

The case for $n > 1$ is proved by induction: Assume then that $\varphi^{(n)}(t) = i^n \int_{-\infty}^{\infty} e^{itx} x^n \mathrm{d}F(x)$. Hence

$$\frac{\varphi^{(n)}(t+h) - \varphi^{(n)}(t)}{h} = i^n \int_{-\infty}^{\infty} e^{itx} \frac{e^{ith} - 1}{h} x^n \mathrm{d}F(x).$$

The rest of the arguments is as in the case $n = 1$. □

Lemma 3.6.2. *Let $p \geq 0$ be an integer. Let*

$$g_p(t) = (-1)^{p+1} \{\cos t - \sum_{j=0}^{p} (-1)^j \frac{t^{2j}}{(2j)!}\}.$$

(i) *It is easy to verify that, for $p \geq 1$, $g_p(t) = \frac{t^{2p}}{(2p)!} - g_{p-1}(t)$.*

(ii) *For all t, $-\infty < t < \infty$, $g_p(t) \geq 0$.* (Consequently, $g_p(t) \leq \frac{t^{2p}}{(2p)!}$) *and*

(iii) $\lim_{t \to 0} \frac{g_p(t)}{t^{2p+2}} = \frac{1}{(2p+2)!}$.

Proof. (ii) It is enough to consider the range $t \geq 0$ since g_p is an even function. It is easily verified that, if $p \geq 1$, then $g_p''(t) = g_{p-1}(t)$. We have $\frac{\mathrm{d}}{\mathrm{d}t} g_1'(t) = g_2''(t) = g_0(t) = 1 - \cos(t) \geq 0$. $g_0(t) > 0$ for all $t \in [0, \infty)$ except at $2k\pi$, $k = 0, 1, 2, \ldots$ g_1' is a continuous function, strictly increasing in $(0, 2\pi)$ and $g_1'(0) = 0$. Hence $g_1'(t) > 0$ for all $t > 0$. This, in turn, implies that g_1 is strictly increasing in $[0, \infty)$. g_1 is a continuous function and $g_1(0) = 0$. Hence $g_1(t) > 0$ in $(0, \infty)$. Starting with g_1 now and arguing on similar lines, we show that g_2 is positive in $(0, \infty)$. This argument can be kept up to establish the claim that $g_p(t) \geq 0$ for all t.

(iii) For $t \neq 0$, we have by Taylor expansion

$$\cos t = \sum_{0}^{2p} (-1)^j \frac{t^{2j}}{(2j)!} + (-1)^{p+1} \frac{t^{2p+2}}{(2p+2)!} \cos(\theta t).$$

The claim follows from this. □

When $\mathbb{E}X^{2p} < \infty$, define $u_p(t) = \mathbb{E}g_p(tX)$.

Theorem 3.6.3. *The following two statements are equivalent.*

(i) $\lim_{t \to 0} \frac{u_p(t)}{t^{2p+2}}$ *exists finitely, equal to, say,* α.

(ii) $\mathbb{E}X^{2p+2} < \infty$. *And then* $\alpha = \frac{1}{(2p+2)!} \mathbb{E}X^{2p+2}$

Proof. Suppose (i) holds. Then

$$\begin{aligned}
\infty > \lim_{t \to 0} \frac{u_p(t)}{t^{2p+2}} &= \lim_{t \to 0} \int_{-\infty}^{\infty} \frac{g_p(tx)}{t^{2p+2}} \mathrm{d}F(x) \\
&\geq \int_{-\infty}^{\infty} \lim_{t \to 0} \frac{g_p(tx)}{t^{2p+2}} \mathrm{d}F(x) \quad \text{(by Fatou's Lemma)} \\
&= \frac{1}{(2p+2)!} \int_{-\infty}^{\infty} x^{2p+2} \mathrm{d}F(x) \quad \text{(by } \mathbf{(3.6.2)}(iii)\text{)}.
\end{aligned}$$

Thus (i) implies (ii).

Suppose now (ii) holds.

$$\lim_{t\to 0}\frac{u_p(t)}{t^{2p+2}} = \lim_{t\to 0}\int_{-\infty}^{\infty}\frac{g_p(tx)}{t^{2p+2}}\mathrm{d}F(x)$$

$$= \int_{-\infty}^{\infty}\lim_{t\to 0}\frac{g_p(tx)}{(tx)^{2p+2}}x^{2p+2}\mathrm{d}F(x)$$

(by the bounded convergence theorem which applies since by parts (ii) and (iii) of **(3.6.2)**

$$\frac{g_p(tx)}{(tx)^{2p+2}} \leq \frac{1}{(2p+2)!} \text{ for all } t \text{ and } x\Big)$$

$$= \frac{1}{(2p+2)!}\int_{-\infty}^{\infty}x^{2p+2}\mathrm{d}F(x) = \alpha. \qquad \square$$

Theorem 3.6.4 (Partial converse to **(3.6.1)**). *If, for $n \geq 1$, $\varphi^{(2n)}(0)$ exists finitely, then* (i) $\mathbb{E}X^{2n} < \infty$ *and* (ii) $\mathbb{E}X^{2n} = (-1)^n\varphi^{(2n)}(0)$.

Proof. It remains for us to establish (i) only since (ii) is a consequence of (i) by **(3.6.1)**. We give the proof first for the case $n = 1$ and then establish the claim for $n > 1$ by the method of induction. Let $u(t)$ be the real part of $\varphi(t)$.

Case $n = 1$. The hypothesis implies that $u''(0)$ exists finitely. This, in turn implies that $u'(t)$ exists finitely for all t near 0. Let us evaluate $u'(0)$. It is equal to $\lim\limits_{t\to 0}\frac{u(t)-1}{t}$. The function under the limit sign is ≤ 0 for $t > 0$ and is ≥ 0 for $t < 0$. Hence $u'(0) = 0$.

$$u''(0) = \lim_{t\to 0}\frac{u'(t)-u'(0)}{t} = \lim_{t\to 0}\frac{u'(t)}{t}.$$

From

$$\frac{1-u(t)}{t^2} = |\frac{u'(\theta t)}{t}| = |\frac{u'(\theta t)}{\theta t}||\theta| \leq |\frac{u'(\theta t)}{\theta t}| \text{ (since } |\theta| \leq 1),$$

we conclude

$$\overline{\lim_{t\to 0}}\frac{1-u(t)}{t^2} \leq |u''(0)| < \infty.$$

We then have

$$\infty > \lim_{t\to 0}\frac{1-u(t)}{t^2} = \frac{1}{2}\lim_{t\to 0}\int_{-\infty}^{\infty}\left(\frac{\sin(\frac{tx}{2})}{\frac{tx}{2}}\right)^2 x^2 \mathrm{d}F(x)$$

$$\geq \frac{1}{2}\int_{-\infty}^{\infty}\lim_{t\to 0}\left(\frac{\sin(\frac{tx}{2})}{\frac{tx}{2}}\right)^2 x^2 \mathrm{d}F(x) \quad \text{(by Fatou's Lemma)}$$

$$= \frac{1}{2}\int_{-\infty}^{\infty} x^2 \mathrm{d}F(x).$$

Thus the finiteness of $\varphi''(0)$ implies $\mathrm{E}X^2 < \infty$, proving the validity of (i).

Case $n \geq 2$. Assume that the claim has been established for $1 \leq n \leq k$. Let $\varphi^{(2k+2)}(0)$ exist finitely. By the induction hypothesis, the existence finitely of $\varphi^{(2k)}(0)$ implies $\mathbb{E}X^{2k} < \infty$. This, in turn, implies (ref. **(3.6.1)**) that $\varphi^{(2k)}(t)$ exists finitely for all t and that

$$\varphi^{(2k)}(t) = (-1)^n \int_{-\infty}^{\infty} e^{itx} x^{2k} \mathrm{d}F(x).$$

Write

$$G_k(x) = \frac{1}{\alpha_{2k}}\int_{-\infty}^{x} y^{2k}\mathrm{d}F(y)$$

where $\alpha_{2k} = \int_{-\infty}^{\infty} y^{2k}\mathrm{d}F(y)$ and note that G_k is a *df*. Its *cf*

$$g_k(t) = \int_{-\infty}^{\infty} e^{itx}\mathrm{d}G_k(x) = \frac{1}{\alpha_{2k}}(-1)^k \varphi^{(2k)}(t).$$

It follows that $g_k''(0)$ exists finitely. Arguing as in the case $n = 1$, we conclude that

$$\int_{-\infty}^{\infty} x^2 \mathrm{d}G_k(x) < \infty$$

or, what is same,

$$\int_{-\infty}^{\infty} x^{2k+2}\mathrm{d}F(x) < \infty. \qquad \square$$

Moments of odd orders

We saw in **(3.6.1)** that if $\mathbb{E}X$ exists finitely, then $\varphi'(0)$ exists finitely. The following example shows that the converse is not true.

Example 3.6.5. Let $\frac{1}{2c} = \sum_{n=2}^{\infty} \frac{1}{n^2 \text{log} n}$. Let $P(X = \pm n) = \frac{c}{n^2 \text{log} n}$ each, $n \geq 2$. Clearly EX^+ =EX^- $= \infty$. Hence EX does not exist. In this case,

$$\varphi(t) = 2c \sum_{n=2}^{\infty} \frac{\cos nt}{n^2 \text{log} n}.$$

We will show that $\varphi'(0)$ exists finitely.

Consider the series

$$\sum_{n=2}^{\infty} \frac{\sin nt}{n \text{log} n}.$$

To establish the convergence pointwise of this series as also to establish its uniform convergence, we need consider only the range $[0, 2\pi]$ for t, since all the terms of the series are periodic functions with period 2π.

Let $\mathrm{u}_r(t) = \sin rt$; $U_k(t) = \sum_{r=1}^{k} \mathrm{u}_r(t)$, $k \geq 1$; and $v_k = \frac{1}{k \text{log} k}$, $k \geq 2$. We note that $v_k > \mathrm{u}_{k+1} > 0$. Then it is easy to verify that

$$\begin{aligned}
\psi_{m,n}(t) &= \sum_{k=m}^{n} \frac{\sin kt}{k \text{log} k} = \sum_{k=m}^{n} v_k \mathrm{u}_k(t) \\
&= \sum_{k=m}^{n-1} U_k(t)(v_k - v_{k+1}) - U_{m-1}(t) v_m + U_n(t) v_n.
\end{aligned}$$

Let $A = \{j\pi, j = 1, 2, ...\}$. For $t \in A$, $U_k(t) = 0$. For other values of t,

$$U_k(t) = \frac{\sin \frac{k}{2} t \sin \frac{k+1}{2} t}{\sin \frac{t}{2}}.$$

Hence for $t \notin A$, $|U_k(t)| \leq \dfrac{1}{|\sin \frac{t}{2}|}$. The convergence pointwise of the series is now immediate since, for $t \notin A$, $|\psi_{m,n}(t)| \leq |\frac{1}{\sin \frac{t}{2}}| 2 v_m \to 0$ as $m \to \infty$.

Denote by $\psi(t)$ the sum of the series $\sum\limits_{n=2}^{\infty} \frac{\sin nt}{n \log n}$.

We will, through the following steps, show that the convergence of the series to ψ is uniform over $[0, 2\pi]$.

Step 1. If $\frac{\pi}{4} \leq t \leq \frac{7\pi}{4}$, $|\psi_{m,n}(t)| \leq \frac{1}{\sin(\frac{\pi}{8})} 2 v_m$. Hence in this region the series converges uniformly.

We record two easily proved results for later use : (i) if $0 \leq \theta \leq 1$ then $\sin\theta > \frac{5\theta}{6}$ and (ii) $\sum\limits_{k=2}^{\infty}(v_k - v_{k+1}) < \infty$.

Step 2. Now we consider the range $I = (0, \frac{\pi}{4})$. Let M be a fixed large integer.

$$\begin{aligned} r_M(t) &= \sum_{M}^{\infty} v_k \mathrm{u}_k(t) = \sum_{M}^{\infty} v_k\{U_k(t) - U_{k-1}(t)\} \\ &= \sum_{M}^{\infty}(v_k - v_{k+1})U_k(t) - v_M U_{M-1}(t). \end{aligned}$$

Hence

$$\begin{aligned} |r_M(t)| &\leq \sum_{M}^{\infty}(v_k - v_{k+1})|U_k(t)| + v_M|U_{M-1}(t)| \\ &\leq \frac{\sum\limits_{M}^{\infty}(v_k - v_{k+1})}{\sin\frac{t}{2}} + \frac{v_M}{\sin\frac{t}{2}} \\ &= \frac{2v_M}{\sin\frac{t}{2}} \leq \frac{24v_M}{5t} < \frac{5v_M}{t} \end{aligned} \tag{3.3}$$

This, we stress, is true for all t in the range.

Let $I_1 = \{t : t \in I, [\frac{1}{t}] + 1 \leq M\}$ and $I_2 = \{t : t \in I, [\frac{1}{t}] + 1 > M\}$. If $t \in I_1$ then $\frac{1}{t} \leq M$ and $|r_M(t)| \leq 5Mv_M \leq \frac{5}{\log M}$.

If $t \in I_2$, then, write

$$N = N(t) = [\frac{1}{t}] + 1, r_M(t) = \sum_{M}^{N-1} v_k \mathrm{u}_k(t) + r_N(t).$$

Hence

$$\begin{aligned} |r_M(t)| &\leq \sum_{M}^{N-1} v_k kt + |r_N(t)| \\ &\leq \text{ (by (3.6.5))} t(N - M)\frac{1}{\log M} + \frac{5v_N(t)}{t} \\ &\leq tN\frac{1}{\log M} + \frac{5Nv_N(t)}{Nt} \leq \frac{2}{\log M} + \frac{5}{\log N} = \frac{7}{\log M}. \end{aligned}$$

Thus for all t in the range $|r_M(t)| \leq \frac{7}{\log M} \to 0$ as $M \to \infty$, proving the uniform convergence of the series to its limit function ψ.

In the region $\frac{7\pi}{4} < t < 2\pi$, the behaviour of $\psi(t)$ is similar to that in I and hence uniform convergence holds for $t \in [0,\ 2\pi]$.

Being the uniform limit of an infinite series of continuous functions, ψ is a continuous function and further

$$\int_0^t \psi(u)\mathrm{d}u = -\sum_2^{\infty} \frac{1}{k^2 \log k}\{\cos kt - 1\}.$$

Thus $\varphi(t) = 1 - 2c\int_0^t \psi(u)\mathrm{d}u$. Since ψ is a continuous function, $\varphi'(t)$ exists finitely for all t and $\varphi'(t) = -2c\psi(t)$. In particular, $\varphi'(0) = 0$.

It must be noted that even though $\varphi'(t)$ exists finitely for all t, the representation $\varphi'(t) = i\int_{-\infty}^{\infty} e^{itx} x\mathrm{d}F(x)$ is not valid. In fact the integral does not exist.

According to **(3.6.1)**, $\mathbb{E}|X|^{2n+1} < \infty \Rightarrow \varphi^{(2n+1)}(0)$ exists finitely. The above example shows that the converse is not true. Hence the condition that $\mathbb{E}X^{2n+1}$ exists finitely is a sufficient condition, and not a necessary condition, for the existence finitely of $\varphi^{(2n+1)}(0)$. These observations raise four questions naturally : (i) What is a necessary and sufficient condition for the existence finitely of $\varphi^{(2n+1)}(0)$? (ii) If $\varphi^{(2n+1)}(0)$ exists finitely but $\mathbb{E}|X|^{2n+1} = \infty$, then to what parameter of the *df* F does $\varphi^{(2n+1)}(0)$ relate? (iii) What is a necessary and sufficient condition, in terms of the *cf* φ, for the finiteness of $\mathbb{E}|X|^{(2n+1)}$? and (iv) If $\varphi^{(2n+1)}(0)$ exists finitely then for what values of $\alpha > 0$ is $\mathbb{E}|X|^{\alpha} < \infty$? Sections **(3.6.7)** and **(3.6.8)** contain the answers.

Lemma 3.6.6. *Below the statements* (i), (ii), (iii) *are equivalent; and the statements* (α),(β),(γ) *are equivalent. As* $r \to \infty$,

(i) $r\int_r^{\infty} dF(x) = o(1)$,

(ii) $r^{-1}\int_0^r x^2 dF(x) = o(1)$,

(iii) $r\int_0^{\infty} \sin^2(\frac{x}{r})dF(x) = o(1)$.

(α) $r\int_{-\infty}^{-r} dF(x) = o(1)$,

(β) $r^{-1}\int_{-r}^{0} x^2 dF(x) = o(1)$,

(γ) $r\int_{-\infty}^{0} \sin^2(\frac{x}{r})dF(x) = o(1)$.

Proof. Let (i) hold. Given $\varepsilon > 0$, use hypothesis to find λ such that $x\{1 - F(x)\} < \varepsilon$ for all $x \geq \lambda$. Using the relation noted in the Theorem in **(2.6.4)**,

$$\lim_{r\to\infty} r^{-1}\int_0^r x^2 \mathrm{d}F(x)$$

$$= \lim_{r\to\infty} r^{-1}\{-r^2(1-F(r)) + 2\int_0^r x(1-F(x))\mathrm{d}x\}$$

$$\leq \lim_{r\to\infty}\{\frac{2}{r}\int_0^\lambda x(1-F(x))\mathrm{d}x + \frac{2}{r}\varepsilon r\} = 2\varepsilon.$$

i.e. (i) implies (ii), since ε is arbitrary.
Suppose now (ii) holds. Given $\varepsilon > 0$, find R large to satisfy $r^{-1}\int_0^r x^2\mathrm{d}F(x) < \varepsilon$ for all $r \geq R$. It follows then that

$$(R2^n)^{-1}\int_0^{R2^n} x^2\mathrm{d}F(x) < \varepsilon \quad \text{for } n = 1, 2, \ldots.$$

Now,

$$R\int_R^\infty \mathrm{d}F(x) = R\sum_{n=1}^\infty \int_{R2^{n-1}}^{R2^n} \mathrm{d}F(x) \leq \sum_{n=1}^\infty \frac{R}{R^2 2^{2n-2}}\int_{R2^{n-1}}^{R2^n} x^2\mathrm{d}F(x)$$

$$\leq \sum_{n=1}^\infty \frac{4}{2^n}\varepsilon = 4\varepsilon.$$

Thus (ii) implies (i).

Again let (ii) hold. Hence (i) holds too.

$$r\int_0^\infty \sin^2(\frac{x}{r})\mathrm{d}F(x) = r\int_0^r \sin^2(\frac{x}{r})\mathrm{d}F(x) + r\int_r^\infty \sin^2(\frac{x}{r})\mathrm{d}F(x)$$

$$\leq r\int_0^r \frac{x^2}{r^2}\mathrm{d}F(x) + r\int_r^\infty \mathrm{d}F(x).$$

The first term on the right is $o(1)$ by (ii) and the second is $o(1)$ by (i). Thus (ii) implies (iii).

Suppose now that (iii) holds. As noted in step 1 under **(3.6.5)**, $\sup\limits_{0\leq\theta<1} \frac{\theta}{\sin\theta} < \frac{6}{5}$. We have

$$r^{-1}\int\limits_0^r x^2 \mathrm{d}F(x) = r\int\limits_0^r \frac{x^2}{r^2}\mathrm{d}F(x) \leq c^2 r\int\limits_0^r \sin^2(\frac{x}{r})\mathrm{d}F(x)$$
$$\leq c^2 r\int\limits_0^\infty \sin^2(\frac{x}{r})\mathrm{d}F(x).$$

Hence (iii) implies (ii).

The proof that (i), (ii) and (iii) are equivalent is complete. That (α), (β) and (γ) are equivalent can be proved on similar lines. □

Theorem 3.6.7. *$\varphi'(0)$ exists finitely iff the following two conditions are satisfied.*

(i) $\tilde{m} = \lim\limits_{r\to\infty} \int\limits_{-r}^{r} x dF(x)$ exists finitely and

(ii) $\lim\limits_{r\to\infty} r\{1 - F(r) + F(-r)\} = 0$.

When $\varphi'(0)$ exists it will be equal to $i\tilde{m}$.

Proof. Let (i) and (ii) hold and let $t > 0$.

$$\frac{\varphi(t)-\varphi(0)}{t} = \int\limits_{-\infty}^{\infty} \frac{e^{itx}-1}{t}\mathrm{d}F(x)$$
$$= -\int\limits_{-\infty}^{\infty} \frac{\sin^2(\frac{tx}{2})}{\frac{t}{2}}\mathrm{d}F(x) + i\int\limits_{-\infty}^{\infty} \frac{\sin tx}{t}\mathrm{d}F(x)$$
$$= -I_1 + iI_2, \quad \text{say.}$$

Write $r = \frac{2}{t}$. Now $I_1 = r\int_{-\infty}^{\infty} \sin^2(\frac{x}{r})\mathrm{d}F(x) = o(1)$ as $t \to 0$ or as $r \to \infty$ by (ii) read with **(3.6.6)**. Setting now $r = \frac{1}{t}$, we have

$$I_2 = r\int\limits_{-\infty}^{\infty} \sin(\frac{x}{r})\mathrm{d}F(x)$$
$$= r\int\limits_{-r}^{r} \sin(\frac{x}{r})\mathrm{d}F(x) + r\int\limits_{r}^{\infty} \sin(\frac{x}{r})\mathrm{d}F(x) + r\int\limits_{-\infty}^{-r} \sin(\frac{x}{r})\mathrm{d}F(x)$$
$$= J_1 + J_2 + J_3, \text{ say.}$$

$$J_1 = r\int_{-r}^{r}\{\sin(\frac{x}{r}) - \frac{x}{r}\}\mathrm{d}F(x) + \int_{-r}^{r} x\mathrm{d}F(x) = J_{1,1}(r) + J_{1,2}(r), \text{ say.}$$

$J_{1,2}(r) \to \tilde{m}$, by (i). From the expansion

$$-\sin\theta + \theta = \theta^3\left(\frac{1}{3!} - \frac{\theta^2}{5!}\right) + \theta^7\left(\frac{1}{7!} - \frac{\theta^2}{9!}\right) + \cdots,$$

we see that if $-1 \le \theta \le 1$ then the terms of this expansion are all non-negative for $0 \le \theta \le 1$ and negative for $-1 \le \theta < 0$. Hence if $-1 \le \theta \le 1$, then $|\sin\theta - \theta| \le \theta^2$. This leads to $|J_{1,1}| \le r\int_{-r}^{r}\frac{x^2}{r^2}\mathrm{d}F(x) = r^{-1}\int_{-r}^{r} x^2\mathrm{d}F(x) = o(1)$ by (ii) read with **(3.6.6)**. Next observe that $|J_2| \le r\int_r^\infty \mathrm{d}F(x) = o(1)$. Similarly $J_3 = o(1)$.

Thus $\lim\limits_{t>0, t\to 0}\frac{\varphi(t)-\varphi(0)}{t}$ exists and is equal to $i\tilde{m}$. On similar lines it can be shown that $\lim\limits_{t<0, t\to 0}\frac{\varphi(t)-\varphi(0)}{t}$ exists and is equal to $i\tilde{m}$. The proof is now complete that (i) and (ii) imply $\varphi'(0)$ exists and is equal to $i\tilde{m}$.

To prove the converse, write $\varphi(t) = u(t) + iv(t)$ and assume $\varphi'(0)$ exists finitely. Hence $u'(0)$ and $v'(0)$ exist finitely. Since u is an even function, $u'(0) = 0$. Thus

$$0 = \lim_{t\to 0}\frac{u(0)-u(t)}{t} = \lim_{t\to 0}\int_{-\infty}^{\infty}\frac{1-\cos tx}{t}\mathrm{d}F(x)$$

$$= \lim_{t\to 0}\int_{-\infty}^{\infty}\frac{2\sin^2(\frac{tx}{2})}{t}\mathrm{d}F(x).$$

Taking $t > 0$ and $r = 2t^{-1}$, this translates to
$r\int_{-\infty}^{\infty}\sin^2(\frac{x}{r})\mathrm{d}F(x) = o(1)$, as $r \to \infty$. Hence by **(3.6.6)**,
$r\{1 - F(r) + F(-r)\} = o(1)$ and $r^{-1}\int_{-r}^{r} x^2\mathrm{d}F(x) = o(1)$. Now,

$$|r\int_r^\infty \sin(\frac{x}{r})\mathrm{d}F(x)| \le r\int_r^\infty \mathrm{d}F(x) = o(1)$$

and similarly

$$r\int_{-\infty}^{-r}\sin(\frac{x}{r})\mathrm{d}F(x) = o(1).$$

Hence, taking $r = t^{-1}$ now,

$$v'(0) = \lim_{r\to\infty} r \int_{-r}^{r} \sin(\frac{x}{r})\mathrm{d}F(x)$$

$$= \lim_{r\to\infty} r\{\int_{-r}^{r} \left(\sin(\frac{x}{r}) - \frac{x}{r}\right) \mathrm{d}F(x)\} + \int_{-r}^{r} x\mathrm{d}F(x).$$

Proceeding as before and using the fact $r^{-1}\int_{-r}^{r} x^2\mathrm{d}F(x) = o(1)$, we see that $\lim_{r\to\infty} \int_{-r}^{r} x\mathrm{d}F(x)$ exists finitely and hence (i) holds. The converse part is now fully established. □

Remarks 3.6.8.
(A) If $\mathbb{E}X$ exists finitely, then trivially $\lim_{r\to\infty}\int_{-r}^{r} x\mathrm{d}F(x)$ exists finitely and is equal to $\mathbb{E}X$. Further $\mathbb{E}|X| < \infty$ implies $\int_{|x|\geq r} |x|\mathrm{d}F(x) = o(1)$. This, in turn, implies that $r\{1 - F(r) + F(-r)\} = o(1)$.

It is desirable to satisfy ourselves that the conditions (i) and (ii) above are independent of each other. Let $P(X = \pm\pi a^n) = \frac{b^n(1-b)}{2b}$ each, $n = 1, 2, \ldots,\ 0 < b < 1$. Then $\varphi(t) = \frac{1-b}{b}\sum_{1}^{\infty} b^n\cos(a^n\pi t)$. For $a > 0$ and $ab > 1 + \frac{3\pi}{2}$, φ is Weierstrass' continuous and nowhere differentiable function. It is easy to see that $\mathbb{E}X$ does not exist and that $\tilde{m} = 0$. Condition (i) is therefore satisfied. We note as $r \to \infty$, $r\{1-F(r)+F(-r)\} \sim 2r\{1-F(r)\}$. Now

$$a^k\{1 - F(\pi a^k)\} = a^k \sum_{n=k+1}^{\infty} \frac{b^n(1-b)}{2b} = a^k(\frac{1}{2}b^k) > \frac{1}{2}(1 + \frac{3\pi}{2})^k \to \infty.$$

Thus condition (ii) is not satisfied.

Let X have frequency function $\frac{c}{x^2\mathrm{log}x}$, for $x \geq e$ and 0 otherwise. $\lim_{r\to\infty}\int_{-r}^{r} x\mathrm{d}F(x) = \infty$. $r\{1 - F(r) + F(-r)\} = r\{1 - F(r)\} \sim \frac{c}{\mathrm{log}r}$ (as $r \to \infty$). Thus condition (ii) is satisfied but not (i). The proof of the theorem shows that when (ii) holds,

$$\lim_{t\to 0} \frac{\varphi(t) - \varphi(0)}{t} = \lim_{r\to\infty} \int_{-r}^{r} x\mathrm{d}F(x) = \infty.$$

(B) Every symmetric *df* satisfies (i). An example of a non-symmetric distribution with mean not existing but satisfying both (i) and (ii) is as follows. Let X have frequency function

$$f(x) = \begin{cases} \frac{c_1}{x^2 \log|x|} & \text{if } x \leq -e; \\ \frac{c_1}{x^2 \log x} + \frac{c_2}{x^3} & \text{if } x \geq e; \\ 0 & \text{elsewhere,} \end{cases}$$

where c_1 and c_2 are such that $2c_1 \int_e^\infty \frac{dx}{x^2 \log x} + \frac{c_2}{2e^2} = 1$. Then

$$\lim_{r\to\infty} \int_{-r}^{r} x f(x)\, dx = c_2 \int_e^\infty \frac{dx}{x^3} = \frac{c_2}{2e^2}.$$

Further,

$$r\{1 - F(r) + F(-r)\} = r\{2 \int_r^\infty \frac{c_1\, dx}{x^2 \log x} + c_2 \int_r^\infty \frac{dx}{x^3}\}$$
$$\leq r\{2c_1 \frac{1}{r \log r} + \frac{c_2}{2r^2}\} \to 0.$$

Further implications of the existence finitely of $\varphi'(0)$ can be found at **(3.6.10)** (ii)

(C) Let $n \geq 1$. Consider now the following proposition extending the result in **(3.6.7)**.

$\varphi^{(2n+1)}(0)$ exists finitely *iff* the following two conditions are satisfied :(i) $\tilde{m}_{2n+1} = \lim_{r\to\infty} \int_{-r}^{r} x^{2n+1}\, dF(x)$ exists finitely and (ii) $\lim_{r\to\infty} r \int_{|x|>r} x^{2n}\, dF(x) = 0$. A proof of this can be constructed as follows, making use of **(3.6.7)**. Let $\varphi^{(2n+1)}(0)$ exist finitely. Hence $\varphi^{(2n)}$ exists and is finite. This implies (ref. **(3.6.4)**) $\alpha_{2n} = \mathbb{E}X^{2n} < \infty$. Define *df* G: $G(x) = \frac{1}{\alpha_{2n}} \int_{-\infty}^{x} x^{2n}\, dF(x)$. If g is its *cf*, then $g(t) = (-1)^n \frac{1}{\alpha_{2n}} \varphi^{(2n)}(t)$ and $g'(0)$ exists finitely. By **(3.6.7)**, it follows that $\lim_{r\to\infty} \int_{-r}^{r} x\, dG(x) = 0$ and $\lim_{r\to\infty} r \int_{|x|>r} dG(x) = 0$.

i.e. $\lim_{r\to\infty} \int_{-r}^{r} x^{2n+1}\, dF(x) = 0$ and $\lim_{r\to\infty} r \int_{|x|>r} x^{2n}\, dF(x) = 0$. i.e. (i) and (ii) hold.

Conversely suppose these conditions hold. Condition (ii) implies $\alpha_{2n} < \infty$. Hence *cf* g can be defined. If *df* G corresponds to g, then (i) and (ii) would read (i)$'$: $\lim_{r\to\infty} \int_{-r}^{r} x\,\mathrm{d}G(x) = 0$ and (ii)$'$: $\lim_{r\to\infty} r\int_{|x|>r} \mathrm{d}G(x) = 0$. Hence by the converse part of **(3.6.7)**, $g'(0)$ exists finitely. This is equivalent to saying $\varphi^{(2n+1)}(0)$ exists finitely.

(D) The *cf* φ of an arbitrary *rv* X possesses the following property : Given $\varepsilon > 0$, an $\eta > 0$ can be found such that $|\varphi(t)| > 1-\varepsilon$ for all t with $|t| \le \eta$. This derives immediately from the continuity of φ at 0. If $\mathbb{E}X = 0$, $\mathbb{E}X^2 = \sigma^2$ and $\mathbb{E}|X|^3 = \beta < \infty$, the following sharper result is possible. Recall $\sigma^3 \le \beta$. Since $\mathbb{E}|X|^3 < \infty$, φ has finite third differential coefficient. Hence $\varphi(t) = \varphi(0) + t\varphi'(0) + \frac{t^2}{2}\varphi''(0) + \frac{t^3}{6}\varphi'''(\theta t)$. This gives $\varphi(\frac{t}{\sigma}) = 1 - \frac{t^2}{2} - i\frac{t^3}{6\sigma^3}\int_{-\infty}^{\infty} e^{i\theta t x}x^3\,\mathrm{d}F(x)$. Hence $|\varphi(\frac{t}{\sigma})| \ge 1 - \frac{t^2}{2} - \frac{|t|^3}{6\sigma^3}\beta$.

For $|t| \le \frac{\sigma^3}{5\beta}$,

$$\begin{aligned}|\varphi(t)| &\ge 1 - \frac{\sigma^6}{50\beta^2} - \frac{\sigma^6\beta}{750\beta^3} = 1 - \frac{\sigma^6}{50\beta^2}\{1+\frac{1}{15}\} \ge 1 - \frac{1}{50}\frac{16}{15}\\ &= \frac{367}{15\times 25} > \frac{360}{15\times 25} = \frac{24}{25}.\end{aligned}$$

Absolute moments of odd orders and of fractional orders

We use the notation and results of **(3.6.2)**. Let $p \ge 0$ be an integer. Let $t > 0$. Let $\lambda_p = \frac{1}{(2p)!}$. Then $\lim_{t\to 0}\frac{g_p(t)}{t^{2p+2}} = \lambda_{p+1}$ and $\lim_{t\to\infty}\frac{g_p(t)}{t^{2p}} = \lambda_p$. For $2p < \beta < 2p+2$, $\frac{g_p(t)}{t^{1+\beta}} \sim \frac{\lambda_{p+1}}{t^{\beta-2p-1}}$ as $t \to 0$. Since $0 < \beta - 2p < 2$, it follows that $\int_0^{\varepsilon}\frac{g_p(t)}{t^{1+\beta}}dt$ exists finitely for every $\varepsilon > 0$. Since, as $t \to \infty$, $\frac{g_p(t)}{t^{1+\beta}} \sim \frac{\lambda_p}{t^{1+\beta-2p}}$ and since $1+\beta-2p > 1$, it follows that $\int_{\varepsilon}^{\infty}\frac{g_p(t)}{t^{1+\beta}}dt$ exists finitely for all $\varepsilon > 0$. Thus $\int_0^{\infty}\frac{g_p(t)}{t^{1+\beta}}dt$ exists finitely. Denote its value by $c_{p,\beta}$.

This constant $c_{p,\beta}$ can be evaluted as follows. We recall the result, obtainable by contour integration (ref. p 108, [T]): for $0 < b < 1$, and $a > 0$,

$$\begin{aligned}\int_0^{\infty} t^{b-1}\cos at\,\mathrm{d}t &= \frac{1}{a^p}\Gamma(b)\cos(\frac{1}{2}b\pi);\\ \int_0^{\infty} t^{b-1}\sin at\,\mathrm{d}t &= \frac{1}{a^p}\Gamma(b)\sin(\frac{1}{2}b\pi). \qquad (3.3)\end{aligned}$$

Now, for $0 < \beta < 1$,

$$c_{0,\beta} = \int_0^{\infty}\frac{g_0(t)}{t^{1+\beta}}\,\mathrm{d}t = \int_0^{\infty}\frac{1-\cos t}{t^{1+\beta}}\,\mathrm{d}t$$

$$
\begin{aligned}
&= \int_0^\infty (1-\cos t)\, \mathrm{d}(-\frac{1}{\beta t^\beta}) = \int_0^\infty \frac{\sin t}{\beta t^\beta}\, \mathrm{d}t \\
&= \frac{1}{\beta}\Gamma(1-\beta)\sin(\frac{1-\beta}{2}\pi) = \frac{1}{\beta}\frac{\Gamma(\beta)\Gamma(1-\beta)}{\Gamma(\beta)}\cos(\frac{\beta\pi}{2}) \\
&= \frac{\pi}{\sin(\beta\pi)}\frac{1}{\Gamma(\beta+1)}\cos(\frac{\beta\pi}{2}) = \frac{\pi}{2}\frac{1}{\Gamma(\beta+1)}\frac{1}{\sin(\frac{\beta\pi}{2})} \\
&= \frac{\pi}{2\Gamma(\beta+1)\cos(\frac{\beta-1}{2}\pi)}.
\end{aligned}
$$

If $\beta = 1$,

$$
c_{0,\beta} = \int_0^\infty \frac{1-\cos t}{t^2}\, \mathrm{d}t = \frac{\pi}{2} = \frac{\pi}{2}\frac{1}{\Gamma(\beta+1)}\frac{1}{\cos(\frac{\beta-1}{2}\pi)}.
$$

If $1 < \beta < 2$,

$$
\begin{aligned}
c_{0,\beta} &= \int_0^\infty \frac{1-\cos t}{t^{1+\beta}}\, \mathrm{d}t = \int_0^\infty (1-\cos t)\, \mathrm{d}(-\frac{1}{\beta t^\beta}) \\
&= \frac{1}{\beta}\int_0^\infty \frac{\sin t}{t^\beta}\, \mathrm{d}t = \frac{1}{\beta(\beta-1)}\int_0^\infty \frac{\cos t}{t^{\beta-1}} \\
&= \frac{1}{\beta(\beta-1)}\Gamma(2-\beta)\cos(\frac{\pi}{2}(2-\beta))(ref.(3.3)) \\
&= \frac{\Gamma(\beta-1)\Gamma(2-\beta)}{\beta(\beta-1)\Gamma(\beta-1)}\sin(\frac{\beta-1}{2}\pi) \\
&= \frac{\pi}{\Gamma(\beta+1)\sin(\overline{\beta-1}\pi)}\sin(\frac{\beta-1}{2}\pi) = \frac{\pi}{2\Gamma(\beta+1)\cos(\frac{\beta-1}{2}\pi)}.
\end{aligned}
$$

Thus for all $\beta,\ 0 < \beta < 2$,

$$
\int_0^\infty \frac{1-\cos t}{t^{1+\beta}}\, \mathrm{d}t = \frac{\pi}{2\Gamma(\beta+1)\cos(\frac{\beta-1}{2}\pi)}.
$$

Let now $p \geq 1$ and let $2p < \beta < 2p+2$. Then after repeatedly carrying out integration by parts and using the just proved result for $\beta \in (0,\ 1)$, we can show that $c_{p,\beta} = \frac{\pi}{2\Gamma(\beta+1)\cos(\frac{\beta+1}{2}\pi)}$.

Theorem 3.6.9. *Let* $2p < \beta < 2p+2$. *Let* $E|X|^{2p} < \infty$. *Then* $E|X|^\beta < \infty$ *iff* $I_\beta = \int\limits_0^\infty \frac{u_p(t)}{t^{1+\beta}} dt < \infty$.

Proof.

$$I_\beta = \int\limits_0^\infty \frac{1}{t^{1+\beta}} \left(\int\limits_{-\infty}^{\infty} g_p(tx) \mathrm{d}F(x) \right) \mathrm{d}t$$

$$= \int\limits_{-\infty}^{\infty} \left(\int\limits_0^\infty \frac{g_p(tx)}{t^{1+\beta}} \mathrm{d}t \right) \mathrm{d}F(x)$$

(change of order of integration is justified since the integrand is non-negative)

$$= \int\limits_{-\infty}^{\infty} \left(\int\limits_0^\infty \frac{g_p(t)}{t^{1+\beta}} \mathrm{d}t \right) |x|^\beta \mathrm{d}F(x) = c_{p,\beta} \mathbb{E}|X|^\beta.$$

□

Remarks 3.6.10.
(i) In particular, with $u(t)$ denoting $\Re(\varphi(t))$, a necessary and sufficient condition for the finiteness of $\mathbb{E}|X|^\beta, 0 < \beta < 2$ is the finiteness of $\int_0^\infty \frac{1-u(t)}{t^{1+\beta}} dt$. When this integral is finite,

$$\mathbb{E}|X|^\beta = c_\beta \int_0^\infty \frac{1-u(t)}{t^{1+\beta}} dt \quad \text{where } c_\beta = \frac{1}{\Gamma(2-\beta)} \frac{\beta(1-\beta)}{\sin((1-\beta)\frac{\pi}{2})}.$$

c_1 is to be taken as $\frac{2}{\pi}$.
(ii) If $\varphi'(0)$ exists finitely, then $\mathbb{E}|X|^\alpha < \infty$ for every $\alpha,\ 0 \le \alpha < 1$. Since result is obvious for $\alpha = 0$, let $0 < \alpha < 1$. The hypothesis implies $u'(0)$ exists finitely. Hence, given $\varepsilon > 0$ there can be found a $\delta,\ 0 < \delta < 1$ such that over $[0,\ \delta]$, $\frac{1-u(t)}{t^{1+\alpha}}$ is integrable since it is dominated by the integrable function $\frac{u'(0)+\varepsilon}{t^\alpha}$. Over $[\delta,\ \infty)$, $\frac{1-u(t)}{t^{1+\alpha}}$ is dominated by the integrable function $\frac{2}{t^{1+\alpha}}$. The stated assertion follows from the finiteness of $\int\limits_0^\infty \frac{1-u(t)}{t^{1+\alpha}} dt$.
(iii) Let $n \ge 0$ be an integer. Using the method developed in **(3.6.8)**(C), it can be shown that the existence finitely of $\varphi^{(2n+1)}(0)$ implies $\mathbb{E}|X|^{2n+\alpha} < \infty$ for every $\alpha,\ 0 \le \alpha < 1$.

Finite series expansion for cfs

Let X be a rv with *cf* φ and with $\mathbb{E}|X|^{n+\delta} < \infty$, $n \geq 0$, an integer and $0 \leq \delta < 1$. Hence $\varphi^{(n)}(u)$ exists finitely for all u and is a continuous function. Let $m_k = \mathbb{E}X^k$. We note $\varphi^{(k)}(0) = i^k m_k$.

Theorem 3.6.11. *(i) For $u \in R$,*

$$\rho_n(u) = \varphi(u) - \sum_0^{n-1} m_k \frac{(iu)^k}{k!} \text{ where } \rho_n(u) = u^n \int_0^1 \frac{(1-t)^{n-1}}{(n-1)!}\varphi^{(n)}(tu)\, dt.$$

(ii) If $\delta = 0$, then

$$\varphi(u) = \sum_0^n m_k \frac{(iu)^k}{k!} + o(u^n),$$

as $u \to 0$. $\varphi(u)$ takes also the form

$$\varphi(u) = \sum_0^{n-1} m_k \frac{(iu)^k}{k!} + O(\frac{\mathbb{E}|X|^n}{n!}|u|^n),$$

true for all $u \in R$.
(iii) If $0 < \delta < 1$, then

$$\varphi(u) = \sum_0^n m_k \frac{(iu)^k}{k!} + a_n(u,\, \delta),$$

valid for all $u \in R$. Here,

$$|a_n(u,\, \delta)| \leq 2^{1-\delta}\mathbb{E}|X|^{n+\delta}|u|^{n+\delta}\frac{1}{(n-1)!}\int_0^1 t^\delta(1-t)^{n-1}\, dt.$$

Proof. (i) On integration by parts,

$$\rho_n(u) = u^{n-1}\frac{(1-t)^{n-1}}{(n-1)!}\varphi^{(n-1)}(tu)\Big|_0^1 + \rho_{n-1}(u)$$
$$= -\frac{u^{n-1}}{(n-1)!}\varphi^{(n-1)}(0) + \rho_{n-1}(u).$$

Repeating this process $n-1$ times we get

$$\rho_n(u) = -\sum_{k=1}^{n-1}\frac{u^k}{k!}\varphi^{(k)}(0) + \rho_1(u) = -\sum_{k=0}^{n-1}\frac{u^k}{k!}\varphi^{(k)}(0) + \varphi(u)$$

which is the desired result.
(ii) Since $\varphi^{(n)}$ is a continuous function, $\varphi^{(n)}(tu) = \varphi^{(n)}(0) + o(1)$ as $u \to 0$. Hence

$$\rho_n(u) = u^n \int_0^1 \frac{(1-t)^{n-1}}{(n-1)!} \{\varphi^{(n)}(0) + o(1)\}\,\mathrm{d}t = \frac{u^n}{n!}\varphi^{(n)}(0) + o(u^n).$$

The first claim follows from this.

The second claim follows since

$$\varphi(u) = \sum_0^{n-1} m_k \frac{(iu)^k}{k!} + \rho_n(u)$$

and since

$$|\rho_n(u)| \leq |u|^n \int_0^1 \frac{(1-t)^{n-1}}{n!} |\varphi^{(n)}(tu)|\,\mathrm{d}t \leq |u|^n \int_0^1 \frac{(1-t)^{n-1}}{(n-1)!} \mathbb{E}|X|^n\,\mathrm{d}t.$$

(iii) Let

$$g(u) = e^{iu} - \sum_{k=0}^{n} \frac{(iu)^k}{k!}.$$

We note $g^{(n)}(u) = i^n\{e^{iu} - 1\}$. Proceeding as in (i) we conclude

$$e^{iu} = \sum_{k=0}^{n-1} \frac{(iu)^k}{k!} + \lambda_n(u) \text{ where } \lambda_n(u) = u^n \int_0^1 \frac{(1-t)^{n-1}}{(n-1)!} g^{(n)}(tu)\,\mathrm{d}t.$$

Hence

$$\varphi(u) = \mathbb{E}e^{iuX} = \sum_{k=0}^{n-1} \frac{(iu)^k}{k!} m_k + \mathbb{E}\lambda_n(uX).$$

We have

$$|\mathbb{E}\lambda_n(uX)| = |\int_R \{u^n x^n \int_0^1 \frac{(1-t)^{n-1}}{(n-1)!}(e^{itux} - 1)\,\mathrm{d}t\}\,\mathrm{d}F(x)|$$

$$\leq \int_0^1 \frac{(1-t)^{n-1}}{(n-1)!} \{\int_R |e^{itux} - 1||ux|^n\,\mathrm{d}F(x)\}\,\mathrm{d}t.$$

Now, we derive a suitable bound for $|e^{iy} - 1|$. We consider two cases. Case 1 : $|\frac{y}{2}| < 1$. In this case, $|e^{iy} - 1| = 2|\sin(\frac{y}{2})| \leq 2\frac{|y|}{2} \leq 2|\frac{y}{2}|^\delta$; Case 2 : $|\frac{y}{2}| \geq 1$.

In this case, $|e^{iy}-1| \le 2 \le 2|\frac{y}{2}|^\delta$. Thus for all y, $|e^{iy}-1| \le 2|\frac{y}{2}|^\delta$. Using this bound, we get

$$|\mathbb{E}\lambda_n(uX)| \le |u|^n \int_0^1 \frac{(1-t)^{n-1}}{(n-1)!}\{\int_R 2^{1-\delta}|x|^n|utx|^\delta \,\mathrm{d}F(x)\}\,\mathrm{d}t. \qquad \square$$

3.7 Characteristic function criteria for uniform integrability

Suppose $F_n \xrightarrow{d} F_0$. Suppose further that $\int_{-\infty}^{\infty}|x|^\beta \,\mathrm{d}F_n(x) < \infty$, $n \ge 0$. In **(2.6.9)** we showed that $\int_{-\infty}^{\infty}|x|^\beta \,\mathrm{d}F_n(x) \to \int_{-\infty}^{\infty}|x|^\beta \,\mathrm{d}F_0(x)$ *iff* the function $|x|^\beta$ is uniformly integrable with respect to the sequence (F_n). In sections **(3.7.1)** and **(3.7.2)** we establish necessary and sufficient conditions, in terms of *cf*s, for the uniform integrability of $|x|^\beta$.

Let $p \ge 0$ be an integer and let $\int_{-\infty}^{\infty}|x|^{2p}\mathrm{d}F_n(x) < \infty$, $n \ge 1$. Let $g_p(t)$ be as defined in **(3.6.2)** and recall

$$u_{p;n}(t) = \mathbb{E}g_p(tX_n) = \int_R g_p(tx)\,\mathrm{d}F_n(x).$$

Theorem 3.7.1. *$|x|^{2p}$ is uniformly integrable with respect to the sequence (F_n) of dfs iff* $\lim\limits_{t\to 0}\sup\limits_n \frac{u_{p;n}(t)}{t^{2p}} = 0$.

Note. We recall the definition of equicontinuity and see that the condition in the theorem is same as the condition that the sequence $\left(\frac{u_{p;n}(t)}{t^{2p}}\right)$ be equicontinuous at 0.

In particular, the equicontinuity of the sequence $(u_{0;n})$ or $\{1-\Re(\varphi_n)\}$ at $t=0$ is a necessary and sufficient condition for the tightness of the sequence (F_n) of dfs. (ref. **(2.3.2)**)

Proof. Suppose $|x|^{2p}$ is uniformly integrable with respect to the F_ns. Given $\varepsilon > 0$, we can find r large such that $\sup\limits_n \int_{|x|\ge r}|x|^{2p}\mathrm{d}F_n(x) < \varepsilon$. It follows that

$$\begin{aligned} q = \sup_n \mu_{p;n} &= \sup_n \int_{-\infty}^{\infty} |x|^{2p}\mathrm{d}F_n(x) \\ &\le \sup_n\{\int_{|x|\le r}|x|^{2p}\mathrm{d}F_n(x) + \int_{|x|\ge r}|x|^{2p}\mathrm{d}F_n(x)\} \\ &\le r^{2p} + \varepsilon < \infty. \end{aligned}$$

Now $\frac{u_{p;n}(t)}{t^{2p}} = I_n(t,\varepsilon) + J_n(t,\varepsilon)$, say, where

$$I_n(t,\varepsilon) = \int\limits_{|tx|\le\varepsilon} \frac{g_p(tx)}{t^{2p}} \mathrm{d}F_n(x) \le \lambda_{p+1}\varepsilon^2 E X_n^{2p} \le \lambda_{p+1} q \varepsilon^2.$$

For all t with $|t| \le \frac{\varepsilon}{r}$,

$$J_n(t,\varepsilon) = \int\limits_{|x|>\frac{\varepsilon}{|t|}} \frac{g_p(tx)}{t^{2p}} \mathrm{d}F_n(x) \le \int\limits_{|x|>\frac{\varepsilon}{|t|}} \lambda_p |x|^{2p} \mathrm{d}F_n(x)$$

$$\le \int\limits_{|x|>r} \lambda_p |x|^{2p} \mathrm{d}F_n(x) < \lambda_p \varepsilon.$$

Since ε is arbitrary, the forward result is established.

Assume now $\limsup\limits_{t\to 0}\limits_{n} \frac{u_{p;n}(t)}{t^{2p}} = 0$. Let $\varepsilon > 0$ be given. Use the result that $\lim\limits_{t\to\infty} \frac{g_p(t)}{t^{2p}} = \lambda_p$ to find r such that $\frac{g_p(t)}{t^{2p}} \ge \frac{1}{2}\lambda_p$ for all t with $|t| > r$. Then the converse part follows from the relations

$$\frac{u_{p;n}(t)}{t^{2p}} = \int\limits_{-\infty}^{\infty} \frac{g_p(tx)}{|tx|^{2p}} |x|^{2p} \mathrm{d}F_n(x) \ge \frac{1}{2}\lambda_p \int\limits_{|tx|>r} |x|^{2p} \mathrm{d}F_n(x)$$

$$\ge \frac{1}{2}\lambda_p \int_{|x|>r} |x|^{2p}\, \mathrm{d}F_n(x) \text{ (since } r < |t|). \qquad \square$$

Theorem 3.7.2. *Let* $2p < \beta < 2p+2$ *and let* $\int_{-\infty}^{\infty} |x|^\beta dF_n(x) < \infty$, $n \ge 1$. *Then* $|x|^\beta$ *is uniformly integrable with respect to the sequence* (F_n) *of dfs iff*

$$\limsup_{\varepsilon\to 0}\limits_{n} \int\limits_0^\varepsilon \frac{u_{p;n}(t)}{t^{1+\beta}} dt = 0.$$

Proof. In **(3.6.9)** we showed that $c_{p,\beta} = \int_0^\infty \frac{g_p(t)}{t^{1+\beta}} \mathrm{d}t < \infty$. Hence an $\varepsilon > 0$ can be found such that $\int_0^{\varepsilon^{-1}} \frac{g_p(t)}{t^{1+\beta}} \mathrm{d}t \ge \frac{1}{2} c_{p,\beta}$. Now,

$$I = \int\limits_0^\varepsilon \frac{u_{p;n}(t)}{t^{1+\beta}} \mathrm{d}t = \int\limits_0^\varepsilon \left(\int\limits_{-\infty}^{\infty} g_p(tx) \mathrm{d}F_n(x) \right) \frac{1}{t^{1+\beta}} \mathrm{d}t$$

$$= \int\limits_{-\infty}^{\infty} \left(\int\limits_0^\varepsilon \frac{g_p(tx)}{t^{1+\beta}} \mathrm{d}t \right) \mathrm{d}F_n(x)$$

(change of order of integration is justified since the integrand is non-negative)

$$= \int_{-\infty}^{\infty} \left(\int_0^{\varepsilon|x|} \frac{g_p(t)}{t^{1+\beta}} \mathrm{d}t \right) |x|^\beta \mathrm{d}F_n(x) \geq \int_{|x|\geq\varepsilon^{-2}} \left(\int_0^{\varepsilon^{-1}} \frac{g_p(t)}{t^{1+\beta}} \mathrm{d}t \right) |x|^\beta \mathrm{d}F_n(x)$$

$$\geq \frac{1}{2} c_{p,\beta} \int_{|x|\geq\varepsilon^{-2}} |x|^\beta \mathrm{d}F_n(x).$$

This inequality implies that the condition is sufficient. Let $\theta = \frac{2p+2-\beta}{2(2p+2)}$.

$$\begin{aligned} I &= \int_{|x|\leq\varepsilon^{-\theta}} \left(\int_0^{\varepsilon|x|} \frac{g_p(t)}{t^{1+\beta}} \mathrm{d}t \right) |x|^\beta \mathrm{d}F_n(x) \\ &\quad + \int_{|x|>\varepsilon^{-\theta}} \left(\int_0^{\varepsilon|x|} \frac{g_p(t)}{t^{1+\beta}} \mathrm{d}t \right) |x|^\beta \mathrm{d}F_n(x) \\ &= J_1 + J_2, \text{ say.} \\ J_1 &\leq \int_{|x|\leq\varepsilon^{-\theta}} \left(\int_0^{\varepsilon|x|} \frac{g_p(t)}{t^{1+\beta}} \mathrm{d}t \right) |x|^\beta \mathrm{d}F_n(x) \leq \int_0^{\varepsilon^{1-\theta}} \frac{g_p(t)}{t^{1+\beta}} \mathrm{d}t \, \varepsilon^{-\beta\theta} \\ &\leq \varepsilon^{-\beta\theta} \lambda_{p+1} \int_0^{\varepsilon^{1-\theta}} \frac{t^{2p+2}}{t^{1+\beta}} \mathrm{d}t = \lambda_{p+1} \frac{1}{2p+2-\beta} \varepsilon^{2p+2-\beta-\theta(2p+2)} \\ &= \frac{\lambda_{p+1}\varepsilon^\delta}{2p+2-\beta} \end{aligned}$$

where $\delta = \frac{1}{2}(2p+2-\beta) > 0$.

$$J_2 \leq c_{p,\beta} \int_{|x|>\varepsilon^{-\theta}} |x|^\beta \mathrm{d}F_n(x).$$

Thus

$$I \leq \frac{\lambda_{p+1}}{2p+2-\beta} \varepsilon^\delta + c_{p,\beta} \int_{|x|>\varepsilon^{-\theta}} |x|^\beta \, \mathrm{d}F_n(x).$$

The necessity of the condition derives from this inequality. □

Theorem 3.7.3. *For $n \geq 0$, let φ_n be the cf of the df F_n. If $\varphi_n(t) \to \varphi_0(t)$ for all t and if the φ_ns are uniformly integrable (with respect to Lebesgue measure), then $F_n'(x) \to F_0'(x)$ for all x.*

Proof. Each F_n is absolutely continuous, by **(3.1.3)**(v). Given $\varepsilon > 0$, find r such that $\sup\limits_{n\geq 0} \int_{|u|\geq r} |\varphi_n(t)|\mathrm{d}t < \varepsilon$. This is possible because the φ_ns are given to be uniformly integrable. Now,

$$2\pi|F_n'(x) - F_0'(x)| = \left|\int_{-\infty}^{\infty} e^{-iux}\varphi_n(u)\mathrm{d}u - \int_{-\infty}^{\infty} e^{-iux}\varphi_0(u)\mathrm{d}u\right|$$

$$\leq \left|\int_{|t|\leq r} e^{-iux}\varphi_n(u)\mathrm{d}u - \int_{|t|\leq r} e^{-iux}\varphi_0(u)\mathrm{d}u\right| + 2\varepsilon.$$

The claim follows now by appealing to the bounded convergence theorem. □

3.8 A uniqueness theorem for moment sequences

The example at **(2.6.7)** makes it clear that the moments of a *df* do not determine the *df*. We present in this section a sufficient condition that a moment sequence, say, (μ_n) (i.e. the sequence of moments of a *df*) should satisfy so that it is the moment sequence of one and only one *df*. Let (μ_n) be the moment sequence of a *df* F. Let φ be the cf of F. Let $\beta_n = \int_{-\infty}^{\infty} |x|^n \mathrm{d}F(x)$. The existence of moments of all orders implies that φ is infinitely differentiable at all t (ref. **(3.6.1)**) and $\varphi^{(n)}(t) = i^n \int_{-\infty}^{\infty} e^{itx}x^n \mathrm{d}F(x)$. Consider the two Taylor series generated by φ:

(i) $\sum\limits_0^{\infty} \frac{t^n}{n!}\varphi^{(n)}(0)$ and (ii) $\sum\limits_0^{\infty} \frac{(t-t_0)^n}{n!}\varphi^{(n)}(t_0),\ t_0 \neq 0.$

The radius r of convergence of series (i) is $\frac{1}{l}$ where $l = \overline{\lim\limits_{n\to\infty}} (\frac{|\mu_n|}{n!})^{\frac{1}{n}}$. Since $|\mu_n| \leq \beta_n$, since $\beta_n^{\frac{1}{n}} \leq \beta_{n+1}^{\frac{1}{n+1}}$ (ref. **(2.6.6)**) and since $\mu_{2n} = \beta_{2n}$, it follows that $l = \overline{\lim\limits_{n\to\infty}} (\frac{\mu_{2n}}{(2n)!})^{\frac{1}{2n}}$. Since $|\varphi^{(n)}(t_0)| \leq \beta_n$, the radius $r(t_0)$ of convergence of series (ii) will be not less than r. Thus series (ii) will have a positive radius $r(t_0)$ of convergence if series (i) does. (That series (i) may not always have a positive radius of convergence is brought out by the example in **(2.6.7)**. For the dfs considered there $\mu_n = \beta_n = e^{\frac{1}{2}n^2}$). Consider now the finite Taylor expansion of φ around t_0 :

$$\varphi(t) = \sum_{k=0}^{n} \frac{(t-t_0)^k}{k!}\varphi^{(k)}(t_0) + R_n(t, t_0)$$

where

$$R_n(t,t_0) = \frac{(t-t_0)^{n+1}}{(n+1)!}\varphi^{(n+1)}(t_0+\theta(t-t_0))$$

for some θ, $|\theta| < 1$.

$$|R_n(t,t_0)| \le \frac{r^{n+1}}{(n+1)!}\beta_{n+1} \to 0$$

if $|t-t_0| < r$. Hence for all t with $|t-t_0| < r$,

$$\varphi(t) = \sum_{n=0}^{\infty} \frac{(t-t_0)^n}{n!}\varphi^{(n)}(t_0).$$

Thus $\varphi(t)$ is determined for all values t (and hence the *df* is determined) if series (i) has a positive radius of convergence.

3.9 Symmetric distribution functions\Symmetric probability measures

A rV $\mathbb{X} \in R^p$ is said to be symmetrically distributed or its *df* $\mathbb{F}$ is symmetric if the *df*s of $\mathbb{X}$ and $-\mathbb{X}$ are the same. Equivalently, $\mu(A) = \mu(-A)$ for every Borel set A where μ is the pm generated by $\mathbb{F}$ on $\mathscr{R}^p$ The equivalent condition in terms of the *cf* φ is : $\varphi(\mathbf{t}) = \varphi(-\mathbf{t})$ (ref. **(2.3.7)**(iii)).

If $p = 1$, this condition on the *df* F translates to : $F(-x) = 1 - F(x-)$ for all x. If the frequency function f exists, then the equivalent condition for symmetry is : $f(x) = f(-x)$. If X is symmetrically distributed, if n is a positive odd integer and if $\theta_n = \mathbb{E}X^n$ exists finitely, then $\theta_n = 0$. The following example shows that the converse is not true.

Let $f_\alpha(x)$ be as in **(2.6.7)**. For $-1 < \alpha, \beta < 1$, define $h_{\alpha,\beta}(x) = \frac{1}{2}f_\alpha(x)$ if $x > 0$ and $= \frac{1}{2}f_\beta(-x)$ if $x < 0$. If $\alpha \neq \beta$, then $h_{\alpha,\beta}(x) \neq h_{\alpha,\beta}(-x)$ for every x. Define *df* $H_{\alpha,\,\beta} : H_{\alpha,\,\beta}(x) = \int_{-\infty}^{x} h_{\alpha,\,\beta}(y)\,\mathrm{d}y$ and note that it is not a symmetric distribution. But for n odd,

$$\begin{aligned}2\int_{-\infty}^{\infty} x^n h_{\alpha,\beta}(x)\mathrm{d}x &= \int_{0}^{\infty} x^n f_\alpha(x)\mathrm{d}x + \int_{-\infty}^{0} x^n f_\beta(-x)\mathrm{d}x \\ &= \int_{0}^{\infty} x^n f_\alpha(x)\mathrm{d}x - \int_{0}^{\infty} x^n f_\beta(x)\mathrm{d}x = e^{\frac{n^2}{2}} - e^{\frac{n^2}{2}} = 0.\end{aligned}$$

3.10 Moments and Laplace transforms

In the case of non-negative rvs, it is natural to use Laplace transforms in place of cfs. We state below parallel results *without proofs* which can be constructed on similar lines. Let F be the *df* of a non-negative rv X and let ψ be its Laplace transform : $\psi(\lambda) = \mathbb{E}e^{-\lambda X}, \lambda > 0$. Define for $s \geq 0$ and non-negative integer k,

$$\gamma_k(s) = (-1)^{k+1}\{e^{-s} - \sum_{j=0}^{k}(-1)^j \frac{s^j}{j!}\} \text{ and, when } \mathbb{E}X^k < \infty,$$

$$\Lambda_k(s) = \mathbb{E}\gamma_k(sX) = (-1)^{k+1}\{\psi(s) - \sum_{j=0}^{k} \frac{s^j}{j!}\psi^{(j)}(0)\}.$$

3.10.1. For $k \geq 1$, an integer, $\psi^{(k)}(0)$ exists finitely *iff* $\mathbb{E}X^k < \infty$.

3.10.2. $\mathbb{E}X^{n+1} < \infty$ *iff* $\mathbb{E}X^n < \infty$ and $\alpha_n = \lim\limits_{s\to 0+} \frac{\Lambda_n(s)}{s^{n+1}}$ exists finitely and then $\alpha_n = \frac{1}{(n+1)!}\mathbb{E}X^{n+1}$.

3.10.3. Let $m \geq 0$ be an integer and let τ be an arbitrary number, $m < \tau < m+1$. Then statement (i) is equivalent to statements (ii) and (iii) where (i) : $\mathbb{E}X^\tau < \infty$, (ii) : $\mathbb{E}X^m < \infty$ and (iii) : $\int\limits_0^\infty \frac{\Lambda_m(s)}{s^{1+\tau}}\mathrm{d}s < \infty$.

Let X_n be a non-negative rv with *df* F_n, with Laplace transform ψ_n, $n \geq 1$ and with finite moment of order k. Define

$$\Lambda_{k,n}(s) = (-1)^{k+1}\{\psi_n(s) - \sum_{j=0}^{k} \frac{s^j}{j!}\psi_n^{(j)}(0)\}.$$

3.10.4. x^k is uniformly integrable with respect to the F_ns *iff*

$$\lim_{s\to 0} \sup_n \frac{\Lambda_{k,n}(s)}{s^k} = 0.$$

3.10.5. Let ψ_n be the Laplace transform of the *df* F_n, $n \geq 0$. Then $F_n \overset{d}{\to} F_0$ *iff* $\psi_n(\lambda) \to \psi_0(\lambda)$ for every $\lambda > 0$.

3.10.6. Let $m \geq 0$ be an integer and let τ be an arbitrary number, $m < \tau < m+1$. Let $\int_0^\infty x^\tau \mathrm{d}F_n(x) < \infty$, $n \geq 1$. Then x^τ is uniformly integrable with respect to the F_ns *iff*

$$\lim_{\varepsilon\to 0} \sup_n \int\limits_0^\varepsilon \frac{\Lambda_m(s)}{s^{1+\tau}}\mathrm{d}s = 0.$$

Definition 3.10.7. A positive valued measurable function L defined on $[0, \infty)$ is said to be slowly varying (*sv*) at infinity *iff* for all $x > 0$, $\lim_{t \to \infty} \frac{L(tx)}{L(t)} = 1$. A positive valued function U is said to be regularly varying at infinity with exponent ρ, $-\infty < \rho < \infty$ *iff* $U(x) \sim x^\rho L(x)$ for some *sv* function L.

The first part of **Q**12 below gives the Karamata representation for sv functions. (ref. Feller Vol . II, p 274). The second part quotes a Tauberian theorem (ref. Feller Vol II, pp 274 and 418) needed to prove **(3.10.8)**.

Q12 Let U be a monotonic non-decreasing function defined on $(0, \infty)$ with $\lim_{x \to \infty} U(x) = \infty$. Let $\omega(\lambda) = \int_0^\infty e^{-\lambda x} \, dU(x)$ exist finitely for every $\lambda > 0$.
(i) U varies slowly at ∞ *iff* it is of the form

$$U(x) = a(x) e^{\int_1^x \frac{\varepsilon(y)}{y} \, dy}$$

where $a(x) \to a < \infty$, $\varepsilon(x) \geq 0$, $\varepsilon(x) \to 0$ as $x \to \infty$ and ε continuously differentiable.
(ii) U regularly varying at ∞ with exponent $\rho \geq 0$ implies and is implied by ω regularly varying at 0 with exponent $-\rho$ and then, as $\tau \to 0$, the relation $\omega(\tau) \sim U(\frac{1}{\tau})\Gamma(\rho + 1)$ holds.

Theorem 3.10.8. *Let rv* $X \geq 0$ *have df* G *and Laplace transform* ψ. *Let* $\mathbb{E}X = \infty$. *Let* $L(x) = \int_0^x \{1 - G(y)\} \, dy$ *be sv at infinity. Then as* $\lambda \to 0$, $1 - \psi(\lambda) \sim \lambda L(\frac{1}{\lambda})$. *Further if, for some* L *slowly varying at* ∞, $\frac{1-\psi(\lambda)}{\lambda} \sim L(\frac{1}{\lambda})$ *then* $\int_0^x \{1 - G(y)\} \, dy \sim L(x)$ *as* $x \to \infty$.

Proof.

$$\frac{1 - \psi(\lambda)}{\lambda} = \int_0^\infty \frac{1 - e^{-\lambda y}}{\lambda} \, dG(y) = \lim_{x \to \infty} \int_0^x \frac{1 - e^{-\lambda y}}{\lambda} \, dG(y)$$

$$= \lim_{x \to \infty} \left\{ - \frac{1 - e^{-\lambda x}}{\lambda} (1 - G(x)) + \int_0^x e^{-\lambda y} (1 - G(y)) \, dy \right\}$$

(by integration by parts (ref. **(2.6.4)**))

$$= \int_0^\infty e^{-\lambda y} \, (1 - G(y)) \, dy$$

$$= \int_0^\infty e^{-\lambda y}\, \mathrm{d}L(y)$$

(property of Radon-Nikodym derivative).

Now L is a sv monotonic non-decreasing function. Further $\mathbb{E}X = \infty$ implies $\lim_{x\to\infty} L(x) = \infty$. Both the forward and converse assertions of the theorem now follow from **Q**12. □

Remark. This theorem will be used in sec. **(4.5)** Let U be monotonic non-decreasing sv function with $\lim_{x\to\infty} U(x) = \infty$ and with Karamata representation $U(x) = a(x)H(x)$, $H(x) = 1, 0 \le x \le 1$, $H(x) = e^{\int_1^x \frac{\varepsilon(y)}{y}}\, \mathrm{d}y$, $x \ge 1$.

Theorem 3.10.9. *(i) Given $\delta > 0$, we can find x_0 such that, in the region $x \ge x_0$, $\frac{H(x)}{x^\delta} \downarrow 0$ as $x \uparrow \infty$. A consequence is : $\frac{U(x)}{x^\delta} \to 0$ as $x \to \infty$ for every $\delta > 0$.*

(ii) For every $\theta > 0$, $\int_1^\infty x^{-\theta}\, dU(x) < \infty$.

Proof. (i) Choose x_0 so $0 \le \epsilon(y) \le \frac{1}{2}\delta$ for all $y \ge x_0$. We then have, for $x \ge x_0$, $\frac{H(x)}{x^\delta}$ = constant $\times \exp(-\int_{x_0}^x \frac{\delta - \epsilon(y)}{y} dy)$. The claim that $\frac{H(x)}{x^\delta} \downarrow$ follows since the integrand is non-negative; that the limit is 0 follows since the integral

$$\int_{x_0}^x \frac{\delta - \varepsilon(y)}{y}\, \mathrm{d}y \ge \int_{x_0}^x \frac{\frac{1}{2}\delta}{y}\, \mathrm{d}y \to \infty \text{ as } x \to \infty.$$

(ii) We note that it is enough to show that $I_a = \int_a^\infty x^{-\theta}\, \mathrm{d}U(x) < \infty$ for some $a \ge 1$. Given θ, set $\delta = \frac{1}{2}\theta$. Choose a large such that $\frac{U(x)}{x^\delta} < 1$ for all $x \ge a$. Let b be a continuity point of U. Then, by integration by parts,

$$I_a = \lim_{b\to\infty} \{\frac{U(x)}{x^\theta}\Big|_a^b + \int_a^b \theta \frac{U(x)}{x^{1+\theta}}\, \mathrm{d}x\}.$$

Thus $I_a < \infty$ if $\int_a^\infty \frac{U(x)}{x^{1+2\delta}}\, \mathrm{d}x < \infty$. But this is true since the integrand is dominated by the integrable function $\frac{1}{x^{1+\delta}}$. □

Chapter 4

Independence

Introduction

In this chapter we define and develop the concept of independence (more precisely, *statistical independence*) for events and for rEs. The concept of independence has to do with classes of events and classes of *rv* / *rE*s. Let $(\Omega, \mathscr{A}, P)$ be a fixed probability space. All events cosidered are members of $\mathscr{A}$; all *rE*s considered are $\mathscr{A}$-measurable mappings.

4.1 Definitions

Definitions and Consequences 4.1.1.

(i) Let $A_\lambda, \lambda \in \Lambda$ be a family of events. i.e., $A_\lambda \in \mathscr{A}, \ \lambda \in \Lambda$. The members of this family are said to be mutually independent or, simply, independent, if

$$P(\bigcap_{j \in J} A_j) = \times_{j \in J} P(A_j) \text{ for every finite} J \subseteq \Lambda \tag{4.1}$$

It is trivial to see that, according to this definition, the null set ϕ and the whole space Ω are independent of every set $A \in \mathscr{A}$.

The need for this involved definition will be clear from the following examples.

(α) Let $\Omega = \{\omega_0, \omega_1, \omega_2, \omega_3\}$. $P(\{\omega_k\}) = \frac{1}{4}$, $k = 0, 1, 2, 3$. Define events $A_i = \{\omega_0, \omega_i\}$, $i = 1, 2, 3$. It is easy to see that, for each i, $P(A_i) = \frac{1}{2}$ and that, if $i \neq j$, $P(A_i \cap A_j) = P(\{\omega_0\}) = \frac{1}{4} = P(A_i)P(A_j)$. But $P(A_1 \cap A_2 \cap A_3) = P(\{\omega_0\}) = \frac{1}{4}$ while $P(A_1)P(A_2)P(A_3) = \frac{1}{8}$. Thus (4.1) is satisfied when any two of the three events are considered but not when $J = \{1, 2, 3\}$. Hence the three events can not be said to be independent.

(β) Let $\Omega = \{w_0, w_1, w_2, w_3, w_4, w_5, w_6, w_7\}$ and $P(\{w_0\}) = \frac{1}{27}$; $P(\{w_k\}) = \frac{4}{27}$, $1 \leq k \leq 6$; $P(\{w_7\}) = \frac{2}{27}$.

Define events $A_1 = \{w_0, w_1, w_2\}$; $A_2 = \{w_0, w_3, w_4\}$; $A_3 = \{w_0, w_5, w_6\}$. Easy computation leads to : $P(A_i) = \frac{1}{3}, i = 1, 2, 3$; $P(A_1 \cap A_2 \cap A_3) = P(\{w_0\}) = \frac{1}{27} = P(A_1)P(A_2)P(A_3)$. But $P(A_1 \cap A_2) = P(\{w_0\}) = \frac{1}{27} \neq P(A_1)P(A_2)$; $P(A_2 \cap A_3) = P(\{w_0\}) = \frac{1}{27} \neq P(A_2)P(A_3)$; $P(A_1 \cap A_3) = P(\{w_0\}) = \frac{1}{27} \neq P(A_1)P(A_3)$. Here the opposite of ($\alpha$) happens : if $J = \{1,\ 2\}$ or $\{2,\ 3\}$ or $\{3,\ 1\}$, then (4.1) is not satisfied while it is satisfied if $J = \{1,\ 2,\ 3\}$. Again the three events can not be said to be independent.
(γ) Two fair coins are tossed. Let A be the event that the first coin comes up heads, B the event that the second coin comes up heads and C the event we obtain exactly one head. Then A and B are, by the nature of the experiment, independent. Hence A and B' are independent and so are A' and B. Since $P(A \cap C) = P(A \cap B') = P(A)P(B') = \frac{1}{2} \times \frac{1}{2}$ and since $P(C) = P(A \cap B') + P(A' \cap B) = \frac{1}{4} + \frac{1}{4} = \frac{1}{2}$, it follows that A and C are independent. Similarly B and C are independent. But since $P(A)P(B)P(C) = \frac{1}{8}$ and $P(A \cap B \cap C) = 0$, $A,\ B,\ C$ are not mutually independent.
(ii) Let T be a fixed index set. For each $t \in T$ let $\mathcal{E}_t$ be a collection or a class of events. Classes $\mathcal{E}_t, t \in T^* \subseteq T$ of this family are said to be independent if, for every choice of $A_t \in \mathcal{E}_t, t \in T^*$, $\{A_t, t \in T^*\}$ is a family of mutually independendent events as per Definition (i). Equivalently, events $A_{t_1}, A_{t_2}, \ldots, A_{t_k}$ are mutually independent for every choice of k and every choice of $t_i \in T^*,\ 1 \leq i \leq k$.
(iii) Consider classes $\mathcal{E}_t, t \in T$ of random elements $X_{t,u}$ all defined on a common probability space $(\Omega,\ \mathscr{A},\ P)$ and taking values in possibly different measurable spaces $(\mathbf{X}_{t,u},\ \mathcal{X}_{t,u})$. These classes are said to be independent if the family of the σ-fields generated by the classes are independent.

In particular if $(\mathbf{X}, \mathcal{X}) = (R^p, \mathscr{R}^p)$ is the common range space for the random vectors $X_1, X_2, \ldots, X_n$ then these would be mutually independent *iff*

$$P(X_1 \leq \mathbf{x}_1, X_2 \leq \mathbf{x}_2, \ldots, X_n \leq \mathbf{x}_n) = \mathbb{F}_1(\mathbf{x}_1)\mathbb{F}_2(\mathbf{x}_2)\ldots\mathbb{F}_n(\mathbf{x}_n)$$

where $\mathbb{F}_j$ is the *df* of X_j, $1 \leq j \leq n$.

It is clear that in a family of events, the members are mutually independent *iff* their indicator functions are mutually independent *rvs*.

Some consequences

($\mathfrak{a}$) **Theorem**. If $X,\ Y$ are independent rvs with finite expected values, then $\mathbb{E}XY$ exists finitely and $\mathbb{E}XY = \mathbb{E}X\mathbb{E}Y$ but the converse is not true.
Proof. If $F_1,\ F_2$ are the dfs of X and Y respectively, then

$$\mathbb{E}|XY| = \iint\limits_{R^2} |x||y| \,\mathrm{d}F(x)\,\mathrm{d}F(y) = \int\limits_R |x| \,\mathrm{d}F(x) \int\limits_R |y| \,\mathrm{d}F(y) < \infty.$$

The relation $\mathbb{E}XY = \mathbb{E}X\mathbb{E}Y$ follows now.

Consider rvs $X,\ Y$ with joint cf.

$$\varphi(t,\ u) = \mathbb{E}e^{i(tX+uY)} = \frac{1}{2}\{e^{-iu-\frac{t^2}{2}} + e^{iu}\}.$$

It is a proper cf, being a convex combination of two cfs : the first being the cf of $(X_1,\ Y_1)$ where $X_1,\ Y_1$ are independent, X_1 having the standard normal distribution and Y_1 degenerate at -1; the second being the cf of $(X_2,\ Y_2)$ where X_2 is degenerate at 0 and Y_2 degenerate at 1. By differentiation at the origin, we see $\mathbb{E}X = 0$, $\mathbb{E}Y = 0$ and $\mathbb{E}XY = 0$. But $X,\ Y$ are not independent. For, if they were, then aguing as above we will have for all $t,\ u \in R$, $\varphi(t,u) = \mathbb{E}e^{itX}\mathbb{E}e^{iuY} = \frac{1}{2}\{e^{-\frac{t^2}{2}} + 1\}\frac{1}{2}\{e^{-iu} + e^{iu}\}$. For t arbitrary and $u = \frac{\pi}{2}$, this relation would read $\frac{1}{2}\{-ie^{-\frac{t^2}{2}} + i\} = 0$ or $e^{t^2} \equiv 1$, which is not possible.

However we have the following

Theorem. Let $X,\ Y$ be measurable mappings of $(\Omega,\ \mathscr{A},\ P)$ into, respectively, the metric spaces $(M_1,\ d_1)$ and $(M_2,\ d_2)$ with Borel σ-fields $\mathcal{m}_i$, $i = 1,\ 2$. Then $(\alpha) \Leftrightarrow (\beta)$ where

(α) : $X,\ Y$ are independent and

(β) : $\mathbb{E}f_1(X)f_2(Y) = \mathbb{E}f_1(X)\mathbb{E}f_2(Y)$ for every choice of the real bounded continuous functions f_i, $i = 1,\ 2$ defined on M_i.

Proof. (α) implies $f_1(X),\ f_2(Y)$ are independent. Hence $(\alpha) \Rightarrow (\beta)$.

Suppose now (β) holds. Let $A_i \in \mathcal{m}_i$ be closed sets in M_1 and M_2 respectively. Define $f_{i,n(.)} = 1 - nd_i(.,\ A_i)$ if $d_i(.,\ A_i) \leq \frac{1}{n}$ and zero otherwise. We note that each $f_{i,n}$, $i = 1,\ 2$; $n = 1,\ 2,\ ..$ is a bounded continuous function and that $\lim\limits_{n\to\infty} f_{i,n}(.) = \chi_{A_i}(.)$. By hypothesis, $\mathbb{E}f_{1,n}(X)f_{2,n}(Y) = \mathbb{E}f_{1,n}(X)\mathbb{E}f_{2,n}(Y)$. In this relation take limit as $n \to \infty$ and conclude that $\mathbb{E}\chi_{A_1}(X)\chi_{A_2}(Y) = \mathbb{E}\chi_{A_1}(X)\mathbb{E}\chi_{A_2}(Y)$ by appeal to the bounded convegence theorem. It follows from this that for closed sets $A_i \subset M_i$, $(X \in A_1)$ and $(Y \in A_2)$ are independent.

Fix $A \in \mathcal{m}_1$. If $P(X \in A) = 0$ then trivially the events $(X \in A)$ and $(Y \in B)$ are independent for every $B \in \mathcal{m}_2$. Let now $A \in \mathcal{m}_1$ be a closed set with $P(X \in A) > 0$. We note $\frac{P(X\in A,\ Y\in .)}{P(X\in A)}$ is a pm on $\mathcal{m}_2$. For arbitrary $B \in \mathcal{m}_2$, which is not a closed set, we will then have, by **(1.9.2)**, $\frac{P(X\in A,\ Y\in B)}{P(X\in A)} =$

$\lim_{n\to\infty} \frac{P(X\in A,\, Y\in B_n)}{P(X\in A)}$, where $B_n \subset B$ is a specially chosen monotonic increasing sequence of closed sets, which is such that $P(Y \in B_n) \to P(Y \in B)$. It follows $P(X \in A,\ Y \in B) = P(X \in A)\frac{P(X\in A,\, Y\in B)}{P(X\in A)} = P(X \in A) \lim_{n\to\infty} \frac{P(X\in A,\, Y\in B_n)}{P(X\in A)} = P(X \in A) \lim_{n\to\infty} P(Y \in B_n) = P(X \in A)P(Y \in B)$. Thus if $A \in \mathfrak{m}_1$ is a closed set, then $(X \in A)$ is independent of the σ-field generated by Y. Now starting with $(Y \in B)$ in this σ-field, and using similar arguments we can show that for every $B \in \mathfrak{m}_2$, $(Y \in B)$ is independent of $(X \in A)$ for every $A \in \mathfrak{m}_1$. This completes the proof. □

(𝔟) **Theorem**. Let random elements X_i defined on a common probability space $(\Omega,\ \mathscr{A},\ P)$ take values in the complete and separable metric space $(M_i,\ d_i,\ m_i, 1 \le i \le k,\ k \ge 2)$. Then $(\alpha_k) \Leftrightarrow (\beta_k)$ where

(α_k) : $X_1,\ X_2,\ \ldots,\ X_k$ are independent and

(β_k) : $\mathbb{E}\prod_{i=1}^{k} g_i(X_i) = \prod_{i=1}^{k} \mathbb{E}g_i(X_i)$ for every choice g_i of real bounded continuous function defined on M_i, $1 \le i \le k$.

Proof.

$(\alpha_k) \Leftrightarrow (g_i(X_i),\ 1 \le i \le k)$ are independent. Hence $(\alpha_k) \Rightarrow (\beta_k)$.

Suppose now (β_k) holds. Define $g_{i,n}(x) = 1 - nd_i(x,\ A_i)$ if $d_i(x_i,\ A_i) \le \frac{1}{n}$ and zero otherwise, $x_i \in M_i,\ A_i \subset M_i$ is a closed set, $1 \le i \le k$. Use hypothesis, take $n \to \infty$ and appeal to the bounded convergence theorem to conclude that the events $(X_i \in A_i),\ 1 \le i \le k$ are independent whenever the A_is are closed sets.

[*Note*. If $k = 2$, the *proof* in part (𝔞) above applies to any metric spaces $M_1,\ M_2$, not necessarily complete and separable ones].

Let the A_is be arbitrary Borel sets. By **(1.9.5)**, Remark (ii), there exist, corresponding to each i, compact (and, hence, closed) sets $(C_{i,n},\ C_{i,n} \uparrow$ as $n \uparrow)$ such that $P(X_i \in A_i,\ 1 \le i \le k) = \lim_{n\to\infty} P(X_i \in C_{i,n},\ 1 \le i \le k) = \lim_{n\to\infty} \prod_{i=1}^{k} P(X_i \in C_{i,n}) = \prod_{i=1}^{k} P(X_i \in A_i)$ as was to be proved. □

(𝔠)**Theorem**. Let $(X_n,\ n \ge 0)$ be measurable mappings from Ω into M_1; let Y be a measurable map from Ω into M_2; let X_n be independent of Y for each $n \ge 1$ and let $\lim_{n\to\infty} X_n(\omega) = X(\omega),\ \omega \in \Omega$. Then X is independent of Y.

Proof. The hypothsis implies $\mathbb{E}f_1(X_n)f_2(Y) = \mathbb{E}f_1(X_n)\mathbb{E}f_2(Y)$ for every choice of real bounded continuous functions f_i defined on M_i, $i = 1,\ 2$. Take limit as $n \to \infty$ and appeal to the bounded convergence theorem to claim $\mathbb{E}f_1(X_0)f_2(Y) = \mathbb{E}f_1(X_0)\mathbb{E}f_2(Y)$. Now appeal to (𝔞) and conclude the proof. □

Theorem 4.1.2. *Let* $\{\mathcal{E}_t, t \in T\}$ *be a family of classes of events. Let* $\{\sigma(\mathcal{E}_t),$

$t \in T\}$ be the family of the σ-fields generated by these classes. Assume that each class $\mathcal{E}_t$ is a π-class. (ref. ***(1.1.1)****) Then the classes $\mathcal{E}_t$ are independent iff the classes $\sigma(\mathcal{E}_t)$ are independent.*

Proof. That independence of the classes $\{\sigma(\mathcal{E}_t)\}$ implies the independence of the classes $\{\mathcal{E}_t\}$ is obvious. To establish implication in the reverse direction, assume that the classes $\{\mathcal{E}_t\}$ are independent. Take two members of the family. Label them $\mathcal{E}_1$ and $\mathcal{E}_2$. Let $\mathcal{C}$ be the collection of all events A such that $P(A \cap B) = P(A)P(B)$ for every $B \in \mathcal{E}_2$. $\mathcal{C} \supseteq \mathcal{E}_1$, by hypothesis. We note : $P(\Omega \cap B) = P(B) = P(\Omega)P(B)$, showing that $\Omega \in \mathcal{C}$. If $A_1, A_2 \in \mathcal{C}$ and if $A_1 \subseteq A_2$, then

$$\begin{aligned} P((A_2 \sim A_1) \cap B) &= P((A_2 \cap B) \sim (A_1 \cap B)) \\ &= P(A_2 \cap B) - P(A_1 \cap B) = P(A_2)P(B) - P(A_1)P(B) \\ &= P(A_2 \sim A_1)P(B). \end{aligned}$$

This implies $A_2 \sim A_1 \in \mathcal{C}$. Next, let $A_j \in \mathcal{C}, j \geq 1, A_j \cap A_k = \phi, j \neq k$ and $B \in \mathcal{E}_2$. Then

$$\begin{aligned} P((\cup_{j\geq 1} A_j) \cap B) &= P(\cup_{j\geq 1}(A_j \cap B)) = \sum_{j=1}^{\infty} P(A_j \cap B) \\ &= \sum_{j=1}^{\infty} P(A_j)P(B) = (\sum_{j=1}^{\infty} P(A_j))P(B) = P(\cup_{j\geq 1} A_j)P(B), \end{aligned}$$

proving that $\mathcal{C}$ is closed under countable disjoint unions. These three properties show that $\mathcal{C}$ is a λ-class. Since it contains $\mathcal{E}_1$, a π-class of events, it follows by **(1.1.2)** that $\mathcal{C} \supseteq \sigma(\mathcal{E}_1)$. Thus $P(A \cap B) = P(A)P(B)$ for every $A \in \sigma(\mathcal{E}_1)$ and $B \in \mathcal{E}_2$. Start now with the collection $\mathcal{D}$ of events B such that $P(A \cap B) = P(A)P(B)$ for every $A \in \sigma(\mathcal{E}_1)$ and$B \in \mathcal{E}_2$ and reason as before to prove that $P(A \cap B) = P(A)P(B)$ for every $A \in \sigma(\mathcal{E}_1)$ and $B \in \sigma(\mathcal{E}_2)$. Since these arguments carry over for a finite number of classes, the proof is complete. □

Existence of independent sequences

In order to make sure that the definition given above of independence of families of rEs is not vacuous we present the following theorem.

Theorem 4.1.3. *For each n, $n = 1, 2, \ldots$ let Ω_n be a complete separable metric space, $\mathscr{A}_n$ be its Borel σ-field and μ_n an arbitrary pm on $\mathscr{A}_n$. Then there exist rEs X_n in Ω_n such that the dm of X_n is μ_n and such that (X_n) is a sequence of independent rEs.*

Proof. Let $(\Omega,\ \mathscr{A},\ P)$ be the product measure space (ref. Remark under **(1.11.6)**). Write $\omega = (\omega_1,\ \omega_2,\ \ldots),\ \omega \in \Omega,\ \omega_n \in \Omega_n$ and $X_n : \Omega \to \Omega_n,\ X_n(\omega) = \omega_n$ and note that for any $B \in \mathscr{A}_n,\ P(X_n \in B) = \mu_n(B)$. Further

$$\begin{aligned} &P(X_k \in B_k,\ 1 \le k \le n) \\ &\quad = (\mu_1 \times \mu_2 \times \cdots \times \mu_n)(B_1 \times B_2 \times \ldots \times B_n) \\ &\quad = \mu_1(B_1)\mu_2(B_2)\ldots\mu_n(B_n) \\ &\quad = P(X_1 \in B_1)P(X_2 \in B_2) \ \ldots\ P(X_n \in B_n). \end{aligned}$$

Hence (X_n) is a sequence of independent rEs with property as stated. □

Tail σ-fields, symmetric events and the 0-1 laws

Definitions 4.1.4. Let $T = \{1,\ 2,\ \ldots\}$. For each $n \in T$, let $(\Omega_n,\ \mathscr{A}_n)$ be a measurable space. Let $(\Omega_T,\ \mathscr{A}_T)$ be product measurable space.

Let $(\Omega,\ \mathscr{A})$ be a measurable space. Let $X_n,\ n \in T$ be a measurable mapping of $(\Omega,\ \mathscr{A})$ into $(\Omega_n,\ \mathcal{A}_n)$. Let $\mathbf{X} = (X_1,\ X_2,\ ..)$. By **(1.4.7)**, $\mathbf{X}$ is a measurable mapping of Ω into Ω_T.

Let $\mathcal{A}_1 = \mathscr{A}_T;\ \mathcal{A}_n = \otimes_{r=1}^{\infty}\hat{\mathscr{A}}_{r,n},\ n \ge 2$ where $\hat{\mathscr{A}}_{r,n} = \{\Omega_r,\ \phi\}$ if $r < n$ and $\mathscr{A}_r$ if $r \ge n$. We note $\mathcal{A}_n$ is a monotonic decreasing sequence of sub σ-fields of $\mathscr{A}_T$.

(i) The tail σ-field $\mathcal{A}$ of the product space is defined as $\mathcal{A} = \cap_n \mathcal{A}_n$.

Let $\mathcal{B}_n = \otimes_{r=1}^{\infty}\tilde{\mathscr{A}}_{r,n}$ where $\tilde{\mathscr{A}}_{r,n} = \mathscr{A}_r$ if $r \le n$ and $\tilde{\mathscr{A}}_{r,n} = \{\Omega_r,\ \phi\}$ if $r > n$.

(ii) We note
that $\mathcal{B}_n$ is a monotonic increasing sequence of sub σ-fields of $\mathscr{A}_T$;
that $\mathcal{B} = \cup_n \mathcal{B}_n$ is a field;
that $\sigma(\mathcal{B}) = \mathscr{A}_T$;
that $\mathbf{X}^{-1}(\mathcal{B}_n)$ is the minimal σ-field in Ω with respect to which the mappings $X_1,\ X_2,\ \ldots,\ X_n$ are measurable; and
that $\mathcal{A}_n$ is the minimal σ-field in Ω with respect to which the mappings X_m, $m \ge n$ are measurable.

(iii) $\mathbf{X}^{-1}(\mathcal{A}) = \cap_n \mathbf{X}^{-1}(\mathcal{A}_n) \subset \mathscr{A}$ is called the tail σ-field of the sequence (X_n).

(iv) Let $(\Omega_n, \mathscr{A}_n) = (U, \mathcal{U})$ for all $n \in T$. Define $A \in \mathcal{U}^T$ to be invariant under n-permutation if it has the following property . $(\mathbf{x} \in A) \Rightarrow (\mathbf{y} \in A)$ if $x_k = y_k, k \ge n+1$ and $(y_1, y_2, \ldots, y_n)$ is a permutation of $(x_1, x_2, \ldots, x_n)$.

Denote the collection of all sets A invariant under n-permutation by $\mathcal{V}_n$. Clearly $\mathcal{V}_n \downarrow$. $\mathcal{V} = \cap_n \mathcal{V}_n \subseteq \mathcal{A}_T$ is called the class of finite permutation invariant measurable subsets or symmetric measurable subsets or, simply, symmetric events of Ω_T.

Consider $\mathcal{V}_n$ for n fixed. Let $A \in \mathcal{V}_n$. Let $\mathbf{x} = (x_1, x_2, \ldots, x_n, ..) \in A^{'}$. If $(y_1, y_2, \ldots, y_n)$ is a permutation of $(x_1, x_2, \ldots, x_n)$ then we claim that $(y_1, y_2, \ldots, y_n, x_{n+1}, x_{n+2}, \ldots) \in A^{'}$. For otherwise, the point will belong to A. But that would imply, since $A \in \mathcal{V}_n$, that $\mathbf{x} \in A$. i.e., A' is invariant under n-permutation. Thus $\mathcal{V}_n$ is closed under complementation. That $\mathcal{V}_n$ is closed under countable unions is obvious. It follows that all the $\mathcal{V}_n$s, and hence $\mathcal{V}$, are σ-fields.

We note that every non-null member of $\mathcal{A}_{n+1}$ is of the form $U^n \times E$ where E is a non-null member of $\otimes_{n+1}^{\infty} \mathscr{A}_n$ and hence is invariant under n-permutation. Hence $\mathcal{A}_{n+1} \subset \mathcal{V}_n$. It follows $\mathcal{A} \subset \mathcal{V}$. i.e., every tail event of the product space U^T is a symmetric event. Suppose (A_n) is a sequence of events such that $A_n \in \mathcal{V}_n$, $n = 1, 2, \ldots$. Since $\mathcal{V}_n \downarrow$, $A_\nu \in \mathcal{V}_n$, $\nu \geq n$. Consequently, $\bigcup_{\nu=n}^{\infty} A_\nu \in \mathcal{V}_n$. It follows now that $A^* = \overline{\lim_{n \to \infty}} A_n = \bigcap_{n=1}^{\infty} \bigcup_{\nu=n}^{\infty} A_\nu \in \bigcap_{n=1}^{\infty} \mathcal{V}_n = \mathcal{V}$. In other words, if $A_n \in \mathcal{V}_n$, $n = 1, 2, \ldots$. then A^* is a symmetric event.

(v) Let $(U, \mathcal{U}) = (R, \mathscr{R})$. Let f_n be the map taking R^n into R : $f_n(x_1, x_2, \ldots, x_n) = x_1 + x_2 + \ldots + x_n$. We note the σ-field $\mathfrak{C}_n = f_n^{-1}(\mathscr{R}) \otimes \mathscr{R} \otimes ..$ is contained in $\mathcal{V}_n$. Since $f_n^{-1}(\mathscr{R}) \otimes \mathscr{R} \supset f_{n+1}^{-1}(\mathscr{R})$, $\mathfrak{C}_n \downarrow$. If $\mathfrak{C}_n \downarrow \mathfrak{C}$, then $\mathfrak{C} \subset \mathcal{V}$. A consequence is that if g is a Borel measurable function defined on the line and if E is a linear Borel set, then $A_n \in \mathcal{V}_n$ where $A_n = \{\mathbf{x} : g(x_1 + x_2 + \cdots + x_n) \in E\}$. In particular, given $a > 0$, $A_n = \{\mathbf{x} : |\sum_{k=1}^{n} x_k| < a\} \in \mathcal{V}_n$. This implies A^* is a symmetric event.

Let $S_n = \sum_1^n X_k$. Then the minimal σ-field with respect to which $S_n, S_{n+1}, \ldots$ are measurable (or, what is same, the minimal σ-field with respect to which $S_n, X_{n+1}, X_{n+2}, \ldots$ are measurable) is $\mathbf{X}^{-1}(\mathfrak{C}_n)$. Hence the tail σ-field of the sequence (S_n) is $\mathfrak{C}$.

From the definition of $\mathcal{A}_{n+1}$ and that of $\mathfrak{C}_n$, the inclusion $\mathcal{A}_n \subset \mathfrak{C}_n$ is obvious. Hence $\mathcal{A} \subset \mathfrak{C}$. i.e., the tail σ-field of the sequence (X_n) is contained in that of the sequence (S_n).

(vi) **Definition.** A σ-field of a probability space is said to be a trivial one if it consists only of events with probability 0 or probability 1.

Let μ_n be a *pm* on $\mathscr{A}_n$, $n \geq 1$ and μ_T the product measure. Then

Theorem 4.1.5. *$\mu_T(A) = 0$ or 1 for every $A \in \mathcal{A}$, the tail σ-field of the product space.*

Proof. Let $A \in \mathcal{A}$ and let $\varepsilon > 0$ given. Since $\sigma(\cup_n \mathcal{B}_n) = \mathscr{A}_T$ and since $A \in \mathscr{A}_T$, Theorem D, p56, [H], as applied to the present situation states: a positive integer N and a set $A_N \in \mathcal{B}_N$ can be found such that

$$\mu_T(A \Delta A_N) < \varepsilon \tag{4.2}$$

This implies

$$|\mu_T(A) - \mu_T(A_N)| < \varepsilon \tag{4.3}$$

Since $A \in \mathcal{A}_m$, $m > N$, the events A and A_N are independent. (This assertion is proved by considering the probability measures of the finite dimensional cylinder sets generating the σ-field $\mathcal{B}_N$ and the measures of those generating $\mathcal{A}_m$, $m > N$). Hence,

$$\begin{aligned}
\mu_T(A) &= \mu_T(A \cap A_N) + \mu_T(A \cap A'_N) \\
&\leq \mu_T(A)\mu_T(A_N) + \mu_T(A \Delta A_N) \\
&\leq \mu_T(A)\{\mu_T(A) + \varepsilon\} + \varepsilon \qquad \text{(by (4.2) and (4.3))} \\
&\leq \{\mu_T(A)\}^2 + \varepsilon\mu_T(A) + \varepsilon.
\end{aligned}$$

Since $\varepsilon > 0$ is arbitrary and since $\mu_T(A) < \infty$, this leads to

$$\mu_T(A) \leq \{\mu_T(A)\}^2. \tag{4.4}$$

Since μ_T is a pm, $\mu_T(A) \leq 1$ and hence $\mu_T(A) \geq \{\mu_T(A)\}^2$. This with (4.4) implies $\mu_T(A) = \{\mu_T(A)\}^2$, leading to $\mu_T(A) = 0$ or 1. □

Corollary 4.1.6. *Let (X_n) be a sequence of independent rEs, X_n taking values in Ω_n with pm μ_n. Then the tail σ-field of the sequence (X_n) is a trivial one.*

Proof. We noted in **(4.1.4)** that the tail σ-field of the X_n sequence is $\mathbf{X}^{-1}(\mathcal{A})$. We note that the dm of $\mathbf{X}$ in Ω_T is μ_T. Since $\mathcal{A}$ is a trivial σ-field under μ_T, the claim follows that $\mathbf{X}^{-1}(\mathcal{A})$ is a trivial σ-field in $(\Omega, \mathscr{A}, P)$. □

The following theorem is only a reformulation of **(4.1.5)**.

4.1.7. (Part 1)
Definition. Let $(\Omega, \mathscr{A}, P)$ be a probability space. Let $(\mathscr{A}_n \subset \mathscr{A})$ be a sequence of σ-fields. Define the σ-field $\mathcal{B}_n = \sigma(\cup_{k \geq n} \mathscr{A}_k)$. Then the tail σ-field of the sequence $(\mathscr{A}_n)$ is defined as $\mathcal{B}_{(\infty)} = \cap_n \mathcal{B}_n$

Given a sequence $(X_n,\ n \geq 1)$ of rVs, define $\mathscr{A}_n,\ n \geq 1$ to be the σ-field generated by X_n. Then, in the above notation, $\mathcal{B}_n$ is the σ-field generated by the elements $X_\nu,\ \nu \geq n$. Given a two sided infinite sequence $(X_n,\ -\infty <$

$n < \infty$), define $\mathscr{A}_n$ as before and define $\mathcal{B}_n$ as the σ-field generated by the elements $X_\nu,\ |\nu| \geq n$.

(Part 2)

Theorem (*Kolmogorov 0-1 law*).

Let $(\Omega,\ \mathscr{A},\ P)$ be a probability space. Let $(\mathscr{A}_n \subset \mathscr{A})$ be a sequence of independent σ-fields. Define the σ-field $\mathcal{B}_n = \sigma(\cup_{k\geq n}\mathscr{A}_k)$. Then the tail σ-field $\mathcal{B}_{(\infty)} = \cap_n \mathcal{B}_n$ is the trivial σ-field, denoted by $\{\Omega, \phi\}$, consisting of all the events with probability 0 and all those with probability 1.

Proof. Denote by $\mathscr{A}^{[n]}$ the σ-field $\sigma(\cup_1^n \mathscr{A}_k)$ and by $\mathscr{A}^{[\infty]}$ the σ-field $\sigma(\cup_1^\infty \mathscr{A}_k)$. Let $A \in \mathcal{B}_{(\infty)}$. We note (i) that $A \in \mathscr{A}^{[\infty]}$ and (ii) that A i.e the class consisting of this single member is (since the σ-fields are independent) independent of $\mathscr{A}^{[n]}$ for every n and hence independent of the field of events $\cup_1^\infty \mathscr{A}^{[n]}$. It now follows by **(4.1.2)** that A is independent of $\mathscr{A}^{[\infty]}$. Thus A is independent of itself. i.e. $P(A) = P(A \cap A) = P(A)P(A)$. Hence $P(A) = 0$ or $P(A) = 1$. □

Corollary. (*Borel 0-1 law*)

If the events A_n are independent and if $A^* = \overline{\lim\limits_{n\to\infty}} A_n$, then $P(A^*) = 0$ or 1.

Proof. By hypothesis the indicator functions χ_ns of the A_ns are mutually independent *rvs*. i.e. $\mathscr{A}_n = \{\Omega, A_n, A_n^{'}, \phi\}$ is a sequence of independent σ-fields. Hence, by the theorem, the tail σ-field of this sequence is the trivial σ-field. We note that $\chi_{A^*} = \overline{\lim\limits_{n\to\infty}}\ \chi_n$ and that $\overline{\lim\limits_{n\to\infty}}\ \chi_n$ is measurable with respect to $\sigma(\cup_{k\geq n}\mathscr{A}_k)$ for every n. Hence χ_{A^*}, being a tail function (i.e. a rv measurable with respect to the tail σ-field), is degenerate. This means that χ_{A^*} is 0 with probability 1 or is 1 with probability 1. Equivalently, $P(A^*) =$ 0 or 1. □

While Borel's $0 - 1$ law says $P(A^*)$ can take only the values 0 and 1, the Lemma that follows develops criteria for each of these two values to be assumed.

Theorem 4.1.8 (Borel-Cantelli Lemma). *Given an arbitrary sequence* (A_n) *of events write* $a_k = P(A_k)$.

(i) If $\sum a_n < \infty$, *then* $P(\overline{\lim\limits_{n\to\infty}} A_n) = 0$;

(ii) if $\sum a_n = \infty$ *and if the* A_n*s are independent, then* $P(\overline{\lim\limits_{n\to\infty}} A_n) = 1$.

Proof. Suppose $\sum_1^\infty a_n < \infty$. Hence $\lim\limits_{m\to\infty} \sum_{k\geq m} a_k = 0$. We then have,

$$P(\overline{\lim_{n\to\infty}} A_n) = \lim_{m\to\infty} P(\cup_{k\geq m} A_k) \leq \lim_{m\to\infty} \sum_{k\geq m} a_k = 0.$$

(ii) We recall that for every $0 \le a \le 1$, $1 - a \le e^{-a}$. We have

$$\begin{aligned} P(\overline{\lim_{n\to\infty}} A_n) &= \lim_{m\to\infty} P(\cup_{k\ge m} A_k) \\ &= \lim_{m\to\infty} \lim_{n\to\infty} P(\cup_{k=m}^{n} A_k) \\ &= \lim_{m\to\infty} \lim_{n\to\infty} \{1 - \times_{k=m}^{n}(1 - a_k)\} \\ &\ge \lim_{m\to\infty} \lim_{n\to\infty} \{1 - e^{-\sum\limits_{k=m}^{n} a_k}\} = 1, \end{aligned}$$

since $\sum_n a_n = \infty$. □

Remark 4.1.9. To quickly see that in part (ii) above, the condition of independence can not be dispensed with, let $\Omega = [0,\ 1]$, $\mathscr{A}$ is the Borel σ-field and P the Lebesgue measure. Let $A_n = [0,\ \frac{1}{n}]$. We note $P(A_n) = \frac{1}{n}$ and $\sum_{n=1}^{\infty} P(A_n) = \infty$. But the A_ns are not mutually independent. For example $P(A_n \cap A_{n+1}) = P(A_{n+1}) = \frac{1}{n+1}$ while $P(A_n)P(A_{n+1}) = \frac{1}{n(n+1)}$. We can not apply part (ii) and claim $P(\overline{\lim_{n\to\infty}} A_n) = 1$. In fact $\overline{\lim_{n\to\infty}} A_n = \{0\}$. Hence $P(\overline{\lim_{n\to\infty}} A_n) = 0$.

Example 1.
To illustrate the application of the above theorem, let (X_n) be a sequence of independent and identically distributed (*iid*) rvs and consider the problem of finding the limit points of the sequence $(\frac{X_n}{\sqrt{2\log n}})$ under the assumption that the distribution of X_1 is standard normal (ref.**(2.1.2)**).

As $x \to \infty$,

$$P(X_1 > x) \sim \frac{q}{x} e^{-\frac{1}{2}x^2} \tag{4.5}$$

where q is an absolute constant. (ref. **(4.4.5)**). Using this we see that for every $\varepsilon > 0$, $\sum_n P\big(X_n > (1+\varepsilon)\sqrt{2\log n}\big) < \infty$. The arbitrariness of ε and Borel-Cantelli lemma help us conclude that a.s. $\overline{\lim_{n\to\infty}} \frac{X_n}{\sqrt{2\log n}} \le 1$. By symmetry, a.s. $\underline{\lim_{n\to\infty}} \frac{X_n}{\sqrt{2\log n}} \ge -1$.

Let $0 < a < 1$. Let $\varepsilon > 0$ be arbitrary but small so that $0 < a - \varepsilon < a < a + \varepsilon < 1$. Then, appealing to the Mean Value theorem, we have, for some $y \in \big((a-\varepsilon)\sqrt{2\log n},\ (a+\varepsilon)\sqrt{2\log n}\big)$,

$$P(a - \varepsilon \le \tfrac{X_n}{\sqrt{2\log n}} \le a + \varepsilon) = \int\limits_{(a-\varepsilon)\sqrt{2\log n}}^{(a+\varepsilon)\sqrt{2\log n}} \frac{1}{\sqrt{2\pi}} e^{-\frac{1}{2}x^2}\, dx = q\sqrt{\log n}\, e^{-\frac{1}{2}y^2}$$

$$\ge q\sqrt{\log n} e^{-\frac{1}{2}(a+\varepsilon)^2 2\log n} = q\sqrt{\log n} \frac{1}{n^{(a+\varepsilon)^2}} \ge \frac{q}{n^{(a+\varepsilon)^2}}.$$

Since $(a+\varepsilon)^2 < 1$, the series $\sum\limits_n P(a - \varepsilon \le \frac{X_n}{\sqrt{2\log n}} \le a + \varepsilon)$ diverges.

Since the X_ns are independent, we conclude, by the second part of the Borel-Cantelli lemma, that for all integers k with $0 < a - \frac{1}{k} < a + \frac{1}{k} < 1$, $P(a - \frac{1}{k} < \frac{X_n}{\sqrt{2 \log n}} < a + \frac{1}{k} \text{ i.o.}) = 1$. Thus for each k large, there exists a set A_k such that $P(A_k) = 1$ and such that for every $\omega \in A_k$, $\frac{X_n(\omega)}{\sqrt{2 \log n}} \in (a - \frac{1}{k},\ a + \frac{1}{k})$ for infinitely many k. If $A = \cap_k A_k$, then $P(A) = 1$ and for every $\omega \in A$, a is a limit point of the sequence $(\frac{X_n(\omega)}{\sqrt{2 \log n}})$. The set $A = A_a$ would depend on the number a. Let A_r correspond to the rational number $r \in E = (-1,\ 0) \cup (0,\ 1)$. Let $A = \cap_r A_r$. Then, since the rationals are countable, $P(A) = 1$. Further for every $\omega \in A$, every rational in $[-1,\ 1]$ is a limit point of the sequence $(\frac{X_n(\omega)}{\sqrt{2 \log n}})$. Since the rationals in E form a dense subset of $[-1,\ 1]$, it follows that for every $\omega \in A$ every point in $[-1,\ 1]$ is a limit point of the sequence $(\frac{X_n(\omega)}{\sqrt{2 \log n}})$. Since $wp1$, $\overline{\lim_{n \to \infty}} \frac{X_n}{\sqrt{2 \log n}} \le 1$ and $\underline{\lim_{n \to \infty}} \frac{X_n}{\sqrt{2 \log n}} \ge -1$, there are no other limit points. To summarise : $wp1$, $[-1,\ 1]$ is the limit set of the sequence $(\frac{X_n(\omega)}{\sqrt{2 \log n}})$.

In the above example independence played a crucial part. In situations where the events A_n are not independent but $\sum_n P(A_n) = \infty$ and we know that $P(A_n\ i.o.)$ is 0 or 1, the technique to decide which one is the correct value varies from problem to problem. In the following example we present an often used technique that helps us show that $P(A_n\ i.o.) = 0$.

Example 2.

Let $X,\ X_1,\ X_2,\ X_3,\ \ldots.$ be *iid* Cauchy variables with common frequency function $\frac{1}{\pi} \frac{1}{1+x^2}$, $-\infty < x < \infty$. We note $P(|X| > x) \sim \frac{2}{\pi x}$ as $x \to \infty$. Let $S_n = \sum_{k=1}^n X_k$, $T_n = \frac{S_n}{n}$ and $\theta_n = (\log \log n)^{-1}$. For any $k \ge 1$,

$$\begin{aligned}
\overline{\lim_{n \to \infty}} |T_n|^{\theta_n} &= \overline{\lim_{n \to \infty}} \left| \frac{S_k}{n} + \frac{S_n - S_k}{n} \right|^{\theta_n} \\
&\le \overline{\lim_{n \to \infty}} \left| \frac{S_k}{n} \right|^{\theta_n} + \overline{\lim_{n \to \infty}} \left| \frac{S_n - S_k}{n} \right|^{\theta_n} = \overline{\lim_{n \to \infty}} \left| \frac{S_n - S_k}{n} \right|^{\theta_n}.
\end{aligned}$$

Again

$$\overline{\lim_{n \to \infty}} \left| \frac{S_n - S_k}{n} \right|^{\theta_n} \le \overline{\lim_{n \to \infty}} \left| \frac{S_n}{n} \right|^{\theta_n} + \overline{\lim_{n \to \infty}} \left| \frac{S_k}{n} \right|^{\theta_n} = \overline{\lim_{n \to \infty}} |T_n|^{\theta_n}.$$

Thus $\xi = \overline{\lim_{n \to \infty}} |T_n|^{\theta_n}$ is measurable with respect to the tail σ-field of the X_n sequence and hence is a constant $a.s.$, by Kolmogorov's $0 - 1$ law. Denote it by α. We note that T_n is distributed as X.

For any $a > 0$, let $A_n = \{|T_n|^{\theta_n} > a\}$. Since $A^* = \{A_n \text{ i.o.}\} = \{\xi > a\}$, it follows that A^* is a tail event of the sequence (X_n) and hence $P(A^*) = 0$ or 1.

We consider the question of proving that $\alpha \leq e$.

For $\varepsilon > 0$ arbitrary, define events $A_n = \{|T_n|^{\theta_n} > e^{1+\varepsilon}\} = \{|S_n| > n(\log n)^{1+\varepsilon}\}$. If we prove that $P(A_n\ i.o.) = 0$, it will follow that $\alpha \leq e$.

Now,

$$P(A_n) = P(|T_n| > (\log n)^{1+\varepsilon}) \sim \frac{2}{(\log n)^{1+\varepsilon}}.$$

Hence $\sum\limits_n P(A_n) = \infty$. Since the A_ns are not independent events, the second part of Borel-Cantelli lemma does not apply and we are not able to conclude if $P(A_n\ i.o.) = 1$ or $= 0$. Finer analysis is called for.

Let $m_r = [e^r] \sim e^r$ as $r \to \infty$. Let

$$B_r = \{\max_{m_r \leq n \leq m_{r+1}} |S_n| > m_r(\log m_r)^{1+\varepsilon}\}.$$

We note that $(A_n \text{ i.o.}) \subseteq (B_r \text{ i.o.})$. For $m_r \leq n \leq m_{r+1}$,

$$\begin{aligned}
&P(|S_{m_{r+1}} - S_n| > \tfrac{1}{2}m_r(\log m_r)^{1+\varepsilon}) \\
&= P(|S_{m_{r+1}-n}| > \tfrac{1}{2}m_r(\log m_r)^{1+\varepsilon}) \\
&= P(|X| > \tfrac{1}{2}\tfrac{m_r}{m_{r+1}-n}(\log m_r)^{1+\varepsilon}) \\
&\leq P(|X| > \tfrac{1}{2}\tfrac{m_r}{m_{r+1}-m_r}(\log m_r)^{1+\varepsilon}) \\
&\leq P(|X| > q(\log m_r)^{1+\varepsilon}) = o(1)
\end{aligned}$$

as $r \to \infty$. Hence $P(|S_{m_{r+1}} - S_n| > \frac{1}{2}m_r(\log m_r)^{1+\varepsilon}) \leq \frac{1}{2}$ for every n, $m_r \leq n \leq m_{r+1}$ and for all r large.

Write $Y_1 = S_{m_r}$, $Y_2 = X_{m_r+1}$, $Y_3 = X_{m_r+2}$, $Y_4 = X_{m_r+3}$, $Y_{m_{r+1}-m_r+1} = X_{m_r+m_{r+1}-m_r} = X_{m_{r+1}}$ and $R_k = \sum_{j=1}^k Y_j$, $1 \leq k \leq m_{r+1} - m_r + 1$. We note that $R_{m_{r+1}-m_r+1} = S_{m_{r+1}}$ and that

$$\begin{aligned}
&\max_{1\leq j\leq m_{r+1}-m_r+1} P(|R_{m_{r+1}-m_r+1} - R_j| > \tfrac{1}{2}m_r(\log m_r)^{1+\varepsilon}) \\
&= \max_{m_r\leq n\leq m_{r+1}} P(|S_{m_{r+1}} - S_n| > \tfrac{1}{2}m_r(\log m_r)^{1+\varepsilon}) \leq \tfrac{1}{2}.
\end{aligned}$$

Hence

$$\begin{aligned}
P(B_r) &= P(\max_{m_r\leq n\leq m_{r+1}} |S_n| > m_r(\log m_r)^{1+\varepsilon}) \\
&= P(\{\max_{1\leq j\leq m_{r+1}-m_r+1} |R_j| > m_r(\log m_r)^{1+\varepsilon}\}) \\
&\leq 2P(|R_{m_{r+1}-m_r+1}| > \tfrac{1}{2}m_r(\log m_r)^{1+\varepsilon}) \\
&\qquad\qquad \text{(by the Skorohod inequality (ref. } \mathbf{(4.4.2)}\text{))} \\
&= 2P(|S_{m_{r+1}}| > \tfrac{1}{2}m_r(\log m_r)^{1+\varepsilon}) \\
&= P(|X| > \tfrac{1}{2}\tfrac{m_r}{m_{r+1}}(\log m_r)^{1+\varepsilon}) \sim \tfrac{q}{(\log m_r)^{1+\varepsilon}} = \tfrac{q}{r^{1+\varepsilon}}.
\end{aligned}$$

This implies that $\sum_r P(B_r) < \infty$. Hence $P(A_n \text{ i.o.}) \leq P(B_r \text{ i.o.}) = 0$. As noted earlier it follows now $\overline{\lim\limits_{n\to\infty}} \left|\frac{S_n}{n}\right|^{\theta_n} \leq e$ or $\alpha \leq e$.

Example 3. Let $(X_k,\ -\infty < k < \infty)$ be a two-sided infinite sequence of iid Bernoulli variables such that $P(X_k = 0) = P(X_k = 1) = \frac{1}{2}$.

Define $l_m = \max\{i \geq 1 : X_m = 1,\ X_{m-1} = 1,\ \dots\ ,\ X_{m-i+1} = 1\}$; $l_m = 0$ if $X_m = 0$. Define $L_n = \max\limits_{1 \leq m \leq n} l_m$. Then

$$\overline{\lim_{n\to\infty}} \frac{L_n}{\log_2 n} = 1 \tag{4.6}$$

where $\log_2$ is logarithm to base 2.

Proof.

Step 1. We note $P(l_m = 0) = P(X_m = 0) = \frac{1}{2}$; and, for $k > 0$, $(l_m = k) = (X_m = 1,\ X_{m-1} = 1,\ \dots,\ X_{m-k+1} = 1, X_{m-k} = 0)$. Hence $P(l_m = k) = \frac{1}{2^{k+1}}$. (Thus the l_ns, $n \geq 1$ are identically distributed). This yields

$$\begin{aligned} P(l_n > (1+\varepsilon)\log_2 n) &= \sum_{k=[(1+\varepsilon)\log_2 n]+1}^{\infty} P(l_n = k) \\ &= \sum_{k=[(1+\varepsilon)\log_2 n]+1}^{\infty} \frac{1}{2^{k+1}} \\ &= \frac{1}{2^{[(1+\varepsilon)\log_2 n]+1}} \leq \frac{1}{2^{1+\varepsilon)\log_2 n}} = \frac{1}{n^{1+\varepsilon}} \end{aligned}$$

This implies $\sum\limits_n P(l_n > (1+\varepsilon)\log_2 n) < \infty$, leading to

$$P(\frac{l_n}{\log_2 n} \leq 1+\varepsilon \text{ for all } n \text{ large}) = 1 \tag{4.7}$$

We conclude from this and the arbitrariness of $\varepsilon > 0$ that $wp1$,

$$\overline{\lim_{n\to\infty}} \frac{l_n}{\log_2 n} \leq 1 \tag{4.8}$$

Step 2. (4.7) and the fact that $\log_2 n \uparrow \infty$ as $n \uparrow \infty$ imply:

$$\text{wp1}, \quad \overline{\lim_{n\to\infty}} \frac{L_n}{\log_2 n} \leq 1. \tag{4.9}$$

Let $k_n = [(1-\varepsilon)\log_2 n]$. To complete the proof it remains to show that $P(\frac{L_n}{\log_2 n} > k_n \text{ for all } n \text{ large}) = 1$ Equivalently, what needs to be shown is that

$$P(L_n \leq k_n \text{ i.o.}) = 0. \tag{4.10}$$

Define $j_n = [\frac{n}{k_n}]$; $C_n = (L_n \le k_n)$;
$D_i = \{X_{(i-1)k_n+1} = X_{(i-1)k_n+2} = \ldots = X_{ik_n} = 1\}$.
Note that $(D_i,\ i = 1,\ 2,\ 3,\ \ldots,\ j_n)$ is a sequence of independent events. If D_i occurs, then $l_{ik_n} \ge k_n$. Consequently $L_n \ge k_n$. Thus $D_i \subset C_n'$. This being true for all i in the range, it follows that $\cup_{i=1}^{j_n} D_i \subset C_n'$. What is same, $C_n \subset \cap_{i=1}^{j_n} D_i'$. Since the events D_i', $1 \le i \le j_n$ are independent and since $P(D_i) = P(D_1)$, $1 \le i \le j_n$, we have

$$\begin{aligned}
P(C_n) \le \{P(D_1')\}^{j_n} &= \{1 - \tfrac{1}{2^{k_n}}\}^{j_n} \le \{1 - \tfrac{1}{2^{(1-\varepsilon)\log_2 n}}\}^{j_n} \\
&= \{1 - \tfrac{1}{n^{1-\varepsilon}}\}^{j_n} \le \{1 - \tfrac{1}{n^{1-\varepsilon}}\}^{\frac{n}{k_n}-1} \\
&\le 2\{1 - \tfrac{1}{n^{1-\varepsilon}}\}^{\frac{n}{k_n}} \text{ (for all } n \text{ large)} \\
&= 2e^{-\frac{n}{k_n}\{-\log_e(1-\frac{1}{n^{1-\varepsilon}})\}} \le 3e^{-\frac{n}{k_n}\frac{1}{n^{1-\varepsilon}}} \\
&\le 3e^{-\frac{n^{\varepsilon}}{(1-\varepsilon)\log_2 n}} \\
&\le 3e^{-2n^{\frac{\varepsilon}{2}}} \text{ (for all } n \text{ large and } \varepsilon < \tfrac{1}{2}) \\
&\le \tfrac{3}{n^2} \text{ (for all } n \text{ large).}
\end{aligned}$$

Since this implies that $\sum_n P(C_n) < \infty$, it follows by an appeal to the Borel-Cantelli lemma that $P(L_n \le k_n \text{ i.o.}) = 0$. Hence $wp1$, $\underline{\lim}_{n\to\infty} \frac{L_n}{\log_2 n} \ge 1 - \varepsilon$. Arguing as before, taking $\varepsilon \downarrow 0$, we conclude that $wp1$,

$$\underline{\lim}_{n\to\infty} \frac{L_n}{\log_2 n} \ge 1 \tag{4.11}$$

(4.9) and (4.11) imply (4.6). □

In **(4.1.5)**, we showed that the tail σ-field in the product measurable space is a trivial σ-field under the product measure. The following theorem shows that σ-fields that are not tail σ-fields too can turn out to be trivial σ-fields under the product measure.

Refer to **(4.1.4)** for notation.

Theorem 4.1.10 (Hewitt-Savage 0-1 law). *Let $T = \{1, 2, \ldots\}$; for each $n \in T$, let $(\Omega_n, \mathcal{A}_n, \mu_n) = (U, \mathcal{U}, \mu)$; let $(\Omega_T, \mathcal{A}_T, \mu_T)$ be the product measure space (ref.* ***(1.11.6)****). Then $\mu_T(A) = 0$ or 1 for every $A \in \mathcal{V}$.*

Proof. Let $A \in \mathcal{V}$. Being a measurable set, $A \in \mathcal{A}_T$. We recall that $\mathcal{A}_T = \sigma(\widetilde{\mathcal{A}}_T)$ where $\widetilde{\mathcal{A}}_T = \cup\{\mathbf{a}_{J;T}\}$. Write $J_N = \{1, 2, \ldots, N\}$. Theorem D, p 56, [H], as applied to the present situation states: *Given $\varepsilon > 0$, an integer N and a set $A_N \in \mathbf{a}_{J_N;T}$ can be found such that*

$$\mu_T(A \Delta A_N) < \varepsilon. \tag{4.12}$$

This implies

$$|\mu_T(A) - \mu_T(A_N)| \le \mu_T(A \Delta A_N) < \varepsilon. \tag{4.13}$$

Define transformation $\mathfrak{T}_N$ from Ω_T to itself as follows.

$$\begin{aligned}&\mathfrak{T}_N(x_1, x_2, \ldots, x_N, \ldots, x_{2N}, x_{2N+1}, \ldots)\\ &\qquad = (x_{N+1}, x_{N+2}, \ldots, x_{2N}, x_1, x_2, \ldots, x_N, x_{2N+1}, x_{2N+2}, \ldots).\end{aligned}$$

This is a one-to-one onto mapping, measurable bothways. We note $\mathfrak{T}_N(A) = A$. Further it preserves all set operations :
$\mathfrak{T}_N(E \cup F) = \mathfrak{T}_N(E) \cup \mathfrak{T}_N(F); (\mathfrak{T}_N(E))' = \mathfrak{T}_N(E')$.
Hence, $\mathfrak{T}_N(A \sim E) = \mathfrak{T}_N(A) \sim \mathfrak{T}_N(E) = A \sim \mathfrak{T}_N(E)$. And then we will have $\mathfrak{T}_N(A\Delta E) = A\Delta\mathfrak{T}_N(E)$.

Since the μ_ns are same and because μ_T is the product measure, it follows that $\mu_T(\mathfrak{T}_N(E)) = \mu_T(E)$ for every measurable set E.

Let $\hat{A}_N = \mathfrak{T}_N(A_N)$. Then

$$\mu_T(A\Delta\hat{A}_N) < \varepsilon \tag{4.14}$$

since

$$\begin{aligned}\mu_T(A\Delta\hat{A}_N) &= \mu_T(\mathfrak{T}_N(A)\Delta\mathfrak{T}_N(A_N))\\ &= \mu_T(\mathfrak{T}_N(A\Delta A_N)) = \mu_T(A\Delta A_N)\end{aligned}$$

and (4.13) applies.

We note that $A_N \in \bigotimes_{r=1}^{\infty} \mathbf{A}_r$ where $\mathbf{A}_r = \mathcal{A}_r, r \leq N$ and $\mathbf{A}_r = \{\Omega_r, \phi\}, r \geq N+1$. We note too that $\hat{A}_N \in \bigotimes_{r=1}^{\infty} \mathbf{B}_r$ where $\mathbf{B}_r = \mathcal{A}_r, N+1 \leq r \leq 2N$ and $\mathbf{B}_r = \{\Omega_r, \phi\}$ for all other values of r. Again since μ_T is the product measure, it follows that, under μ_T, A_N and $\hat{A}_N$ are mutually independent. Further $\mu_T(A_N) = \mu_T(\hat{A}_N)$, since the μ_ns are all same.

We have,

$$\begin{aligned}\mu_T(A_N) &= \mu_T(A_N \cap \hat{A}_N) + \mu_T(A_N \cap \hat{A}_N')\\ &= \mu_T(A_N)\mu_T(\hat{A}_N) + \mu_T(A_N \cap \hat{A}_N' \cap A) + \mu_T(A_N \cap \hat{A}_N' \cap A')\\ &\qquad \text{(using the independence of } A_N \text{ and } \hat{A}_N)\\ &\leq \mu_T(A_N)\mu_T(\hat{A}_N) + \mu_T(\hat{A}_N' \cap A) + \mu_T(A_N \cap A')\\ &\leq \{\mu_T(A_N)\}^2 + \mu_T(\hat{A}_N\Delta A) + \mu_T(A_N\Delta A)\\ &\leq \{\mu_T(A_N)\}^2 + 2\varepsilon \text{ (by (4.12) and (4.14))}\\ &\leq \{\mu_T(A) + \varepsilon\}^2 + 2\varepsilon \text{ (by (4.13)).}\end{aligned}$$

Since $\varepsilon > 0$ is arbitrary, this gives $\mu_T(A) \leq \{\mu_T(A)\}^2$. This, together with the obvious inequality $\mu_T(A) \geq \{\mu_T(A)\}^2$ imply $\mu_T(A) = \{\mu_T(A)\}^2$. This is possible *iff* $\mu_T(A) = 0$ or 1. □

Remark 4.1.11. Let $(U, \mathscr{U}) = (R, \mathscr{R})$. Let the X_ns be *iid* variables. Then $P(A) = 0$ or 1 for every A in the tail σ-field of the sequence (S_n).

Proof. Let μ_n be the pm induced in $\mathscr{R}$ by X_n. Then the dm of $\mathbf{X}$ on $\mathscr{R}^T$ is the product measure μ_T. In **(4.1.4)** we noted that the tail σ-field of the sequence (S_n) is $\mathbf{X}^{-1}(\mathfrak{C})$. By the theorem, $\mathcal{V}$ is a trivial σ-field under μ_T. Since $\mathfrak{C} \subset \mathcal{V}$, it follows that $\mathfrak{C}$ is a trivial σ-field under μ_T. Hence $\mathbf{X}^{-1}(\mathfrak{C})$ is a trivial σ-field under P. □

Examples of symmetric events : $A_1 = \{\omega : S_n(\omega) > \lambda\sqrt{n}\ i.o\}$ and $A_2 = \{\omega : |S_n(\omega)| < \varepsilon\ i.o.\}$ where (S_n) is the partial sum sequence of the X_ns. Hence $P(A_i) = 0$ or 1, $i = 1,\ 2$.

If $\sum\limits_{n=1}^{\infty} P(A_n) < \infty$ for some sequence (A_n) of events, then by the first part of the Borel-Cantelli Lemma, $P(A^*) = 0$ where $A^* = \overline{\lim\limits_{n\to\infty}} A_n$. If the A_ns considered here are not independent, the second part of the Borel-Cantelli Lemma does not apply. The following Lemmas prepare the ground to prove that, under certain conditions, $\sum_{n=1}^{\infty} P(A_n) = \infty$ implies $P(A^*) = 1$.

Lemma 4.1.12. *Let (A_n) be an arbitrary sequence of events, with $P(A_1) > 0$. Then for every n,*

$$\alpha_n \geq 1 \textit{ where } \alpha_n = \frac{\sum\limits_{j=1}^{n}\sum\limits_{k=1}^{n} P(A_j \cap A_k)}{\{\sum\limits_{j=1}^{n} P(A_j)\}^2}.$$

Proof. Let χ_j be the indicator function of the event A_j. Let $\xi_n = \sum_{k=1}^{n} \chi_k$. Then $\int_\Omega \xi_n \, \mathrm{d}P = \sum_{k=1}^{n} P(A_k)$ and $\int_\Omega \xi_n^2 \, \mathrm{d}P = \sum_{j=1}^{n}\sum_{k=1}^{n} P(A_j \cap A_k)$. Thus the Lemma follows if $(\int_\Omega \xi_n \mathrm{d}P)^2 \leq \int_\Omega \xi_n^2 \mathrm{d}P$, which is true by the Hölder inequality (ref. **(2.6.5)**). □

Lemma 4.1.13. *Let the A_js, the α_ns and the ξ_ns be as in the last Lemma and let $\underline{\lim\limits_{n\to\infty}}\ \alpha_n = c$. (By the last Lemma, $c \geq 1$). Further let $\beta_n \to \infty$ where $\beta_n = \sum_{k=1}^{n} P(A_k)$. Then $P(A^*) \geq \frac{1}{c}$, where $A^* = \overline{\lim\limits_{n\to\infty}} A_n$.*

Proof.
Step 1. Fix $\varepsilon > 0$, small. Define events $C_k = \{\omega : \xi_k > \varepsilon\beta_k\}$; $B_n = \cup_{k=n}^{\infty} C_k$; and $D_n = \cup_{k=n}^{\infty} A_k$. Note that $D_n \downarrow A^*$ and that $B_n \downarrow$. Let $B = \lim_n B_n$. Claim $B \subseteq A^*$. To establish the claim, we argue as follows. Suppose B occurs. We must show that A^* occurs. i.e. we must show that each D_n occurs, $n \geq 1$. If this is not true, then we can conclude, because of the monotonicity of the sequence (D_n), that no D_n occurs after some stage. i.e. there exists N such that $D_n^{'}$ occurs for all $n \geq N$. But $D_n^{'} = \cap_{k=n}^{\infty} A_k^{'}$. Hence $D_n^{'}$ occurring implies that of $A_k^{'}$ for all $k \geq n$. This in turn implies that $\xi_k = \xi_n$ for all $k \geq n$. Thus $0 \leq \xi_k \leq n$ for all $k \geq n$. Since $\beta_n \to \infty$, it follows that C_k does not occur for all k large, implying that B_n does not occur for all n. Hence B can not occur, a contradiction to our assumption. The truth of the claim is established. Thus $P(A^*) \geq P(B)$.

Step 2. Let f_n denote the indicator function of B_n. Now,

$$\begin{aligned} P(B_n) &= \int_{\Omega} f_n^2 \, \mathrm{d}P \\ &= \int_{\Omega} f_n^2 \, \mathrm{d}P \int_{\Omega} \xi_n^2 \, \mathrm{d}P \Big(\int_{\Omega} \xi_n^2 \, \mathrm{d}P \Big)^{-1} \\ &\geq \Big(\int_{\Omega} f_n \xi_n \, \mathrm{d}P \Big)^2 \Big(\int_{\Omega} \xi_n^2 \, \mathrm{d}P \Big)^{-1} \end{aligned}$$

using **(2.6.5)**(iii) with $p = q = 2$.
Now, $\int_{\Omega} (1 - f_n)\xi_n \, \mathrm{d}P = \int_{B_n^{'}} \xi_n \, \mathrm{d}P \leq \varepsilon\beta_n$. Using this,

$$\begin{aligned} \int_{\Omega} f_n \xi_n \, \mathrm{d}P &= \int_{\Omega} \xi_n \, \mathrm{d}P - \int_{\Omega} (1 - f_n)\xi_n \, \mathrm{d}P = \beta_n - \int_{B_n^{'}} \xi_n \, \mathrm{d}P \\ &\geq \beta_n - \varepsilon\beta_n. \end{aligned}$$

Hence
$$\begin{aligned} P(A^*) \geq P(B) = \lim_n P(B_n) &\geq \overline{\lim_{n\to\infty}} (1 - \varepsilon)^2 \beta_n^2 \Big(\int_{\Omega} \xi_n^2 \, \mathrm{d}P \Big)^{-1} \\ &= \overline{\lim_{n\to\infty}} (1 - \varepsilon)^2 \tfrac{1}{\alpha_n} = (1 - \varepsilon)^2 \, \tfrac{1}{c}. \end{aligned}$$
$\varepsilon > 0$ being arbitrary, the claim follows. □

Corollary 4.1.14. *If the A_ns are pairwise independent and if* $\sum\limits_{n=1}^{\infty} P(A_n) = \infty$, *then* $P(A^*) = 1$. *This is because in this case* $\alpha_n = 1$ *for all* n *and hence* $c = 1$.

The following theorem is of interest not only because of the result it reports but also because the proof makes use of **(4.1.13)**.

Theorem 4.1.15. *Let* (X_n) *be a sequence of iid rvs and let* $S_n = \sum_{k=1}^{n} X_k$, *let* $\varepsilon > 0$ *be arbitrary and small but fixed and let* $A_n = \{\omega : |S_n(\omega)| < \varepsilon\}$. *Then* $P(A^*) = 1$ *iff* $\sum_{n=1}^{\infty} P(A_n) = \infty$, *where* $A^* = \overline{\lim\limits_{n\to\infty}} A_n$.

Proof. Necessity follows from the Borel-Cantelli lemma.
Let α_n, β_n be as in the last two sections. As noted earlier, A^* is a symmetric event and hence $P(A^*) = 0$ or $P(A^*) = 1$. It is therefore enough to show that $P(A^*) > 0$ when $\beta_n \to \infty$.
Step 1. Write $S_0 = 0$. Given j an arbitrary integer, define *rv* $T_j(\omega) = \inf\{n : n \geq 0, j\varepsilon \leq S_n(\omega) < (j+1)\varepsilon\}$. Note that T_j may take the value ∞; that $(T_j = r, j\varepsilon \leq S_i < (j+1)\varepsilon)$ can be a non-null event only if $r \leq i$; that the event $(T_j = r)$ depends only on the variables $X_1, X_2, \ldots, X_r$; that therefore the events $(T_j = r)$ and $(|S_i - S_r| < \varepsilon)$ are mutually independent for $r \leq i$. Now,

$$\begin{aligned}
\sum_{i=0}^{N} P(j\varepsilon \leq S_i < (j+1)\varepsilon) &= \sum_{i=0}^{N}\sum_{r=0}^{i} P(T_j = r, j\varepsilon \leq S_i < (j+1)\varepsilon) \\
&\leq \sum_{i=0}^{N}\sum_{r=0}^{i} P(T_j = r,\ |S_i - S_r| < \varepsilon) \\
&= \sum_{i=0}^{N}\sum_{r=0}^{i} P(T_j = r)P(|S_i - S_r| < \varepsilon) \\
&\leq \sum_{i=0}^{N}\sum_{r=0}^{i} P(T_j = r)P(|S_{i-r}| < \varepsilon) \\
&= \sum_{r=0}^{N} P(T_j = r) \sum_{k=0}^{N-r} P(|S_k| < \varepsilon) \\
&\leq \sum_{r=0}^{N} P(T_j = r) \sum_{k=0}^{N} P(|S_k| < \varepsilon) \\
&\leq \sum_{k=0}^{N} P(|S_k| < \varepsilon).
\end{aligned}$$

Step 2. Writing the above inequality for $j = -2, -1, 0, 1$ and adding, we get

$$\sum_{i=0}^{N} P(|S_i| < 2\varepsilon) \leq 4 \sum_{i=0}^{N} P(|S_i| < \varepsilon).$$

Consider

$$\begin{aligned}
\Gamma_n &= \sum_{j=1}^{n-1} \sum_{k=j+1}^{n} P(A_j \cap A_k) \\
&= \sum_{j=1}^{n-1} \sum_{k=j+1}^{n} P(|S_j| < \varepsilon, |S_k| < \varepsilon) \\
&\leq \sum_{j=1}^{n-1} \sum_{k=j+1}^{n} P(|S_j| < \varepsilon, |S_k - S_j| < 2\varepsilon) \\
&= \sum_{j=1}^{n-1} \sum_{k=j+1}^{n} P(|S_j| < \varepsilon) P(|S_k - S_j| < 2\varepsilon) \\
&= \sum_{j=1}^{n-1} \sum_{k=j+1}^{n} P(|S_j| < \varepsilon) P(|S_{k-j}| < 2\varepsilon) \\
&= \sum_{j=1}^{n-1} \sum_{r=1}^{n-j} P(|S_j| < \varepsilon) P(|S_r| < 2\varepsilon) \\
&\leq (\sum_{j=0}^{n} P(|S_j| < 2\varepsilon))^2 \\
&\leq 16(\sum_{j=0}^{n} P(|S_j| < \varepsilon))^2 \\
&= 16\beta_n^2.
\end{aligned}$$

Hence $\underline{\lim}_{n\to\infty} \alpha_n = \underline{\lim}_{n\to\infty} \frac{2\Gamma_n}{\beta_n^2} \leq 32$. Now appeal to **(4.1.13)** to get, $P(A^*) \geq \frac{1}{32}$ and the proof is complete. □

Let us discuss two special sequences.
(i) Let the X_ns be iid standard normal variables. Then

$$P(|S_n| < \varepsilon) = P(|\xi| < \frac{\varepsilon}{\sqrt{n}}) = \frac{2}{\sqrt{n}} \frac{1}{\sqrt{2\pi}} e^{-\frac{1}{2}u^2}$$

for some u, $|u| < \frac{1}{\sqrt{n}}$. Hence

$$\sum_{n=1}^{\infty} P(|S_n| < \varepsilon) = \infty \quad \text{since } P(|S_n| < \varepsilon) \geq \frac{q}{\sqrt{n}} \text{ all } n \text{ large.}$$

Thus, in this case, $P(A^*) = 1$, whatever $\varepsilon > 0$ be.

(ii) Let the X_ns be *iid* symmetric Bernoulli variables, taking the values 1 and -1. Let $\varepsilon < \frac{1}{2}$. Then we claim $\sum_{n=1}^{\infty} P(|S_n| < \varepsilon) = \infty$. For, $P(|S_n| < \varepsilon) = P(S_n = 0) = 0$ if n is odd and $P(S_{2n} = 0) = \binom{2n}{n}\frac{1}{2^{2n}} \sim \frac{q}{\sqrt{n}}$. Hence, by the theorem, $P(S_n = 0 \text{ i.o.}) = 1$.

[*A different proof of* (ii) *can be given based on* (S_n) *being a Markov Chain with state space* $\{\pm n,\ n = 0,\ 1,\ 2,\ ...\}$ *and with stationary transition probabilities, ref.* ***(7.2.4)*** *and* ***(7.4.2)***]

4.2 Moments and uniform integrability of sums

In **(2.6.8)** and **(2.6.9)** we defined uniform integrability and studied its role in the convergence of moments under weak convergence of *df*s. In the present section, we study the mutual implication of uniform integrability of the sum of two sequences of rvs and that of the the individual sequences.

Let $\mathbb{X}_1,\ \mathbb{X}_2$ be two rEs in a separable metric space. Let the dms of $\mathbb{X}_1,\ \mathbb{X}_2$ and $(\mathbb{X}_1,\ \mathbb{X}_2)$ be respectively $\mu_1,\ \mu_2$ and μ. It is trivial to see that $\mathbb{X}_1$ and $\mathbb{X}_2$ are independent *iff* $\mu = \mu_1 \times \mu_2$.

Hence the *cf* of the sum of two independent random p-vectors is the product of their individual *cf*s. For details refer **(4.2.2)** below.

In particular, let $X_i,\ 1 \leq i \leq n$ be n standard normal variables. Each X_i^2 is χ^2 distributed with 1 degree of freedom and has *cf* $\frac{1}{(1-2it)^{\frac{1}{2}}}$ (ref. **(3.2)**(x)). Hence the *cf* of $\sum_1^n X_i^2$ is $(1-2it)^{-\frac{n}{2}}$. Thus $\sum_1^n X_i^2$ is χ^2 distributed with n degrees of freedom.

Let $X,\ Y$ be two real independent variables. Let $\mathbb{E}|X| < \infty$, $\mathbb{E}|Y| < \infty$. Then by an application of Fubini's theorem we get : $\mathbb{E}|XY| < \infty$ and $\mathbb{E}XY = \mathbb{E}X\mathbb{E}Y$.

If $X,\ Y$ are any two square integrable rvs, then by **(2.6.5)** $\mathbb{E}|XY| < \infty$.

Definition 4.2.1. Two square integrable rvs $X,\ Y$ defined on the same probability space are said to be orthogonal if $\mathbb{E}XY = 0$; they are said to be uncorrelated if $X - \mathbb{E}X$ and $Y - \mathbb{E}Y$ are orthogonal. Equivalently, X and Y are uncorrelated if $\mathbb{E}XY = \mathbb{E}X\mathbb{E}Y$.

It is readily seen that if $X,\ Y$ have zero means, then they are orthogonal *iff* they are uncorrelated. As remarked in para 2 above, independent zero mean rvs are orthogonal. That the converse is not true is brought out by the following example.

Let $X,\ Y$ be two $iid\ rvs$ with $P(X = 1) = \frac{1}{2} = P(X = -1)$. Let $\xi = X+Y$ and $\eta = X-Y$. We note that $\xi,\ \eta$ are not independent. For $P(\xi = 2) = P(X = 1,\ Y = 1) = \frac{1}{4}$ and $P(\eta = 2) = P(X = 1,\ Y = -1) = \frac{1}{4}$ while $P(\xi = 2,\ \eta = 2) = P(X = 1, Y = 1,\ X = 1,\ Y = -1) = 0$. That ξ and η are uncorrelated follows from the relations $\mathbb{E}\xi = 0 = \mathbb{E}\eta$ and $\mathbb{E}\xi\eta = \mathbb{E}(X^2 - Y^2) = 0$.

Theorem 4.2.2. *Let* $\mathbb{X}_1,\ \mathbb{X}_2$ *be independent* R^p*-valued* rVs *with dfs* $\mathbb{F}_1 and \mathbb{F}_2$, *pms* μ_1 *and* μ_2 *and* cfs φ_1 *and* φ_2 *and let* $\mathbb{X} = \mathbb{X}_1 + \mathbb{X}_2$ *have df* $\mathbb{F}$ *and cf* φ. *Then* (i) $\varphi(\boldsymbol{t}) = \varphi_1(\boldsymbol{t})\varphi_2(\boldsymbol{t})$ *for all* $\boldsymbol{t} \in R^p$. (ii) $\mathbb{F}(\boldsymbol{x}) = \int_{R^p} \mathbb{F}_1(\boldsymbol{x} - \boldsymbol{y})d\mathbb{F}_2(\boldsymbol{y})$.

Proof.
(i) $\varphi(\mathbf{t}) = \mathrm{E}e^{i\mathbf{t}'(\mathbb{X}_1+\mathbb{X}_2)} = \int_{R^{2p}} e^{i\mathbf{t}'(\mathbf{x}+\mathbf{y})}\mathrm{d}(\mu_1 \times \mu_2)(\mathbf{x},\ \mathbf{y})$
$= \int_{R^p} e^{i\mathbf{t}'\mathbf{x}}\mathrm{d}\mu_1(\mathbf{x}) \int_{R^p} e^{i\mathbf{t}'\mathbf{y}}\mathrm{d}\mu_2(\mathbf{y}) = \varphi_1(\mathbf{t})\varphi_2(\mathbf{t})$.
(ii) We have already shown (ref. **(1.12.1)** and **(2.3.10)**) that $\varphi_1\varphi_2$ is the *cf* of $\mathbb{F}$. This combined with the just proved result that the *cf* of $\mathbb{X}$ is $\varphi_1\varphi_2$ and an appeal to the uniqueness theorem (ref. **(2.3.4)**) completes the proof. □

Theorem 4.2.3. *Let* $X,\ Y$ *be independent* rvs *and let* $Z = X + Y$ *be their sum. Then*
(i) *(Symmetrization inequality).*
If Y *has median* 0, *then* $P(|Z| \geq t) \geq \frac{1}{2}P(|X| \geq t)$.
This result together with **(2.6.4)** *implies* $\mathbb{E}|X|^\alpha \leq 2\mathbb{E}|Z|^\alpha,\ \alpha > 0$.
(ii) *for* $\alpha > 0$, $\mathbb{E}|Z|^\alpha < \infty$ *iff* $\mathbb{E}|X|^\alpha < \infty$ *and* $\mathbb{E}|Y|^\alpha < \infty$ *and*
(iii) *Let* $X,\ Y$ *have finite mean values. Then* $var(X+Y) = var(X)+var(Y)$, *where for a* $rv\ U$, $var(U) =$ *variance of* $U = \mathbb{E}(U - \mathbb{E}U)^2$. *This equality is in the sense that if either side is finite then so is the other and two sides are equal.*

Proof.
(i) From

$$(|Z| \geq t) = (X + Y \geq t) \cup (X + Y \leq -t)$$
$$\supseteq (X \geq t,\ Y \geq 0) \cup (X \leq -t,\ Y \leq 0),$$

we have

$$\begin{aligned}P(|Z| \geq t) &\geq P(X \geq t,\ Y \geq 0) + P(X \leq -t,\ Y \leq 0)\\ &= P(X \geq t)P(Y \geq 0) + P(X \leq -t)P(Y \leq 0)\\ &\geq \frac{1}{2}\{P(X \geq t) + P(X \leq -t)\}\\ &= \frac{1}{2}P(|X| \geq t).\end{aligned}$$

(ii) By **(2.6.5)** we have $|Z|^\alpha \leq c_\alpha\{|X|^\alpha + |Y|^\alpha\}$ where $c_\alpha = \max(1,\ 2^{\alpha-1})$. The sufficiency of the condition follows from this inequality.

Let $\mathbb{E}|Z|^\alpha < \infty$. Let $X_1 = X - m_1$ and $Y_1 = Y - m_2$ where $m_1,\ m_2$ are medians respectively of X and Y. Since $\mathbb{E}|Z|^\alpha < \infty$ implies $\mathbb{E}|Z - m_1 - m_2|^\alpha < \infty$ and since it is enough to show that $\mathbb{E}|X - m_1|^\alpha < \infty$ and $\mathbb{E}|Y - m_2|^\alpha < \infty$, we may take, without loss of generality, $m_1 = 0,\ m_2 = 0$. By (i) $P(|X| \geq t) \leq 2P(|Z| \geq t)$. This implies $\int_0^\infty x^{\alpha-1}P(|X| \geq t)\ dt \leq 2\int_0^\infty x^{\alpha-1}P(|Z| \geq t)\ dt < \infty$. This is equivalent to (ref. **(2.6.4)**) $\mathbb{E}|X|^\alpha < \infty$. Similarly $\mathbb{E}|Y|^\alpha < \infty$.

(iii) For, we note first that $\mathbb{E}(X + Y) = \mathbb{E}X + \mathbb{E}Y$. Write $\xi = X - \mathbb{E}X$ and $\eta = Y - \mathbb{E}Y$.

Suppose $var(Z)$ is finite. Then $var(X+Y) = var(\xi+\eta) = \mathbb{E}(\xi+\eta)^2 = \mathbb{E}(\xi^2 + \eta^2 + 2\xi\eta) = \mathbb{E}\xi^2 + \mathbb{E}\eta^2 + 2\mathbb{E}\xi\mathbb{E}\eta = \mathbb{E}\xi^2 + \mathbb{E}\eta^2$.

Suppose now the right side is finite. Then by part (ii), $\mathbb{E}\xi^2 < \infty$ and $\mathbb{E}\eta^2 < \infty$. Now the earlier argument applies and the proof is complete. □

4.2.4. For some $\alpha > 0$ let each of the sequences $(|X_n|^\alpha)$ and $(|Y_n|^\alpha)$ be uniformly integrable. Then $(|X_n + Y_n|^\alpha)$ is a uniformly integrable sequence (no assumption of independence is made).

Proof. Write $A_n = (|X_n| \geq \frac{t}{2})$; $B_n = (|Y_n| \geq \frac{t}{2})$. Since $(|X_n|^\alpha)$ and $(|Y_n|^\alpha)$ are uniformly integrable sequences,

$$\lim_{t\to\infty} \sup_n \int_{A_n} |X_n|^\alpha\ dP = 0 \text{ and } \lim_{t\to\infty} \sup_n \int_{B_n} |Y_n|^\alpha\ dP = 0. \tag{4.15}$$

From the definition of uniform integrability,

$$\lim_{t\to\infty}(\frac{t}{2})^\alpha \sup_n P(A_n) = \lim_{t\to\infty}(\frac{t}{2})^\alpha \sup_n P(|X_n| \ge \frac{t}{2})$$
$$\le \lim_{t\to\infty} \sup_n \int\limits_{|X_n|\ge\frac{t}{2}} |X_n|^\alpha \,\mathrm{d}P(\omega) = 0 \quad (4.16)$$

Similarly,

$$\lim_{t\to\infty} t^\alpha \sup_n P(B_n) = 0. \quad (4.17)$$

Further,

$$\int\limits_{B_n} |X_n|^\alpha \,\mathrm{d}P = \int\limits_\Omega |X_n|^\alpha \chi_{B_n} \,\mathrm{d}P$$
$$= \int\limits_{A_n} |X_n|^\alpha \chi_{B_n} \,\mathrm{d}P + \int\limits_{A_n'} |X_n|^\alpha \chi_{B_n} \,\mathrm{d}P$$
$$\le \int\limits_{A_n} |X_n|^\alpha \,\mathrm{d}P + (\frac{t}{2})^\alpha P(B_n). \quad (4.18)$$

Similarly,

$$\int\limits_{A_n} |Y_n|^\alpha \,\mathrm{d}P \le \int\limits_{B_n} |Y_n|^\alpha \,\mathrm{d}P + (\frac{t}{2})^\alpha P(A_n). \quad (4.19)$$

We have

$$\int\limits_{|X_n+Y_n|\ge t} |X_n+Y_n|^\alpha \,\mathrm{d}P \le \int\limits_{A_n\cup B_n} |X_n+Y_n|^\alpha \,\mathrm{d}P$$
$$\le \int\limits_\Omega 2^\alpha\{|X_n|^\alpha + |Y_n|^\alpha\}\{\chi_{A_n} + \chi_{B_n}\} \,\mathrm{d}P$$
$$= 2^\alpha \int\limits_{A_n} |X_n|^\alpha \,\mathrm{d}P + 2^\alpha \int\limits_{B_n} |Y_n|^\alpha \,\mathrm{d}P$$
$$+ 2^\alpha \int\limits_{B_n} |X_n|^\alpha \,\mathrm{d}P + 2^\alpha \int\limits_{A_n} |Y_n|^\alpha \,\mathrm{d}P$$
$$\le 2^{\alpha+1} \int\limits_{A_n} |X_n|^\alpha \,\mathrm{d}P + 2^{\alpha+1} \int\limits_{B_n} |Y_n|^\alpha \,\mathrm{d}P$$
$$+ t^\alpha P(A_n) + t^\alpha P(B_n).$$

(by (4.17) and (4.18))

The uniform integrability of $(|X_n+Y_n|^\alpha)$ follows now from (4.15)–(4.19). □

To prove a converse to the above result, we first note that, without further conditions, the uniform integrability alone of $(|X_n+Y_n|^\alpha)$ does not imply that of the sequences $(|X_n|^\alpha)$ and $(|Y_n|^\alpha)$ as is clear from taking $Y_n = -X_n$.

(i) The condition of independence of X_n and Y_n is natural to impose. That the condition of independence is not sufficient is brought out by the following examples.

(ii) Let $\xi_n,\ \eta_n$ be independent standard normal variables. Let $X_n = \xi_n - n$ and $Y_n = \eta_n + n$. $X_n + Y_n$ is distributed centered normal with variance 2. Hence $|X_n + Y_n|^\alpha$ is trivially uniformly integrable. But this does not imply even the weaker result $\sup\limits_n \mathrm{E}|X_n|^\alpha = c_1 < \infty$. For, if c_1 is finite, then (X_n) is a tight sequence since $P(|X_n| \geq \lambda) \leq \frac{c_1}{\lambda^\alpha}$ independent of n. (ref. text under **(2.6.8)**). But this is not possible because, whatever $\lambda > 0$ may be, $P(|X_n| \geq \lambda) \geq P(\xi_n \leq n - \lambda)$ which tends to 1 as $n \to \infty$. Also for the above sequence $\sup\limits_n |median X_n|,\ \sup\limits_n |median Y_n|$ are not finite. These sup values would be finite if $(|X_n|^\alpha),\ (|Y_n|^\alpha)$ are uniformly integrable sequences.

These considerations lead to the following converse to **(4.2.4)**.

Theorem 4.2.5. *For each $n \geq 1$, let $X_n,\ Y_n$ be independent variables. Further let s_n be a median of X_n and t_n a median of Y_n. Let $\alpha > 0$.*
(i) *If $(X_n + Y_n)$ is a tight sequence, then so are the sequences $(X_n + t_n)$ and $(Y_n + s_n)$.*
(ii) *If the sequence $(|X_n + Y_n|^\alpha)$ is uniformly integrable, then so are the sequences $(|X_n + t_n|^\alpha)$ and $(|Y_n + s_n|^\alpha)$.*
(iii) *If $\sup\limits_n |s_n| = s$ and $\sup\limits_n |t_n| = t$ are finite and if the sequence $(|X_n+Y_n|^\alpha)$ is uniformly integrable, then each of the two sequences $(|X_n|^\alpha)$ and $(|Y_n|^\alpha)$ is uniformly integrable.*

Proof. (i) By **(4.2.3)**,

$$
\begin{aligned}
P(|X_n + Y_n| \geq x) &= P(|X_n + t_n + Y_n - t_n| \geq x) \\
&\geq \frac{1}{2} P(|X_n + t_n| \geq x)
\end{aligned}
\tag{4.20}
$$

The claim regarding the sequence $(X_n + t_n)$ follows from this. Tightness of the sequence $(Y_n + s_n)$ can be argued on similar lines.

(ii) The hypothesis is equivalent to:
$\lim\limits_{r\to\infty} \sup\limits_n \int_r^\infty x^{\alpha-1} P(|X_n + Y_n| \geq x)\mathrm{d}x = 0$. This implies, by (4.20),
$\lim\limits_{r\to\infty} \sup\limits_n \int_r^\infty x^{\alpha-1} P(|X_n + t_n| \geq x)\mathrm{d}x = 0$. This is equivalent to the uniform integrability of the sequence $(|X_n + t_n|^\alpha)$. Similarly, sequence $(|Y_n + s_n|^\alpha)$ can be shown to be uniformly integrable.

(iii) Using **(2.6.5)**(iv), we have

$$\lim_{r\to\infty} \sup_n \int_{|X_n|>r} |X_n|^\alpha \, dP \le \lim_{r\to\infty} \sup_n \int_{|X_n+t_n|>r-2t} 2^\alpha\{|X_n+t_n|^\alpha + t^\alpha\} \, dP = 0,$$

since $(|X_n + t_n|^\alpha)$ is a uniformly integrable sequence.
Similar step leads to proving the uniform integrability of $(|Y_n|^\alpha)$. □

4.3 Modes of convergence

Convergence with $wp1$ or almost surely ($a.s.$) of a sequence of rvs was defined at **(1.5.1)**. Convergence in distribution was defined at **(2.1.1)**. Weak convergence of pms in metric spaces was defined at **(2.4)**. Convergence of mean of order 1 was defined at **Q**2. In this section we introduce additional modes of convergence and study the mutual implications of the various modes.

Let $(M,\ d)$ be a separable metric space and let $\mathcal{D}$ be its Borel σfield. Then (ref. **(1.2.2)**) the Borel σ-field of the product space $M \times M$ endowed with the product topology is the product σ-field $\mathcal{D} \otimes \mathcal{D}$. Since $d(.\ ,\ .)$ is a continuous function in the product topology, it is $\mathcal{D} \otimes \mathcal{D}$-measurable. Hence if the rEs $X,\ Y$ take values in M then $d(X,\ Y)$ is a *rv*.

Definitions 4.3.1. Let X_n, $n \ge 0$ be rEs in a separable metric space $(M,\ d)$. Let $Y_n = d(X_n,\ X_0)$.

(i) We say that $\lim_{n\to\infty} X_n = X_0$ $a.s.$ or $wp1$ ($X_n \overset{wp1}{\longrightarrow} X_0$) if the limit of the real sequence (Y_n) is 0 $wp1$. (ref. **(1.5.1)**).

(ii) We say X_n converges to X_0 in probability ($X_n \overset{pr}{\to} X_0$) if $\lim_{n\to\infty} P(Y_n > \varepsilon) = 0$ for every $\varepsilon > 0$.

Note. If $X_n \overset{pr}{\to} X$ and if $X_n \overset{pr}{\to} Y$, then $wp1$, $X = Y$. For, $P(|X - Y| > \varepsilon) \le P(|X_n - X| > \frac{\varepsilon}{2}) + P(|X_n - Y| > \frac{\varepsilon}{2}) \to 0$. i.e., $P(|X - Y| > \varepsilon) = 0$. Hence, for each $k \ge 1$, there exists a set A_k such that $P(A_k) = 1$ and such that $(\omega \in A_k) \Rightarrow (|X(\omega) - Y(\omega)| \le \frac{1}{k})$. Let $A = \cap_k A_k$ We note that $P(A) = 1$ and that on A, $X = Y$.

(iii) We say X_n converges to X_0 in mean of order $r > 0$ ($X_n \overset{r\text{th mean}}{\to} X_0$) if $\mathbb{E}Y_n^r < \infty$ for each n and $\lim_{n\to\infty} \mathbb{E}Y_n^r = 0$.

(iv) We say X_n converges to X_0 completely ($X_n \overset{c}{\to} X_0$) if, for every $\varepsilon > 0$, $\sum_n P(Y_n > \varepsilon) < \infty$.

Remarks 4.3.2.
(i) $(X_n \overset{wp1}{\longrightarrow} X_0) \Leftrightarrow (\sup_{k \geq n} d(X_k,\ X_0) \overset{pr}{\to} 0)$.

Proof.

$$\begin{aligned}(X_n \overset{wp1}{\longrightarrow} X_0) &\Leftrightarrow (Y_n \overset{wp1}{\longrightarrow} 0) \Leftrightarrow (P(Y_n > \varepsilon \text{ i.o.}) = 0 \text{ for every } \varepsilon > 0)\\ &\Leftrightarrow (\lim_{n\to\infty} P(\cup_{k \geq n}\{Y_k > \varepsilon\}) = 0 \text{ for every } \varepsilon > 0)\\ &\Leftrightarrow (\lim_{n\to\infty} P(\cup_{k \geq n}(Y_k > \varepsilon)) = 0 \text{ for every } \varepsilon > 0)\\ &\Leftrightarrow (\lim_{n\to\infty} P(\sup_{k\geq n} Y_k > \varepsilon) = 0 \text{ for every } \varepsilon > 0)\\ &\Leftrightarrow (\lim_{k \geq n} Y_k \overset{pr}{\to} 0), \text{ as was to be proved.}\end{aligned}$$

Note. An obvious corollary to the above result is:
$(X_n \overset{wp1}{\longrightarrow} X_0) \Rightarrow (X_n \overset{pr}{\to} X_0)$.

(ii) $X_n \overset{pr}{\to} X_0 \not\Rightarrow X_n \overset{wp1}{\longrightarrow} X_0$.

Proof (Example).
Let (X_n) be a sequence of *independent rvs* with $P(X_n = 0) = 1 - \frac{1}{n}$ and $P(X_n = 1) = \frac{1}{n}$. For $0 < \varepsilon < 1$, $P(|X_n| > \varepsilon) = P(X_n = 1) = \frac{1}{n} \to 0$. Hence $X_n \overset{pr}{\to} 0$. Now, $A = \{\omega : \lim_{n\to\infty} X_n(\omega) \text{ exists}\}$ is a tail event of the independent sequence (X_n). Hence $P(A) = 1$ or 0. Suppose $P(A) = 1$. It follows the limit function $X = \lim_n X_n$ is necessarily a constant *wp1*, being measurable with respect to the tail σ-field. Since it has already been shown that $X_n \overset{pr}{\to} 0$, it follows by part (i) that the constant is necessarily 0. Thus $\left(P(A) = 1\right) \Rightarrow \left(X_n \overset{a.s.}{\longrightarrow} 0\right)$. But this is not possible. For, by the Borel-Cantelli lemma applied to independent events, $X_n \overset{a.s.}{\longrightarrow} 0$ *iff* $\sum_n P(|X_n| > \varepsilon) < \infty$ for every $\varepsilon > 0$. This condition is not satisfied for $0 < \varepsilon < 1$, since for such ε, $\sum_n P(|X_n| > \varepsilon) = \sum_n P(X_n = 1) = \sum_n \frac{1}{n} = \infty$. This proves the stated assertion.

In fact applying Borel-Cantelli for independent events it is easy to show that *wp1*, $\overline{\lim_{n\to\infty}}\, X_n = 1,\ \underline{\lim_{n\to\infty}}\, X_n = 0$. □

This topic is pursued in section **(4.5)**

(iii) $\left(X_n \overset{pr}{\to} X_0\right) \Rightarrow \left(X_n \overset{d}{\to} X_0\right)$.
Proof. The hypothesis implies that for each $\varepsilon > 0$, $P(d(X_n,\ X_0) > \varepsilon) \to 0$. Let μ_n be the distribution in M of $X_n,\ n \geq 0$. Let $C \subseteq M$ be a closed set.

Let $C_k = \{x : x \in M,\ d(x,\ C) \leq \frac{1}{k}\}$. We note that $C_k \downarrow C$.

$$\begin{aligned}\mu_n(C) &= P(X_n \in C)\\ &\leq P(X_n \in C,\ d(X_n,\ X_0) \leq \frac{1}{k}) + P(d(X_n,\ X_0) > \frac{1}{k})\\ &\leq P(X_0 \in C_k) + P(d(X_n,\ X_0) > \frac{1}{k})\\ &= \mu_0(C_k) + P(d(X_n,\ X_0) > \frac{1}{k}).\end{aligned}$$

Hence $\overline{\lim_{n\to\infty}}\ \mu_n(C) \leq \mu_0(C_k)$. Letting $k \to \infty$, we conclude $\overline{\lim_{n\to\infty}}\ \mu_n(C) \leq \mu_0(C)$. This is same as $\mu_n \overset{d}{\to} \mu_0$ or $X_n \overset{d}{\to} X_0$. □

In the special case $M = R^p$, a proof of the weaker result, $(X_n \overset{\text{a.s.}}{\to} X_0) \Rightarrow (X_n \overset{\text{d}}{\to} X_0)$, on the following lines is possible. By the bounded convergence theorem $X_n \overset{a.s.}{\longrightarrow} X_0$ implies $\int_\Omega e^{it'X_n}\ dP(\omega) \to \int_\Omega e^{it'X_0}\ dP(\omega)$. i.e. $\lim_{n\to\infty} \varphi_n(\mathbf{t}) = \varphi_0(\mathbf{t})$ for every $\mathbf{t} \in R^p$, where the φ s are the corresponding cfs. This implies (ref. **(2.3.5)**) that $X_n \overset{d}{\to} X$.

(iv) $\left(X_n \overset{d}{\to} X_0\right) \nRightarrow \left(X_n \overset{pr}{\to} X_0\right)$. Also $\left(X_n \overset{d}{\to} X_0\right) \nRightarrow \left(X_n \overset{a.s.}{\longrightarrow} X_0\right)$
Example. Let $\Omega = [0,\ 1]$; let $\mathscr{A}$ be the Borel σ-field of Ω. With Lebesgue measure on $\mathscr{A}$ for P, $(\Omega,\ \mathscr{A},\ P)$ is a probability measure space. Define $\xi(\omega) = \omega$ and note that ξ is uniformly distributed on $[0,\ 1]$. Define $\eta(\omega) = 1 - \omega$ and note that η is also uniformly distributed on $[0,\ 1]$. Let $X_{2n}(\omega) = \xi(\omega)$ and $X_{2n-1}(\omega) = \eta(\omega)$ for all ω and for all $n \geq 1$. Since X_ns are identically distributed, it follows trivially that $X_n \overset{d}{\to} X_1$.

For ε small, $P(\{\omega : |X_{2n}(\omega) - X_1(\omega)| > \varepsilon\}) = P(\{\omega : |1-2\omega| > \varepsilon\}) = 1 - \varepsilon$, proving the first assertion.

Let $\omega \neq \frac{1}{2}$, (so $\omega \neq 1 - \omega$). Then $wp1$, $\lim_{n\to\infty} X_{2n+1}(\omega) = 1 - \omega$ and $\lim_{n\to\infty} X_{2n}(\omega) = \omega$. Hence $wp1$, $\lim_{n\to\infty} X_n(\omega)$ does not exist. By ((**4.3.2**)(i)) and by the first assertion, the second assertion follows. □

However the following are true
(v) If $X_0 = c\ wp1$. c, a constant, $c \in M$ then $(X_n \overset{d}{\to} X_0) \Rightarrow (X_n \overset{pr}{\to} X_0)$.

Proof. Given $\varepsilon > 0$, let S denote the closed sphere in M with center c and radius ε. Trivially $P(X_0 \in Bd(S)) = 0$. The hypothesis implies $P(X_n \in S) \to P(X_0 \in S)$. i.e. $P(X_n \in S) \to 1$. Now $P(d(X_n,\ X_0) > \varepsilon) = P(d(X_n,\ c) > \varepsilon) = P(X_n \notin S) = 1 - P(X_n \in S) \to 0$. □
(vi) **Theorem**. *Let $X_n,\ n \geq 0$ be rvs defined on a probability measure space*

$(\Omega, \mathscr{A}, P)$ and let $X_n \xrightarrow{d} X_0$. Then there exists a probability measure space $(U, \mathscr{U}, \mu)$ and rvs Y_n, $n \geq 0$ defined on it such that Y_n has the same df F_n as X_n and such that a.s.(μ) $\lim_{n\to\infty} Y_n(u) = Y_0(u)$.

Proof. For $0 < u < 1$, let $G_n(u) = \inf\{x : F_n(x) \geq u\}$. It follows from this that $F_n(x) \geq u$ *iff* $G_n(u) \leq x$. Define probability space $(U, \mathscr{U}, \mu)$ where $U = (0,\ 1)$, $\mathscr{U}$ is the Borel σ-field of U and μ is the Lebesgue measure. Define, on U, rvs : $Y_n(u) = G_n(u)$. Then the *df* of Y_n is F_n since $\mu(u : Y_n(u) \leq x) = \mu(u : G_n(u) \leq x) = \mu(u : F_n(x) \geq u) = \mu(u : u \leq F_n(x)) = F_n(x)$.

Since the F_ns are right continuous, there corresponds, for each n a unique number $x_{n;u}$, depending on u, such that $F_n(x_{n;u}) \geq u$ but $F_n(x_{n;u}-\varepsilon) \leq u$ for every $\varepsilon > 0$. By definition, $G_n(u) = x_{n;u}$. $F_n \xrightarrow{d} F_0$ implies $\overline{\lim_{n\to\infty}} F_n(y) \leq F_0(y)$ for every y. Consider appropriate subsequences wherever necessary in the following steps. Suppose $x_{n;u} \to \infty$ as $n \to \infty$. Let λ be a continuity point of F_0 such that $F_0(\lambda) > \frac{u+3}{4}$. For all n large,say $n \geq N_1$, $x_{n;u} > \lambda$. Since $F_n(\lambda) \to F_0(\lambda)$, there exists N_2 such that $n \geq N_2$ implies $F_n(\lambda) \geq F_0(\lambda) - \frac{1-u}{4}$. Let $N = \max(N_1,\ N_2)$. We then have for all $n \geq N$, $u \geq F_n(x_{n;u} - \varepsilon) \geq F_n(\lambda) \geq F_0(\lambda) - \frac{1-u}{4} > \frac{u+3}{4} - \frac{1-u}{4} = \frac{u+1}{2}$, a contradiction. Hence the sequence $(x_{n;u})$, for u fixed, is bounded above. That the sequence is bounded below can be similarly argued. Thus $(x_{n;u})$ is a bounded sequence.

For u fixed , to show that $Y_n(u) \to Y_0(u)$ with Lebesgue measure1 is to show that $x_{n;u} \to x_{0;u}$. Suppose the subsequence $(x_{m;u})$ converges to y. Choose $\varepsilon > 0$ arbitrary but subject to $y + \varepsilon$, $y - 2\varepsilon$ being continuity points of F_0. Then for all m large, $y - 2\varepsilon \leq x_{m;u} - \varepsilon < x_{m;u} \leq y + \varepsilon$. It follows now $u \leq \overline{\lim_{m\to\infty}} F_m(x_{m;u}) \leq \overline{\lim_{m\to\infty}} F_m(y+\varepsilon) = F_0(y+\varepsilon)$. By the right continuity of F_0, this yields $u \leq F_0(y)$. Again, $u \geq \overline{\lim_{m\to\infty}} F_m(x_{m;u} - \varepsilon) \geq \overline{\lim_{m\to\infty}} F_m(y - 2\varepsilon) = F_0(y - 2\varepsilon)$. Combining the two results, $F_0(y-) \leq u \leq F_0(y)$. This implies $y = x_{0;u}$. This being true for every convergent subsequence $(x_{m;u})$, it follows that $\lim_{n\to\infty} x_{n;u}$ exists and is equal to $x_{0;u}$.

Note. The construction above for the case of the real line does not extend even to R^2, since the arguments rest heavily on the linear order of the line.

(vii) **Definition**. A sequence (X_n) of rEs in M is said to be fundamental or Cauchy in probability if for every $\varepsilon > 0$, $P(d(X_n,\ X_m) > \varepsilon) \to 0$ as $m,\ n \to \infty$.

(viii) **Theorem**. Let M satisfy the additional condition of being complete. Let (X_n) be a sequence of rEs in M, Cauchy in probability. Then (X_n) admits of a subsequence converging $wp1$ (to say, X). Further, the sequence (X_n) converges in probability to X. The limit rE is unique in the sense that if another subsequence converges $wp1$ to, say Y, then P(X = Y) = 1.

Conversely if every subsequence $(X_{n(k)})$ of (X_n) has a further subsequence $(X_{n(k(r))})$ converging to X $wp1$, then $X_n \overset{pr}{\to} X$.

Proof. We are given that for every $\varepsilon > 0$, $P(d(X_n, X_m) > \varepsilon) \to 0$ as $m, n \to \infty$. Hence there exists n_1 such that $P(d(X_n, X_m) > \frac{1}{2}) < \frac{1}{2}$ for all $m, n \geq n_1$. Again there would exist $n_2 > n_1$ such that $P(d(X_n, X_m) > \frac{1}{2^2}) < \frac{1}{2^2}$ for all $m, n \geq n_2$. In this way construct the infinite sequence $n_1 < n_2 < \ldots$. and consider the sequence (X_{n_k}). Define $A_k = \{d(X_{n_{k+1}}, X_{n_k}) > \frac{1}{2^k}\}$ and note that $P(A_k) < \frac{1}{2^k}$. If $B_r = \cup_r^\infty A_j$, then we note that $B_r \downarrow$ and that $P(B_r) \leq \sum_r^\infty P(A_j) < \sum_r^\infty \frac{1}{2^j} = \frac{1}{2^{r-1}}$. Hence $P(B) = 0$ where $B = \cap_r B_r$. We will show that on B', X_{n_k} converges. Given $\varepsilon > 0$ take r sufficiently large such that $\frac{1}{2^{r-1}} < \varepsilon$. Let $\omega \in B'$. Hence $\omega \in B'_m$ for all m large, say, $m \geq r$. Let $r < j < k$. Then

$$d(X_{n_j}(\omega), X_{n_k}(\omega)) \leq \sum_{s=j}^{k-1} d(X_{n_s}(\omega), X_{n_{s+1}}(\omega))$$
$$\leq \sum_{s=j}^{k-1} \frac{1}{2^s} < \sum_{s=r}^{\infty} \frac{1}{2^s} = \frac{1}{2^{r-1}} < \varepsilon.$$

Thus for each $\omega \in B'$, $(X_{n_k}(\omega))$ is a Cauchy sequence in the complete metric space M and, hence, is convergent. Denote its limit by $X(\omega)$. Since $P(B') = 1$, we conclude $X_{n_k} \overset{wp1}{\longrightarrow}$ to some rE X.

Given $\varepsilon > 0$ we can find N such that for all $n_k \geq N, n \geq N$,

(a) $P(d(X_{n_k}, X_n) > \varepsilon) < \varepsilon$ and (b) $P(d(X_{n_k}, X) > \varepsilon) < \varepsilon$.

(a) is possible by hypothesis. (b) is possible since $X_{n_k} \overset{wp1}{\longrightarrow} X$ and hence $X_{n_k} \overset{pr}{\to} X$ (Ref. (i) above). That $X_n \overset{pr}{\to} X$ follows now since

$$P(d(X_n, X) > 2\varepsilon) \leq P(d(X_n, X_{n_k}) + d(X_{n_k}, X) > 2\varepsilon)$$
$$\leq P(d(X_n, X_{n_k}) > \varepsilon) + P(d(X_{n_k}, X) > \varepsilon)$$
$$< 2\varepsilon.$$

The assertion regarding the uniqueness of X follows from the Note under **(4.3.1)**(ii).

Now to prove the converse part, assume the conditions hold. If possible

$$\text{let } X_n \text{ not converge to } X \text{ in probability.} \tag{4.21}$$

Since $\varepsilon < \varepsilon'$ implies $P(d(X_n, X) > \varepsilon) \geq P(d(X_n, X) > \varepsilon')$, (4.21) means that there exists ε_0 such that whatever $\varepsilon < \varepsilon_0$ may be, $\overline{\lim_{n\to\infty}} P(d(X_n, X) >$

$\varepsilon) = a = a(\varepsilon) > 0$. Hence there exists a subsequence (n_k) such that $\lim_{k\to\infty} P(d(X_{n_k}, X) > \varepsilon) = a$. By hypothesis there exists a subsequence $(n_{k(r)})$ such that $X_{n(k(r))} \xrightarrow{wp1} X$. This implies, by (i) that $X_{n(k(r))} \xrightarrow{pr} X$. i.e. $P(d(X_{n(k(r))}, X) > \varepsilon) \to 0$ and also to a. This contradiction establishes the converse. □

(ix) A sequence (X_n) of rvs is Cauchy in probability *iff* it converges in probability.

Proof. If (X_n) is Cauchy in probability, then that it converges in probability follows from (viii).

Conversely, let $X_n \xrightarrow{pr} X_0$. Hence for each $\varepsilon > 0$, $\lim_{n\to\infty} P(|X_n - X_0| > \varepsilon) = 0$. That (X_n) is Cauchy in probability follows now since

$$\begin{aligned} P(|X_m - X_n| > \varepsilon) &= P(|X_m - X_0 + X_0 - X_n| > \varepsilon) \\ &\le P(|X_m - X_0| > \frac{\varepsilon}{2}) + P(|X_n - X_0| > \frac{\varepsilon}{2}) \to 0 \end{aligned}$$

as $m, n \to \infty$.

Note. If (X_n) is a monotonic sequence of rvs and if $X_n \xrightarrow{pr} X$, then $X_n \xrightarrow{wp1} X$.

Proof. By (viii), there exists a subsequence (X_{n_k}) of (X_n) converging to X $wp1$. Since the sequence (X_n) is monotonic, the claim follows. □

(x) Let $r > 0$. $X_n \xrightarrow{a.s.} X_0$ or $X_n \xrightarrow{pr} X_0$ and $\mathbb{E}\xi_n^r < \infty$, $n \ge 1 \not\Rightarrow \left(X_n \xrightarrow{r^{th} mean} X_0\right)$.

Proof (Example). Let $P(Y_n = e^n) = \frac{1}{n^2}$, $P(Y_n = 0) = 1 - \frac{1}{n^2}$. Then, for every $\varepsilon > 0$, $\sum_n P(|Y_n| > \varepsilon) = \sum_n \frac{1}{n^2} < \infty$. Hence, by the Borel-Cantelli Lemma $Y_n \to 0$ $wp1$. But $\mathbb{E}|Y_n|^r = \frac{e^{nr}}{n^2} \to \infty$. □

(xi) Let $r > 0$. Then $\left(X_n \xrightarrow{r^{th}\ mean} X_0\right) \Rightarrow \left(X_n \xrightarrow{pr} X_0\right)$.

Proof. The claim follows since

$P(\xi_n > \varepsilon) = \int_{\xi_n > \varepsilon} \mathrm{d}P(\omega) \le \frac{1}{\varepsilon^r} \int_{\xi_n > \varepsilon} \xi_n^r \, \mathrm{d}P(\omega) \le \frac{1}{\varepsilon^r}\mathbb{E}\xi_n^r \to 0.$ □

(xii) Let $r > 0$; $(X_n \xrightarrow{r^{th} mean} X_0) \not\Rightarrow (X_n \xrightarrow{a.s.} X_0)$.

Proof (Example). Let (X_n) be a sequence of independent rvs with $P(X_n = n^{\frac{1}{2r}}) = \frac{1}{n}$, $P(X_n = 0) = 1 - \frac{1}{n}$. Since $\mathrm{E}|X_n|^r = \frac{1}{\sqrt{n}} \to 0$, it follows that $X_n \xrightarrow{r^{th} mean} 0$. Since (X_n) is a sequence of independent variables, X_n will converge to 0 $wp1$ *iff* $\sum_n P(|X_n| > \varepsilon) < \infty$. But $\sum_n P(|X_n| > \varepsilon) = \sum_n \frac{1}{n} = \infty$ for every $\varepsilon > 0$. □

(xiii) If (M, d) is a complete metric space, then a sufficient condition for the sequence (X_n) to converge $wp1$ is the convergence of the series

$\sum\limits_n P\big(d(X_n,\ X_{n+1}) > \varepsilon_n\big)$ for some sequence (ε_n) of positive numbers with $\sum\limits_n \varepsilon_n < \infty$.

Proof. Define events $A_n = \{d(X_n,\ X_{n+1}) > \varepsilon_n\}$. The hypothesis that $\sum_n P(A_n) < \infty$ implies that there exists a set E with $P(E) = 1$, that to each $\omega \in E$ there corresponds an integer $N = N(\omega)$ and that $d\big(X_n(\omega),\ X_{n+1}(\omega)\big) \leq \varepsilon_n$ for all $n \geq N$. If $\omega \in E$ and if $N \leq m < n$, then $d(X_m,\ X_n) \leq \sum_{r=m}^{n-1} d\big(X_r(\omega),\ X_{r+1}(\omega)\big) \leq \sum_{r=m}^{n-1} \varepsilon_r$ which tends to zero as $m \to \infty$. This shows that for each $\omega \in E$, $(X_n(\omega))$ is a Cauchy sequence in M. Since, M is a complete metric space, $\lim\limits_{n\to\infty} X_n(\omega)$ exists. □

(xiv) $(X_n \overset{c}{\to} X_0) \Rightarrow (X_n \overset{wp1}{\longrightarrow} X_0)$ but $(X_n \overset{wp1}{\longrightarrow} X_0) \not\Rightarrow (X_n \overset{c}{\to} X_0)$.

Proof. The first assertion is obvious.

The following example serves to validate the second assertion. Let $\Omega = [0,\ 1]$, $\mathscr{A}$ the Borel σ-field and P the Lebesgue measure. Define for $n \geq 1$,

$$X_n(\omega) = \begin{cases} 1 & \text{if } 0 \leq \omega \leq \frac{1}{n}, \\ 0 & \text{if } \frac{1}{n} < \omega \leq 1 \text{ and } X_0 \equiv 0. \end{cases}$$

We note $\lim\limits_{n\to\infty} X_n(\omega) = X_0(\omega)$ for all ω except $\omega = 0$. Thus $X_n \overset{wp1}{\longrightarrow} X_0$. But $\sum\limits_n P(|X_n| > \varepsilon) = \sum\limits_n P(X_n = 1) = \sum\limits_n \frac{1}{n} = \infty$. □

(xv) If g is a continuous mapping of M into a separable metric space $(S,\ \rho)$, then $(X_n \overset{pr}{\to} X_0) \Rightarrow g(X_n) \overset{pr}{\to} g(X_0)$.

Proof. We recall that if $X,\ Y$ are rE in M, then the separability of M ensures that $d(X,\ Y)$ is a rv. Since g is a continuous mapping, $g(X),\ g(Y)$ are rEs in S. Since $(S,\ \rho)$ is separable , $\rho(g(X),\ g(Y))$ is a rv. Since g is a continuous function we can, given $\varepsilon > 0$, find a δ such that $\rho(g(x),\ g(y)) < \varepsilon$ whenever $d(x,\ y) < \delta$. Since $X_n \overset{pr}{\to} X_0$ we can find $N = N(\delta,\ \varepsilon)$ such that $P(A_n) < \varepsilon$ for all $n \geq N$ where $A_n = \{\omega : d(X_n(\omega),\ X_0(\omega)) \geq \delta\}$. Thus $\{\omega : \rho(g(X_n(\omega)),\ g(X_0(\omega))) < \varepsilon\} \supseteq A'_n$ for all $n \geq N$. Hence $P(\rho(g(X_n),\ g(X_0)) > \varepsilon) \leq P(A_n) < \varepsilon$. □

(xvi) Let $(M,\ d)$, $(S,\ \rho)$ and g be as in (xiv) above. Then
$(X_n \overset{d}{\to} X_0) \Rightarrow (g(X_n) \overset{d}{\to} g(X_0))$.

Proof. Let G be a ρ-open set in S.

Now $g(X_n) \overset{d}{\to} g(X_0)$ if $\underline{\lim}_{n\to\infty} P(g(X_n) \in G) \geq P(g(X_0) \in G)$. i.e. if $\liminf\limits_{n\to\infty} P(X_n \in g^{-1}(G)) \geq P(X_0 \in g^{-1}(G))$. But this is true since $g^{-1}(G)$ is d-open set in M and since $X_n \overset{d}{\to} X_0$. □

(xvii) Let $X_n,\ Y_n,\ n \geq 0$ be random elements in a separable metric space $(M,\ d)$. Let $X_n \overset{d}{\to} X_0$. Let $d(X_n,\ Y_n) \overset{pr}{\to} 0$. Then $Y_n \overset{d}{\to} X_0$.

Proof. Let $C \subseteq M$ be a closed subset. Given integer $k \geq 1$, define $C^{(k)} = \{u : u \in M,\ d(u,\ C) \leq \frac{1}{k}\}$ and note that $C \subset C^{(k)}$, that $C^{(k)}$ is a closed set and that $C^{(k)} \downarrow C$ as $k \uparrow \infty$. Now,

$$\begin{aligned}\overline{\lim_{n\to\infty}}\, P(Y_n \in C) &\leq \overline{\lim_{n\to\infty}}\, P(Y_n \in C,\ d(Y_n,\ X_n) > \frac{1}{k}) \\ &\quad + \overline{\lim_{n\to\infty}}\, P(Y_n \in C,\ d(Y_n,\ X_n) \leq \frac{1}{k}) \\ &\leq \overline{\lim_{n\to\infty}}\, P(d(Y_n,\ X_n) > \frac{1}{k}) + \overline{\lim_{n\to\infty}}\, P(X_n \in C^{(k)}) \\ &= \overline{\lim_{n\to\infty}}\, P(X_n \in C^{(k)}) \\ &\leq P(X_0 \in C^{(k)})\end{aligned}$$

(since $C^{(k)}$ is a closed set and since $X_n \overset{d}{\to} X_0$). Let $k \to \infty$ and get $\overline{\lim_{n\to\infty}}\, P(Y_n \in C) \leq P(X_0 \in C)$. This being true for every closed set C, the desired result follows. □

(xviii) If $(X_n),\ (Y_n)$ are sequences of rvs, if $X_n \overset{d}{\to} X_0$ and if $Y_n \overset{pr}{\to} 0$, then $X_n + Y_n \overset{d}{\to} X_0$.

Proof. Given $\varepsilon > 0$, we have

$$\begin{aligned}P(X_n + Y_n \leq x) &= P(X_n + Y_n \leq x,\ |Y_n| \leq \varepsilon) \\ &\quad + P(X_n + Y_n \leq x,\ |Y_n| > \varepsilon) \\ &\leq P(X_n \leq x + \varepsilon) + P(|Y_n| > \varepsilon),\end{aligned}$$

leading to $\overline{\lim_{n\to\infty}}\, P(X_n + Y_n \leq x) \leq P(X_0 \leq x + \varepsilon)$. Now take $\varepsilon \to 0$ keeping $x + \varepsilon$ to be a continuity point of the *df* F_0 of X_0 and conclude $\overline{\lim_{n\to\infty}}\, P(X_n + Y_n \leq x) \leq P(X_0 \leq x)$. Again,

$$\begin{aligned}P(X_n + Y_n \leq x) &\geq P(X_n + Y_n \leq x,\ |Y_n| \leq \varepsilon) \\ &\geq P(X_n \leq x - \varepsilon,\ |Y_n| \leq \varepsilon) \\ &\geq P(X_n \leq x - \varepsilon) - P(|Y_n| > \varepsilon),\end{aligned}$$

leading to

$$\underline{\lim_{n\to\infty}}\, P(X_n + Y_n \leq x) \geq \underline{\lim_{n\to\infty}}\, P(X_n \leq x - \varepsilon).$$

Now take $\varepsilon \to 0$ keeping $x - \varepsilon$ to be a continuity point of F_0 and conclude $\underline{\lim_{n\to\infty}}\, P(X_n + Y_n \leq x) \geq P(X_0 \leq x)$. The two inequalities imply the desired result. □

4.4 Inequalities. Set II

Theorem 4.4.1 (Tchebichev inequality). *Let X be a rv with $\mathbb{E}|X|^\alpha < \infty$ for some $\alpha > 0$. Then for every $\lambda > 0$, $P(|X| \geq \lambda) \leq \frac{\mathbb{E}|X|^\alpha}{\lambda^\alpha}$.*

Proof

$$\mathbb{E}|X|^\alpha = \int_\Omega |X(\omega)|^\alpha \, \mathrm{d}P(\omega) \geq \int_{|X|\geq\lambda} |X(\omega)|^\alpha \, \mathrm{d}P(\omega)$$

$$\geq \lambda^\alpha \int_{|X|\geq\lambda} \mathrm{d}P(\omega) = \lambda^\alpha P(|X| \geq \lambda). \qquad \square$$

Theorem 4.4.2 (Skorohod inequality). *Let X_r, $1 \leq r \leq n$ be n mutually independent rvs. Let $S_r = \sum_{k=1}^{r} X_k$, $1 \leq r \leq n$. Let $0 < a < \lambda$ be arbitrary numbers.*

(i) Let $\max_{1\leq j\leq n} P(S_n - S_j \geq a) \leq p < 1$. Then

$$P(\max_{1\leq j\leq n} S_j \geq \lambda) \leq \frac{1}{1-p} P(S_n > \lambda - a).$$

(ii) Let $\max_{1\leq j\leq n} P(|S_n - S_j| \geq a) \leq p < 1$. Then

$$P(\max_{1\leq j\leq n} |S_j| \geq \lambda) \leq \frac{1}{1-p} P(|S_n| > \lambda - a).$$

Proof. (i) For $n = 1$, the inequality would read

$$P(X_1 \geq \lambda) \leq \frac{1}{1-p} P(X_1 \geq \lambda - a)$$

and is trivially true. Let $n \geq 2$. Define events $A_1 = \{X_1 \geq \lambda\}$, $A_j = \{S_r < \lambda,\ 1 \leq r \leq j-1,\ S_j \geq \lambda\}$, $2 \leq j \leq n$. We note that the A_js are mutually exclusive and that

$$B_n = \cup_1^n A_j = \{\max_{1\leq j\leq n} S_j \geq \lambda\}.$$

Now, by hypothesis,

$$P(A_j) \leq \frac{1}{1-p} P(S_n - S_j < a) P(A_j)$$

(since the events are independent)

$$= \frac{1}{1-p} P(S_n - S_j < a,\ A_j)$$

$$\leq \frac{1}{1-p} P(S_n > \lambda - a,\ A_j).$$

Adding up over $j \leq n$, we get

$$P(\max_{1 \leq j \leq n} S_j \geq \lambda) \leq \tfrac{1}{1-p} P(S_n > \lambda - a,\ B_n)$$
$$\leq \tfrac{1}{1-p} P(S_n > \lambda - a).$$

(ii) The arguments are similar to those in (i). The details are as under.

For $n = 1$, the inequality would read

$$P(|X_1| \geq \lambda) \leq \frac{1}{1-p} P(|X_1| \geq \lambda - a)$$

and is trivially true. Let $n \geq 2$. Define events $A_1 = \{|X_1| \geq \lambda\}$, $A_j = \{|S_r| < \lambda,\ 1 \leq r \leq j-1,\ |S_j| \geq \lambda\}$, $2 \leq j \leq n$. We note that the A_js are mutually exclusive and that $B_n = \cup_1^n A_j = \{\max_{1 \leq j \leq n} |S_j| \geq \lambda\}$. Now,

$$P(A_j) \leq \frac{1}{1-p} P(|S_n - S_j| < a) P(A_j)$$

$$= \frac{1}{1-p} P(|S_n - S_j| < a,\ A_j) \quad \text{(the events being independent)}$$

$$\leq P(|S_n| > \lambda - a,\ A_j).$$

Adding up over $j \leq n$, we get

$$P(\max_{1 \leq j \leq n} |S_j| \geq \lambda) \leq \tfrac{1}{1-p} P(|S_n| > \lambda - a,\ B_n)$$
$$\leq \tfrac{1}{1-p} P(|S_n| > \lambda - a).$$

□

Theorem 4.4.3 (Kolmogorov-Hājeck-Rēnyi inequality). *Let X_r, $1 \leq r \leq n$ be n mutually independent rvs with $\mathbb{E}X_r = 0$ and $\mathbb{E}X_r^2 = \sigma_r^2 < \infty$ and let (c_n) be a sequence of positive numbers with $c_{n+1} \leq c_n$, $n \geq 1$. Then for every $\lambda > 0$,*

$$P(\max_{1 \leq r \leq n} c_r |S_r| \geq \lambda) \leq \frac{1}{\lambda^2} \sum_1^n c_r^2 \sigma_r^2$$

where $S_r = \sum_1^r X_k$. *If additionally* $|X_n| \le c$ *for all* n *and for some* $c > 0$, *then*

$$P(\max_{1\le r\le n} |S_r| \ge \lambda) \ge 1 - \frac{(\lambda + c)^2}{\sum_1^n \sigma_r^2}.$$

Proof. We record first that for $n = 1$, the first inequality is the same as the Tchebichev inequality with $\alpha = 2$. Let $n \ge 2$.

Let $s_r^2 = \sum_1^r \sigma_r^2 = \mathbb{E}S_r^2$. Define events : $A_1 = \{c_1|S_1| \ge \lambda\}$, $A_r = \{c_k|S_k| < \lambda,\ 1 \le k \le r-1,\ c_r|S_r| \ge \lambda\}$, $2 \le r \le n$. We note that the n events $A_1, A_2, \ldots, A_n$ are mutually exclusive and that $B_n = \cup_1^n A_r = \{\max_{1\le r\le n} c_r|S_r| \ge \lambda\}$. Denote by χ_r the indicator function of the event A_r.

The following identity will be useful in the sequel. Let $a_i > 0$, $b_i^2 = \sum_{j=1}^i a_j^2$, $1 \le i \le n$. Then

$$\sum_1^n c_r^2 a_r^2 = \sum_1^{n-1} b_k^2(c_k^2 - c_{k+1}^2) + c_n^2 b_n^2. \tag{4.22}$$

Thus

$$\begin{aligned}
\alpha_n &= \sum_1^n c_r^2\sigma_r^2 = \sum_1^{n-1} s_k^2(c_k^2 - c_{k+1}^2) + c_n^2 s_n^2 \\
&= \sum_1^{n-1}(c_k^2 - c_{k+1}^2)\int_\Omega S_k^2(\omega)\,\mathrm{d}P(\omega) + c_n^2\int_\Omega S_n^2(\omega)\,\mathrm{d}P(\omega) \\
&\ge \sum_1^{n-1}(c_k^2 - c_{k+1}^2)\int_{B_k} S_k^2(\omega)\,\mathrm{d}P(\omega) + c_n^2\int_{B_n} S_n^2(\omega)\,\mathrm{d}P(\omega) \\
&= \sum_1^{n-1}(c_k^2 - c_{k+1}^2)\sum_{j=1}^k\int_{A_j}(S_k(\omega) - S_j(\omega) + S_j(\omega))^2\,\mathrm{d}P(\omega) \\
&\quad + c_n^2\sum_{r=1}^n\int_{A_r}(S_n(\omega) - S_r(\omega) + S_r(\omega))^2\,\mathrm{d}P(\omega) \\
&\ge \sum_1^{n-1}(c_k^2 - c_{k+1}^2)\sum_{j=1}^k\int_{A_j}\{2(S_k(\omega) - S_j(\omega))S_j(\omega) + (S_j(\omega))^2\}\,\mathrm{d}P(\omega)
\end{aligned}$$

$$+ c_n^2 \sum_{r=1}^{n} \int_{A_r} \{2(S_n(\omega) - S_r(\omega))S_r(\omega) + (S_r(\omega))^2\} \, \mathrm{d}P(\omega).$$

For each k and for $r < k$, $S_k - S_r$ is independent of the variables S_r and χ_r. Further $\mathbb{E}(S_k - S_r) = 0$. These facts help us see

$$\begin{aligned}
\alpha_n &\geq \sum_{1}^{n-1}(c_k^2 - c_{k+1}^2) \sum_{j=1}^{k} \int_{A_j} \{S_j(\omega)\}^2 \, \mathrm{d}P(\omega) \\
&\quad + c_n^2 \sum_{r=1}^{n} \int_{A_r} \{S_r(\omega)\}^2 \, \mathrm{d}P(\omega) \\
&\geq \sum_{1}^{n-1}(c_k^2 - c_{k+1}^2) \sum_{j=1}^{k} \int_{A_j} \frac{\lambda^2}{c_j^2} \, \mathrm{d}P(\omega) \; + c_n^2 \sum_{r=1}^{n} \int_{A_r} \frac{\lambda^2}{c_r^2} \, \mathrm{d}P(\omega) \\
&\geq \lambda^2 \sum_{1}^{n-1}(c_k^2 - c_{k+1}^2) \sum_{j=1}^{k} \frac{1}{c_j^2} P(A_j) + \lambda^2 c_n^2 \sum_{r=1}^{n} \frac{1}{c_r^2} P(A_r),
\end{aligned}$$

leading to

$$\begin{aligned}
\frac{1}{\lambda^2}\alpha_n &\geq \sum_{1}^{n-1}(c_k^2 - c_{k+1}^2) \sum_{j=1}^{k} \frac{1}{c_j^2} P(A_j) + c_n^2 \sum_{r=1}^{n} \frac{1}{c_r^2} P(A_r) \\
&= \sum_{1}^{n-1}(c_k^2 - c_{k+1}^2) \sum_{j=1}^{k} a_j^2 + c_n^2 \sum_{r=1}^{n} a_r^2 \\
&= \sum_{1}^{n-1}(c_k^2 - c_{k+1}^2) b_k^2 + c_n^2 b_n^2
\end{aligned}$$

(where $a_j^2 = \frac{1}{c_j^2} P(A_j)$ and $b_r^2 = \sum_{1}^{r} a_j^2$)

$$\begin{aligned}
&= \sum_{1}^{n} c_r^2 a_r^2 \text{ (appealing to the identity at (4.22))} \\
&= \sum_{1}^{n} P(A_j) = P(B_n).
\end{aligned}$$

Now we establish the other inequality. The inequality is obvious for $n = 1$ since the right side is negative. Let $n \geq 2$. Assume $c_n = 1$ for all n and take the A_rs defined accordingly.

Let I_r denote the indicator function of B'_r. We note that $B'_{k-1} = B'_k \cup A_k$ and hence that $S_k I_k + S_k \chi_k = S_k I_{k-1} = S_{k-1} I_{k-1} + X_k I_{k-1}$. Since $B'_k \cap A_k = \phi, \mathbb{E} S_k I_k \chi_k = 0$; since $S_{k-1} I_{k-1}$ and X_k are mutually independent and since $\mathbb{E} X_k = 0, \mathbb{E} S_{k-1} I_{k-1} X_k = 0$. Taking the expected values of the squares of the two extreme expressions, we get

$$\mathbb{E}(S_k I_k)^2 + \mathbb{E}(S_k \chi_k)^2 = \mathbb{E}(S_{k-1} I_{k-1})^2 + \mathbb{E}(X_k I_{k-1})^2.$$

Since

$$|S_k \chi_k| \leq |S_{k-1} \chi_k| + |X_k \chi_k| \leq \lambda \chi_k + c \chi_k,$$

we have

$$\mathbb{E}(S_k \chi_k)^2 \leq (\lambda + c)^2 \mathbb{E} \chi_k = (\lambda + c)^2 P(A_k) \leq (\lambda + c)^2 P(A_k).$$

Also $\mathbb{E}(X_k I_{k-1})^2 = \sigma_k^2 P(B'_{k-1}) \geq \sigma_k^2 P(B'_n)$. Thus

$$\mathbb{E}(S_k I_k)^2 + (\lambda + c)^2 P(A_k) \geq \mathbb{E}(S_{k-1} I_{k-1})^2 + \sigma_k^2 P(B'_n).$$

These inequalities for $k = 1, 2, \ldots, n$ yield, when summed,

$$\mathbb{E}(S_n I_n)^2 + (\lambda + c)^2 P(\cup_1^n A_k) \geq P(B'_n) \sum_1^n \sigma_k^2.$$

Or,

$$\mathbb{E}(S_n I_n)^2 + (\lambda + c)^2 P(B_n) \geq P(B'_n) \sum_1^n \sigma_k^2.$$

Since $|S_n| I_n \leq \lambda I_n$, we have $\mathbb{E}(S_n I_n)^2 \leq \lambda^2 P(B'_n)$. Hence

$$\lambda^2 P(B'_n) + (\lambda + c)^2 P(B_n) \geq P(B'_n) \sum_{k=1}^n \sigma_k^2. \tag{4.23}$$

It follows from this

$$P(B_n) \geq \frac{\sum\limits_{k=1}^n \sigma_k^2 - \lambda^2}{\sum\limits_{k=1}^n \sigma_k^2 + (\lambda+c)^2 - \lambda^2} = 1 - \frac{(\lambda+c)^2}{\sum\limits_{k=1}^n \sigma_k^2 + (\lambda+c)^2 - \lambda^2}$$
$$\geq 1 - \frac{(\lambda+c)^2}{\sum\limits_{k=1}^n \sigma_k^2}.$$

□

Theorem 4.4.4 (Levy inequality). *Suppose* X_r, $1 \leq r \leq n$ *are* n *mutually independent* rvs *and* $S_r = \sum\limits_{k=1}^r X_k$. *Let* med *stand for median (ref.* **(2.5.1)***).*

Then for every $\lambda > 0$,
(i) $P\big(\max_{1\le r\le n}\{S_r - med(S_r - S_n)\} \ge \lambda\big) \le 2P(S_n \ge \lambda)$.
(ii) $P\big(\max_{1\le r\le n}|S_r - med(S_r - S_n)| \ge \lambda\big) \le 2P(|S_n| \ge \lambda)$.

Proof. (i) For $n = 1$, what needs to be proved is $P(X_1 \ge \lambda) \le 2P(X_1 \ge \lambda)$ which is obvious. Let $n \ge 2$. Let $S_0 = 0$. Write $S_r^* = \max_{1\le k\le r}\{S_k - med(S_k - S_n)\}$. Define events

$$A_1 = \{S_1 - med(S_1 - S_n) \ge \lambda\};$$
$$A_r = \{S_{r-1}^* < \lambda,\ S_r - med(S_r - S_n) \ge \lambda\},\ 2 \le r \le n.$$

The events A_rs are mutually exclusive and $\cup_1^n A_r = B_n = \{S_n^* \ge \lambda\}$. Let $C_r = \{S_n - S_r - med(S_n - S_r) \ge 0\}$. We note that C_r depends only on the variables $X_k,\ r+1 \le k \le n$, that $P(C_r) \ge \frac{1}{2}$ and that, for each r, A_r and C_r are mutually independent. Now

$$\begin{aligned} A_r \cap C_r = \{&S_{r-1}^* < \lambda,\ S_r - med(S_r - S_n) \ge \lambda,\\ &S_n - S_r - med(S_n - S_r) \ge 0\} \subseteq \{S_{r-1}^* < \lambda,\ S_n \ge \lambda\}. \end{aligned}$$

Thus

$$\begin{aligned} \frac{1}{2}P(S_n^* \ge \lambda) = \frac{1}{2}P(B_n) &= \sum_1^n \frac{1}{2}P(A_r)\\ &\le \sum_1^n P(A_r \cap C_r)\\ &\le \sum_1^n P(S_{r-1}^* < \lambda, S_r - med(S_r - S_n) \ge \lambda,\ S_n \ge \lambda)\\ &= P(S_n^* \ge \lambda,\ S_n \ge \lambda) = P(S_n \ge \lambda). \end{aligned}$$

(ii) Replace each X_i in (i) by $-X_i$ to get :
(i′) $P\big(\min_{1\le r\le n}\{S_r - med(S_r - S_n)\} \le -\lambda\big) \le 2P(S_n \le -\lambda)$.
Now, (i) and (i′) lead to (ii). □

Theorem 4.4.5 (Inequalities relating to the normal df). *Let* Φ *denote the standard normal df:*

$$\Phi(x) = \int_{-\infty}^{x} \frac{1}{\sqrt{2\pi}} e^{-\frac{1}{2}y^2}\, dy.$$

Let $(X_k,\ 1 \le k \le n)$ *be* n *iid* $\mathcal{N}(0,\ 1)$ *variables. Write* $S_n = \sum_1^r X_k$, $1 \le r \le 1;\ S_n^* = \max_{1\le k\le n} S_k;\ S_n^{**} = \max_{1\le k\le n} |S_k|$. *Then*
(i) $1 - \Phi(x) \le \frac{1}{\sqrt{2\pi}}\frac{1}{x}e^{-\frac{1}{2}x^2},\ x > 0.$
(ii) $1 - \Phi(x) \ge \frac{1}{\sqrt{2\pi}}\left(\frac{1}{x} - \frac{1}{x^3}\right)e^{-\frac{1}{2}x^2},\ x > 0.$
(iii) If $-\infty < a < b < \infty$, *then*

$$(b-a)e^{-\frac{\alpha}{2}} \le \sqrt{2\pi}\{\Phi(b) - \Phi(a)\} \le b - a,$$

where $\alpha = \max(a^2,\ b^2)$.
(iv) $P(S_n^* \ge b) \le 2P(S_n \ge b)$.
(v) $P(S_n^{**} \ge b) \le 2P(|S_n| \ge b)$.

(i) We note

$$\frac{1}{x}e^{-\frac{1}{2}x^2} = -\int_x^\infty \mathrm{d}\left(\frac{1}{y}e^{-\frac{1}{2}y^2}\right) = \int_x^\infty e^{-\frac{1}{2}y^2}\left(1 + \frac{1}{y^2}\right)\,\mathrm{d}y.$$

Hence

$$\frac{1}{x}e^{-\frac{1}{2}x^2} \ge \int_x^\infty e^{-\frac{1}{2}y^2}\,\mathrm{d}y = \sqrt{2\pi}\{1 - \Phi(x)\}.$$

(ii) We note

$$\int_x^\infty e^{-\frac{1}{2}y^2}\left(1 - \frac{3}{y^4}\right)\,\mathrm{d}y = -\int_x^\infty \mathrm{d}\left(e^{-\frac{1}{2}y^2}\left\{\frac{1}{y} - \frac{1}{y^3}\right\}\right) = e^{-\frac{1}{2}x^2}\left(\frac{1}{x} - \frac{1}{x^3}\right).$$

Rewriting,

$$\sqrt{2\pi}\{1 - \Phi(x)\} = \int_x^\infty e^{-\frac{1}{2}y^2}\frac{3}{y^4}\,\mathrm{d}y + e^{-\frac{1}{2}x^2}\left(\frac{1}{x} - \frac{1}{x^3}\right) \ge e^{-\frac{1}{2}x^2}\left(\frac{1}{x} - \frac{1}{x^3}\right).$$

(iii) The proof is immediate on observing that if $y \in [a,\ b]$, then $e^{-\frac{1}{2}\alpha} \le e^{-\frac{1}{2}y^2} \le 1$.
(iv) Taking $\lambda = b$ and $a = 0$ in **(4.4.2)**(i) and noting the consequence that $p = \frac{1}{2}$, the desird result is obtained.
(v) Applying the result in (iv) to the variables $-X_k$ we get $P(\min_{1\le k\le n} \le -b) \le 2P(S_n \le -b)$. The claim follows since

$$(S_n^{**} \ge b) \subset (\max_{1\le k\le n} S_k \ge b) \cup (\min_{1\le k\le n} S_k \le -b)$$

and since

$$P(S_n \geq b) + P(S_n \leq -b) = P(|S_n| \geq b). \qquad \square$$

4.5 Continuation of (4.3.2)(viii) and (4.3.2)(ix)

Even though it is true that a sequence (X_n) of rvs converging in probability to a rv X admits of a subsequence, say, (X_{n_k}) converging $wp1$ to X, it is not always easy, except in simple cases, to explicitly describe the sequence (n_k).

For example let (X_n) be a sequence of non-negative iid rvs with common df G and with $\mathbb{E}X_1 = \infty$. Let $S_n = \sum_1^n X_j$. Let L be sv where $L(x) = \int_0^x \{1 - G(y)\}\, dy$. Write $\psi(\lambda) = \int_0^\infty e^{-\lambda x}\, dG(x)$, $\lambda > 0$. We note L is a monotonic non-decreasing continuous function. By integration by parts (on the lines of **(2.6.3)**) we get

$$\frac{1-\psi(\lambda)}{\lambda} = \int_0^\infty e^{-\lambda y}\{1 - G(y)\}\, dy = \int_0^\infty e^{-\lambda y}\, dL(y) \sim L(\frac{1}{\lambda}) \text{ as } \lambda \downarrow 0$$

(by **Q**10, using the hypothesis that L is sv). We use Karamata representation for L and conclude that $\lim_{x\to\infty} \frac{L(x)}{x} = 0$. Define a_n asymptotically to satisfy the relation $\frac{L(a_n)}{a_n} \sim \frac{1}{n}$. So defined, a_n will be of the form $nl(n)$ where $l(n) \to \infty$ and l is sv.

As $n \to \infty$, $n \log \psi(\frac{\lambda}{a_n}) \sim -n\{1 - \psi(\frac{\lambda}{a_n})\} \sim -n\frac{\lambda}{a_n} L(\frac{a_n}{\lambda})$ $\sim -\lambda \frac{n}{a_n} L(a_n) \sim -\lambda$. Thus the Laplace transform $\psi^n(\frac{\lambda}{a_n})$ of $\frac{S_n}{a_n}$ converges to $e^{-\lambda}$. This implies $\frac{S_n}{a_n} \xrightarrow{d} X_0 (\equiv 1)$ (ref. **(3.10.5)**). This, in turn, implies $\frac{S_n}{a_n} \xrightarrow{pr} 1$ (ref. **(4.3.2)**(v)). The question before us now is : to find a subsequence (n_k) tending to infinity not very fast and such that $\frac{S_{n_k}}{a_{n_k}} \xrightarrow{wp1} 1$. We will find subsequences under the following growth condition (4.25) on G.

Step 1. Denote by $\mathfrak{L}$ the σ-finite measure generated on R by the monotonic function L and note that $\mathfrak{L}$ is absolutely continuous with respect to Lebesgue measure, with $1 - G(y)$ for one version of the Radon-Nikodym derivative. Apply formula for change of variables (ref. **(2.3.6)**) to recognise that

$$\int_1^x \frac{1}{L(t)}\, d\mathfrak{L}(t) = \int_{L(1)}^{L(x)} \frac{1}{y}\, dy = \log \frac{L(x)}{L(1)}.$$

Thus

$$\frac{L(x)}{L(1)} = e^{\log \frac{L(x)}{L(1)}} = e^{\int_1^x \frac{1}{L(y)}\, d\mathcal{L}(y)} = e^{\int_1^x \frac{1-G(y)}{L(y)}\, dy} = e^{\int_1^x \frac{r(y)}{y}\, dy}$$

where $r(y) = \frac{y\{1-G(y)\}}{L(y)}$. For $\lambda > 1$, $\lim_{x\to\infty} \frac{L(\lambda x)}{L(x)} = 1$ since L is sv. Hence

$$\lim_{x\to\infty} \int_{x}^{\lambda x} \frac{r(y)}{y}\, dy = 0.$$

i.e. $\lim_{x\to\infty} \int_{x}^{\lambda x} \frac{1-G(y)}{L(y)}\, dy = 0$, leading to $\lim_{x\to\infty} \frac{1-G(\lambda x)}{L(\lambda x)}(\lambda - 1)x = 0$.
Hence $r(x) \to 0$. Thus in the Karamata representation for L, "ε" $= r$.
Step 2. By partial integration

$$\begin{aligned} M(x) &= \int_0^x y\, dG(y) = -x\{1 - G(x)\} + L(x) \\ &= -r(x)L(x) + L(x). \end{aligned}$$

Hence

$$M(x) \sim L(x) \tag{4.24}$$

We assume

$$r(x)(\log x) = o(1) \text{ as } x \to \infty. \tag{4.25}$$

(4.25) implies that there exists $u > 1$ such that for all $x > u$, $r(x) < \frac{1}{2\log x}$.

Here and in the rest of the book, q with or without suffix, will denote a constant, possibly with different values at different occurrences.

$$L(x) \le q_1 e^{\int_u^x \frac{1}{2y\log y}\, dy} = q_2 e^{\frac{1}{2}(\log\log x)} = q_2(\log x)^{\frac{1}{2}}. \tag{4.26}$$

Further

$$I = \int_a^\infty \frac{dL(x)}{L(x)(\log x)^{2\delta}} < \infty. \tag{4.27}$$

for every $\delta > 0$, since for all large values of a,

$$I \le \int_a^\infty \frac{r(x)}{x(\log x)^{2\delta}}\, dx \le \int_a^\infty \frac{dx}{x(\log x)^{1+2\delta}}$$

Assume $\delta < \frac{1}{2}$. We note (4.27) is equivalent to

$$\int_a^\infty \frac{x\, dG(x)}{L(x)(\log x)^{2\delta}} < \infty. \tag{4.28}$$

Step 3. Let $f(x) = \frac{x}{L(x)(\log x)^{2\delta}}$. We note f is a monotonic increasing function of x. Then, as $x \to \infty$,

$$f^{-1}(x) \sim xL(x)(\log x)^{2\delta}. \tag{4.29}$$

This will be true if

$$L(xL(x)(\log x)^{2\delta}) \sim L(x) \tag{4.30}$$

i.e. if $\lim_{x\to\infty} J(x) = 0$ where $J(x) = \int_x^{xL(x)(\log x)^{2\delta}} \frac{r(y)}{y}\, dy$. Now

$$\begin{aligned} 2J(x) &\leq \int_x^{xL(x)(\log x)^{2\delta}} \frac{1}{y\log y}\, dy \\ &= \log \frac{\log x + \log L(x) + 2\delta \log\log x}{\log x} \to 0 \qquad \text{as } x \to \infty. \end{aligned}$$

Hence (4.29) is true.

On the same lines it can be shown that $L(a_n) \sim L(n)$.

Define $\lambda_n = f^{-1}(n)$. We note that $\lambda_n \sim nL(n)(\log n)^{2\delta}$ and that $L(\lambda_n) \sim L(n)$.

Step 4. Define $Y_n = X_n$ if $X_n \leq \lambda_n$ and $Y_n = 0$ otherwise. Hence

$$\begin{aligned} \sum_n P(Y_n \neq X_n) &= \sum_n P(X_n > \lambda_n) = \sum_n \{1 - G(\lambda_n)\} \\ &\leq \sum_n \int_n^{n+1} \{1 - G(f^{-1}(t))\}\, dt \\ &= \int^{\infty} \{1 - G(f^{-1}(t))\}\, dt < \infty \end{aligned}$$

if (ref. **(2.3.6)**) $\int^{\infty} \{1 - G(t)\}\, df(t) < \infty$. By integration by parts we see that this integral is finite if $\int_0^{\infty} f(x)\, dG(x) < \infty$ which is true by (4.28). Hence by Borel-Cantelli lemma, $P(X_n \neq Y_n \ i.o.) = 0$. This implies that if

$$T_n = \sum_1^n Y_k \quad \text{and if} \quad \frac{T_{n_k}}{a_{n_k}} \xrightarrow{wp1} 1, \quad \text{then} \quad \frac{S_{n_k}}{a_{n_k}} \xrightarrow{wp1} 1.$$

$$\mathbb{E}Y_n = \int_0^{\lambda_n} x\, dG(x) = M(\lambda_n) \sim L(\lambda_n) \sim L(n).$$

Hence $\mathbb{E}T_n = \sum_1^n \mathbb{E}Y_k \sim \sum_1^n L(k) \sim nL(n) \sim a_n$. i.e. $\mathbb{E}(\frac{T_n}{a_n}) \to 1$. We note (Y_n) is a sequence of mutually independent variables. Hence

$$\begin{aligned}
varT_n &= \sum_1^n varY_k \\
&\leq \sum_1^n \mathbb{E}Y_k^2 = \sum_1^n \int_0^{\lambda_k} y^2 \, \mathrm{d}G(y) \\
&= \sum_1^n \big[-\lambda_k^2\{1-G(\lambda_k)\} + 2\int_0^{\lambda_k} y\{1-G(y)\} \, \mathrm{d}y \big] \\
&\leq 2\sum_1^n \int_0^{\lambda_k} y\{1-G(y)\} \, \mathrm{d}y \\
&\leq 2\sum_1^n \int_0^{\lambda_k} r(y)L(y) \, \mathrm{d}y \\
&\leq \sum_1^n L(\lambda_k) \int_0^{\lambda_k} r(y) \, \mathrm{d}y.
\end{aligned}$$

For all k sufficiently large, $\int_0^{\lambda_k} r(y) \, \mathrm{d}y \leq q\frac{\lambda_k}{\log \lambda_k}$ where the constant q is independent of k. Here we make use of (4.25) and the fact that

$$\int^x \frac{\mathrm{d}y}{\log y} \sim \frac{x}{\log x}.$$

We note that as $k \to \infty$,

$$\frac{\lambda_k}{\log \lambda_k} \sim \frac{kL(k)}{(\log k)^{1-2\delta}}.$$

Hence

$$varT_n \leq q\frac{n^2L^2(n)}{(\log n)^{1-2\delta}} \sim \frac{a_n^2}{(\log n)^{1-2\delta}}.$$

Thus $var\frac{T_n}{a_n} \leq \frac{q}{(\log n)^{1-2\delta}}$. If

$$n_k = [k^{k^{\frac{1}{1-2\delta}}(\log k)^{\frac{1-\delta}{(1-2\delta)^2}}}],\ k \geq 3,$$

then

$$\sum_k var\left(\frac{T_{n_k}}{a_{n_k}}\right) \le \sum_k \frac{q}{k(\log k)^{\frac{1-\delta}{1-2\delta}}} < \infty.$$

Thus

$$\sum_k P\left(\left|\frac{T_{n_k} - \mathbb{E}T_{n_k}}{a_{n_k}}\right| \ge \varepsilon\right) \le \text{(by ((**4.4.1**))} \sum_k \frac{1}{\varepsilon^2} var\left(\frac{T_{n_k}}{a_{n_k}}\right) < \infty.$$

This being true for every $\varepsilon > 0$, we conclude, appealing to Borel-Cantelli lemma, that $\frac{T_{n_k} - \mathbb{E}T_{n_k}}{a_{n_k}} \xrightarrow{wp1} 0$. Since $\frac{\mathbb{E}T_n}{a_n} \to 1$, it follows that $\frac{T_{n_k}}{a_{n_k}} \xrightarrow{wp1} 1$. □

4.6 Infinite series of independent random variables.

Let (X_n) be a sequence of independent rvs. Let $S_n = \sum_1^n X_k$. The set A of points of convergence of the series $\sum_n X_n$ is a tail event of the sequence (X_n). Hence $P(A) = 0$ or 1.

Theorem 4.6.1. *(i) Let* $\mathbb{E}X_n = 0$; $\mathbb{E}X_n^2 = \sigma_n^2 < \infty$. *If* $\sum_1^\infty \sigma_n^2 < \infty$ *then* $\sum_n X_n(\omega)$ *converges* $wp1$.
(ii) If the $X_n s$ *are uniformly bounded and if* $\sum_n X_n$ *converges* $wp1$, *then the two series* $\sum_n \mathbb{E}X_n$ *and* $\sum_n varX_n$ *are convergent.*

Proof. Consider an arbitrary infinite series $\sum_n a_n$ of real numbers. Write $s_n = \sum_1^n a_k$. Then the series converges *iff* given $\varepsilon > 0$ there exists $N = N(\varepsilon)$ such that $|s_{n+k} - s_n| < \varepsilon$ for all $k \ge 1$ and all $n \ge N$. i.e. *iff* $t_n = \sup_{k\ge1} |s_{n+k} - s_n| \to 0$. Suppose $\inf_n t_n = 0$. This implies that, given $\varepsilon > 0$, an integer N can be found such that $t_N < \varepsilon$. i.e. $\sup_{k\ge1} |s_{N+k} - s_N| < \varepsilon$. Thus $\inf_n t_n = 0$ is a sufficient for the series to converge. That the condition is also necessary is easily established.

Let $\alpha_n(\omega) = \sup_{k\ge1} |S_{n+k}(\omega) - S_n(\omega)|$. Let $\alpha = \inf_n \alpha_n$. By the reasoning above, the series $\sum_n X_n$ will converge $wp1$ if $\alpha = 0$ $wp1$. If $T_{m;n} = \max_{1\le k\le m} |S_{n+k} - S_n|$ then $T_{m;n}(\omega) \uparrow \alpha_n(\omega)$ as $m \uparrow \infty$. Hence $T_{m;n} \xrightarrow{d} \alpha_n$. Thus if ε is a continuity point of the *df* of α_n, then

$$P(\alpha_n \geq \varepsilon) = \lim_{m\to\infty} P(T_{m;n} \geq \varepsilon) \leq \lim_{m\to\infty} \tfrac{1}{\varepsilon^2} \sum_{k=n+1}^{n+m} \sigma_k^2 \text{ (ref. } \mathbf{(4.4.3)}\text{)}$$
$$= \tfrac{1}{\varepsilon^2} \sum_{n+1}^{\infty} \sigma_k^2.$$

This inequality extends to all ε since a *df* is right continuous. Since for all n, $\alpha \leq \alpha_n$, we have

$$P(\alpha \geq \varepsilon) \leq \overline{\lim_{n\to\infty}}\, P(\alpha_n \geq \varepsilon) \leq \lim_{n\to\infty} \frac{1}{\varepsilon^2} \sum_{\nu=n+1}^{\infty} \sigma_\nu^2 = 0.$$

That $\alpha = 0\ wp1$ follows now since $\varepsilon > 0$ is arbitrary.
(ii) Let $\sup_n |X_n| \leq c\ wp1$ and let $\sum_n X_n$ converge $wp1$. Let (X_n') be an independent copy of the sequence (X_n). Write $Z_n = X_n - X_n'$ and note that $|Z_n| \leq 2c$, $\sum_n Z_n$ converges $wp1$, $\mathbb{E}Z_n = 0$ and $\tau_n^2 = var Z_n = 2\sigma_n^2$. Let $V_n = \sum_{k=1}^{n} Z_k$ and $b_n = \sup_{k\geq 1} |V_{n+k} - V_n|$. Then $\sum_n Z_n$ converges $wp1$ *iff* $b_n \stackrel{pr}{\to} 0$. We have, by the inequality in the second half of **(4.4.3)**,

$$\begin{aligned} P(b_n > \varepsilon) &= \lim_{m\to\infty} P(\max_{1\leq k\leq m} |V_{n+k} - V_n| > \varepsilon) \\ &\geq \lim_{m\to\infty} 1 - \frac{(\varepsilon + 2c)^2}{\sum_{k=n+1}^{n+m} \tau_k^2} \\ &= 1 - \frac{(\varepsilon + 2c)^2}{\sum_{k=n+1}^{\infty} \tau_k^2}. \end{aligned}$$

If $\sum_n \tau_n^2 = \infty$, we will have $P(b_n > \varepsilon) \geq 1$. But $\lim_{n\to\infty} P(b_n > \varepsilon) = 0$, since $b_n \stackrel{pr}{\to} 0$. Hence $\sum_n \tau_n^2 < \infty$. This implies $\sum_n \sigma_n^2 < \infty$. Let now $Y_n = X_n - \mathbb{E}X_n$. Then (Y_n) is a sequence of mutually independent rvs, $\mathbb{E}Y_n = 0$ and $\sum_n var Y_n = \sum_n \sigma_n^2 < \infty$. Hence by (i), $\sum_n Y_n$ converges $wp1$. Since by hypothesis $\sum_n X_n$ converges $wp1$. it follows that $\sum_n \mathbb{E}X_n$ is a convergent series. □

Definition 4.6.2. Two sequences (X_n) and (Y_n) of rvs are said to be equivalent in probability if $\sum_n P(X_n \neq Y_n) < \infty$.

This definition implies, by the Borel-Cantelli lemma, that $P(X_n \neq Y_n \text{ i.o}) = 0$. i.e., there exists a set A with $P(A) = 1$ such that if $\omega \in A$, then $X_n(\omega) = Y_n(\omega)$ for all n large, the largeness depending on ω. We note that if the two sequences (X_n) and (Y_n) are equivalent in probability, then the series $\sum_n X_n$ converges *iff* $\sum_n Y_n$ does.

Let X_n have standard Cauchy distribution. i.e., it has frequency function $\frac{1}{\pi}\frac{1}{1+x^2}$, $-\infty < x < \infty$. Let $Y_n = X_n$ if $|X_n| \leq n^2$ and zero otherwise. Then $P(X_n \neq Y_n) = P(|X_n| > n^2) = \frac{2}{\pi}\int_{n^2}^{\infty} \frac{1}{1+x^2}\, dx \leq \frac{2}{\pi n^2}$. Hence the two sequences (X_n) and (Y_n) are equivalent in probability. It is worthy of note that the X_ns do not possess even the mean while the Y_ns possess moments of all orders.

(ii) **Remark.** Let (X_n) be a sequence of independent variables. Let $X_1 \equiv 0$; $P(X_n = n) = P(X_n = -n) = \frac{1}{n^3}$; $P(X_n = 0) = 1 - \frac{2}{n^3}$, $n \geq 2$. We note that $\mathbb{E}X_n = 0$ and that $\sum_n varX_n = \sum_n \frac{2}{n} = \infty$. For any $c > 0$, $P(|X_n| \geq c) = \frac{2}{n^2}$ for all $n \geq N = N(c)$. Hence $\sum_n P(|X_n| \geq c) < \infty$. i.e. $P(X_n \neq 0 \ i.o.) = 0$. What is same, $P(X_n = 0 \text{ eventually}) = 1$. Hence $\sum_n X_n$ converges $wp1$.

This example does not violate **(4.6.1)**(i) since the conditions stated there are sufficient but not necessary. Again, this example does not violate **(4.6.1)**(ii)) since the X_ns are not uniformly bounded.

An example to enhance understanding of the theorem

Let (X_n) be a sequence of independent rvs with $P(X_n = \pm\frac{1}{n^2}) = \frac{1}{2}(1 - \frac{1}{n^2})$ each and $P(X_n = \pm n) = \frac{1}{2n^2}$ each. We note that $\mathbb{E}X_n = 0$ and $var(X_n) \sim 1$. Hence part (i) of the theorem does not apply to the sequence (X_n) and we are not able to say if $\sum_n X_n$ converges $wp1$ or with probability 0. Let $Y_n = X_n$ if $X_n = \pm\frac{1}{n^2}$ and 0 if $X_n = \pm n$. Then (Y_n) is a sequence of independent variables, $\mathbb{E}Y_n = 0$, $\mathbb{E}Y_n^2 \sim \frac{1}{n^4}$ and hence $\sum_n var(Y_n) < \infty$. Hence by part (i) of the theorem, $\sum_n Y_n$ converges $wp1$ Since $\sum_n P(X_n \neq Y_n) = \sum_n \frac{1}{n^2} < \infty$ the sequences (X_n) and (Y_n) are equivalent. Hence $\sum_n X_n$ converges $wp1$.

Theorem 4.6.3 (Kolmogorov's Three Series Theorem). *Let (X_n) be a sequence of independent rvs.*

($\mathfrak{p}$) $P(\sum_{n=1}^{\infty} X_n$ converges$) = 0$ or 1.

($\mathfrak{q}$) Let $c > 0$ be an arbitrary but a fixed number and let $A_n = \{\omega : |X_n(\omega)| > c\}$. Define $Y_n = X_n$ if $|X_n| \leq c$ and $Y_n = 0$ otherwise. Let $\mathbb{E}Y_n = \alpha_n$ and $varY_n = \beta_n^2$. If (i) $\sum_n P(A_n) < \infty$, (ii) $\sum_n \alpha_n$ is convergent and (iii) $\sum_n \beta_n^2 < \infty$ then $\sum_n X_n$ converges $wp1$.

($\mathfrak{r}$) Conversely if $\sum_n X_n$ converges $wp1$, then the above three series converge for every $c > 0$.

Proof. ($\mathfrak{p}$) Write $S_n = \sum_{k=1}^{n} X_k$. Then $\lim_{n\to\infty} S_n$ exists finitely *iff* (S_n) is a

Cauchy sequence. The event (S_n) is a Cauchy sequence i.e., the event that $\lim_{m,\, n\to\infty}(S_m - S_n) = 0$ is clearly a tail event of the independent variables X_n, $n \geq 1$. Hence, by the zero-one law, the claim follows.
q) By (i), $\sum_n P(X_n \neq Y_n) < \infty$. This implies by Borel-Cantelli lemma that $P(X_n = Y_n \text{ eventually}) = 1$. Hence it is enough to show that $\sum_n Y_n$ converges $wp1$. Since $\sum_n \alpha_n$ is a convergent series, it is enough to show that $\sum_n Z_n$ converges $wp1$ where $Z_n = Y_n - \alpha_n$. We note (Z_n) is a sequence of mutually independent rvs with $\mathbb{E}Z_n = 0$ and $\sum_n var Z_n = \sum_n \beta_n^2 < \infty$, Theorem **(4.6.1)** applies and the desired result follows.
(r) Suppose now $\sum_n X_n$ converges $wp1$ Hence $X_n \overset{wp1}{\longrightarrow} 0$. Since the X_ns are mutually independent this is equivalent to saying that $\sum_n P(|X_n| > c) < \infty$ for every $c > 0$. Fix c. Define the Y_ns as above and note that $\sum_n Y_n$ converges $wp1$ The convergence of $\sum_n \alpha_n$ and that of $\sum_n \beta_n^2$ follow now from **(4.6.1)**(ii). □

Remarks 4.6.4.
(0) It must be recorded that the theorem does not impose any moment condition on the X_n variables.
(i) Let X_n be a sequence of *iid* symmetric Bernoulli variables. Consider the series $\sum_{n=1}^{\infty} \frac{X_n}{n}$. Take $c = 1$ in the theorem and note that $Y_n = X_n$, that $\alpha_n = 0$, that $A_n = \phi$, that $\beta_n^2 = \frac{1}{n^2}$ and that all the three series of the theorem are convrgent. Hence the series $\sum_{n=1}^{\infty} \frac{X_n}{n}$ converges $wp1$. But the series $\sum_{n=1}^{\infty} \frac{X_n}{\sqrt{n}}$ converges $wp0$, since the third condition of the theorem is violated. For the same reason, the series $\sum_{n=1}^{\infty} X_n$ converges $wp\ 0$.
(ii) Let (X_n) be a sequence of independent normal variables with zero means and variances (σ_n^2). Then $\sum\limits_{n=1}^{\infty} X_n$ converges to a finite valued variable $wp1$ *iff* $\sum\limits_{n=1}^{\infty} \sigma_n^2 < \infty$.
Proof.
Let $S_n = \sum\limits_{k=1}^{n} X_k$ and $s_n^2 = \sum\limits_{k=1}^{n} \sigma_k^2$. If the series converges $wp1$ to the rv X, then $S_n \overset{d}{\to} X$. Examining the cf $e^{-\frac{1}{2}s_n^2 t^2}$ of S_n, we conclude that the weak convergence of S_n implies $\lim\limits_{n\to\infty} s_n^2 = \sum\limits_{n=1}^{\infty} \sigma_n^2 < \infty$.

Suppose now $\sum\limits_{n=1}^{\infty} \sigma_n^2 < \infty$. For any $c > 0$, $\sum\limits_{n} P(|X_n| > c) \leq \sum\limits_{n} \frac{1}{c^2}\mathbb{E}X_n^2 = \frac{1}{c^2}\sum\limits_{n} \sigma_n^2 < \infty$. Thus condition (i) of the theorem is satisfied. The other two conditions are easily verified. Hence the converse part follows.

(iii) Often convergence to a finite valued variable $wp0$ is described as divergence $wp1$. To understand the usage of this word better, let us have a closer look at the last mentioned series. Let $S_n = \sum_{k=1}^{n} X_k$. For $\lambda > 0$, arbitrary, $P(S_n > \lambda \text{ i.o.}) = 1$. A quick way to prove the validity of this claim is to appeal to the theorem at **(6.9.2)** with $\varphi(n) = \sqrt{2 \log \log n}$. By that theorem, $P(S_n > \sqrt{n} \text{ i.o.}) = 1$. Since $(S_n > \lambda) \supset (S_n > \sqrt{n})$, it follows that $P(S_n > \lambda \text{ i.o.}) = 1$. Since $\lambda > 0$ is arbitrary, we can conclude that $wp1$ $\overline{\lim\limits_{n\to\infty}} S_n = \infty$. Similarly, $wp1$ $\underline{\lim\limits_{n\to\infty}} S_n = -\infty$. Again, for every ε, $0 < \varepsilon < 1$, $P(|S_n| < \varepsilon) = P(S_n = 0)$, which is equal to 0 if n is odd and $P(S_{2n} = 0) = \binom{2n}{n}\frac{1}{2^{2n}} \sim \frac{c}{\sqrt{n}}$. Hence $\sum_n P(|S_n| < \varepsilon) = \infty$. This implies (ref. **(4.1.15)**) $P(|S_n| < \varepsilon \text{ i.o.}) = 1$. This being true for every $\varepsilon > 0$, it follows that $wp1$, 0 is a limit point of the sequence (S_n). Thus when $\sum_n X_n$ is said to diverge it means only that it converges to a finite valued variable with probability 0 and not that $\lim\limits_n S_n$ exists $wp1$ and equals $+\infty$ or $-\infty$. We have already noted in **(4.1.15)** that the last result $P(S_n = 0 \text{ i.o.}) = 1$ can be obtained from the theory of Markov chains. By the same theory (ref. **(7.3.6)**), we know $P(S_n = k \text{ i.o.}) = 1$ for every $k = \pm 1, \pm 2, \ldots\ldots$ These being countable in number, we conclude that $wp1$, $\{\pm\infty, \pm k,\ k = 0,\ 1,\ \ldots.\}$ is the limit set of the series $\sum\limits_n X_n$. Thus it is amply clear that the usage of the expression *diverges* (wp1) is not in conformity with its usage in the theory of infinite series.

(iv) The discussion below will further enhance our understanding. Let X_n be the iid Bernoulli variables discussed above. Let (S_n) denote the partial sum sequence corresponding to the series $\sum_n X_n$ and (T_n) that correspnding to the series $\sum_n (X_n - \theta)$ where θ is unique positive number such that $e^\theta = \mathbb{E}e^{X_1} = \frac{1}{2}(e + e^{-1})$. Both are series of iid variables. Hence, by the Three Series Theorem criteria, each of them converges to a finite valued variable with probability 0. So both are often described as divergent $wp1$. The case of (S_n) has been exhaustively discussed above. With the use of some sophisticated tools, we will show, in section **(6.4.2)**, that

$$T_n \stackrel{wp1}{\longrightarrow} -\infty. \tag{4.31}$$

Let (X_n) be a sequence of independent rvs. Let $S_n = \sum_1^n X_k$. In general (ref. **(4.3.2)**(ii) and (iv)) convergence in distribution does not imply convergence in probability and convergence in probability does not imply convergence $a.s.$ But we show in the next two sections that S_n converging in distribution implies that it converges $wp1$. Let φ_n be the *cf* of X_n.

Theorem 4.6.5. *$(S_n \stackrel{d}{\to} S) \Rightarrow (S_n \stackrel{pr}{\to} S)$.*

Proof. By the hypothesis there exists a *cf* φ such that $\prod_1^\infty \varphi_k(t) = \varphi(t)$ for every $t \in R$. Let $[-a,\ a]$ be an interval in which φ does not vanish. Hence, for $t \in [-a,\ a]$, $\lim \prod_m^n \varphi_k(t) = 1$ as $m,\ n \to \infty$. Or, what is same, $S_n - S_m \stackrel{d}{\to} 0$ as $m,\ n \to \infty$. This implies S_n converges in probability (ref.**(4.3.2)**(v) and **(4.3.2)**(xvi)). Denote the limit variable by S^*. Since convergence in probability always implies convergence in distribution (ref. **(4.3.2)**(iii)), $S^* = S\ wp1$. Hence $S_n \stackrel{pr}{\to} S$, since $P(|S_n - S| > 2\varepsilon) = P(|S_n - S^* + S^* - S| > 2\varepsilon) \leq P(|S_n - S^*| + |S - S^*| > 2\varepsilon) \leq P(|S_n - S^*| > \varepsilon) + P(|S - S^*| > \varepsilon) = P(|S_n - S^*| > \varepsilon) \to 0$. □

Theorem 4.6.6. *$(S_n \stackrel{pr}{\to} S) \Rightarrow (S_n \stackrel{wp1}{\longrightarrow} S)$.*

Proof. Fix $\frac{1}{2} > \varepsilon > 0$, the target error.

The proof will be done if we show (ref.**(4.3.2)**(xii)) $\sup_{j \geq n} |S_j - S| \stackrel{pr}{\to} 0$. Since $\sup_{j \geq n} |S_j - S| \leq \sup_{j \geq n} |S_j - S_n| + |S_n - S|$ and since $|S_n - S| \stackrel{pr}{\to} 0$, it is enough to show that $\sup_{j \geq n} |S_j - S_n| \stackrel{pr}{\to} 0$. i.e. we must find $N = N(\varepsilon)$ such that $P(A) < \varepsilon$ for all $n \geq N$ where $A = A(n) = \{\sup_{j \geq n} |S_j - S_n| > \varepsilon\}$.

Define event $A_k = A_{k,n} = \{\max_{n \leq j \leq k} |S_j - S_n| > \varepsilon\}$ and note that $A_k \uparrow A$ as $k \uparrow \infty$. Thus it is enough to show that, for every n larger than some N and for all $k > n$, $P(\max_{n \leq j \leq k} |S_j - S_n| > \varepsilon) < \varepsilon$.

The hypothesis implies (ref. remark under **(4.3.2)**(xvi)) that the sequence (S_n) is Cauchy in probability. There exists then $M = M(\varepsilon)$ such that $P(|S_n - S_m| > \varepsilon) < \varepsilon$ for all $n,\ m \geq M$. Let $M \leq n < j < k$. Taking the independent variables as $S_{n+1},\ X_{n+2},\ X_{n+3},\ \ldots,\ X_k$ in Skorohod inequality (ref.**(4.4.2)**) we get

$$P(\max_{n \leq j \leq k} |S_j - S_n| \geq 2\varepsilon) \leq \frac{1}{1-\varepsilon} P(|S_k - S_n| \geq \varepsilon) \leq \frac{\varepsilon}{1-\varepsilon} \leq 2\varepsilon.$$

The proof is completed by taking $N = M$. □

4.7 The weak law of large numbers

Let (X_n) be a sequence of independent *rvs*; (S_n) their partial sums. In **(4.5)** we established the limit in probability for the properly normed partial sums of a class of positive valued variables. Such a limit theorem is called a weak law of large numbers. Formally,

Definition 4.7.1. The weak law of large numbers (*WLLN*) is said to hold or to exist for the sequence (X_n) or the sequence (S_n) is said to be stable in probability if there exist real numbers a_n and $b_n \uparrow \infty$ such that $\frac{S_n}{b_n} - a_n \stackrel{pr}{\to} 0$.

Attention is drawn to the fact that the limit in probability, if it exists, can not be a non-degenerate rv Z. For, by **(4.3.2)**(viii) there exists a subsequence (n_k) such that $\frac{S_{n_k}}{b_{n_k}} - a_{n_k} \stackrel{wp1}{\longrightarrow} Z$. We note that the limit is measurable with respect to the tail σ-field of the X_n- sequence. Hence, by the Kolmogorov's *zero-one* law, Z is a constant $a.s.$ This constant can be absorbed in the constant a_n and the $WLLN$ would read as stated.

The burden of the theorems below is to describe explicitly the norming constants a_n and b_n in special cases or to determine the conditions the X_ns must satisfy so that a regularly varying b_n would be the right choice.

Theorem 4.7.2. *Let the X_ns be iid with common df F and common cf φ. In order that there exist constants a_n and regularly varying b_n with exponent $\alpha \geq 1$ such that $\frac{S_n}{b_n} - a_n \stackrel{pr}{\to} 0$ it is necessary and sufficient that $\lim_{r\to\infty} rP(|X_1| > b_r) = 0$. When the condition is satisfied a_n can be taken to be*

$$\frac{n}{b_n}\int_{-b_n}^{b_n} x\, dF(x).$$

Proof. Necessity. Let (Y_n) be an independent copy of the X_n sequence and let $T_n = \sum_1^n Y_k$. We note $\sum_{k=r}^m (X_k - Y_k)$, being the sum of independent symmetric variables, is symmetrically distributed. Hence $med\{\sum_{k=1}^m (X_k - Y_k)\} = 0$, The hypothesis implies $\frac{S_n - T_n}{b_n} \stackrel{pr}{\to} 0$. We have

$$|X_k - Y_k| = |S_k - T_k - (S_{k-1} - T_{k-1})| \leq |S_k - T_k| + |S_{k-1} - T_{k-1}|$$

where $S_0 = 0 = T_0$. This implies

$$\max_{1\leq k\leq n} |X_k - Y_k| \leq 2 \max_{1\leq k\leq n} |S_k - T_k|$$

leading to

$$\begin{aligned} P(\frac{1}{b_n}\max_{1\leq k\leq n} |X_k - Y_k| > \varepsilon) &\leq P(\frac{2}{b_n}\max_{1\leq k\leq n} |S_k - T_k| > \varepsilon) \\ &\leq 2P(|S_n - T_n| > \frac{b_n\varepsilon}{2}) \\ &\quad \text{(by Levy inequality, ref. (4.4.4)(ii))} \\ &\to 0. \end{aligned}$$

Define $\theta_n = P(|X_1 - Y_1| > b_n\varepsilon)$. Since

$$\begin{aligned}P(\frac{1}{b_n}\max_{1\le k\le n}|X_k - Y_k| > \varepsilon) &= 1 - P(\frac{1}{b_n}\max_{1\le k\le n}|X_k - Y_k| \le \varepsilon)\\ &= 1 - \{P(|X_1 - Y_1| \le b_n\varepsilon)\}^n\\ &= 1 - (1-\theta_n)^n\end{aligned}$$

we conclude that $(1-\theta_n)^n \to 1$. This implies $n\theta_n \to 0$. Thus $nP(|X_1 - Y_1| > b_n\varepsilon) \to 0$ for every $\varepsilon > 0$ and hence for $\varepsilon = 1$. With this the proof of necessity is complete for symmetric rvs.

Let m be a median of X_1. Define events $A = \{X_1 - m \ge b_r,\ Y_1 - m \le 0\}$ and $B = \{X_1 - m \le -b_r,\ Y_1 - m \ge 0\}$. We note $A,\ B$ are mutually exclusive. Further $A \subset (X_1 - Y_1 \ge b_r)$ and hence $A \subset (|X_1 - Y_1| \ge b_r)$. Similarly $B \subset (|X_1 - Y_1| \ge b_r)$. Thus $(|X_1 - Y_1| \ge b_r) \supseteq A \cup B$. Hence

$$\begin{aligned}P(|X_1 - Y_1| \ge b_r) &\ge P(A) + P(B)\\ &= P(X_1 - m \ge b_r)P(Y_1 \le m)\\ &\quad + P(X_1 - m \le -b_r)P(Y_1 \ge m)\\ &= \frac{1}{2}\{P(X_1 - m \ge b_r) + P(X_1 - m \le -b_r)\}\\ &= \frac{1}{2}P(|X_1 - m| \ge b_r) \ge \frac{1}{2}P(|X_1| \ge b_r + |m|).\end{aligned}$$

It follows that $(\lim_{r\to\infty} rP(|X_1 - Y_1| \ge b_r) = 0) \Rightarrow (\lim_{r\to\infty} rP(|X_1| \ge b_r) = 0)$ since $b_r \to \infty$ and m is finite. We note that in the proof above of the necessity part, the assumption that b_n is regularly varying is not required.

Sufficiency. Let the condition hold.

Write $b_n = n^\alpha l(n)$, l sv. We note that if $\beta \ge 0$ and L sv, then $\sum_1^n k^\beta L(k) \le qn^{\beta+1}L(n)$.

Define $Z_k = Z_{k,n} = X_k$ if $-b_n < X_k \le b_n$ and $Z_k = 0$ otherwise. We note that, for n fixed, the Z_ks are *iid*. Let $V_n = \sum_1^n Z_k$ and $\alpha_n = \mathbb{E}\frac{V_n}{b_n} = \frac{n}{b_n}\mathbb{E}Z_1 = \frac{n}{b_n}\int_{(-b_n,\, b_n]} x \,\mathrm{d}F(x)$;

$$\begin{aligned}&P(|\frac{S_n}{b_n} - \alpha_n| > \varepsilon)\\ &\quad = P(|\frac{S_n}{b_n} - \alpha_n| > \varepsilon,\ S_n = V_n) + P(|\frac{S_n}{b_n} - \alpha_n| > \varepsilon,\ S_n \ne V_n)\\ &\quad \le P(|\frac{V_n}{b_n} - \alpha_n| > \varepsilon) + P(S_n \ne V_n)\end{aligned}$$

$$\leq \frac{1}{b_n^2\varepsilon^2} var(V_n) + P(X_k \neq Z_k \text{ for some } k,\ 1 \leq k \leq n)$$

$$\leq \frac{n}{b_n^2\varepsilon^2}\mathbb{E}Z_1^2 + \sum_1^n P(X_k \neq Z_k)$$

$$\leq \frac{n}{b_n^2\varepsilon^2}\mathbb{E}Z_1^2 + nP(|X_1| \geq b_n).$$

$$\mathbb{E}Z_1^2 \leq \int\limits_{|x|\leq b_n} x^2 \,\mathrm{d}F(x) = \sum_1^n \int\limits_{b_{k-1}<|x|\leq b_k} x^2 \,\mathrm{d}F(x)$$

$$\leq \sum_1^n b_k^2 P(b_{k-1} < |X_1| \leq b_k)$$

$$= \sum_1^n (k^\alpha l(k))^2 P(b_{k-1} < |X_1| \leq b_k)$$

$$\leq q\sum_1^n \sum_{j=1}^k j^{2\alpha-1} l^2(j) P(b_{k-1} < |X_1| \leq b_k)$$

$$= q\sum_{j=1}^n j^{2\alpha-1} l^2(j) \sum_{k=j}^n P(b_{k-1} < |X_1| \leq b_k)$$

$$= q\sum_{j=1}^n j^{2\alpha-1} l^2(j) P(b_{j-1} < |X_1| \leq b_n)$$

$$\leq q\sum_{j=1}^n j^{2\alpha-1} l^2(j) P(|X_1| > b_{j-1})$$

$$\leq q\sum_{j=1}^n j^{2\alpha-2} l^2(j) jP(|X_1| > b_j).$$

Given $\delta > 0$, use hypothesis to find N such that $jP(|X_1| > b_j) < \delta$ for all $j \geq N$. We have, for all n large,

$$P(|\frac{S_n}{b_n} - \alpha_n| > \varepsilon) \leq \frac{n}{b_n^2\varepsilon^2}\{q_1 + \sum_N^n j^{2\alpha-2} l^2(j)\delta\} + nP(|X_1| \geq b_n)$$

$$\leq o(1) + q\delta\frac{n}{b_n^2} n^{2\alpha-1} l^2(n) \leq q\delta. \qquad \square$$

Remarks 4.7.3.

(i) The case $b_n = n$ merits special mention and admits of a simple proof based on theorems we have already established.

Let (Y_n) be an indepedent copy of the X_n- sequence and $\psi(t) = |\varphi(t)|^2$ Let $T_n = \sum_{j=1}^n Y_j$, $U_n = S_n - T_n$ and G the *df* of U_1. Denote by m a median of F.

Necessity of the condition.
$(\frac{S_n}{n} - a_n \overset{pr}{\to} 0) \Rightarrow (\frac{U_n}{n} \overset{pr}{\to} 0)$. Now,

$$\begin{aligned}
\frac{U_n}{n} \overset{pr}{\longrightarrow} 0 &\Leftrightarrow \psi^n(\frac{t}{n}) \to 1 \text{ for all } t \\
&\Leftrightarrow n\{1 - \psi(\frac{t}{n})\} \to 0 \Leftrightarrow \psi'(0) \text{ exists finitely (and} = 0) \\
&\Leftrightarrow \{nP(|U_1| \geq n) \to 0\} \text{ (ref. } \mathbf{(3.6.7)}).
\end{aligned}$$

For r sufficiently large so $|m| < \frac{r}{2}$, we have
$$P(|U_1| \geq r) \geq \tfrac{1}{2}P(|X_1 - m| \geq r) \geq \tfrac{1}{2}P(|X_1| \geq \tfrac{3r}{2}).$$
From this the necessity of the condition follows.

Proof of the sufficiency of the condition will be on the earlier lines. Let

$$\lim_{n\to\infty} nP(|X_1| > n) = 0 \tag{4.32}$$

With the same notation as above,

$$\begin{aligned}
\mathbb{E}Z_1^2 &\leq \int\limits_{|x|\leq n} x^2 \,\mathrm{d}F(x) = \sum_{k=1}^{n} \int\limits_{k-1<|x|\leq k} x^2 \,\mathrm{d}F(x) \\
&\leq \sum_{k=1}^{n} k^2 P(k-1 < |X_1| \leq k) \\
&\leq q \sum_{k=1}^{n} \sum_{j=1}^{k} jP(k-1 < |X_1| \leq k) \\
&\leq q \sum_{j=1}^{n} j \sum_{k=j}^{n} P(k-1 < |X_1| \leq k) \\
&\leq q \sum_{j=1}^{n} jP(|X_1| > j-1) \\
&= q \sum_{j=0}^{n-1} (j+1)P(|X_1| > j) \\
&\leq q \sum_{j=1}^{n} jP(|X_1| > j) + q \sum_{j=0}^{n} P(|X_1| > j).
\end{aligned}$$

The first sum is $o(n)$ by (4.32). The limit, as $n \to \infty$, of second sum divided by n is the Cesaro limit of the sequence $(P(|X_1| > n))$ which tends to zero. Hence the second sum too is $o(n)$.

Since

$$P(|\frac{S_n}{n} - \alpha_n| > \varepsilon) \leq \frac{1}{n\varepsilon^2}\mathbb{E}Z_1^2 + nP(|X_1| > n) = o(1),$$

the proof of the sufficiency part is complete.
(ii) If $\frac{S_n}{n} - a_n \overset{pr}{\to} 0$, then $\mathbb{E}|X_1|^\alpha < \infty,\ 0 \leq \alpha < 1$.
Proof. The hypothesis implies (4.32). Let $0 \leq \alpha < 1$. We note $\mathbb{E}|X_1|^\alpha < \infty$ *iff* $I = \int\limits_0^\infty x^{\alpha-1}P(|X_1| \geq x)\ dx$ exists finitely. Over the region $0 < x < 1$ the integrand is bounded by the integrable function $x^{\alpha-1}$. Over the region $[1, \infty)$ appeal to (4.32) and conclude that the integrand is bounded by $x^{\alpha-1}\frac{1}{x}$. i.e. by $\frac{1}{x^{2-\alpha}}$ which is integrable since $2 - \alpha > 1$. The claim follows now. □
(iii) Let F have frequency function $\frac{1}{2|x|^2}$ if $|x| \geq 1$ and 0 elsewhere. In this case $\int_{-n}^{n} x\ dF(x) = 0$ but $nP(|X| \geq n) = 1$. Hence WLLN *with norming constant n* does not hold. With b_n such that $n = o(b_n)$, WLLN holds. i.e. $\frac{S_n}{b_n} \overset{pr}{\to} 0$.

If F has frequency function $\frac{c}{y^2 \log |y|}$ for $|y| \geq e$ and zero elsewhere, then

$$nP(|x| \geq n) = 2cn \int\limits_n^\infty \frac{1}{y^2 \log y}\ dy \sim qn\frac{1}{n \log n} \to 0.$$

Hence in this case $WLLN$ holds : $\frac{S_n}{n} \overset{pr}{\to} 0$.
(iv) Suppose $\frac{S_n}{n} \overset{pr}{\to} 0$. Does complete convergence hold? (ref. **(4.3.1)**(iv) and **(4.3.1)**(xiii)). If not, what additional condition needs to be satisfied so that $\sum_n P(|\frac{S_n}{n}| > \varepsilon) < \infty$ for every $\varepsilon > 0$? The proof that a necessary and sufficient condition for complete convergence is that $\mathbb{E}X_1^2 < \infty$ is beyond the scope of this book. A proof that complete convergence holds if $\mathbb{E}|X_1|^{2+\delta} < \infty$ for some $\delta > 0$ will be given in the next chapter. (ref. **(5.3.8)**)

Here we record some consequences of the complete convergence of $\frac{S_n}{n}$ to 0. Since $|X_n| = |S_n - S_{n-1}| \leq |S_n| + |S_{n-1}|$, it follows that $\sum_n P(|X_n| > n) < \infty$. This implies (ref. **(2.6.4)**) that $\mathbb{E}|X_1| < \infty$. Let $\mathbb{E}X_1 = \mu$. The general term $P(|X_n| \geq n) = P(|X_1| \geq n)$ is a monotonic decreasing function of n. Hence the convergence of the series implies $nP(|X_1| \geq n) \to 0$. This together with the fact that $\lim\limits_{n\to\infty} \int\limits_{-n}^{n} x\ dF(x) = \mu$, implies by **(4.7.2)** that $\frac{S_n}{n} - \mu$

$\xrightarrow{pr} 0$. Since it has already been noted (ref. **(4.3.2)**(xiii) and **(4.3.2)**(i)) that $\frac{S_n}{n} \xrightarrow{pr} 0$, we conclude $\mu = 0$.

(v) Let X_1 be positive valued and let $L(x) = \int_0^x \{1 - F(y)\}\, dy$ be sv. Let $M(x) = \int_0^x y \, dF(y)$. Define b_n asymptotically from the condition $\frac{L(b_n)}{b_n} \sim \frac{1}{n}$. In **(4.5)** we showed that $\frac{S_n}{n} - 1 \xrightarrow{pr} 0$. The proof was by Laplace transform, using some intrinsic properties of sv functions. An alternative proof can be constructed as follows using **(4.7.2)**. According to this $\frac{S_n}{b_n} - \frac{n}{b_n} \int_0^{b_n} x \, dF(x) \xrightarrow{pr} 0$ if $nP(|X_1| \geq b_n) \to 0$. In Step 1 of **(4.5)** it was shown that $\lim_{x\to\infty} \frac{x\{1-F(x)\}}{L(x)} = 0$. Hence for the choice of b_n as above, $\frac{b_n\{1-F(b_n)\}}{L(b_n)} \to 0$. i.e. $nP(X_1 \geq b_n) \to 0$. Since $\frac{n}{b_n} \int_0^{b_n} x \, dF(x) = \frac{n}{b_n} M(b_n) \sim \frac{n}{b_n} L(b_n) \to 1$, it follows then $\frac{S_n}{b_n} \xrightarrow{pr} 1$.

(vi) For a variational sequence of positive variables we have a WLLN as follows.

Theorem Let (X_n) be a sequence of independent non-negative rvs with dfs (F_n) and positive, finite means (μ_n). Write $s_n = \sum_1^n \mu_k$ and for each $\varepsilon > 0$ let

$$\frac{1}{s_n} \sum_1^n \int_0^{\varepsilon s_n} x \, dF_k(x) \to 1$$

or, equivalently

$$\frac{1}{s_n} \sum_1^n \int_{\varepsilon s_n}^{\infty} x \, dF_k(x) = o(1).$$

Then $\frac{S_n}{s_n} \xrightarrow{pr} 1$ where $S_n = \sum_1^n X_j$.

Proof. Define $Y_{j,n} = X_j$ if $X_j \leq s_n$ and equal to zero otherwise, $1 \leq j \leq n$. Write $Y_n = \sum_1^n Y_{j,n}$. Let $\varepsilon > 0$, small, be given. Then

$$P(|S_n - Y_n| > \varepsilon s_n) \leq P\,(X_j > \varepsilon s_n \text{ for at least one } j,\ 1 \leq j \leq n)$$
$$\leq \sum_1^n \int_{\varepsilon s_n}^{\infty} dF_j(x) \leq \frac{1}{\varepsilon s_n} \sum_1^n \int_{\varepsilon s_n}^{\infty} t \, dF_k(t) = o(1).$$

i.e. $\left|\frac{S_n}{s_n} - \frac{Y_n}{s_n}\right| \xrightarrow{pr} 0$. Now

$$var\left(\frac{Y_n}{s_n}\right) = \frac{1}{s_n^2} var(Y_n) = \frac{1}{s_n^2}\sum_{j=1}^{n} var(Y_{j,n}) \leq \frac{1}{s_n^2}\sum_{j=1}^{n} \mathbb{E}Y_{j,n}^2$$

$$= \frac{1}{s_n^2}\sum_{j=1}^{n}\int_0^{s_n} t^2\, dF_j(t) = \frac{1}{s_n^2}\sum_{k=1}^{n}\int_0^{\varepsilon s_n} t^2\, dF_k(t) + \frac{1}{s_n^2}\sum_{k=1}^{n}\int_{\varepsilon s_n}^{s_n} t^2\, dF_k(t)$$

$$\leq \frac{\varepsilon}{s_n}\sum_{1}^{n}\int_0^{\varepsilon s_n} t\, dF_k(t) + \frac{1}{s_n}\sum_{1}^{n}\int_{\varepsilon s_n}^{\infty} t\, dF_k(t) \leq \varepsilon + o(1).$$

It follows from this that $\frac{Y_n - EY_n}{s_n} \xrightarrow{pr} 0$. Since $EY_n = \sum_1^n \int_0^{s_n} u\, dF_k(u) \sim s_n$ we can conclude that $\frac{Y_n}{s_n} \xrightarrow{pr} 1$. This together with the earlier result that $\frac{S_n}{s_n} - \frac{Y_n}{s_n} \xrightarrow{pr} 0$ implies $\frac{S_n}{s_n} \xrightarrow{pr} 1$. □

(vii) Let (X_n) be a sequence of *iid* variables and let $S_n = \sum_1^n X_k$. Then, unless the variables are degenerate at 0, for no $\beta,\ 0 < \beta \leq \frac{1}{2}$ can $\frac{S_n}{n^\beta} \xrightarrow{pr} 0$.

Given $\beta,\ \frac{1}{2} < \beta \leq 1$ a sequence of *iid* variables can always be found such that $\frac{S_n}{n^\beta} \xrightarrow{pr} 0$.

Proof. If possible let $\frac{S_n}{n^\beta} \xrightarrow{pr} 0$ for some $\beta \in (0,\ \frac{1}{2}]$. Let (Y_n) be an independent copy of the sequence (X_n). Let g be the *cf* of $X_1 - Y_1$ and let $V_n = \sum_1^n (X_k - Y_k)$. The hypothesis implies $\frac{V_n}{n^\beta} \xrightarrow{pr} 0$. Hence $\frac{V_n}{n^{\frac{1}{2}}} \xrightarrow{pr} 0$. This can be written $\frac{V_{[u]}}{u^{\frac{1}{2}}} \xrightarrow{pr} 0$ as $u \to \infty$. This is equivalent to $g^u(\frac{t}{u^{\frac{1}{2}}}) \to 1$ for every t. Or, $\lim_{u\to\infty} u\{1 - g(\frac{1}{\sqrt{u}})\} = 0$. Or, $\lim_{t\to 0} \frac{1-g(t)}{t^2} = 0$. But this is equivalent to $var(X_1 - Y_1) = 0$. That would imply X_1 is degenerate at 0.

Let $\beta \in (\frac{1}{2},\ 1)$ be given. Choose $\beta_1,\ \frac{1}{2} < \beta_1 < \beta < 1$. Let $\theta = 1 + \frac{1}{\beta_1}$. We note $2 < \theta < 3$. Let *rv* X have frequency function $\frac{\theta-1}{2|x|^\theta},\ |x| \geq 1$. If g is its *cf*, $1 - g(t) = 2\int_1^\infty \frac{\theta-1}{2}\frac{1-\cos tx}{x^\theta}\, dx$. i.e.

$$\frac{1-g(t)}{|t|^{\theta-1}} = (\theta - 1)\int_t^\infty \frac{1-\cos x}{x^\theta}\, dx \to (\theta-1)\int_0^\infty \frac{1-\cos x}{x^\theta}\, dx, \text{ as } t \to 0.$$

The integral is finite and positive, since $\frac{1-\cos x}{x^\theta} \sim \frac{1}{2x^{\theta-2}}$ as $x \to 0$ and $0 <$

$\theta - 2 < 1$. Thus $\frac{1-g(t)}{|t|^{\theta-1}} \to q,\ q > 0$. Hence

$$\lim_{n\to\infty} \frac{(n^{\beta_1})^{\theta-1}}{|t|^{\theta-1}}\{1 - g(\frac{t}{n^{\beta_1}})\} = q.$$

Or,

$$\lim_{n\to\infty} n\{1 - g(\frac{t}{n^{\beta_1}})\} = q|t|^{\theta-1}.$$

Or, $g^n(\frac{t}{n^{\beta_1}}) \to e^{-q|t|^{\theta-1}}$, a continuous function. This is equivalent to $\frac{S_n}{n^{\beta_1}}$ converging weakly to a non-degenerate rv. Since $\beta > \beta_1$, it follows $\frac{S_n}{n^{\beta}} \overset{pr}{\to} 0$.

In **(4.7.5)** below we take a closer look at the convergence $\frac{S_n}{n} \overset{pr}{\to} 0$ in some more detail. This convergence is, by definition, $\lim\limits_{n\to\infty} a_n(\varepsilon) = 0$ for every $\varepsilon > 0$ where $a_n(\varepsilon) = P(|\frac{S_n}{n}| > \varepsilon)$. We saw earlier that this implies the finiteness of $E|X|^{\theta}, 0 \le \theta < 1$ but $E|X_1|$ may be infinite. (The example at **(3.6.5)** illustrates this point) It therefore seems reasonable to expect that $E|X_1|$ may be finite if $a_n(\varepsilon) \to 0$ fast enough. The proof that it is indeed so is presented in sec. **(4.7.5)** for symmetric variables and in sec. **(4.7.6)** for general variables. Now for the proof of **(4.7.5)** we need the following lemma.

Lemma 4.7.4. *Define $a_n = P(|\frac{S_n}{n}| > 1)$ and $b_n = nP(|X_1| > n)$. Then $b_{2n} = O(a_n)$.*

Proof. Define $B_{k,n} = \{\omega : |S_n - X_k| < n\}$; $C_{k,n} = \{\omega : |X_k| > 2n\}$ and note that $P(C_{k,n}) = \frac{b_{2n}}{2n}$ and that $P(B_{k,n}) = P(B_{1,n})$ and $P(C_{k,n}) = P(C_{1,n})$. Choose N such that, for $n \ge N$, $P(B_{1,n}) > \frac{1}{2}$, which is possible because $\frac{S_n}{n} \to 0$ in probabilty. Observe that (i) the $C_{j,n}$s, $1 \le j \le n$ are independent events (ii) $B_{k,n}$ and $C_{k,n}$ are independent events for each k and (iii) $D_{k,n} = B_{k,n} \cap C_{k,n} \subseteq A_n = \{\omega : |S_n| > n\}$ for each k. Now

$$a_n = P(A_n) \ge P(\cup_1^n D_{k,n}) = P(D_{1,n}) + \sum_{j=2}^{n} P(\{\cap_1^{j-1} D'_{r,n}\} \cap D_{j,n})$$

$$\ge P(D_{1,n}) + \sum_{j=2}^{n} P(\{\cap_1^{j-1} C'_{r,n}\} \cap D_{j,n})$$

$$= P(D_{1,n}) + \sum_{j=2}^{n} \{P(D_{j,n}) - P(D_{j,n} \cap \cup_1^{j-1} C_{r,n})\}$$

$$\ge P(B_{1,n})P(C_{1,n}) + \sum_{j=2}^{n} \{P(D_{j,n}) - P(C_{j,n} \cap \cup_1^{j-1} C_{r,n})\}$$

$$= P(B_{1,n})P(C_{1,n}) + \sum_{j=2}^{n}\{P(B_{j,n})P(C_{j,n}) - P(C_{j,n})P(\cup_1^{j-1}C_{r,n})\}$$

$$= P(B_{1,n})P(C_{1,n}) + \sum_{j=2}^{n} P(C_{j,n})\{P(B_{j,n}) - P(\cup_1^{j-1}C_{r,n})\}$$

$$\geq P(B_{1,n})P(C_{1,n}) + \sum_{j=2}^{n} P(C_{j,n})\{P(B_{j,n}) - \sum_{r=1}^{j-1} P(C_{r,n})\}$$

$$\geq P(B_{1,n})P(C_{1,n}) + \sum_{j=2}^{n} P(C_{1,n})\{P(B_{1,n}) - nP(C_{1,n})\}$$

$$\geq nP(C_{1,n})\{\frac{1}{2} - \frac{1}{4}\} = \frac{b_{2n}}{8}. \qquad \square$$

Theorem 4.7.5. *If the X_n's are iid symmetric variables and if $\sum_{n=1}^{\infty}\frac{1}{n}P(|\frac{S_n}{n}| > \varepsilon) < \infty$ for every $\varepsilon > 0$, then $\frac{S_n}{n} \to 0$ in probability. Further $E|X_1| < \infty$. And then $EX_1 = 0$.*

Proof. If the claim is not true, there would exist $\eta > 0$ such that $\overline{\lim_{n\to\infty}} P(|\frac{S_n}{n}| > 2\eta) = c > 0$. A subsequence (n_k) can be found such that $\lim_{k\to\infty} P(|\frac{S_{n_k}}{n_k}| > 2\eta) = c$. There exists then $K = K(\eta)$ such that for all $k \geq K$, $P(|\frac{S_{n_k}}{n_k}| > 2\eta) \geq \frac{c}{2}$. i.e., $P(\frac{S_{n_k}}{n_k} > 2\eta) \geq \frac{c}{4}$, by reason of symmetry. Taking a further subsequence if necessary, assume $n_{k+1} > 2n_k$. For $n_k < n \leq 2n_k$, we have

$$P(S_n > n\eta) = P(S_{n_k} + \sum_{n_k+1}^{n} X_j > n\eta) \geq P(S_{n_k} \geq n\eta, \sum_{n_k+1}^{n} X_j \geq 0)$$

$$= P(S_{n_k} > n\eta)P(\sum_{n_k+1}^{n} X_j \geq 0) \text{ (using independence)}$$

$$\geq \frac{1}{2}P(S_{n_k} > n\eta) \text{ (reason : symmetry of the variables)}$$

$$\geq \frac{1}{2}P(S_{n_k} > 2n_k\eta) \geq \frac{c}{8}.$$

This leads to:

$$\sum_{n=1}^{\infty}\frac{1}{n}P(|\frac{S_n}{n}| > \eta) \geq \sum_{k=1}^{\infty}\sum_{n=n_k+1}^{2n_k}\frac{1}{n}P(|\frac{S_n}{n}| > \eta)$$

$$\geq \sum_{k=1}^{\infty}\sum_{n=n_k+1}^{2n_k}\frac{1}{2n_k}\frac{c}{8} \geq \sum_{k=1}^{\infty}\frac{c}{16} = \infty.$$

This contradiction establishes that $\frac{S_n}{n} \to 0$ in probability.

By **(4.7.4)**, it now follows that $P(|X_1| > n) = O(\frac{1}{n}P(|\frac{S_n}{n}| > 1))$. Hence $\sum_{n=1}^{\infty} P(|X_1| > n) < \infty$, which is equivalent to $E|X_1| < \infty$. By **(4.7.3)**(i), $\varphi'(0)$ exists and is equal to 0. But when $\mathbb{E}|X_1| < \infty$, $\varphi'(0) = i\mathbb{E}X_1$. Hence $\mathbb{E}X_1 = 0$. □

Theorem 4.7.6. *Let (X_n) be an iid sequence and let there exist some number λ such that $\sum_{n=1}^{\infty} \frac{1}{n}P(|S_n - n\lambda| > n\varepsilon) < \infty$ for every $\varepsilon > 0$. Then $\mathbb{E}|X_1| < \infty$ and $\mathbb{E}X_1 = \lambda$.*

Proof. Let (Y_n) be an independent copy of the X_n sequence. Write $Z_n = X_n - Y_n$; $T_n = \sum_{j=1}^{n} Y_j$; $V_n = S_n - T_n = \sum_{j=1}^{n} Z_j$. Now $P(|V_n| > n\varepsilon) = P(|(S_n - \lambda n) - (T_n - \lambda n)| > n\varepsilon) \leq P(|S_n - \lambda n| > \frac{1}{2}n\varepsilon) + P(|T_n - \lambda n| > \frac{1}{2}n\varepsilon) = 2P(|S_n - \lambda n| > \frac{1}{2}n\varepsilon)$. This inequality and the hypothesis imply $\sum_{n=1}^{\infty} \frac{1}{n}P(|V_n| > n\varepsilon) < \infty$ for every $\varepsilon > 0$. Theorem **(4.7.5)** applies and we have $\mathbb{E}|Z_1| < \infty$. Hence $\mathbb{E}|X_1| < \infty$. Suppose $\mathbb{E}X_1 = \mu \neq \lambda$. Let $\varepsilon < \frac{1}{2}|\lambda - \mu|$. Now, let φ be the *cf* of X_1. We claim $\frac{S_n}{n} \overset{pr}{\to} \mu$. This would be true if the *cf* $\varphi^n(\frac{t}{n})$ of $\frac{S_n}{n}$ converges to $e^{i\mu t}$ for each $t \in R$ (ref. **(4.3.2)**(v) and **(2.3.5)**). Now, $\lim_{n\to\infty} \varphi^n(\frac{t}{n}) = e^{i\mu t}$ if $\lim_{n\to\infty}\{-n(1 - \varphi(\frac{t}{n}))\} = i\mu t$. i.e., if $-\lim_{n\to\infty} \frac{1-\varphi(\frac{t}{n})}{\frac{t}{n}} = i\mu$. i.e., if $\varphi'(0) = i\mu$, which is true (ref. **(3.6.1)**). This implies $P(|\frac{S_n}{n} - \mu| < \varepsilon) \geq \frac{1}{2}$ for all n large. Since $\{\omega : |\frac{S_n(\omega)}{n} - \mu| < \varepsilon)\}$ $\subseteq \{\omega : |[|\frac{S_n(\omega)}{n} - \lambda| - |\lambda - \mu|]| < \varepsilon\} \subseteq \{\omega : -\varepsilon + |\lambda - \mu| < |\frac{S_n}{n}(\omega) - \lambda|\}$ $\subseteq \{\omega : |\frac{S_n(\omega)}{n} - \lambda| > \varepsilon\}$, we have, for all n large, $\frac{1}{2} \leq P(|\frac{S_n}{n} - \mu| < \varepsilon) \leq P(|\frac{S_n}{n} - \lambda| > \varepsilon)$. This violates the convergence of $\sum_{n=1}^{\infty} P(|S_n - n\lambda| > n\varepsilon)$. Hence $\lambda = \mu = \mathbb{E}X_1$. □

Theorem 4.7.7. *Let (X_n) be an iid sequence with $\mathbb{E}|X_1| < \infty$ and $\mathbb{E}X_1 = \mu$. Then for every $\varepsilon > 0$,*

$$\sum_{n=1}^{\infty} \frac{1}{n}P(|\frac{S_n}{n} - \mu| > \varepsilon) < \infty.$$

Proof. It is clear that with no loss of generality we may take $\mu = 0$. Define *rvs* $X_{j,n}$, $1 \leq j \leq n$ thus:

$$X_{j,n} = \begin{cases} X_j & \text{if } |X_j| < n \\ 0 & \text{if } |X_j| \geq n \end{cases}$$

For each n fixed, the $X_{j,n}$s are independent and identically distributed. Also $EX_{j,n} = EX_{1,n} \to EX_1 = 0$. Considering the variables $\frac{X_j}{\varepsilon}$ in place of the

X_js, we see that it is enough to show that $\sum_{n=1}^{\infty} \frac{1}{n} P(|\frac{S_n}{n}| > 1) < \infty$. Now,

$$\{|S_n| > n\} = \{|S_n| > n, |X_k| \geq n \text{ for some } k, 1 \leq k \leq n\} \\ \cup \{|S_n| > n, |X_k| < n \text{ for all } k, 1 \leq k \leq n\}.$$

Hence

$$\begin{aligned} P(|S_n| > n) &\leq P(|X_k| \geq n \text{ for some } k, 1 \leq k \leq n) \\ &\quad + P(|\sum_{k=1}^{n} X_{k,n}| > n) \\ &\leq nP(|X_1| \geq n) + P(|\sum_{k=1}^{n} X_{k,n}| > n). \end{aligned}$$

Since $E|X_1| < \infty$, it follows that $\sum_{n=1}^{\infty} P(|X_1| \geq n) < \infty$. Hence it remains to show that $\sum_{n=1}^{\infty} \frac{1}{n} P(|\sum_{k=1}^{n} X_{k,n}| > n) < \infty$. Now,

$$\begin{aligned} P(|\sum_{k=1}^{n} X_{k,n}| > n) &= P(|\sum_{k=1}^{n} \{X_{k,n} - \mathbb{E}X_{k,n} + \mathbb{E}X_{k,n}\}| > n) \\ &\leq P(|\sum_{k=1}^{n} (X_{k,n} - \mathbb{E}X_{k,n})| + |\sum_{k=1}^{n} \mathbb{E}X_{k,n}| > n) \\ &\leq P(|\sum_{k=1}^{n} X_{k,n} - \mathbb{E}X_{k,n}| + n|\mathbb{E}X_{1,n}| > n) \\ &\leq P(|\sum_{k=1}^{n} X_{k,n} - \mathbb{E}X_{k,n}| + \frac{1}{2}n > n) \\ &\qquad \text{(for all } n \text{ large; since } \mathbb{E}X_{1,n} \to 0) \\ &= P(|\sum_{k=1}^{n} X_{k,n} - \mathbb{E}X_{k,n}| > \frac{1}{2}n) \\ &\leq \frac{4}{n^2} \sum_{k=1}^{n} \mathbb{E}(X_{k,n} - EX_{k,n})^2 \text{ (due to independence)} \\ &\leq \frac{4}{n^2} \sum_{k=1}^{n} \mathbb{E}X_{k,n}^2 = \frac{4}{n} \mathbb{E}X_{1,n}^2. \end{aligned}$$

Thus we have to show that

$$\sum_{n=1}^{\infty} \frac{1}{n^2} \mathbb{E}X_{1,n}^2 < \infty. \tag{4.33}$$

This series is

$$= \sum_{n=1}^{\infty} \frac{1}{n^2} \sum_{k=1}^{n} \int_{k-1 \le |X(\omega)| < k} X^2(\omega) dP(\omega)$$

$$\le \sum_{n=1}^{\infty} \frac{1}{n^2} \sum_{k=1}^{n} k^2 P(k-1 \le |X| < k)$$

$$= \sum_{n} n^2 P(n-1 \le |X| < n) \{ \sum_{k=n}^{\infty} \frac{1}{k^2} \}.$$

This series is convergent *iff*

$$\sum_{n} nP(n-1 \le |X| < n) < \infty. \tag{4.34}$$

Setting $b_k = P(|X_1| \ge k)$, $k \ge 0$, the sum of the first n terms of the series in (4.34) is $\sum_{k=1}^{n} k(b_{k-1} - b_k) = \sum_{k=0}^{n-1} b_k - nb_n$.

Now, $\mathbb{E}|X| < \infty$ implies $\sum_n b_n = \sum_n P(|X| \ge n) < \infty$. This, in turn, implies $\lim_{n \to \infty} nP(|X| \ge n) = 0$, since the nth term $P(|X| \ge n)$ of the series monotonically decreases to 0. Thus (4.34), and hence (4.33) are proved. □

4.8 The strong law of large numbers

In **(4.7.1)**, we defined stability in probability or the $WLLN$. Parallel to that we say:

Definition 4.8.1. The strong law of large numbers (*SLLN*) is said to hold or to exist for the sequence (X_n) or the sequence (S_n) is said to be stable $wp1$ if there exist real numbers a_n and $b_n \uparrow \infty$ such that $\frac{S_n}{b_n} - a_n \xrightarrow{wp1} 0$.

Needless to add that, where possible, one should prove the $SLLN$ instead of the $WLLN$. To see that it is not always possible ref. **(4.8.5)**.

Let (τ_n) be a sequence of independent non-negative rvs. Let *df* of τ_n be V_n and let $\mathbb{E}\tau_n = \alpha_n < \infty$. Write $T_n = \sum_1^n \tau_k$, $\beta_n = \sum_1^n \alpha_k$. Assume

$$\inf_n \frac{\beta_n}{n} > 0. \tag{4.35}$$

We assume further that there exists a *df* V on $[0, \infty)$ with

$$\int_0^{\infty} x \, \mathrm{d}V(x) < \infty. \tag{4.36}$$

and that

$$\sup_n \int_{x\geq r} \mathrm{d}V_n(x) = O\Big(\int_{x\geq r} \mathrm{d}V(x)\Big) \tag{4.37}$$

$$\sup_n \int_{x\geq r} x\,\mathrm{d}V_n(x) = O\Big(\int_{x\geq r} x\,\mathrm{d}V(x)\Big) \tag{4.38}$$

We note, for future applications, that (4.37) is equivalent to saying that there exists an absolute constant A such that for all r,

$$\sup_n \int_{x\geq r} \mathrm{d}V_n(x) \leq A \int_{x\geq r} \mathrm{d}V(x).$$

(4.38) has a similar meaning.
If $u(r) = \int_{x\geq r} x\,\mathrm{d}V(x)$, then we note that $u(t) \downarrow 0$ as $t \uparrow \infty$ and that, for every choice of $a > 0$ and $c > 1$,

$$(4.36) \Leftrightarrow \sum_n u(an) < \infty \Leftrightarrow \sum_n c^n u(ac^n) < \infty \tag{4.39}$$

$$(4.38) \Rightarrow \sup_n \alpha_n < \infty. \tag{4.40}$$

This together with (4.35) implies that there exist numbers $\beta < \alpha$ such that for all n

$$0 < \beta \leq \frac{n}{\beta_n} \leq \alpha < \infty \tag{4.41}$$

Since $\alpha_n > 0$, $\beta_n < \beta_{n+1}$, $n \geq 1$. By (4.35), $\beta_n \to \infty$. This, together with (4.40), implies that

$$\lim_{n\to\infty} \frac{\beta_{n+1}}{\beta_n} = 1 \tag{4.42}$$

Theorem 4.8.2. $\frac{T_n}{\beta_n} \stackrel{wp1}{\longrightarrow} 1.$

Proof. Define $Y_j = \tau_j$ if $\tau_j \leq \beta_j$ and $Y_j = 0$ if $\tau_j > \beta_j$. We have

$$\begin{aligned}\sum_{\nu=1}^{\infty} P(\tau_\nu \neq Y_\nu) &= \sum_{\nu=1}^{\infty} \int_{x>\beta_\nu} dV_\nu(x) \\ &\leq \sum_{\nu=1}^{\infty} \int_{x\geq \frac{\nu}{\alpha}} dV_\nu(x) (\text{by (4.41)}) \\ &\leq A \sum_{\nu=1}^{\infty} \int_{x\geq \frac{\nu}{\alpha}} dV(x) (\text{by (4.37)}) \\ &\leq A \sum_{1}^{\infty} u(\frac{\nu}{\alpha}) < \infty \ (\text{by (4.39)}),\end{aligned}$$

where A is a positive constant. Thus, by Borel-Cantelli lemma, $P(\tau_\nu \neq Y_\nu\ i.o.) = 0$. Let $\xi_n = \sum_1^n Y_j$, then the theorem follows if we show that $\frac{\xi_n}{\beta_n} \xrightarrow{a.s.} 1$. This we proceed to establish.

Let t be an arbitrary positive number. Let $J = [t\alpha]$ and $n > J$. We have

$$\begin{aligned}0 \leq \beta_n - \mathbb{E}\xi_n &= \sum_{1}^{n} \int_{x\geq\beta_j} x\, dV_j(x) \leq \sum_{1}^{n} \int_{x\geq \frac{j}{\alpha}} x\, dV_j(x) \\ &\leq A \sum_{1}^{n} \int_{x\geq \frac{j}{\alpha}} x\, dV(x) \\ &= A \sum_{1}^{J} \int_{x\geq \frac{j}{\alpha}} x\, dV(x) + A \sum_{J+1}^{n} \int_{x\geq \frac{j}{\alpha}} x\, dV(x) \\ &\leq q(J) + A \sum_{J+1}^{n} \int_{x\geq t} x\, dV(x) \\ &\leq q(J) + nA \int_{x\geq t} x\, dV(x).\end{aligned}$$

Divide throughout by β_n and use (4.41) to get

$$0 \leq 1 - \frac{\mathbb{E}\xi_n}{\beta_n} \leq \frac{q(J)}{\beta_n} + A\alpha \int_{x\geq t} x\, dV(x).$$

Now let $n \to \infty$ and then let $t \to \infty$ to see that $\frac{\mathbb{E}\xi_n}{\beta_n} \to 1$.

Given $c \in (1,\ 2)$, let (n_k) be the the sequence of integers such that

$$\beta_{n_k} \leq c^k < \beta_{n_k+1} \tag{4.43}$$

Set $n_0 = 0$. We note that $n_1 \geq 1$, that $n_k < n_{k+1}$ for all k and that $n_k \to \infty$ as $k \to \infty$. Use (4.42) and (4.43) to conclude that

$$\beta_{n_k} \sim c^k \tag{4.44}$$

Next we show that $\frac{\xi_{n_k}}{\beta_{n_k}} \overset{wp1}{\longrightarrow} 1$.

Write $a_k = \int\limits_{x>\frac{k}{\beta}} dV_s(x)$. For $s \geq 3$, we have

$$\begin{aligned}
\mathbb{E}Y_s^2 &= \int\limits_{x\leq\beta_s} x^2\, dV_s(x) \leq \int\limits_{x\leq\frac{s}{\beta}} x^2\, dV_s(x) = \sum_1^s \int\limits_{\frac{k-1}{\beta}<x\leq\frac{k}{\beta}} x^2\, dV_s(x) \\
&\leq \frac{1}{\beta^2}\sum_{k=1}^s k^2 \int\limits_{\frac{k-1}{\beta}<x\leq\frac{k}{\beta}} dV_s(x) = \frac{1}{\beta^2}\sum_{k=1}^s k^2(a_{k-1}-a_k) \\
&= \frac{1}{\beta^2} + \frac{1}{\beta^2}\{\sum_{k=1}^{s-1} a_{k-1}(k^2-(k-1)^2) - s^2 a_s\} \\
&\leq \frac{1}{\beta^2} + \frac{1}{\beta^2}\sum_{k=1}^{s-1} 2k a_{k-1} \leq \frac{1}{\beta^2} + A\frac{2}{\beta^2}\sum_{k=2}^s k \int\limits_{x\geq\frac{k-1}{\beta}} dV(x) \\
&= q_1 + q_2 \sum_{k=2}^s k \int\limits_{x\geq\frac{k}{2\beta}} dV(x) \leq q_1 + q_2\sum_{k=2}^s k u(\frac{k}{2\beta}).
\end{aligned}$$

For all n large,

$$\begin{aligned}
var\xi_n &= \sum_{\nu=1}^n varY_\nu \leq \sum_{\nu=1}^n \mathbb{E}Y_\nu^2 \leq \sum_{\nu=1}^n \{q_1 + q_2\sum_{k=1}^\nu ku(\frac{k}{2\beta})\} \\
&\leq nq_3 + nq_2\sum_{k=1}^n ku(\frac{k}{2\beta}).
\end{aligned}$$

Hence

$$var\frac{\xi_{n_k}}{\beta_{n_k}} \le q_3\frac{n_k}{\beta_{n_k}^2} + q_2\frac{n_k}{\beta_{n_k}^2}\sum_{j=1}^{n_k} ju(\frac{j}{2\beta}) \le \frac{q_4}{c^k} + \frac{q_5}{c^k}\sum_{r=0}^{k-1}\sum_{j=n_r+1}^{n_{r+1}} ju(\frac{j}{2\beta})$$

$$\le \frac{q_4}{c^k} + \frac{q_5}{c^k}\sum_{r=0}^{k-1} n_{r+1}u(\frac{n_r}{2\beta})(n_{r+1}-n_r)$$

$$\le \frac{q_4}{c^k} + \frac{q_5}{c^k}\sum_{r=0} k-1c^r u(\frac{c^r}{4})c^r \text{ (for all } k \text{ sufficiently large).}$$

Thus that $(\sum_{k=1}^{\infty} var\frac{\xi_{n_k}}{\beta_{n_k}} < \infty)$ would follow if we show that

$$\sum_{k=1}^{\infty}\frac{1}{c^k}\sum_{r=1}^{k} c^{2r}u(\frac{c^r}{4}) < \infty.$$

Set $g_n = c^n u(\frac{c^n}{4})$ and $h_n = \frac{1}{c^n}$. Then by (4.39), $\sum_{n=1}^{\infty} g_n < \infty$; $\sum_{n=1}^{\infty} h_n < \infty$, since $c > 1$. All the terms of both the series are positive. Hence their Cauchy product $\sum_{n=1}^{\infty} j_n < \infty$ where $j_n = \sum_{t=1}^{n} g_t h_{n-t} = \sum_{t=1}^{n} c^t u(\frac{c^t}{4})\frac{1}{c^{n-t}}$. This shows that $\sum_1^{\infty} var(\frac{\xi_{n_k}}{\beta_{n_k}}) < \infty$; which implies, by Borel-Cantelli lemma and Tchebichev inequality, that

$$P(|\frac{\xi_{n_k}}{\beta_{n_k}} - \frac{\mathbb{E}\xi_{n_k}}{\beta_{n_k}}| > \varepsilon \; i.o.) = 0 \text{ for every } \varepsilon > 0.$$

i.e. $\frac{\xi_{n_k}}{\beta_{n_k}} - \frac{\mathbb{E}\xi_{n_k}}{\beta_{n_k}} \stackrel{wp1}{\longrightarrow} 0$. Since it has already been established that $\mathbb{E}\frac{\xi_n}{\beta_n} \to 1$, it follows now $\frac{\xi_{n_k}}{\beta_{n_k}} \stackrel{wp1}{\longrightarrow} 1$. Since the sequences (τ_n) and (Y_n) are equivalent, it follows from this that $\frac{T_{n_k}}{\beta_{n_k}} \stackrel{wp1}{\longrightarrow} 1$.

Given n, find k such that $n_k \le n < n_{k+1}$. The τ_js being non-negative, we have, $\frac{\beta_{n_k}}{\beta_{n_{k+1}}}\frac{T_{n_k}}{\beta_{n_k}} \le \frac{T_n}{\beta_n} \le \frac{\beta_{n_{k+1}}}{\beta_{n_k}}\frac{T_{n_{k+1}}}{\beta_{n_{k+1}}}$. Refer to (4.44) and let $k \to \infty$ to get $\frac{1}{c} \le \underline{\lim}_{n\to\infty} \frac{T_n}{\beta_n} \le \overline{\lim}_{n\to\infty} \frac{T_n}{\beta_n} \le c \; wp1$ Letting $c \to 1$, the proof is completed. □

Remarks 4.8.3. (i) In case the τ_ns in the theorem are identically distributed with $\mathbb{E}\tau_n = \mu < \infty$, then $\frac{T_n}{n} \stackrel{wp1}{\longrightarrow} \mu$, since the assumptions (4.35)–(4.38) are satisfied, taking for V the common *df* of the τ_ns.

(ii) If the τ_ns are *iid* with $\mathbb{E}\tau_1 = \alpha$ and if $\mathbb{E}\tau_1^2 < \infty$ then the proof that $\frac{T_{n_k}}{n_k}$

$\xrightarrow{a.s.} \alpha$ follows immediately from

$$\sum_k P(|T_{n_k} - n_k\alpha| > n_k\varepsilon) \leq \sum_k \frac{1}{\varepsilon^2 n_k^2} var(T_{n_k}) = \frac{1}{\varepsilon^2 n_k} var(\tau_1)$$
$$\leq \sum_k \frac{q}{c^k} < \infty.$$

Given a real number x, denote $\max(x,\ 0)$ by x^+ and denote $\max(-x,\ 0)$ by x^-. We note that $x^+ \geq 0,\ x^- \geq 0,\ x = x^+ - x^-$ and $|x| = x^+ + x^-$.

Theorem 4.8.4 (Kolmogorov). *Let (X_n) be a sequence of iid rvs. Write $S_n = \sum_1^n X_k$.*
(i) *If $\mathbb{E}|X_1| < \infty$, and $\mathbb{E}X_1 = \mu$, then $\frac{S_n}{n} \xrightarrow{wp1} \mu$.*
(ii) *If $\lim_{n\to\infty} \frac{S_n}{n}$ exists finitely $wp1$, then $\mathbb{E}|X_1| < \infty$ and $\lim_{n\to\infty} \frac{S_n}{n} = \mathbb{E}X_1\ wp1$*
(iii) *If $\mathbb{E}X_1^+ < \infty$ and $\mathbb{E}X_1^- = \infty$, then $\frac{S_n}{n} \xrightarrow{wp1} -\infty$;*
If $\mathbb{E}X_1^+ = \infty$ and $\mathbb{E}X_1^- < \infty$, then $\frac{S_n}{n} \xrightarrow{wp1} \infty$.

Proof. (i) $\mathbb{E}|X_1| < \infty \Rightarrow \mathbb{E}X_1^+ < \infty$ and $\mathbb{E}X_1^- < \infty$. Set $\tau_n = X_n^+, n \geq 1$ and note that the τ_ns are *iid* with common *df*, say, G and with common mean $\mathbb{E}\tau_n = \alpha_1 < \infty$. If $\alpha_1 = 0$, the τ_ns are $0\ wp1$. Let $\alpha_1 > 0$. Then by **(4.8.3)**(i), $\frac{X_1^+ + X_2^+ + \ldots + X_n^+}{n\alpha_1} \xrightarrow{wp1} 1$. Similarly, when $\mathbb{E}X_1^- = \alpha_2 > 0$, we get $\frac{X_1^- + X_2^- + \ldots + X_n^-}{n\alpha_2} \xrightarrow{wp1} 1$. Hence

$$\frac{S_n}{n} = \frac{X_1^+ + X_2^+ + \ \ldots\ + X_n^+}{n} - \frac{X_1^- + X_2^- + \ \ldots\ + X_n^-}{n}$$
$$\xrightarrow{wp1} (\alpha_1 - \alpha_2) = \mu.$$

(ii) The hypothesis implies $\frac{X_n}{n} \xrightarrow{wp1} 0$. Since the X_ns are independent, this is equivalent to $\sum_n P(|X_n| \geq n) < \infty$. i.e. $\sum_n P(|X_1| \geq n) < \infty$. Hence $\mathbb{E}|X_1| < \infty$ (ref. **(2.6.4)**). Now the result in (i) applies and we conclude $\frac{S_n}{n} \xrightarrow{wp1} \mathbb{E}X_1$.

(iii) Let $\mathbb{E}X_1^- < \infty$ and $\mathbb{E}X_1^+ = \infty$. By (i), $\frac{X_1^- + X_2^- + \ldots + X_n^-}{n} \xrightarrow{wp1} \alpha_2$.

For $\lambda > 0$, arbitrary, define $Y_n = X_n^+$ if $X_n^+ < \lambda$ and $Y_n = 0$ if $X_n^+ \geq \lambda,\ n \geq 1$. We note that (Y_n) is a sequence of non-negative *iid rvs*, that $X_n^+ \geq Y_n$, that $\mathbb{E}Y_n \leq \lambda$ and that, as $\lambda \to \infty$, $\mathbb{E}Y_1 \to \mathbb{E}X_1^+ = \infty$. By (i), $\frac{1}{n}(Y_1 +$

$Y_2 + \ldots + Y_n) \xrightarrow{wp1} \mathbb{E}Y_1$. It follows now

$$\lim_{\lambda\to\infty} \varliminf_{n\to\infty} \frac{X_1^+ + X_2^+ + \ \ldots \ + X_n^+}{n}$$
$$\geq \lim_{\lambda\to\infty} \lim_{n\to\infty} \frac{1}{n}(Y_1 + Y_2 + \ldots + Y_n)$$
$$= wp1 \lim_{\lambda\to\infty} \mathbb{E}Y_1 = \infty.$$

Thus in this case $\frac{S_n}{n} \xrightarrow{wp1} \infty$.

On similar lines it can be shown that $\frac{S_n}{n} \xrightarrow{wp1} -\infty$ if $\mathbb{E}X_1^- = \infty$ and $\mathbb{E}X_1^+ < \infty$. □

Remark Suppose $\mathbb{E}X_1 = 0$. Hence $\frac{S_n}{n} \xrightarrow{wp1} 0$. The question arises : is it true that $\frac{S_n}{n^\theta} \xrightarrow{wp1} 0$ for $0 < \theta < 1$? In sec. **(6.9.2)**, we will show that, when $\mathbb{E}X_1^2 < \infty$, $wp1 \varlimsup_{n\to\infty} \frac{|S_n|}{\sqrt{2n \log\log n}} \leq 1$. It will follow from this result that under the additional condition that $\mathbb{E}X_1^2 < \infty$, $\frac{S_n}{n^\theta} \xrightarrow{wp1} 0$ for $\frac{1}{2} < \theta < 1$.

4.8.5. *An example of a sequence of iid variables X_n for which the $WLLN$ holds but not the $SLLN$.*

Let X_n have frequency function $f(x) = \frac{c}{x^2 \log|y|}$ if $|y| \geq e$ and zero elsewhere. $S_n = \sum_{k=1}^n X_k$. We showed in **(4.7.3)**(iii) that $\frac{S_n}{n} \xrightarrow{pr} 0$. Now we will show that the $SLLN$ does not hold for this sequence. i.e., $\frac{S_n}{n} \xrightarrow{wp1} 0$ is not possible. For, if this is true, then we will have : $\frac{X_n}{n} \xrightarrow{wp1} 0$. Since the X_ns are independent variables, Borel-Cantelli lemma can be invoked to conclude that $\sum_{n=1}^\infty P(|X_n| > n) < \infty$. i.e., $\mathbb{E}|X| < \infty$. But $\mathbb{E}|X| = 2\int_e^\infty \frac{c}{x \log x}\, dx = \infty$.

More can be said about the long time behaviour of $(\frac{S_n}{n})$.

We note that $\mathbb{E}X_1^+ = \infty = \mathbb{E}X_1^-$ and that $\mathbb{E}X_1$ does not exist. By Kolmogorov zero-one law, $\varlimsup_{n\to\infty} \frac{S_n}{n}$ and $\varliminf_{n\to\infty} \frac{S_n}{n}$ are constants, finite or infinite. They are negatives of each other, since the *df* of X_1 is symmetric.

Since $\mathbb{E}|X_1| = \infty$, $\sum_n P(|X_1| > \lambda n) = \infty$ for every $\lambda \geq 0$. Hence $\varlimsup_{n\to\infty} \frac{|X_n|}{n} = \infty$. This leads to

$$\infty = \varlimsup_{n\to\infty} \frac{|X_n|}{n} = \varlimsup_{n\to\infty} \frac{|S_n - S_{n-1}|}{n} \leq \varlimsup_{n\to\infty} \frac{|S_n|}{n} + \varlimsup_{n\to\infty} \frac{|S_{n-1}|}{n}.$$

Hence $\varlimsup_{n\to\infty} \frac{|S_n|}{n} = \infty$. Since the variables are symmetric, this implies $\varlimsup_{n\to\infty} \frac{S_n}{n} = \infty = -\varliminf_{n\to\infty} \frac{S_n}{n}$. It follows that for any constant c, $P(\lim_{n\to\infty} \frac{S_n}{n} = c) = 0$.

Theorem 4.8.6. *If the iid rvs (X_n) are non-negative, if $S_n = \sum_1^n X_k$ and if $\frac{S_n}{n} \overset{pr}{\to} \theta$, θ finite, then $\frac{S_n}{n} \overset{wp1}{\longrightarrow} \theta$.*

Proof. Since X_1 is non-negative, $\mathbb{E}X_1$ is well defined, finite or infinite. If $\mathbb{E}X_1 = \infty$, then by **(4.8.4)**(i) $\frac{S_n}{n} \overset{wp1}{\longrightarrow} \infty$. But the hypothesis implies that for sequence (n_k) of integers $\frac{S_{n_k}}{n_k} \overset{wp1}{\longrightarrow} \theta$, leading to a contradiction. Hence necessarily $\mathbb{E}X_1 < \infty$. Now by the SLLN, $\frac{S_n}{n} \overset{wp1}{\longrightarrow} \mathbb{E}X_1$. It follows that $\mathbb{E}X_1 = \theta$. □

The following Lemma is needed to prove **(4.8.8)** where we establish a $SLLN$ for a variational sequence of independent rvs.

Lemma 4.8.7. *Let $a_n \geq 0$, $n \geq 1$. Then $(\sum\limits_n \frac{a_n}{n^2} < \infty) \Rightarrow (\lim\limits_{m\to\infty} \frac{1}{m^2} \sum\limits_1^m a_k =$ $0)$.*

Proof. Given $\varepsilon > 0$, find M such that $\sum_{M+1}^\infty \frac{a_k}{k^2} < \varepsilon$. Then for all m large, $m > M$, such that $\frac{1}{m^2} \sum_1^M a_k < \varepsilon$, we have

$$\begin{aligned}\tfrac{1}{m^2} \textstyle\sum_1^m a_k &= \tfrac{1}{m^2} \textstyle\sum_1^M a_k + \tfrac{1}{m^2} \textstyle\sum_{M+1}^m a_k \\ &\leq \varepsilon + \textstyle\sum_{M+1}^m \tfrac{a_k}{k^2} \leq \varepsilon + \textstyle\sum_{M+1}^\infty \tfrac{a_k}{k^2} \leq 2\varepsilon.\end{aligned}$$ □

Theorem 4.8.8. *Let (X_n) be a sequence of independent rvs with $\mathbb{E}X_n = 0$ and $\mathbb{E}X_n^2 = \sigma_n^2 < \infty$ such that $\sum_n \frac{\sigma_n^2}{n^2} < \infty$. Let $S_n = \sum_1^n X_k$. Then $\frac{S_n}{n}$ $\overset{wp1}{\longrightarrow} 0$.*

Proof. Proving the claim is equivalent to proving

$$\lim_{k\to\infty} P(\sup_{n\geq k} |\frac{S_n}{n}| > \varepsilon) = 0 \text{ for every } \varepsilon > 0.$$

Let $Y_1 = S_k$, $Y_j = X_{k+j-1}$, $j \geq 2$. Let $T_n = \sum_{\nu=1}^n Y_\nu$ and note $T_n = S_{k+n-1}$. Thus $\max\limits_{k\leq n\leq k+r} |\frac{S_n}{n}| = \max\limits_{1\leq\nu\leq r+1} |\frac{T_\nu}{k+\nu-1}|$. Hence

$$\begin{aligned}P(\sup_{n\geq k} |\frac{S_n}{n}| > \varepsilon) &= \lim_{r\to\infty} P(\max_{k\leq n\leq k+r} |\frac{S_n}{n}| > \varepsilon) \\ &= \lim_{r\to\infty} P(\max_{1\leq\nu\leq r+1} |\frac{T_\nu}{k+\nu-1}| > \varepsilon) \\ &\leq \lim_{r\to\infty} \frac{1}{\varepsilon^2}\{\frac{\mathbb{E}Y_1^2}{k^2} + \sum_{j=2}^{r+1} \frac{\sigma_{k+j-1}^2}{(k+j-1)^2}\} \quad (\text{by } \mathbf{(4.4.3)})\end{aligned}$$

$$= \lim_{r\to\infty} \frac{1}{\varepsilon^2}\{\frac{1}{k^2}\sum_{j=1}^{k}\sigma_j^2 + \sum_{k+1}^{k+r}\frac{\sigma_j^2}{j^2}\}$$

$$= \frac{1}{\varepsilon^2}\{\frac{1}{k^2}\sum_{j=1}^{k}\sigma_j^2 + \sum_{k+1}^{\infty}\frac{\sigma_j^2}{j^2}\}.$$

The desired result follows by using the hypothesis, by appealing to **(4.8.7)** and letting $k \to \infty$. □

A second proof.

Let $T_n = \sum_1^n \frac{X_j}{j}$. Since $\sum_n \frac{\sigma_n^2}{n^2} < \infty$, it follows (ref. **(4.6.1)**) that $T = \lim_{n\to\infty} T_n = \sum_n \frac{X_n}{n}$ exists finitely $a.s.$ We note $nT_n = nT_{n-1} + X_n$. Hence $\frac{1}{n}S_n = T_n - \frac{1}{n}\sum_{j=1}^{n-1} T_j$. Since $\lim\limits_{n\to\infty} T_n = T$ exists finitely $a.s.$, the Cesaro limit $\lim\limits_{n\to\infty} \frac{1}{n}\sum_{j=1}^{n-1} T_j$ exists finitely $a.s.$ and is equal to T. That $\frac{1}{n}S_n \xrightarrow{a.s.} 0$ follows now.

Chapter 5

The Central Limit Theorem and its Ramifications

Let (X_n) be a sequence of independent rvs and $S_n = \sum_{r=1}^{n} X_r$, $n \geq 1$, their partial sums. Let ξ stand for a standard normal variable (**(3.2)**(x)). In this chapter we prove that, under certain conditions, there exist constants $b_n > 0$ and a_n such that $\frac{S_n}{b_n} - a_n \stackrel{d}{\to} \xi$. When this happens, we say that the central limit theorem (*CLT*) holds for the X_ns. We investigate further conditions that will ensure the convergence of the moments of $\frac{S_n}{b_n} - a_n$ to the corresponding ones of ξ. Then we illustrate the Erdös-Kac invariance principle by finding the limiting distribution of $\max\limits_{1 \leq k \leq n} |S_k|$, suitably normed. Finally we establish the functional form or the Donsker form of the *CLT*.

5.1 Some auxiliary results

The inequalities derived in this section are crucial in proving in the next section that the weak limit, if it exists, of the sums of infinitesimal rvs is necessarily infinitely divisible.

Definitions 5.1.1.
Definition 1. Two sequences (Y_n) and (Z_n) are said to be equivalent in law if $P(Y_n \neq Z_n) \to 0$.
Remark. If the two sequences (Y_n) and (Z_n) are equivalent in law and if, say, $Y_n \stackrel{d}{\to} Y$ then $Z_n \stackrel{d}{\to} Y$. For,

$$\begin{aligned} P(Y_n \leq x) &= P(Y_n \leq x,\ Y_n = Z_n) + P(Y_n \leq x,\ Y_n \neq Z_n) \\ &= P(Z_n \leq x,\ Y_n = Z_n) + P(Y_n \leq x,\ Y_n \neq Z_n) \\ &= P(Z_n \leq x) - P(Z_n \leq x,\ Y_n \neq Z_n) + P(Y_n \leq x,\ Y_n \neq Z_n). \end{aligned}$$

The claim follows from this since the absolute value of each of the last two terms does not exceed $P(Y_n \neq Z_n)$ which is $o(1)$. □

Definition 2. For each $n \geq 1$ and integers $k_n \to \infty$, let $X_{i,n}$, $1 \leq i \leq k_n$ be k_n independent rvs with *df* $F_{i,n}$ and with *cf* $\varphi_{i,n}$. Further let the variables be infinitesimal. i.e.

$$\lim_{n\to\infty} \max_{1\leq k\leq k_n} P(|X_{i,n}| > \varepsilon) = 0 \text{ for every } \varepsilon > 0.$$

Let $S_n = \sum_{k=1}^{k_n} X_{k,n}$, $\varphi_n(t) = \prod_{k=1}^{k_n} \varphi_{k,n}(t)$. We note φ_n is the *cf* of S_n. In Sec. **(5.2)** we show that the weak limit, if it exists, of S_n properly normed is necessarily infinitely divisible. The needed auxiliary results are presented now.

Theorem 5.1.2. *The following three statements are equivalent:*
(i) The variables $(X_{k,n},\ 1 \leq k \leq k_n)$ *are infinitesimal.*
(ii) $\max_{1\leq k\leq k_n} |1 - \varphi_{k,n}(t)| \to 0$ *uniformly over every bounded t-interval.*
iii) $\max_{1\leq k\leq k_n} \int_R \frac{x^2}{1+x^2}\, dF_{k,n}(x) \to 0.$

Proof. Let (i) hold. Let $|t| \leq b$. Now,

$$\begin{aligned}
\max_{1\leq k\leq k_n} |1 - \varphi_{k,n}(t)| &= \max_{1\leq k\leq k_n} \left|\int_R (1 - e^{itx})\, \mathrm{d}\, F_{k,n}(x)\right| \\
&\leq \max_{1\leq k\leq k_n} \int_{|x|\leq\varepsilon} |1 - e^{itx}|\, \mathrm{d}\, F_{k,n}(x) \\
&\quad + \max_{1\leq k\leq k_n} \int_{|x|\geq\varepsilon} |1 - e^{itx}|\, \mathrm{d}\, F_{k,n}(x) \\
&\leq \varepsilon|t| + 2\sup_k P(|X_{k,n}| \geq \varepsilon) \\
&\leq \varepsilon b + 2\sup_k P(|X_{k,n}| \geq \varepsilon).
\end{aligned}$$

That (i) $\Rightarrow$ (ii) follows from this.

Suppose (ii) holds. Given $u > 0$, $\delta > 0$ find $N = N(u, \delta)$ such that

$$\max_{|t|\leq u} \max_{1\leq k\leq k_n} |1 - \varphi_{k,n}(t)| < \frac{\delta}{2u}.$$

We recall a classical result : for every $\varepsilon > 0$, the improper Riemann integral $\int_0^\infty \frac{\sin \varepsilon x}{x}\, \mathrm{d}\, x = \frac{\pi}{2}$. Now, using **(3.1.2)**,

$$P(|X_{k,n}| > \varepsilon) \leq 1 - \{F_{k,n}(\varepsilon) - F_{k,n}(-\varepsilon)\}$$

$$
\begin{aligned}
&= 1 - \frac{1}{2\pi} \lim_{u\to\infty} \int_{-u}^{u} \frac{e^{it\varepsilon} - e^{-it\varepsilon}}{it} \varphi_{k,n}(t)\, \mathrm{d}t \\
&= \frac{1}{\pi} \lim_{u\to\infty} \int_{-u}^{u} \frac{\sin \varepsilon t}{t}\, \mathrm{d}t - \frac{1}{\pi} \lim_{u\to\infty} \int_{-u}^{u} \frac{\sin \varepsilon t}{t} \varphi_{k,n}(t)\, \mathrm{d}t \\
&= \frac{1}{\pi} \lim_{u\to\infty} \int_{-u}^{u} \frac{\sin \varepsilon t}{t} \{1 - \varphi_{k,n}(t)\}\, \mathrm{d}t \\
&\leq \frac{1}{\pi} \lim_{u\to\infty} \int_{-u}^{u} \left|\frac{\sin \varepsilon t}{t}\right| |1 - \varphi_{k,n}(t)|\, \mathrm{d}t \\
&\leq \frac{\varepsilon}{\pi} \lim_{u\to\infty} \int_{-u}^{u} \max_{1\leq k\leq k_n} |1 - \varphi_{k,n}(t)|\, \mathrm{d}t < \frac{\varepsilon}{\pi}\delta.
\end{aligned}
$$

Since $\delta > 0$ is arbitrary, we conclude that (ii) $\Rightarrow$ (i). This completes the proof that (i) $\Leftrightarrow$ (ii).

Suppose (iii) holds. We have

$$
\begin{aligned}
P(|X_{k,n}| \geq \varepsilon) = \int_{|x|\geq\varepsilon} \mathrm{d}\, F_{k,n}(x) &= \int_{|x|\geq\varepsilon} \frac{1+x^2}{x^2} \frac{x^2}{1+x^2}\, \mathrm{d}\, F_{k,n}(x) \\
&\leq \frac{1+\varepsilon^2}{\varepsilon^2} \int_{R} \frac{x^2}{1+x^2}\, \mathrm{d}\, F_{k,n}(x).
\end{aligned}
$$

That (iii) $\rightarrow$ (i) follows from this. (i) $\Rightarrow$ (iii) since, for every $\varepsilon > 0$, we have

$$
\begin{aligned}
\int_{R} \frac{x^2}{1+x^2}\, \mathrm{d}\, F_{k,n}(x) &= \int_{|x|\leq\varepsilon} \frac{x^2}{1+x^2}\, \mathrm{d}\, F_{k,n}(x) \\
&\qquad + \int_{|x|\geq\varepsilon} \frac{x^2}{1+x^2}\, \mathrm{d}\, F_{k,n}(x) \\
&\leq \varepsilon^2 + \int_{|x|\geq\varepsilon} \mathrm{d}\, F_{k,n}(x).
\end{aligned}
$$

□

(iv) Remark. If the variables $(X_{k,n},\ 1 \leq k \leq k_n)$ are infinitesimal, then $\lim_{n\to\infty} \max_{1\leq k\leq k_n} |m_{k,n}| = 0$ where $m_{k,n}$ is a median of $X_{k,n}$.

Proof. The hypothesis implies that given $\varepsilon > 0$, we can find $N = N(\varepsilon)$ such that for all $n \geq N$, $\max\limits_{1 \leq k \leq k_n} P(|X_{k,n}| > \varepsilon) < \frac{1}{2}$. Hence

$$\min_{1 \leq k \leq k_n} P(|X_{k,n}| \leq \varepsilon) > \frac{1}{2}.$$

This implies $-\varepsilon \leq m_{k,n} \leq \varepsilon$, $1 \leq k \leq k_n$, $n \geq N$, which is the desired result. □

(v) Remark. If the variables $(X_{k,n},\ 1 \leq k \leq k_n)$ are infinitesimal, then

$$\lim_{n \to \infty} \sup_{1 \leq k \leq k_n} |\alpha_{k,n}| = 0$$

where

$$\alpha_{k,n} = \int\limits_{|x| \leq \tau} |x|^r \, \mathrm{d} F_{k,n}(x), \quad \tau > 0,\ r > 0$$

arbitrary.

Proof. If $\lim\limits_{n \to \infty} \sup\limits_{1 \leq k \leq k_n} |\alpha_{k,n}| \neq 0$, taking a subsequence, if necessary, assume that the limit is $c > 0$. Hence we can find a subsequence $(F_{j_n,n})$ such that

$$\lim_{n \to \infty} \int\limits_{|x| \leq \tau} |x|^r \, \mathrm{d} F_{j_n,n}(x) = c.$$

The hypothesis implies $F_{j_n,n}$ converges weakly to the distribution degenerate at 0. Hence by **(2.4.3)**(v),

$$\lim_{n \to \infty} \int\limits_{|x| \leq \tau} |x|^r \, \mathrm{d} F_{j_n,n}(x) = 0,$$

a contradiction. The claim follows. □

Lemma 5.1.3. (i) *Let (u_n), (v_n) be sequences of complex numbers such that (u_n) is a bounded sequence and such that $u_n - v_n \to 0$. Then $e^{u_n} - e^{v_n} \to 0$.*
(ii) *For $b > 0$ the function f is bounded away from zero where*

$$f(x) = (1 - \frac{\sin bx}{bx}) \frac{1 + x^2}{x^2}, \ x \in R.$$

i.e. $\inf\limits_{x \in R} f(x) = c(b) > 0$.
(iii) *Let F be an arbitrary df on R. For t a real number, define df F^t : $F^t(x) =$*

$F(t+x)$. Let $\tau > 0$, θ be arbitrary real numbers subject to $\tau > |\theta|$. Let $a = \int_{|x|\le\tau} x\, dF(x)$. Then there exists $q = q(\tau,\ \theta)$ such that

$$\int_R \frac{x^2}{1+x^2}\, dF^a(x) \le q \int_R \frac{x^2}{1+x^2}\, dF^\theta(x).$$

q is an absolute constant, independent of the df F.
(iv) Let F be an arbitrary df with cf φ. Let $\theta,\ \tau,\ a,\ F^a$ be as in (iii). Then for any $b > 0$, there corresponds a constant $q_1 = q(\theta,\ \tau,\ b)$, independent of the df F, such that

$$\int_R \frac{x^2}{1+x^2}\, dF^a(x) \le q_1 \int_0^b (1-|\varphi(t)|^2)\, dt.$$

(v) Let df F be arbitrary and $\tau > 0$ be arbitrary. Suppose $|a| < \frac{\tau}{2}$ where $a = \int_{|x|<\tau} x\, dF(x)$. Then

$$\left|\int_{|x|<\tau} x\, dF(x+a)\right| \le 2\tau \int_{|x|\ge\frac{\tau}{2}} dF(x+a).$$

Proof. (i) Let $|u_n| \le q_1$. Hence $|e^{u_n}| \le e^{q_1} = q$, say. We then have
$$|e^{u_n} - e^{v_n}| = |e^{u_n}||1 - e^{v_n - u_n}| \le q|1 - e^{v_n-u_n}| \to 0.$$
(ii) With the understanding that $f(0) = \lim_{x\to 0} f(x) = \frac{b^2}{6}$, we note that f is positive and continuous everywhere. Hence for some $\delta > 0$, $\inf_{|x|\le\delta} f(x) > 0$, In the region $\delta \le |x| \le \frac{\pi}{2b}$, $f(x) \ge 1 - \frac{\sin bx}{bx} = 1 - \frac{\sin b|x|}{b|x|}$, a monotonic decreasing function of $|x|$. Hence in this region, $f(x) \ge 1 - \frac{2}{\pi} > 0$. In the region $|x| \ge \frac{\pi}{2b}$, $f(x) \ge 1 - \frac{|\sin bx|}{b|x|} \ge 1 - \frac{1}{b|x|} \ge 1 - \frac{2}{\pi} > 0$. The claim is obvious from these values.

(iii) We note that $\int_{|x|\le\tau}(x-a)\, \mathrm{d}F(x) = a\int_{|x|>\tau} \mathrm{d}F(x)$. We note also $(x-a)^2 = (x-\theta)^2 + 2(\theta-a)(x-a) + (\theta-a)^2 = (x-\theta)^2 + 2(\theta-a)(x-a+a-\theta) + (\theta-a)^2 \le (x-\theta)^2 + 2(x-a)(\theta-a)$. Now,

$$\int_R \frac{x^2}{1+x^2}\, \mathrm{d}F^a(x) = \int_R \frac{(x-a)^2}{1+(x-a)^2}\, \mathrm{d}F(x)$$
$$= \int_{|x|\le\tau} \frac{(x-a)^2}{1+(x-a)^2}\, \mathrm{d}F(x) + \int_{|x|>\tau} \frac{(x-a)^2}{1+(x-a)^2}\, \mathrm{d}F(x)$$
$$\le \int_{|x|\le\tau} (x-a)^2\, \mathrm{d}F(x) + \int_{|x|>\tau} \mathrm{d}F(x)$$

$$\leq \int\limits_{|x|\leq\tau} (x-\theta)^2 \, \mathrm{d}F(x) + 2|a-\theta| \, \Big| \int\limits_{|x|\leq\tau} (x-a) \, \mathrm{d}F(x)\Big| + \int\limits_{|x|>\tau} \mathrm{d}F(x)$$
$$= \int\limits_{|x|\leq\tau} (x-\theta)^2 \, \mathrm{d}F(x) + 2|a-\theta||a| \int\limits_{|x|>\tau} \mathrm{d}F(x) + \int\limits_{|x|>\tau} \mathrm{d}F(x)$$
$$\leq \int\limits_{|x|\leq\tau} (x-\theta)^2 \, \mathrm{d}F(x) + (4\tau^2+1) \int\limits_{|x|>\tau} \mathrm{d}F(x)$$
$$\text{(since } |a|\leq\tau,\ |\theta|\leq\tau)$$
$$= I_1 + (4\tau^2+1)I_2, \quad \text{say.}$$

$$I_1 = \int\limits_{|x|\leq\tau} \tfrac{(x-\theta)^2}{1+(x-\theta)^2}\{1+(x-\theta)^2\} \, \mathrm{d}F(x)$$

$$\leq \{1+(\tau+|\theta|)^2\} \int\limits_{|x+\theta|\leq\tau} \tfrac{x^2}{1+x^2} \, \mathrm{d}F^{\theta}(x).$$

Noting that, by the hypothesis on τ, $x-\theta$ does not vanish in the region $\{|x|>\tau\}$, we have,

$$I_2 = \int\limits_{|x|>\tau} \tfrac{(x-\theta)^2}{1+(x-\theta)^2}\tfrac{1+(x-\theta)^2}{(x-\theta)^2} \, \mathrm{d}F(x)$$
$$\leq \tfrac{1+(\tau+|\theta|)^2}{(\tau-|\theta|)^2} \int\limits_{|x+\theta|>\tau} \tfrac{x^2}{1+x^2} \, \mathrm{d}F^{\theta}(x).$$

It is clear from these that a q can be found such that

$$\int\limits_R \tfrac{x^2}{1+x^2} \, \mathrm{d}F^{a}(x) \leq q\{ \int\limits_{|x+\theta|\leq\tau} \tfrac{x^2}{1+x^2} \, \mathrm{d}F^{\theta}(x) + \int\limits_{|x+\theta|>\tau} \tfrac{x^2}{1+x^2} \, \mathrm{d}F^{\theta}(x)\}$$
$$= q\int\limits_R \tfrac{x^2}{1+x^2} \, \mathrm{d}F^{\theta}(x).$$

It is clear too that q does not depend on the *df* F.

(iv) Let X, Y be two *iid rvs* with common *df* F and *cf* φ. Let G be the *df* of $X-Y$. We note that G is a symmetric *df* with *cf* $|\varphi|^2$. We have ,

$$\int\limits_0^b (1-|\varphi(t)|^2) \, \mathrm{d}t = \int\limits_0^b \{\int\limits_R (1-\cos tx) \, \mathrm{d}G(x)\} \, \mathrm{d}t$$
$$= b\int\limits_R \{1-\tfrac{\sin bx}{bx}\} \, \mathrm{d}G(x)$$
$$= b\int\limits_R \{1-\tfrac{\sin bx}{bx}\}\tfrac{1+x^2}{x^2}\tfrac{x^2}{1+x^2} \, \mathrm{d}G(x)$$
$$\geq \mathrm{bc}\,(b)\int\limits_R \tfrac{x^2}{1+x^2} \, \mathrm{d}G(x),$$

by (ii) above. Thus the claim is established for symmetric distributions.

To prove the result without the assumption of symmetry, we note first (ref.

(2.3.6)) that for any *df* F,

$$\int_{-\infty}^{0} \frac{x^2}{1+x^2}\,\mathrm{d}\,F(x) = -\int_{0}^{\infty} \frac{x^2}{1+x^2}\,\mathrm{d}\,F(-x).$$

Trivially

$$\int_{0}^{\infty} \frac{x^2}{1+x^2}\,\mathrm{d}\,F(x) = -\int_{0}^{\infty} \frac{x^2}{1+x^2}\,\mathrm{d}\,(1-F(x)).$$

Hence

$$\int_{R} \frac{x^2}{1+x^2}\,\mathrm{d}\,F(x) = -\int_{0}^{\infty} \frac{x^2}{1+x^2}\,\mathrm{d}\,P(|X| \geq x).$$

Let θ be a median of F. Let F^{θ} be the *df* of $X - \theta$. Zero will be a median of F^{θ}. We have, by integration by parts. (ref. **(2.6.3)**,**(2.6.4)**),

$$\begin{aligned}
\int_{R} \frac{x^2}{1+x^2}\,\mathrm{d}\,F^{\theta}(x) &= -\frac{x^2}{1+x^2} P(|X-\theta| \geq x)\Big|_0^{\infty} \\
&\qquad + \int_{0}^{\infty} P(|X-\theta| \geq x)\,\mathrm{d}\,(\frac{x^2}{1+x^2}) \\
&\leq 2\int_{0}^{\infty} P(|X-Y| \geq x)\,\mathrm{d}\,(\frac{x^2}{1+x^2}) \text{ (ref. } \mathbf{(4.2.3)}\text{(i))} \\
&= 2\frac{x^2}{1+x^2} P(|X-Y| \geq x)\Big|_0^{\infty} \\
&\qquad - 2\int_{0}^{\infty} \frac{x^2}{1+x^2}\,\mathrm{d}\,P(|X-Y| \geq x) \\
&= -2\int_{0}^{\infty} \frac{x^2}{1+x^2}\,\mathrm{d}\,P(|X-Y| \geq x) = 2\int_{-\infty}^{\infty} \frac{x^2}{1+x^2}\mathrm{d}G(x)
\end{aligned}$$

This implies that (iv) is true for arbitrary *df*s with 0 for a median.

Let a, τ be as in (iii) with $\tau > |\theta|$. The stated result in (iv) follows now since, by (iii),

$$\int_{R} \frac{x^2}{1+x^2}\,\mathrm{d}\,F^{a}(x) \leq q(\tau,\,\theta) \int_{R} \frac{x^2}{1+x^2}\,\mathrm{d}\,F^{\theta}(x).$$

(v)

$$\Big|\int_{|x|<\tau} x\,\mathrm{d}F(x+a)\Big| \le \Big|\int_{|x|<\tau} x\,\mathrm{d}F(x+a) - \int_{|x+a|<\tau} x\,\mathrm{d}F(x+a)\Big|$$
$$+\Big|\int_{|x+a|<\tau} x\,\mathrm{d}F(x+a)\Big| = A+B, \text{ say}$$

Assume $a>0$. Argument will be similar when $a<0$. We note

$$\int_{|x-(-a)|<\tau} x\,\mathrm{d}F(x+a) = \int_{|x|<\tau} x\,\mathrm{d}F(x+a) - \int_{-a+\tau}^{\tau} x\,\mathrm{d}F(x+a)$$
$$+\int_{-a-\tau}^{-\tau} x\,\mathrm{d}F(x+a).$$

Hence

$$\begin{aligned}
A &\le \int_{-a+\tau}^{\tau} |x|\,\mathrm{d}F(x+a) + \int_{-a-\tau}^{-\tau} |x|\,\mathrm{d}F(x+a)\\
&\le \int_{\frac{1}{2}\tau}^{\tau} |x|\,\mathrm{d}F(x+a) + \int_{-\frac{3}{2}\tau}^{-\tau} |x|\,\mathrm{d}F(x+a)\\
&\le \int_{\frac{1}{2}\tau\le|x|\le\frac{3}{2}\tau} |x|\,\mathrm{d}F(x+a)\\
&\le \frac{3\tau}{2}\int_{|x|\ge\frac{\tau}{2}} \mathrm{d}F(x+a).
\end{aligned}$$

Next,

$$B = \Big|\int_{|x|<\tau} (x-a)\,\mathrm{d}F(x)\Big| = \Big|a - a\int_{|x|<\tau} \mathrm{d}F(x)\Big| \le |a|\int_{|x|\ge\tau} \mathrm{d}F(x)$$
$$\le \frac{\tau}{2}\int_{|x|\ge\tau} \mathrm{d}F(x) = \frac{\tau}{2}\int_{|x+a|\ge\tau} \mathrm{d}F(x+a) \le \frac{\tau}{2}\int_{|x|\ge\frac{\tau}{2}} \mathrm{d}F(x+a).$$

In the case $a=0$, which will happen if F is the *df* of a symmetric *rv*, the L.H.S. is zero and hence the claimed inequality is obvious. □

The next result we establish helps us show in **(5.2.1)** that sequence (φ_n) converges to a *cf iff* a certain accompanying sequence of *id cf*s converges to a *cf*, paving the way for showing further that the limit *cf* of φ_n, if it exists, will necessarily be *id*.

Theorem 5.1.4. *Let* τ, a, F^a *be as in* **(5.1.3)**(iii).
Let $a_{k,n} = \int_{|x|\le\tau} x\, d\,F_{k,n}(x)$. *If* φ_0 *is a cf and if* $|\varphi_n(t)| \to |\varphi_0(t)|$ *then there exists* $q > 0$ *such that for all* n,

$$\sum_{k=1}^{k_n} \int_R \frac{x^2}{1+x^2}\, d\,F_{k,n}^{a_{k,n}}(x) \le q.$$

Proof. By **(5.1.3)**(iv),

$$\int_R \frac{x^2}{1+x^2}\, \mathrm{d}\,F_{k,n}^{a_{k,n}}(x) \le q \int_0^b \{1 - |\varphi_{k,n}(t)|^2\}\, \mathrm{d}\,t,$$

q being independent of k and n. Hence

$$\begin{aligned}\sum_{k=1}^{k_n} \int_R \frac{x^2}{1+x^2}\, \mathrm{d}\,F_{k,n}^{a_{k,n}}(x) &\le q \sum_{k=1}^{k_n} \int_0^b \{1 - |\varphi_{k,n}(t)|^2\}\, \mathrm{d}\,t \\ &= q \int_0^b \sum_{k=1}^{k_n} \{1 - |\varphi_{k,n}(t)|^2\}\, \mathrm{d}\,t.\end{aligned}$$

Since the convergence of $|\varphi_n|$ to $|\varphi_0|$ is necessarily uniform over bounded t-intervals, there can be found $b > 0$ such that $\inf_{|t|\le b} |\varphi_n(t)| > \frac{1}{2}$ for every $n \ge 0$. Since $|\varphi_n(t)| = \prod_{k=1}^{k_n} |\varphi_{k,n}(t)|$, we have, for each t with $|t| \le b$, for each k, $1 \le k \le k_n$ and for each $n \ge 1$, $|\varphi_{k,n}(t)| > \frac{1}{2}$. When $|\varphi_{k,n}(t)| > 0$, we have $1 - |\varphi_{k,n}(t)|^2 \le -\log|\varphi_{k,n}(t)|^2 = -2\log|\varphi_{k,n}(t)|$. Hence, by **(5.1.3)**(iv),

$$\begin{aligned}\sum_{k=1}^{k_n} \int_R \frac{x^2}{1+x^2}\, \mathrm{d}\,F_{k,n}^{a_{k,n}}(x) &\le q \int_0^b \sum_{k=1}^{k_n} \{-2\log|\varphi_{k,n}(t)|\}\, \mathrm{d}\,t \\ &= q \int_0^b \{-2\log|\varphi_n(t)|\}\, \mathrm{d}\,t.\end{aligned}$$

Since $-\log|\varphi_n(t)| \to -\log|\varphi_0(t)|$ uniformly over $[0,\ b]$ and since $\inf\limits_{|t|\le b} |\varphi_n(t)| > \frac{1}{2}$, a number c exists such that

$$\sum_{k=1}^{k_n} \int_R \frac{x^2}{1+x^2}\, \mathrm{d}\, F_{k,n}^{a_{k,n}}(x) \le c \int_0^b \{-\log|\varphi_0(t)|\}\, \mathrm{d}t \le c_2 \qquad \square$$

5.2 Weak limit of sums of infinitesimal variables

Theorem 5.2.1. *Let $X_{k,n}$, $\varphi_{k,n}$, φ_n be as defined in* **(5.1.1)**. *Let $a_{k,n}$, $F_{k,n}^{a_{k,n}}$ be as defined in* **(5.1.4)**. *Let*

$$\psi_n(t) = \sum_{k=1}^{k_n} \{ita_{k,n} + \int_R (e^{itx} - 1)\, d\, F_{k,n}^{a_{k,n}}(x)\}$$

Then for each $t \in R$, $\varphi_n(t) - e^{\psi_n(t)} \to 0$.

Proof. Let $\varphi_{k,n}^*(t) = \int_R e^{itx}\, \mathrm{d}\, F_{k,n}^{a_{k,n}}(x)$. Let $\beta_{k,n} = \beta_{k,n}(t) = \varphi_{k,n}^*(t) - 1$. Since $(X_{k,n} - a_{k,n},\ 1 \le k \le k_n)$ are infinitesimal (ref. **(5.1.3)**(v)), it follows that (ref. **(5.1.2)**) $\max\limits_{1\le k\le k_n} |\beta_{k,n}(t)| \to 0$ uniformly over every bounded t-intervals $I_\lambda = \{|t| \le \lambda\}$. Hence for all n large, for all $t \in I_\lambda$ and for all k in the range $1 \le k \le k_n$, $|\beta_{k,n}(t)| < \frac{1}{2}$. This implies $1 + \beta_{k,n}(t) \ne 0$ for all n large, all k, $1 \le k \le k_n$ and for all $t \in I_\lambda$

Let $B_n(t) = it\alpha_n + \sum\limits_{k=1}^{k_n} \log(1 + \beta_{k,n}(t)) = it\alpha_n + C_n(t)$, say, where log represents the principal value of the logarithm. It is clear $e^{B_n(t)} = \varphi_n(t)$

We note $\psi_n(t)$ can be written $\psi_n(t) = it\alpha_n + \sum\limits_{k=1}^{k_n} \beta_{k,n}(t) = it\alpha_n + A_n(t)$, say.

Fix $t \in I_\lambda$. If we show (i) that $(|C_n(t)|)$ is a bounded sequence and (ii) that $|A_n(t) - C_n(t)| \to 0$, then by **(5.1.3)** (i), it will follow that $e^{A_n(t)} - e^{C_n(t)} \to 0$. This is equivalent to the stated claim that $e^{\psi_n(t)} - \varphi_n(t) \to 0$, since

$$\varphi_n(t) = \prod_1^{k_n} \varphi_{k,n}(t) = \prod_1^{k_n} e^{ita_{k,n}} \mathbb{E} e^{it(X_{k,n} - a_{k,n})} = \prod_1^{k_n} e^{ita_{k,n}} \varphi_{k,n}^*(t) = e^{B_n(t)}$$

We proceed to establish (i) and (ii).

$$|A_n(t) - C_n(t)|$$
$$\le \sum_1^{k_n} |\log(1 + \beta_{k,n}) - \beta_{k,n}|$$
$$\le \sum_1^{k_n} \sum_{r=2}^{\infty} \tfrac{1}{r} |\beta_{k,n}|^r \le \tfrac{1}{2} \sum_1^{k_n} \tfrac{|\beta_{k,n}|^2}{1 - |\beta_{k,n}|} \le \sum_{k=1}^{k_n} |\beta_{k,n}|^2$$

$\leq \max\limits_{1\leq k\leq k_n} |\beta_{k,n}| \sum\limits_{k=1}^{k_n} |\beta_{k,n}|.$

Similarly, $|C_n(t)| \leq \sum\limits_{k=1}^{k_n} |\beta_{k,n}|$

We have,

$$|\beta_{k,n}(t)| = |\int\limits_{|x|<\tau} (e^{itx}-1)\,\mathrm{d}\,F_{k,n}^{a_{k,n}}(x) + \int\limits_{|x|\geq\tau} (e^{itx}-1)\,\mathrm{d}\,F_{k,n}^{a_{k,n}}(x)|$$

$$\begin{aligned}
&= |\int\limits_{|x|<\tau} (e^{itx}-1-itx)\,\mathrm{d}\,F_{k,n}^{a_{k,n}}(x) + it\int\limits_{|x|<\tau} x\,\mathrm{d}\,F_{k,n}^{a_{k,n}}(x) \\
&\quad + \int\limits_{|x|\geq\tau} (e^{itx}-1)\,\mathrm{d}\,F_{k,n}^{a_{k,n}}(x)| \\
&\leq \frac{1}{2}|t|^2 \int\limits_{|x|<\tau} x^2\,\mathrm{d}\,F_{k,n}^{a_{k,n}}(x) + |t||\int\limits_{|x|<\tau} x\,\mathrm{d}\,F_{k,n}^{a_{k,n}}(x)| \\
&\quad + 2\int\limits_{|x|\geq\tau} \mathrm{d}\,F_{k,n}^{a_{k,n}}(x) \\
&= \frac{t^2}{2}u_{k,n} + |t|v_{k,n} + 2w_{k,n}, \text{ say.}
\end{aligned}$$

Now, by **(5.1.4)**,

$$\sum_{k=1}^{k_n} u_{k,n} \leq (1+\tau^2)\sum_{k=1}^{k_n}\int\limits_{|x|<\tau} \frac{x^2}{1+x^2}\,\mathrm{d}\,F_{k,n}^{a_{k,n}}(x) \leq (1+\tau^2)q$$

$$\begin{aligned}
\sum_{k=1}^{k_n} v_{k,n} &= \sum_{k=1}^{k_n}|\int\limits_{|x|<\tau} x\,\mathrm{d}\,F_{k,n}^{a_{k,n}}(x)| \leq 2\tau\sum_{k=1}^{k_n}\int\limits_{|x|\geq\frac{\tau}{2}} \mathrm{d}\,F_{k,n}^{a_{k,n}}(x) \\
&\leq 2\frac{\tau^2+4}{\tau}\sum_{k=1}^{k_n}\int\limits_{|x|\geq\frac{\tau}{2}} \frac{x^2}{1+x^2}\,\mathrm{d}\,F_{k,n}^{a_{k,n}}(x) \\
&\leq 2\frac{\tau^2+4}{\tau}q \text{ (by \textbf{(5.1.3)}(v) and \textbf{(5.1.4)})}.
\end{aligned}$$

Similarly, $\sum_{k=1}^{k_n} w_{k,n} \leq \frac{\tau^2+1}{\tau^2}q$. Collecting these results, we have

$$C_n(t) \leq \tilde{q}(t,\,\tau) \text{ and } |C_n(t)-A_n(t)| \leq q_1(t,\,\tau)\max_{1\leq k\leq k_n}|\beta_{k,n}| \to 0,$$

as was to be proved. □

Remarks 5.2.2.
(i) $e^{\psi_n(t)}$ is of the form $e^{i\alpha_n t}\prod_{k=1}^{k_n} e^{g_{k,n}(t)-1}$, where $g_{k,n}$ is a *cf* and $\alpha_n = \sum_{k=1}^{k_n} a_{k,n}$. Hence e^{ψ_n} is an *id cf* (ref. **(2.3.11)**, **(2.3.10)** and **(3.3.2)**). The weak limit, if it exists, of e^{ψ_n} is *id* (ref. **(3.3.2)**). Hence if $\varphi_n \to \varphi_0$, then φ_0 is an *id cf*.
(ii) If the ψ_ns are defined as above and if $e^{\psi_n} \to e^{\psi}$ then $\varphi_n \to e^{\psi}$.
(iii) Suppose that the $X_{k,n}$s are symmetrically distributed. Then φ_n converges to a *cf iff* e^{ψ_n} converges to a *cf* where

$$\psi_n(t) = \sum_{k=1}^{k_n} \int_R (\cos tx - 1)\, \mathrm{d}\, F_{k,n}(x).$$

(iv) On the same lines as in **(5.2.1)** the following version can be proved.

Let φ_n be the *cf* of $\sum_{k=1}^{k_n} X_{k,n} - a_n$ and let

$$\psi_n(t) = -ita_n + \sum_{k=1}^{k_n} \{ita_{k,n} + \int_R (e^{itx} - 1)\, \mathrm{d}\, F_{k,n}^{a_{k,n}}(x)\}.$$

If the weak limit φ of φ_n exists then for each $t \in R$, $\varphi_n(t) - e^{\psi_n(t)} \to 0$.

5.3 *CLT* (Lindeberg)

Let (X_n), (Y_n), $n \geq 1$ be two independent sequences of independent *rvs*. Let X_n and Y_n have a common *df* F_n, and common *cf* φ_n. When the X_ns are *iid*, write φ for φ_1. Let $\mathbb{E}X_n = 0$ and $\mathbb{E}X_n^2 = \sigma_n^2 \neq 0$. Let $s_n^2 = \sum_{i=1}^{n} \sigma_i^2$ and $S_n = \sum_{i=1}^{n} X_i$. Clearly (s_n) is a monotonic increasing sequence. Let ξ denote a standard normal variable. Define $Z_n = \frac{1}{\sqrt{2}}(X_n - Y_n)$ and note that $\mathbb{E}Z_n = 0$ and $\mathbb{E}Z_n^2 = \sigma_n^2$. Denote by G_n the *df* of Z_n. Let m_n be a median of X_n and let H_n denote the *df* of $X_n - m_n$. Consider the following statements:
(A) $\frac{S_n}{s_n} \xrightarrow{d} \xi$.
(B) $\max\limits_{1\leq k\leq n} \sigma_k^2 = o(s_n^2)$.
(C) For an $r \geq 2$ and for every $\varepsilon > 0$,

$$g_n(r,\, \varepsilon) \to 0 \text{ where } g_n(r,\, \varepsilon) = \frac{1}{s_n^r}\sum_{i=1}^{n} \int_{|x|\geq\varepsilon s_n} |x|^r\, \mathrm{d}\, F_i(x).$$

The condition $\lim\limits_{n\to\infty} g_n(2,\, \varepsilon) = 0$ for every $\varepsilon > 0$ is called the Lindeberg-Feller condition.

(D) For an $r \geq 2$ and for every $\varepsilon > 0$,

$$g_n^*(r,\, \varepsilon) \to 0 \text{ where } g_n^*(r,\, \varepsilon) = \frac{1}{s_n^r} \sum_{i=1}^{n} \int\limits_{|x| \geq \varepsilon s_i} |x|^r \, \mathrm{d}\, F_i(x).$$

(E$_1$) For a $p > 2$, let $\mathbb{E}|X_k|^p < \infty$, $1 \leq k \leq n$ and let $h_n(p) \to 0$ where $h_n(p) = \frac{1}{s_n^p} \sum_{k=1}^{n} \mathbb{E}|X_k|^p$.

The condition $\lim\limits_{n\to\infty} h_n(3) = 0$ is called the Liapounov condition.

(E$_2$) For a $p > 2$, let $\mathbb{E}|X_k|^p < \infty$, $1 \leq k \leq n$ and let $h_n^*(p) \to 0$ where $h_n^*(p) = \frac{1}{s_n^p} \sum_{k=1}^{n} \mathbb{E}|Z_k|^p$.

(F) For every $\varepsilon > 0$, $\hat{g_n}(2,\, \varepsilon) \to 0$ where

$$\hat{g_n}(2,\, \varepsilon) = \frac{1}{s_n^2} \sum_{i=1}^{n} \int\limits_{|x| \geq \varepsilon s_n} x^2 \, \mathrm{d}\, G_i(x).$$

(G) For every $\varepsilon > 0$, $\tilde{g_n}(2,\, \varepsilon) \to 0$ where

$$\tilde{g_n}(2,\, \varepsilon) = \frac{1}{s_n^2} \sum_{i=1}^{n} \int\limits_{|x| \geq \varepsilon s_n} x^2 \, \mathrm{d}\, H_i(x).$$

Theorem 5.3.1. (i) *(B)* $\Rightarrow$ *($s_n \to \infty$).*
(ii) *(C)* $\Rightarrow$ *(B) and (G)* $\Rightarrow$ *(B).*
(iii) *(D)* $\Rightarrow$ *(C).*
(iv) *(C)* $\Rightarrow$ *(D).*
(v) *(E$_1$)* $\Rightarrow$ *(C) when* $2 \leq r \leq p$, $p > 2$.
(C) $\Rightarrow$ *(E$_1$) if* $2 < r = p$.
(vi) *When* $r = 2$, *(G)* $\Leftrightarrow$ *(C).*
(vii) *When* $r = 2$, *(F)* $\Leftrightarrow$ *(C).*
(viii) *(A) and (B)* $\Rightarrow$ *(C) with* $r = 2$.
(ix) *(C)* $\Rightarrow$ *(A) and (B).*
(x) *For every postive integer k, (B) implies that, as* $n \to \infty$,
$s_n^{2k} - s_{n-1}^{2k} \sim k s_{n-1}^{2k-2} \sigma_n^2$
(xi) *(E$_1$)* $\Leftrightarrow$ *(E$_2$)*
(xii) *The following consequence of* (ix) *will be needed in* ***(5.6.1)***.

Let $\tau(n) < \eta(n) < n$ *be integers such that* $\frac{s_{\tau(n)}^2}{s_n^2} \to \alpha$, $\frac{s_{\eta(n)}^2}{s_n^2} \to \beta$, $\alpha < \beta$. *Then* $\frac{S_{\eta(n)} - S_{\tau(n)}}{s_n} \xrightarrow{d} \xi\sqrt{\beta - \alpha}$.

Note. (viii) and (ix) *constitute the Central Limit Theorem for independent variables with finite variances and satisfying the Lindeberg-Feller condition.*

(viii) and (ix) *hold also if the* $X_k = X_{k,n}$, $1 \le k \le n$, *depend on* n.

Proof (i) We have noted $s_n \uparrow$. If $\lim_{n\to\infty} s_n < \infty$, then (B) would imply $\sigma_n = 0$ for all n, contrary to assumption.

(ii) If (C) holds with $r > 2$, it holds with $r = 2$. Hence

$$\begin{aligned}\frac{1}{s_n^2}\max_{1\le k\le n}\sigma_k^2 &= \frac{1}{s_n^2}\int\limits_{|x|<\varepsilon s_n} x^2 \,\mathrm{d}\,F_k(x) + \int\limits_{|x|\ge\varepsilon s_n} x^2 \,\mathrm{d}\,F_k(x)\\ &\le \frac{1}{s_n^2}\{\varepsilon^2 s_n^2 + g_n(2,\,\varepsilon)s_n^2\} = \varepsilon^2 + o(1).\end{aligned}$$

The desired result follows by first letting $n \to \infty$ and then letting $\varepsilon \to 0$.

To prove the second part, we need the result that for any *rv* X, $\mathbb{E}(X-a)^2$ is least when $a = \mathbb{E}X$. The proof for this claim is by finding the minimum by differentiation.

$$\begin{aligned}\frac{1}{s_n^2}\max_{1\le k\le n}\sigma_k^2 &\le \frac{1}{s_n^2}\max_{1\le k\le n}\mathbb{E}(X-m_k)^2\\ &\le \frac{1}{s_n^2}\int\limits_{|x-m_k|<\varepsilon s_n}(x-m_k)^2\,\mathrm{d}\,F_k(x)\\ &\qquad + \frac{1}{s_n^2}\int\limits_{|x-m_k|\ge\varepsilon s_n}(x-m_k)^2\,\mathrm{d}\,F_k(x)\\ &\le \frac{1}{s_n^2}\{\varepsilon^2 s_n^2 + \tilde{g_n}(2,\,\varepsilon)s_n^2\} = \varepsilon^2 + o(1).\end{aligned}$$

The desired result follows by first letting $n \to \infty$ and then letting $\varepsilon \to 0$.

(iii) Since (s_n) is monotonic increasing $g_n(r,\,\varepsilon) \le g_n^*(r,\,\varepsilon)$. Hence (D) $\Rightarrow$ (C).

(iv) Let now (C) hold with some $r \ge 2$. Fix $\varepsilon > 0$. We must show that $g_n^*(r,\,\varepsilon) \to 0$ as $n \to \infty$. Fix target error η, $0 < \eta < \varepsilon$. Find $k = k(\eta; n)$

such that $s_k \le \frac{\eta^2}{\varepsilon} s_n < s_{k+1}$. This is possible because $s_n \uparrow$. Now,

$$\begin{aligned}
g_n^*(r,\varepsilon) &= \frac{1}{s_n^r}\sum_{j=1}^{k}\int\limits_{|x|>\epsilon s_j} |x|^r \,\mathrm{d}\,F_j(x) + \frac{1}{s_n^r}\sum_{j=k+1}^{n}\int\limits_{|x|\ge \epsilon s_j} |x|^r \,\mathrm{d}\,F_j(x) \\
&\le \frac{1}{s_n^r}\sum_{j=1}^{k}\int\limits_{R} |x|^r \,\mathrm{d}\,F_j(x) + \frac{1}{s_n^r}\sum_{j=k+1}^{n}\int\limits_{|x|>\eta^2 s_n} |x|^r \,\mathrm{d}\,F_j(x) \\
&\le \frac{1}{s_n^r}\sum_{j=1}^{k}\int\limits_{R} |x|^r \,\mathrm{d}\,F_j(x) + g_n(r,\eta^2).
\end{aligned}$$

If $r = 2$, the first term on the right side equals

$$\frac{1}{s_n^2}\sum_{j=1}^{k}\int\limits_{R} x^2 \,\mathrm{d}\,F_j(x) = \frac{s_k^2}{s_n^2} \le \frac{\eta^4}{\varepsilon^2}(\text{by the choice of } k) \le \eta^2;$$

if $r > 2$, this term is

$$\begin{aligned}
&\frac{1}{s_n^r}\sum_{j=1}^{k}\int\limits_{|x|<\eta s_n} |x|^r \,\mathrm{d}\,F_j(x) + \frac{1}{s_n^r}\sum_{j=1}^{k}\int\limits_{|x|\ge \eta s_n} |x|^r \,\mathrm{d}\,F_j(x) \\
&\le \frac{1}{s_n^r}\sum_{j=1}^{k}\int\limits_{|x|<\eta s_n} |x|^r \,\mathrm{d}\,F_j(x) + g_n(r,\ \eta) \\
&\le \frac{(\eta s_n)^{r-2}}{s_n^r}\sum_{j=1}^{k}\int\limits_{R} x^2 \,\mathrm{d}\,F_j(x) + g_n(r,\ \eta) \le \eta^{r-2} + g_n(r,\ \eta).
\end{aligned}$$

Thus $g_n^*(r,\ \varepsilon) \le \eta^{\min(2,\ r-2)} + 2g_n(r,\ \eta^2)$. The claim $\lim\limits_{n\to\infty} g_n^*(r,\ \varepsilon) = 0$ follows from this by taking $n \to \infty$ and then taking $\eta \to 0$, since $\lim\limits_{n\to\infty} g_n(r,\ \eta^2) = 0$ and η is arbitrary.

(v) Let (E_1) hold with some $p > 2$ and let $\varepsilon > 0$ be arbitrary.

$$\begin{aligned}
g_n(r,\ \varepsilon) &= \frac{1}{s_n^r}\sum_{k=1}^{n}\int\limits_{|x|\ge \varepsilon s_n} \frac{|x|^p}{|x|^{p-r}} \,\mathrm{d}\,F_k(x) \\
&\le \frac{1}{\varepsilon^{p-r} s_n^p}\sum_{k=1}^{n}\int\limits_{R} |x|^p \,\mathrm{d}\,F_k(x) \\
&= \frac{1}{\varepsilon^{p-r}} h_n(p).
\end{aligned}$$

Hence (E_1) $\Rightarrow$ (C), when $2 \le r \le p$, $p > 2$.

Suppose now (C) is true for some $r > 2$. Given $\varepsilon > 0$,

$$h_n(r) = \frac{1}{s_n^r} \sum_{k=1}^{n} \int_{|x|<\varepsilon s_n} |x|^r \, \mathrm{d}\, F + g_n(r,\, \varepsilon) \le \frac{(\varepsilon s_n)^{r-2} s_n^2}{s_n^r} + g_n(r,\, \varepsilon).$$

That $\lim_{n\to\infty} h_n(r) = 0$ follows by taking the limit of the rightside first as $n \to \infty$ and then as $\varepsilon \to 0$. Thus (C) $\Rightarrow$ (E_1) if $2 < r = p$.

(vi) Before we proceed to prove (vi), we record:

If X has mean 0, median m and variance σ^2 then $|m| \le \sqrt{2}\sigma$.

Suppose $m > 0$. Then

$$\sigma^2 = \int_R x^2 \, \mathrm{d}\, F(x) \ge \int_{[m,\, \infty)} x^2 \, \mathrm{d}\, F(x) \ge m^2 P(X \ge m) \ge \frac{1}{2} m^2.$$

If $m < 0$, the proof that $-m \le \sigma\sqrt{2}$ follows by considering the variable $-X$.

Now we proceed to the proof of (vi).

We note that (C) and (G) together imply (B). By (B),

$$\max_{1\le k\le n} |m_k| \le \sqrt{2} \max_{1\le k\le n} \sigma_k \le \frac{\varepsilon s_n}{2} \text{ for all } n \text{ sufficiently large.}$$

Now

$$\begin{aligned}
\tilde{g}_n(2,\, \varepsilon) &= \frac{1}{s_n^2} \sum_{k=1}^{n} \int_{|x|\ge\varepsilon s_n} x^2 \, \mathrm{d}\, F_k(x + m_k) \\
&= \frac{1}{s_n^2} \sum_{k=1}^{n} \int_{|x-m_k|\ge\varepsilon s_n} (x - m_k)^2 \, \mathrm{d}\, F_k(x) \\
&\ge \frac{1}{s_n^2} \sum_{k=1}^{n} \int_{|x|\ge 2\varepsilon s_n} (x - m_k)^2 \, \mathrm{d}\, F_k(x) \\
&\ge \frac{1}{s_n^2} \sum_{k=1}^{n} \{ \int_{|x|\ge 2\varepsilon s_n} x^2 \, \mathrm{d}\, F_k(x) - 2\, m_k \int_{|x|\ge 2\varepsilon s_n} x \, \mathrm{d}\, F_k(x) \}.
\end{aligned}$$

We note

$$2|m_k| \int\limits_{|x|\geq 2\varepsilon s_n} |x| \,\mathrm{d}\, F_k(x) \leq \varepsilon s_n \int\limits_{|x|\geq 2\varepsilon s_n} |x| \,\mathrm{d}\, F_k(x)$$

$$\leq \frac{1}{2} \int\limits_{|x|\geq 2\varepsilon s_n} x^2 \,\mathrm{d}\, F_k(x).$$

Hence

$$\tilde{g_n}(2,\ \varepsilon) \geq \frac{1}{2} \sum_{k=1}^{n} \int\limits_{|x|\geq 2\varepsilon s_n} x^2 \,\mathrm{d}\, F_k(x) = \frac{1}{2} g_n(2,\ 2\varepsilon)$$

for all n sufficiently large.

Similar arguments lead to $\tilde{g_n}(2,\ \varepsilon)$

$$\leq \frac{1}{s_n^2} \sum_{k=1}^{n} \Big\{ \int\limits_{|x|\geq \frac{1}{2}\varepsilon s_n} x^2 \,\mathrm{d}\, F_k(x) + 2\,|m_k| \int\limits_{|x|\geq \frac{1}{2}\varepsilon s_n} |x| \,\mathrm{d}\, F_k(x)$$

$$+ m_k^2 \int\limits_{|x|\geq \frac{1}{2}\varepsilon s_n} \mathrm{d}\, F_k(x) \Big\} \leq 4 g_n(2,\ \tfrac{1}{2}\varepsilon)$$

for all n large. Take limit, as $n \to \infty$, in the resulting inequality $\frac{1}{2} g_n(2,\ 2\varepsilon) \leq \tilde{g_n}(2,\ \varepsilon) \leq 4 g_n(2,\ \frac{1}{2}\varepsilon)$, valid for all n large, and complete the proof.

(vii) By integration by parts

$$\hat{g_n}(2,\varepsilon) \leq \frac{1}{s_n^2} \sum_{k=1}^{n} \{\varepsilon^2 s_n^2 P(|X_k - Y_k| \geq \sqrt{2}\varepsilon s_n)$$

$$+ \int\limits_{\varepsilon s_n}^{\infty} 2x P(|X_k - Y_k| \geq \sqrt{2}x) \,\mathrm{d}\, x\}$$

$$\leq \frac{1}{s_n^2} \sum_{k=1}^{n} \{\varepsilon^2 s_n^2 2P(|X_k| \geq \frac{1}{2}\sqrt{2}\varepsilon s_n)$$

$$+ \int\limits_{\varepsilon s_n}^{\infty} 4x P(|X_k| \geq \frac{1}{2}\sqrt{2}x) \,\mathrm{d}\, x\}$$

(since X_k and Y_k are identically distributed)

$$\leq \frac{1}{s_n^2}\sum_{k=1}^{n}\{\varepsilon^2 s_n^2 2P(|X_k| \geq \frac{1}{2}\varepsilon s_n) + \int_{\varepsilon s_n}^{\infty} 4xP(|X_k| \geq \frac{1}{2}x)\,\mathrm{d}\,x\}$$

$$\leq \frac{1}{s_n^2}\sum_{k=1}^{n}\{2\varepsilon^2 s_n^2 \int_{|x|\geq\frac{1}{2}\varepsilon s_n} \mathrm{d}\,F_k(x) + 8\int_{\frac{1}{2}\varepsilon s_n}^{\infty} 2xP(|X_k| \geq x)\,\mathrm{d}\,x\}$$

$$\leq \frac{1}{s_n^2}\sum_{k=1}^{n}\{8\int_{|x|\geq\frac{1}{2}\varepsilon s_n} x^2\,\mathrm{d}\,F_k(x) + 8\int_{|x|\geq\frac{1}{2}\varepsilon s_n} x^2\,\mathrm{d}\,F_k(x)\}$$

$$= 16 g_n(2, \frac{1}{2}\varepsilon).$$

Again

$$\hat{g_n}(2, \varepsilon) \geq \frac{1}{s_n^2}\sum_{k=1}^{n}\varepsilon^2 s_n^2 P(|X_k - Y_k| \geq \sqrt{2}\varepsilon s_n)$$
$$+ \frac{1}{s_n^2}\sum_{k=1}^{n}\int_{\varepsilon s_n}^{\infty} 2xP(|X_k - Y_k| \geq \sqrt{2}x)\,\mathrm{d}\,x$$
$$\geq \frac{1}{s_n^2}\sum_{k=1}^{n}\varepsilon^2 s_n^2 \frac{1}{2}P(|X_k - m_k| \geq \sqrt{2}\varepsilon s_n)$$
$$+ \frac{1}{s_n^2}\sum_{k=1}^{n}\int_{\varepsilon s_n}^{\infty} 2x\frac{1}{2}P(|X_k - m_k| > \sqrt{2}x)\,\mathrm{d}\,x \text{ (ref. \textbf{(4.2.3)}(i))}$$
$$\geq \frac{1}{4}\tilde{g_n}(2, \varepsilon\sqrt{2}) \geq \frac{1}{8}g_n(2, 2\sqrt{2}\varepsilon)$$

by the first part of (vi) above.

These two inequalities lead to the desired result.

Proof of (viii) and (ix)

Define $X_{k,n} = \frac{X_k}{s_n}$, $1 \leq k \leq n$ and note that they are infinitesimal since $\lim_{n\to\infty}\max_{1\leq k\leq n} P(|X_{k,n}| > \varepsilon) \leq \lim_{n\to\infty}\max_{1\leq k\leq n}\frac{\sigma_k^2}{\varepsilon^2 s_n^2}$ (ref. **(4.4.1)**) $= 0$ by (B).

We establish the claim first by assuming that the X_ks are symmetrically distributed.

Let (A) and (B) hold. By **(5.2.1)**, the fact that $\frac{S_n}{s_n} \xrightarrow{d} \xi$ implies $\psi_n(t) = \sum_{k=1}^{n}\int_R(\cos tx - 1)\,\mathrm{d}\,F_k(xs_n) \to -\frac{t^2}{2}$ or, what is same, for t fixed, $\frac{t^2}{2} +$

$\psi_n(t) = o(1)$ as $n \to \infty$. This result is true for all t. Given $\varepsilon > 0$, choose and fix a t such that $\frac{t^2}{2} > \frac{2}{\varepsilon^2}$.

Now, $-\psi_n(t) = I_1 + I_2$ where

$$I_1 = \sum_{k=1}^{n} \int_{|x|<\varepsilon s_n} (1 - \cos(\frac{tx}{s_n}))\, \mathrm{d}\, F_k(x)$$

and

$$I_2 = \sum_{k=1}^{n} \int_{|x|\geq\varepsilon s_n} (1 - \cos(\frac{tx}{s_n}))\, \mathrm{d}\, F_k(x).$$

We have

$$I_1 \leq \frac{t^2}{2s_n^2} \sum_{k=1}^{n} \int_{|x|<\varepsilon s_n} x^2\, \mathrm{d}\, F_k(x) = \frac{t^2}{2}\{1 - g_n(2,\, \varepsilon)\}$$

and

$$I_2 \leq 2 \sum_{k=1}^{n} \int_{|x|\geq\varepsilon s_n} \mathrm{d}\, F_k(x) \leq \frac{2}{\varepsilon^2} g_n(2,\, \varepsilon),$$

leading to

$$I_1 + I_2 \leq \frac{t^2}{2} - \{\frac{t^2}{2} - \frac{2}{\varepsilon^2}\} g_n(2,\, \varepsilon).$$

Hence

$$0 < (\frac{t^2}{2} - \frac{2}{\varepsilon^2}) g_n(2,\, \varepsilon) < \frac{t^2}{2} - (I_1 + I_2) = \frac{t^2}{2} + \psi_n(t) = o(1)$$

as $n \to \infty$. That (C) is true follows immediately from this.

In the reverse direction, assume (C) holds. That this implies (B) has already been established in (ii). Again by **(5.2.1)**, $\frac{S_n}{s_n} \xrightarrow{d} \xi$ will follow if we show $\psi_n(t) \to -\frac{t^2}{2}$ for every $t \in R$. We recall the well known result $\lim_{\theta\to 0} \frac{\sin\theta}{\theta} = 1$. Hence given $\varepsilon > 0$, we can find a $\delta > 0$ such that $\frac{\sin\theta}{\theta} > 1 - \varepsilon$ for all $\theta,\ |\theta| \leq \delta$. Fix $t > 0$. Choose δ_1 such that $t\delta_1 \leq 2\delta$. Now

$$-\psi_n(t) = \sum_{k=1}^{n} \int_{|x|<\delta_1 s_n} (1 - \cos(\frac{tx}{s_n}))\, \mathrm{d}\, F_k(x) + \alpha_n$$

where

$$\alpha_n = \sum_{k=1}^{n} \int_{|x|\geq\delta_1 s_n} (1-\cos(\frac{tx}{s_n}))\, \mathrm{d}\, F_k(x) \leq \frac{2}{\delta_1^2} g_n(2,\ \delta_1) = o(1),$$

by hypothesis. Consider

$$\sum_{k=1}^{n} \int_{|x|<\delta_1 s_n} (1-\cos(\tfrac{tx}{s_n}))\mathrm{d}F_k(x) = \tfrac{t^2}{2}\tfrac{1}{s_n^2} \sum_{k=1}^{n} \int_{|x|<\delta_1 s_n} \lambda(t,\ x)x^2 \mathrm{d}F_k(x)$$

where

$$\lambda(t,\ x) = \frac{\sin^2(\frac{tx}{2s_n})}{\frac{t^2x^2}{4s_n^2}}.$$

Since $\lambda(t,\ x) \leq 1$ for all t and x and since $\lambda(t,\ x) \geq 1-\varepsilon$ when $|x| < \delta_1 s_n$, it follows that

$$\begin{aligned}(1-\varepsilon)\frac{t^2}{2} \lim_{n\to\infty} \frac{1}{s_n^2} \sum_{k=1}^{n} \int_{|x|<\delta_1 s_n} x^2\, \mathrm{d}\, F_k(x) \\ \leq \varliminf_{n\to\infty} \{-\psi_n(t)\} \leq \varlimsup_{n\to\infty} \{-\psi_n(t)\} \\ \leq \lim_{n\to\infty} \frac{t^2}{2}\frac{1}{s_n^2} \sum_{k=1}^{n} \int_{|x|<\delta_1 s_n} x^2\, \mathrm{d}\, F_k(x).\end{aligned}$$

Using hypothesis that (C) holds, take limit as $n \to \infty$ and then as $\varepsilon \to 0$ and conclude that $-\lim_{n\to\infty} \psi_n(t) = \frac{t^2}{2}$.

With this, the proof is complete that if the X_ks are symmetrically distributed then (A) & (B) $\Leftrightarrow$ (C).

Now we extend result to the variables X_k which are not necessarily symmetrically distributed. Let $Y_n,\ Z_n,\ G_n$ be as defined in Sec. **(5.3)**. Let $T_n = \sum_{k=1}^n Y_k$ and $V_n = \sum_{k=1}^n Z_k$.

Suppose (A) & (B) hold for the X_k. Hence (A) & (B) hold for the symmetric variables Z_k when $r = 2$. By what has been proved above for symmetric variables it follows that (F) holds. This implies (C) by (vii).

Suppose now (C) holds. This implies (F), when $r = 2$, by (vii). By the result for symmetric variables this, in turn, implies that (B) holds and that $\frac{V_n}{s_n} \overset{d}{\to} \xi$. We must show that $\frac{S_n}{s_n} \overset{d}{\to} \xi$. Since $P(|\frac{S_n}{s_n}| \geq \lambda) \leq \frac{1}{\lambda^2}\mathbb{E}(\frac{S_n}{s_n})^2 = \frac{1}{\lambda^2}$, we see, by choosing λ large, that $(\frac{S_n}{s_n})$ is a tight sequence. So is the sequence $(\frac{T_n}{s_n})$. Let $(\frac{S_{n_r}}{s_{n_r}})$ be a weakly convergent subsequence, converging to, say, ζ_1. This implies that there exists ζ_2 such that $\frac{T_{n_r}}{s_{n_r}} \overset{d}{\to} \zeta_2$. By (B), $(\frac{X_k}{s_{n_r}},\ 1 \leq k \leq$

$n_r)$ and $\left(\frac{Y_k}{s_{n_r}},\ 1 \le k \le n_r\right)$ are infinitesimal. Hence the weak limits ζ_1 and ζ_2 are necessarily id (ref. **(5.2.2)**(i)). We note $\zeta_1,\ \zeta_2$ are *iid* and that, because S_n and T_n are mutually independent, $\frac{V_{n_r}}{s_{n_r}} = \frac{1}{\sqrt{2}}\{\frac{S_{n_r}}{s_{n_r}} - \frac{T_{n_r}}{s_{n_r}}\} \overset{d}{\to} \frac{1}{\sqrt{2}}(\zeta_1 - \zeta_2)$. Hence $\frac{1}{\sqrt{2}}(\zeta_1 - \zeta_2)$ is distributed as ξ. By Cramér's theorem (ref. **(3.4.6)**), $\zeta_1,\ \zeta_2$ are normal variables, say, $\mathcal{N}(\mu,\ \sigma^2)$. Since $1 = var\xi = var(\frac{\zeta_1-\zeta_2}{\sqrt{2}}) = \sigma^2$, we get $\sigma = 1$. Since $\frac{V_n}{s_n} \overset{d}{\to} \xi$ and since $\mathbb{E}(\frac{V_n}{s_n})^2 = 1 = \mathbb{E}\xi^2$, it follows (ref. **(2.6.9)**(iii)) that $(\frac{V_n}{s_n})^2$ is uniformly integrable. We have (ref. proof of (**(5.3.1)**(vi)) $|median(\frac{S_n}{s_n})| \le \sqrt{2var(\frac{S_n}{s_n})} = \sqrt{2}$ and, similarly, $|median(\frac{T_n}{s_n})| \le \sqrt{2}$. These facts and the uniform integrability of $(\frac{S_n}{s_n} - \frac{T_n}{s_n})^2$ imply (ref. **(4.2.5)**) that $\left((\frac{S_n}{s_n})^2\right)$ is a uniformly integrable sequence. Since $\frac{S_{n_r}}{s_{n_r}} \overset{d}{\to} \xi + \mu$ and since $\left((\frac{S_{n_r}}{s_{n_r}})^2\right)$ is uniformly integrable, $\mathbb{E}\frac{S_{n_r}}{s_{n_r}} \to \mathbb{E}(\xi + \mu)$ (ref. **(2.6.9)**(ii)). Since $\mathbb{E}\frac{S_{n_r}}{s_{n_r}} = 0$ and $\mathbb{E}\xi = 0$, we conclude $\mu = 0$. Thus every weakly convergent subsequence of the tight sequence $(\frac{S_n}{s_n})$ converges to the $\mathcal{N}(0,\ 1)$ *df*. Hence $\frac{S_n}{s_n} \overset{d}{\to} \xi$. i.e. (A) holds.

(x) For $k = 1$ we have the obvious relation $s_n^2 - s_{n-1}^2 = \sigma_n^2$.
For $k \ge 2$,

$s_n^{2k} = (s_n^2)^k = (s_{n-1}^2 + \sigma_n^2)^k = s_{n-1}^{2k} + ks_{n-1}^{2k-2}\sigma_n^2 + \sum_{j=2}^{k} \binom{k}{j} s_{n-1}^{2k-2j}\sigma_n^{2j}$. Now,

$$\frac{s_{n-1}^{2k-2j}\sigma_n^{2j}}{s_{n-1}^{2k-2}\sigma_n^2} = \left(\frac{\sigma_n^2}{s_{n-1}^2}\right)^{j-1} \sim \left(\frac{\sigma_n^2}{s_n^2}\right)^{j-1} = o(1),$$

by (B). Hence $s_n^{2k} - s_{n-1}^{2k} \sim ks_{n-1}^{2k-2}\sigma_n^2$.

(xi) By the inequality **(2.6.5)**(iv), $\mathbb{E}|Z_k|^p \le 2^p\mathbb{E}|X_k|^p$. This implies that $(E_1) \Rightarrow (E_2)$.

We proceed to prove implication in the reverse direction. We note first that, by (v), (E_2) implies (C) for the variables (Z_n). Hence (B) holds by (ii). Let m_k be a median of X_k. We then get by **(4.2.3)**, $\mathbb{E}|X_k - m_k|^p \le 2(\sqrt{2})^p\mathbb{E}|Z_k|^p$. Hence $\lim_{n\to\infty} \frac{1}{s_n^p}\sum_{k=1}^n \mathbb{E}|X_k - m_k|^p = 0$. Now $\mathbb{E}|X_k|^p \le \mathbb{E}(|X_k - m_k| + |m_k|)^p \le$ (by **(2.6.5)**(iv)) $2^{p-1}\{\mathbb{E}|X_k - m_k|^p + |m_k|^p\}$. Hence $(E_2) \Rightarrow (E_1)$ if $\theta_n = \frac{1}{s_n^p}\sum_{k=1}^n |m_k|^p \to 0$. But this is true since $\theta_n \le$ (ref. proof of **(5.3.1)**(vi))

$$\frac{1}{s_n^p}\sum_{k=1}^n (\sigma_k\sqrt{2})^p \le q\frac{1}{s_n^p}\max_{1\le k\le n}\sigma_k^{p-2}\sum_{k=1}^n(\sigma_k)^2 = q\left(\frac{\max_{1\le k\le n}\sigma_k^2}{s_n^2}\right)^{\frac{p-2}{2}} \to 0.$$

(xii) Since $\frac{S_{\eta(n)}}{s_{\eta(n)}} \overset{d}{\to} \xi$ and since $s_{\eta(n)}^2 \sim \beta s_n^2$, we have $\frac{S_{\eta(n)}}{s_n} \overset{d}{\to} \xi\sqrt{\beta}$. Similarly

$\frac{S_{\tau(n)}}{s_n} \xrightarrow{d} \xi\sqrt{\alpha}$. We now use independence to conclude

$$\frac{S_{\eta(n)} - S_{\tau(n)}}{s_n} \xrightarrow{d} \xi\sqrt{\beta - \alpha}.$$

Remarks 5.3.2.

(i) If the mean zero X_ks are *iid* with finite variance, then they satisfy Lindeberg-Feller condition.

(ii) If (C) does not hold, then either (A) or (B) or both must fail to hold. If (C) does not hold but (B) holds then necessarily (A) fails to hold. i.e. it is not true that $\frac{S_n}{s_n} \xrightarrow{d} \xi$. What does this mean? The following examples help enhance our understanding of the possibilities.

For $j = 1, 2, \ldots$, let $X_j = \pm j^2$ with probability $\frac{1}{12j^2}$ each; $X_j = \pm j$ with probability $\frac{1}{12}$ each; $X_j = 0$ with probability $1 - \frac{1}{6} - \frac{1}{6j^2}$. We note $\mathbb{E}X_j = 0$, $\mathbb{E}X_j^2 = \frac{j^2}{3}$, $var(S_n) \sim \frac{n^3}{9}$. Let $Y_j = X_j$ if $|X_j| \leq j$ and $Y_j = 0$ otherwise. Let $T_n = \sum_{j=1}^{n} Y_j$. $t_n^2 = varT_n \sim \frac{n^3}{18}$. Given $\varepsilon > 0$, take $n > \frac{72}{\varepsilon^2}$ and note that $\int_{|x|>\varepsilon t_n} x^2 \,\mathrm{d}P(Y_j \leq x) = 0,\ 1 \leq j \leq n$. i.e. Lindebeg-Feller condition holds for the sequence (Y_n). Hence $\frac{T_n}{t_n} \xrightarrow{d} \xi$. We note that $\sum_{j=1}^{\infty} P(X_j \neq Y_j) = \sum_{j=1}^{\infty} \frac{1}{12j^2} < \infty$. Hence $(\frac{T_n}{t_n} \xrightarrow{d} \xi) \Rightarrow (\frac{S_n}{t_n} \xrightarrow{d} \xi)$. Thus $\frac{S_n}{s_n} \xrightarrow{d} \frac{1}{\sqrt{2}}\xi$. We can infer from this or prove directly that (C) does not hold for the X_ks. Yet there exist constants t_n such that $\frac{S_n}{t_n} \xrightarrow{d} \xi$. Thus for the X_ns, neither (C) nor (A) holds but (B) holds.

(iii) For each $n \geq 1$, let Y_n be a $\mathcal{N}(0,\ 1)$ variable and suppose that variable Z_n is such that $P(Z_n = 0) = 1 - \frac{1}{n^2}$ and over the region $\{|x| \geq n^2\}$ it has frequency function $\frac{1}{2|x|^2}$. All variables are assumed to be mutually independent. Define $X_n = Y_n + Z_n$. We note that $\mathbb{E}X_n$ does not exist, that $\sum_n P(X_n \neq Y_n) = \sum_n P(Z_n \neq 0) = \sum_n \frac{1}{n^2} < \infty$ and that $\frac{\sum_1^n Y_k}{\sqrt{n}}$ has $\mathcal{N}(0,\ 1)$ distribution. Hence limiting distribution of $\frac{\sum_1^n X_k}{\sqrt{n}}$ is the same as that of $\frac{\sum_1^n Y_k}{\sqrt{n}}$ which is $\mathcal{N}(0,\ 1)$.

It may be noted that this does not contradict **(5.3.2)**(ii) since the X_ns are not identically distributed.

(iv) *A direct proof of the Central limit theorem for iid variables with finite variances*

Let $(X_n,\ n \geq 1)$ be a sequence of *iid* variables with $\mathbb{E}X_1 = 0,\ \mathbb{E}X_1^2 = 1$. Let $T_n = \frac{S_n}{\sqrt{n}} = \sum_{\nu=1}^{n} \frac{X_\nu}{\sqrt{n}}$. Let (n_k) be an integer sequence with $n_k \uparrow \infty$.

Then the variables $(\frac{X_\nu}{\sqrt{n_k}},\ 1 \le \nu \le n_k)$ are infinitesimal. Suppose the limiting distribution of the sequence (T_{n_k}) exists. It will be necessarily infinitely divisible (ref. **(5.2.2)**). If G_n is the distribution of $\frac{S_n}{\sqrt{n}}$ and G is the limiting distribution of G_{n_k}, then $1 = \int_R x^2 \,\mathrm{d}\, G_{n_k}(x) \ge \int_{-\lambda}^{\lambda} x^2 \,\mathrm{d}\, G_{n_k}(x)$, for every $\lambda > 0$. Choose $\lambda,\ -\lambda$ to be continuity points of G, take limit as $n \to \infty$ and appeal to the definition of weak convergence of *df*s to get $1 \ge \int_{-\lambda}^{\lambda} x^2 \,\mathrm{d}\, G(x)$. Now allow $\lambda \to \infty$ and conclude $\int_R x^2 \,\mathrm{d}\, G(x) \le 1$. Let the limit *cf* be $e^{\psi(t)}$ where, using the Kolmogorov canonical form for an infinitely divisible law (ref. **(3.4.5)**),

$$\Re(\psi(t)) = -\frac{a^2t^2}{2} - \int\limits_{|x|>0} \frac{1-\cos tx}{x^2} \,\mathrm{d}\, K(x).$$

If $\varphi(t)$ is the *cf* of X_1, then

$$n_k\{1 - \Re(\varphi(\frac{t}{\sqrt{n_k}}))\} \to \frac{a^2t^2}{2} + \int_{|x|>0} \frac{1-\cos tx}{x^2} \,\mathrm{d}\, K(x).$$

The limit of $n_k\{1-\Re(\varphi(\frac{t}{\sqrt{n_k}}))\}$ is also equal to $\frac{t^2}{2} \times$ the variance of $X_1 = \frac{t^2}{2}$. Hence, for all t,

$$\frac{1}{2}t^2 = \frac{a^2t^2}{2} + \int\limits_{|x|>0} \frac{1-\cos tx}{x^2} \,\mathrm{d}\, K(x).$$

Divide by t^2 throughout, allow $t^2 \to \infty$ and appeal to the bounded convergence theorem to conclude $1 = a^2$. This implies

$$\int_{|x|>0} \frac{1-\cos tx}{x^2} \,\mathrm{d}\, K(x) = 0.$$

Since the integrand is non-negative, this implies that the K-measure of $R \sim \{0\} = 0$. And then the *cf* of G is that of a standard normal variable. Thus every convergent subsequence of the sequence (T_n) converges to the standard normal distribution. Since $P(|T_n| > t) \le \frac{1}{t^2}\mathbb{E}T_n^2 = \frac{1}{t^2}$, it follows that sequence (T_n) is tight. Hence $\frac{S_n}{\sqrt{n}} \xrightarrow{d} \xi$.

(v) The version of Lindeberg - Feller CLT stated below can be established by following closely the steps in the proofs of (**(5.3.1)**, (viii) and (ix)).

For each $n \ge 1$, let $(X_{k,n},\ 1 \le k \le k_n)$ be independent rvs with dfs $F_{k,n}$, with zero means and finite variances $\sigma_{k,n}^2$. Let $S_n = \sum_{k=1}^{k_n} X_{k,n}$ and

$s_n^2 = \sum_{k=1}^{k_n} \sigma_{k,n}^2$. Then $(\mathfrak{A})$ and $(\mathfrak{B}) \Leftrightarrow (\mathfrak{C})$ where
$(\mathfrak{A})$: $\frac{S_n}{s_n} \xrightarrow{d} \xi$
$(\mathfrak{B})$: $\max\limits_{1\le k\le k_n} \sigma_{k,n}^2 = o(s_n^2)$ and
$(\mathfrak{C})$: $\frac{1}{s_n^2}\sum\limits_{k=1}^{k_n} \int\limits_{|x|>\varepsilon} x^2 \,\mathrm{d}\, F_{k,n}(x) = o(1)$ for every $\varepsilon > 0$.

An illustration.

For each n fixed, $n = 1, 2, ...$, let $(X_{n,m}, 1 \le m \le n)$ be *iid* Bernoulli variables, with $P(X_{n,m} = 1) = \frac{c}{n^\alpha} = 1 - P(X_{n,m} = 0)$, $c > 0$, $0 < \alpha < 1$. Define $Y_{n,m} = n^{-\frac{1}{2}(1-\alpha)}(X_{n,m} - \frac{c}{n^\alpha})$ and $S_n = \sum\limits_{m=1}^{n} Y_{n,m}$. We note $\mathbb{E}Y_{n,m} = 0$ and $\mathbb{E}Y_{n,m}^2 = \frac{c}{n}(1 - \frac{c}{n^\alpha})$.
$s_n^2 = \sum\limits_{m=1}^{n} var(Y_{n,m}) = \sum\limits_{m=1}^{n} \frac{c}{n}(1-\frac{c}{n^\alpha}) = c(1-\frac{c}{n^\alpha}) \to c$. Hence $\frac{\max\limits_{1\le m\le n} \sigma_{n,m}^2}{s_n^2} +$ $o(1)$, proving that condition $(\mathfrak{B})$ is satisfied. For all $\varepsilon > 0$,

$$\lim_{n\to\infty} \frac{1}{s_n^2} \sum_{m=1}^{n} \int\limits_{|x|>\varepsilon s_n} x^2 \mathrm{d}P(Y_{n,m} \le x) = 0 \textit{ iff}$$

$\lim\limits_{n\to\infty} n \int\limits_{|x|>\varepsilon s_1} x^2\mathrm{d}P(Y_{n,1} \le x) = 0$. i.e., *iff* $\lim\limits_{n\to\infty} n \int\limits_{|x|>\varepsilon c(1-c)} x^2\mathrm{d}P(Y_{n,1} \le x) = 0$. We note $P(|Y_{n,1}| > a) = 0$ for any $a > 0$ for all n sufficiently large, largeness depending on a. Hence condition $(\mathfrak{C})$ is satified and we conclude $Z_n = \frac{S_n - cn^{1-\alpha}}{n^{\frac{1}{2}(1-\alpha)}} \xrightarrow{d} \sqrt{c}\xi$ or $Z_n = \frac{S_n - np_n}{\sqrt{np_n}} \xrightarrow{d} \xi$. □

(vi) The weak convergence $(\frac{S_n}{B_n} \xrightarrow{d} \xi)$ in the CLT can not be strengthened to convergence in probabilty. i.e., never can $(\frac{S_n}{B_n} \xrightarrow{pr} \xi)$ be true.
Proof. Let

$$\frac{S_n}{B_n} \xrightarrow{d} \xi \tag{5.1}$$

Case 1. The X_ns are assumed to be *iid* with zero means. If $EX_1^2 < \infty$, then take $\mathbb{E}X_1^2 = 1$. We then have $B_n = \sqrt{n}$. When the variance is infinite, $B_n = \sqrt{n l(n)}$ where $l(n)$ is a sv function with $\lim\limits_{n\to\infty} l(n) = \infty$ ref.**(5.4.2)**. For both the sub cases, we note $B_{2n} \sim \sqrt{2}B_n$. If $\frac{S_n}{B_n} \xrightarrow{pr} \xi$, then $\frac{S_{2n}}{B_{2n}} \xrightarrow{pr} \xi$ or, $\frac{S_{2n}}{B_n} \xrightarrow{pr} \sqrt{2}\xi$. Hence $\frac{S_{2n}-S_n}{B_n} \xrightarrow{pr} (\sqrt{2}-1)\xi$, leading to $\frac{S_{2n}-S_n}{B_n} \xrightarrow{d} (\sqrt{2}-1)\xi$. But, since the variables are identically distributed, this implies that $\frac{S_n}{B_n} \xrightarrow{d} (\sqrt{2}-1)\xi$, contradicting (5.1).
Case 2. Let the independent vaiables (X_n) have zero means and finite variances (σ_n^2) and let Lindeberg-Feller condition be satisfied. Hence

$$\max_{1\le k\le n} \sigma_k^2 = o(s_n^2) \tag{5.2}$$

and with $B_n^2 = s_n^2 = \sum_{k=1}^n \sigma_k^2$, (5.1) holds. We recall that (5.2) implies

$$s_n \to \infty \text{ and } s_n \sim s_{n+1}. \tag{5.3}$$

Fix $\lambda > 1$. Consider the sequence $\frac{s_{n+\nu}}{s_n}$, $\nu = 1, 2, \ldots$. Since $s_{n+k} \to \infty$ as $k \to \infty$, we can find $\alpha(n) > n$ such that $\frac{s^2_{\alpha(n)}}{s_n^2} \le \lambda < \frac{s^2_{\alpha(n)+1}}{s_n^2}$ or $\frac{s^2_{\alpha(n)}}{s_n^2} \le \lambda < \frac{s^2_{\alpha(n)}+\sigma^2_{\alpha(n)+1}}{s_n^2}$. Since $\frac{\sigma^2_{\alpha(n)+1}}{s_n^2} \le \lambda \frac{\sigma^2_{\alpha(n)+1}}{s^2_{\alpha(n)}} \to 0$ (ref. (5.2) and (5.3)), it follows that

$$s^2_{\alpha(n)} \sim \lambda s_n^2. \tag{5.4}$$

If $\frac{S_n}{s_n} \xrightarrow{pr} \xi$, then we will have $\frac{S_{\alpha(n)}}{s_{\alpha(n)}} \xrightarrow{pr} \xi$ or, by (5.4), $\frac{S_{\alpha(n)}}{s_n} \xrightarrow{pr} \sqrt{\lambda}\xi$. Hence $\frac{S_{\alpha(n)}-S_n}{s_n} \xrightarrow{pr} (\sqrt{\lambda}-1)\xi$ which implies

$$\frac{S_{\alpha(n)} - S_n}{s_n} \xrightarrow{d} (\sqrt{\lambda}-1)\xi. \tag{5.5}$$

Since $s^2_{\alpha(n)} - s_n^2 \sim (\lambda - 1)s_n^2$, the Lindeberg-Feller condition holds for the variables X_ν, $n+1 \le \nu \le \alpha(n)$, if

$$u_n = \frac{1}{s_n^2} \sum_{k=n+1}^{\alpha(n)} \int_{|x|>\varepsilon s_n} x^2 \, \mathrm{d} F_k(x) = o(1).$$

But this is true since $u_n = O(g_n(2, \varepsilon))$ (ref. Sec. **(5.3)** for notation). Hence by part (v) above,

$$\frac{S_{\alpha(n)} - S_n}{\sqrt{s^2_{\alpha(n)} - s_n^2}} \xrightarrow{d} \xi. \text{ i.e.}, \frac{S_{\alpha(n)} - S_n}{\sqrt{\lambda - 1}s_n} \xrightarrow{d} \xi. \text{ i.e.}, \frac{S_{\alpha(n)} - S_n}{s_n} \xrightarrow{d} \sqrt{\lambda - 1}\xi,$$

contradicting (5.5), since, for $\lambda > 1$, $\sqrt{\lambda} - 1 \ne \sqrt{\lambda - 1}$. The stated claim stands proved.

Convergence of moments in Lindeberg-Feller CLT

Follow the notation of section **(5.3)**. For a positive integer r let $\mathbb{E}X_k^{2r} < \infty$, $k = 1, 2, \ldots$. This implies (ref. **(2.6.5)**(v)) that $\mathbb{E}S_n^{2r} < \infty$, $n = 1, 2, \ldots$. Suppose Lindeberg-Feller condition holds. Hence $\frac{S_n}{s_n} \xrightarrow{d} \xi$. We now investigate condition(s) that will ensure, for $r > 1$, $\mathbb{E}(\frac{S_n}{s_n})^{2r} \to \mathbb{E}\xi^{2r}$.

Let $r \ge 2$ be an integer. Consider the following 3 statements (i), (ii) & (iii) where, as $n \to \infty$,

(i) : (C) holds; (ii) : $(\mathbb{E}(\frac{S_n}{s_n})^{2r} \to \mathbb{E}\xi^{2r})$ and (iii) : $(h_n(2r) \to 0)$. The theorem below investigates the mutual implications of the three statements.

Theorem 5.3.3. (i) & (ii) $\Leftrightarrow$ (iii)

Proof. We first show that if $2 < t < r$, then

$$(\lim_{n\to\infty} h_n(r) = 0) \Rightarrow (\lim_{n\to\infty} h_n(t) = 0). \tag{5.6}$$

Proof of (5.6).

Let $2 < t < r$ and let $h_n(r) \to 0$.

Let $\varepsilon > 0$, the target error be given. Write $h_n(t) = h_{n,1}(t) + h_{n,2}(t)$.

$$h_{n,1}(t) = \frac{1}{s_n^t}\sum_{k=1}^{n}\int_{|X_k|\le\varepsilon s_n} |X_k|^t \,\mathrm{d}P \le \frac{(\varepsilon s_n)^{t-2}}{s_n^t}\sum_{k=1}^{n}\int_{\Omega} |X_k|^2 \,\mathrm{d}P$$
$$= \varepsilon^{t-2}$$

$$h_{n,2}(t) = \frac{1}{s_n^t}\sum_{k=1}^{n}\int_{|X_k|>\varepsilon s_n} |X_k|^t \,\mathrm{d}P \tag{5.7}$$

We note that the kth term in the summation would be zero if $P(|X_k| > \varepsilon s_n) = 0$. Let $A = \{k : P(|X_k| > \varepsilon s_n) > 0\}$.

$$h_n(t) = \frac{1}{s_n^t}\sum_{k\in A} P(|X_k| > \varepsilon s_n)\{\int_{|X_k|>\varepsilon s_n} \frac{1}{P(|X_k| > \varepsilon s_n)}|X_k|^t \,\mathrm{d}P\}$$
$$= \frac{1}{s_n^t}\sum_{k\in A} P(|X_k| > \varepsilon s_n)\{\int_{|X_k|>\varepsilon s_n} |X_k|^t \,\mathrm{d}Q\}$$

where Q is a probabilty measure on $\{|X_k| > \varepsilon s_n\}$

$$\le \frac{1}{s_n^t}\sum_{k\in A} P(|X_k| > \varepsilon s_n)\{\int_{|X_k|>\varepsilon s_n} \frac{1}{P(|X_k| > \varepsilon s_n)}|X_k|^r \,\mathrm{d}P\}^{\frac{t}{r}}$$
$$\le \frac{1}{s_n^t}\sum_{k\in A}\{P(|X_k| > \varepsilon s_n)\}^{1-\frac{t}{r}}\{\int_{\Omega} |X_k|^r \,\mathrm{d}P\}^{\frac{t}{r}}$$
$$= \frac{1}{s_n^t}\sum_{k\in A}\{P(|X_k| > \varepsilon s_n)\}^{1-\frac{t}{r}}\{\mathbb{E}|X_k|^r\}^{\frac{t}{r}}$$
$$\le \frac{1}{s_n^t}\sum_{k\in A}\{\frac{1}{\varepsilon^r s_n^r}\mathbb{E}|X_k|^r\}^{1-\frac{t}{r}}\{\mathbb{E}|X_k|^r\}^{\frac{t}{r}}$$
$$= \frac{1}{\varepsilon^{r-t}}\frac{1}{s_n^r}\sum_{k\in A}\mathbb{E}|X_k|^r$$

$$= \frac{1}{\varepsilon^{r-t}} h_n(r).$$

Thus $h_n(t) \le \varepsilon^{t-2} + \frac{1}{\varepsilon^{r-t}} h_n(r)$. Now let $n \to \infty$ and then let $\varepsilon \to 0$ to see that $h_n(t) \to 0$.

Proof of the theorem.

That (iii) implies (i) follows from **(5.3.1)**(v).

To show that (iii) implies (ii), let us first consider the case $r = 2$. Then $\mathbb{E}S_k^4 - \mathbb{E}S_{k-1}^4 = \mathbb{E}(S_{k-1} + X_k)^4 - \mathbb{E}S_{k-1}^4 = 6s_{k-1}^2\sigma_k^2 + \mathbb{E}X_k^4$. Set $S_0 = 0$ and sum over $k = 1, 2, \ldots, n$ to get $\mathbb{E}S_n^4 = 6\sum_{k=2}^n s_{k-1}^2\sigma_k^2 + \sum_{k=1}^n \mathbb{E}X_k^4 = 6u_n + v_n$, say. $u_n \sim$ (by **(5.3.1)**(x)) $\sum_{k=2}^n \frac{1}{2}\{s_k^4 - s_{k-1}^4\} \sim \frac{1}{2}s_n^4$ and $v_n = o(s_n^4)$, by hypothesis. Hence $\mathbb{E}(\dfrac{S_n}{s_n})^4 \sim 6 \times \frac{1}{2} = 3 = \mathbb{E}\xi^4$. Thus if $r = 2$, then (iii) implies (i) and (ii).

Let now $r \ge 3$. The method of proof is by induction on r.

The induction hypothesis consists of two parts: $(\mathfrak{a})$: $h_n(2r) \to 0$ and $(\mathfrak{b})$: $(\lim_{n\to\infty} h_n(2r-2) = 0) \Rightarrow (\mathbb{E}(\frac{S_n}{s_n})^{2r-2} \to \mathbb{E}\xi^{2r-2})$.

The convergence $\mathbb{E}(\frac{S_n}{s_n})^{\overline{2r-1}} \to \mathbb{E}\xi^{\overline{2r-1}}$ implies (ref. **(2.6.8)** and **(2.6.9)**) $\mathbb{E}|\frac{S_n}{s_n}|^s \to \mathbb{E}|\xi|^s$, $s \le 2r-2$.

We note $\mathbb{E}|\xi|^\theta > 0$ for all $\theta \ge 0$. Hence there exists an integer $N = N(r)$ such that $\mathbb{E}|\frac{S_n}{s_n}|^\nu \le 2\mathbb{E}|\xi|^\nu$ for all $n \ge N$ and for $\nu = 1, 2, \ldots, 2r-2$. We have

$$\begin{aligned}
\mathbb{E}S_n^{2r} &= \sum_{j=1}^n \mathbb{E}\{S_j^{2r} - S_{j-1}^{2r}\} = \sum_{j=1}^n \mathbb{E}\{(S_{j-1} + X_j)^{2r} - S_{j-1}^{2r}\} \\
&= \sum_{j=1}^n \mathbb{E}\{\sum_{m=1}^{2r}\binom{2r}{m} S_{j-1}^{2r-m}X_j^m\} = \sum_{j=1}^n\{\sum_{m=1}^{2r}\binom{2r}{m}\mathbb{E}S_{j-1}^{2r-m}\mathbb{E}X_j^m\} \\
&= \sum_{j=1}^n \mathbb{E}X_j^{2r} + \sum_{j=1}^n\sum_{m=2}^{2r-2}\binom{2r}{m}\mathbb{E}S_{j-1}^{2r-m}\mathbb{E}X_j^m \\
&= \sum_{j=1}^n \mathbb{E}X_j^{2r} + \sum_{j=1}^n\binom{2r}{2}\mathbb{E}S_{j-1}^{2r-2}\mathbb{E}X_j^2 + \sum_{j=1}^n\sum_{m=3}^{2r-2}\mathbb{E}S_{j-1}^{2r-m}\mathbb{E}X_j^m \\
&= a_n + b_n + c_n, \text{ say.}
\end{aligned}$$

By the induction hypothesis $(\mathfrak{a})$, $a_n = o(s_n^{2r})$. For $m = 2, 3, \ldots, 2r-2$, we have, by induction hypothesis $(\mathfrak{b})$ and (5.6),

$$|c_n| \le \sum_{j=1}^N\sum_{m=3}^{2r-2}\mathbb{E}|S_{j-1}|^{2r-m}\mathbb{E}|X_j|^m + \sum_{m=3}^{2r-2}\sum_{j=N+1}^n 2s_{j-1}^{2r-m}\mathbb{E}|X_j|^m$$

$$\leq q + \sum_{m=3}^{2r-2} \sum_{j=1}^{n} 2 s_n^{2r-m} \mathbb{E}|X_j|^m \leq q + \sum_{m=3}^{2r-2} 2 s_n^{2r-m} \sum_{j=1}^{n} \mathbb{E}|X_j|^m$$

$$\leq q + 2 s_n^{2r} \sum_{m=3}^{2r-2} h_n(m) = o(s_n^{2r}).$$

As $j \to \infty$, $\binom{2r}{2} \mathbb{E} S_{j-1}^{2r-2} \mathbb{E} X_j^2 \sim \binom{2r}{2} s_{j-1}^{2r-2} \mathbb{E}\xi^{2r-2} \sigma_j^2 \sim \binom{2r}{2} \mathbb{E}\xi^{2r-2} \frac{1}{r} \{s_j^{2r} - s_{j-1}^{2r}\}$ (by **(5.3.1)**(x)).

Hence

$$b_n \sim \binom{2r}{2} \mathbb{E}\xi^{2r-2} \sum_{j=1}^{n} \frac{1}{r} \{s_j^{2r} - s_{j-1}^{2r}\} \sim \frac{1}{r} \binom{2r}{2} \mathbb{E}\xi^{2r-2} s_n^{2r}.$$

Thus

$$\mathbb{E}\Big(\frac{S_n}{s_n}\Big)^{2r} \sim \frac{1}{r} \binom{2r}{2} \mathbb{E}\xi^{2r-2} = \mathbb{E}\xi^{2r}.$$

Conversely, suppose (i) & (ii) hold for some $r \geq 2$. Let (Y_n) be an independent copy of the X_n sequence. Let $T_n = \sum_{k=1}^{n} Y_k$. Write $Z_n = \frac{1}{\sqrt{2}}(X_n - Y_n)$ and $V_n = \sum_{k=1}^{n} Z_k$. By **(5.3.1)**(vii), (i) implies $\frac{S_n}{s_n} \xrightarrow{d} \xi$ and $\frac{T_n}{s_n} \xrightarrow{d} \xi$. These weak convergences imply $\frac{V_n}{s_n} \xrightarrow{d} \xi$. The weak convergence $\frac{S_n}{s_n} \xrightarrow{d} \xi$ and (ii) imply (ref. **(2.6.9)**) the uniform integrability of the sequence $(\frac{S_n}{s_n})^{2r}$. Similarly the sequence $(\frac{T_n}{s_n})^{2r}$ is uniformly integrable. Hence by **(4.2.4)** $(\frac{V_n}{s_n})^{2r}$ is a uniformly integrable sequence. This fact and the fact that $\frac{V_n}{s_n} \xrightarrow{d} \xi$ imply (ref. **(2.6.9)**) $\mathbb{E}(\frac{V_n}{s_n})^{2r} \to \mathbb{E}\xi^{2r}$. Hence (ref. **(2.6.8)** and **(2.6.9)**) $\mathbb{E}(\frac{V_n}{s_n})^j \to \mathbb{E}\xi^j$, $j = 1, 2, \ldots, 2r$. Now,

$$\mathbb{E}V_n^{2r} = \sum_{m=1}^{n} (\mathbb{E}V_m^{2r} - \mathbb{E}V_{m-1}^{2r}) = \sum_{m=1}^{n} \sum_{j=1}^{r} \binom{2r}{2j} \mathbb{E}V_{m-1}^{2r-2j} \mathbb{E}Z_m^{2j}$$

$$= \binom{2r}{2} \sum_{m=1}^{n} \mathbb{E}V_{m-1}^{2r-2} \sigma_m^2 + \sum_{j=2}^{r} \binom{2r}{2j} \sum_{m=1}^{n} \mathbb{E}V_{m-1}^{2r-2j} \mathbb{E}Z_m^{2j}$$

The first term on the rightside is asymptotically equal to

$$\binom{2r}{2} \sum_{m=1}^{n} s_{m-1}^{2r-2} \mathbb{E}\xi^{2r-2} \sigma_m^2$$

$$\sim \text{(by } \mathbf{(5.3.1)}\text{(x))} \binom{2r}{2} \mathbb{E}\xi^{2r-2} \sum_{m=1}^{n} \frac{1}{r} (s_m^{2r} - s_{m-1}^{2r})$$

$$\sim (2r-1) \mathbb{E}\xi^{2r-2} s_n^{2r} \sim s_n^{2r} \mathbb{E}\xi^{2r}.$$

Since $\mathbb{E}(\frac{V_n}{s_n})^{2r} \to \mathbb{E}\xi^{2r}$, this shows that

$$\frac{1}{s_n^{2r}} \sum_{j=2}^{r} \binom{2r}{2j} \sum_{m=1}^{n} \mathbb{E}V_{m-1}^{2r-2j} \mathbb{E}Z_m^{2j} \to 0.$$

All the terms of this sum are non-negative. In particular the term corresponding to $j = r$ tends to 0. i.e. $\frac{1}{s_n^{2r}} \sum\limits_{m=1}^{n} \mathbb{E}Z_m^{2r} = o(1)$. i.e. ($E_2$) holds with $p = 2r$. Hence, by **(5.3.1)**(xi), (E_1) holds with $p = 2r$. i.e. (iii) holds. □

Extension of **(5.3.3)** to moments of odd orders and of fractional orders require proving first many *hilfsatz* which is beyond the scope of this book. We satisfy ourselves proving

Theorem 5.3.4. *Let (C) hold with "r" = 2, let* $\mathbb{E}\big(\frac{S_n}{s_n}\big)^{2r} \to \mathbb{E}\xi^{2r}$, *let* $\overline{\lim\limits_{n\to\infty}}\, h_n(2r+2) \neq 0$ *and let* $\sup\limits_n \mathbb{E}\big|\frac{S_n}{s_n}\big|^{\alpha} = \hat{q}_\alpha < \infty$, *for some* α, $2r < \alpha \leq 2r + 2$, $r \geq 1$, *a fixed integer. Then for every* β, $0 < \beta < \alpha$, $\mathbb{E}\big|\frac{S_n}{s_n}\big|^{\beta} \to \mathbb{E}|\xi|^{\beta}$.

Proof. Since (C) holds, $\frac{S_n}{s_n} \stackrel{d}{\to} \xi$. This together with the assumption that $\mathbb{E}\big(\frac{S_n}{s_n}\big)^{2r} \to \mathbb{E}\xi^{2r}$, or equivalently, $\mathbb{E}|\frac{S_n}{s_n}|^{2r} \to \mathbb{E}|\xi|^{2r}$ implies (ref. **(2.6.9)** and **(2.6.8)**) $\mathbb{E}\big|\frac{S_n}{s_n}\big|^{\beta} \to \mathbb{E}|\xi|^{\beta}$, $0 \leq \beta \leq 2r$. Hence we need to consider only $2r < \beta < \alpha$. Let $q_\alpha = \max\{2, \hat{q}_\alpha\}$. The assumption $\sup\limits_n \mathbb{E}\big|\frac{S_n}{s_n}\big|^{\alpha} \leq q_\alpha < \infty$ implies (ref. **(2.6.6)**) $\sup\limits_{0\leq\lambda\leq\alpha} \sup\limits_n \mathbb{E}\big|\frac{S_n}{s_n}\big|^{\lambda} \leq q_\alpha$. Denote by ψ_n the real part of the *cf* of $\frac{S_n}{s_n}$. By **(3.6.9)**, there exists a constant $c_{r,\beta}$ such that

$$c_{r,\beta}\mathbb{E}\Big|\frac{S_n}{s_n}\Big|^{\beta} = \int_0^{\infty} (-1)^{r+1} \frac{\psi_n(t) - \sum\limits_{j=0}^{r} (-1)^j \frac{t^{2j}}{(2j)!}\mathbb{E}(\frac{S_n}{s_n})^{2j}}{t^{1+\beta}}\, \mathrm{d}t.$$

Appeal to **(3.6.11)**(iii) to conclude that, for $t > 0$, the integrand is bounded by $q(\alpha)\mathbb{E}\big|\frac{S_n}{s_n}\big|^{\alpha} t^{\alpha-\beta-1} \leq q_1(\alpha)t^{\alpha-\beta-1}$. Since $|\alpha - \beta - 1| < 1$, $t^{\alpha-\beta-1}$ is integrable over $(0, 1)$. Over $[1, \infty)$ the integrand is bounded by $\frac{4q_\alpha}{t^{\beta+1-2r}}$. Since $\beta - 2r > 0$, $\int_1^\infty \frac{4q_\alpha}{t^{\beta+1-2r}}\, \mathrm{d}t < \infty$. Hence the bounded convergence theorem (ref. **Q**2) applies and we get, by **(3.6.9)**,

$$\begin{aligned} c_{r,\beta} \lim_{n\to\infty} \mathbb{E}\big|\tfrac{S_n}{s_n}\big|^{\beta} &= \int_0^{\infty} (-1)^{r+1} \frac{e^{-\frac{1}{2}t^2} - \sum\limits_{0}^{r}(-1)^j \frac{t^{2j}}{(2j)!}\mathbb{E}\xi^{2j}}{t^{1+\beta}}\, \mathrm{d}t \\ &= c_{r,\beta}\mathbb{E}|\xi|^{\beta}. \end{aligned}$$

□

Example 5.3.5. In this section we present an example to show that the theorem just proved is not vacuous.

Let $(X_n,\ n \geq 1)$ be a sequence of independent symmetric variables. $X_n = \pm n^2$ with probability $\frac{1}{12n^4}$ each; $= \pm n$ with probability $\frac{1}{12}$ each; $= 0$ with probability $1 - \frac{1}{6} - \frac{1}{6n^4}$, $n \geq 1$.
$S_n = \sum_{k=1}^{n} X_k$; $s_n^2 = \mathbb{E}S_n^2$. $\mathbb{E}|X_n|^\nu \sim \frac{n^\nu}{6}$ if $2 \leq \nu < 4$; $\mathbb{E}|X_n|^4 \sim \frac{n^4}{3}$; $\mathbb{E}|X_n|^\nu \sim \frac{n^{2\nu-4}}{6}$ if $\nu > 4$. Hence $s_n^2 \sim \frac{n^3}{18}$ and $\frac{\Gamma_n^{(\nu)}}{s_n^\nu} = \frac{1}{s_n^\nu}\sum_{k=1}^{n} \mathbb{E}|X_k|^\nu \to 0$ if $2 < \nu < 6$ and for $\nu = 6$, the limit is 108. Since $\frac{\Gamma_n^{(3)}}{s_n^3} \to 0$, it follows by Lyapunoff's theorem that $\frac{S_n}{s_n} \stackrel{d}{\to} \xi$ where ξ is a standard normal variable. Setting $X_0 \equiv 0 \equiv S_0$, we have

$$\begin{aligned}
\lim_{n\to\infty} \mathbb{E}(\frac{S_n}{s_n})^4 &= \lim_{n\to\infty} \frac{18^2}{n^6} \sum_{k=1}^{n} \mathbb{E}\{S_k^4 - S_{k-1}^4\} \\
&= \lim_{n\to\infty} \frac{18^2}{n^6} \sum_{k=1}^{n} \mathbb{E}\{4S_{k-1}^3 X_k + 6S_{k-1}^2 X_k^2 + 4S_{k-1}X_k^3 + X_k^4\} \\
&= \lim_{n\to\infty} \frac{18^2}{n^6} \sum_{k=1}^{n} \{6\mathbb{E}S_{k-1}^2 EX_k^2 + EX_k^4\} \\
&= \lim_{n\to\infty} \frac{18^2}{n^6} \sum_{k=1}^{n} \{6\frac{(k-1)^3}{18}(\frac{1}{6} + \frac{k^2}{6}) + \frac{k^4}{3}\} \\
&= \lim_{n\to\infty} \frac{18^2}{n^6} \sum_{k=1}^{n} \frac{k^5}{18} = 3 = \mathbb{E}\xi^4.
\end{aligned}$$

$$\begin{aligned}
\lim_{n\to\infty} E(\frac{S_n}{s_n})^6 &= \lim_{n\to\infty} \frac{18^3}{n^9} \mathbb{E}\sum_{k=1}^{n} \{S_k^6 - S_{k-1}^6\} \\
&= \lim_{n\to\infty} \frac{18^3}{n^9} \mathbb{E}\sum_{k=1}^{n} \{15S_{k-1}^4 X_k^2 + 15S_{k-1}^2 X_k^4 + X_k^6\} \\
&= \lim_{n\to\infty} \frac{18^3}{n^9} \sum_{k=1}^{n} \{15 \times 3\frac{k^6}{18^2}\frac{k^2}{6} + 15\frac{(k-1)^3}{18}\frac{k^4}{3} + \frac{k^8}{6}\} \\
&= \lim_{n\to\infty} \frac{18^3}{n^9} \sum_{k=1}^{n} \{15 \times 3\frac{k^8}{18^2 \times 6} + 15\frac{k^7}{18 \times 3} + \frac{k^8}{6}\} \\
&= \lim_{n\to\infty} \frac{18^3}{n^9} \{\frac{15 \times 3}{6 \times 18^2}\frac{n^9}{9} + \frac{n^9}{9 \times 6}\} = 123.
\end{aligned}$$

But $\mathbb{E}\xi^6 = 15$.

The proof that $\mathbb{E}\left|\frac{S_n}{s_n}\right|^\beta \to \mathbb{E}|\xi|^\beta$, $0 \le \beta < 6$ is on the same lines as in the theorem.

5.3.6. *Convergence of moments in CLT for iid variables with finite variances.*

Let (X_n) be an iid sequence with $\mathbb{E}X_1 = 0$, $\mathbb{E}X_1^2 = 1$ and common *df* F and common *cf* ψ. As usual $S_n = \sum_1^n X_k$. Let $r > 1$ be an integer. If $\mathbb{E}|X_1|^{2r} < \infty$, then trivially, $h_n(2r) \to 0$. Hence by **(5.3.3)**, $\mathbb{E}(\frac{S_n}{\sqrt{n}})^{2r} \to \mathbb{E}\xi^{2r}$.

Let $\mathbb{E}|X_1|^\beta < \infty$. If $\beta > 2$ is an odd integer, then following the steps in **(5.3.5)** and using **(5.3.3)** where necessary, it can be shown that $\mathbb{E}(\frac{S_n}{\sqrt{n}})^\beta \to 0 \ (= \mathbb{E}\xi^\beta)$. If $\beta > 2$ is a non-integer, then, for reasons stated earlier (refer para preceding **(5.3.4)**) we will not be proving the convergence of absolute moments of order β. However to provide the proof promised in **(4.7.3)**(iii), a truncated result (sufficient for our purpose) is established in the following theorem when $2 < \beta < 4$.

Theorem 5.3.7. *If* $\mathbb{E}|X_1|^\beta < \infty$, $2 < \beta < 4$, *then* $\mathbb{E}|\frac{S_n}{\sqrt{n}}|^\alpha \to \mathbb{E}|\xi|^\alpha$ *for every* $0 < \alpha < \beta$.

Proof. We need consider only the case $2 < \alpha < \beta$.
We note $\psi_n(t) = \psi^n(\frac{t}{\sqrt{n}})$ is the cf of $\frac{S_n}{\sqrt{n}}$. Hence (ref. **(3.6.9)**)

$$c_{1,\beta}\mathbb{E}|\frac{S_n}{\sqrt{n}}|^\beta = \int_0^\infty \frac{u_n(t)}{t^{1+\beta}}\,dt$$

where
$u_n(t) = (-1)\{\Re(\psi_n(t)) - \sum_{j=0}^1 \frac{(-1)^j}{(2j)!}t^{2j}\mathbb{E}(\frac{S_n}{\sqrt{n}})^{2j}\} = 1 - \Re(\psi_n(t)) - \frac{t^2}{2}$.

The cf ψ of X_1 admits of the following expansion in the neighborhood of the origin (ref. **(3.6.11)** : $\psi(t) = 1 - \frac{t^2}{2!} + O(|t|^\beta)$. Hence for t fixed and as $n \to \infty$, $1 - \Re\psi(\frac{t}{\sqrt{n}}) = \frac{t^2}{2n} + O((\frac{|t|}{\sqrt{n}})^\beta)$, the constant in the order term depending only on the moments of X_1. This leads to

$$\psi^n(\frac{t}{\sqrt{n}}) = \left(1 - (1 - \psi(\frac{t}{\sqrt{n}}))\right)^n = 1 - \frac{t^2}{2} + O\left(\frac{|t|^\beta}{n^{\frac{\beta}{2}-1}}\right).$$

Thus for all t, the non-negative function $u_n(t) \le A|t|^\beta$ where A is a constant. Hence over $0 < t < 1$,

$$\frac{u_n(t)}{t^{1+\alpha}} = \frac{1 - \Re\psi_n(t) - \frac{t^2}{2}}{t^{1+\alpha}} \le \frac{A}{t^{1+\alpha-\beta}}.$$

Since $-1 < 1+\alpha-\beta < 1$, $\frac{1}{t^{1+\alpha-\beta}}$ is integrable over $(0,\ 1)$. We note (ref. **(3.6.3)**) $u_n(t) = \mathbb{E}g_1(t\frac{S_n}{\sqrt{n}}) \leq \mathbb{E}(t\frac{S_n}{\sqrt{n}})^2 \leq t^2$. Hence $\frac{u_n(t)}{t^{1+\alpha}} \leq \frac{1}{t^{\alpha-1}}$. Since $\alpha - 1 > 1$, $\frac{1}{t^{\alpha-1}}$ is integrable over $[1,\ \infty)$. It follows now by the bounded convergence theorem that

$$c_{1,\alpha} \lim_{n\to\infty} \mathbb{E}|\frac{S_n}{\sqrt{n}}|^\alpha = \lim_{n\to\infty} \int_0^\infty \frac{u_n(t)}{t^{1+\alpha}} dt = \int_0^\infty \lim_{n\to\infty} \frac{1 - \Re\psi_n(t) - \frac{t^2}{2}}{t^{1+\alpha}} dt$$

$$= \int_0^\infty \frac{1 - e^{-\frac{1}{2}t^2} - \frac{t^2}{2}}{t^{1+\alpha}} dt = c_{1,\alpha}\mathbb{E}|\xi|^\alpha. \qquad \square$$

Theorem 5.3.8. (Continuation of **(4.7.3)**(iii)).
Let (X_n) be a sequence of iid variables with $\mathbb{E}X_1 = 0$, $\mathbb{E}X_1^2 = 1$ and $\mathbb{E}|X_1|^\beta < \infty$ for some $\beta > 2$. Then $\frac{S_n}{n}$ converges to 0 completely.

Proof. By **(5.3.6)**, $\mathbb{E}|\frac{S_n}{\sqrt{n}}|^\alpha \to \mathbb{E}|\xi|^\alpha$ for every α, $2 < \alpha < \beta$. Since

$$P(|\frac{S_n}{n}| > \varepsilon) \leq \frac{1}{\varepsilon^\alpha}\mathbb{E}|\frac{S_n}{n}|^\alpha \sim \frac{1}{\varepsilon^\alpha}\frac{\mathbb{E}|\xi|^\alpha}{n^{\frac{\alpha}{2}}}$$

and since $\frac{\alpha}{2} > 1$, the convergence of $\sum_n P(|\frac{S_n}{n}| > \varepsilon)$ follows. $\square$

5.4 *CLT* for iid variables without variance and a converse to the *CLT*

5.4.1. As motivation for investigating the possibility of proving a CLT for rvs without variance, we consider the following example.

Let the *df* F of X have frequency function $\frac{1}{|x|^3}$, $|x| \geq 1$. We note $\mathbb{E}|X|^\theta < \infty$, $0 \leq \theta < 2$ and $\mathbb{E}X^2 = \infty$. Let φ be its *cf*. We note $\varphi(t) = \varphi(-t)$. We have, for $t > 0$,

$$1 - \varphi(t) = 2\int_1^\infty \frac{1-\cos tx}{x^3} dx = 2t^2 \int_t^\infty \frac{1-\cos u}{u^3} du.$$

For t small,

$$\frac{1-\varphi(t)}{2t^2} = \int_t^1 \frac{1-\cos u}{u^3} du + \int_1^\infty \frac{1-\cos u}{u^3} du$$

$$= \int_t^1 \frac{1-\cos u}{u^3}\,\mathrm{d}u + q \text{ (where } q \text{ is a constant free from } t)$$

$$= \int_t^1 \frac{1}{2u}\,\mathrm{d}u + \int_t^1 \frac{1-\cos u - \frac{1}{2}u^2}{u^3}\,\mathrm{d}u + \alpha$$

$$= \frac{1}{2}(-\log t) + \beta(t) + \alpha, \text{ say.}$$

Since $\lim_{t\to 0}\beta(t) = \beta < \infty$, $1-\varphi(t) \sim -t^2 \log t$ as $t \to 0+$. For $t > 0$ fixed this asymptotic value gives

$$\lim_{n\to\infty} n\{1-\varphi(\frac{t}{\sqrt{2n\log n}})\} = -\lim_{n\to\infty} n\frac{t^2}{2n\log n}\log\frac{t}{\sqrt{2n\log n}} = \frac{t^2}{2}.$$

This is equivalent to : $\varphi^n(\frac{t}{\sqrt{2n\log n}}) \to e^{-\frac{1}{2}t^2}$. Since φ is an even function, this is true for all $t \in R$. If then $X_n, n \geq 1$ are independent copies of X and if $S_n = \sum_1^n X_k$, we have $\frac{S_n}{\sqrt{2n\log n}} \stackrel{d}{\to} \xi$.

It is instructive to note that even though $\mathbb{E}X^2 = \infty$, the variance defining function $U(x) = \int_{-x}^{x} y^2\,\mathrm{d}F(y) \sim 2\log x$ is tending to infinity with x as a sv function. It is reasonable to conclude from this that a CLT probably holds whenever U is sv. Through what follows, we show that this indeed is the case.

Let X, Y, X_n, Y_n, $n \geq 1$ be *iid rvs* with common *df* F and common characteristic function φ. Let G denote the *df* of $Z = \frac{X-Y}{\sqrt{2}}$ and g its *cf*. Let $Z_n = \frac{X_n - Y_n}{\sqrt{2}}$; $S_n = \sum_1^n X_k$; $T_n = \sum_1^n Y_k$ and $V_n = \sum_1^n Z_k$. For $x \geq 0$, define $U(x) = \int_{-x}^{x} y^2 dF(y)$. $EX^2 < \infty$ iff $\lim_{x\to\infty} U(x)$ is finite. We assume $\lim_{x\to\infty} U(x) = \infty$ and that U is slowly varying at ∞. We recall that such a function U can be written $U(x) = a(x)H(x)$ where $\lim_{x\to\infty} a(x) = a > 0$, finite, exists and H has the representation $H(x) = 1$ for $0 \leq x \leq 1$ and $H(x) = \exp(\int_1^x \frac{\epsilon(y)}{y} dy)$ for $x \geq 1$, $\epsilon(y) \geq 0$, $\epsilon(y) \to 0$ as $y \to \infty$, ϵ a continuously differentiable function. Recall $\frac{H(x)}{x^2}$ is a continuous function, strictly monotonically decreasing over x exceeding some x_0, with limit 0 as $x \to \infty$. Define $b(t)$ to be the unique value c satisfying $\frac{aH(c)}{c^2} = \frac{1}{t}$ and note that $b(t) \uparrow \infty$ as $t \uparrow \infty$. Write $b_n = b(n)$ and note $\frac{aH(b_n)}{b_n^2} = \frac{1}{n}$ and $\frac{U(b_n)}{b_n^2} \sim \frac{1}{n}$. If $\mathfrak{a}(x) = \frac{aH(x)}{x^2}$, then $b_n = \mathfrak{a}^{-1}(\frac{1}{n})$. In part (i) of the following theorem, it is shown that $\mathbb{E}|X| < \infty$ if U is sv. Write $\mathbb{E}X = \mu$.

Theorem 5.4.2. (i) $E|X|^\nu < \infty,\ 0 < \nu < 2$.
(ii) $b_n \uparrow \infty$.

(iii) $\frac{b_n}{\sqrt{n}} \to \infty$.
(iv) *If* $b_n = \sqrt{n}l(n)$ *then* $l(n) \uparrow \infty$ *and* l *is sv.*
(v) *For* a *a real number, let* F^a *be the df of* $X - a$ *and let* $V(x) = \int_{-x}^{x} y^2 dF^a(y)$. *Then* $V(x) \sim U(x)$.
(vi) $W(x) \sim U(x) \sim aH(x)$ *where* $W(x) = \int_{|y| \le x} y^2 dG(y)$, *whenever* W *or* U *is sv.*
(vii) *Let* $Q(c) = \frac{c^2 \int_{|x|>c} dF(x)}{U(c)}$, $c > 0$. *Then* U *is sv, iff* $Q(c) \to 0$ *as* $c \to \infty$.

Proof. (i) Let $a > 0$;

$$\mathbb{E}|X|^\nu = \int_R |x|^\nu \, \mathrm{d}F(x) < \infty \text{ if } \int_{|x|>a} |x|^\nu \, \mathrm{d}F(x) < \infty.$$

i.e. if $\int_a^\infty \frac{1}{x^{2-\nu}} \, \mathrm{d}U(x) < \infty$. But this is true by **(3.10.9)**.
(ii) That b_t is strictly monotonic increasing follows from $\frac{H(x)}{x^2}$ being a strictly monotonic decreasing function. Since $b_n \uparrow$, $\lim_{n\to\infty} b_n$ exists. If the limit is finite, then $aH(b_n) \sim \frac{b_n^2}{n}$ would imply $H \equiv 0$.
(iii) By (ii), $b_n \to \infty$. Hence $H(b_n) \to \infty$. From $aH(b_n) \sim \frac{b_n^2}{n}$, the claim follows.
(iv) Let

$$r_1(x) = \frac{x^2}{H(x)} = e^{\int_1^x \frac{2-\varepsilon(y)}{y} \, \mathrm{d}y}.$$

Since r_1 is a strictly increasing continuous function, it has a unique inverse r_2. We recognise that $b_t \sim r_2(t)$. On differentiation, the relation $r_1\big(r_2(x)\big) = x = r_2\big(r_1(x)\big)$ gives $r_1'\big(r_2(x)\big)r_2'(x) = 1 = r_2'\big(r_1(x)\big)r_1'(x)$. Use the relation $r_1'(x) = r_1(x)\frac{2-\varepsilon(x)}{x}$ to get
$r_1'\big(r_2(x)\big) = r_1\big(r_2(x)\big)\frac{2-\varepsilon\big(r_2(x)\big)}{r_2(x)} = x\frac{2-\varepsilon\big(r_2(x)\big)}{r_2(x)}$. Using first the relation $r_1'\big(r_2(x)\big)r_2'(x) = r_2'\big(r_1(x)\big)r_1'(x)$ and then $r_1'\big(r_2(x)\big)r_2'(x) = 1$, we get

$$r_2'\big(r_1(x)\big) = \frac{x\frac{2-\varepsilon\big(r_2(x)\big)}{r_2(x)}r_2'(x)}{r_1'(x)} = \frac{x\frac{2-\varepsilon\big(r_2(x)\big)}{r_2(x)}\frac{1}{r_1'\big(r_2(x)\big)}}{r_1'(x)} = \frac{1}{r_1'(x)}.$$

Hence

$$\frac{r_2'\big(r_1(x)\big)r_1(x)}{r_2\big(r_1(x)\big)} = \frac{r_1(x)}{xr_1'(x)} = \frac{1}{2-\varepsilon(x)}.$$

Writing $x = r_2(t)$ in this relation, we get

$$\frac{r_2'(t)}{r_2(t)}t = \frac{1}{2-\varepsilon_1(t)} = \frac{1}{2}\frac{1}{1-\frac{1}{2}\varepsilon_1(t)} = \frac{1}{2} + \varepsilon_2(t)), \text{ say.}$$

Hence

$$r_2(x) = r_2(1)e^{\int_1^x \frac{\frac{1}{2}+\varepsilon_2(t)}{t}\,\mathrm{d}t} = r_2(1)x^{\frac{1}{2}}e^{\int_1^x \frac{\varepsilon_2(t)}{t}\,\mathrm{d}t}$$

Thus $\frac{b_t}{\sqrt{t}} \sim \frac{r_2(t)}{\sqrt{t}}$ is sv if $\varepsilon_2(t) \to 0$ as $t \to \infty$. But this is true since $\varepsilon(t) \to 0$ and $r_2(t) \sim b_t \to \infty$ and since $1 + \varepsilon_2(t) = \frac{4-\varepsilon(r_2(t))}{4-2\varepsilon(r_2(t))}$. Since $\varepsilon_2 \geq 0$, it follows that $l(n) \uparrow$. That $l(n) \to \infty$ was already shown in (iii).
(v) $V(x) = \int_{|X|\leq x}(X-a)^2 dP = \int_{|X|\leq x} X^2 dP - 2a\int_{|X|\leq x} X dP + a^2 = U(x) + O(1)$. (The second term is $O(1)$ since it is bounded by $2\mathbb{E}|X|$ which is finite by (i))
(vi) Choose c large so $U(y) \leq y$ for all $y \geq c$. Now,

$$\begin{aligned}
2W(c) &= \int_{|X-Y|\leq c} (X-Y)^2 dP \\
&= \int_{|X-Y|\leq c} X^2\,\mathrm{d}P + \int_{|X-Y|\leq c} Y^2 dP - 2\int_{|X-Y|\leq c} XY dP \\
&= I_1(c) + I_2(c) - 2I_3(c), \textit{say}.
\end{aligned}$$

$|I_3(c)| \leq \int |X||Y| dP = E|X|E|Y|$ since X and Y are mutually independent. Hence $I_3(c) = O(1)$ as $c \to \infty$.

$$\begin{aligned}
I_1(c) &= \iint\limits_{|x-y|\leq c} x^2\,\mathrm{d}F(x)\,\mathrm{d}F(y) \leq \iint\limits_{|x|\leq c+|y|} x^2 dF(x) dF(y) \\
&= \int_{-\infty}^{\infty} U(c+|y|) dF(y) \\
&= \int_{|y|\leq c} U(c+|y|) dF(y) + \int_{|y|>c} U(c+|y|) dF(y) \\
&\leq U(2c) + \int_{|y|>c} U(2|y|) dF(y) \leq U(2c) + 2\int_{-\infty}^{\infty} |y| dF(y).
\end{aligned}$$

Hence $\overline{\lim_{c\to\infty}} \frac{I_1(c)}{U(2c)} \leq 1$. By reason of symmetry, $I_2(c) = I_1(c)$.
Hence $\overline{\lim_{c\to\infty}} \frac{2W(c)}{U(2c)} \leq 2$. Again $I_1(c) \geq \int_{(|X|\leq\frac{c}{2})\cap(|Y|\leq\frac{c}{2})} X^2 dP = P(|Y| \leq \frac{c}{2})U(\frac{c}{2})$, leading to $\underline{\lim_{c\to\infty}} \frac{2W(c)}{U(\frac{c}{2})} \geq 2$. Since U is sv, $U(c) \sim U(2c) \sim U(\frac{c}{2})$. It follows $W(c) \sim U(c)$. If W is sv , then we will have $W(c) \sim W(\frac{c}{2}) \sim W(2c)$, leading to $U(c) \sim W(c)$.

(vii) Suppose $Q(c) \to 0$. For $\lambda > 1$,

$$\begin{aligned} U(\lambda c) &= U(c) + \int_{c\le|x|\le\lambda c} x^2 dF(x) \le U(c) + \lambda^2 c^2 \int_{|x|>c} dF(x) \\ &= U(c) + \lambda^2 Q(c) U(c). \end{aligned}$$

Hence $\overline{\lim}_{c\to\infty} \frac{U(\lambda c)}{U(c)} \le 1$. U being a non-decreasing function, $\frac{U(\lambda c)}{U(c)} \ge 1$. Hence U is $s.v.$ Conversely, let U be *sv*.
$U(3c) = U(c) + \int_{c<|x|\le 3c} x^2 dF(x)$. Since U is *sv* this implies

$$\frac{\int_{c<|x|\le 3c} x^2 dF(x)}{U(c)} \to 0$$

as $c \to \infty$. From this, we have $\frac{c^2}{U(c)} \int_{c<|x|\le 3c} dF(x) \to 0$. Given $\epsilon > 0$ we can then find c_1 such that for all $c \ge c_1$,

$$\frac{c^2}{U(c)} \int_{c<|x|\le 3c} dF(x) < \epsilon. \tag{5.8}$$

Since $1 \le \frac{U(3c)}{U(c)} \to 1$, we can find a c_2 such that $\frac{U(3c)}{U(c)} < 2$ for all $c \ge c_2$. Define $c_0 = \max(c_1, c_2)$ and take $c > c_0$. In (5.8), replacing c by $3^k c$, we get

$$\frac{3^{2k} c^2}{U(3^k c)} \int_{3^k c<|x|\le 3^{k+1}c} \mathrm{d}\, F(x) < \epsilon.$$

$$\begin{aligned} \text{i.e.,} \quad \frac{c^2}{U(c)} \int_{3^k c<|x|\le 3^{k+1}c} \mathrm{d}\, F(x) &\le \frac{\epsilon}{3^{2k}} \frac{U(3^k c)}{U(c)} = \frac{\epsilon}{3^{2k}} \prod_{j=0}^{k-1} \frac{U(3^{j+1}c)}{U(3^j c)} \\ &\le \frac{\epsilon}{3^{2k}} 2^k = \left(\frac{2}{9}\right)^k \epsilon. \end{aligned}$$

Summing up over k,

$$\frac{c^2}{U(c)} \sum_{k=0}^{\infty} \int_{3^k c<|x|\le 3^{k+1}c} dF(x) < \frac{9}{7}\varepsilon.$$

i.e., $\frac{c^2}{U(c)} \int_{|x|>c} dF(x) < 3\epsilon$ for all $c \ge c_0$. i.e $Q(c) \to 0$ as $c \to \infty$. □

In **(5.4.1)**, we promised to show that, if U is sv , the CLT holds. We use the properties of the functions V, W and Q observed in **(5.4.2)** to establish this claim. Further, we prove the convergence of the absolute moments of order $\alpha < 2$ of the partial sums.

Theorem 5.4.3. (i) *Recall g is the cf of Z. If W is sv then as $t \to 0$,*

$$1 - g(t) \sim \frac{t^2}{2} W\left(\frac{1}{|t|}\right) \sim \frac{t^2}{2} a H\left(\frac{1}{|t|}\right).$$

Note. On the same lines it can be shown that $\lim\limits_{t \to 0} \frac{1 - \Re(\varphi(t))}{t^2 U(\frac{1}{t})} = \frac{1}{2}$.
(ii) *If W is sv, then $\frac{V_n}{b_n} \xrightarrow{d} \xi$ and $\mathbb{E}|\frac{V_n}{b_n}|^\beta \to \mathbb{E}|\xi|^\beta$, $0 \le \beta < 2$.*
(iii) *If U is sv, then $\frac{S_n}{b_n} \xrightarrow{d} \xi$ and $\mathbb{E}|\frac{S_n}{b_n}|^\beta \to \mathbb{E}|\xi|^\beta$, $0 \le \beta < 2$.*
(iv) (*In* **(5.6.1)**, *the following simple consequence of* (iii) *will be used. Compare with* **(5.3.1)**(xii))
Let U be sv. Let $0 < \alpha < \beta < 1$. Then $\frac{S_{[\beta n]} - S_{[\alpha n]}}{b_n} \xrightarrow{d} \xi\sqrt{\beta - \alpha}$.

Proof. (i) Since g is an even function, it is enough to prove result for $t > 0$. Define $\frac{1 - \cos tx}{x^2}$ to be $\frac{1}{2}t^2$ for $x = 0$ and note that, so defined, the resulting function is continuous in x for every t. We have,

$$1 - g(t) = \int_{-\infty}^{\infty} (1 - \cos tx)\, \mathrm{d}G(x) = \int_{-\infty}^{\infty} \frac{1 - \cos tx}{x^2} x^2\, \mathrm{d}G(x)$$
$$= \int_{0-}^{\infty} \frac{1 - \cos tx}{x^2}\, \mathrm{d}W(x).$$

Or,

$$\frac{1 - g(t)}{t^2 W(\frac{1}{t})} = \int_{[0,\,\infty)} \frac{1 - \cos ty}{t^2 y^2 W(\frac{1}{t})}\, \mathrm{d}W(y)$$
$$= \int_{[0,\,\alpha]} \frac{1 - \cos ty}{t^2 y^2 W(\frac{1}{t})}\, \mathrm{d}W(y) + \int_{(\alpha,\,\infty)} \frac{1 - \cos ty}{t^2 y^2 W(\frac{1}{t})}\, \mathrm{d}W(y)$$
$$= \int_{[0,\,\alpha]} \frac{1 - \cos y}{y^2}\, \mathrm{d}W_t(y) + \int_{(\alpha,\,\infty)} \frac{1 - \cos y}{y^2}\, \mathrm{d}W_t(y)$$
$$= I_1(t) + I_2(t), \text{ say,}$$

where $\alpha > 1$ is arbitrary and $W_t(y) = \frac{W(\frac{y}{t})}{W(\frac{1}{t})}$.

The integral $I_2(t)$ is dominated by the finite integral

$$\int_{(\alpha,\,\infty)} \frac{1}{y^2}\, \mathrm{d}W_t(y) = \lim_{r \to \infty} \int_{(\alpha+,\, r]} y^{-2}\, \mathrm{d}W_t(y).$$

We appeal to **(2.6.3)** which applies since y^{-2} is continuously differentiable in $(\alpha,\ r]$ and see that

$$I_2(t) \le \lim_{r\to\infty} \Big[y^{-2}W_t(y)\Big|_\alpha^r + 2\int_\alpha^r W_t(y)\frac{1}{y^3}\,\mathrm{d}\,y\Big]$$

$$= \lim_{r\to\infty}\Big[\frac{1}{r^2}W_t(r) - \frac{1}{\alpha^2}W_t(\alpha)\Big] + 2\int_\alpha^\infty \frac{W(\lambda y)}{W(\lambda)}\frac{1}{y^3}\,\mathrm{d}\,y$$

where $\lambda = \frac{1}{t} \to \infty$ as $t \to 0$. Thus $I_2(\frac{1}{\lambda}) \le -\frac{W(\lambda\alpha)}{\alpha^2 W(\lambda)} + I_3(\lambda)$, say, where $I_3(\lambda) = 2\int_\alpha^\infty \frac{W(\lambda y)}{W(\lambda)}\frac{1}{y^3}\,\mathrm{d}\,y$. Since W is *sv*,

$$W(x) \sim b e^{\int_1^x \frac{\varepsilon_1(y)}{y}\,\mathrm{d}\,y} \quad \text{as } x \to \infty,$$

with $\varepsilon_1(y) \to 0$ as $y \to \infty$. Find λ_0 such that $\varepsilon_1(y) \le \frac{1}{2}$ for all $y \ge \lambda_0$ and such that

$$\frac{1}{2}b e^{\int_1^x \frac{\varepsilon_1(y)}{y}\,\mathrm{d}\,y} \le W(x) \le 2b e^{\int_1^x \frac{\varepsilon_1(y)}{y}\,\mathrm{d}\,y} \quad \text{for all } x > \lambda_0.$$

We have

$$I_3(\lambda) \le \int_\alpha^\infty e^{\int_\lambda^{\lambda y} \frac{1}{2z}\,\mathrm{d}\,z}\frac{1}{y^3}\,\mathrm{d}\,y \le \int_\alpha^\infty \frac{1}{y^2}\,\mathrm{d}\,y = \frac{1}{\alpha}.$$

Thus

$$I_2(t) \le -\frac{W(\lambda\alpha)}{\alpha^2 W(\lambda)} + \frac{1}{\alpha}.$$

Let $\lambda \to \infty$ and then $\alpha \to \infty$ to conclude that $\lim_{t\to 0} I_2(t) = 0$.

Now,

$$I_1(t) = \frac{W(\lambda\alpha)}{W(\lambda)}\int_{[0,\ \alpha]} \frac{1-\cos y}{y^2}\,\mathrm{d}\,W_\lambda^*(y)$$

where $W_\lambda^*(y)$ is a *df* with $W_\lambda^*(0) = 0$ and $W_\lambda^*(\alpha) = 1$. Hence for every $\lambda_n \to \infty$, $\lim_{n\to\infty} W_{\lambda_n}^*(y) = W^*(y)$ where $W^*(0) = 0$ and $W^*(y) = 1,\ y > 0$. Since $\lim_{\lambda\to\infty}\frac{W(\lambda\alpha)}{W(\lambda)} = 1$, this implies $\lim_{n\to\infty} I_1(t_n) = \frac{1}{2}$. i.e., $\lim_{t\to 0}\frac{1-g(t)}{t^2}W(\frac{1}{t}) = \frac{1}{2}$, which is the desired result.

The last asymptotic relation follows from **(5.4.2)**(vi)

(ii) For $t \neq 0$, fixed,

$$\lim_{n\to\infty} n\{1 - g(\frac{t}{b_n})\} = \lim_{n\to\infty} n\frac{t^2}{2b_n^2} W\left(\frac{b_n}{|t|}\right) \qquad \text{(by (i))}$$

$$= \lim_{n\to\infty} n\frac{t^2}{2b_n^2} W(b_n) = \lim_{n\to\infty} n\frac{t^2}{2b_n^2} U(b_n) = \frac{t^2}{2}.$$

We note that for t fixed and n large, $g(\frac{t}{b_n})$ is close to unity. Hence

$$\lim_{n\to\infty} n \log g(\frac{t}{b_n}) = \lim_{n\to\infty} -n\{1 - g(\frac{t}{b_n})\} = -\frac{t^2}{2}.$$

Thus $\lim_{n\to\infty} g^n(\frac{t}{b_n}) = e^{-\frac{t^2}{2}}$ which is equivalent to the desired result.

Let $0 < \beta < 2$ and let $t > 0$. By **(5.4.2)**(i), $\mathbb{E}|Z|^\beta < \infty$. Hence (ref. **(4.2.3)**(ii)) $\mathbb{E}|\frac{V_n}{b_n}|^\beta < \infty$. By **(3.6.9)**,

$$\mathbb{E}|\frac{V_n}{b_n}|^\beta = \frac{1}{c_{0,\beta}} I_n \text{ where } I_n = \int_0^\infty \frac{1 - g^n(\frac{t}{b_n})}{t^{1+\beta}}\, \mathrm{d}t.$$

Find $q > 0$ small such that $1 - g(t) \le 2\frac{t^2}{2} aH(\frac{1}{t})$ for all $t \in (0,\ q]$. This is possible by (i). Choose $\eta \in (0,\ 2 - \beta)$. This implies $|\beta + \eta - 1| < 1$. Then for $t \in (0, 1]$ and for all n large such that $\frac{1}{b_n} < q$ and such that $\varepsilon(y) < \eta$ if $y > b_n$, we have

$$\frac{1 - g^n(\frac{t}{b_n})}{t^{1+\beta}} \le \frac{n\{1 - g(\frac{t}{b_n})\}}{t^{1+\beta}} \le n\frac{\frac{t^2}{b_n^2} aH(\frac{b_n}{t})}{t^{1+\beta}} = a\frac{nH(b_n)}{b_n^2}\frac{H(\frac{b_n}{t})}{H(b_n)}\frac{1}{t^{\beta-1}}$$

$$= O(1)\frac{1}{t^{\beta-1}} e^{\int_{b_n}^{\frac{b_n}{t}} \frac{\varepsilon(y)}{y}\,\mathrm{d}y} = O(1)\frac{1}{t^{\beta+\eta-1}}.$$

Thus over $(0,\ 1]$, the integrand in I_n is bounded, for all large n by the integrable function $\frac{q}{t^{\beta+\eta-1}}$. Over $[1,\ \infty)$ the integrand is bounded, for all n, by the integrable function $\frac{2}{t^{1+\beta}}$. Taking limits and appealing to the bounded convegence theorem we then get

$$\lim_{n\to\infty} \mathbb{E}|\frac{V_n}{b_n}|^\beta = \frac{1}{c_{0,\beta}} \int_0^\infty \frac{1 - e^{-\frac{t^2}{2}}}{t^{1+\beta}}\, \mathrm{d}t = \mathbb{E}|\xi|^\beta$$

(ref. **(3.6.9)**).

(iii) U sv implies, by **(5.4.2)**(vi), that W is sv. Hence, by (ii), $\frac{V_n}{b_n} \xrightarrow{d} \xi$ and $\mathbb{E}|\frac{V_n}{b_n}| \to \mathbb{E}|\xi|$. This implies (ref. **(2.6.9)**(iii)) that $(|\frac{V_n}{b_n}|)$ is a uniformly integrable sequence. Let θ_n be a median of $\frac{S_n}{b_n}$. Writing $\sqrt{2}\frac{V_n}{b_n} = (\frac{S_n}{b_n} - \theta_n) - (\frac{T_n}{b_n} - \theta_n)$ and appealing to **(4.2.5)**, we conclude that $(|\frac{S_n}{b_n} - \theta_n|)$ is a uniformly integrable sequence and, as a consequence (ref. **(2.6.8)**), is a tight sequence. Let $n_r \to \infty$ and let $\frac{S_{n_r}}{b_{n_r}} - \theta_{n_r} \xrightarrow{d} \zeta_1$ and, similarly, let $\frac{T_{n_r}}{b_{n_r}} - \theta_{n_r} \xrightarrow{d} \zeta_2$. Since the X_ks and the Y_ks are identically distributed and since $b_n \to \infty$, it follows that $(\frac{X_k}{b_{n_r}}, \frac{Y_k}{b_{n_r}}, 1 \leq k \leq n_r)$ are infinitesimal. Hence (ref. **(5.2.2)**) ζ_1, ζ_2 are id. We note ζ_1, ζ_2 are independent. Hence $\frac{V_{n_r}}{b_{n_r}} \xrightarrow{d} \frac{1}{\sqrt{2}}(\zeta_1 - \zeta_2)$. Since ζ_1, ζ_2 are *iid* and since $\frac{1}{\sqrt{2}}(\zeta_1 - \zeta_2)$ is distributed as ξ, it follows by **(3.4.6)**(iii) that ζ_1, ζ_2 are normally distributed, say, $\mathcal{N}(\mu, \sigma^2)$. We have $1 = var(\xi) = var(\frac{\zeta_1 - \zeta_2}{\sqrt{2}}) = \sigma^2$. Use uniform integrability of $(|\frac{S_{n_r}}{b_{n_r}} - \theta_{n_r}|)$ and the weak convergence $\frac{S_{n_r}}{b_{n_r}} - \theta_{n_r} \xrightarrow{d} \xi + \mu$ to conclude (ref. **(2.6.9)**) that $\mathbb{E}\{\frac{S_{n_r}}{b_{n_r}} - \theta_{n_r}\} \to \mathbb{E}(\xi + \mu)$. i.e. $\theta_{n_r} \to -\mu$. Again, by (**(2.5.2)**), $0 = median(\frac{S_{n_r}}{b_{n_r}} - \theta_{n_r}) \to median(\xi + \mu) = \mu$. Hence $\theta_{n_r} \to 0$ and then $\frac{S_{n_r}}{b_{n_r}} \xrightarrow{d} \xi$. Thus given a subsequence $\frac{S_m}{b_m}$ of $\frac{S_n}{b_n}$, we consider the tight sequence $(\frac{S_m}{b_m} - \theta_m)$ and arrive at a further subsequence, say, $\frac{S_\nu}{b_\nu}$ converging weakly to ξ. Hence $\frac{S_n}{b_n} \xrightarrow{d} \xi$ (ref. **(2.4.3)**(iv)).

We have shown in (ii) that $\mathbb{E}|\frac{V_n}{b_n}|^\beta \to \mathbb{E}|\xi|^\beta$. Since for each n, S_n and T_n are independent and since $\theta_n \to 0$, it follows (ref. **(4.2.5)**) that $(|\frac{S_n}{b_n}|^\beta)$ is uniformly integrable. This together with the fact $\frac{S_n}{b_n} \xrightarrow{d} \xi$ implies (ref. **(2.6.9)**(ii)) that $\mathbb{E}|\frac{S_n}{b_n}|^\beta \to \mathbb{E}|\xi|^\beta$.

We give below an alternate proof of the first part of (iii) above. The proof rests on truncation suitably of the variables X_n and showing that the Lindeberg-Feller condition applies to the truncated variables.

The argument in the proof of (iii) shows that it is sufficient to prove the theorem for symmetrically distributed variables (X_n). For $\nu = 1, 2, \ldots, n$, define $Y_{\nu,n} = X_\nu$if $|X_\nu| \leq b_n$ and 0 otherwise and note that these are n *iid* symmetrically distribted variables. Let $Y_n = \sum_{\nu=1}^n Y_{\nu,n}$. We note var $Y_n = n\int_{|x|\leq b_n} x^2 dF(x) = nU(b_n) \sim b_n^2$. Now, The Lindeberg-Feller normal convergence criterion holds for the sequence $(Y_{\nu,n}, \nu = 1, 2, \ldots, n)$ if $\frac{1}{b_n^2}\sum_{\nu=1}^n \int_{|Y_{\nu,n}|>\epsilon b_n} Y_{\nu,n}^2 dP \to 0$ for every $\epsilon > 0$; i.e if $\frac{n}{b_n^2}\int_{\epsilon b_n \leq |x| \leq b_n} x^2 dF(x)$ $\to 0$ for every $\epsilon > 0$; i.e if $\frac{1}{U(b_n)}\{U(b_n) - U(\epsilon b_n)\} \to 0$, which is true. We thus have $P(\frac{Y_n}{b_n} \leq x) \to \Phi(x)$. Now, $P(S_n \neq Y_n) \leq \sum\limits_{\nu=1}^{n} P(X_\nu \neq Y_{\nu,n}) =$

$nP(|X| > b_n) = n \int\limits_{|x|>b_n} dF(x) = \frac{n}{b_n^2} b_n^2 \int\limits_{|x|>b_n} dF(x) \sim \frac{b_n^2}{U(b_n)} \int\limits_{|x|>b_n} dF(x)$
$= Q(b_n) \to 0$, since U is *sv* (ref. **(5.4.2)**(vii)). That $P(\frac{S_n}{b_n} \le x) \to \Phi(x)$ follows now.

(iv) Since the variables are identically distributed, the distribution of $S_{[\beta n]} - S_{[\alpha n]}$ is the same as that of $S_{[\beta n]-[\alpha n]}$. (iii) now gives $\frac{S_{[\beta n]-[\alpha n]}}{b_{[\beta n]-[\alpha n]}} \xrightarrow{d} \xi$. Since $\frac{b_{[\beta n]-[\alpha n]}}{b_n} \to \sqrt{\beta - \alpha}$ (ref. **(5.4.2)**(iv))), the claim follows. □

Theorem 5.4.4. *(A converse to the CLT). Let (X_n) be a sequence if iid variables with common df F and common cf φ. Let $S_n = \sum\limits_{k=1}^{n} X_k$, $n = 1, 2, \ldots$.. Let ξ denote a standard normal variable. Let (b_n) be a sequence of positive numbers. Then,*

(i) $\frac{S_n}{b_n} \xrightarrow{d} \xi$ *implies* $\underline{\lim}_{n\to\infty} \frac{b_n}{\sqrt{n}} > 0$

(ii) $\frac{S_n}{b_n} \xrightarrow{d} \xi$ *and* $\underline{\lim}_{n\to\infty} \frac{b_n}{\sqrt{n}} = \lambda$, $0 < \lambda < \infty$ *imply* $\lim_{n\to\infty} \frac{b_n}{\sqrt{n}} = \lambda$, $\mathbb{E}X_1^2 < \infty$, $\mathbb{E}X_1 = 0$ *and* $\mathbb{E}X_1^2 = \lambda^2$.

(iii) $\frac{S_n}{b_n} \xrightarrow{d} \xi$ *and* $\lim_{n\to\infty} \frac{b_n}{\sqrt{n}} = \infty$ *imply* $U(x) = \int\limits_{|y|\le x} y^2\, dF(y) \to \infty$ *as* $x \to \infty$ *and is a slowly varying (sv) function.*

Proof.

Let (Y_n) be an independent copy of the X_n-sequence. Let $Z_n = \frac{1}{\sqrt{2}}(X_n - Y_n)$ and $T_n = \sum\limits_{k=1}^{n} Z_k$. Let V be the *df* of Z_n and v its *cf*. We note $v \ge 0$, since $v(t) = |\varphi(\frac{t}{\sqrt{2}})|^2$.

(i) The hypothesis that $\frac{S_n}{b_n} \xrightarrow{d} \xi$ implies $\frac{T_n}{b_n} \xrightarrow{d} \xi$.

Being a symmetric variable Z_1, if degenerate, will have to be zero $wp1$. But this is not possible since $\frac{T_n}{b_n} \xrightarrow{d} \xi$. Hence the variables Z_n can not be degenerate.

Now, if possible let $\underline{\lim}_{n\to\infty} \frac{b_n}{\sqrt{n}} = 0$. Hence there can be found a sequence $(n_k \uparrow \infty)$ such that $\frac{b_{n_k}}{\sqrt{n_k}} \to 0$. This implies $\frac{T_{n_k}}{\sqrt{n_k}} \xrightarrow{d} 0$. i.e., $v^{n_k}(\frac{t}{\sqrt{n_k}}) \to 1$ for all t. Either there exists an infinite subsequence (m_r) of (n_k) such that $v(\frac{1}{\sqrt{m_r}}) = 1$ or $0 \le v(\frac{1}{\sqrt{m_r}}) < 1$. In the first case we have trivially $m_r\{1 - v(\frac{1}{\sqrt{m_r}})\} = 0$ and in the second case we will have $m_r\{1 - v(\frac{1}{\sqrt{m_r}})\} \to 0$ as $r \to \infty$, since $v^{m_r}(\frac{1}{\sqrt{m_r}}) \to 1$.

By Fatou's Lemma, we have in both the cases,

$$0 = \lim_{r\to\infty} m_r\{1 - v(\tfrac{1}{\sqrt{m_r}})\} = \lim_{r\to\infty} \int_{-\infty}^{\infty} \frac{1-\cos(\frac{x}{\sqrt{m_r}})}{\frac{1}{m_r}}\,\mathrm{d}V(x)$$
$$\geq \int_{-\infty}^{\infty} \lim_{r\to\infty} \frac{1-\cos(\frac{x}{\sqrt{m_r}})}{\frac{1}{m_r}}\,\mathrm{d}V(x)$$
$$= \tfrac{1}{2}\int_{-\infty}^{\infty} x^2\,\mathrm{d}V(x),$$

a contradiction, since the variables Z_n are non-degenerate. The claim follows that necessarily $\underline{\lim}_{n\to\infty} \frac{b_n}{\sqrt{n}} > 0$.

(ii) Since $\underline{\lim}_{n\to\infty} \frac{b_n}{\sqrt{n}} = \lambda$, a subsequence $(n_k \uparrow \infty)$ exists such that $\lim_{n\to\infty} \frac{b_{n_k}}{\sqrt{n_k}} = \lambda$. The hypothesis implies $\frac{T_n}{b_n} \overset{d}{\to} \xi$. Hence $\frac{T_{n_k}}{\lambda\sqrt{n_k}} \overset{d}{\to} \xi$. i.e., $v^{n_k}(\frac{t}{\lambda\sqrt{n_k}}) \to e^{-\frac{t^2}{2}}$. This implies that for all k sufficiently large, $0 \leq v(\frac{1}{\lambda\sqrt{n_k}}) < 1$. Hence $n_k\{1 - v(\frac{1}{\lambda\sqrt{n_k}})\} \to \frac{1}{2}$. As before we get, by appealing to Fatou's Lemma, $\frac{1}{2} \geq \int_{-\infty}^{\infty} \frac{x^2}{2\lambda^2}\,\mathrm{d}P(Z_1 \leq x)$. i.e., $\mathbb{E}Z_1^2 < \infty$. Hence $\mathbb{E}|Z_1| < \infty$. Since the Z_ns are symmetric variables, $\mathbb{E}Z_1 = 0$. Let $\mathbb{E}Z_1^2 = a^2$, $a > 0$. The Lindeberg-Feller condition applies to the sequence (Z_n) (ref. **(5.3.2)**(i)) and we conclude that $\frac{T_n}{a\sqrt{n}} \overset{d}{\to} \xi$. Comparing this result with what is already noted, namely, $\frac{T_n}{b_n} \overset{d}{\to} \xi$, we conclude $b_n \sim a\sqrt{n}$ (ref. **(2.1.14)**). This together with $b_{n_k} \sim \lambda\sqrt{n_k}$ gives $a = \lambda$ and then $b_n \sim \lambda\sqrt{n}$. Now, $\mathbb{E}Z_1^2 < \infty$ implies (ref. **(4.2.3)**) $\mathbb{E}X_1^2 < \infty$. Let $\mathbb{E}X_1 = \theta$ and let $\mathbb{E}(X_1 - \theta)^2 = \sigma^2$. Then, as before, by **(5.3.2)**(i), $\frac{S_n - n\theta}{\sigma\sqrt{n}} \overset{d}{\to} \xi$. This, together with the hypothesis $\frac{S_n}{\lambda\sqrt{n}} \overset{d}{\to} \xi$ implies $\theta = 0$ and $\sigma = \lambda$ (ref. **(2.1.14)**).

(iii) First, we claim that the hypothesis implies $U(x) \to \infty$ as $x \to \infty$ i.e., $\mathbb{E}X_1^2 = \infty$. For, if $\mathbb{E}X_1^2 < \infty$ and if $var X_1 = \sigma^2$, then $\mathbb{E}Z_1^2$ would be σ^2. That would imply $\frac{T_n}{\sigma\sqrt{n}} \overset{d}{\to} \xi$. This, together with the fact, derived from the hypothesis, that $\frac{T_n}{b_n} \overset{d}{\to} \xi$ imply $b_n \sim \sigma\sqrt{n}$, a contradiction. Hence $U(x) \to \infty$.

Step 1.

Define $W(x) = \int_{|y|\leq x} y^2\,\mathrm{d}V(y)$. Since by **(5.4.2)** (vi), $U(x) \sim W(x)$ as $x \to \infty$, it is enough if we show W is sv. By **(5.4.2)**(vii), this is equivalent to proving

$$\lim_{c\to\infty} Q(c) = 0 \text{ where } Q(c) = \frac{c^2 \int_{|x|>c} \mathrm{d}V(x)}{\int_{|x|<c} x^2\,\mathrm{d}V(x)} \tag{5.9}$$

This we proceed to establish.

Step 2.

In this step we show that the hypothesis $\frac{S_n}{b_n} \xrightarrow{d} \xi$ implies $G_n \xrightarrow{d} G_0$ where $G_n(x) = \int\limits_{(-\infty,\, x]} n\frac{y^2}{1+y^2}\, \mathrm{d}V(yb_n),\ n \geq 1$ and $G_0(x) = 0$ if $x < 0$ and $G_0(x) = 1$ if $x \geq 1$.

The hypothesis implies, as argued in part (ii),

$$n\{v(\frac{t}{b_n}) - 1\} \to -\frac{t^2}{2} \tag{5.10}$$

Now,

$$\begin{aligned}
n\{v(\tfrac{t}{b_n}) - 1\} &= n \int\limits_{-\infty}^{\infty} \{e^{itx} - 1\}\, \mathrm{d}V(xb_n) \\
&= \int\limits_{-\infty}^{\infty} \{e^{itx} - 1 - \tfrac{itx}{1+x^2}\}\, \mathrm{d}nV(xb_n) \text{ (since } V \text{ is a symmetric } df) \\
&= \int\limits_{-\infty}^{\infty} \{e^{itx} - 1 - \tfrac{itx}{1+x^2}\}\tfrac{1+x^2}{x^2}\, \mathrm{d}G_n(x)
\end{aligned}$$

where the integrand $K(t,\, x) = \{e^{itx} - 1 - \frac{itx}{1+x^2}\}\frac{1+x^2}{x^2}$ is to be taken as $-\frac{t^2}{2}$ at $x = 0$. Also $-\frac{t^2}{2} = \int\limits_{-\infty}^{\infty} K(t,\, x)\, \mathrm{d}G_0(x)$.

We note G_n is a right continuous non-decreasing function with

$$G_n(-\infty) = 0, G_n(\infty) < \infty$$

$$G_n(0) - G_n(0-) = \int\limits_{\{0\}} n\frac{y^2}{1+y^2}\, \mathrm{d}V(xb_n) = 0 \tag{5.11}$$

and, by (5.10), for all t,

$$\psi_n(t) = \int\limits_{-\infty}^{\infty} K(t,\, x)\, \mathrm{d}G_n(x) \to \int\limits_{-\infty}^{\infty} K(t,\, x)\, \mathrm{d}G_0(x) = \psi_0(t) \tag{5.12}$$

Now,

$$\begin{aligned}
\tfrac{1}{\tau}\int_0^{\tau}\{-\Re(\int\limits_{-\infty}^{\infty} K(t,\, x)\, \mathrm{d}G_n(x))\}\, \mathrm{d}t &= \tfrac{1}{\tau}\int\limits_{0}^{\tau}\{\int\limits_{-\infty}^{\infty} (1 - \cos tx)\tfrac{1+x^2}{x^2}\, \mathrm{d}G_n(x)\}\, \mathrm{d}t \\
&= \int\limits_{-\infty}^{\infty} (1 - \tfrac{\sin \tau x}{\tau x})\tfrac{1+x^2}{x^2}\, \mathrm{d}G_n(x) \\
&\geq \int\limits_{-\infty}^{\infty} (1 - \tfrac{|\sin \tau x|}{\tau x})\, \mathrm{d}G_n(x)
\end{aligned}$$

$$\geq \int_{|\tau x| \geq \frac{1}{2}} dG_n(x).$$

Since the *cf* e^{ψ_n} converges pointwise to the *cf* e^{ψ_0}, the convergence is uniform over bounded intervals. Hence $\psi_n(t) \to \psi_0(t)$ uniformly over bounded intervals. Consequently, $\int_0^\tau \{-\psi_n(t)\}\,dt \to \int_0^\tau \{-\psi_0(t)\}\,dt = \frac{\tau^3}{6}$. Given $\varepsilon > 0$ choose $\tau^2 \leq 6\varepsilon$. We have,

$$\int_{|x| \geq \frac{1}{2\tau}} dG_n(x) \leq \frac{1}{\tau} \int_0^\tau \{-\psi_n(t)\}\,dt \to \frac{1}{\tau} \int_0^\tau \{-\psi_0(t)\}\,dt$$

$$= \frac{1}{\tau} \int_0^\tau \frac{t^2}{2}\,dt = \frac{\tau^2}{6} < \varepsilon,$$

showing that the sequence (G_n) of measures is tight. By **(2.2.2)**, it admits of a subsequence (G_{n_k}) conveging weakly to, say, G^*. We note that for t fixed, $K(t,\,x)$ is a continuous function of x. Further, $\sup_{x \in R} |K(t,\,x)| \leq 6t^2 + 2|t| + 4$ (ref. **(3.4.1)**, end of Step 1). Since $K(t,\,x)$ is a bounded continuous function of x, it follows that $\int_{-\infty}^{\infty} K(t,\,x)\,dG_{n_k}(x) \to \int_{-\infty}^{\infty} K(t,\,x)\,dG^*(x)$. Hence $\int_{-\infty}^{\infty} K(t,\,x)\,dG^*(x) = \int_{-\infty}^{\infty} K(t,\,x)\,dG_0(x)$. From this we conclude (ref. **(3.4.3)**) $G^* = G_0$. Since this is true of every convergent subsequence of (G_n), it follows that $G_n \xrightarrow{d} G_0$. i.e., $G_n(\infty) \to G_0(\infty)$ and $G_n(x) \to G_0(x)$ at all the continuity points x of G_0. i.e., at all $x \neq 0$. □

Define $M_n(u) = \int_{(-\infty,\,u]} \frac{1+x^2}{x^2}\,dG_n(x),\ u < 0$

and $N_n(u) = -\int_{[u,\,\infty)} \frac{1+x^2}{x^2}\,dG_n(x),\ u > 0$. Then,

Step 2.

M_n (and, similarly, N_n) is right continuous.

Proof.

The claim follows by an appeal to the bounded convergence theorem, since $M_n(u) = \int_{-\infty}^{\infty} \frac{1+x^2}{x^2} \chi_{(-\infty,\,u]}\,dG_n(x)$, since $\chi_{(-\infty,v]} \downarrow \chi_{(-\infty,u]}$ as $v \downarrow u$.

Step 3.

Every discontinuity point of M_n is a discontinuity point of G_n and every discontinuity point of G_n in the range $(-\infty,\,0)$ is a discontinuity point of M_n. [A similar statement is valid for N_n].

Proof. For $-\infty < u_0 < 0$,

$M_n(u_0) - M_n(u_0-) = \int_{\{u_0\}} \frac{1+x^2}{x^2}\,dG_n(x) = \frac{1+u_0^2}{u_0^2}\{G_n(u_0) - G_n(u_0-)\}$. This proves the claim for M_n.

Step 4.

$(G_n \overset{d}{\to} G_0) \Leftrightarrow (M_n \overset{d}{\to} M_0$ and $N_n \overset{d}{\to} N_0)$.

Proof.

We recall $G_0(x) = 0$ if $x < 0$ and $G(x) = 1$ if $x \geq 0$ and hence 0 is a point of discontinuity of G_0.

Let $G_n \overset{d}{\to} G_0$. Let u be a continuity point of M_0. Hence it is a continuity point of G_0. Now, since $(-\infty,\ u]$ is a continuity set of G_0 and since $\frac{1+x^2}{x^2}$ is a bounded continuous function in $(-\infty,\ u^*)$, $u < u^* < 0$, it follows by Remark ((**2.4.3**)(v)) that $\int\limits_{(-\infty,\ u]} \frac{1+x^2}{x^2}\, dG_n(x) \to \int\limits_{(-\infty,\ u]} \frac{1+x^2}{x^2}\, dG_0(x)$. i.e., $M_n(u) \to M_0(u)$. Similarly, $N_n(u) \to N_0(u)$ at all the continuity points u of N_0.

To prove the converse of this result, let u be a continuity point of M_0 and let $M_n \overset{d}{\to} M_0$. These assumptions imply u is a continuity point of G_0 and that $G_n(u) = \int\limits_{(-\infty,\ u]} \frac{x^2}{1+x^2}\, dM_n(x) \to \int\limits_{(-\infty,\ u]} \frac{x^2}{1+x^2}\, dM_0(x) = G_0(u)$, again appealing to ((**2.4.3**)(v)). Thus for every continuity point $u < 0$, $G_n(u) \to G_0(u)$. Starting with $N_n \overset{d}{\to} N_0$, similar arguments lead to : for every continuity point $u > 0$, $G_n(u) \to G_0(u)$.

Step 5.

For $\varepsilon > 0$,

$$\begin{aligned}
1 &= \int_{(-\varepsilon,\,\varepsilon)} \mathrm{d}G_0(x) \\
&= \lim_{n\to\infty} \int_{(-\varepsilon,\,\varepsilon)} \mathrm{d}G_n(x) \\
&= \lim_{n\to\infty} \Big\{ \int_{(-\varepsilon,\,0)} \mathrm{d}G_n(x) + \int_{\{0\}} \mathrm{d}G_n(x) + \int_{(0,\,\varepsilon)} \mathrm{d}G_n(x) \Big\} \\
&= \lim_{n\to\infty} \Big\{ \int_{(-\varepsilon,\,0)} \mathrm{d}G_n(x) + \int_{(0,\,\varepsilon)} \mathrm{d}G_n(x) \Big\} \text{ (ref. (5.11))} \\
&= \lim_{n\to\infty} \Big\{ \int_{(-\varepsilon,\,0)} \frac{x^2}{1+x^2}\, \mathrm{d}M_n(x) + \int_{(0,\,\varepsilon)} \frac{x^2}{1+x^2}\, \mathrm{d}N_n(x) \Big\}
\end{aligned} \tag{5.13}$$

(since u is the lower limit of the integral defining $N_n(u)$)

$$\geq \frac{1}{1+\varepsilon^2} \overline{\lim_{n\to\infty}} \Big\{ \int_{(-\varepsilon,\,0)} x^2\, \mathrm{d}M_n(x) + \int_{(0,\,\varepsilon)} x^2\, \mathrm{d}N_n(x) \Big\} \tag{5.14}$$

We also have from (5.13),

$$1 \leq \underline{\lim}_{n\to\infty} \Big\{ \int_{(-\varepsilon,\,0)} x^2\, \mathrm{d}M_n(x) + \int_{(0,\,\varepsilon)} x^2\, \mathrm{d}N_n(x) \Big\}.$$

Thus

$$\begin{aligned}
\lim_{\varepsilon\to 0} \tfrac{1}{1+\varepsilon^2} \overline{\lim_{n\to\infty}} \Big\{ \int_{(-\varepsilon,\,0)} x^2\, \mathrm{d}M_n(x) + \int_{(0,\,\varepsilon)} x^2\, \mathrm{d}N_n(x) \Big\} &\leq 1 \\
\leq \lim_{\varepsilon\to 0} \underline{\lim}_{n\to\infty} \Big\{ \int_{(-\varepsilon,\,0)} x^2\, \mathrm{d}M_n(x) + \int_{(0,\,\varepsilon)} x^2\, \mathrm{d}N_n(x) \Big\}
\end{aligned}$$

Or,

$$\begin{aligned}
\lim_{\varepsilon\to 0} \overline{\lim_{n\to\infty}} \Big\{ \int_{(-\varepsilon,\,0)} x^2\, \mathrm{d}M_n(x) + \int_{(0,\,\varepsilon)} x^2\, \mathrm{d}N_n(x) \Big\} &\leq 1 \\
\leq \lim_{\varepsilon\to 0} \underline{\lim}_{n\to\infty} \Big\{ \int_{(-\varepsilon,\,0)} x^2\, \mathrm{d}M_n(x) + \int_{(0,\,\varepsilon)} x^2\, \mathrm{d}N_n(x) \Big\}
\end{aligned}$$

i.e.,

$$\begin{aligned}
\lim_{\varepsilon\to 0} \overline{\lim_{n\to\infty}} \Big\{ \int_{(-\varepsilon,\,0)} x^2\, \mathrm{d}M_n(x) + \int_{(0,\,\varepsilon)} x^2\, \mathrm{d}N_n(x) \Big\} &= 1 \\
= \lim_{\varepsilon\to 0} \underline{\lim}_{n\to\infty} \Big\{ \int_{(-\varepsilon,\,0)} x^2\, \mathrm{d}M_n(x) + \int_{(0,\,\varepsilon)} x^2\, \mathrm{d}N_n(x) \Big\}
\end{aligned}$$

i.e.,

$$\lim_{\varepsilon\to 0}\overline{\lim_{n\to\infty}}\, n\Big\{\int_{(-\varepsilon,\,0)} x^2\,\mathrm{d}V(xb_n)+\int_{(0,\,\varepsilon)} x^2\,\mathrm{d}V(xb_n)\Big\}=1$$
$$=\lim_{\varepsilon\to 0}\underline{\lim_{n\to\infty}}\, n\Big\{\int_{(-\varepsilon,\,0)} x^2\,\mathrm{d}V(xb_n)+\int_{(0,\,\varepsilon)} x^2\,\mathrm{d}V(xb_n)\Big\}$$

i.e.,

$$\lim_{\varepsilon\to 0}\overline{\lim_{n\to\infty}}\,\frac{n}{b_n^2}\int_{|x|<\varepsilon b_n} x^2\,\mathrm{d}V(x)=1=\lim_{\varepsilon\to 0}\underline{\lim_{n\to\infty}}\,\frac{n}{b_n^2}\int_{|x|<\varepsilon b_n} x^2\,\mathrm{d}V(xb_n) \tag{5.15}$$

Step 6.

By Step 1 and Step 4, $M_n(u)\to M_0(u)\equiv 0$ for all $u<0$ and $N_n(u)\to N_0(u)\equiv 1$ for all $u>0$. i.e.,

$$nV(xb_n)\to 0,\ -\infty<x<0 \text{ and } n\{1-V(xb_n)\}\to 0,\ 0<x<\infty \tag{5.16}$$

(5.16) is equivalent to

$$\lim_{n\to\infty} n\int_{|y|>xb_n}\mathrm{d}V(y)=0,\ \text{for every } x>0 \tag{5.17}$$

Step 7.

$A(\varepsilon)=\overline{\lim_{n\to\infty}}\frac{n}{b_n^2}\int_{|x|<\varepsilon b_n} x^2\,\mathrm{d}V(x)$ is independent of ε. (And, similarly, $B(\varepsilon)=\underline{\lim_{n\to\infty}}\frac{n}{b_n^2}\int_{|x|<\varepsilon b_n} x^2\,\mathrm{d}V(x)$ is independent of ε). i.e., we claim $A(\varepsilon)=A(\eta)$ for every $\varepsilon>0$ and $\eta>0$. The claim follows since

$$\Big|\frac{n}{b_n^2}\int_{|x|<\varepsilon b_n} x^2\,\mathrm{d}V(x)-\frac{n}{b_n^2}\int_{|x|<\eta b_n} x^2\,\mathrm{d}V(x)\Big|=\frac{n}{b_n^2}\int_{\eta b_n\le|x|<\varepsilon b_n} x^2\,\mathrm{d}V(x)$$
$$\le n\varepsilon^2\int_{\eta b_n\le|x|<\varepsilon b_n}\mathrm{d}V(x)$$
$$\le n\varepsilon^2\int_{\eta b_n\le|x|<\varepsilon b_n}\mathrm{d}V(x)$$
$$\le n\varepsilon^2\int_{|x|\ge\eta b_n}\mathrm{d}V(x)$$
$$=o(1)\ \text{(by (5.17))}.$$

Consequently,

$$\lim_{n\to\infty}\frac{n}{b_n^2}\int_{|x|<\varepsilon b_n} x^2\,\mathrm{d}V(x)=1 \tag{5.18}$$

(5.17) together with (5.18) give $\lim_{n\to\infty} \dfrac{n \int_{|x|>\varepsilon b_n} dV(x)}{\frac{n}{b_n^2} \int_{|x|<\varepsilon b_n} x^2\, dV(x)} = 0.$

Set $\varepsilon b_n = c$ and conclude $\lim_{c\to\infty} Q(c) = 0.$

With this the proof of assertion (iii) is complete. □

5.5 Erdös-Kac invariance principle

To explain the invariance principle, which led to the formulation of functional CLT (ref. sec. **(5.6)**), let us start by having a fresh and close look at the CLT. Let (X_n) be a sequence of zero mean independent rvs with partial sums (S_n). Let $\frac{S_n}{b_n} \xrightarrow{d} \xi$ (ref. **(5.3.1)**(viii), **(5.3.1)**(ix) and **(5.4.3)**). The beautiful aspect of the CLT we wish to draw pointed attention to is that whatever be the underlying distributions of the X_ns, provided certain conditions are satisfied, the sequence $(\frac{S_n}{b_n})$ has a weak limit and is the same (namely, the standard normal distribution). Now S_n is a functional of the first n partial sums. It therefore is natural to ask whether other functionals like $Y_n = Y_n(S_1, S_2, \ldots, S_n) = \max_{1\le k\le n} S_k$, $\zeta_n = \zeta_n(S_1, S_2, \ldots, S_n) = \max_{1\le k\le n} |S_k|$, $V_n = V_n(S_1, S_2, \ldots, S_n) = \sum_{k=1}^n S_k^2$, properly normed, will have limiting distributions and if the limit distribution would be the same, once the CLT holds for the X_ns. The pleasant answer to this question is *yes* and this property is called the *invariance principle*. The evaluation of the limit distribution can be done by choosing for the X_ns some convenient distributions, like symmetric Bernoulli, standard normal etc. Below we illustrate the invariance principle by discussing the sequence (ζ_n) described above.

Let F_n be the *df* of X_n. We consider two different cases but present a unified treatment: (i) The X_ns possess finite variances and satisfy the Lindeberg-Feller condition and (ii) The X_ns are *iid* with common *df* F and satisfy the growth condition that $U(x) = \int_{|y|\le x} y^2 \, dF(y)$ is sv at ∞. Follow the notation in Sec. **(5.3)** or in **(5.4.1)** as is appropriate. Let B_n stand for s_n in case (i) and, in case (ii), let it stand for the solution of $\frac{aH(B_n)}{B_n^2} = \frac{1}{n}$. We know $\frac{S_n}{B_n} \xrightarrow{d} \xi$. Let sequence (Y_n) be an independent copy of the sequence (X_n). Write $T_n = \sum_{k=1}^n Y_k$ and note $\frac{T_n}{B_n} \xrightarrow{d} \xi$. Let $\widetilde{\zeta}_n = \max_{1\le k\le n} |T_k|$. The method consists of two parts. First show that the weak limit of $\frac{\zeta_n}{B_n}$ exists *iff* the weak limit of $\frac{\widetilde{\zeta}_n}{B_n}$ exists and the two limits are the same. Then choose Y_n conveniently and evaluate the limit.

To carry out this plan, we need the following Lemma.

Lemma 5.5.1. *There exists a constant c such that* (i) *for all* $n \geq 1$, $\mathbb{E}|\frac{S_n}{B_n}| \leq c$ *and* (ii) *such that for all* $\lambda > 0$ *and for all* $n \geq 1$, $P(\zeta_n \geq \lambda) \leq 2P(|S_n| \geq \lambda - 2cB_n)$.

Proof. In Case (i), we have, by **(2.6.6)**, $\mathbb{E}|\frac{S_n}{B_n}| \leq \left(\mathbb{E}(\frac{S_n}{B_n})^2\right)^{\frac{1}{2}} = 1$. In case (ii), we have from **(5.4.3)**(iii) $\mathbb{E}|\frac{S_n}{B_n}| \to \mathbb{E}|\xi|$. By **(2.6.8)** it then follows that $c_1 = \sup_{1 \leq n < \infty} \mathbb{E}|\frac{S_n}{B_n}| < \infty$. Let c stand for 1 in case (i) and for c_1 in case (ii).

For any rv Y with $\mathbb{E}|Y| < \infty$, $P(|Y| \geq 2\mathbb{E}|Y|) \leq \frac{1}{2}$.

Hence

$$-2\mathbb{E}|Y| \leq \text{med}(Y) \leq 2\mathbb{E}|Y|.$$

Now

$$\begin{aligned}\{|S_k - \text{med}(S_k - S_n)| \leq u\} &\subseteq \{|S_k| \leq u + |\text{med}(S_k - S_n)|\} \\ &\subseteq \{|S_k| \leq u + 2\mathbb{E}|S_k - S_n|\}.\end{aligned}$$

In case (i), $\mathbb{E}|S_k - S_n| \leq \left(\mathbb{E}(S_k - S_n)^2\right)^{\frac{1}{2}} \leq s_n = cs_n$. In case (ii), $\mathbb{E}|S_k - S_n| \leq cB_n$. Thus in both the cases

$$\begin{aligned}&\{\max_{1\leq k\leq n} |S_k - \text{med}(S_k - S_n)| \leq u\} \subseteq \{\max_{1\leq k\leq n} |S_k| \leq u + 2cB_n\} \\ \text{or } &\{\max_{1\leq k\leq n} |S_k - \text{med}(S_k - S_n)| \geq u\} \supseteq \{\max_{1\leq k\leq n} |S_k| \geq u + 2cB_n\}.\end{aligned}$$

Use this inclusion relation in Levy's inequality (ref. **(4.4.4)**): for every $u > 0$

$$P\left(\max_{1\leq k\leq n} |S_k - \text{med}(S_k - S_n)| \geq u\right) \leq 2P(|S_n| \geq u)$$

to obtain $P(\max_{1\leq k\leq n} |S_k| \geq u + 2cB_n) \leq 2P(|S_n| \geq u)$.

Writing $\lambda = u + 2cB_n$ we arrive at the desired result. □

Theorem 5.5.2. *Limit distribution of* $\frac{\zeta_n}{B_n}$ *exists iff that of* $\frac{\tilde{\zeta}_n}{B_n}$ *exists and then the two limit distributions are the same.*

Proof. Write $D_n(x) = P(\zeta_n \leq xB_n) = P(\max_{1\leq k\leq n} |S_k| \leq xB_n), x > 0$. Fix an integer k and determine $n_j, j = 0, 1, 2, \ldots, k$ to satisfy $\frac{B^2_{n_j}}{B^2_n} \leq \frac{j}{k} < \frac{B^2_{n_j+1}}{B^2_n}$. In case (i), (B) of Sec. **(5.3)** holds. Hence $B^2_{n_j} \sim \frac{j}{k} B^2_n$. In case (ii), n_j can be taken to be $[j\frac{n+1}{k}]$. Since in this case $B_n = \sqrt{n}l(n)$, where l is sv, it follows

$\frac{B_{n_j}^2}{B_n^2} \sim \frac{j}{k}$. Let c have the same meaning as in **(5.5.1)**. Let $\epsilon > 0$ be arbitrary but satisfying $\frac{1}{\epsilon} > c$. Choose $k > \frac{4}{\varepsilon^3} > 4c^3$.

Write $D_{n,k}(x) = P(\max_{1\leq j\leq k} |S_{n_j}| \leq xB_n)$ and note that $D_n(x) \leq D_{n,k}(x)$ for all n, k and x.

Define $E_{1,n} = \{|S_1| > xB_n\}$; for $r = 2, \ldots, n$, $E_{r,n} = \{|S_q| \leq xB_n, 1 \leq q \leq r-1, |S_r| > xB_n\}$. The n events are mutually exclusive and $\cup_{r=1}^{n} E_{r,n} = \{\max_{1\leq r\leq n} |S_r| > xB_n\}$. Hence $1 - D_n(x) = P(\cup_{r=1}^n E_{r,n}) = \sum_{r=1}^n P(E_{r,n})$. For $n_\nu \leq r < n_{\nu+1}$, we have $P(E_{r,n}) = u_{r,n} + v_{r,n}$ where $u_{r,n} = P(E_{r,n}, |S_{n_{\nu+1}} - S_r| \geq \epsilon B_n)$ and $v_{r,n} = P(E_{r,n}, |S_{n_{\nu+1}} - S_r| < \epsilon B_n)$.

Since the events involved are independent,

$$u_{r,n} = P(E_{r,n})P(|S_{n_{\nu+1}} - S_r| \geq \epsilon B_n).$$

In case (i),

$$u_{r,n} \leq P(E_{r,n})\frac{1}{s_n^2\varepsilon^2}\{s_{n_{\nu+1}}^2 - s_{n_\nu}^2\} \leq P(E_{r,n})\frac{1}{k\varepsilon^2} < \varepsilon P(E_{r,n}).$$

Thus in this case $\sum_{r=1}^n u_{r,n} \leq \varepsilon$. In case (ii),

$$\begin{aligned}
u_{r,n} &\leq P(E_{r,n})P(\max_{n_\nu\leq r\leq n_{\nu+1}} |S_{n_{\nu+1}} - S_r| \geq \epsilon B_n) \\
&= P(E_{r,n})P(\max_{1\leq r\leq n_{\nu+1}-n_\nu} |S_r| \geq \epsilon B_n) \\
&\qquad \text{(since the variables are identically distributed)} \\
&\leq P(E_{r,n})P(\max_{1\leq r\leq [\frac{2n}{k}]} |S_r| \geq \epsilon B_n) \\
&\leq P(E_{r,n})2P(|S_{[\frac{2n}{k}]}| \geq \epsilon B_n - 2cB_{[\frac{2n}{k}]}) \\
&\leq P(E_{r,n})2P(|S_{[\frac{2n}{k}]}| \geq B_{[\frac{2n}{k}]}(\epsilon\frac{B_n}{B_{[\frac{2n}{k}]}} - 2c) \quad \text{(by (5.5.1))} \\
&\leq 2P(E_{r,n})P(|S_{[\frac{2n}{k}]}| \geq B_{[\frac{2n}{k}]}(\epsilon\sqrt{\frac{k}{2}} - 2c)) \\
&\qquad \text{for all } n \text{ large} \quad \text{(largeness depending only on } k) \\
&\leq 2P(E_{r,n})P(|S_{[\frac{2n}{k}]}| \geq B_{[\frac{2n}{k}]}\frac{\epsilon\sqrt{k}}{2}) \\
&\leq 2P(E_{r,n})\frac{\mathbb{E}|\frac{S_{[\frac{2n}{k}]}}{B_{[\frac{2n}{k}]}}|}{\frac{\varepsilon\sqrt{k}}{2}} \leq 2P(E_{r,n})\frac{2c}{\varepsilon\sqrt{k}} \leq P(E_{r,n})2c\sqrt{\varepsilon}.
\end{aligned}$$

We have $\sum_{r=2}^{n} u_{r,n} \leq 2c\sqrt{\epsilon}$, since the $E_{r,n}$'s are mutually exclusive. Thus in both the cases $\sum_{r=1}^{n} u_{r,n} \leq 2c\sqrt{\varepsilon}$.
For r with $n_\nu \leq r < n_{\nu+1}$ and $\nu \leq k-1$,

$$\begin{aligned} v_{r,n} &\leq P(\max_{1\leq l\leq r-1} |S_l| \leq xB_n, |S_r| > xB_n, |S_{n_{\nu+1}}| > \overline{x-\epsilon}\, B_n) \\ &\leq P(\max_{1\leq l\leq r-1} |S_l| \leq xB_n, |S_r| > xB_n, \max_{1\leq \nu\leq k} |S_{n_\nu}| > \overline{x-\epsilon}\, B_n) \\ &= P(E_{r,n}, \max_{1\leq \nu\leq k} |S_{n_\nu}| > \overline{x-\epsilon}\, B_n). \end{aligned}$$

Summing up we get, since the $E_{r,n}$'s are mutually exclusive, $\sum_{r=1}^{n} v_{r,n} \leq P(\max_{1\leq \nu\leq k} |S_{n_\nu}| > \overline{x-\epsilon} B_n)$. Thus $1 - D_n(x) \leq P(|S_1| > xB_n) + 4\epsilon + 1 - D_{n,k}(x-\epsilon) < 5\epsilon + 1 - D_{n,k}(x-\epsilon)$ for all $n \geq n_1(k,\epsilon,x)$. This leads to $D_{n,k}(x-\epsilon) - 5\epsilon \leq D_n(x) \leq D_{n,k}(x)$. Since $\frac{1}{B_n}(S_{n_1}, S_{n_2}, \ldots, S_{n_k})$ converges in distribution to $\frac{1}{\sqrt{k}}(R_1, R_2, \ldots, R_k)$ where $R_j = \sum_{v=1}^{j} \xi_\nu$, $1 \leq j \leq k$, the ξ_νs being *iid* standard normal variables, we have

$$\begin{aligned} P(\max_{1\leq j\leq k} |R_j| \leq \overline{x-\epsilon}\sqrt{k}) - 5\epsilon &= \lim_{n\to\infty} D_{n,k}(x-\epsilon) - 5\epsilon \leq \underline{\lim}_{n\to\infty} D_n(x) \\ &\leq \overline{\lim}_{n\to\infty} D_n(x) \leq \lim_{n\to\infty} D_{n,k}(x) = P(\max_{1\leq j\leq k} |R_j| \leq x\sqrt{k}). \end{aligned}$$

Similarly

$$\begin{aligned} P(\max_{1\leq j\leq k} |R_j| \leq \overline{x-\epsilon}\sqrt{k}) - 5\varepsilon \leq \underline{\lim}_{n\to\infty} \widetilde{D}_n(x) &\leq \overline{\lim}_{n\to\infty} \widetilde{D}_n(x) \\ &\leq P(\max_{1\leq j\leq k} |R_j| \leq x\sqrt{k}). \end{aligned}$$

where $\widetilde{D}_n(x) = P(\widetilde{\zeta}_n \leq x)$.
Combining the two inequalities, we have

$$\overline{\lim}_{n\to\infty} \widetilde{D}_n(x) - 5\epsilon \leq \underline{\lim}_{n\to\infty} D_n(x+\varepsilon) \tag{5.19}$$

and

$$\overline{\lim}_{n\to\infty} D_n(x-\varepsilon) \leq \underline{\lim}_{n\to\infty} \widetilde{D}_n(x) + 5\epsilon \tag{5.20}$$

Similarly,

$$\overline{\lim}_{n\to\infty} D_n(x) - 5\epsilon \leq \underline{\lim}_{n\to\infty} \widetilde{D}_n(x+\varepsilon) \tag{5.21}$$

and

$$\overline{\lim}_{n\to\infty} \widetilde{D}_n(x-\varepsilon) \leq \underline{\lim}_{n\to\infty} D_n(x) + 5\epsilon \tag{5.22}$$

Suppose $\widetilde{D}_n \overset{d}{\to} \Lambda$. Let x be a continuity point of Λ. Choose ε_r, $r = 1, 2, \ldots$, $\varepsilon_r \downarrow 0$ such that $x + \varepsilon_r$, $x - \varepsilon_r$, $r \geq 1$ are continuity points of Λ. This is always possible since the continuity points of a *df* is the complement of a countable subset of the real line. From (5.21), $\overline{\lim_{n\to\infty}} D_n(x) \leq \Lambda(x+\varepsilon_r)+5\varepsilon_r$. Now let $r \to \infty$ to get

$$\overline{\lim_{n\to\infty}} D_n(x) \leq \Lambda(x) \tag{5.23}$$

Now, (5.22) yields : $\Lambda(x - \varepsilon_r) \leq \underline{\lim}_{n\to\infty} D_n(x) + 5\varepsilon_r$. Letting now $r \to \infty$, we get

$$\Lambda(x) \leq \underline{\lim}_{n\to\infty} D_n(x) \tag{5.24}$$

Thus $\lim_{n\to\infty} D_n(x) = \Lambda(x)$ at all the continuity points x of Λ. Or, what is same, $D_n \overset{d}{\to} \Lambda$.

On similar lines, we can show that $(D_n \overset{d}{\to} \Lambda) \Rightarrow (\widetilde{D}_n \overset{d}{\to} \Lambda)$. □

In the case of symmetric Bernoulli variables Y_k the limiting distribution Λ of $\frac{1}{\sqrt{n}}\widetilde{\zeta}_n$ has been found to be

$$\Lambda(x) = \frac{4}{\pi}\sum_{m=0}^{\infty}\frac{(-1)^m}{2m+1}e^{-\frac{(2m+1)^2\pi^2}{8x^2}}, \; x > 0. \tag{5.25}$$

Corollary 5.5.3. *In both case* (i) *and case* (ii), *the limiting distribution of* $\frac{\zeta_n}{B_n}$ *is* Λ.

Theorem 5.5.4. *Let* (X_n) *be iid* $\mathcal{N}(0, 1)$ *variables. Let* $\alpha > 0$ *be arbitrary. Let* S_n, ζ_n *be as already defined in* **(5.5)**. *Then*
(i) $\mathbb{E}e^{\alpha\zeta_n} < \infty$, $n \geq 1$ *and*
(ii) $\int_0^\infty e^{\alpha x}\, d\,\Lambda(x) < \infty$ *and* $\mathbb{E}e^{\alpha\frac{\zeta_n}{\sqrt{n}}} \to \int_0^\infty e^{\alpha x}\, d\,\Lambda(x)$.

Proof. Integration by parts (ref. **(2.6.3)**) gives

$$\int_0^b e^{\alpha x}\,\mathrm{d}\,P(\zeta_n \leq x) = e^{\alpha b}P(\zeta_n \leq b) - \alpha\int_0^b e^{\alpha x}P(\zeta_n \leq x)\,\mathrm{d}\,x$$

$$= e^{\alpha b}P(\zeta_n \leq b) - \alpha\int_0^b e^{\alpha x}\,\mathrm{d}\,x + \alpha\int_0^b e^{\alpha x}P(\zeta_n \geq x)\,\mathrm{d}\,x$$

$$= -e^{\alpha b}P(\zeta_n \geq b) + 1 + \alpha\int_0^b e^{\alpha x}P(\zeta_n \geq x)\,\mathrm{d}\,x.$$

Clearly

$$\begin{aligned} e^{\alpha b}P(\zeta_n \geq b) &\leq 2e^{\alpha b}P(|S_n| \geq b)(\text{by } \mathbf{(4.4.5)}(\text{v})) \\ &= 2e^{\alpha b}P(|\xi| \geq \frac{b}{\sqrt{n}}) = 4e^{\alpha b}P(\xi \geq \frac{b}{\sqrt{n}}) \\ &\leq 4e^{\alpha b}\frac{1}{\sqrt{2\pi}}\frac{\sqrt{n}}{b}e^{-\frac{1}{2}\frac{b^2}{n}} \ (\text{by } \mathbf{(4.4.5)}(\text{i})) \\ &\to 0 \text{ as } b \to \infty. \end{aligned}$$

On similar simplification, we see that the integrand

$$e^{\alpha x}P(\zeta_n \geq x) \leq 4e^{\alpha x}\frac{1}{\sqrt{2\pi}}\frac{\sqrt{n}}{x}e^{-\frac{1}{2}\frac{x^2}{n}}$$

which is Lebesgue integrable over $[1, \infty)$. Over $[0, 1]$, the integrand is bounded by the integrable function $e^{\alpha x}$.

Hence $\int_0^\infty e^{\alpha x}P(\zeta_n \leq x)\,\mathrm{d}\,x < \infty$ and claim (i) follows.

(ii) We note $\int_0^\infty e^{\alpha x}P(\xi \geq x)\,\mathrm{d}\,x < \infty$. Arguing as in (i), we get

$$\begin{aligned} \int_0^b e^{\alpha x}\,\mathrm{d}\,P(\zeta_n \leq x\sqrt{n}) &\leq 1 + \alpha\int_0^b e^{\alpha x}P(\zeta_n \geq x\sqrt{n})\,\mathrm{d}\,x \\ &\leq 1 + \alpha\int_0^b e^{\alpha x}2P(|S_n| \geq x\sqrt{n})\,\mathrm{d}\,x \\ &= 1 + \alpha\int_0^b e^{\alpha x}2P(|\xi| \geq x)\,\mathrm{d}\,x. \end{aligned}$$

Thus

$$\sup_{n\geq 1}\int_0^\infty e^{\alpha x}\,\mathrm{d}\,P(\zeta_n \leq x\sqrt{n})\,\mathrm{d}\,x < \infty \tag{5.26}$$

Further, since $P(\zeta_n \leq x\sqrt{n}) \to \Lambda(x)$,

$$\int_0^b e^{\alpha x}\,\mathrm{d}\,P(\zeta_n \leq x\sqrt{n}) \to \int_0^b e^{\alpha x}\,\mathrm{d}\,\Lambda(x).$$

Hence we can find $N = N(b)$ such that

$$\int_0^b e^{\alpha x}\,\mathrm{d}\,\Lambda(x) \leq \int_0^b e^{\alpha x}\,\mathrm{d}\,P(\zeta_n \leq x\sqrt{n}) + 1 \text{ for all } n \geq N.$$

It follows that

$$\int_0^b e^{\alpha x}\, \mathrm{d}\,\Lambda(x) \leq 1 + \sup_{n\geq 1} \int_0^b e^{\alpha x}\, \mathrm{d}\, P(\zeta_n \leq x\sqrt{n})$$

$$\leq 1 + \sup_{n\geq 1} \int_0^\infty e^{\alpha x}\, \mathrm{d}\, P(\zeta_n \leq x\sqrt{n}) < \infty \text{ (by (5.26))}.$$

Letting $b \to \infty$, we get the desired result : $\int_0^\infty e^{\alpha x}\, \mathrm{d}\,\Lambda(x) < \infty$.

Since $e^{\alpha x} P(\zeta_n \geq x\sqrt{n}) \leq 2e^{\alpha x} P(|\xi| \geq x)$, an integrable function, bounded convergence theorem applies and we get

$$\int_0^\infty e^{\alpha x}\, \mathrm{d}\, P(\zeta_n \leq x\sqrt{n}) = 1 + \alpha \int_0^\infty e^{\alpha x} P(\zeta_n \geq x\sqrt{n})\, \mathrm{d}\, x$$

$$\to 1 + \alpha \int_0^\infty e^{\alpha x}\{1 - \Lambda(x)\}\, \mathrm{d}\, x = \int_0^\infty e^{\alpha x}\, \mathrm{d}\,\Lambda(x). \qquad \square$$

5.6 Functional CLT or Donsker form of CLT

The notaion in **(5.5)** and in **(2.7)** will be followed. Let $T = \{t_1, t_2, \ldots\}$ be an arbitrary but fixed countable dense subset of $[0, 1]$.

Let $x(t)$, $0 \leq t \leq 1$ be a family of random variables such that for each ω, $x(\cdot, \omega) \in C$. Define random element x taking values in C : $x(\omega)$ is that member of C which, at t takes the value $x(t, \omega)$. If $a \in C$ and if $S = \{x : \rho(x, a) \leq r\}$ then $\{\omega : x(\omega) \in S\} = \{\omega : \rho(x(\omega), a) \leq r\} =$
$= \{\omega : \sup_{0\leq t\leq 1} |x(t, \omega) - a(t)| \leq r\} = \{\omega : \lim_{n\to\infty} \max_{1\leq j\leq n} |x(t_j, \omega) - a(t_j)| \leq r\}$,
since T is dense in $[0, 1]$. This shows that $x^{-1}S$ is a measurable set. Since the spheres generate the σ-field $\mathscr{C}$, it follows that $x^{-1}\mathscr{C} \subseteq \mathscr{A}$. The pm μ generated by x or the distribution μ of x is given by $\mu(A) = P(x^{-1}(A))$, $A \in \mathscr{C}$.

Set $S_0(\omega) \equiv 0$. For $0 \leq t \leq 1$, define $x_n(t, \omega)$ as follows. $x_n(\frac{B_k^2}{B_n^2}, \omega) = \frac{S_k(\omega)}{B_n}$ and x_n is linear elsewhere. We note that for ω fixed, $x_n(., \omega) \in C$ and for each t fixed $x_n(t, .)$ is a rv. Denote by μ_n the distribution of x_n. x_n is called the partial sum process of the variables (X_k). The functional central limit theorem consists in proving that

$$\text{there exists a } pm\ \mu \text{ on } \mathscr{C} \text{ such that } \mu_n \xrightarrow{d} \mu. \tag{5.27}$$

Here is how we explain why this result is called the functional central limit theorem or the $FCLT$. Suppose (5.27) holds. Then (ref. **(2.4.1)**(vii))

$$\mu_n f^{-1} \xrightarrow{d} \mu f^{-1} \tag{5.28}$$

for every real continuous function defined on C. If, for $x \in C$, we define $f(x) = \sup_{0 \le t \le 1} |x(t)|$, then f can be shown to be a continuous function (ref. **(5.6)**). Since $\mu_n f^{-1}$ is the pm of $f(x_n)$ i.e. of $\frac{1}{B_n} \max_{1 \le k \le n} |S_k|$ i.e. of $\frac{\zeta_n}{B_n}$, (5.28) implies that the limit distribution of $\frac{\zeta_n}{B_n}$ exists. Similarly we can claim the existence of the limit distributions of various functionals of $\{S_1, S_2, \ldots, S_n\}$ that can be expressed as the continuous functional of x_n. For this reason the label $FCLT$ seems justified and the title of this section could therefore also be "Convergence in distribution of the partial sum process sequence".

Theorem 5.6.1. *The finite dimensional distributions of the sequence (μ_n) have weak limits which are multivariate normal distributions.*

Proof. Let $\pi_{t_1, t_2, \ldots, t_k}$ denote the projection mapping from C into R^k : $\pi_{t_1, t_2, \ldots, t_k} x = \big(x(t_1), x(t_2), \ldots, x(t_k)\big)$. Corresponding to $0 < t_1 < t_2 < \ldots < t_k \le 1$, the finite dimensional distribution or marginal of μ_n is the multivariate distribution of the vector variable $\big(x_n(t_1), x_n(t_2), \ldots, x_n(t_k)\big)$. Now if n is taken to be sufficiently large then no two of the k numbers $t_1, t_2, \ldots, t_k$ will lie in an interval of the type $(\frac{B_j^2}{B_n^2}, \frac{B_{j+1}^2}{B_n^2}]$. Hence corresponding to t_i, $i = 1, 2, \ldots, k$, we can find, when n is sufficiently large, $j_i(n)$ such that

$$\frac{B^2_{j_i(n)}}{B_n^2} < t_i \le \frac{B^2_{j_i(n)+1}}{B_n^2}.$$

We note $\frac{B^2_{j_i(n)}}{B_n^2} \to t_i$. Now,

$$x_n(t_i) = \frac{S_{j_i(n)}}{B_n} + \alpha_{i,n} \text{ where } \alpha_{i,n} = \frac{\frac{X_{j_i(n)+1}}{B_n}}{\frac{B^2_{j_i(n)+1} - B^2_{j_i(n)}}{B_n^2}} \Big(t_i - \frac{B^2_{j_i(n)}}{B_n^2}\Big).$$

Hence $|\alpha_{i,n}| \le |\frac{X_{j_i(n)+1}}{B_n}|$.
In case (i),

$$P(|\frac{X_{j_i(n)+1}}{B_n}| > \varepsilon) \le \frac{1}{\varepsilon^2 s_n^2} var(X_{j_i(n)+1}) \le \frac{1}{\varepsilon^2 s_n^2} \max_{1 \le k \le n} \sigma_k^2 = o(1).$$

In case (ii), the distribution of $X_{j_i(n)+1}$ does not depend on n and $B_n \to \infty$. Thus in both cases $\alpha_{i,n} \overset{pr}{\to} 0$. This implies that the limiting distributions of $\big(x_n(t_1), x_n(t_2), \ldots, x_n(t_k)\big)$ and $\frac{1}{B_n}\big(S_{j_1(n)}, S_{j_2(n)}, \ldots, S_{j_k(n)}\big)$ are same.

Let $\xi_1, \xi_2, \ldots, \xi_k$ be *iid* standard normal vaiables. Set $S_{j_0(n)} = 0$ and $t_0 = 0$. Now $\frac{S_{j_1(n)}}{B_n} \stackrel{d}{\to} \sqrt{t_1}\xi_1$; for $i \geq 2$,

$$\frac{1}{B_n} S_{j_i(n)} = \frac{1}{B_n} \sum_{r=1}^{i} \{S_{j_r(n)} - S_{j_{r-1}(n)}\} \stackrel{d}{\to} \sum_{r=1}^{i} \sqrt{t_r - t_{r-1}} \xi_r$$

(ref. **(5.3.1)**(xii) and **(5.4.3)**(iv)). Thus $\big(x_n(t_1), x_n(t_2), \ldots, x_n(t_k)\big)$ converges in distribution to the multivariate normal vector variable $(\eta_1, \eta_2, \ldots, \eta_k)$ where $\mathbb{E}\eta_r = 0$ and for $r \leq s$, $\mathbb{E}\eta_r\eta_s = t_r$. □

Remark 5.6.2. We note that these limit distributions, for every choice of k and every choice of $t_1, t_2, \ldots, t_k$ determine unique *pms*, say, $\mu_{t_1,t_1,\ldots,t_k}$ on the sub σ-fields $\mathscr{C}_{t_1,t_2,\ldots,t_k} = \pi^{-1}_{t_1,t_2,\ldots,t_k}\mathscr{R}^k \subset \mathscr{C}$ and that the $\mu_{t_1,t_1,\ldots,t_k}$s form a consistent family. Hence, by **(2.7.9)** they determine a unique *pm*, say, μ on $\mathscr{C}$, called the standard Wiener measure, such that μ restricted to $\mathscr{C}_{t_1,t_2,\ldots,t_k}$ is $\mu_{t_1,t_1,\ldots,t_k}$.

The weak convergence of the marginals of μ_n can be written:

$$\mu_n \pi^{-1}_{t_1,t_2,\ldots,t_k} \stackrel{d}{\to} \mu\pi^{-1}_{t_1,t_2,\ldots,t_k} \tag{5.29}$$

To prove **(5.6.4)**, where we establish the $FCLT$, we need the following Lemma.

Lemma 5.6.3. *Let Λ be an index set. Then the family $(\mu_\alpha, \alpha \in \Lambda)$ of pms on the Borel σ-field $\mathscr{C}$ of C is tight if for every pre-assigned $\varepsilon > 0$ and $\eta > 0$, a number $c > 0$ can be found such that for all α, $\mu_\alpha\{x : \delta_c(x) \leq \varepsilon\} > 1 - \eta$ where* (ref. Sec. **(2.7)**)

$$\delta_c(x) = \sup_{\substack{0 \leq t,\, u \leq 1 \\ |t-u| \leq c}} |x(t) - x(u)|.$$

Proof. Recall (ref. Sec. **(2.7)**) that the only constant function in C, as defined in **(2.7)**, is the one identically zero. Denote by **0** this zero function.

Assume, with no loss of generality, (i) that all the μ_αs are distinct and (ii) that $\mu_\alpha(\{\mathbf{0}\}) < 1$, $\alpha \in \Lambda$. We note that it is enough to prove the tightness of this collection of measures satisfying the condition of the theorem, since we might only be possibly omitting one single tight measure μ with $\mu(\{\mathbf{0}\}) = 1$.

Suppose the condition is satisfied. Given η find $c(\nu)$ such that

$$\mu_\alpha\{x : 0 < \delta_{c(\nu)}(x) \leq \frac{1}{\nu}\} > 1 - \frac{\eta}{2^{\nu+1}} \tag{5.30}$$

for all α. Choose the $c(\nu)$s to be strictly decreasing. This is possible since $\delta_c(x) < \delta_{c'}(x)$ if $c < c'$. If $c(\nu) \downarrow c > 0$ then taking limit as $\nu \to \infty$ in the above inequality, we get $\mu_\alpha(\{\mathbf{0}\}) \geq 1$, a contradiction of our assumtion. Hence $c(\nu) \downarrow 0$. Define $A_\nu = \{x : \delta_{c(\nu)}(x) \leq \frac{1}{\nu}\}$ and $K = \cap_1^\infty A_\nu$. That K is a compact set follows from **(2.7.1)**. That the family (μ_α) is tight follows now since

$$\mu_\alpha(K') \leq \sum_{\nu=1}^{\infty} \mu_\alpha(A'_\nu) \leq \sum_{1}^{\infty} \frac{\eta}{2^{\nu+1}} < \eta. \qquad \square$$

Theorem 5.6.4. *Let x_n, μ_n be as in* **(5.6)**. *Then*
(i) (μ_n) *is a tight sequence and*
(ii) $\mu_n \overset{d}{\to} \mu$ *where μ is the (unique) pm referred to in* **(5.6.2)**.

Proof. (i) Since (C, ρ) is a complete separable metric space, every finite collection of pms on $\mathscr{C}$ is tight. Hence the tightness of the sequence (μ_n) will follow (ref. **(5.6.3)**) if, given $\varepsilon > 0$, $\eta > 0$, we can find a $c > 0$ such that $\mu_n\{x : \delta_c(x) \leq \varepsilon\} > 1 - \eta$ for all n large. Given ε and η, choose and fix integer $k > \frac{256}{\varepsilon^4 \eta}$. From **(5.5.1)**(i), we know that there exists a number τ such that $\mathbb{E}|\frac{S_n}{B_n}| \leq \tau$ for all n. There is no loss in generality in assuming that ε, η are small enough to satisfy $\frac{2}{\varepsilon\sqrt{\eta}} > \tau$. Let n_j, $0 \leq j \leq k$ be as in **(5.5.2)**. We recall $t_j = \frac{B_{n_j}^2}{B_n^2} \sim \frac{j}{k}$. Let n be large, $n > 2k$. Set $c = c_k = \frac{1}{k}$ and note $t_j - t_{j-1} \leq \frac{2}{k} = 2c$ for all n large. Given $0 < t < u < 1$, $|t - u| < c$, there corresponds a j such that $t_j \leq t < u \leq t_{j+1}$, $0 \leq j \leq k-1$ or $t_j \leq t \leq t_{j+1} \leq u \leq t_{j+2}$, $0 \leq j < k-2$.

In the first case $|x(t) - x(u)| \leq |x(t) - x(t_{j+1})| + |x(u) - x(t_{j+1})|$. Hence in this case, with t, u as above, $(|x(t) - x(u)| \geq 3\varepsilon) \Rightarrow (|x(t) - x(t_{j+1})| \geq \varepsilon)$ or $(|x(u) - x(t_{j+1})| \geq \varepsilon)$. Thus

$$(|x(t) - x(u)| \geq 3\varepsilon) \Rightarrow \sup_{t_j \leq t \leq t_{j+1}} |x(t) - x(t_{j+1})| \geq \varepsilon.$$

In the second case $|x(t) - x(u)| \leq |x(t) - x(t_{j+1}) + |x(t_{j+1}) - x(t_{j+2})| + |x(t_{j+2}) - x(u)|$. Hence in this case, $(|x(t) - x(u)| \geq 3\varepsilon) \Rightarrow$ $\{|x(t) - x(t_{j+1})| \geq \varepsilon\} \cup \{|x(t_{j+1}) - x(t_{j+2})| \geq \varepsilon\} \cup \{|x(t_{j+2}) - x(u)| \geq \varepsilon\}$

Thus for t, u as above

$$\{|x(t) - x(u)| \geq 3\varepsilon\} \subset \{\sup_{t_j \leq t \leq t_{j+1}} |x(t) - x(t_{j+1})| \geq \varepsilon\}$$
$$\cup \{\sup_{t_{j+1} \leq t \leq t_{j+2}} |x(t) - x(t_{j+1})| \geq \varepsilon\}.$$

These considerations lead to

$$\{x : \delta_c(x) \geq 3\varepsilon\} \subset \bigcup_{j=0}^{k-1} \{x : \sup_{t_j \leq t \leq t_{j+1}} |x(t) - x(t_{j+1})| \geq \varepsilon\}.$$

Hence

$$\begin{aligned}\mu_n\{x : \delta_c(x) \geq 3\varepsilon\} &\leq \sum_{j=0}^{k-1} \mu_n\{x : \sup_{t_j \leq t \leq t_{j+1}} |x(t) - x(t_{j+1})| \geq \varepsilon\} \\ &= \sum_{j=0}^{k-1} P(\max_{n_j \leq r < n_{j+1}} |S_{n_{j+1}} - S_r| \geq \varepsilon B_n) \\ &= \sum_{j=0}^{k-1} a_{j,n}, \text{ say.}\end{aligned}$$

We consider the two cases separately as in **(5.5.2)**.
Case (i): Lindeberg-Feller condition holds. We have,

$$P(|S_{n_{j+1}} - S_r| \geq \frac{\varepsilon}{2} s_n) \leq \frac{4\mathbb{E}(S_{n_{j+1}} - S_r)^2}{\varepsilon^2 s_n^2} \leq \frac{4}{\varepsilon^2} \frac{s_{n_{j+1}}^2 - s_{n_j}^2}{s_n^2} \to \frac{4}{\varepsilon^2} \frac{1}{k}.$$

This implies that for all n large and for all r, $n_j \leq r < n_{j+1}$, $P(|S_{n_{j+1}} - S_r| \geq \frac{\varepsilon}{2} s_n) \leq \frac{1}{2}$. Hence by **(4.4.2)**, it follows that for all n large, $a_{j,n} < 2P(|S_{n_{j+1}} - S_{n_j}| \geq \frac{\varepsilon}{2} s_n)$.

Now we appeal to **(5.3.1)**(xii) $\frac{1}{s_n} |S_{n_{j+1}} - S_{n_j}| \overset{d}{\to} \frac{|\xi|}{\sqrt{k}}$. Since the convergence is uniform, i.e., since $\sup_x \left|P(|S_{n_{j+1}} - S_{n_j}| \leq x s_n) - P(|\xi| \leq x\sqrt{k})\right| \to 0$, $P(|S_{n_{j+1}} - S_{n_j}| \geq \frac{\varepsilon}{2} s_n) \leq P(|\xi| \geq \frac{\sqrt{k}\varepsilon}{2}) + \frac{\eta}{k}$.
Case (ii) : X_ks are iid with $H(x) \to \infty$ and is sv. In this case we can take $n_j = [\frac{jn}{k}]$. With the n_js so defined, it is readily verified that $\frac{B_{n_j}^2}{B_n^2} \sim \frac{j}{k}$. We note

$$\begin{aligned}a_{j,n} &= P(\max_{1 \leq r < n_{j+1} - n_j} |S_r| \geq \varepsilon B_n) \leq P(\max_{1 \leq r \leq [\frac{n}{k}]+2} |S_r| \geq \varepsilon B_n) \\ &\leq \text{ (by } \mathbf{(5.5.1)}) 2P(|S_{[\frac{n}{k}]+2}| \geq \varepsilon B_n - 2\tau B_{[\frac{n}{k}]+2}) \\ &= 2P\big(|S_{[\frac{n}{k}]+2}| \geq B_{[\frac{n}{2}]+2}(\varepsilon \frac{B_n}{B_{[\frac{n}{k}]+2}} - 2\tau)\big) \\ &\leq 2P\big(|S_{[\frac{n}{k}]+2}| \geq B_{[\frac{n}{2}]+2}(\frac{\varepsilon}{2}\sqrt{k} - 2\tau)\big), \text{ for all } n \text{ large.}\end{aligned}$$

We then have, by CLT, for all n large

$$a_{j,n} \leq 2\{P(|\xi| \geq \frac{\varepsilon}{2}\sqrt{k} - 2\tau) + \frac{\eta}{k}\} \leq 2\{P(|\xi| \geq \frac{\varepsilon}{4}\sqrt{k}) + \frac{\eta}{k}\}.$$

Thus

$$\mu_n\{x : \delta_c(x) \geq 3\varepsilon\} \leq \sum_{j=0}^{k-1} 2\{P(|\xi| \geq \frac{\varepsilon}{4}\sqrt{k}) + \frac{\eta}{k}\}$$
$$\leq 2k\{\frac{3 \times 256}{k^2\varepsilon^4} + \frac{\eta}{k}\}\text{by Tchebichev inequality with fourth moment}$$
$$\leq 7\eta \text{ for all } n \text{ sufficiently large, say, } n > N.$$

By **(2.7.4)**, there exists $\bar{c} > 0$ such that $\mu_\nu\big(x : \delta_{\bar{c}}(x) \geq \varepsilon\big) \leq \eta,\ 1 \leq \nu \leq N$. If $c^* = \min(c,\ \bar{c})$, then $\mu_n\big(x : \delta_{c^*}(x) \geq 3\varepsilon\big) \leq 4\eta$ for all n. With this the proof that (μ_n) is a tight sequence is complete.

(ii) In **(5.6.2)** we showed that $\mu_n\pi^{-1}_{t_1,t_2,\ldots,t_k} \overset{d}{\to} \mu\pi^{-1}_{t_1,t_2,\ldots,t_k}$ for every choice of $k \geq 1$ and every choice of $t_1,\ t_2, \ldots,\ t_k$ in $[0,\ 1]$. This fact together with the tightness of the sequence (μ_n) implies the desired result (ref. **(2.7.10)**). □

Some continuous functionals on C

$f_i,\ i = 1,\ 2,\ 3$ are continuous functionals on C where, for $x \in C$,

(i) $f_1(x) = \max\limits_{0\leq t\leq 1} x(t)$

(ii) $f_2(x) = \max\limits_{0\leq t\leq 1} |x(t)| = \|x\|$

(iii) $f_3(x) = \int_0^1 |x(t)|^\alpha \,\mathrm{d}t,\ \alpha > 0.$

Proof. Let $\|x_n - x\| \to 0$. Hence, given $\varepsilon > 0$, there exists then $N = N(\varepsilon)$ such that for all $t \in [0,\ 1]$ and all $n \geq N$,

$$x(t) - \varepsilon \leq x_n(t) \leq x(t) + \varepsilon. \tag{5.31}$$

We must show that $f_i(x_n) \to f_i(x),\ i = 1,\ 2,\ 3.$

(i) (5.31) yields $\max\limits_{0\leq t\leq 1} x(t) - 2\varepsilon \leq \max\limits_{0\leq t\leq 1} x_n(t) \leq \max\limits_{0\leq t\leq 1} x(t) + \varepsilon$. It follows $|\max\limits_{0\leq t\leq 1} x_n(t) - \max\limits_{0\leq t\leq 1} x(t)| \leq 2\varepsilon$ for all $n \geq N$. This is equivalent to $f_1(x_n) \to f_1(x)$.

(ii) From the inequalities $\|x_n\| \leq \|x_n - x\| + \|x\|$ and from the hypothesis $\|x_n - x\| \to 0$ we conclude $\overline{\lim\limits_{n\to\infty}} \|x_n\| \leq \|x\|$. Similarly the inequality $\|x\| \leq \|x - x_n\| + \|x_n\|$ leads to $\|x\| \leq \underline{\lim\limits_{n\to}} \|x_n\|$. Hence $\|x_n\| \to \|x\|$. i.e. $f_2(x_n) \to f_2(x)$.

(iii) Since $|x_n(t)| \leq |x_n(t) - x(t)| + |x(t)| \leq |x(t)| + \varepsilon$ for all n large, it follows by the bounded convergence theorem $\int_0^1 |x_n(t)|^\alpha \,\mathrm{d}t \to \int_0^1 |x(t)|^\alpha \,\mathrm{d}t$. i.e $f_3(x_n) \to f_3(x)$. □

Remark 5.6.5. If x_n is as in **(5.6)**, then
(i) $f_1(x_n(., \omega)) = \frac{1}{B_n} \max\limits_{1 \le k \le n} S_k(\omega)$.
(ii) $f_2(x_n(., \omega)) = \frac{1}{B_n} \max\limits_{1 \le k \le n} |S_k(\omega)|$.

Theorem 5.6.6. *Let the X_ns and the B_ns be as in* **(5.5)** *and let μ be as in* ***(5.6.2)****. Then the weak limit of*

$$\frac{1}{B_n^{2+\frac{\alpha}{2}}} \sum_{k=1}^{n} (B_k^2 - B_{k-1}^2)|S_k|^{\alpha} \text{ exists and the limit is } \mu f_3^{-1}.$$

Proof. Define $y_n(t) = \frac{S_{k-1}}{B_n}$ for $\frac{B_{k-1}^2}{B_n^2} \le t < \frac{B_k^2}{B_n^2}$, $k = 1, 2, \dots, n$; $y_n(1) = \frac{S_n}{B_n}$. The sample functions of this process are not continuous. To handle this we introduce the metric space D of real functions x defined on $[0, 1]$ such that either $x \in C$ or x is a right continuous *step function* with finite number of steps. Endow D with the uniform metric: for $x, y \in D$,

$$d(x, y) = \|x - y\| = \sup_{0 \le t \le 1} |x(t) - y(t)|.$$

Under this metric D is separable. Further, y_n induces a pm ν_n on the Borel σ-field $\mathscr{D}$ of (D, d). (Proof of this assertion is similar to the arguments for x in Sec. **(5.6)**).

We have $Z_n = \|x_n - y_n\| \le \frac{1}{B_n} \max\limits_{1 \le k \le n} |X_k|$ where x_n is as in Sec. **(5.6)**.

$Z_n \overset{pr}{\to} 0$ if $P(\max\limits_{1 \le k \le n} |X_k| > \varepsilon B_n) \to 0$. For this, it is sufficient if $\alpha_n = \sum\limits_{k=1}^{n} P(|X_k| > \varepsilon B_n) \to 0$.

Case 1. Let the X_ks satisfy the Lindeberg-Feller condition. In this case $B_n = s_n$. Then

$$\alpha_n = \sum_{k=1}^{n} \int_{|x| > \varepsilon s_n} d F_k(x) \le \sum_{k=1}^{n} \int_{|x| \ge \varepsilon s_n} \frac{x^2}{\varepsilon^2 s_n^2} d F_k(x) \to 0.$$

Case 2. The X_ks are iid with mean zero and variance 1. In this case $B_n = \sqrt{n}$. We recall $\mathbb{E}X_1^2 < \infty$ is equivalent to $\sum\limits_{n=1}^{\infty} nP(|X_1| \ge n) < \infty$. This implies $\lim\limits_{n \to \infty} n^2 P(|X_1| \ge n) = 0$. Since $\alpha_n = nP(|X_1| \ge \varepsilon\sqrt{n})$, it follows, after some simple adjustments, that $\alpha_n \to 0$.

Case 3. The X_ks are iid, $\mathbb{E}X_1 = 0$, $\mathbb{E}X_1^2 = \infty$, and $U(x) = \int_{-x}^{x} y^2 \, \mathrm{d}F(y)$ is sv. In this case, $\frac{U(B_n)}{B_n^2} \sim \frac{1}{n}$. Also we have (ref. **(5.4.2)**)

$$\lim_{c \to \infty} \frac{c^2 \int_{|x| \geq c} \mathrm{d}F(x)}{U(c)} \to 0.$$

These two relations give $nP(|X_1| \geq \varepsilon B_n) \to 0$. i.e. $\alpha_n \to 0$.

Since $y_n = y_n - x_n + x_n$ and since $\|y_n - x_n\| \overset{pr}{\to} 0$, it follows (ref. **(4.3.2)**(xvii)) that the limit distribution of y_n exists and is the same as that of x_n. This, together with **(5.6.4)**, implies that $\nu_n \overset{d}{\to} \mu$. We note f_3 on $(D,\ d)$ is a continuous function. It now follows by **(4.3.2)**(xvi) that the limiting distribution of $f_3(x_n)$ is the same as that of $f_3(y_n)$, which is μf_3^{-1}. Since

$$f_3(y_n) = \int_0^1 |y_n(t)|^\alpha \, \mathrm{d}t = \sum_{k=1}^{n} \int_{\frac{B_{k-1}^2}{B_n^2}}^{\frac{B_k^2}{B_n^2}} |\frac{S_k}{B_n}|^\alpha \, \mathrm{d}t$$

$$= \frac{1}{B_n^\alpha} \sum_{k=1}^{n} |S_k|^\alpha \frac{B_k^2 - B_{k-1}^2}{B_n^2},$$

the desired result follows.

In the case (ii), $f_3(y_n) = \frac{1}{n^{1+\frac{\alpha}{2}}} \sum_{k=1}^{n} |S_k|^\alpha$. □

Theorem 5.6.7. *Define $f_4(x)$ = Lebesgue measure of the linear set $\{t : x(t) > 0\}$. f_4 is not a continuous functional.* .

Proof. Let $x_n(0) = 0 = x_n(1)$, $x_n(\frac{1}{2}) = \frac{1}{n}$ and x_n is linear elsewhere. Let $x \equiv 0$. Clearly $\|x_n - x\| \to 0$. But $f_4(x_n) \nrightarrow f_4(x)$ since $f_4(x_n) = \mathcal{L}\{t : x_n(t) = 0\} = 0$ whereas $f_4(x) = 1$, $\mathcal{L}$ denoting Lebesgue measure. □

Theorem 5.6.8. *f_4 is continuous at all points x which are such that $\mathcal{L}\{t : x(t) = 0\} = 0$, $\mathcal{L}$ denoting Lebesgue measure*

Proof. If $x \in C$ with $\mathcal{L}\{t : x(t) = 0\} = 0$. Let $\|x_n - x\| \to 0$. Hence, given $k \geq 1$, an integer, we can find $N(k)$ such that $x(t) - \frac{1}{k} \leq x_n(t) \leq x(t) + \frac{1}{k}$ for all $t \in [0,\ 1]$ and for all $n \geq N(k)$. Define $A_n = \{t : x_n(t) > 0\}$; $C_k = \{t : x(t) + \frac{1}{k} > 0\}$ and $D_k = \{t : x(t) - \frac{1}{k} > 0\}$ and note that, for all $n \geq N(k)$, $D_k \subseteq A_n \subseteq C_k$. Hence, letting letting $n \to \infty$, we get $\mathcal{L}(D_k) \leq \underline{\lim}_{n\to\infty} \mathcal{L}(A_n) \leq \overline{\lim}_{n\to\infty} \mathcal{L}(A_n) \leq \mathcal{L}(C_k)$. Now, we note $D_k \uparrow \{t : x(t) > 0\}$ and

$C_k \downarrow \{t : x(t) \geq 0\}$. Since $\mathcal{L}\{t : x(t) = 0\} = 0$, we get, by letting $k \to \infty$, $\lim_{n\to\infty} \mathcal{L}(A_n)$ exists and is equal to $\mathcal{L}\{t : x(t) > 0\}$. i.e. $f_4(x_n) \to f_4(x)$. □

Theorem 5.6.9. *Let the pm μ be the Wiener measure described in* **(5.6.2)**. *Then $\mu(A) = 1$ where $A = \{x : x \in C,\ \mathcal{L}(t : x(t) = 0) = 0\}$.*

Proof. Consider the product space $[0,\ 1] \times C$ endowed with the product σ-field and the product measure $\nu = \mathcal{L} \times \mu$. Let $E = \{(t,\ x) : x(t) = 0\}$. The x-section E_x of E is $\{t : x(t) = 0\}$. The t-section E_t of E is $\{x : x(t) = 0\}$ and $\mu(E_t) = 0$ since, for t fixed, $x(t)$ has, under μ, a continuous distribution function (in fact a zero mean normal *df* with variance t). By Fubini's theorem we have $\int_0^1 \mu(E_t)\ \mathrm{d}\,\mathcal{L}(t) = \int_C \mathcal{L}(E_x)\ \mathrm{d}\,\mu(x)$. Hence $\int_C \mathcal{L}(E_x)\ \mathrm{d}\,\mu(x) = 0$. Since the integrand is non-negative, this implies that the μ-measure of the following set is 1 : the set of all $x \in C$ with the property $\mathcal{L}\{t : x(t) = 0\} = 0$. □

Remark 5.6.10. In view of **(5.6.7)**, the theorem implies that there exists $E \in \mathscr{C}$ such that $\mu(E) = 1$ and such that f_4 is continuous on E.

Theorem 5.6.11. *Let μ_n and μ be as in* **(5.6.4)**. *Then $\mu_n f_i^{-1} \overset{d}{\to} \mu f_i^{-1}$, $i = 1,\ 2,\ 3,\ 4$.*

Proof. Recall $\mu_n \overset{d}{\to} \mu$. Since $f_i,\ i = 1,\ 2,\ 3$ are continuous functionals on C, the claim follows immediately from **(2.4.2)**.

In **(5.6.4)** we have shown that (μ_n) is a tight sequence and that $\mu_n \overset{d}{\to} \mu$. In **(5.6.9)** it was proved that f_4 is continuous over a set of μ-measure 1. Hence **(2.4.6)** applies and the desired result follows for f_4. □

Definition 5.6.12. Define $\psi(x) = 1$ if $x > 0$ and $\psi(x) = 0$ otherwise. Let $T_n = \sum_{k=1}^{n} \psi(S_k)$ and note that the number of positive terms among $\{S_1,\ S_2, \ldots,\ S_n\}$ is T_n.

Theorem 5.6.13. *The limiting distribution of $f_4(x_n)$ is the same as that of $\frac{T_n}{n}$ in the case where the X_ns are iid zero mean unit variance rvs.*

Proof. Let $x_n,\ \mu_n,\ \mu$ be as in **(5.6)** and let $y_n,\ \nu_n$ be as in **(5.6.6)**. In **(5.6.6)** we showed μ_n and ν_n have the same weak limits. Hence $\nu_n \overset{d}{\to} \mu$. Since f_4 is continuous over a set of μ-measure 1, **(2.4.6)** applies and we conclude $\nu_n f_4^{-1}$ and $\mu_n f_4^{-1}$ have the same weak limit. Thus the limiting distribution of $f_4(x_n)$ is the same as that of $f_4(y_n)$. Since, $f_4(y_n)$ is precisely $\frac{T_n}{n}$, the proof is complete. □

5.7 Wiener measure and representation for a Brownian Motion Process

Recall the unique pm μ referred to in **(5.6.2)** is called the standard Wiener measure. Consider the probability space $(\Omega,\ \mathscr{A},\ \mu)$ where $\Omega = C$ and $\mathscr{A} = \mathscr{C}$. We note each $\omega \in \Omega$ is a continuous function. For each $t \in [0,\ 1]$, define $x_t(\omega) = \omega(t) = \pi_t(\omega)$. Since π_t is a continuous function under the ρ-metric and since $\mathscr{C}$ is the ρ-Borel σ-field, it follows that x_t is a rv. The joint distribution of $(x(t_1),\ x(t_2), \ldots ,\ x(t_k))$ is $\mu\pi^{-1}_{t_1,t_2,\ldots,t_k}$ which, according to **(5.6.1)**, is multivariate normal with zero mean vector and covariances given by $\mathbb{E}x(t_i)x(t_j) = \min(t_i,\ t_j)$. This shows that if $t_1 < t_2 < \ldots < t_k$, then $x(t_1),\ x(t_2) - x(t_1),\ x(t_3) - x(t_2), \ldots ,\ x(t_k) - x(t_{k-1})$ are independent variables with zero mean values and variances $t_1,\ t_2 - t_1,\ t_3 - t_2, \ldots ,\ t_k - t_{k-1}$. Further note that the process x_t has all its sample functions continuous. i.e. the functions $x_t(\omega)$ of t for ω fixed are continuous.

Definitions 5.7.1. (i) A family $\{x_t,\ t \in T\}$ of random elements, all defined on the same probability space $(\Omega,\ \mathscr{A},\ P)$ is called a stochastic process or a random process. The x_ts may take values in possibly different measurable spaces $(\Omega_t,\ \mathscr{A}_t)$. The pms $\mu_{t_1,\ t_2,\ \ldots,\ t_k}$ induced in $\bigotimes_{i=1}^{k} \mathscr{S}_{t_i}$ by the vector element $(x_{t_1},\ x_{t_2},\ \ldots,\ x_{t_k})$, for various choices of $k \geq 1$ and various choices of $t_1,\ t_2,\ \ldots,\ t_k \in T$ are called the finite dimensional distributions of the process. That they satisfy the consistency conditions (ref. **(1.11)**) is obvious.

Suppose we start with a consistent family $\mu_{J:T}$, $J \subset T$, J finite, of pms on $\mathbf{a}_{J:T}$ (ref. **(1.2)** for notation). We ask : does there exist a stochastic process whose finite dimensional distributions are the corresponding members of this family? The answer is *yes iff* these $\mu_{J:T}$s determine a pm μ on the product σ-field $\mathscr{A}_T$. For, denoting by $u = (u_t)$ a generic element of Ω_T, define the family of rE $x_t,\ t \in T\ :\ x_t(u) = u(t)$. The *fdd*s of this process are obviously the $\mu_{J:T}$s. A sufficient condition for the existence of μ is contained in the theorem **(1.11.1)**.

When $(\Omega_t,\ \mathscr{A}_t) = (R,\ \mathscr{R})$, (x_t) is called a real process. Since R is a complete separable metric space, real processes with prescribed *fdd*s always exist.

A (real) process is said to have independent increments if T is a linear interval and if for all $k \geq 2$, and $t_1,\ t_2, \ldots,\ t_k \in T$, $x(t_1),\ x(t_2) - x(t_1),\ x(t_3) - x(t_2), \ldots ,\ x(t_k) - x(t_{k-1})$ are independent, where $t_1 < t_2 < \ldots < t_k$.

(ii) A standard Brownian Motion Process (BMP) is a process B i.e $(B(t),$

$0 \leq t \leq a$) or $(B(t),\ 0 \leq t < \infty)$ with the following properties:

(p_1) $B(0) = 0$.

(p_2) All its finite dimensional distributions are multivariate normal with zero mean vectors and variance of $B(t) - B(u)$ is $|t - u|$ or, equivalently, with $\mathbb{E}B(t)B(u) = \min(t,\ u)$. (This property implies that B is a process with independent increments)

(p_3) With probability 1 the sample functions of B are continuous.

If $\mathfrak{N} \in \mathscr{A}$ is the exceptional set such that $P(\mathfrak{N}) = 0$ and $B(.,\ \omega)$ is a continuous function for every $\omega \in \mathfrak{N}'$, then define probability space $(\Omega_1,\ \mathscr{A}_1,\ P_1)$: $\Omega_1 = \Omega \cap \mathfrak{N}',\ \mathscr{A}_1 = \mathscr{A} \cap \mathfrak{N}'$ and $P_1(A \cap \mathfrak{N}') = P(A),\ A \in \mathscr{A}$. The process B now restricted to Ω_1 possesses properties (p_1) and (p_2) and has all its sample functions continuous. Hence forward, we will assume that BMP processes considered have all their sample functions continuous.

For a more useful definition of BMP processes, refer to **(6.7.4)**.

The construction in **(5.7)** proves that standard Brownian Motion processes on the interval $[0,\ 1]$ exist.

We raise the question if property (p_3) is a consequence of or independent of p_2. To answer this question, let x be the process constructed in **(5.7)** on the probability space $(C,\ \mathscr{C},\ \mu)$ where μ is the Wiener pm. Let $I = [0,\ 1]$; $\mathscr{I}$ be the Borel σ-field of I and $\mathcal{L}$ the Lebesgue measure on I. For $t \in [0,\ 1]$ and $v \in I$, define $y(0;\ v) = 0$ for all v; for $t > 0$, $y(t;\ v) = 1$ if $t = v$ and zero otherwise. Let $(\Omega,\ \mathscr{A},\ P)$ be the product probability space of the spaces $(C,\ \mathscr{C},\ \mu)$ and $(I,\ \mathscr{I},\ \mathcal{L})$. For $u \in C,\ v \in I$ and $t \in [0,\ 1]$ define $z(t;\ \omega) = z(t;\ (u,\ v)) = u(t) + y(t;\ v)$. All the sample functions of the process y and, hence, all the sample functions of the process z are discontinuous. For each t, $P(z(t,\ \omega) = u(t)) = 1$ since $P((u,\ v)\ : y(t,\ v) = 0) = \mathcal{L}(v\ : v \neq t) = 1$. It follows that the finite dimensional distributions of the process z are the same as the corresponding ones of the process u. This shows that p_2 does not imply p_3.

However if a process satisfies the condition of separability (ref. **(5.7.2)**) for definition) then p_2 implies p_3. This claim will be established in the sequel.

In **(1.1.1)** we defined separability of a subset of R^T with respect to a countable dense set of T.

Definitions 5.7.2. (i) Let D be a countable dense subset of T. A stochasic process $x = \{x(t),\ t \in T\}$ on $(\Omega,\ \mathscr{A},\ P)$ is said to be D-separable if there exists $N \in \mathscr{A}$ such that $P(N) = 0$ and such that for every $\omega \notin N$, the sample path $x(.\ ,\ \omega)$ is D-separable, according to the definition **(1.1.1)**. We say the process is separable if it is D-separable for some countable dense subset $D \subset T$.

(ii) A (real) stochastic process $x = \{x(t),\ t \in T\}$ is said to be stochastically continuous or continuous in probability at $t \in T$ if $x(s) \stackrel{pr}{\rightarrow} x(t)$ as $s \rightarrow t$. It is said to be stochastically continuous if it is so at every $t \in T$.

Theorem 5.7.3. *Let a process x be separable and stochastically continuous. Then x is separable with respect to every countable dense subset of T.*

Proof. Let D be a countable dense subset of T and let x be D-separable. Hence there exists $N_1 \in \mathscr{A}$ such that $P(N_1) = 0$ and such that for each $\omega \notin N_1$ and each $t \in T$, there exist $t_n \in D,\ t_n \rightarrow t$ and $x(t_n,\ \omega) \rightarrow x(t,\ \omega)$. Now let $S = \{s_j\}$ be any other countable dense subset of T. If $s_j \rightarrow t$ then, by hypothesis, $x(s_j) \stackrel{pr}{\rightarrow} x(t)$. Hence there exists a subsequence $(x(s_{j'}))$ and $N(t) \in \mathscr{A}$ such that $P(N(t)) = 0$ and such that for every $\omega \notin N_t$, $x(s_{j'},\ \omega) \rightarrow x(t)$. Write $N_2 = \cup_n N(t_n)$ and $N = N_1 \cup N_2$. We note $P(N) = 0$. Let $\omega \notin N$. Fix target error $\varepsilon > 0$. We have for every $n \geq 1$, $s_{n,j} \rightarrow t_n$ and $x(s_{n,j},\ \omega) \rightarrow x(t_n,\ \omega)$ as $j \rightarrow \infty$. Hence we can find $\nu(n) \uparrow \infty$ such that $|s_{n,j} - t_n| < \frac{1}{2^n}$ and $|x(s_{n,j}) - x(t_n)| < \frac{1}{2^n}$ for all $j \geq \nu(n)$. Thus $s_{n,\nu(n)} - t_n \rightarrow 0$ and $x(s_{n,\nu(n)}) - x(t_n) \rightarrow 0$. These convergences and the facts $t_n \rightarrow t$ and $x(t_n,\ \omega) \rightarrow x(t,\ \omega)$ imply $s_{n,\nu(n)} \rightarrow t$ and $x(s_{n,\nu(n)},\ \omega) \rightarrow x(t)$. That the process x is separable with respect to S follows now. (ref. definition **(5.7.2)**(i)) □

Theorem **(2.7.9)** has the following enhanced version.

Theorem 5.7.4. *With probability* 1*, the sample functions of a separable process $x(t,\ \omega), t \in T = [0,\ \infty)$ defined on a pobability space $(\Omega,\ \mathscr{A},\ P)$ will be continuous if $\mathbb{E}|x(t) - x(s)|^\beta \leq c|t - s|^{1+\alpha}$ for some positive constants $c,\ \alpha,\ \beta$.*

Proof. The steps are essentially the same as in **(2.7.9)** but the reasoning uses a slightly different language. Let us first consider the case $T = [0,\ 1]$.

$$P(|x(t) - x(s)| > \varepsilon) \leq \varepsilon^{-\beta}\mathbb{E}|x(t) - x(s)|^\beta \leq \varepsilon^{-\beta}c|t - s|^{1+\alpha} \qquad (5.32)$$

Hence $x(s) \stackrel{pr}{\rightarrow} x(t)$ as $s \rightarrow t$. i.e. the process is stochasically continuous. Hence, by **(5.7.3)**, x is separable with respect to $D = \cup_n D_n$ where $D_n = \{\frac{k}{2^n},\ 1 \leq k \leq 2^n\}$. and this implies that there exists $\Omega^* \in \mathscr{A}$ with $P(\Omega^*) = 1$ such that for each $\omega \in \Omega^*$ and each $t \in T$, we have : a sequence (t_n) in D is available with the properties that $t_n \rightarrow t$ and $x(t_n,\ \omega) \rightarrow x(t,\ \omega)$.

Taking $\varepsilon = 2^{-\gamma n}$, $\gamma < \frac{\alpha}{\beta}$, $s = \frac{k-1}{2^n}$, and $t = \frac{k}{2^n}$ in (5.32) we have

$$\sum_{n=1}^{\infty} P\{\max_{1\leq k\leq 2^n} |x(\tfrac{k}{2^n}) - x(\tfrac{k-1}{2^n})|$$

$\geq 2^{-\gamma n}\} \leq \sum_{n=1}^{\infty} \sum_{k=1}^{2^n} P(|x(\frac{k}{2^n}) - x(\frac{k-1}{2^n})| \geq 2^{-\gamma n})$

$\leq \sum_{n=1}^{\infty} 2^{n\gamma\beta} 2^n c 2^{-n(1+\alpha)} = c \sum_{n=1}^{\infty} 2^{-n(\alpha-\beta\gamma)} < \infty$ since $\alpha - \beta\gamma > 0$.

Hence by Borel-Cantelli lemma, there exists Ω^{**} with $P(\Omega^{**}) = 1$ such that there exists a positive integer valued rv N with the property that for each $\omega \in \Omega^{**}$,

$$\max_{1 \leq k \leq 2^n} |x(\frac{k}{2^n}, \omega) - x(\frac{k-1}{2^n}, \omega)| \leq 2^{-\gamma n} \text{ for all } n \geq N(\omega). \tag{5.33}$$

Or

$$\sup_{\substack{s,t \in D_N \\ |t-s| \leq 2^{-N}}} |x(t, \omega) - x(s, \omega)| \leq 2^{-\gamma N}, \text{ for } n = N, \omega \in \Omega^{**}.$$

Now consider

$$v_{N+1} = \sup_{\substack{s,t \in D_{N+1} \\ |t-s| \leq 2^{-N}}} |x(t) - x(s)|.$$

We note $s < t$ have to be of the form $s = \frac{j}{2^{N+1}}$ and $t = \frac{k}{2^{N+1}}$, $j < k$. Define $s' = s$ if $s \in D_N$; $s' = \frac{j+1}{2^{N+1}}$ if $s \notin D_N$. Define $t' = t$ if $t \in D_N$; $t' = \frac{k-1}{2^{N+1}}$ if $t \notin D_N$. Hence $|x(t, \omega) - x(s, \omega)| \leq |x(t, \omega) - x(t', \omega)| + |x(t', \omega) - x(s', \omega)| + |x(s, \omega) - x(s', \omega)| \leq 2^{-\gamma(N+1)} + 2^{-\gamma N} + 2^{-\gamma(N+1)} \leq 3 \times 2^{-\gamma(N+1)}$. This argument can be kept up to get, for every $n \geq N$,

$$\sup_{\substack{s,t \in D_n \\ |t-s| \leq 2^{-N}}} |x(t) - x(s)| \leq 2 \sum_{j=N}^{n} 2^{-\gamma j}.$$

Thus we have

$$\sup_{\substack{s,t \in D \\ |t-s| \leq 2^{-N}}} |x(t, \omega) - x(s, \omega)| \leq 2 \sum_{j=N}^{\infty} 2^{-\gamma j}.$$

The same argument gives, for every $n > N$,

$$\sup_{\substack{s,t \in D \\ |t-s| \leq 2^{-n}}} |x(t, \omega) - x(s, \omega)| \leq 2 \sum_{j=n}^{\infty} 2^{-\gamma j}.$$

Define $\tilde{\Omega} = \Omega^* \cap \Omega^{**}$ and note that $P(\tilde{\Omega}) = 1$. Let $\omega \in \tilde{\Omega}$. Given $\varepsilon > 0$, find $n > N$ such that $2 \sum_{j=n}^{\infty} 2^{-\gamma j} < \varepsilon$. Let $t, s \in T$, $|t - s| \leq \frac{1}{2^{n+2}}$ be arbitrary. Since $\omega \in \Omega^*$, there exist sequences $(t_k = t_k(\omega))$, $(s_k = s_k(\omega))$ in D converging respectively to t and s. Choose k large such that $|t_k - t| < \frac{1}{2^{n+2}}$, $|s_k - s| < \frac{1}{2^{n+2}}$. Then $|t_k - s_k| \leq |t_k - t| + |t - s| + |s - s_k| < \frac{1}{2^n}$. Hence $|x(t_k, \omega) - x(s_k, \omega)| \leq 2\sum_{j=n}^{\infty} 2^{-\gamma j} < \varepsilon$. Now $|x(t, \omega) - x(s, \omega)| = \lim_{k\to\infty} |x(t_k, \omega) - x(s_k, \omega)| < \varepsilon$, establishing that for every $\omega \in \tilde{\Omega}$, $x(., \omega)$ is uniformly continuous on T.

Let $T = [0, \infty)$. For k a positive integer, proceeding as above, we can show that there exists $\Omega_k \in \mathscr{A}$ such that $P(\Omega_k) = 1$ and such that for each $\omega \in \Omega_k$, $x(., \omega)$ is continuous over $[0, k]$. Let $\bar{\Omega} = \cap_k \Omega_k$. Then $P(\bar{\Omega}) = 1$ and for every $\omega \in \bar{\Omega}$, $x(., \omega)$ is continuous over $[0, \infty)$. □

Remarks 5.7.5. (i) we have established more than what the theorem states. Let $\omega \in \tilde{\Omega}$. Consider the ratio $v_n = \frac{|x(t,\, \omega) - x(s,\, \omega)|}{|t-s|^\gamma}$ for $|t - s| \leq 2^{-N}$. By (5.33) this ratio, for $|t - s| = 2^{-N}$, is ≤ 1. If $|t - s| < 2^{-N}$, find $n \geq N$ such that $2^{-(n+1)} \leq |t - s| < 2^{-n}$. We have already shown that

$$|x(t, \omega) - x(s, \omega)| \leq 2\sum_{j=n}^{\infty} 2^{-\gamma j}.$$

Hence

$$\frac{|x(t, \omega) - x(s, \omega)|}{|t - s|^\gamma} \leq \frac{|x(t, \omega) - x(s, \omega)|}{2^{-\gamma(n+1)}} \leq \frac{2\sum\limits_{j=n}^{\infty} 2^{-\gamma j}}{2^{-\gamma(n+1)}} = 2\frac{2^\gamma}{1 - 2^{-\gamma}}.$$

Summarizing, what we have established is, putting $h(\omega) = 2^{-N(\omega)}$,

$$P\left[\omega : \sup_{\substack{0<|t-s|<h(\omega)\\ 0\leq s,t\leq 1}} \frac{|x(t,\omega) - x(s,\omega)|}{|t - s|^\gamma} \leq \frac{2^{\gamma+1}}{1 - 2^{-\gamma}}\right] = 1 \tag{5.34}$$

$$\sup_{\substack{0<|t-s|\geq h(\omega)\\ 0\leq s,t\leq 1}} \frac{|X_t(\omega) - X_s(\omega)|}{|t - s|^\gamma} \leq \frac{2\|x\|(\omega)}{(h(\omega))^\gamma} = M_1(\omega), \text{ say.}$$

Let $M(\omega) = \max(\frac{2^{\gamma+1}}{1-2^{-\gamma}}, M_1(\omega))$, leading to

$$\sup_{s,t\in[0,\, 1],\, s\neq t} \frac{|x(t,\, \omega) - x(s,\, \omega)|}{|t - s|^\gamma} \leq M(\omega)\ wp1.$$

And then each sample path corresponding to $\omega \in \Omega^{**}$ is said to be γ-Hölder continuous. For a formal definition of this property and Hölder spaces refer to section **(5.10)**.

(ii) In **(5.7.1)**(ii) we defined a standard BMP as a process satisfying the conditions p_1, p_2, and p_3 listed there and we promised to show that if the process is separable then p_2 implies p_3. Here is the proof of this claim.

Let $B = \{B(t),\ t \in T = [0,1]\}$ be a separable process satisfying condition p_2 of **(5.7.1)**(ii). From p_2, we get $\mathbb{E}|B(s) - B(t)|^4 = 3|t-s|^2$. Hence Theorem **(5.7.4)**, with $\beta = 4,\ \alpha = 1$ and $c = 3$, applies to B and the claim that p_3 is true follows.

Again, $\mathbb{E}|B(s) - B(t)|^{2n} = c_n|t-s|^n$. Hence B satisfies (5.34) with $\gamma < \frac{n-1}{2n} = \frac{1}{2} - \frac{1}{2n}$. This is true for every $n \geq 1$. Hence B satisfies (5.34) for every $\gamma < \frac{1}{2}$.

5.8 Brownian Motion Process on $[0,\ \infty)$

(i) By a Brownian Motion Process (BMP) on the half line $[0,\ \infty)$, we understand a stochastic process with time parameter set $[0,\ \infty)$ and satisfying the conditions (p_1), (p_2) and (p_3) listed in sec. **(5.7.1)**(ii). We shall now prove the existence of such a process.

We recall that the existence of a BMP B on the interval $[0,\ 1]$ was established in sec. **(5.7)**.

Let $B_0,\ B_1, \ldots$ be a sequence of independent Brownian Motion Processes on $[0,\ 1]$. Since each B_r can be thought of (ref. Sec. **(5.6)**) as a random element in the complete separable metric space $C = C[0,\ 1]$, the existence of such an independent sequnce is assured to us by sec. **(4.1.3)**.

Define $B(t) = B_{[t]}(t - [t]) + \sum_{0 \leq i < [t]} B_i(1), t \geq 0$. Then (p_1), (p_2) and (p_3) are easily verified to show that B is a standard BMP on $[0,\ \infty)$.

(ii) Another example of a process satisfying the condition of the theorem **(5.7.4)** is contained in the following

Theorem. Let (X_n) be a sequence of zero mean unit variance iid variables. Denote their partial sums by $S_n,\ n \geq 1$. Define process $x_n : x_n(\frac{k}{n}) = \frac{S_k}{\sqrt{n}},\ 0 \leq k \leq n$ and x_n linear elsewhere on $[0,\ 1]$.

Assume $\tau_p = \mathbb{E}|X_1|^{2p} < \infty$ for some $p,\ 1 < p < 2$.

Then there exists a constant λ independent of n such that

$$\mathbb{E}|x_n(t) - x_n(s)|^{2p} \leq \lambda|t-s|^p, \quad 0 \leq s,\ t \leq 1.$$

Proof. By **(5.3.6)**, $\mathbb{E}|\frac{S_n}{\sqrt{n}}|^{2p} \to \mathbb{E}|\xi|^{2p}$. Hence there exists a constant c such that $\sup_n \mathbb{E}|\frac{S_n}{\sqrt{n}}|^{2p} \leq c$ for all $n \geq 1$. We note $\tau_p \leq c$.

If, for some k, $s = \frac{k}{n}$ and $t = \frac{k+1}{n}$ then

$$\mathbb{E}|x_n(t) - x_n(s)|^{2p} = \mathbb{E}|\frac{X_{k+1}}{\sqrt{n}}|^{2p} \leq \frac{c}{n^p} = c|t-s|^p.$$

Now consider the case $\frac{k}{n} \leq s < t \leq \frac{k+1}{n}$. In this case, $x_n(t) - x_n(s) = \frac{X_{k+1}(t-s)}{\sqrt{n}}$. Hence

$$\mathbb{E}|x_n(t) - x_n(s)|^{2p} = \frac{\mathbb{E}|X_1|^{2p}(t-s)^{2p}}{(\sqrt{n})^{2p}} \leq c|t-s|^p.$$

If $s < t$ lie in different intervals, then there exist $k < r$ such that $\frac{k}{n} \leq s < \frac{k+1}{n} \leq \frac{r}{n} \leq t < \frac{r+1}{n}$. In this case
$|x_n(t) - x_n(s)| \leq |x_n(t) - x_n(\frac{r}{n})| + |x_n(\frac{r}{n}) - x_n(\frac{k+1}{n})| + |x_n(\frac{k+1}{n}) - x_n(s)|$.
Appealing to the inequality **(2.6.5)**(iii), we have

$$\begin{aligned}
&\mathbb{E}|x_n(t) - x_n(s)|^{2p} \\
&\leq 2^{4p-2}\{\mathbb{E}|x_n(t) - x_n(\frac{r}{n})|^{2p} + \mathbb{E}|x_n(\frac{r}{n}) - x_n(\frac{k+1}{n})|^{2p} \\
&\qquad + \mathbb{E}|x_n(\frac{k+1}{n}) - x_n(s)|^{2p}\} \\
&\leq 2^{4p-2}\{c|t - \frac{r}{n}|^p + \mathbb{E}|\frac{X_{k+2} + \cdots + X_r}{\sqrt{n}}|^{2p} + c|\frac{k+1}{n} - s|^p\} \\
&\leq 2^{4p-2}\{2c|t-s|^p + \mathbb{E}|\frac{S_{r-k-1}}{\sqrt{r-k-1}} \frac{\sqrt{r-k-1}}{\sqrt{n}}|^{2p}\} \\
&\leq 2^{4p-2}\{2c|t-s|^p + c(\frac{r-k-1}{n})^p\} \\
&\leq 2^{4p-2}c\{2|t-s|^p + |t-s|^p\} \leq c_1|t-s|^p, \text{ say.}
\end{aligned}$$

Thus for all $0 \leq s < t \leq 1, \mathbb{E}|x_n(t) - x_n(s)|^{2p} \leq \lambda|t-s|^p$ where the constant λ does not depend on n. □

5.9 Some basic properties of a Brownian Motion Process

Let B be a standard BMP on $T = [0, \infty)$. Let $s, t \in T$. Then ,

(i)$Y_1(t) = -B(t), t \geq 0$ is a standard BMP.

This will follow if we verify condition p_2 of **(5.7.1)**(ii). p_2 holds since the distribution of $(-B(t_1),\ -B(t_2), \dots,\ -B(t_k))$ is the same as that of $(B(t_1),\ B(t_2), \dots,\ B(t_k))$.

Remark. We note that this has the implication $-\max_{0 \le s \le t}\{-B(s)\}$ and $\min_{0 \le s \le t} B(s)$ are identically distributed.

(ii) $Y_2(t) = B(t+s) - B(s),\ t \ge 0,\ s$ fixed, is a standard BMP, since the distribution of $(B(t_1+s) - B(s),\ B(t_2+s) - B(s), \dots, B(t_k+s) - B(s))$ is the same as that of $B(t_1),\ B(t_2), \dots,\ B(t_k))$.

(iii) $Y_3(t) = \sqrt{a}B(\frac{t}{a}),\ a > 0$ is a standard BMP.

We note that $Y_3(0) = 0$, that the joint distribution of $(Y_3(t_1),\ Y_3(t_2), \dots, Y_3(t_k))$ is a zero mean multivariate normal vector with $cov(\sqrt{a}B(\frac{t_i}{a}), \sqrt{a}B(\frac{t_j}{a})) = \min(t_i,\ t_j)$. Thus Y_3 is a standard BMP.

Corollary. $\sqrt{a} \max_{0 \le t \le 1} |B(t)|$ is distributed as $\sqrt{a} \max_{0 \le t \le a} |B(\frac{t}{a})|$. i.e. as $\max_{0 \le t \le a} |Y_3(t)|$.

(iv) Let $Z = \max_{0 \le t \le 1} B(t)$ and $x \ge 0$. Then

$$P(Z \ge x) \le 2P(B(1) \ge x).$$

Proof.

For $x = 0$, the right side has the value unity and hence the inequality is trivially true.

Let $x > 0$. Let $B_n = \max_{0 \le k \le 2^n} B(\frac{k}{2^n})$ and note that $B_n \le B_{n+1}$ and that, since the sample functions of B are continuous, $B_n \uparrow Z$. Hence $B_n \xrightarrow{d} Z$. Let $0 < y \le x$ be an arbitrary continuity point of the *df* of Z. We have $P(Z \ge x) \le P(Z \ge y) = P(Z > y) = \lim_{n \to \infty} P(B_n > y)$. Now

$$B(\frac{k}{2^n}) = \sum_{j=1}^{k} \{B(\frac{j}{2^n}) - B(\frac{j-1}{2^n})\},$$

the sum of k *iid* normal variables. If X_r in **(4.4.4)** is taken as $B(\frac{r}{2^n}) - B(\frac{r-1}{2^n}))$, then $B(\frac{k}{2^n}) = \sum_{j=1}^{k} X_j = S_k$. We have

$$\begin{aligned} P(Z > y) &\le \lim_{n \to \infty} P(B_n > y) = \lim_{n \to \infty} P(\max_{0 \le k \le 2^n} B(\frac{k}{2^n}) > y) \\ &= \lim_{n \to \infty} P(\max_{0 \le k \le 2^n} S_k > y) \\ &\le 2P(B(1) > y) \end{aligned}$$

(by **(4.4.4)**(i)), remembering $medX_j = 0$).

Thus, $P(Z \geq x) \leq \lim_{y \uparrow x} 2P(B(1) > y) = 2P(B(1) > x) = 2P(B(1) \geq x)$, since the *df* of $B(1)$ is continuous.

(v) $P(\min_{0 \leq t \leq 1} B(t) \leq -x) \leq 2P(B(1) \leq -x), \ x \geq 0.$

This result follows from (iv) above (ref. Remark under (i)).

(vi) For $x \geq 0, \ P(\max_{0 \leq t \leq 1} |B(t)| \geq x) \leq 2P(|B(1)| \geq x)$. Since

$$\{\max_{0 \leq t \leq 1} |B(t)| \geq x\} = \{\max_{0 \leq t \leq 1} B(t) \geq x\} \cup \{-\min_{0 \leq t \leq 1} B(t) \geq x\},$$

the claim follows from (iv) and (v).

(vii) $\frac{B(t)}{t} \xrightarrow{wp1} 0$ as $t \to \infty$, or, $tB(\frac{1}{t}) \xrightarrow{wp1} 0$ as $t \to 0$.

Proof. For n a positive integer

$$P(|\frac{1}{n}B(n)| > \varepsilon) = P(|\frac{B(n)}{\sqrt{n}}| > \varepsilon\sqrt{n}) \leq \frac{3}{\varepsilon^4 n^2}.$$

Hence $\sum_n P(|\frac{B(n)}{n}| > \varepsilon) < \infty$. By the Borel-Cantelli lemma and the abitrariness of ε, it now follows $\frac{B(n)}{n} \xrightarrow{wp1} 0$.

Now $P(\max_{n \leq s \leq n+1} |B(s) - B(n)| \geq n\varepsilon) =$ (ref. (ii) above) $P(\max_{0 \leq t \leq 1} |B(t)| \geq n\varepsilon) \leq 2P(|B(1)| \geq n\varepsilon) \leq \frac{2}{n^2\varepsilon^2}$. Hence $\sum_n P(\max_{n \leq t \leq n+1} |\frac{B(t)-B(n)}{n}| \geq \varepsilon) < \infty$. By the Borel-Cantelli lemma and by the arbitrariness of ε, it follows that $\max_{n \leq s \leq n+1} |\frac{B(s)-B(n)}{n}| \xrightarrow{wp1} 0$.

Let $[t] = n$. Then $\lim_{t \to \infty} \frac{B(t)}{t} = \lim_{n \to \infty} \{\frac{B(n)}{n} + \frac{B(t)-B(n)}{n}\}$. Since $|\frac{B(t)-B(n)}{n}| \leq \max_{n \leq t \leq n+1} \frac{1}{n}|B(t) - B(n)| \xrightarrow{wp1} 0$, the claim follows.

(viii) The processes $\{Y_4(t), t \geq 0\}, \{Y_5(t), t \geq 0\}\}$ and $\{Y_6(t), 0 \leq t \leq 1\}$ are standard BMPs where $Y_4(t) = tB(\frac{1}{t}), \ t > 0$ and $Y_4(0) = 0$; $Y_5(t) = (1+t)B(\frac{t}{1+t}) - tB(1)$, $Y_6(t) = B(1) - B(1-t)$.

Proof. The sample function continuity of Y_4 over $(0, \ \infty)$ derives from that of B. That Y_4 is continuous at 0 follows from (vii). $(Y_4(t_1), Y_4(t_2), \ldots, Y_4(t_k))$ having a multivariate normal distribution is immediate from the similar property of B. Since $\text{cov}(Y_4(t_i), \ Y_4(t_j)) = t_i t_j \text{cov}(B(\frac{1}{t_i}), B(\frac{1}{t_j}))$

$= t_i t_j \min(\frac{1}{t_i}, \ \frac{1}{t_j}) = \min(t_i, \ t_j)$, for $t_i > 0, \ t_j > 0$,

the proof is complete.

Now we study the process Y_5. Clearly $Y_5(0) = 0$. The continuity of the sample paths of Y_5 follow trivially from that of B. The vector $\mathbb{V} =$

$\left(B(\frac{t_1}{1+t_1}),\ B(\frac{t_2}{1+t_2}),\ldots,\ B(\frac{t_k}{1+t_k}),\ B(1)\right)'$ has a multivariate normal distribution since B is a BMP. That $\Lambda = \left(Y_5(t_1),\ Y_5(t_2),\ldots,\ Y_5(t_k)\right)'$ has multivariate normal distribution follows now since $\Lambda = \mathbb{A}\mathbb{V}$ where $\mathbb{A}$ is the $k \times (k+1)$ matrix with entries $a_{i,j} = 0$, for $i \neq j,\ 1 \le i,\ j \le k$; $a_{j,j} = 1 + t_j,\ 1 \le j \le k$; $a_{j,k+1} = t_j$. Now for $s < t$

$$\begin{aligned}
&\operatorname{cov}(Y_5(s),\ Y_5(t)) \\
&\quad = \mathbb{E}\{(1+s)B(\frac{s}{1+s}) - sB(1)\}\{(1+t)B(\frac{t}{1+t}) - tB(1)\} \\
&\quad = (1+s)(1+t)\mathbb{E}B(\frac{s}{1+s})B(\frac{t}{1+t}) \\
&\qquad - (1+s)t\mathbb{E}B(\frac{s}{1+s})B(1) - (1+t)s\mathbb{E}B(\frac{t}{1+t})B(1) \\
&\qquad\quad + st\mathbb{E}B^2(1) \\
&\quad = s(1+t) - st - st + st = s.
\end{aligned}$$

The claim about Y_5 follows.

That $Y_6(0) = 0$ is obvious. The sample function continuity of the process derives directly from that of the process B. Let $0 < u_p = 1 - t_p < u_{p-1} = 1 - t_{p-1} < \cdots < u_1 = 1 - t_1 < 1$. Then the joint distribution of $(Y_6(u_p), Y_6(u_{p-1}), \ldots, Y_6(u_1))$ is zero-mean multivariate normal since the joint distribution of $(B(t_1), B(t_2), \ldots, B(t_p), B(1))$ is zero-mean multivariate normal. Further, if $0 < s < t < 1$, then $\mathbb{E}Y_6(s)Y_6(t) = \mathbb{E}\{B^2(1) - B(1)B(1-s) - B(1)B(1-t) + B(1-s)B(1-t)\} = 1 - (1-s) - (1-t) + (1-t) = s$. That Y_6 is a standard BMP follows.

(ix) With probability 1, the sample functions of B are monotone in no interval.

Proof. If possible let there exist an interval in which B is monotone. Hence a finite closed interval exists in which B is monotone. We use (ii) and (iii) to say that this interval can be taken to be $[0,\ 1]$. Again it is enough to show that $P(A) = 0$ where

$$A = \{\omega : B(t,\ \omega) \text{ is non-decreasing in } [0,\ 1]\}.$$

Let $A_n = \{\omega : B(\frac{i+1}{n},\ \omega) \ge B(\frac{i}{n},\ \omega),\ 0 \le i \le n-1\}$. Then $A \subset A_n,\ n \ge 1$. Since $B(\frac{i+1}{n}) - B(\frac{i}{n}),\ 0 \le i \le n-1$ are independent symmetric variables, $P(A_n) = \frac{1}{2^n}$. This implies $P(A) = 0$. Similarly $B(.,\ \omega)$ is non-increasing $wp1$.

(x) Quadratic variation. Let $\Delta_m = \{0 = t_0 < t_1 < \ldots < t_m = t\}$ be a partition of $[0,\ t]$. Define

$$\|\Delta_m\| = \max_{1 \le k \le m}(t_k - t_{k-1}) \text{ and } V_t^{(p)}(\Delta_m) = \sum_{k=1}^{m} |B(t_k) - B(t_{k-1})|^p$$

and note that $V_t^{(p)}$ is a rv. As $m \to \infty$ and $\|\Delta_m\| \to 0$, if $V_t^{(p)}(\Delta_m)$ tends to a limit and if the limit is the same for all partitions then the limit rv is called the pth variation of B. The limit may be in probability or $wp1$ or in mean of some order etc. If Δ_n is a refinement of Δ_m the inequality $V_t^{(p)}(\Delta_m) \le V_t^{(p)}(\Delta_n)$ holds when $p = 1$ but not when $p > 1$.

For the BMP B, $V_t^{(2)}(\Delta_m) \to t$ in mean of order 2, as $\|\Delta_m\| \to 0$.
Proof.

$$
\begin{aligned}
&\mathbb{E}\Big(\sum_{k=1}^{m}(B(t_k) - B(t_{k-1}))^2 - t\Big)^2 \\
&= t^2 - 2t\mathbb{E}\sum_{k=1}^{m}(B(t_k) - B(t_{k-1}))^2 + \mathbb{E}\Big\{\sum_{k=1}^{m}(B(t_k) - B(t_{k-1}))^2\Big\}^2 \\
&= t^2 - 2t\sum_{k=1}^{m}(t_k - t_{k-1}) + \mathbb{E}\sum_{k=1}^{m}\{B(t_k) - B(t_{k-1})\}^4 \\
&\quad + \mathbb{E}\sum_{\substack{1 \le j,k \le m \\ j \ne k}}\{B(t_k) - B(t_{k-1})\}^2\{B(t_j) - B(t_{j-1})\}^2 \\
&= t^2 - 2t^2 + 3\sum_{k=1}^{n}(t_k - t_{k-1})^2 + \sum_{\substack{1 \le j,k \le m \\ j \ne k}}(t_k - t_{k-1})(t_j - t_{j-1}) \\
&= -t^2 + 2\sum_{k=1}^{m}(t_k - t_{k-1})^2 + \Big\{\sum_{k=1}^{m}(t_k - t_{k-1})\Big\}^2 \\
&= 2\sum_{k=1}^{m}(t_k - t_{k-1})^2 \le 2\|\Delta_m\|\sum_{k=1}^{m}(t_k - t_{k-1}) \\
&= 2\|\Delta_m\|t \to 0 \text{ as } \|\Delta_m\| \to 0.
\end{aligned}
$$

Remark. If $\sum_m \|\Delta_m\| < \infty$, then, by the Borel-Cantelli lemma, $V_t^{(2)}(\Delta_m) \xrightarrow{wp1} t$.

(xi) $wp1$, the sample functions of a BMP are of bounded variation in no interval.

Proof. If possible let there exist a set $A \in \mathscr{A}$ with $P(A) > 0$ and an interval $[a,\ b]$ such that for each $\omega \in A$, $B(.,\ \omega)$ is a function of bounded variation over this interval. By suitable translation and scaling (ref. (ii) and (iii) above) we may take this interval as $[0,\ 1]$. Consider a partition of $[0,\ 1]$ such that $\sum_m \|\Delta_m\| < \infty$. By (x) there exists then $E \in \mathscr{A}$ with $P(E) = 1$ such that $V_1^{(2)}(\Delta_m)(\omega) \to 1$ for every $\omega \in E$. Let $A_1 = E \cap A$ and note that

$P(A_1) = P(A)$. Let $\omega \in A_1$. The function $B(., \omega)$, being continuous, is uniformly continuous over $[0, 1]$. Hence given $\varepsilon > 0$, we can find a $\delta > 0$ such that $|B(t, \omega) - B(s, \omega)| < \varepsilon$ whenever $|t - s| < \delta$. Since (by hypothesis) $B(., \omega)$ is a function of bounded variation, there exists $M(\omega)$ such that $\sum_{j=1}^{n} |B(u_j, \omega) - B(u_{j-1}, \omega)| \leq M(\omega)$ for all partitions $\{0 = u_0 < u_1 < ... u_n = 1\}$. For the partition through $\{\frac{k}{m}, 0 \leq k \leq m\}$ and for all m large such that $\|\Delta_m\| < \delta$, we have

$$V_1^{(2)}(\Delta_m)(\omega) = \sum_{k=1}^{m} |B(\frac{k}{m}, \omega) - B(\frac{k-1}{m})|^2 \leq \varepsilon M(\omega).$$

Thus $\lim_{m\to\infty} V_1^{(2)}(\Delta_m)(\omega) = 0$, a contradiction.

(xii) $wp1$, $Y = \int_0^\infty \frac{|B(t)|}{1+t^2}\, \mathrm{d}t < \infty$.

Proof. By Fubini's theorem,

$$\mathbb{E}Y = \mathbb{E}\int_0^\infty \frac{|B(t)|}{1+t^2}\,\mathrm{d}t = \int_0^\infty \frac{1}{1+t^2}\mathbb{E}|B(t)|\,\mathrm{d}t$$

$$\leq \int_0^\infty \frac{1}{1+t^2}\{\mathbb{E}|B(t)|^2\}^{\frac{1}{2}} = \int_0^\infty \frac{c\sqrt{t}}{1+t^2}\,\mathrm{d}t < \infty.$$

This implies that Y is finite valued $wp1$ and that is the desired result.

This result can also be stated thus : if $E = \{x : x \in C, \int_0^\infty \frac{|x(t)|}{1+t^2}\,\mathrm{d}t < \infty\}$, then there exists $A \in \mathscr{A}$ such that $P(A) = 1$ and for $\omega \in A$, $B(., \omega) \in E$ or $\mu(E) = 1$ where μ is the (Wiener) measure induced in $\mathscr{C}$ by B.

(xiii) $wp1$, $\overline{\lim}_{t\to\infty} \frac{B(t)}{\sqrt{t}} = \infty$ and $\underline{\lim}_{t\to\infty} \frac{B(t)}{\sqrt{t}} = -\infty$.

Proof. We note that it is enough to show that $wp1$, $\overline{\lim}_{n\to\infty} \frac{B(n)}{\sqrt{n}} = \infty$ and $\underline{\lim}_{n\to\infty} \frac{B(n)}{\sqrt{n}} = -\infty$ where the limits are taken along positive integers.

Let $X_r = B(r) - B(r-1)$. For $r = 1, 2, \ldots, (X_r)$ is a sequence of *iid rvs* and $(B(n))$ the corresponding partial sum sequence. Let $A_{k,n} = (B(n) > k\sqrt{n})$, $k \geq 1$ an arbitrary integer . By Remark **(4.1.11)**(iii), it follows that $P(A^{(k)}) = 0$ or 1 where $A^{(k)} = (A_{k,n}$ occurs for infinitely many $n)$. Since $A^{(k)} = \lim_{n\to\infty} \cup_{r=n}^\infty A_{k,r}$ and since $\cup_{r=n}^\infty A_{k,r}$, $n \geq 1$ is a monotonic sequence,

$$P(A^{(k)}) = \lim_{n\to\infty} P(\cup_{r=n}^\infty A_{k,r}) \geq \overline{\lim_{n\to\infty}}\, P(A_{k,n}) = P(B(1) > k) > 0,$$

we conclude $P(A^{(k)}) = 1$. Since $\{\omega : \overline{\lim}_{n\to\infty} \frac{B(n)}{\sqrt{n}} = \infty\} = \cap_k A^{(k)}$, the claim follows.

Proof that $\varliminf_{n\to\infty} \frac{B(n)}{\sqrt{n}} = -\infty$ is on similar lines.

The property just established implies that $B(t)$ (remember $B(.)$ is a continuous function) crosses 0 for arbitrarily large values of t. If we use the BMP $tB(\frac{1}{t})$, we see that a BMP crosses 0 for arbitrarily small values of t.

The result that $wp1$, $\lim_{t\to\infty} \frac{B(t)}{t} = 0$ established in (vii) and the result that $wp1$, $\varlimsup_{t\to\infty} \frac{B(t)}{\sqrt{t}} = \infty$ raise the following question. For which function $a(t)$ (necessarily tending to infinity with t) will we have $0 < \varlimsup_{t\to\infty} \frac{B(t)}{a(t)} < \infty$? We will answer this question in the next chapter.

We record the following useful consequence. We have shown that $wp1$, $\varlimsup_{t\to\infty} \frac{B(t)}{\sqrt{t}} = \infty$. This implies that, $wp1$, $\varlimsup_{t\to\infty} \frac{tB(\frac{1}{t})}{\sqrt{t}} = \infty$, since $tB(\frac{1}{t})$ is a standard BMP. i.e. $wp1$, $\varlimsup_{t\to 0} \frac{B(t)}{\sqrt{t}} = \infty$. We also have $wp1$, $\varlimsup_{t\to 0} \frac{B(t+s)-B(s)}{\sqrt{t}} = \infty$.

Definition. A function f defined on an interval I, finite or infinite, is called Hölder continuous of order $\gamma > 0$ at a point $t_0 \in I$ if there exists a constant c (which may depend on t_0) such that $|f(t) - f(t_0)| \le c|t - t_0|^\gamma$ for all t in a neighbourhood, however small, of t_0.

The property that $wp1$, $\varlimsup_{t\to 0} \frac{B(t+s)-B(s)}{\sqrt{t}} = \infty$ implies that $wp1$, at no point s, the paths of B are Hölder continuous of order $\frac{1}{2}$.

(xiv) *Definition.* The limit, if it exists, of $V_1^{(1)}(\Delta_m)$ as $\|\Delta_m\| \to 0$ is called the total variation of B. The limit may be under any mode of convergence.

We prove now: The total variation of B is infinite $wp1$.

Proof. Let target error $\varepsilon > 0$ be given. For each $\omega \in \Omega$, $B(., \omega)$ is a uniformly continuous function. Hence there exists a $\delta = \delta(\omega) > 0$ such that for every partition Δ_m with $\|\Delta_m\| < \delta$, then

$$V_1^{(2)}(\Delta_m)(\omega) \le \varepsilon V_1^{(1)}(\Delta_m)(\omega). \tag{5.35}$$

In this section we denote by Δ_m the partition of $[0, 1]$ through the division points $\{\frac{k}{2^m}, 0 \le k \le 2^m\}$. We note that $\|\Delta_m\| = \frac{1}{2^m}$ and that $\sum_m \|\Delta_m\| < \infty$. By (x), there exists then E with $P(E) = 1$ such that $V_1^{(2)}(\Delta_m)(\omega) \to 1$ as $m \to \infty$ for every $\omega \in E$. Hence by (5.35), $\lim_{m\to\infty} V_1^{(2)}(\Delta_m)(\omega) \le \varepsilon \varliminf_{m\to\infty} V_1^{(1)}(\Delta_m)(\omega)$. i.e., $\varliminf_{m\to\infty} V_1^{(1)}(\Delta_m)(\omega) \ge \frac{1}{\varepsilon}$. The claim follows from this since $\varepsilon > 0$ is arbitrary.

(xv) Let μ be the standard Wiener measure induced in C. For $\lambda \ne 0$, define measure ν on $\mathscr{C}$: $\nu_\lambda(A) = \mu(\lambda A)$. Then, for $|\lambda| \ne 1$, $\nu_\lambda \perp \mu$.

Proof. Observe that $\nu_1 \equiv \nu_{-1}$.

Let Δ_m be as in (xiv). Let $E = \{x : V_1^{(2)}(\Delta_m)(x) \to 1\}$ as $m \to \infty$. Then by (x), $\mu(E) = 1$. By the definition of E the function which is identically 0 does not belong to E and hence does not belong to λE, $|\lambda| \neq 1$. We claim $E \cap \lambda E = \phi$. For, if possible let $x \in E \cap \lambda E$. $x \in E$ implies $V_1^{(2)}(\Delta_m)(x) \to 1$. Since $x \in \lambda E$, there exists $y \in E$ such that $x = \lambda y$. This implies $1 = \lim\limits_{m\to\infty} V_1^{(2)}(\Delta_m)(x) = \lim\limits_{m\to\infty} V_1^{(2)}(\Delta_m)(\lambda y) = \lambda^2 \lim\limits_{m\to\infty} V_1^{(2)}(\Delta_m)(y) = \lambda^2$, a contradiction. Thus, for $|\lambda| \neq 1$, $\nu_\lambda(E) = \mu(\lambda E) = 0$. $\nu_\lambda(\frac{1}{\lambda}E) = \mu(E) = 1$. Hence $\nu_\lambda \perp \mu$.

In the last section we considered the transformation $\mathcal{T}$ of C into C : $\mathcal{T}(x) = \lambda x, |\lambda| \neq 1$ and showed that $\nu = \mu\mathcal{T}^{-1} \perp \mu$. In the next section we study a transformation T complementary to $\mathcal{T}$:

(xvi) Given $f \in C$ with a continuous differential coefficient f', define transformation $T = T_f$ taking C into itself : $T(x) = x - f$. Hence $T^{-1}(x) = x + f$. Define *pm* ν : $\nu = \mu T^{-1}$. We prove $\nu \ll \mu$.

In this section and the next, the expressions $wp1$ *or 'in probability', are to be understood as referring to measure space* $(C, \mathscr{C}, \mu)$.

Step 1. Let g be an arbitrary continuous function defined on $[0, 1]$. g will be automatically uniformly continuous on $[0, 1]$. Let $t_{j,n}$ be an arbitrary point in the interval $[\frac{j-1}{2^n}, \frac{j}{2^n}]$, $j = 1, 2, \ldots, 2^n$, $n = 1, 2, \ldots$. Let $g_n = \frac{1}{2^n}\sum_{j=1}^{n} g(t_{j,n})$. Then (g_n) is a Cauchy sequence and hence is convergent.

Proof. Given $\varepsilon > 0$, choose $\mathbf{m}$ such that $|g(t) - g(t')| < \varepsilon$ whenever $|t - t'| < \frac{1}{2^{\mathbf{m}}}$. This is possible since g is uniformly continuous. Let $\mathbf{m} < m < n$. Now

$$\begin{aligned} g_n - g_m &= \frac{1}{2^n}\sum_{j=1}^{2^n} g(t_{j,n}) - \frac{1}{2^m}\sum_{j=1}^{2^m} g(t_{j,m}) \\ &= \frac{1}{2^m}\sum_{j=1}^{2^m}\frac{1}{2^{n-m}}\sum_{r=1}^{2^{n-m}} g(t_{2^{n-m}(j-1)+r,\, n}) - \frac{1}{2^m}\sum_{j=1}^{2^m} g(t_{j,m}) \\ &\frac{1}{2^m}\sum_{j=1}^{2^m}\frac{1}{2^{n-m}}\sum_{r=1}^{2^{n-m}}\Big\{g(t_{2^{n-m}(j-1)+r,\, n}) - g(t_{j,m})\Big\}. \end{aligned}$$

All the numbers $t_{2^{n-m}(j-1)+r,\, n}$, $1 \le r \le 2^{n-m}$ lie in the interval $[\frac{j-1}{2^m}, \frac{j}{2^m}]$. Hence

$$|g_n - g_m| \le \frac{1}{2^m}\sum_{j=1}^{2^m}\frac{1}{2^{n-m}}\sum_{r=1}^{2^{n-m}} \varepsilon = \varepsilon.$$

That (g_n) is a Cauchy sequence, and hence a convergent sequence, of numbers follows from this.

Note. Since $|g_n| \leq \|g\|$ and since g is a continuous function, there exists $t \in [0,\ 1]$ such that $g(t) = \lim_{n\to\infty} g_n$.

Step 2. Let g denote an arbitrary continuous function defined on $[0,\ 1]$. When the underlying measure on $\mathscr{C}$ is μ, consider the sequence of rvs (α_n) where $\alpha_n(x) = \sum_{j=1}^{2^n} g(t_{j,n})\Big(x(\frac{j}{2^n}) - x(\frac{j-1}{2^n})\Big)$ ($t_{j,n} \in (\frac{j-1}{2^n},\ \frac{j}{2^n})$) being an arbitrary point. We note $\alpha_n \in L^2(\mu)$. Let us examine if this sequence has a limit in the norm topology of $L^2(\mu)$.

$$\mathbb{E}\alpha_n^2 = \sum_{j=1}^{2^n} \left(g(t_j)\right)^2 \frac{1}{2^n} \to \int_0^1 \left(g(t)\right)^2 \mathrm{d}t \tag{5.36}$$

since g^2, being a continuous function, is Riemann integrable. Let $1 \leq m < n < \infty$.

$$\begin{aligned}
\mathbb{E}\alpha_m\alpha_n &= \mathbb{E}\Big[\sum_{j=1}^{2^m}\{(x(\frac{j}{2^m}) - x(\frac{j-1}{2^m}))g(t_{j,m})\}\times \\
&\quad \{\sum_{j=1}^{2^n}(x(\frac{j}{2^n}) - x(\frac{j-1}{2^n}))g(t_{j,n})\}\Big] \\
&= \mathbb{E}\Big[\sum_{j=1}^{2^m}\{(x(\frac{j}{2^m}) - x(\frac{(j-1)}{2^m}))g(t_{j,m})\}\times \\
&\quad \{\sum_{j=1}^{2^m}(\sum_{r=1}^{2^{n-m}}(x(\frac{(j-1)}{2^m} + \frac{r}{2^n}) - x(\frac{(j-1)}{2^m} + \frac{r-1}{2^n}))\times \\
&\quad g(t_{2^{n-m}(j-1)+r,n})\}\Big] \\
&= \sum_{j=1}^{2^m} g(t_{j,m}) \times \Lambda_j, \text{ say, where}
\end{aligned}$$

$$\begin{aligned}
\Lambda_j &= \mathbb{E}\Big[\Big(x(\frac{j}{2^m}) - x(\frac{(j-1)}{2^m})\Big)\times \\
&\quad \sum_{r=1}^{2^{n-m}}\Big\{\Big(x(\frac{2^{n-m}(j-1)+r}{2^n}) - x(\frac{2^{n-m}(j-1)+r-1}{2^n})\Big)\times \\
&\quad \Big(g(t_{2^{n-m}(j-1)+r,\ n})\Big)\Big\}\Big]
\end{aligned}$$

$$= \sum_{r=1}^{2^{n-m}} g(t_{2^{n-m}(j-1)+r,\ n}) \times$$

$$\times\ \mathbb{E}\Big\{\big(x(\frac{2^{n-m}j}{2^n}) - x(\frac{2^{n-m}(j-1)}{2^n})\big) \times \big(x(\frac{2^{n-m}(j-1)+r}{2^n})$$

$$- x(\frac{2^{n-m}(j-1)+r-1}{2^n})\big)\Big\}$$

$$= \sum_{r=1}^{2^{n-m}} g(t_{2^{n-m}(j-1)+r,\ n}) \times$$

$$\times\ \mathbb{E}\big(x(\frac{2^{n-m}(j-1)+r}{2^n}) - x(\frac{2^{n-m}(j-1)+r-1}{2^n})\big)^2$$

$$= \sum_{r=1}^{2^{n-m}} g(t_{2^{n-m}(j-1)+r,\ n}) \frac{1}{2^n}$$

(the other terms in the product being zero each)

Hence

$$\lim_{m\to\infty}\lim_{n\to\infty} \mathbb{E}\alpha_m\alpha_n$$

$$= \lim_{m\to\infty}\lim_{n\to\infty} \sum_{j=1}^{2^m} g(t_{j,m}) \frac{1}{2^m} \Big\{ \sum_{r=1}^{2^{n-m}} g(t_{2^{n-m}(j-1)+r,\ n}) \frac{1}{2^{n-m}} \Big\}$$

$$= \lim_{m\to\infty} \frac{1}{2^m} \sum_{j=1}^{2^m} g(t_{j,m}) \lim_{n\to\infty} \Big\{ \frac{1}{2^{n-m}} \sum_{r=1}^{2^{n-m}} g(t_{2^{n-m}(j-1)+r,\ n}) \Big\}.$$

All the points $t_{2^{n-m}(j-1)+r,n}$, $r = 1, 2, \ldots, 2^{n-m}$ lie in the interval $[\frac{j-1}{2^m}, \frac{j}{2^m}]$. By Step 1 above, there exists $t'_{j,m}$ such that

$$\lim_{n\to\infty} \Big\{ \frac{1}{2^{n-m}} \sum_{r=1}^{2^{n-m}} g(t_{2^{n-m}(j-1)+r,\ n}) \Big\} = g(t'_{j,m}).$$

We have

$$\lim_{m\to\infty}\lim_{n\to\infty} \mathbb{E}\alpha_m\alpha_n = \lim_{m\to\infty} \frac{1}{2^m} \sum_{j=1}^{2^m} g(t_{j,m}) g(t'_{j,m}) = \int_0^1 g^2(t)\, \mathrm{d}t \qquad (5.37)$$

(5.36) and (5.37) imply (α_n) is a Cauchy sequence. Since $L^2(\mu)$ is a complete metric space, there exists $\alpha \in L^2(\mu)$ to which the sequence (α_n) converges in norm.

Step 3. Henceforth in this section, $f \in C$ is a fixed element and is assumed to have a continuous differential coefficient f'.

Let $\mathcal{D}_n = \{\frac{j}{2^n},\ 0 \le j \le 2^n\}$. We note that $\mathcal{D}_n \subset \mathcal{D}_{n+1}$ and that $\bigcup_{n=1}^{\infty} \mathcal{D}_n$ is a countable dense subset of $[0,\ 1]$. Let $\pi_{\mathcal{D}_n}$ be the projection map of C into R^{2^n}. Denote by $\mathfrak{C}_n$ the σ-field $\pi_{\mathcal{D}_n}^{-1}(\mathscr{R}^n)$. We note that $\mathfrak{C}_n \uparrow$ and that $\sigma(\bigcup_{n=1}^{\infty} \mathfrak{C}_n) = \mathscr{C}$. Let $\mu_n,\ \nu_n$ denote the restrictions respectively of $\mu,\ \nu$ to $\mathfrak{C}_n$. The rV $\pi_{\mathcal{D}_n}$ has a non-singular multivariate normal *df*, both under μ and ν with common variance-covariance matrix. The mean vector under μ is the null vector. We note $\mathbb{E}x(t)$ under measure ν is

$$\int_C \pi_t(x)\, \mathrm{d}\,\nu(x) = \int_C \pi_t(Tx)\, \mathrm{d}\,\mu(x) \quad \text{(refer } \textbf{(2.3.6)}\text{)}$$

$$= \int_C \{x(t) - f(t)\}\, \mathrm{d}\,\mu(x) = 0 - f(t) = -f(t).$$

Hence the mean vector under ν is $(-f(\frac{j}{2^n}),\ j = 0,\ 1, \ldots,\ 2^n)$. The Radon-Nikodym (R-N) derivative $\frac{\mathrm{d}\,\nu_n}{\mathrm{d}\,\mu_n}$ of ν_n with respect to μ_n would be given by the ratio of the frequency functions of the two normal distributions : $\frac{\mathrm{d}\,\nu_n}{\mathrm{d}\,\mu_n}(x) = e^{-H_n(x)}$ where $H_n(x) = a_n(x) - b_n(x)$, say, with

$$a_n(x) = \tfrac{1}{2} \sum_{j=1}^{2^n} 2^n \Big\{x(\tfrac{j}{2^n}) - x(\tfrac{j-1}{2^n}) + f(\tfrac{j}{2^n}) - f(\tfrac{j-1}{2^n})\Big\}^2 \text{ and}$$

$$b_n(x) = \tfrac{1}{2} \sum_{j=1}^{2^n} \Big\{x(\tfrac{j}{2^n}) - x(\tfrac{j-1}{2^n})\Big\}^2,$$

leading to

$$\begin{aligned} H_n(x) &= 2^{n-1} \sum_{j=1}^{2^n} \Big\{2(x(\frac{j}{2^n}) - x(\frac{j-1}{2^n}))(f(\frac{j}{2^n}) - f(\frac{j-1}{2^n})\\ &\qquad + \Big(f(\frac{j}{2^n}) - f(\frac{j-1}{2^n})\Big)^2\Big\}\\ &= u_n(x) + v_n, \qquad \text{say.}\end{aligned}$$

$$v_n = 2^{n-1} \sum_{j=1}^{2^n} \Big\{f(\frac{j}{2^n}) - f(\frac{j-1}{2^n})\Big\}^2 = 2^{n-1} \sum_{j=1}^{2^n} \Big(f'(t_{j,n})\Big)^2 \frac{1}{2^{2n}}$$

(mean value theorem is applied; $t_{j,n} \in [\frac{j-1}{2^n}, \frac{j}{2^n}]$)

$$= \frac{1}{2}\sum_{j=1}^{2^n} \Big(f'(t_{j,n})\Big)^2 \frac{1}{2^n} \to v(= v_f) = \frac{1}{2}\int_0^1 \Big(f'(t)\Big)^2 \, \mathrm{d}\, t$$

Now, $u_n(x) = \sum_{j=1}^{2^n} f'(t_{j,n})\big(x(\frac{j}{2^n}) - x(\frac{j-1}{2^n})\big)$ (as in the last step).
By Step 2, $\lim\limits_{n\to\infty} u_n = u = u_f$ exists in $L^2(\mu)$. Necessarily, u is $wp1$ finite valued. Convergence in $L^2(\mu)$ implies $u_n \overset{pr}{\to} u$. There exists then a subsequence (n_r) such that $u_{n_r} \overset{wp1}{\longrightarrow} u$. It follows $H_{n_r} \overset{wp1}{\longrightarrow} u+v$. Thus if $Z_r(x) = \frac{\mathrm{d}\,\nu_{n_r}}{\mathrm{d}\,\mu_{n_r}}(x)$ and if $Z(x)(= Z_f(x)) = e^{-(u(x)+v)}$, then $Z_r \overset{wp1}{\longrightarrow} Z$. Since u, being an element of $L^2(\mu)$, is finite valued $wp1$, it follows that $Z(x)$ is positive valued and finite valued $wp1$.
Step 4. Let $A \in \mathfrak{C}_k$ and let $r \geq k$. We recall $\mathfrak{C}_n \uparrow$. Hence $A \in \mathfrak{C}_k$ for all $r \geq k$. Now, $\nu(A) = \nu_{n_r}(A) = \int_A \mathrm{d}\,\nu_{n_r}(x) = \int_A \frac{\mathrm{d}\,\nu_{n_r}}{\mathrm{d}\,\mu_{n_r}}(x)\ \mathrm{d}\,\mu_{n_r}(x) = \int_A Z_r(x)\ \mathrm{d}\,\mu_{n_r}(x) = \int_A Z_r(x)\ \mathrm{d}\,\mu(x)$ (since μ_{n_r}, ν_{n_r} are the restrictions of μ, ν to $\mathfrak{C}_r$). If limit can be taken inside the integral sign, we will have $\nu(A) = \int\limits_A Z(x)\ \mathrm{d}\,\mu(x)$. Since the bounded convergence theorem may not be applicable for the sequence (Z_n), we argue as follows.

Take limit as $r \to \infty$ and appeal to Fatou's lemma to get

$$\nu(A) \geq \int\limits_A \varliminf_{r\to\infty} Z_r(x)\,\mathrm{d}\,\mu(x) = \int\limits_A Z(x)\,\mathrm{d}\,\mu(x). \tag{5.38}$$

To get the inequality in the reverse direction, we consider first the set $B_r(\lambda) = A \cap (Z_r < \lambda)$ where $\lambda > 0$ is arbitrary. We note $B_r(\lambda) \in \mathfrak{C}_{n_r}$. Now,

$$\begin{aligned}
\nu(A) &= \lim_{\lambda\to\infty} \nu(B_r(\lambda)) \quad \text{(since } Z_r \text{ is finite valued)}\\
&= \lim_{\lambda\to\infty} \nu_{n_r}(B_r(\lambda))\\
&= \lim_{\lambda\to\infty} \int\limits_{A\cap(Z_r<\lambda)} \mathrm{d}\,\nu_{n_r}(x)\\
&= \lim_{\lambda\to\infty} \int\limits_A Z_r(x)\chi_{(Z_r<\lambda)}\ \mathrm{d}\,\mu(x)\\
&\leq \lim_{\lambda\to\infty} \int\limits_A \min\{Z_r(x),\ \lambda\}\ \mathrm{d}\,\mu(x)
\end{aligned}$$

$$\leq \lim_{\lambda\to\infty} \lim_{r\to\infty} \int_A \min\{Z_r(x),\ \lambda\}\ \mathrm{d}\,\mu(x)$$

$$\leq \lim_{\lambda\to\infty} \int_A \min\{Z(x),\ \lambda\}\ \mathrm{d}\,\mu(x)$$

(by the bounded convergence theorem)

$$= \int_A Z(x)\ \mathrm{d}\,\mu(x). \tag{5.39}$$

(5.38) and (5.39) help us conclude that $\nu(A) = \int_A Z(x)\ \mathrm{d}\,\mu(x)$, for $A \in \cup_{k=1}^{\infty}\mathfrak{C}_k$. Since ν is a probability measure, it follows that $\int_C Z(x)\ \mathrm{d}\,\mu(x) < \infty$. If $\mathfrak{C} \subset \mathscr{C}$ is the collection of all sets A with $\nu(A) = \int_A Z(x)\ \mathrm{d}\,\mu(x)$, then $\cup_{k=1}^{\infty}\mathfrak{C}_k \subset \mathfrak{C}$. Further, if $A_n \in \mathfrak{C},\ n \geq 1,$ and $A_n \uparrow A$, then

$$\begin{aligned}
\nu(A) &= \lim_{n\to\infty} \nu(A_n) = \lim_{n\to\infty} \int_{A_n} Z(x)\ \mathrm{d}\,\mu(x) \\
&= \lim_{n\to\infty} \int_C Z(x)\chi_{A_n}(x)\ \mathrm{d}\,\mu(x) \\
&= \int_C Z(x)\chi_A(x)\ \mathrm{d}\,\mu(x) \text{ (by the bounded convergence theorem)} \\
&= \int_A Z(x)\ \mathrm{d}\,\mu(x).
\end{aligned}$$

This shows that $\mathfrak{C}$ is a monotone class. Hence (ref. Theorem D, p56, [H]) $\sigma(\cup_{k=1}^{\infty}\mathfrak{C}_k) \subset \mathfrak{C}$. With this the proof is complete that for every $A \in \mathscr{C}$, $\nu(A) = \int_A Z(x)\ \mathrm{d}\,\mu(x)$. That $\nu \ll \mu$ and that Z is the R-N derivative of ν with respect to μ follow from this. □

(xvii) The property of the Wiener measure μ that we establish in this section is that its support (ref. Sec. **(1.8.2)** is the whole of C.

From the definition of support of a pm, it follows that the support of μ would be C if we show that the μ-measure of every open subset of C is positive. Showing this is equivalent to showing that every open sphere has positive μ-measure. It is clearly enough to show that open spheres with centers at the points of a dense subset have positive μ-measures. The collection $\mathcal{P}$ of all polynomials is a well known dense subset of C. The collection $\mathcal{S}$ of all functions $f \in C$ with a continuous differential coefficient contains $\mathcal{P}$ and hence is a dense subset of C. If A_r denotes the open sphere with center at the null element and radius r, then it is enough to show that $\mu(A_r + f) > 0$ for every $f \in \mathcal{S}$. By (xvi), $\mu(A_r + f) = \nu(A_r) = \int_{A_r} Z_f(x)\, \mathrm{d}\,\mu(x)$. Since Z_f is positive valued $wp1$, it follows from this that it is enough to show that $\mu(A_r) > 0$ for every $r > 0$. This we proceed to do now.

If there exists r such that $\mu(A_r) = 0$, then it would follow that $\mu(A_s + f) = 0$ for all $f \in \mathcal{S}$ and for all $s \leq r$. These spheres $A_s + f,\ f \in \mathcal{S},\ 0 < s \leq r$ constitute a neighborhood system for the toplogy of $(C,\ \rho)$. Since C is separable, this implies that the μ-measure of every open subset of C is zero. That the μ-measure of every set in $\mathscr{C}$ is zero will follow from this since μ is an outer regular regular measure (ref. **(1.9.2)**). Hence, necessarily, every open subset of C has positive μ-measure. That C is the support of μ follows now. □

Remark. The proof of the claim in (xvii) was by arriving at a contradiction if $\mu(A_r) = 0$ for some $r > 0$. That $\mu(A_r) > 0$ for all $r > 0$ can be established directly if we make use of the fact that the *df* G of $\|x\|$ or $\max\limits_{0 \leq t \leq 1} |x(t)|$ under μ is:

$$G(u) = \frac{4}{\pi} \sum_{k=0}^{\infty} \frac{(-1)^k}{2k+1} e^{-\frac{(2k+1)\pi^2}{8u^2}}, \quad u > 0 \text{ (ref. (5.25))}.$$

Hence,

$$\mu(A_r) = G(r) \geq \frac{4}{\pi}\{e^{-\frac{\pi^2}{8r^2}} - \frac{1}{3} e^{-\frac{9\pi^2}{8r^2}}\} > 0 \text{ for every } r > 0.$$

(xviii) For a function f, define $D^* f(t) = \overline{\lim\limits_{h \downarrow 0}} \frac{f(t+h) - f(t)}{h}$ and $D_* f(t) = \underline{\lim\limits_{h \downarrow 0}} \frac{f(t+h) - f(t)}{h}$.

Fix $t_0 \geq 0$. Then $wp1$, $D^* B(t_0) = +\infty$ and $D_* B(t_0) = -\infty$.
Proof. Since $B(t + t_0) - B(t_0),\ t \geq 0$ is a standard BMP we see that it is enough to prove the claim for $t_0 = 0$.

$$D^* B(0) \geq \overline{\lim_{n \to \infty}} \frac{B(\frac{1}{n}) - B(0)}{\frac{1}{n}} = \overline{\lim_{n \to \infty}}\, n B(\frac{1}{n}) = \overline{\lim_{n \to \infty}} \sqrt{n} \frac{B(n)}{\sqrt{n}}$$

(since the two processes $(\frac{B(t)}{\sqrt{t}},\ t > 0)$ and $(\sqrt{t} B(\frac{1}{t}),\ t > 0)$ are both standard BMPs). Hence $D^* B(0)$ is infinite since (ref. (xiii)) $\overline{\lim\limits_{n \to \infty}} \frac{B(n)}{\sqrt{n}} = \infty\ wp1$.

Similar steps lead to the second assertion.

5.9.1 Convention

In view of **(5.9)**(xiii) above, we agree that the BMP B is defined on $(\Omega, \mathscr{A}, P)$ and $\overline{\lim_{n\to\infty}}\ B(n, \omega) = \infty$; $\underline{\lim_{n\to\infty}}\ B(n, \omega) = -\infty$ for all $\omega \in \Omega$.

5.10 Hölder continuous spaces

Definition 5.10.1. Let $0 < \alpha \leq 1$. We define the Hölder space $H_\alpha = H_\alpha[0, 1]$ to be the space of functions x vanishing at 0 such that

$$\|x\|_\alpha = \sup_{0\leq s,t\leq 1,\ s\neq t} \frac{|x(t) - x(s)|}{|t - s|^\alpha} < \infty \tag{5.40}$$

Clearly $H_\alpha \subset C$. Define the Hölderian modulus of continuity of x:

$$\omega_\alpha(x;\ \delta) = \sup_{0<|t-s|<\delta} \frac{|x(t) - x(s)|}{|t - s|^\alpha} \tag{5.41}$$

and the subspace H_α^0 of H_α by

$$x \in H_\alpha^0 \Leftrightarrow x \in H_\alpha \text{ and } \lim_{\delta\to 0} \omega_\alpha(x,\ \delta) = 0. \tag{5.42}$$

We list some characteristic properties of these spaces.

Q10 (i) $\|.\|_\alpha$ as defined above qualifies to be a norm function.
(ii) under this norm, H_α is a non-separable Banach space and H_α^0 is a separable Banach space.
(iii) H_α^0 is a closed subspace of H_α.
Let $0 < \alpha_1 < \alpha_2 < 1$ arbitrary.
(iv) Trivially $H_{\alpha_2} \subset H_{\alpha_1}$. Further if $x \in H_{\alpha_2}^0$ then $\|x\|_{\alpha_1} \leq \|x\|_{\alpha_2}$.
(v) H_{α_2} endowed with the norm $\|.\|_{\alpha_1}$ is a separable Banach space, embedded topologically in H_{α_1}.
We will call H_α^0 the α-Hölder continuous space.
(vi) All polygonal functions belong to H_α^0, $0 < \alpha < 1$.
Only the function identically zero belongs to H_1^0.
(vii) $wp1$ the sample functions of a BMP lie in H_γ^0 for every γ, $0 < \gamma < \frac{1}{2}$.
To see this fix γ. Let $\gamma < \theta < \frac{1}{2}$. Since, $wp1$,

$$\sup_{\substack{0\leq s,\ t\leq 1\\ s\neq t}} \frac{|x(t,\ \omega) - x(s,\ \omega)|}{|t - s|^\theta} \leq M_\theta(\omega) \quad \text{(ref. ((\textbf{5.7.5}))}$$

it follows that $\omega(x(., \omega), \delta) \leq M_\theta(\omega)|t-s|^{\theta-\gamma} \leq M_\theta(\omega)\delta^{\theta-\gamma} \to 0$ as $\delta \to 0$.

(viii) Since $(H^0_\alpha, \|.\|_\alpha)$ is a Banach space, $\|.\|_\alpha$ is a continuous functional on H^0_α. Hence $\{x : x \in H^0_\alpha, \|x\|_\alpha \leq c\}$ is a closed set for every $c > 0$.

We have the following facts about compact sets in H^0_α.

(ix) Let $\mathbb{Q}$ be a bounded subset of $H^0_{\alpha_2}$. i.e. $\sup_{x\in\mathbb{Q}} \|x\|_{\alpha_2} < \infty$. Then $\mathbb{Q}$ is precompact in $H^0_{\alpha_1}$. i.e. $\mathbb{Q}$ has compact closure in $H^0_{\alpha_1}$. This property and that in (viii) imply that $\{x : x \in H^0_{\alpha_1}, \|x\|_{\alpha_2} \leq c\}$ is a compact subset of $H^0_{\alpha_1}$ for every $c > 0$.

(x) Let $\mathcal{M}$ be a family of *pms* on the Borel σ-field of $(H^0_\alpha, \|\cdot\|_\alpha)$, $0 < \alpha < 1$. Then this family is tight if for every $\varepsilon > 0$ there exists $\eta > 0$ with $\alpha + \eta < 1$ and $c < \infty$ such that $\mu\{x : x \in H^0_{\alpha_1}, \|x\|_{\alpha+\eta} \leq c\} > 1 - \varepsilon$ for every $\mu \in \mathcal{M}$.

5.10.2. It is instructive to present in alternative form certain results already proved in this chapter : The set of all functions $x \in C$ with the following properties is of full Wiener measure.

(i) x is monotone in no interval (ref. **(5.9)**(ix))

(ii) x is of unbounded variation in every interval (ref. **(5.9)**(x))

(iii) x is holder continuous of every order γ, $0 < \gamma < \frac{1}{2}$ (ref. **(5.7.5)**).

Chapter 6

The law of the iterated logarithm

6.1 Introduction

For B, a standard BMP, we showed in Sec. **(5.9)** that $\frac{B(t)}{t} \stackrel{wp1}{\longrightarrow} 0$, as $t \to \infty$, that $wp1$ $\overline{\lim_{t\to\infty}} \frac{B(t)}{\sqrt{t}} = \infty$ and $\underline{\lim_{t\to\infty}} \frac{B(t)}{\sqrt{t}} = -\infty$. Then we raised the question : For which function $a(t)$ (necessarily tending to infinity with t) will we have $-\infty < \underline{\lim_{t\to\infty}} \frac{B(t)}{a(t)} \leq \overline{\lim_{t\to\infty}} \frac{B(t)}{a(t)} < \infty$. In this chapter we answer this question as applied not only to a BMP but also, as applied to the partial sum sequence (S_n) of iid variables (X_n).

Consider a simple symmetric random walk . i.e. Consider the partial sum sequence (S_n) where the X_ns are iid variables taking the values ± 1 with probability $\frac{1}{2}$ each. The values of S_n can be thought of as the position of a particle moving, say, every second one step to the right or one step to the left with probability $\frac{1}{2}$ each. By the CLT the position of the particle, after a long time from the start, will be described nearly well by a zero mean normal distribution with variance n. Thus after n seconds, n large, the particle will be distant from the starting point $1.96\sqrt{n}$ steps or more with probability 0.05. i.e. in repeated experiments, each stretching over a long time, the proportion of times the particle would have travelled $1.96\sqrt{n}$ steps or more would be 5%. If the random walk (S_n) is observed over an infinite length of time, then by the $SLLN$, we can claim that even though $|S_n|$ may assume values larger than n it can happen only a finite number of times. As will be seen shortly, the law of the iterated logarithm (LIL) is a refinement of the $SLLN$.

The LIL as applied to the partial sums (S_n) of the sequence (X_n) of iid variables with $\mathbb{E}X_1 = 0$ and $\mathbb{E}X_n^2 = 1$ would read:

$$wp1, \qquad \overline{\lim_{n\to\infty}} \frac{S_n}{\sqrt{2n \log\log n}} = 1 \text{ and } \underline{\lim_{n\to\infty}} \frac{S_n}{\sqrt{2n \log\log n}} = -1. \qquad (6.1)$$

It describes the magnitude of the fluctuations of the sequence (S_n). The name *Law of the Iterated Logarithm* derives from the rarely occurring norming constant $\log\log n$.

A strengthened version of (6.1) is:

$$wp1,\ [-1,\ 1] \text{ is the limit set of the sequence } \left(\frac{S_n}{\sqrt{2n\log\log n}}\right). \tag{6.2}$$

The phrase 'limit set' needs to be defined.

Sequence (a_n) in $(M,\ d)$ has a for a limit point if there exists a subsequence (a_{n_r}) converging to a as $r \to \infty$. Or, equivalently, for each $\varepsilon > 0,\ d(a_n,\ a) < \varepsilon$ for infinitely many n. A subset $E \subseteq M$ is called the limit set of a sequence (a_n) if none of its subsequences converge to an element outside E and every element $x \in E$ is the limit of one of its subsequences. i.e., E is the set of all the limit points of the convergent subsequences of (a_n). We define the limit points of a sequence $(\mathbb{Z}_n)$ of random elements in $(M,\ d)$ *only for those sequences whose tail σ-field is the trivial σ-field.* Any convergent subsequence will then converge not to a random element but to a non-random point in M. Equivalently, we define the the limit points of the sequence $(\mathbb{Z}_n)$ only if every one of its convergent sub sequences converges $wp1$ to a non-random element of M. Since $\{\omega\ :\ d(\mathbb{Z}_n(\omega),\ a) < \varepsilon \text{ i.o.}\}$ is a tail event of the $\mathbb{Z}_n$ sequence, its probability is either zero or 1. This is so for every $a \in M$ and every $\varepsilon > 0$.

(i) Definition.

An element $a \in M$ will be called a limit point of the sequence $(\mathbb{Z}_n)$ if there exists a set A_a such that $P(A_a) = 1$ and such that $\omega \in A_a$ implies $d(\mathbb{Z}_{n_r}(\omega),\ a) \to 0$ for some subsequence $(n_r = n_r(\omega,\ a))$. Or, equivalently, for every $\varepsilon > 0$, there exists a set $A_{a,\varepsilon}$ with $P(A_{a,\varepsilon}) = 1$ such that if $\omega \in A_{a,\varepsilon}$, then $d(\mathbb{Z}_n(\omega),\ a) < \varepsilon$ for infinitely many n. Taking $A_a = \cap_{k=1}^{\infty} A_{a,\frac{1}{k}}$ and noting that $P(A_a) = 1$, we can say a is a limit point of the sequence $(\mathbb{Z}_n)$ if there exists set A with $P(A) = 1$ such that if $\omega \in A$, then $d(\mathbb{Z}_n(\omega),\ a) < \varepsilon$ for infinitely many n for every $\varepsilon > 0$.

The limit set of a sequence $(\mathbb{Z}_n)$ of random elements in M is said to be the set E if it contains all its limit points and if there exists a set $A \in \mathscr{A}$ such that $P(A) = 1$ and such that for every $\omega \in A$, the sequence $(\mathbb{Z}_n(\omega))$ of elements in M has E for its limit set.

(ii) Note.

It is not enough that each point of E is a limit point of the sequence (Z_n) for E to be called the limit set of the sequence. The existence of the set A described above is essential. Since each point $a \in E$ is a limit point, we have sets A_a, $a \in E$ with $P(A_a) = 1$. But if E is uncountable, $A = \cap_{a \in E} A_a$ may not be a measurable set.

(iii)**Theorem**.

The limit set A of a sequence $(\mathbb{Z}_n)$ is always a closed set.

Proof. For, if possible, let $a_n \in A$, $a_n \to a$, $a \notin A$. Given $\varepsilon > 0$, we can find N such that $a_N \in A$ and $d(a_N, a) < \varepsilon$. Since $\left| d(\mathbb{Z}_n(\omega), a) - d(\mathbb{Z}_n(\omega), a_N) \right| \leq d(a_N, a) < \varepsilon$, we have : $(d(\mathbb{Z}_n(\omega), a_N) < \varepsilon) \Rightarrow (d(\mathbb{Z}_n(\omega), a) < 2\varepsilon)$. Hence $(d(\mathbb{Z}_n(\omega), a_N) < \varepsilon \text{ i.o.})) \Rightarrow (d(\mathbb{Z}_n(\omega), a) < 2\varepsilon \text{ i.o.})$. Since a_N is a limit point of the $\mathbb{Z}_n$-sequence, $P(d(\mathbb{Z}_n(\omega), a_N) < \varepsilon \text{ i.o.})) = 1$. This implies $P(d(\mathbb{Z}_n(\omega), a) < 2\varepsilon \text{ i.o.})) = 1$. This being true fo every $\varepsilon > 0$, we conclude that a is a limit point of the $\mathbb{Z}_n$-sequence. i.e., $a \in A$. Thus A is a closed set. □

(iv)**Theorem**. If each point of $\bar{E}$ is a limit point of the sequence $(\mathbb{Z}_n)$, if no point outside $\bar{E}$ is its limit point and if $\bar{E}$ is separable, then $\bar{E}$ is the limit set of the sequence.

Proof. We must produce a set A with $P(A) = 1$ such that, given $\varepsilon > 0$ and $a \in E$, for every $\omega \in A$, the inequality $d(\mathbb{Z}_n(\omega), a) < \varepsilon$ is valid for infinitely many n. We argue as follows.

Let S be a separability set for $\bar{E}$. To each $s \in S$ there corresponds a set A_s such that $P(A_s) = 1$ and such that $\omega \in A_s$ implies s is a limit point of the sequence $(\mathbb{Z}_n(\omega))$. Let $A = \cap_{s \in S} A_s$, so $P(A) = 1$. $\omega \in A$ would then imply that for each $\varepsilon > 0$ and each $s \in S$, $d(\mathbb{Z}_n(\omega), s) < \frac{\varepsilon}{2}$ for infinitely many n. Let $a \in E \sim S$. Given $\varepsilon > 0$, we can find $s_a \in S$ such that $d(s_a, a) < \frac{\varepsilon}{2}$ for all n large. Since $d(\mathbb{Z}_n(\omega), a) \leq d(\mathbb{Z}_n(\omega), s_a) + d(s_a, a)$, $d(\mathbb{Z}_n(\omega), s_a) < \frac{\varepsilon}{2}$ implies $d(\mathbb{Z}_n(\omega), a) < \varepsilon$. It follows from this that $d(\mathbb{Z}_n(\omega), a) < \varepsilon$ for infinitely many n, as was to be proved. □

(v) Examples.

Let Y be a rv such that $P(Y = 0) = \frac{1}{2} = P(Y = 1)$ and let $Y_n = Y$, $n \geq 1$. We note that for each ω, $Y_n(\omega)$ is a convergent sequence. Yet this sequence does not come under the purview of our definition of limit set since the tail σ-field of the sequence, being the σ-field generated by Y, is not the trivial σ- field. We can not say 1 and 0 are the limit points of the sequence. Since $P(Y_n = 1 \text{ i.o.}) = P(Y = 1) \neq 1$, 1 is not a limit point. Similarly 0 is not a limit point.

Let (X_n) be a sequence of independent rvs with $P(X_n = 1) = \frac{1}{2} - \frac{1}{(n+1)^2} = P(X_n = 2)$ and $P(X_n = 3) = \frac{2}{(n+1)^2}$, $n \geq 1$. Use independence and the divergence of the series $\sum_{n=1}^{\infty} P(X_n = 1)$ and $\sum_{n=1}^{\infty} P(X_n = 2)$, to conclude that $P(X_n = 1 \text{ i.o.}) = 1 = P(X_n = 2 \text{ i.o.})$. Thus 1 and 2 are limit points of the sequence (X_n). 3 is not a limit point (even though the value 3 is assumed by each X_n with positive probability), since $\sum_{n=1}^{\infty} P(X_n = 3) < \infty$, so $P(X_n = 3 \text{ i.o.}) = 0$. It follows that $\{1, 2\}$ is the limit set of the sequence.

Let the X_ns be *iid rvs*, $S_n = \sum_1^n X_k$ and let $a_n \to \infty$. Define the polygonal function $x_n : x_n(\frac{i}{n}) = \frac{S_i}{a_n}$, $0 \le i \le n$; x_n linear elsewhere on $[0, 1]$, $S_0 \equiv 0$. For $k \ge 1$, define the polygonal function $x_{n,k} : x_{n,k}(0) = 0$, $x_{n,k}(t) = x_n(t)$ for $t \ge \frac{k}{n}$ and linear over $[0, \frac{k}{n}]$ and note that x_n, $x_{k,n}$ are random elements in C. Since $a_n \to \infty$, $\|x_n - x_{n,k}\| \to 0$ as $n \to \infty$. This shows that the limit of a convergent subsequence of the x_n sequence is measurable with respect to the σ-field generated by S_ν, $\nu \ge k$. Since this is true for every $k \ge 1$, it follows that the limit of every convergent subsequence of the x_n sequence is measurable with respect to the tail σ-field of the S_n sequence, which we know is the trivial σ-field. This claim is valid since this tail σ-field is a sub σ- field of the σ-field $\mathcal{V}$ of symmetric events and $\mathcal{V}$ is a trivial σ-field under the product measure μ_T when the X_k's are identically distributed. (ref. **(4.1.4)**(v) and **(4.1.10)**).

(vi) Parallel to the functional central limit theorem ($FCLT$) there is a versatile version of the LIL, called the functional LIL ($FLIL$) or Strassen form of the LIL which we describe now. Let (X_n) be a sequence of *iid rvs* with zero means and unit variances. (S_n) is their partial sum sequence. Define on the parameter space $[0, 1]$ the process x_n as in part (v) above, with a_n replaced by $\sqrt{2n \log\log n}$. The $FLIL$ states:

$$wp1, \text{ the limit set of the sequence } (x_n) \text{ is the set } K_1 \tag{6.3}$$

where, for $a > 0$,

$$K_a = \{x \in C,\ x \text{ absolutely continuous and } \int_0^1 \left(\frac{\mathrm{d}x(t)}{\mathrm{d}t}\right)^2 \mathrm{d}t \le a^2\} \tag{6.4}$$

Proofs of (6.1), (6.2) and (6.3) will be given in the following sections.

Clearly (6.2) $\Rightarrow$ (6.1).

Suppose (6.3) has been proved. To claim that this implies (6.2), the argument is as follows. The function f defined on C through $f(x) = x(1)$ is a continuous functional. Hence the limit set of $(x_n(1))$ i.e. of ($V_n = \frac{S_n}{\sqrt{2n \log\log n}}$) is

$f(K)$, the image of K under f. Let us evaluate $f(K)$. If $x \in K$, then

$$|x(1)| = |x(1) - x(0)| = |\int_0^1 \frac{\mathrm{d}\,x(t)}{\mathrm{d}\,t}\,\mathrm{d}\,t|$$

$$\leq \int_0^1 |\frac{\mathrm{d}\,x(t)}{\mathrm{d}\,t}|\,\mathrm{d}\,t = \int_0^1 |\frac{\mathrm{d}\,x(t)}{\mathrm{d}\,t}| \times 1\,\mathrm{d}\,t$$

$$\leq \left(\int_0^1 (\frac{\mathrm{d}\,x(t)}{\mathrm{d}\,t})^2\,\mathrm{d}\,t\right)^{\frac{1}{2}} \left(\int_0^1 1^2\,\mathrm{d}\,t\right)^{\frac{1}{2}} \quad \text{(ref. **(2.6.5)**(iii))} \leq 1.$$

The function $y_1(t) = t$ belongs to K and $f(y_1) = y_1(1) = 1$. Similarly the function $y_2(t) = -t$ belongs to K and $f(y_2) = -1$. These results imply: $wp1$, $\underline{\lim}_{n\to\infty} V_n = -1$ and $\overline{\lim}_{n\to\infty} V_n = 1$. i.e. $1, -1 \in f(K)$ and $f(K) \subseteq [-1, 1]$. Now if $x, y \in K$ and if $\alpha \geq 0, \beta \geq 0, \alpha+\beta = 1$, then $\alpha x+\beta y \in K$. For, let $x, y \in K$. We have

$$\int_0^1 (\frac{\mathrm{d}\,(\alpha x(t) + \beta y(t))}{\mathrm{d}\,t})^2\,\mathrm{d}\,t$$

$$\leq \alpha^2 \int_0^1 (\frac{\mathrm{d}\,x(t)}{\mathrm{d}\,t})^2\,\mathrm{d}\,t + \beta^2 \int_0^1 (\frac{\mathrm{d}\,y(t)}{\mathrm{d}\,t})^2\,\mathrm{d}\,t$$

$$+ 2\alpha\beta \int_0^1 |\frac{\mathrm{d}\,x(t)}{\mathrm{d}\,t}||\frac{\mathrm{d}\,y(t)}{\mathrm{d}\,t}|\,\mathrm{d}\,t$$

$$\leq \alpha^2 + \beta^2 + 2\alpha\beta \left(\int_0^1 (\frac{\mathrm{d}\,x(t)}{\mathrm{d}\,t})^2\,\mathrm{d}\,t\right)^{\frac{1}{2}}$$

$$\left(\int_0^1 (\frac{\mathrm{d}\,y(t)}{\mathrm{d}\,t})^2\,\mathrm{d}\,t\right)^{\frac{1}{2}} \quad \text{(ref. **(2.6.5)**(iii))}$$

$$\leq \alpha^2 + \beta^2 + 2\alpha\beta = (\alpha + \beta)^2 = 1.$$

This shows that $\alpha x + \beta y \in K$. Let y_1, y_2 be the earlier defined functions. Then $\alpha y_1 + \beta y_2 \in K$. This implies $f(\alpha y_1 + \beta y_2) \in [-1, 1]$. i.e. $\alpha - \beta \in$

$[-1,\ 1]$. Now, $\{\alpha - \beta\ :\ \alpha \geq 0,\ \beta \geq 0,\ \alpha + \beta = 1\} = \{2\alpha - 1\ :\ 0 \leq \alpha \leq 1\} = [-1,\ 1]$. i.e., $f(K) = [-1,\ 1]$. Thus (6.3) $\Rightarrow$ (6.2).

It is obvious now that we should aim to prove (6.3). We will give, in **(6.2.11)**, a direct proof of a more general version of (6.3).

The method of proof we have chosen rests on what is called the Skorohod representation which, given a *df* F, consists in finding a rv τ such that $B(\tau)$ has *df* F where, as usual, B is a standard BMP. Proof of the existence of τ and the (much needed) study of the process $B(\tau + t),\ t \geq 0$ lean on the concept and basic properties of conditional expectation. In the circumstances our plan of proof is as follows.

In the rest of the present section i.e in **(6.1.1)-(6.1.3)** we study the functions φ which are non-decreasing and which will take the place of $\sqrt{2 \log \log n}$. We study too the set K_a (ref. **(6.1.4)**). In section **(6.2)**, FLIL for the BMP is established. i.e. we show that (6.3) is true when $x_n(t)$ is replaced by $\frac{B(nt)}{\sqrt{n\varphi(n)}}$ with K suitably changed. In **(6.3)** and **(6.5)**, we explain the concept of Conditional Expectation and establish some of its basic properties. Many useful theorems are proved though they are not relevant to the study of Skorohod representation and to that of the process $B(\tau + t)$. In **(6.6)** we study Skorohod representation and consequences. In **(6.9.2)** we prove (6.3) (Bulinskii version).

6.1.1. Let φ be an arbitrary non-decreasing function defined over $(0,\ \infty)$ with $\lim_{n\to\infty} \varphi(n) = \infty$. Given $c > 1,\ \theta \geq 1$, write $n_k = [c^{k^\theta}]$ and define

$$I(\varphi,\ r,\ \theta,\ c) = \sum_{k=0}^{\infty} e^{\frac{-r\varphi^2(n_k)}{2}}. \tag{6.5}$$

Lemma 6.1.2. *For* $r,\ \theta$ *fixed,* $r > 0,\ \theta > 0$*, if* $I(\varphi,\ r,\ \theta,\ c_0) < \infty$ *for some* $c_0 > 1$*, then* $I(\varphi,\ r,\ \theta,\ c) < \infty$ *for all* $c > 1$.

Proof. Let $n_k = [c^{k^\theta}]$ and $m_k = [c_0^{k^\theta}]$.

If $c > c_0$ then the finiteness of $I(\varphi,\ r,\ \theta,\ c)$ follows trivially from the relation $I(\varphi,\ r,\ \theta,\ c) \leq I(\varphi,\ r,\ \theta,\ c_0)$. Let now $1 < c < c_0$. Let j be the least integer such that $c^{j^\theta} \geq c_0$. We note that $n_{jk} \geq m_k$. We have

$$I(\varphi,\ r,\ \theta,\ c) = \sum_{k=0}^{\infty} \sum_{q=jk}^{j(k+1)-1} e^{-\frac{r}{2}\varphi^2(n_q)} \leq \sum_{k=0}^{\infty} j e^{-\frac{r}{2}\varphi^2(n_{jk})}$$

$$\leq j \sum_{k=0}^{\infty} e^{-\frac{r}{2}\varphi^2(m_k)} = jI(\varphi,\ r,\ \theta,\ c_0) < \infty.$$

□

6.1.3. Let $I(\varphi,\ r,\ \theta,\ c)$ be as defined above. Let

$$r^2(\theta) = r^2(\varphi,\ \theta) = \inf\{r > 0 : I(\varphi,\ r,\ \theta,\ c) < \infty\}. \tag{6.6}$$

$r^2(1)$ will be written R^2. For $r \le R$, let

$$\theta(r) = \theta(\varphi,\ r) = \inf\{\theta \ge 1 : I(\varphi,\ r,\ \theta,\ c) < \infty\}. \tag{6.7}$$

We note $\theta(R) = 1$.

The following assertions are easily verified.
(i) If $\varphi(n) = \sqrt{2\log n}$, then $R = 0$.
(ii) If $\varphi(n) = \sqrt{2\log\log n}$, then $R = 1$.
(iii) If $\varphi(n) = \sqrt{2\log\log\log n}$, then $R = \infty$.

Taking $\varphi(n) = \sqrt{2\log\log n}$ and $\theta = 1$ it may be noted that $\sum\limits_k e^{-\frac{R}{2}\varphi^2(n_k)}$ is not a convergent series.

If $\overline{\lim\limits_{n\to\infty}} \dfrac{\varphi(n)}{\sqrt{2\log\log n}} = b < \infty$, then $R > 0$;

if $\underline{\lim\limits_{n\to\infty}} \dfrac{\varphi(n)}{\sqrt{2\log\log n}} = a > 0$, then $R < \infty$.

Proof. The hypothesis implies that given $\varepsilon > 0$,

$$\frac{\varphi^2(n)}{2\log\log n} \le b^2 + \varepsilon$$

for all large n. Hence

$$\sum_k e^{-\frac{r}{2}\varphi^2(c^k)} \ge \sum_k e^{-\frac{r}{2}(b^2+\varepsilon)2\log\log c^k} = \sum_k \frac{1}{(k\log c)^{r(b^2+\varepsilon)}}.$$

Hence $\sum_k e^{-\frac{r}{2}\varphi^2(c^k)}$ will diverge if $r < \frac{1}{b^2}$. Thus $R \ge \frac{1}{b} > 0$. The assertion $R < \infty$ can be proved similarly.

That the converse is not true is brought out by the following example. Let $\varphi(n) = \sqrt{2\log\log n}$ if $n = 2^k$ for $k = 1,\ 2, \ldots$, log being to base 2. Define

$$\varphi(2^{2^{2^{2k}}} + j) = \varphi(2^{2^{2^{2k}}})\log k\ (\sim q2^{\frac{k}{2}}\log k) \text{ for } 1 \le j \le m_k$$

where m_k is the least j such that

$$\varphi(2^{2^{2^{2k}}})\log k \le \sqrt{2\log\log(2^{2^{2^{2k}}} + j)}.$$

Such a $j > 1$ exists since for $j = 2^{2^{2^{2k+2}}} - 2^{2^{2^{2k}}}$ the value is asymptotically $2^{k+\frac{3}{2}}$. We have

$$e^{-\frac{r}{2}\varphi^2(2^k)} = \frac{1}{(\log 2^k)^r}.$$

Hence $R = 1$ but

$$\lim_{k\to\infty} \frac{\varphi(2^{2^{2^{2k}}}+1)}{\sqrt{2\log\log(2^{2^{2^{2k}}}+1)}} = \infty.$$

In what follows, R is assumed to be finite and positive, a sufficient condition for which is:

$$0 < \varliminf_{n\to\infty} \frac{\varphi(n)}{\sqrt{2\log\log n}} \leq \varlimsup_{n\to\infty} \frac{\varphi(n)}{\sqrt{2\log\log n}} < \infty. \tag{6.8}$$

□

6.1.4. In **(6.9.2)** we will be proving (6.3). The proof does not use the information that K_1 is a closed set. Thus **(6.9.2)** and **(6.1)**(iii) will imply that K_1 is a closed set. Since K_1 is a totally specified subset of C, a direct proof of this property, purely as an excercise in Analysis, should be possible. For such a proof ref. **(2.7.2)** and **(2.7.3)**)

6.2 Strassen LIL for Brownian Motion Processes: Bulinskii version

Let

$$f_n(t) = f_n(t, \omega) = \frac{B(nt,\,\omega)}{\sqrt{n}\varphi(n)}, \quad 0 \leq t \leq 1.$$

To be able to talk about the limit points of the sequence (f_n), we first show that the tail σ-field of the sequence is the trivial σ-field. Let $B_k(t) = B(kt) - B(\overline{k-1}t)$, $1 \leq k \leq n$, $B(0) \equiv 0$, and note $B(nt) = \sum_{k=1}^{n} B_k(t)$. Thus the process $B(nt)$ is the sum of n independent standard BMPs on $[0,\ 1]$. Each process $B_k(t)$ is a rE in C. Hence the tail σ-field of the sequence (f_n) is the trivial σ-field (ref. (**(4.1.7)**)). The limit points of the sequence (f_n) will therefore be non-random. They will, of course, be members of C.

Let R be as defined in **(6.1.3)**. As already stated, our object in the present section is to prove that K_R is the limit set of the sequence (f_n). This consists in showing (i) that $P(f_n \notin K_R \text{ i.o.}) = 0$. (This is done in **(6.2.7)**) and (ii) that $wp1$, each point in K_R is a limit point of the f_n sequence. (This is done in

(6.2.8)). The exceptional set may depend on the limit point). Then complete the proof by invoking the separability of the set K_R.

Now, to carryout this plan, we first approximate the continuous function f_n by the more manageable polygonal functions $\hat{f}_{n,m}$ which coincide with f_n at the points $t = \frac{i}{m}$, $1 \leq i \leq m$ and are linear elsewhere. Explicitly, $\hat{f}_{n,m}(\frac{i}{m}) = f_n(\frac{i}{m})$ at $t = \frac{i}{m}$, $1 \leq i \leq m$ and linear elsewhere on $[0, 1]$. This is done in **(6.2.4)**. In **(6.2.3)** we obtain a useful bound for $P(\hat{f}_{n,m} \notin K_R)$, n, m fixed. Using **(6.2.3)** and **(6.2.4)**, we find a bound for $P(f_n \notin K_{R,\varepsilon})$ where $K_{a;\varepsilon}$ denotes the ε-neighbourhood of K_a under the uniform metric : $K_{a;\varepsilon} = \{x \ : \ x \in C, \ \rho(x, \ K_a) < \varepsilon\}$.

Lemma 6.2.1. *Let $0 < a < b < c$ be three arbitrary numbers. Then there exist $\alpha > 0, \ \beta > 0$ such that $\frac{1}{b} = \frac{\alpha}{a} + \frac{\beta}{c}$ and such that $\alpha + \beta = 1$.*

Proof. We note that $\frac{b}{a} > 1$ and $\frac{b}{c} < 1$. Hence the lines $x + y = 1$ and $\frac{b}{a}x + \frac{b}{c}y = 1$ intersect at a point $(\alpha, \ \beta)$ in the first quadrant. These numbers α and β are the desired ones. □

Lemma 6.2.2. *Let $\alpha \geq 0, \ \beta \geq 0, \ \alpha+\beta = 1$. If $x, y \in K_a$, then $\alpha x+\beta y \in K_a$. If $u, \ v \in K_{a;\varepsilon}$, then $\alpha u + \beta v \in K_{a;\varepsilon}$.*

Proof. For proof that $\alpha x + \beta y \in K_a$ refer to **(6.1)** where a was taken as equal to 1.

Suppose now $u, \ v \in K_{a;\varepsilon}$. Hence there exist $x, \ y \in K_a$ such that $\|u - x\| < \varepsilon$ and $\|v - y\| < \varepsilon$. By the earlier part, $\alpha x + \beta y \in K_a$. Since $\|(\alpha u + \beta v) - (\alpha x + \beta y)\| \leq \alpha\|u - x\| + \beta\|v - y\| < \alpha\varepsilon + \beta\varepsilon = \varepsilon$, it follows that $\alpha u + \beta v \in K_{a;\varepsilon}$. □

Theorem 6.2.3. *Let $s > 1$. Then $P(\frac{1}{s}\hat{f}_{n,m} \notin K_R) \leq q_m e^{-\frac{1}{2}sR^2\varphi^2(n)}$.*

Proof. We note that over $\frac{i}{m} \leq t \leq \frac{i+1}{m}$,

$$\frac{\mathrm{d}\,\hat{f}_{n,m}(t)}{\mathrm{d}t} = m\{\frac{B(\frac{n(i+1)}{m}) - B(\frac{ni}{m}))}{\sqrt{n}\varphi(n)}\} = \frac{m}{\sqrt{n}\varphi(n)}Y_i$$

where Y_i, $0 \leq i \leq m - 1$ are *iid* normal variables with zero means and with common variance $\frac{n}{m}$. Hence

$$\int_0^1 (\frac{\mathrm{d}\hat{f}_{n,m}}{\mathrm{d}t})^2 \,\mathrm{d}t = \sum_{i=1}^{m} \int_{\frac{i-1}{m}}^{\frac{i}{m}} (\frac{\mathrm{d}\hat{f}_{n,m}}{\mathrm{d}t})^2 \,\mathrm{d}t$$

$$= \sum_{i=1}^{m} \frac{m}{n\varphi^2(n)} Y_{i-1}^2 = \sum_{i=1}^{m} \frac{1}{\varphi^2(n)} \xi_i^2$$

where the ξs are iid standard normal variables. Thus if χ_m^2 denotes a χ^2 variable (ref. **(4.2)**) with m degrees of freedom, then

$$\begin{aligned} P(\frac{1}{s}\hat{f}_{n,m} \notin K_R) &= P\left(\int_0^1 (\frac{1}{s}\frac{\mathrm{d}\hat{f}_{n,m}(t)}{\mathrm{d}t})^2 \,\mathrm{d}t > R^2 \right) \\ &= P(\chi_m^2 > s^2R^2\varphi^2(n)) \\ &\sim q_m (s^2R^2\varphi^2(n))^{\frac{m}{2}} e^{-\frac{1}{2}s^2R^2\varphi^2(n)} \\ &= q_m (s^2R^2\varphi^2(n))^{\frac{m}{2}} e^{-\frac{1}{2}(s^2-s)R^2\varphi^2(n)} e^{-\frac{1}{2}sR^2\varphi^2(n)} \\ &\le q_m e^{-\frac{1}{2}sR^2\varphi^2(n)} \end{aligned}$$

for all n large, largeness depending on m. □

Theorem 6.2.4.

$$P(\|f_n - \hat{f}_{n,m}\| > \varepsilon) \le qme^{-\frac{\varepsilon^2}{4}m\varphi^2(n)}.$$

Proof

$$\begin{aligned} &P(\|f_n - \hat{f}_{n,m}\| > \varepsilon) \le \sum_{i=1}^{m} P(\sup_{\frac{i-1}{m} \le t \le \frac{i}{m}} |f_n(t) - \hat{f}_{n,m}(t)| > \varepsilon) \\ &\le \sum_{i=1}^{m} P(\sup_{\frac{i-1}{m} \le t \le \frac{i}{m}} \max\{|f_n(t) - f_n(\tfrac{i-1}{m})|; \\ &\qquad\qquad |f_n(t) - f_n(\tfrac{i}{m})|\} > \varepsilon) \\ &= \sum_{i=1}^{m} P(\max\{ \sup_{\frac{i-1}{m} \le t \le \frac{i}{m}} |f_n(t) - f_n(\tfrac{i-1}{m})|; \\ &\qquad\qquad \sup_{\frac{i-1}{m} \le t \le \frac{i}{m}} |f_n(t) - f_n(\tfrac{i}{m})|\} > \varepsilon) \\ &\le \sum_{i=1}^{m} P(\sup_{\frac{i-1}{m} \le t \le \frac{i}{m}} |f_n(t) - f_n(\tfrac{i-1}{m})| > \varepsilon) \end{aligned}$$

$$
\begin{aligned}
&\qquad\qquad + \sum_{i=1}^{m} P\Big(\sup_{\frac{i-1}{m}\le t\le \frac{i}{m}} |f_n(t) - f_n(\tfrac{i}{m})|\} > \varepsilon\Big) \\
&\le \sum_{i=1}^{m} P\Big(\sup_{\frac{i-1}{m}\le t\le \frac{i}{m}} |f_n(t) - f_n(\tfrac{i-1}{m})| > \varepsilon\Big) \\
&\qquad + \sum_{i=1}^{m} P(|f_n(\tfrac{i-1}{m}) - f_n(\tfrac{i}{m})| > \tfrac{\varepsilon}{2}) \\
&\qquad\qquad + \sum_{i=1}^{m} P\Big(\sup_{\frac{i-1}{m}\le t\le \frac{i}{m}} |f_n(t) - f_n(\tfrac{i-1}{m})|\} > \tfrac{\varepsilon}{2}\Big) \\
&= \sum_{i=1}^{m} P(|f_n(\tfrac{i-1}{m}) - f_n(\tfrac{i}{m})| > \tfrac{\varepsilon}{2}) \\
&\qquad\qquad + 2\sum_{i=1}^{m} P\Big(\sup_{\frac{i-1}{m}\le t\le \frac{i}{m}} |f_n(t) - f_n(\tfrac{i}{m})|\} > \tfrac{\varepsilon}{2}\Big) \\
&= \sum_{i=1}^{m} P(|f_n(\tfrac{i-1}{m}) - f_n(\tfrac{i}{m})| > \tfrac{\varepsilon}{2}) \\
&\qquad + 2\sum_{i=1}^{m} \quad P\Big(\sup_{\frac{i-1}{m}\le t\le \frac{i}{m}} |B(nt) - B(\tfrac{n(i-1)}{m})| > \sqrt{n}\varphi(n)\tfrac{\varepsilon}{2}\Big) \\
&= \sum_{i=1}^{m} P(|B(\tfrac{ni}{m}) - B(n\tfrac{i-1}{m}))| > \tfrac{\varepsilon}{2}\sqrt{n}\varphi(n)) \\
&\qquad + 2\sum_{i=1}^{m} P\Big(\sup_{\frac{n(i-1)}{m}\le t\le \frac{ni}{m}} |B(t) - B(\tfrac{n(i-1)}{m})| > \sqrt{n}\varphi(n)\tfrac{\varepsilon}{2}\Big) \\
&= mP(|\xi| > \tfrac{\varepsilon}{2}\sqrt{m}\varphi(n)) + 2\sum_{i=1}^{m} P\Big(\sup_{0\le t\le \frac{n}{m}} |B(t)| > \sqrt{n}\varphi(n)\tfrac{\varepsilon}{2}\Big) \\
&= mP(|\xi| > \tfrac{\varepsilon}{2}\sqrt{m}\varphi(n)) + 2\sum_{i=1}^{m} P\Big(\sup_{0\le t\le 1} |B(t)| > \sqrt{m}\varphi(n)\varepsilon\Big) \\
&\le mP(|\xi| > \tfrac{\varepsilon}{2}\sqrt{m}\varphi(n)) + 2mP\Big(\sup_{0\le t\le 1} |B(t)| > \sqrt{m}\varphi(n)\tfrac{\varepsilon}{2}\Big) \\
&\le mP(|\xi| > \tfrac{\varepsilon}{2}\sqrt{m}\varphi(n)) + 4mP\big(|\xi| > \sqrt{m}\varphi(n)\tfrac{\varepsilon}{2}\big) \\
&= 5mP\big(|\xi| > \sqrt{m}\varphi(n)\tfrac{\varepsilon}{2}\big) \\
&\le qme^{-\frac{\varepsilon^2}{4}m\varphi^2(n)} \qquad \text{(ref. \textbf{(5.9)}(vi)).}
\end{aligned}
$$

□

Theorem 6.2.5. *There exists $s = s(\varepsilon) > 1$ and close to 1 such that*

$$P(f_n \notin K_{R;\varepsilon}) \le qe^{-\frac{sR^2\varphi^2(n)}{2}}.$$

Proof. We note, for $s > 1$,

$$\{f_n \notin K_{R;\varepsilon}\} = \{\{f_n \notin K_{R;\varepsilon},\ \frac{1}{s}\hat{f}_{n,m} \in K_R\}$$

$$
\begin{aligned}
&\cup \{f_n \notin K_{R;\varepsilon},\ \frac{1}{s}\hat{f}_{n,m} \notin K_R\} \\
&\subset \{f_n \notin K_{R;\varepsilon},\ \frac{1}{s}\hat{f}_{n,m} \in K_R\} \cup \{\frac{1}{s}\hat{f}_{n,m} \notin K_R\} \\
&\subset \{\frac{1}{s}\hat{f}_{n,m} \in K_R,\ \|f_n - \frac{1}{s}\hat{f}_{n,m}\| > \varepsilon\} \cup \{\frac{1}{s}\hat{f}_{n,m} \notin K_R\} \\
&= \{\frac{1}{s}\hat{f}_{n,m} \notin K_R\} \\
&\quad \cup \{\frac{1}{s}\hat{f}_{n,m} \in K_R,\ \|f_n - \hat{f}_{n,m} + \hat{f}_{n,m} - \frac{1}{s}\hat{f}_{n,m}\| > \varepsilon\} \\
&\subset \{\frac{1}{s}\hat{f}_{n,m} \notin K_R\} \cup \{\frac{1}{s}\hat{f}_{n,m} \in K_R,\ \|f_n - \hat{f}_{n,m}\| > \frac{\varepsilon}{2}\} \\
&\quad \cup \{\frac{1}{s}\hat{f}_{n,m} \in K_R,\ \|(1 - \frac{1}{s})\hat{f}_{n,m}\| > \frac{\varepsilon}{2}\}.
\end{aligned}
$$

Choose $s = s(\varepsilon)$ close to 1 such that $(1 - \frac{1}{s})sR = (s-1)R < \frac{1}{2}\varepsilon$. It then follows $\{\frac{1}{s}\hat{f}_{n,m} \in K_R,\ \|(1-\frac{1}{s})\hat{f}_{n,m}\| > \frac{\varepsilon}{2}\} = \phi$, since $\frac{1}{s}\hat{f}_{n,m} \in K_R$ implies $\|\hat{f}_{n,m}\| \le sR$.

We thus have

$$
\begin{aligned}
P(f_n \notin K_{R;\varepsilon}) &\le P(\frac{1}{r}\hat{f}_{n,m} \notin K_R) + P(\|f_n - \hat{f}_{n,m}\| > \frac{\varepsilon}{2}) \\
&\le q_m e^{-\frac{1}{2}sR^2\varphi^2(n)} + mqe^{-\frac{\varepsilon^2}{2}m\varphi^2(n)} \le qe^{-\frac{sR^2\varphi^2(n)}{2}}. \square
\end{aligned}
$$

Remark 6.2.6. From the definition of R, we note that $\sum_k e^{-\frac{sR^2\varphi^2(n_k)}{2}} < \infty$, since $s > 1$. Hence, it follows that $wp1,\ f_{n_k} \in K_{R;\varepsilon}$ eventually. i.e. there exists a set A of probability 1 such that to each $\omega \in A$ there corresponds an integer $\lambda = \lambda(\varepsilon,\ \omega,\ c)$ with the property that $f_{n_k} \in K_{R,\varepsilon}$ for all $k \ge \lambda$.

Theorem 6.2.7. $P(f_n \notin K_R\ i.o.) = 0$. *Or, what is same,* $wp1,\ f_n \in K_R$ *eventually.*

Proof. Let $n_k \le n \le n_{k+1}$. This implies $\varphi(n_k) \le \varphi(n) \le \varphi(n_{k+1})$. Hence (ref. **(6.2.1)**) there exist $\alpha_k \ge 0,\ \beta_k \ge 0$ such that $\alpha_k + \beta_k = 1$ and $\frac{1}{\varphi(n)} = \frac{\alpha_k}{\varphi(n_k)} + \frac{\beta_k}{\varphi(n_{k+1})}$. Define $f^*_{n_k} = \alpha_k f_{n_k} + \beta_k f_{n_{k+1}}$ and note that (ref. **(6.2.2)**) $wp1,\ f^*_{n_k} \in K_{R;\varepsilon}$ eventually since the sequence (f_{n_k}) does so (ref. **(6.2.6)**).

Suppose we show that $wp1,\ \max\limits_{n_k \le n \le n_{k+1}} \|f_n - f^*_{n_k}\| \le \varepsilon$ eventually, it will follow that $wp1,\ f_n \in K_{R;\varepsilon}$ eventually. The following steps are to achieve this. We have

$$
f_n(t) - f^*_{n_k}(t) \doteq \frac{B(nt)}{\sqrt{n}\varphi(n)} - \alpha_k \frac{B(n_k t)}{\sqrt{n_k}\varphi(n_k)} - \beta_k \frac{B(n_{k+1}t)}{\sqrt{n_{k+1}}\varphi(n_{k+1})}
$$

$$= \frac{B(nt)}{\sqrt{n}}\{\frac{\alpha_k}{\varphi(n_k)} + \frac{\beta_k}{\varphi(n_{k+1})}\} - \alpha_k \frac{B(n_k t)}{\sqrt{n_k}\varphi(n_k)} - \beta_k \frac{B(n_{k+1}t)}{\sqrt{n_{k+1}}\varphi(n_{k+1})}$$
$$= \frac{\alpha_k}{\varphi(n_k)}\{\frac{B(nt)}{\sqrt{n}} - \frac{B(n_k t)}{\sqrt{n_k}}\} + \frac{\beta_k}{\varphi(n_{k+1})}\{\frac{B(nt)}{\sqrt{n}} - \frac{B(n_{k+1}t)}{\sqrt{n_{k+1}}}\}$$
$$= \frac{\alpha_k}{\sqrt{n_k}\varphi(n_k)}\{B(nt) - B(n_k t)\} + \alpha_k B(nt)\{\frac{1}{\sqrt{n}} - \frac{1}{\sqrt{n_k}}\}\frac{1}{\varphi(n_k)}$$
$$+ \frac{\beta_k}{\sqrt{n_{k+1}}\varphi(n_{k+1})}\{B(nt) - B(n_{k+1}t)\}$$
$$+ \beta_k B(nt)\{\frac{1}{\sqrt{n}} - \frac{1}{\sqrt{n_{k+1}}}\}\frac{1}{\varphi(n_{k+1})}.$$

Hence $P(\max_{n_k \le n \le n_{k+1}} \|f_n - f^*_{n_k}\| > \varepsilon) \le p_1 + p_2 + p_3 + p_4$, where

$p_1 = P(\max_{n_k \le n \le n_{k+1}} \sup_{0 \le t \le 1} |B(nt) - B(n_k t)| > \frac{\varepsilon}{4\alpha_k}\sqrt{n_k}\varphi(n_k))$

$\le P(\max_{n_k \le n \le n_{k+1}} \sup_{0 \le t \le 1} |B(nt) - B(n_k t)| > \frac{\varepsilon}{4}\sqrt{n_k}\varphi(n_k))$

$\le P(\max_{0 \le m \le n_{k+1}-n_k} \sup_{0 \le t \le 1} |B(mt + n_k t) - B(n_k t)| > \frac{\varepsilon}{4}\sqrt{n_k}\varphi(n_k))$

$= P(\max_{0 \le m \le n_{k+1}-n_k} \sup_{0 \le t \le 1} |B(mt)| > \frac{\varepsilon}{4}\sqrt{n_k}\varphi(n_k))$

$= P(\sup_{0 \le s \le n_{k+1}-n_k} |B(s)| > \frac{\varepsilon}{4}\sqrt{n_k}\varphi(n_k))$

$= P\left(\sup_{0 \le t \le 1} |B(t)| > \frac{\varepsilon}{4}\sqrt{\frac{n_k}{n_{k+1}-n_k}}\varphi(n_k)\right)$

$\le P\left(\sup_{0 \le t \le 1} |B(t)| > \frac{\varepsilon}{8}\sqrt{\frac{1}{c-1}}\varphi(n_k)\right)$ (for all k large, with $n_k = [c^k]$)

$\le q e^{-\frac{1}{2}\frac{\varepsilon^2}{64}\frac{1}{c-1}\varphi^2(n_k)}$

$\le q e^{-\frac{1}{2}rR^2\varphi^2(n_k)}$, with $r = r(\varepsilon) = \frac{\varepsilon^2}{64(c-1)R^2}$.

Now, taking c close to 1 we assume $r > 1$.

$$p_3 \le P\left(\max_{n_k \le n \le n_{k+1}} \sup_{0 \le t \le 1} |B(nt) - B(n_{k+1}t)| > \frac{\varepsilon}{4}\sqrt{n_{k+1}}\varphi(n_{k+1})\right)$$
$$\le p_{3,1} + p_{3,2}.$$
$$p_{3,1} = P\left(\max_{n_k \le n \le n_{k+1}} \sup_{0 \le t \le 1} |B(nt) - B(n_k t)| > \frac{\varepsilon}{8}\sqrt{n_{k+1}}\varphi(n_{k+1})\right)$$
$$\le P\left(\max_{n_k \le n \le n_{k+1}} \sup_{0 \le t \le 1} |B(nt) - B(n_k t)| > \frac{\varepsilon}{8}\sqrt{n_k}\varphi(n_k)\right)$$

$\le q e^{-\frac{1}{2}r(\varepsilon)R^2\varphi^2(n_k)}$ (using the bound for p_1).

$$\begin{aligned}
p_{3,2} &= P\left(\sup_{0\le t\le 1}|B(n_k t)-B(n_{k+1}t)| > \frac{\varepsilon}{8}\sqrt{n_{k+1}}\varphi(n_{k+1})\right)\\
&\le P\left(\sup_{0\le t\le 1}|B(t)| > \frac{\varepsilon}{8}\sqrt{\frac{n_k}{n_{k+1}-n_k}}\varphi(n_k)\right)\\
&\le qe^{-\frac{1}{2}r(\varepsilon)R^2\varphi^2(n_k)}.\\
p_2 &= P\left(\max_{n_k\le n\le n_{k+1}}\sup_{0\le t\le 1}\alpha_k|B(nt)|\{\frac{1}{\sqrt{n_k}}-\frac{1}{\sqrt{n}}\} > \frac{\varepsilon}{4}\varphi(n_k)\right)\\
&\le P\left(\sup_{0\le t\le 1}|B(t)|\{\sqrt{\frac{n_{k+1}}{n_k}}-1\} > \frac{\varepsilon}{4}\varphi(n_k)\right)\\
&\le P\left(\sup_{0\le t\le 1}|B(t)| > \frac{\varepsilon}{2\sqrt{c-1}}\varphi(n_k)\right) \text{ (for all } k \text{ large)}\\
&\le e^{-\frac{1}{2}rR^2\varphi^2(n_k)}
\end{aligned}$$

where $r = r(\varepsilon)$ is as chosen above.

p_4 can be similarly estimated. Collecting all the above results we conclude

$$\sum_k P(\max_{n_k\le n\le n_{k+1}}\|f_n - f^*_{n_k}\| > \varepsilon) < \infty$$

which is equivalent to the desired result. i.e., $f_n \in K_{R,\varepsilon}$.

Taking $\varepsilon = \frac{1}{N}$ and taking $N \to \infty$, we complete the proof that $wp1,\ f_n \in K_R$ eventually. □

Theorem 6.2.8. *Every $x \in K_R$ is, $wp1$, a limit point of the sequence (f_n).*

Proof. Let $x \in K_R$ and let $\varepsilon > 0$ be given. Then the claim would follow if we show that the probability is 1 that a properly chosen subsequence (f_{n_k}) is in $\{x\}_\varepsilon$ infinitely often, where $\{x\}_\varepsilon$ is the open ε-sphere around x.

Let $m \ge 2$ be an integer. Write $u_i = x(\frac{i}{m}) - x(\frac{i-1}{m}),\ 1 \le i \le m$. Choose m sufficiently large so that

$$\sup_{\substack{0\le s,t\le 1\\ |t-s|\le \frac{1}{m}}}|x(t)-x(s)| < \frac{\varepsilon}{4}.$$

This is possible since x is uniformly continuous over $[0,\ 1]$. Hence $\max_{1\le i\le m}|u_i| < \frac{\varepsilon}{2}$. We note m depends on x.

Define $\hat{x}(\frac{i}{m}) = x(\frac{i}{m})$, $1 \le i \le m$ define it to be linear elsewhere and note that $\|x - \hat{x}\| < \frac{\varepsilon}{2}$.

We note that $\int\limits_0^1 \left(\frac{d\hat{x}(t)}{dt}\right)^2 dt < R^2$ implies

$$R^2 > \sum_{i=1}^{m} \int\limits_{\frac{i-1}{m}}^{\frac{i}{m}} \left(\frac{d\hat{x}(t)}{dt}\right)^2 dt$$

$$= \sum_{i=1}^{m} \int\limits_{\frac{i-1}{m}}^{\frac{i}{m}} \left(\frac{x(\frac{i}{m}) - x(\frac{i-1}{m})}{\frac{1}{m}}\right)^2 dt = m\sum_{i=1}^{m} u_i^2. \tag{6.9}$$

Since $\|f_n - x\| \le \|f_n - \hat{x}\| + \|x - \hat{x}\| \le \|f_n - \hat{x}\| + \frac{\varepsilon}{2}$, it is enough to show that $P(\|f_{n_k} - \hat{x}\| < \varepsilon\ i.o.) = 1$ for every $\varepsilon > 0$ for a suitably chosen subsequence (n_k). We take $n_k = [c^k]$, $c > 1$ and choose c more appropriately later. With no more condition on c we showed earlier (ref. **(6.2.4)**) that $P(\|f_{n_k} - \hat{f}_{n_k,m}\| > \varepsilon) \le qme^{-\frac{1}{4}\varepsilon^2 m\varphi^2(n_k)}$. Now choose m large so $\frac{\varepsilon^2}{4}m > R^2$. Hence $\sum_k P(\|f_{n_k} - \hat{f}_{n_k,m}\| > \varepsilon) < \infty$. This implies that $wp1$, $\|f_{n_k} - \hat{f}_{n_k,m}\| \le \varepsilon$ eventually. To prove theorem it is therefore enough to show that $P(\|\hat{f}_{n_k,m} - \hat{x}\| < \varepsilon\ i.o.) = 1$.

Now

$$E_{n,m} = \{\|\hat{f}_{n,m} - \hat{x}\| < \varepsilon\}$$

$$= \{\max_{1\le i\le m} |f_n(\frac{i}{m}) - x(\frac{i}{m})| < \varepsilon\} = \bigcap_{i=1}^{m} E_{n,m,i}$$

where

$$E_{n,m,i} = \{|f_n(\frac{i}{m}) - x(\frac{i}{m})| < \varepsilon\}.$$

Define $A_{n,m,i} = \left\{\left|f_n(\frac{i}{m}) - f_n(\frac{i-1}{m}) - u_i\right| < \delta_i\right\}$ and $A_{n,m} = \cap_{i=1}^{m} A_{n,m,i}$. The δ_is are positive and chosen arbitrarily subject to the condition $\sum\limits_{i=1}^{m} \delta_i < \varepsilon$.

We note that, for n, m fixed, the m events $(A_{n,m,i})$ are independent.

For each i, $A_{n,m} \subset \cap_{j=1}^{i} A_{n,m,j}$. i.e. $-\delta_j < f_n(\frac{j}{m}) - f_n(\frac{j-1}{m}) - u_j < \delta_j$, $1 \le j \le i$. Adding these inequalities, $-\varepsilon < f(\frac{i}{m}) - x(\frac{i}{m}) < \varepsilon$. Thus we see $A_{n,m} \subset E_{n,m,i}$, $1 \le i \le m$. Hence $A_{n,m} \subset E_{n,m}$. From this will follow that $P(\overline{\lim\limits_{n\to\infty}} E_{n,m}) = 1$ if we show $P(\overline{\lim\limits_{n\to\infty}} A_{n,m}) = 1$. This will follow if we show $P(\overline{\lim\limits_{k\to\infty}} A_{n_k,m}) = 1$. In turn, this will follow if we show that, for m fixed, the sequence $(A_{n_k,m})$ is one of independent events and that $\sum\limits_k P(A_{n_k,m}) = \infty$.

Assume $c > m$ and that c is an integer. The random variables determining the event $A_{n_k,m}$ are the m independent variables constituting the, say, k^{th} block: $B(c^k \frac{i}{m}) - B(c^k \frac{i-1}{m})$, $1 \le i \le m$. The $A_{n_k,m}$s will be independent events if any two random variables, chosen one each from two different blocks are independent. Let $k < r$. Consider two such random variables: $B(c^k \frac{i}{m}) - B(c^k \frac{i-1}{m})$ and $B(c^r \frac{j}{m}) - B(c^r \frac{j-1}{m})$. The independence of these two variables would follow from the independent increments property of a BMP if $c^k i < c^r$. Now, $c^k i \le c^k m < c^r$ if $c^{r-k} > m$. This is true since $r - k \ge 1$ and $c > m$, by choice.

We have

$$P(A_{n,m,i}) = P\left\{\left|\frac{B(\frac{ni}{m})}{\sqrt{n}\varphi(n)} - \frac{B(\frac{n(i-1)}{m})}{\sqrt{n}\varphi(n)} - u_i\right| < \delta_i\right\}$$

$$= P\left\{|\xi - \sqrt{m}\varphi(n)u_i| < \sqrt{m}\varphi(n)\delta_i\right\}$$

$$\ge \frac{\sqrt{m}}{\sqrt{2\pi}}\varphi(n)2\delta_i e^{-\frac{1}{2}m\varphi^2(n)\max\{(u_i+\delta_i)^2,\ (u_i-\delta_i)^2\}}$$

(ref. **(4.4.5)**)

$$\ge e^{-\frac{1}{2}m\varphi^2(n)\max\{(u_i+\delta_i)^2,(u_i-\delta_i)^2\}}$$

for all n large satisfying $\frac{2\delta_i\sqrt{m}\varphi(n)}{\sqrt{2\pi}} > 1$. (We note $a_i = \max\{(u_i+\delta_i)^2,\ (u_i - \delta_i)^2\} = (u_i + \delta_i)^2$ or $(u_i - \delta_i)^2$ according as $u_i > 0$ or $u_i < 0$. Clearly $a_i \le (|u_i| + \delta_i)^2$). This gives $P(A_{n,m}) = \prod_1^m P(A_{n,m,i}) \ge e^{-\frac{1}{2}\lambda\varphi^2(n)}$ where $m\sum_{i=1}^m(|u_i| + \delta_i)^2 = \lambda$, say. Given the u_is and m, we use (6.9) to find δ_is such that $\lambda < R^2$. We thus have $P(A_{n,m}) \ge e^{-\frac{1}{2}\lambda\varphi^2(n)}$. Hence $\sum_k P(A_{n_k,m}) \ge \sum_k e^{-\frac{1}{2}\lambda\varphi^2(n_k)} = \infty$, since $\lambda < R^2$. □

Lemma 6.2.9. *K_R is is a closed subset of C and is separable.*

Proof. Take $\alpha = 2$ and $a = R$ in **(6.1.4)** and conclude K_R is a closed set. Since $(C,\ \rho)$ is a separable metric space, every one of its subsets is separable. For this result and a special separability set for $K_{\alpha,a}$, ref. **(6.1.4)**. □

We are now in a position to identify the almost sure limit set of the sequence (f_n).

Theorem 6.2.10. *$wp1$ the limit set of the sequence (f_n) in C is K_R where $f_n(\omega)$ is the function $\frac{B(nt,\ \omega)}{\sqrt{n}\varphi(n)}$.*

Proof. By **(6.2.7)** $wp1$ the limit set is contained in K_R. By **(6.2.9)** there exists a countable dense subset of K_R. By **(6.2.8)**, every $x \in K_R$ is a limit point. Now the claim follows by **(6.1)**(iv). □

Corollary 6.2.11. *$wp1$, the limit set **A** of the sequence $(\frac{B(n)}{\sqrt{n}\varphi(n)})$ is $[-R,\ R]$.*

Proof. Arguments are same as in **(6.1)**(vi). Details are omitted. □

Remark 6.2.12. In order to prove the assertion in the Corollary above, it should not be necessary to establish first the theorem at **(6.2.10)** which is the culmination of **(6.2.1)** through **(6.2.9)**. For this reason we give below

A direct proof of (6.2.11)

First we show that $wp1$, $\overline{\lim}_{n\to\infty} \frac{B(n)}{\sqrt{n}\varphi(n)} \leq R$. i.e., we show that, given $\delta > 0$, $P\big(B(n) > (R+\delta)\sqrt{n}\varphi(n) i.o\big) = 0$. There is no loss in generality in assuming $\delta < \frac{R}{4}$. Write $B^*(n) = \max_{1\leq k\leq n} B(k)$. Define $\sqrt{c} = 1 + \frac{\delta}{8R}$. For $k = 1, 2, \ldots$, let $n_k = [c^k]$. If $n_{k-1} \leq n < n_k$, then clearly $\big(B(n) > (R+\delta)\sqrt{n}\varphi(n)\big) \subset \big(B^*(n_k) > (R+\delta)\sqrt{n_{k-1}}\varphi(n_{k-1})\big)$. Hence

$$\big(B(n) > (R+\delta)\sqrt{n}\varphi(n) \text{for infinitely many } n\big) \subset \big(B^*(n_k) > (R+\delta)\sqrt{n_{k-1}}\varphi(n_{k-1}) \text{for infinitely many } k\big).$$

Thus $P\big(B(n) > (R+\delta)\sqrt{n}\varphi(n)$ for infinitely many $n\big)$ would be zero if we show $\sum_k P\big(B^*(n_k) > (R+\delta)\sqrt{n_{k-1}}\varphi(n_{k-1})\big) < \infty$ and appeal to Borel Cantelli lemma. We note $n_k = [c^k] \leq cc^{k-1} \leq c(n_{k-1}+1)$. We note too that for all k large,

$$\sqrt{\frac{n_{k-1}}{c(n_{k-1}+1)}} \geq \frac{1-\frac{\delta}{2R}}{1+\frac{\delta}{8R}} > 1 - \frac{5\delta}{8R}.$$

Hence

$$\begin{aligned}
&P\big(B^*(n_k) > (R+\delta)\sqrt{n_{k-1}}\varphi(n_{k-1})\big)\\
&\quad\leq P\Big(B^*(n_k) > (R+\delta)\sqrt{\frac{n_{k-1}}{c(n_{k-1}+1)}}\sqrt{n_k}\varphi(n_{k-1})\Big)\\
&\quad\leq P\big(B^*(n_k) > (R+\delta)(1-\frac{5\delta}{8R})\sqrt{n_k}\varphi(n_{k-1})\big)\\
&\quad\leq P\big(B^*(n_k) > (R+\frac{7\delta}{32})\sqrt{n_k}\varphi(n_{k-1})\big)\\
&\quad\leq P\big(\max_{0\leq s\leq n_k} B(s) > (R+\frac{7\delta}{32})\sqrt{n_k}\varphi(n_{k-1})\big)\\
&\quad= P\big(\max_{0\leq s\leq 1} B(s) > (R+\frac{7\delta}{32})\varphi(n_{k-1})\big) \leq e^{-\frac{1}{2}(R+\frac{7\delta}{32})^2\varphi^2(n_{k-1})}
\end{aligned}$$

That $\sum_k P\big(B^*(n_k) > (R+\delta)\sqrt{n_{k-1}}\varphi(n_{k-1})\big) < \infty$ follows now. With this the proof is complete that $\overline{\lim}_{n\to\infty} \frac{B(n)}{\sqrt{n}\varphi(n)} \leq R$.

By symmetry, $wp1$, $\underline{\lim}_{n\to\infty} \frac{B(n)}{\sqrt{n}\varphi(n)} \geq -R$.

Let $0 < r < R$. Find $\theta \geq 1,\ \theta = \theta(r)$ as defined in **(6.1.3)**. Let $n_k = \left[c^{k^\theta}\right]$. Let $\delta > 0$ be arbitrary but small. We have

$$\sum_k P(B(n_k) > \overline{r+\delta}\sqrt{n_k}\varphi(n_k)) \leq \sum_k e^{-\frac{1}{2}(r+\delta)^2\varphi^2(n_k)} < \infty.$$

Hence

$$wp1, \qquad \overline{\lim_{k\to\infty}} \frac{B(n_k)}{\sqrt{n_k}\varphi(n_k)} \leq r$$

Similarly,

$$wp1, \qquad \underline{\lim_{k\to\infty}} \frac{B(n_k)}{\sqrt{n_k}\varphi(n_k)} \geq -r \tag{6.10}$$

This implies that there exists a set A^* such that $P(A^*) = 1$ and such that $\omega \in A^*$ implies $\frac{B(n_k)}{\sqrt{n_k\varphi(n_k)}} \geq -2r$ for all k large, largeness depending on ω.

Let $Z_k = B(n_k) - B(n_{k-1})$ and note that (Z_k) is a sequence of independent zero mean normal variables with variances $(n_k - n_{k-1})$. Since $\sqrt{\frac{n_k}{n_k - n_{k-1}}} \to 1+$, it follows that for all k sufficiently large, $\sqrt{\frac{n_k}{n_k-n_{k-1}}} \leq \frac{2r-\delta}{2r-2\delta}$. For all k large we have,

$$\begin{aligned} P(Z_k > \overline{r-\delta}\sqrt{n_k}\varphi(n_k)) &\geq P(B(1) > (r - \frac{1}{2}\delta)\varphi(n_k)) \\ &\geq e^{-\frac{1}{2}(r^2 - \frac{r\delta}{4})\varphi^2(n_k)}. \end{aligned}$$

Hence $\sum_k P(Z_k > \overline{r-\delta}\sqrt{n_k}\varphi(n_k)) = \infty$. Since the Z_ks are independent, this implies that there exists $A_1(\delta)$ such that $P(A_1(\delta)) = 1$ and such that for every $\omega \in A_1(\delta), B(n_k,\ \omega) - B(n_{k-1},\ \omega) > \overline{r-\delta}\sqrt{n_k}\varphi(n_k)$ for infinitely many k. Let $A(\delta) = A^* \cap A_1(\delta)$. We note $P(A(\delta)) = 1$. Let $\omega \in A(\delta)$. Hence for infinitely many k,

$$\begin{aligned} B(n_k,\ \omega) &> (r-\delta)\sqrt{n_k}\,\varphi(n_k) + B(n_{k-1},\ \omega) \\ &> (r-\delta)\sqrt{n_k}\,\varphi(n_k) - 2r\sqrt{n_{k-1}}\,\varphi(n_{k-1}) \\ &= (r-\delta)\sqrt{n_k}\,\varphi(n_k)\{1 - \frac{2r}{r-\delta}\sqrt{\frac{n_{k-1}}{n_k}}\,\frac{\varphi(n_{k-1})}{\varphi(n_k)}\} \\ &> (r-\delta)\sqrt{n_k}\,\varphi(n_k)\{1 - \frac{2r}{r-\delta}\frac{\delta}{2r}\} \\ &> (r-2\delta)\sqrt{n_k}\,\varphi(n_k). \end{aligned}$$

Thus $wp1$,

$$\overline{\lim_{k\to\infty}} \frac{B(n_k)}{\sqrt{n_k}\varphi(n_k)} \geq r.$$

This, together with (6.2) gives that $wp1$, r is a limit point of $\frac{B(n)}{\sqrt{n\varphi(n)}}$. That $wp1$ $-r$ is a limit point follows by symmetry. Since r is arbitrary, we conclude that every point in $[-R,\ R]$ is $wp1$ a limit point. Since the rationals in this interval are countable, there exists a set A such that $P(A) = 1$ and such that $\omega \in A$ implies every rational number in $[-R,\ R]$ is a limit point of the sequence $(\frac{B(n,\ \omega)}{\sqrt{n\varphi(n)}})$. Since the rationals are dense, it follows that if $\omega \in A$, then every point in $[-R,\ R]$ is a limit point of the sequence $(\frac{B(n,\ \omega)}{\sqrt{n\varphi(n)}})$.

6.3 Conditioning

The notion of conditional probability of an event A, given the outcome of observing a random variable is the key to the study of dependent rvs. Guided by the intuitive evaluation of the frequency of occurrence of an event A in repeated trials where event B occurs, it seems natural to define $P(A|B)$, the conditional probability of the event A given that the event B has occured, to be

$$P(A|B) = \frac{P(A \cap B)}{P(B)} \tag{6.11}$$

provided $P(B) > 0$. This definition seems to be appropriate also when we note that A is independent of B *iff* $P(A|B) = P(A)$. i.e. *iff* the information that B has occured does not change the probability that A occurs. However this definition fails to make sense in every situation. For example, if $X,\ Y$ are rvs representing the height and weight of individuals in a population, it is reasonable to want to know the weight distribution among individuals with a specified height. In other words, we may want to know the conditional probability $P(Y \leq b|X = a)$. If X is thought of as a continuous rv then $P(X = a) = 0$. In this case the above definition does not help. Thus we need a more inclusive definition.

To arrive at a definition which will apply also when $P(B) = 0$, we examine (6.11) again and note that it contains no information about the probability of A given that B does not occur. To fill this gap, we consider defining the conditional probability of A conditioned on the σ-field $\mathcal{B}$ generated by the set B : $\mathcal{B} = \{\Omega,\ \phi,\ B,\ B'\}$. We define $P(A|\mathcal{B})$ as the rv Y where $Y(\omega) = P(A|B)\chi_B(\omega) + P(A|B')\chi_{B'}(\omega)$. Here $P(A|.)$ is as per (6.11). If $0 < P(B) < 1$, then $Y(\omega) = P(A|B)$ if $\omega \in B$ and $Y(\omega) = P(A|B')$ if $\omega \in B'$. If, say, $P(B') = 0$ the second term is undefined (or, $P(A|B')$ may be assigned an arbitrary value) and then $Y = P(A|B)$ $wp1$. This definition easily extends to any finite or countably infinite partition $\{B_1,\ B_2,\ ...\}$ of Ω

: $Y(\omega) = \sum_i P(A|B_i)\chi_{B_i}(\omega)$. We note Y is $\mathcal{B}$- measurable where $\mathcal{B}$ is the σfield generated by the collection $\{B_i,\ i \geq 1\}$..

If Z is a discrete rv such that $B_k = (Z = a_k)$, $\cup_k B_k = \Omega$ then $\mathcal{B}$ is generated by Z and we write, also, $P(A|Z)$ for $P(A|\mathcal{B})$. Since Y is $\mathcal{B}$-measurable there exists (ref. **(1.4.5)**) a Borel-measurable function g defined on the line such that $Y = g(Z)$.

In tune with the unconditional situation (where $P(A) = \mathbb{E}\chi_A$) we write $\mathbb{E}(\chi_A|\mathcal{B})$ for $P(A|\mathcal{B})$ and refer to it as the conditional expectation of χ_A given $\mathcal{B}$.

Our object is to extend, in a consistent manner, this definition to arbitrary σ-fields $\mathcal{B} \subset \mathscr{A}$ which are not necessarily generated by a single discrete rv and to more general rvs X in place of the characteristic function χ_A of a set A.

Let $Z,\ a_k,\ B_k$ be as before. We note $P(B_k) > 0,\ k \geq 1$. In the matter of extending the definition of conditional expectation to more general variables X it is natural to examine first the case where X is discrete valued. Let $A_k = (X = x_k),\ k = 1,\ 2, \ldots;\ X = \sum_1^\infty x_k \chi_{A_k}$. Let $\mathbb{E}|X| < \infty$. It seems reasonable to define

$$E(X|Z) = \sum_{k=1}^{\infty} x_k P(A_k|Z) = \sum_{k=1}^{\infty} x_k (\sum_{j=1}^{\infty} P(A_k|B_j)\chi_{B_j}).$$

For the definition to be acceptable, we must establish that the defining series is convergent. Now,

$$\sum_{k=1}^{\infty} |x_k| (\sum_{j=1}^{\infty} \frac{P(A_k \cap B_j)}{P(B_j)} \chi_{B_j}) \leq (\sum_{k=1}^{\infty} |x_k| P(A_k))(\sum_{j=1}^{\infty} \frac{1}{P(B_j)} \chi_{B_j})$$
$$= \mathbb{E}|X| \sum_{j=1}^{\infty} \frac{1}{B_j} \chi_{B_j}.$$

For each ω this series has exactly one non-zero term and hence is convergent. Thus the defining series is absolutely convergent.

We study the above definition for properties to motivate ourselves to define conditional expectation in the general case. We note
(i) $E(X|Z) = E(X|\mathcal{B})$ is $\mathcal{B}$-measurable and hence is a Borel measurable function $g(Z)$ of Z (ref. **(1.4.5)**).

(ii)

$$\int_{B_j} g(Z)\,\mathrm{d}P = \int_{B_j} \{\sum_{k=1}^{\infty} x_k P(A_k|B_j)\chi_{B_j}\}\,\mathrm{d}P$$
$$= \sum_{k=1}^{\infty} x_k P(A_k|B_j) \int_{B_j} \chi_{B_j}\,\mathrm{d}P.$$

Integration term by term justified since the nth term of the series in absolute value is bounded by the nth term of the convergent series $\sum_k |x_k| P(A_k)$. And then

$$\int_{B_j} g(Z)\,\mathrm{d}P = \sum_{k=1}^{\infty} x_k P(A_k|B_j)P(B_j)$$
$$= \sum_{k=1}^{\infty} x_k \int_{B_j} \chi_{A_k}\,\mathrm{d}P = \int_{B_j} (\sum_{k=1}^{\infty} x_k \chi_{A_k})\,\mathrm{d}P$$

(again using the same argument, in the reverse direction). Thus $\int_{B_j} g(Z)\,\mathrm{d}P = \int_{B_j} X\,\mathrm{d}P$. Since this is true for every B_j we have $\int_E g(Z)\,\mathrm{d}P = \int_E X\,\mathrm{d}P$ for every $E \in \mathcal{B}$.

These properties lead us to adopt the following definition in the general case.

Definition 6.3.1. Let $\mathbb{E}|X| < \infty$ and let $\mathcal{B} \subseteq \mathscr{A}$ be an arbitrary sub σ-field. Then $E(X|\mathcal{B})$, the conditional expectation of X relative to or given $\mathcal{B}$, is defined to be any $\mathcal{B}$-measurable rv Y satisfying, for every $B \in \mathcal{B}$, the equality

$$\int_B Y\,\mathrm{d}P = \int_B X\,\mathrm{d}P. \tag{6.12}$$

$E(\chi_A|\mathcal{B})$, $A \in \mathscr{A}$ is called the conditional probability on $\mathscr{A}$ relative to or given $\mathcal{B}$.

Two questions arising out of this definition are answered in the following theorem.

Theorem 6.3.2.
(i) *Existence. Let $\mathcal{B}$ be an arbitrary sub σ-field of $\mathscr{A}$ and let X be an arbitrary rv with $\mathbb{E}|X| < \infty$. Then there exists a conditional expected value Y of X.*
(ii) *Uniqueness. Any two $\mathcal{B}$-measurable rvs satisfying* (6.12) *are equal $wp1$.*

Proof. (i) Let us consider first the case $X \geq 0$. Define $\mu(A) = \int_A X \, \mathrm{d} P$, $A \in \mathscr{A}$. We note that, since $\mathbb{E}X < \infty$, μ is a finite measure and that $\mu << P$. Restricting μ and P to $\mathcal{B}$ and applying Radon-Nikodym theorem (ref. (**Q**3)) we know there exists a non-negative finite valued $\mathcal{B}$-measurable function f such that $\mu(A) = \int_A f(\omega) \, \mathrm{d} P(\omega)$, $A \in \mathcal{B}$. Since f satisfies (6.12), it qualifies to be $E(X|\mathcal{B})$. Thus the existence of $E(X|\mathcal{B})$ is established when $X \geq 0$.

When X is not necessarily non-negative, write $X = X^+ - X^-$ and note that $\mathbb{E}(X^+|\mathcal{B}) - \mathbb{E}(X^-|\mathcal{B})$ is a $\mathcal{B}$-measurable function and that it satisfies (6.12). It is natural then to define $E(X|\mathcal{B})$ as equal to $E(X^+|\mathcal{B}) - E(X^-|\mathcal{B})$.

(ii) Let the $\mathcal{B}$-measurable rvs $Y_1, \; Y_2$ satisfy (6.12). Hence for every $A \in \mathcal{B}$, $\int_A \{Y_1(\omega) - Y_2(\omega)\} \, \mathrm{d} P(\omega) = 0$. Taking $A = \{\omega : Y_1(\omega) > Y_2(\omega)\}$ we conclude from this that $P(A) = 0$. Similarly, $P\{\omega : Y_1(\omega) < Y_2(\omega)\} = 0$. Hence $P\{\omega : Y_1(\omega) = Y_2(\omega)\} = 1$. □

Any $\mathcal{B}$-measurable rv satisfying (6.12) is called a version of $\mathbb{E}(X|\mathcal{B})$

Remarks 6.3.3. (i) **Conditioning on** rvs.
Suppose $\{X_t, \; t \in T\}$ is a family of rvs defined on $(\Omega, \; \mathscr{A})$ and let $\mathcal{B} \subset \mathscr{A}$ be the minimal σ-field with respect which these variables are measurable. i.e., $\mathcal{B} = \sigma(X_t, \; t \in T)$. Then by $E(X|X_t, \; t \in T)$ we understand $E(X|\mathcal{B})$.
(ii) By $E(X|Y = y)$ we understand $g(y)$ where $g(Y) = E(X|Y)$. We ask : Let $A \in \mathscr{A}$. To find $P(A|Z = z)$, is it necessary to find the function g first and then evaluate it at z or can we attempt to find $P(A|Z = z)$ by evaluating $\lim_{n \to \infty} P(A|z - \frac{1}{n} \leq Z \leq z + \frac{1}{n})$ when z is a possible value of Z but $P(Z = z) = 0$? z is said to be a possible value if $P(Z \in J) > 0$ for every open interval J containing z. But this limit can not be guaranteed to exist. However, this natural approach can be fine tuned to ensure the existence of the limit for most values of z. The details are given in **(6.4.1)**.
(iii) Example 1. Recall $E(X|Z)$ is a Borel function $g(.)$ of Z. Thus if Z is a discrete rv, then $g(Z)$ is a discrete rv even if X is a continuous variable. This point is brought home by the following example.
Let X have frequency function $\frac{3}{2x^2}$, $1 \leq x \leq 3$ and $Z = [X]$. We note $P(Z = 1) = \frac{3}{4}$ and $P(Z = 2) = \frac{1}{4}$. If $B = (Z = 1)$ then $B' = (Z = 2)$. We have $\int_B E(X|Z) \, \mathrm{d} P = \int_B X \, \mathrm{d} P = \int_{(1 \leq X < 2)} X \, \mathrm{d} P = \frac{3}{2} \log 2$; $\int_{B'} X \, \mathrm{d} P = \frac{3}{2}\{\log 3 - \log 2\}$. Define $g(Z) = 2 \log 2$ if $Z = 1$ and $g(Z) = 6 \log(\frac{3}{2})$ if $Z = 2$. It is easily verified that, so defined, $g(Z)$ is a version of $E(X|Z)$.
(iv) Finding "$g(Z)$". Example 2.
Let (X, Y) have joint frequency function $f(x, \; y)$. Let $C \in \mathscr{R}$. We proceed to find $g(X) = P(Y \in C|X)$. Let f_1 be the marginal frequency function of

$X : f_1(x) = \int\limits_R f(x, u)\,\mathrm{d}u$. Define $g(x) = \frac{\lambda(x)}{f_1(x)}$ if $f_1(x) > 0$ and $g(x) = 0$ if $f_1(x) = 0$ where $\lambda(x) = \int\limits_C f(x, u)\,\mathrm{d}u$. We claim $g(X)$ is a version of $P(C|X)$.

The measurability of $\lambda(.)$ follows from integration theory (ref. **Q**8). We note $0 \leq \lambda(x) \leq f_1(x)$. $\{x : g(x) = 0\} = [\{x : f_1(x) = 0\} \cup \{x : \lambda(x) = 0\}] \cap \{x : f_1(x) > 0\}$, a measurable set. $\{x : 0 < g(x) \leq t\} = \{x : \lambda(x) > 0\} \cap \{x : tf_1(x) - \lambda(x) \geq 0\}$, a measurable set. Thus g is a Borel measurable function. It remains to show that g satisfies (6.12). Let $A \in X^{-1}(\mathscr{R})$. Let $B \in \mathscr{R}$ such that $A = X^{-1}(B)$. We have

$$\begin{aligned}
\int\limits_{X^{-1}(B)} g(X)\,\mathrm{d}P &= \int\limits_{\Omega} g(X)\chi_{(X^{-1}(B))}\,\mathrm{d}P \\
&= \int\limits_{R} g(x)\chi_B(x)\,\mathrm{d}P(X \leq x) \\
&= \int\limits_{B} g(x) f_1(x)\,\mathrm{d}x \\
&= \int\limits_{B\cap(f_1(x)>0)} \{\int\limits_{C} f(x, u)\,\mathrm{d}u\}\,\mathrm{d}x \\
&= \int\limits_{C}\{\int\limits_{B\cap(f_1(x)>0)} f(x, u)\,\mathrm{d}x\}\,\mathrm{d}u \\
&= \int\limits_{C}\{\int\limits_{B} f(x, u)\,\mathrm{d}x\}\,\mathrm{d}u \\
&= \iint\limits_{B\times C} f(x, u)\,\mathrm{d}x\,\mathrm{d}u \\
&= P(X \in B,\ Y \in C) = P(A \cap (Y \in C)),
\end{aligned}$$

proving that (6.12) is satisfied.

(v) Example 3. We need first to introduce some definitions.

Let X, Y be two square integrable rvs. Let $\mathbb{E}X = m_1$, $\mathbb{E}(X - m_1)^2 = \sigma_1^2$; $\mathbb{E}Y = m_2$, $\mathbb{E}(Y - m_2)^2 = \sigma_2^2$. The covariance $\sigma_{1,2}$ between X and Y is defined to be $\mathbb{E}(X - m_1)(Y - m_2)$ and the correlation coefficient ρ between them is defined by the formula $\rho = \frac{\sigma_{1,2}}{\sigma_1\sigma_2}$. X, and Y are said to be uncorrelated if $\rho = 0$.

Let now (X, Y) have the bivariate normal distribution with zero means, variances σ_1^2, σ_2^2 and correlation coefficient ρ. Then $E(Y|X) = \rho X\ wp1$.

We note the bivariate frequency function f of (X, Y) is:

$$f(x, y) = \frac{1}{2\pi\sigma_1\sigma_2\sqrt{1-\rho^2}} e^{-\frac{1}{2(1-\rho^2)}\left(\frac{x^2}{\sigma_1^2} - 2\rho\frac{xy}{\sigma_1\sigma_2} + \frac{y^2}{\sigma_2^2}\right)},$$

$-\infty < x,\ y < \infty$. Let $\mathcal{B}$ be the σ-field generated by $X : \mathcal{B} = X^{-1}(\mathscr{R})$. To

$B \in \mathcal{B}$ there corresponds $\widehat{B} \in \mathscr{R}$ such that $B = X^{-1}(\widehat{B})$. Now,

$$\int\limits_{B} Y \,\mathrm{d}P = \int\limits_{\widehat{B}\times R} yf(x,\, y)\,\mathrm{d}x\,\mathrm{d}y$$

$$= \frac{1}{\sigma_1\sqrt{2\pi}} \int\limits_{\widehat{B}} e^{-\frac{1}{2}\frac{x^2}{\sigma_1^2}}\,\mathrm{d}x \frac{1}{\sigma_2\sqrt{2\pi}\sqrt{1-\rho^2}} \times$$

$$\int\limits_{R} ye^{-\frac{1}{2(1-\rho^2)}(\frac{y}{\sigma_2}-\rho\frac{x}{\sigma_1})^2}\,\mathrm{d}y$$

$$= \int\limits_{\widehat{B}} \rho\frac{\sigma_2}{\sigma_1}x\frac{1}{\sigma_1\sqrt{2\pi}}e^{-\frac{x^2}{2\sigma_1^2}}\,\mathrm{d}x = \rho\frac{\sigma_2}{\sigma_1}\int\limits_{B} X\,\mathrm{d}P.$$

This being true for all $B \in \mathcal{B}$, it follows that $E(Y|X) = \rho\frac{\sigma_2}{\sigma_1}X\ wp1$.

If in the above $\mathbb{E}X = m_1$ and $\mathbb{E}Y = m_2$, then $E(Y|X) = m_2 + \rho\frac{\sigma_2}{\sigma_1}(X - m_1)\ wp1$.

As an application of the above result, let $X,\ Y$ be two iid standard normal variables. Define $Z = X - Y$. $(Y,\ Z)$ is a bivariate zero mean normal vector with $\sigma_1^2 = 1,\ \sigma_2^2 = 2$ and $\rho = -\frac{1}{\sqrt{2}}$. Hence $E(Y|Z) = -\frac{1}{2}Z\ wp1$.

(vi) Example 4. Let $(\mathbf{U},\ \mathcal{U})$ =$(\mathbf{X} \times \mathbf{Y},\ \mathcal{X} \otimes \mathcal{Y})$ be the product measurable space of the measurable spaces $(\mathbf{X},\ \mathcal{X})$, $(\mathbf{Y},\ \mathcal{Y})$. Let $g(x,\ y)$ be a real Borel measurable function defined on $\mathbf{U}$. Let $X,\ Y$ be measurable mappings of the probability space$(\Omega,\ \mathscr{A},\ P)$ into $\mathbf{X}$ and $\mathbf{Y}$ respectively. We note $g(X,\ Y)$ is a real rv. Let $\mu,\ \nu$ be the pms induced respectively by X and Y on $\mathcal{X}$ and $\mathcal{Y}$. Let $\mathscr{A}_1$ be a sub σ-field of $\mathscr{A}$ such that Y is measurable with respect to it and X is independent of it (and hence independent of Y).
Then, for every $D \in \mathcal{U}$,

$$P((X,\ Y) \in D|\mathscr{A}_1) = P((X,\ Y) \in D|Y)\ \ wp1 \tag{6.13}$$

Proof of this claim will be given in **(6.4)** (xiii).
(vii) Example 5. Suppose Ω is the set $\{(x,\ y) : \frac{x^2}{a^2} + \frac{y^2}{b^2} \le 1\},\ a,\ b > 0$. $\mathscr{A}$ is the Borel σ-field of its subsets. P is the pm determined by the frequency function $f(x,\ y) = \frac{p}{\pi ab}$ if $(x,\ y) \in \Omega \cap (x < 0)$ and $f(x,\ y) = \frac{q}{\pi ab}$ if $(x,\ y) \in \Omega \cap (x \ge 0),\ p,\ q \ge 0,\ p + q = 1$. Let us find $E(X|Y)$ and $E(X^2|Y)$.

Let $E(X|Y) = g(Y)$. Let B be a linear Borel set.

$$J_1(B) = \int\limits_{Y\in B} g(Y)\,\mathrm{d}P = \int\limits_{B}\{\int\limits_{R} f(x,\ y)\,\mathrm{d}x\}g(y)\,\mathrm{d}y$$

$$= \int_B \{ \int_{-a\sqrt{1-\frac{y^2}{b^2}}}^{0} \frac{p}{\pi ab}\,\mathrm{d}\,x + \int_0^{a\sqrt{1-\frac{y^2}{b^2}}} \frac{q}{\pi ab}\,\mathrm{d}\,x\} g(y)\,\mathrm{d}\,y$$

$$= \frac{1}{\pi ab}\int_B a\sqrt{1-\frac{y^2}{b^2}}\,g(y)\,\mathrm{d}\,y.$$

$$J_2(B) = \int_B \{ \int_{-a\sqrt{1-\frac{y^2}{b^2}}}^{a\sqrt{1-\frac{y^2}{b^2}}} x f(x,y)\,\mathrm{d}\,x\}\,\mathrm{d}\,y$$

$$= \frac{1}{\pi ab}\int_B \{ \int_{-a\sqrt{1-\frac{y^2}{b^2}}}^{0} xp\,\mathrm{d}\,x + \int_0^{a\sqrt{1-\frac{y^2}{b^2}}} xq\,\mathrm{d}\,x\}\,\mathrm{d}\,y$$

$$= \frac{a^2}{2\pi ab}\int_B (1-\frac{y^2}{b^2})(q-p)\,\mathrm{d}\,y.$$

Since $J_1(B) = J_2(B)$ for all Borel sets B (ref. (6.12)), it follows that $a\sqrt{1-\frac{y^2}{b^2}}g(y) = \frac{a^2}{2}(q-p)(1-\frac{y^2}{b^2})$ for all y except on a set of μ_2-measure 0 where μ_2 is the probability measure of Y. Thus one version of $g(y) = \frac{a}{2}(q-p)\sqrt{1-\frac{y^2}{b^2}}$ for all y with $|y| \leq b$.

Let $h(Y) = E(X^2|Y)$.

$J_3(B) = \int\limits_{Y\in B} h(Y)\,\mathrm{d}\,P = \int\limits_B\{\int\limits_R f(x,\ y)\,\mathrm{d}\,x\}h(y)\,\mathrm{d}\,y$

$= \frac{1}{\pi ab}\int\limits_B a\sqrt{1-\frac{y^2}{b^2}}h(y)\,\mathrm{d}\,y.$

$$J_4(B) = \int_B\{ \int_{-a\sqrt{1-\frac{y^2}{b^2}}}^{a\sqrt{1-\frac{y^2}{b^2}}} x^2 f(x,\ y)\,\mathrm{d}\,x\}\,\mathrm{d}\,y$$

$$= \frac{1}{\pi ab}\int_B\{ \int_{-a\sqrt{1-\frac{y^2}{b^2}}}^{0} x^2p\,\mathrm{d}\,x + \int_0^{a\sqrt{1-\frac{y^2}{b^2}}} x^2q\,\mathrm{d}\,x\}\,\mathrm{d}\,y$$

$$= \frac{a^3}{3\pi ab}\int_B (1-\frac{y^2}{b^2})^{\frac{3}{2}}(q+p)\,\mathrm{d}\,y.$$

Since $J_3(B) = J_4(B)$ for all Borel sets B (ref. (6.12)), we have, arguing as before, $h(y) = \frac{a^2}{3}(1 - \frac{y^2}{b^2})$.

The Martingale concept and examples

Let T be the index set standing for $\{0,\ 1, \ldots,\ n\}$ or $\{0,\ 1,\ 2, \ldots\}$ or $[0,\ \infty)$. Given a stochastic process $(X_t,\ t \in T)$, denote by $\mathcal{A}_t$ the σ-field generated by the variables $X_s,\ 0 \le s \le t$. We note $\mathcal{A}_s \subset \mathcal{A}_t$ if $s < t$. The σ-fields $(\mathcal{A}_t,\ t \in T)$ constitute what is called the natural filtration of the process. The process $(X_t,\ t \in T)$ is said to be a martingale with respect to its natural filtration (or, simply, a martingale) if $\mathbb{E}|X_t| < \infty,\ t \in T$ and if for every $s,\ t \in T,\ s < t,\ E(X_t|\mathcal{A}_s) = X_s\ wp1$.

(viii) Martingales formed from a BMP $(B(t),\ t \ge 0)$:

$B_i(t),\ t \ge 0,\ 1 \le i \le 7$ are martingales where (α) $B_1(t) = B(t)$; (β) $B_2(t) = B^2(t) - t$; (γ) $B_3(t) = t^2 B(t) - 2\int\limits_0^t sB(s)ds$; (δ) $B_4(t) = B^3(t) - 3tB(t)$; (ε) $B_5(t) = B^4(t) - 6tB^2(t) + 3t^2$; (η) $B_6(t) = e^{\lambda B(t) - \frac{\lambda^2}{2}t},\ \lambda \in R$; (θ) $B_7(t) = B^{(1)}(t)B^{(2)}(t)$ where $B^{(i)}, i = 1, 2$ are independent standard Brownian Motion Processes and (ξ) the complex valued process $M(t) = e^{\{i\lambda B(t) + \frac{\lambda^2}{2}t\}}$

Arguments are similar for all the processes $B_i, 1 \le i \le 7$. We present the details for B_3, B_6 and $M(t)$

Proof of (γ)

$$
\begin{aligned}
&E\{t^2 B(t) - 2\int\limits_0^t uB(u)du|\mathcal{A}_s\} \\
&= t^2 B(s) - 2E\{\int\limits_0^s uB(u)du|\mathcal{A}_s\} - 2E\{\int\limits_s^t uB(u)du|\mathcal{A}_s\} \\
&= t^2 B(s) - 2\int\limits_0^s uB(u)du - 2B(s)\int\limits_s^t udu \\
&\qquad - 2E\{\int\limits_s^t u[B(u) - B(s)]|\mathcal{A}_s\} \\
&= t^2 B(s) - 2\int\limits_0^s uB(u)du - (t^2 - s^2)B(s)
\end{aligned}
$$

$$- 2\mathbb{E}\{\int_s^t u[B(u) - B(s)]\mathrm{d}u\} \text{ (use independence)}$$

$$= s^2 B(s) - \int_0^s uB(u)\mathrm{d}u \; wp1,$$

proving that B_3 is a martingale.

Proof of (η)

We recall that for a standard normal variable ξ, $\mathbb{E}e^{t\xi} = e^{\frac{t^2}{2}}$. Now,

$$\begin{aligned} E\{e^{\lambda B(t)-\frac{\lambda^2}{2}t}|\mathcal{A}_s\} &= e^{-\frac{1}{2}\lambda^2 t}E\{e^{\lambda(B(t)-B(s))}e^{\lambda B(s)}|\mathcal{A}(s)\} \\ &= e^{-\frac{1}{2}\lambda^2 t}e^{\lambda B(s)}\mathbb{E}e^{\lambda(B(t)-B(s))}wp1 \\ &= e^{-\frac{1}{2}\lambda^2 t+\lambda B(s)}e^{(t-s)\frac{\lambda^2}{2}}\; wp1, \end{aligned}$$

proving that B_6 is a martingale.

Proof of (ξ)

Let $0 \le s < t$.

$$\begin{aligned} E\{M(t)|\mathcal{A}(s)\} &= E\big[e^{i\lambda(B(t)-B(s))+i\lambda B(s)+\frac{t-s}{2}\lambda^2+\frac{s}{2}\lambda^2}|\mathcal{A}(s)\big] \\ &= M(s)e^{\frac{t-s}{2}\lambda^2}e^{i\lambda(B(t)-B(s))} \\ &= M(s), \text{ since } B(t)-B(s) \text{ is independent of } \mathcal{A}(s). \end{aligned}$$

6.4 Properties of conditional expectation

Recall notation : $(\Omega,\ \mathscr{A},\ P)$ is the basic probability space. $\mathcal{B},\ \mathcal{B}_n$, $n \ge 1$ are sub σ-fields. $X,\ Y,\ X_n,\ n \ge 1$ are $\mathscr{A}$-measurable rvs with finite expected values. We will write, say, $E(X|\mathcal{B}) = Z$ without adding $wp1$ or without saying that Z is a version of the conditional expectation. These are meant to be understood.

(i) For $a,\ b \in R$, $E(aX + bY|\mathcal{B}) = aE(X|\mathcal{B}) + bE(Y|\mathcal{B})$.

Proof. Fix a version of the rightside. It is $\mathcal{B}$-measurable. Further, for $B \in \mathcal{B}$,

$$\int_B \{aE(X|\mathcal{B}) + bE(Y|\mathcal{B})\}\,\mathrm{d}P$$

$$= a\int_B E(X|\mathcal{B})\,\mathrm{d}P + b\int_B E(Y|\mathcal{B})\,\mathrm{d}P$$

$$= a\int_B X \,\mathrm{d}P + b\int_B Y \,\mathrm{d}P \text{ by the definition of conditional expectation}$$

$$= \int_B \{aX + bY\}\,\mathrm{d}P.$$

Thus (6.12) is verified.

(ii) If $X \geq 0$ $wp1$, then $Z = E(X|\mathcal{B}) \geq 0$ $wp1$; If $X \leq Y$ $wp1$, then $E(X|\mathcal{B}) \leq E(Y|\mathcal{B})$ $wp1$.

Proof. Let $B = \{\omega : Z(\omega) < 0\}$, so $B \in \mathcal{B}$. By definition, $\int_B Z \,\mathrm{d}P = \int_B X \,\mathrm{d}P = 0$. Hence $P(B) = 0$. The second assertion follows from the fact that $Y - X \geq 0$ and from an application of (i).

(iii) If X is $\mathcal{B}$-measurable, then $E(X|\mathcal{B}) = X$; if X is independent of $\mathcal{B}$, then $E(X|\mathcal{B}) = \mathbb{E}X$.

Proof. The first assertion is obvious and follows immediately from the definition of conditional expectation and **(6.3.2)**(ii).

The constant function, identically equal to $\mathbb{E}X$, is $\mathcal{B}$-measurable. Further for $B \in \mathcal{B}$,

$$\int_B E(X|\mathcal{B})\,\mathrm{d}P = \int_B X \,\mathrm{d}P = \int_\Omega X\chi_B \,\mathrm{d}P = \int_\Omega X \,\mathrm{d}P \times \int_\Omega \chi_B \mathrm{d}P$$

(because of independence)

$$= \mathbb{E}XP(B) = \int_B \mathbb{E}X \,\mathrm{d}P.$$

This being true for all $B \in \mathcal{B}$, the result follows.

Thus if $\mathbb{E}X = 0$ and if rv Y is independent of X, then $g(Y) = 0$ $wp1$ where $g(Y) = E(X|Y)$. That $g(Y) = 0$ $wp1$ can happen even when X, Y are not independent variables is brought home by the following example. Let Y, Z have the joint frequency function

$$f(y, z) = \frac{1}{2\pi}\frac{|y|}{z^2}e^{-\frac{1}{2}y^2\left(\frac{1+z^2}{z^2}\right)}, \quad -\infty < y,\ z\ (z \neq 0)\ < \infty;\ f(y, 0) = 0$$

We note Y has the frequency function $\frac{1}{\sqrt{2\pi}}e^{-\frac{1}{2}y^2}$, $-\infty < y < \infty$. Hence $\mathbb{E}|Y| < \infty$. Let $g(Z) = E(Y|Z)$. Let $\mathcal{B}$ be the σ-field generated by Z. Let $B \in \mathcal{B}$. Hence there exists a linear Borel set $\widehat{B}$ such that $B = Z^{-1}(\widehat{B}) = \{\omega : Z(\omega) \in \widehat{B},\ Y(\omega) \in (-\infty, \infty)\}$. We have, by the definition of conditional expectation,

$$\int_B g(Z)\,\mathrm{d}P = \int_B Y \,\mathrm{d}P = \int_{(-\infty,\, \infty)\times\widehat{B}} yf(y, z)\,\mathrm{d}y\,\mathrm{d}z = 0.$$

Since this is true for every $B \in \mathcal{B}$, it follows that $g(Z) = 0$ $wp1$.

Remark. The above result can be put in the following familiar setting. Let X, Y be iid standard normal variable. Let $Z = \frac{Y}{X}$ if $X \neq 0$ and $Z = 0$ if $X = 0$. Then the joint frequency function of Y and Z is precisely $f(y, z)$ described above. It follows now $E(Y|\frac{Y}{X}) = 0$ $wp1$. This may be compared with **(6.3.3)**(v) where we showed that $E(Y|X - Y) = -\frac{1}{2}(X - Y)$ $wp1$.

(iv) If $Z = E(X|\mathcal{B})$, then $|Z| \leq E(|X||\mathcal{B})$, $\mathbb{E}|Z| \leq \mathbb{E}|X|$ and $\mathbb{E}(Z) = \mathbb{E}X$.

Proof. Write $X = X^+ - X^-$. Then $Z = Z_1 - Z_2$ where $Z_1 = E(X^+|\mathcal{B})$ and $Z_2 = E(X^-|\mathcal{B})$. Hence $|Z| \leq Z_1 + Z_2 = E(X^+|\mathcal{B}) + E(X^-|\mathcal{B}) = E(X^+ + X^-|\mathcal{B})$(by (i)) $= E(|X||\mathcal{B})$.

By (ii), $Z_1 \geq 0$, $Z_2 \geq 0$. We have $\mathbb{E}Z_1 = \int_\Omega Z_1 \,\mathrm{d}P = \int_\Omega X^+ \,\mathrm{d}P$. Similar result holds for Z_2. Since $\mathbb{E}X^\pm < \infty$, it follows that $\mathbb{E}|Z| \leq \mathbb{E}Z_1 + \mathbb{E}Z_2 = \mathbb{E}X^+ + \mathbb{E}X^- = \mathbb{E}(X^+ + X^-) = \mathbb{E}|X| < \infty$. This implies $\mathbb{E}Z$ exists finitely and then $\mathbb{E}Z = \mathbb{E}Z_1 - \mathbb{E}Z_2 = \mathbb{E}X^+ - \mathbb{E}X^- = \mathbb{E}X$.

(v) Monotone convergence theorem. Let $0 \leq X_n \uparrow X$. Then $E(X_n|\mathcal{B}) \uparrow E(X|\mathcal{B})$ $wp1$.

Proof. [Note. In the non-conditional situation, *it is not assumed that* $\mathbb{E}X < \infty$ and we are able to claim $\mathbb{E}X_n \uparrow \mathbb{E}X$. But here we have not defined $E(X|\mathcal{B})$ when $\mathbb{E}X = \infty$. Hence the theorem is proved under the condition $\mathbb{E}X < \infty$]

By (ii), $E(X_n|\mathcal{B}) \uparrow$. Let $Z = \lim_{n\to\infty} E(X_n|\mathcal{B})$. Then, by the monotone convergence theorem (ref. **Q**2(i)), we have, for any $B \in \mathcal{B}$,

$$\int_B X_n \,\mathrm{d}P = \int_\Omega X_n \chi_B \,\mathrm{d}P \to \int_\Omega X \chi_B \,\mathrm{d}P = \int_B X \,\mathrm{d}P.$$

Also, by (**Q**2(i)), $\int_B E(X_n|\mathcal{B}) \to \int_B Z \,\mathrm{d}P$. Now, by the definition of conditional expectation, for any $B \in \mathcal{B}$,

$$\begin{aligned}\int_B Z \,\mathrm{d}P &= \lim_{n\to\infty} \int_B E(X_n|\mathcal{B}) = \lim_{n\to\infty} \int_B X_n \,\mathrm{d}P \\ &= \int_B X \,\mathrm{d}P = \int_B E(X|\mathcal{B}) \,\mathrm{d}P.\end{aligned}$$

Hence $Z = \lim_{n\to\infty} E(X_n|\mathcal{B}) = E(X|\mathcal{B})$ $wp1$.

Note If $X_n \downarrow X$, (the X_ns not necessarily non-negative), then applying the above theorem to the sequence $(X_1 - X_n)$ we have : $E(X_n|\mathcal{B}) \downarrow E(X|\mathcal{B})$ $wp1$.

(vi) Suppose $\mathbb{E}|Y| < \infty$ and $\mathbb{E}|UY| < \infty$. Suppose further that U is $\mathcal{B}$-measurable. Then $E(UY|\mathcal{B}) = UE(Y|\mathcal{B})$.

Proof. We note $UE(Y|\mathcal{B})$ is $\mathcal{B}$-measurable. We will complete the proof by showing

$$\int_B UE(Y|\mathcal{B})\,\mathrm{d}P = \int_B UY\,\mathrm{d}P \text{ for every } B \in \mathcal{B}. \tag{6.14}$$

This is done in stages.

Case 1. Let $U = \chi_A$ for some $A \in \mathcal{B}$. Then

$$\begin{aligned}\int_B \chi_A E(Y|\mathcal{B})\,\mathrm{d}P &= \int_{A\cap B} E(Y|\mathcal{B})\,\mathrm{d}P\\ &= \int_{A\cap B} Y\,\mathrm{d}P \text{ (since } A\cap B \in \mathcal{B})\\ &= \int_B \chi_A Y\,\mathrm{d}P = \int_B UY\,\mathrm{d}P.\end{aligned}$$

Thus (6.14) is verified in this case.

Case 2. Let U be a non-negative simple function. Such a function can be written $U = \sum_{r=1}^m a_r\chi_{B_r}$ where the a_rs are positive, $B_r \in \mathcal{B}$ and the B_rs are mutually exclusive. We note $\mathbb{E}|UY| < \infty$ and

$$\begin{aligned}\int_B UY\,\mathrm{d}P &= \int_B \{\sum_{r=1}^m a_r\chi_{B_r}Y\}\,\mathrm{d}P = \sum_{r=1}^m a_r \int_B \chi_{B_r}Y\,\mathrm{d}P\\ &= \sum_{r=1}^m a_r \int_{B\cap B_r} Y\,\mathrm{d}P\\ &= \sum_{r=1}^m a_r \int_{B\cap B_r} E(Y|\mathcal{B})\,\mathrm{d}P = \sum_{r=1}^m a_r \int_B \chi_{B_r}E(Y|\mathcal{B})\,\mathrm{d}P\\ &= \int_B \sum_{r=1}^m a_r\chi_{B_r}E(Y|\mathcal{B})\,\mathrm{d}P = \int_B UE(Y|\mathcal{B})\,\mathrm{d}P.\end{aligned}$$

Thus (6.14) holds in this case too.

Case 3. Let $U \geq 0$. We can find non-negative $\mathcal{B}$-measurable simple functions $U_n \uparrow U$.

Case 3a. Additionally, let $Y \geq 0$. We have

$$\begin{aligned}\int_B E(UY|\mathcal{B}) &= \int_B UY\,\mathrm{d}P = \int_B \lim_{n\to\infty} U_nY\,\mathrm{d}P\\ &= \lim_{n\to\infty}\int_B U_nY\,\mathrm{d}P\end{aligned}$$

(by Monotone Convergence theorem for integrals)

$$= \lim_{n\to\infty} \int_B E(U_n Y|\mathcal{B})\, \mathrm{d}P \text{ (ref. definition } \textbf{(6.3.1)}\text{)}$$

$$= \lim_{n\to\infty} \int_B U_n E(Y|\mathcal{B})\, \mathrm{d}P \text{ (ref. case 2)}$$

$$= \int_B \lim_{n\to\infty} U_n E(Y|\mathcal{B})\, \mathrm{d}P$$

(again by Monotone Convergence theorem for integrals)

$$= \int_B UE(Y|\mathcal{B})\, \mathrm{d}P.$$

Since $B \in \mathcal{B}$ is arbitrary, this shows $\mathbb{E}(UY|\mathcal{B}) = U\mathbb{E}(Y|\mathcal{B})$.

Case 3b. $U \geq 0$, Y not necessarily non-negative. We note $\mathbb{E}|UY| < \infty$ implies $\mathbb{E}|U_nY| < \infty$, $\mathbb{E}|U_nY^+| < \infty$ and $\mathbb{E}|U_nY^-| < \infty$. Hence

$$\int_B UY\, \mathrm{d}P = \int_B U(Y^+ - Y^-)\, \mathrm{d}P$$

$$= \int_B UY^+\, \mathrm{d}P - \int_B UY^-\, \mathrm{d}P$$

(by the addition rule for convergent integrals)

$$= \int_B UE(Y^+|\mathcal{B})\, \mathrm{d}P - \int_B UE(Y^-|\mathcal{B})\, \mathrm{d}P \text{ (ref. case 3a)}$$

$$= \int_B U\{E(Y^+|\mathcal{B}) - E(Y^-|\mathcal{B})\}\, \mathrm{d}P$$

$$= \int_B UE(Y^+ - Y^-|\mathcal{B})\, \mathrm{d}P \text{ (by (i))} = \int_B UE(Y|\mathcal{B})\, \mathrm{d}P,$$

completing the proof in this case.

Case 4. Let *rvs* U, Y be arbitrary but subject to the stated conditions. Then

$$\int_B UY\, \mathrm{d}P = \int_B (U^+Y - U^-Y)\, \mathrm{d}P$$

$$= \int_B U^+Y\, \mathrm{d}P - \int_B U^-Y\, \mathrm{d}P$$

(since the integrals are finite)

$$= \int_B E(U^+Y|\mathcal{B})\, \mathrm{d}P - \int_B E(U^-Y|\mathcal{B})\, \mathrm{d}P$$

(ref. definition **(6.3.1)**)

$$= \int_B U^+ E(Y|\mathcal{B})\,\mathrm{d}P - \int_B U^- E(Y|\mathcal{B})\,\mathrm{d}P \text{ (by Case 3a)}$$

$$= \int_B (U^+ - U^-) E(Y|\mathcal{B})\,\mathrm{d}P = \int_B U E(Y|\mathcal{B})\,\mathrm{d}P.$$

Thus under the stated conditions, $E(UY|\mathcal{B}) = UE(Y|\mathcal{B})$. □

(vii) Let $\mathcal{B}_1 \subset \mathcal{B}_2$. Let $E(X|\mathcal{B}_i) = X_i,\ i = 1,\ 2$. Then

(α_1) $E\{X_1|\mathcal{B}_2\} = X_1\ wp1$.

(α_2) $E\{X_2|\mathcal{B}_1\} = X_1\ wp1$.

(β) $E(X_2|X_1) = X_1\ wp1$

(γ) If $\mathcal{B} \subset \mathcal{B}_1$ and if X_1 is $\mathcal{B}$-measurable, then $E(X|\mathcal{B}) = X_1\ wp1$.

(δ) $\{\omega : X_1(\omega) \neq X_2(\omega)\} \in \mathcal{B}_2$.

Proof. We appeal repeatedly to the definition of conditional expectation.

(α_1) Since, for every $B \in \mathcal{B}_2$, $\int_B E(X_1|\mathcal{B}_2)\,\mathrm{d}P = \int_B X_1\,\mathrm{d}P$ and since both $E(X_1|\mathcal{B}_2)$ and X_1 are $\mathcal{B}_2$-measurable, it follows that $E(X_1|\mathcal{B}_2) = X_1\ wp1$.

(α_2) Since, for every $B \in \mathcal{B}_1$,

$$\int_B E(X_2|\mathcal{B}_1)\,\mathrm{d}P = \int_B X_2\,\mathrm{d}P = \int_B E(X|\mathcal{B}_2)\,\mathrm{d}P = \int_B X\,\mathrm{d}P$$
$$= \int_B E(X|\mathcal{B}_1)\,\mathrm{d}P = \int_B X_1\,\mathrm{d}P.$$

Since this is true for all $B \in \mathcal{B}_1$ and since both $E(X_2|\mathcal{B}_1)$ and X_1 are $\mathcal{B}_1$-measurable, it follows that $E(X_2|\mathcal{B}_1) = X_1\ wp1$.

(β) Let $\mathcal{C}$ be the σ-field generated by X_1. We note that $\mathcal{C} \subset \mathcal{B}_1$. If $B \in \mathcal{C}$, then $\int_B E(X|\mathcal{C})\,\mathrm{d}P = \int_B X\,\mathrm{d}P = \int_B E(X|\mathcal{B}_1)\,\mathrm{d}P$ (since $B \in \mathcal{B}_1$) $= \int_B X_1\,\mathrm{d}P$. Hence $E(X|\mathcal{C}) = X_1$. We have , $E(X_2|X_1) = E(X_2|\mathcal{C}) = E(X|\mathcal{C})$ (by (α_2)) $= X_1\ wp1$.

(γ) Let $B \in \mathcal{B}$. We have

$$\int_B E(X|\mathcal{B})\,\mathrm{d}P = \int_B X\,\mathrm{d}P = \int_B E(X|\mathcal{B}_1)\,\mathrm{d}P = \int_B X_1\,\mathrm{d}P.$$

Both X_1 and $E(X|\mathcal{B})$ are $\mathcal{B}$-measurable and the above equality holds for all $B \in \mathcal{B}$. Hence $E(X|\mathcal{B}) = X_1\ wp1$.

(δ) The measurability assertion follows since both X_1 and X_2 are $\mathcal{B}_2$-measurable. □

(viii) If $\mathbb{E}X^2 < \infty$, then $Y^2 \le E(X^2|\mathcal{B})\ wp1$ and $\mathbb{E}Y^2 \le \mathbb{E}X^2$ where $Y = E(X|\mathcal{B})$.

Proof. From the identity $x^2 = y^2 + 2y(x-y) + (x-y)^2$, we have, for all $x,\ y,\ x^2 \geq y^2 + 2y(x-y)$. Set $x = X$ and $y = Y = E(X|\mathcal{B})$, fixing a version of Y. We have $X^2(\omega) \geq Y^2(\omega) + 2Y(\omega)(X(\omega) - Y(\omega))$ for all $\omega \in \Omega$. Suppose $\mathbb{E}|Y(X-Y)| < \infty$. Then $E\left(Y(X-Y)|\mathcal{B}\right) = YE\left((X-Y)|\mathcal{B}\right)$ (by (vi)) $= 0\ wp1$ (by (i), remembering $\mathbb{E}|Y| < \infty$ and hence $E(X - Y|\mathcal{B}) = E(X|\mathcal{B}) - Y$). Thus if $\mathbb{E}Y^2 < \infty$ and if $\mathbb{E}|Y(X - Y)| < \infty$, then the desired inequality follows by conditioning both sides of the relation $X^2 \geq Y^2 + 2Y(X-Y)$ on $\mathcal{B}$ and appealing to (ii).

Since we do not know that $\mathbb{E}|Y(X-Y)| < \infty$, we proceed as follows.

Define $B_m = \{\omega : |Y(\omega)| \leq m\}$ and note that $B_m \in \mathcal{B}$. By the Radon-Nikodym theorem, Y is finite valued, $B_m \uparrow \Omega$. We have $X^2\chi_{B_m} \geq Y^2\chi_{B_m} + Y(X-Y)\chi_{B_m}$. All terms in this inequality have finite expected values. Hence $E(X^2\chi_{B_m}|\mathcal{B}) \geq E(Y^2\chi_{B_m}|\mathcal{B}) + 2E(Y(X-Y)\chi_{B_m}|\mathcal{B})\ wp1$. Now $wp1,\ E(Y(X-Y)\chi_{B_m}|\mathcal{B}) = Y\chi_{B_m}E(X-Y|\mathcal{B}) = 0$. Hence $wp1$, $E(X^2\chi_{B_m}|\mathcal{B}) \geq E(Y^2\chi_{B_m}|\mathcal{B})$. Or $\chi_{B_m}\ E(X^2|\mathcal{B}) \geq Y^2\chi_{B_m}$. Let $m \to \infty$ and complete the proof of the first part.

The second part follows immediately from (iv) above,
$\mathbb{E}Y^2 \leq \mathbb{E}\left(E(X^2|\mathcal{B})\right) = EX^2$. □

(ix) Let $X_k,\ 1 \leq k \leq n$ be rvs such that $\mathbb{E}X_k = 0,\ \mathbb{E}X_k^2 = \sigma_k^2$ and $\mathbb{E}X_jX_k = 0,\ j \neq k$. Write $S_j = \sum_{r=1}^{j} X_r,\ 1 \leq j \leq n$. We note $\mathbb{E}S_j = 0,\ \mathbb{E}S_j^2 = s_j^2 = \sum_{r=1}^{j}\sigma_r^2,\ 1 \leq j \leq n$. If

$$E(X_j|X_1,\ X_2, \ldots,\ X_{j-1}) = 0,\ j > 1 \tag{6.15}$$

then $P(\max\limits_{1\leq j\leq n} |S_j| \geq \lambda) \leq \frac{s_n^2}{\lambda^2}$.

Proof. We note that if the X_ks are zero mean independent rvs, then (6.15) is satisfied and the inequality is precisely Kolmogorov inequality (ref. **(4.4.3)**). Thus the present Lemma extends Kolmogorov inequality to uncorrelated rvs satisfying (6.15). The proof is on the same lines as for Kolmogorov inequality.

The following consequence of (6.15) will be used in the proof. Let $\mathcal{B}_r$ be the σ field generated by the variables $X_\nu,\ 1 \leq \nu \leq r$ and note that (6.15) is equivalent to $E(X_j|\mathcal{B}_{j-1}) = 0$. Then for all $j \geq r+1$, $E(X_j|\mathcal{B}_r) = 0$. This claim is true since by (vii), $E(X_j|\mathcal{B}_r) = \mathbb{E}\left(E(X_j|\mathcal{B}_{j-1})|\mathcal{B}_r\right) = 0$.

We note first that for $n = 1$, this inequality is the same as the Tchebichev inequality (ref. **(4.4.1)** with $\alpha = 2$). Let $n \geq 2$.

Define events: $A_1 = \{|S_1| \geq \lambda\}$, $A_r = \{|S_k| < \lambda,\ 1 \leq k \leq r-1,\ |S_r| \geq \lambda, \},\ 2 \leq r \leq n$. We note that the n events $A_1,\ A_2, \ldots,\ A_n$ are mutually exclusive and that $B_n = \cup_1^n A_r = \{\max\limits_{1\leq r\leq n} |S_r| \geq \lambda\}$. Denote by χ_r the indicator function of the event A_r. Thus

$$\begin{aligned} s_n^2 &= \int_\Omega S_n^2(\omega)\,\mathrm{d}\,P \geq \int_{B_n} S_n^2(\omega)\,\mathrm{d}\,P(\omega) \\ &= \sum_{r=1}^{n} \int_{A_r} (S_n(\omega) - S_r(\omega) + S_r(\omega))^2\,\mathrm{d}\,P(\omega) \\ &\geq \sum_{r=1}^{n} \int_{A_r} \{2(S_n(\omega) - S_r(\omega))S_r(\omega) + (S_r(\omega))^2\}\,\mathrm{d}\,P(\omega) \\ &= \sum_{r=1}^{n} \int_{A_r} S_r^2\,\mathrm{d}\,P, \end{aligned}$$

since

$$\begin{aligned} \int_{A_r} (S_n - S_r)S_r\,\mathrm{d}\,P &= \mathbb{E}((S_n - S_r)S_r\chi_{A_r}) \\ &= \mathbb{E}(E((S_n - S_r)S_r\chi_{A_r}|\mathcal{B}_r)) \text{ (by (vii) above)} \\ &= S_r\chi_{A_r}E(S_n - S_r|\mathcal{B}_r) \end{aligned}$$

(using the fact $S_r\chi_{A_r}$ is $\mathcal{B}_r$-measurable and applying (vi) above)

$$= S_r\chi_{A_r} \sum_{s=r+1}^{n} E(X_s|\mathcal{B}_r)\text{(by (i) above)} = 0.$$

Hence

$$s_n^2 \geq \sum_{r=1}^{n} \int_{A_r} S_r^2\,\mathrm{d}\,P \geq \lambda^2 \sum_{r=1}^{n} P(A_r) = \lambda^2 P(B_n),$$

which is the desired result. □

(x) Let $\mathcal{B}_n \uparrow$. Let $Y_n = E(X|\mathcal{B}_n)$. Then for all $\lambda > 0$,

$$P(\max_{1\leq k\leq n} |Y_k| > \lambda) \leq \frac{2\mathbb{E}|X|}{\lambda} \tag{6.16}$$

and

$$P(\sup_{k\geq 1} |Y_k| > \lambda) \leq \frac{2\mathbb{E}|X|}{\lambda}. \tag{6.17}$$

Proof. We note that for $n = 1$, this inequality follows from the Tchebichev inequality (ref. **(4.4.1)** with $\alpha = 1$), since $\mathbb{E}Y_1 = \mathbb{E}X$. Let $n \geq 2$.

We consider first the case $X \geq 0$. This implies $Y_n \geq 0\ wp1,\ n \geq 1$.

Define events for $2 \le k \le n$,

$$E_1 = \{\omega : Y_1(\omega) > \lambda\}; \; E_k = \{\omega : \max_{1\le j\le k-1} Y_j(\omega) \le \lambda, \; Y_k(\omega) > \lambda\}.$$

We note $E_1, E_2, \ldots, E_n$ are mutually exclusive and $E = \cup_1^n E_k = \{\omega : \max_{1\le j\le n} Y_j > \lambda\}$. We note that for each k, the variables $Y_1, Y_2, \ldots, Y_k$ are $\mathcal{B}_k$- measurable. Hence $E_k \in \mathcal{B}_k$, $1 \le k \le n$. Now, $\mathbb{E}X = \int_\Omega X \,\mathrm{d}P \ge \int_E X \,\mathrm{d}P = \sum_{k=1}^n \int_{E_k} X \,\mathrm{d}P = \sum_{k=1}^n \int_{E_k} E(X|\mathcal{B}_k) \,\mathrm{d}P = \sum_{k=1}^n \int_{E_k} Y_k \,\mathrm{d}P$
$\ge \lambda \sum_{k=1}^n P(E_k) = \lambda P(E)$. This establishes (6.16) when $X \ge 0$.

When X is not necessarily non-negative but $\mathbb{E}|X| < \infty$, Write $X = X^+ - X^-$, $Y_j = E(X|\mathcal{B}_j) = \xi_j - \eta_j$, say, where $\xi_j = E(X^+|\mathcal{B}_j)$ and $\eta_j = E(X^-|\mathcal{B}_j)$. From this, $|Y_j| \le \xi_j + \eta_j$; $\max_{1\le j\le n} |Y_j| \le \max_{1\le j\le n} \xi_j + \max_{1\le j\le n} \eta_j$. Hence

$$\begin{aligned} P(\max_{1\le j\le n} |Y_j| > \lambda) &\le P(\max_{1\le j\le n} \xi_j > \frac{\lambda}{2}) + P(\max_{1\le j\le n} \eta_j > \frac{\lambda}{2}) \\ &\le \frac{\mathbb{E}X^+}{\frac{\lambda}{2}} + \frac{\mathbb{E}X^-}{\frac{\lambda}{2}} = 2\frac{\mathbb{E}|X|}{\lambda}. \end{aligned}$$

Let $A_n = \{\omega : \max_{1\le k\le n} |Y_k(\omega)| > \lambda\}$. We note $A_n \uparrow A = \{\omega : \sup_{k\ge 1} |Y_k(\omega)| > \lambda\}$. Hence $P(A) = \lim_{n\to\infty} P(A_n) \le \lim_{n\to\infty} 2\frac{\mathbb{E}|X|}{\lambda}$, establishing (6.17). □

(xi) Let $\mathcal{B}_n \uparrow$. Let $\mathcal{B} = \sigma(\cup_n \mathcal{B}_n)$, the minimal σ-field containing the $\mathcal{B}_n$s. If $\mathbb{E}|X| < \infty$, then
(α) $E(X|\mathcal{B}_n) \stackrel{wp1}{\longrightarrow} E(X|\mathcal{B})$ and
(β) $E(X|\mathcal{B}_n) \to E(X|\mathcal{B})$ in L_1-norm.

Proof. We first establish the claim (α) under the assumtion $\mathbb{E}X^2 < \infty$. This is done through steps 1 to 5 below. In step 6, the claims (α) and (β), as stated, are proved.

Step 1. Write $X_n = E(X|\mathcal{B}_n)$, fixing a particular version. By property (viii) above,

$$\mathbb{E}X_n^2 \le \mathbb{E}X^2 < \infty. \tag{6.18}$$

Let $m < n$. We note X_m is $\mathcal{B}_m$-measurable. Hence the σ-field generated by X_m is contained in $\mathcal{B}_m$. It then follows from property (vii) above, $E(X_n|X_m) = X_m \; wp1$. We now prove

$$E(X_n|X_r, \; 1 \le r \le n-1) = X_{n-1} \; wp1, \; n \ge 2 \tag{6.19}$$

Let $\mathcal{B}_n^*$ be the σ-field generated by $X_r,\ 1 \le r \le n$. We note $\mathcal{B}_n^* \subset \mathcal{B}_n$. We have $wp1,\ E(X_{n+1}|X_r,\ 1 \le r \le n) = E(X_{n+1}|\mathcal{B}_n^*) = E\left(E(X|\mathcal{B}_{n+1})|\mathcal{B}_n^*\right)$ $= E(X|\mathcal{B}_n^*)$ (by property (vii))$= E(X|\mathcal{B}_n)$ (again by property (vii)) $= X_n$.

Step 2. Write $Y_1 = X_1,\ Y_n = X_n - X_{n-1},\ n \ge 2$. Let $\mathbb{E}X = m$. Then $Y_1 - m,\ Y_2,\ Y_3,\ Y_4, \ldots$. is a zero mean orthogonal sequence of rvs and $\sum_{k=1}^{\infty} var(Y_k) < \infty$.

To see this, we note $\mathbb{E}X_n = m,\ n \ge 1$. Hence $\mathbb{E}(Y_1 - m) = 0$. If $n \ge 2$, $\mathbb{E}Y_n = \mathbb{E}(X_n - X_{n-1}) = \mathbb{E}X - \mathbb{E}X = 0$. For $2 \le m < n$, $\mathbb{E}Y_mY_n =$ $\mathbb{E}X_mX_n - \mathbb{E}X_mX_{n-1} - \mathbb{E}X_{m-1}X_n + \mathbb{E}X_{m-1}X_{n-1}$. Now for $n \ge m$, $\mathbb{E}X_mX_n = \mathbb{E}\left(E(X_mX_n|X_m)\right) = \mathbb{E}\left(X_mE(X_n|X_m)\right) = \mathbb{E}X_m^2$. Hence, for $2 \le m < n$, $\mathbb{E}Y_mY_n = \mathbb{E}X_m^2 - \mathbb{E}X_m^2 - \mathbb{E}X_{m-1}^2 + \mathbb{E}X_{m-1}^2 = 0$. For $n \ge 2$, $\mathbb{E}(Y_1 - m)Y_n = \mathbb{E}\{(X_1 - m)(X_n - X_{n-1})\} = \mathbb{E}X_1X_n - \mathbb{E}X_1X_{n-1} - m\mathbb{E}X_n + m\mathbb{E}X_{n-1} = 0$. The claim of orthogonality follows.

To prove the finiteness of the infinite series, we first note from step 1 that $\mathbb{E}X_n^2 \le \mathbb{E}X^2 < \infty$. Now,

$$\begin{aligned}\sum_{k=1}^{\infty} var(Y_k) &= \lim_{n\to\infty} \sum_{k=1}^{n} var(Y_k) = \lim_{n\to} var(\sum_{k=1}^{n} Y_k) \\ &= \lim_{n\to\infty} var(X_n + m) \\ &\le \lim_{n\to\infty} \mathbb{E}X_n^2 \le \mathbb{E}X^2 < \infty.\end{aligned}$$

Step 3. For $n \ge 2$, $wp1$,

$$\begin{aligned}E(Y_n|Y_1,\ Y_2, \ldots,\ Y_{n-1}) &= E(Y_n|X_1,\ X_2, \ldots,\ X_{n-1}) \\ &= E(X_n - X_{n-1}|X_1,\ X_2, \ldots,\ X_{n-1}) \\ &= X_{n-1} - X_{n-1} = 0.\end{aligned}$$

Hence the sequence (Y_n) satisfies (6.15) and its conclusion applies.

Step 4. $wp1$, $\lim_{n\to\infty} X_n = X_\infty$ exists finitely.

To prove this, let $\mathbb{E}Y_j^2 = \sigma_j^2,\ j \ge 1$. Given $\varepsilon > 0$,

$$\begin{aligned}&\lim_{n\to\infty} P(\sup_{m\ge n} |X_m - X_n| > \varepsilon) \\ &= \lim_{n\to\infty} P(\lim_{k\to\infty} \max_{n\le m\le n+k} |X_m - X_n| > \varepsilon) \\ &= \lim_{n\to\infty} \lim_{k\to\infty} P(\max_{n\le m\le n+k} |X_m - X_n| > \varepsilon) \\ &= \lim_{n\to\infty} \lim_{k\to\infty} P(\max_{n\le m\le n+k} |Y_{n+1} + Y_{n+2} + \cdots + Y_m| > \varepsilon) \\ &\le \lim_{n\to\infty} \lim_{k\to\infty} \frac{\sum_{j=n+1}^{n+k} \sigma_j^2}{\varepsilon^2} \text{ (by (ix))}\end{aligned}$$

$$= 0 \text{ (since } \sum_{j=2}^{n} \sigma_j^2 = \mathbb{E}(\sum_{j=2}^{n} Y_j)^2 \leq \mathbb{E}X^2, \text{ ref. Step 2)}$$

Hence $X_\infty = \lim_{n\to\infty} X_n$ exists finitely $wp1$.

Step 5. $wp1$, $X_\infty = \mathbb{E}(X|\mathcal{B})$. i.e., $\lim_{n\to\infty} \mathbb{E}(X|\mathcal{B}_n) = \mathbb{E}(X|\mathcal{B})\ wp1$.

To prove this, recall that the discussion in Step 4 shows that (X_n) is a Cauchy sequence of elements in the Hilbert space $L_2 = L_2(\Omega, \mathcal{B}, P)$ of square integrable functions defined on Ω and $\mathcal{B}$-measurable. Hence there exists $Z \in L_2$ such that $\lim_{n\to\infty} \mathbb{E}(X_n - Z)^2 = 0$. We record two implications of this:

$$\lim_{n\to\infty} \mathbb{E}|X_n - Z| = 0 \tag{6.20}$$

and, by **(4.3.2)**(viii) and **(4.3.2)**(x), there exists a subsequence (n_k) such that $X_{n_k} \stackrel{wp1}{\longrightarrow} Z$. Hence (ref. Step 4) $Z = X_\infty\ wp1$.

Let $\mathcal{F} = \cup_n \mathcal{B}_n$, so $\mathcal{B} = \sigma(\mathcal{F})$. Let $B \in \mathcal{F}$. Hence there exists N such that $B \in \mathcal{B}_n$ for all $n \geq N$. Then (6.20), which can be written $\lim_{n\to\infty} \mathbb{E}|X_n - X_\infty| = 0$, implies $\lim_{n\to\infty} \int_B X_n \,\mathrm{d}P = \int_B X_\infty \,\mathrm{d}P$ since

$$|\int_B X_n \,\mathrm{d}P - \int_B X_\infty \,\mathrm{d}P| \leq \int_B |X_n - X_\infty| \,\mathrm{d}P \leq \mathbb{E}|X_n - X_\infty| \to 0.$$

But

$$\int_B X_n \,\mathrm{d}P = \int_B E(X|\mathcal{B}_n) \,\mathrm{d}P = \int_B X \,\mathrm{d}P = \int_B E(X|\mathcal{B}) \,\mathrm{d}P.$$

Thus

$$\int_B E(X|\mathcal{B}) \,\mathrm{d}P = \int_B X_\infty \,\mathrm{d}P, \quad B \in \mathcal{F}. \tag{6.21}$$

It remains to prove $X_\infty = E(X|\mathcal{B})\ wp1$. We note X_∞ is $\mathcal{B}$-measurable, since each X_n is so. $E(X|\mathcal{B})$ is obviously $\mathcal{B}$-measurable. Hence, the required result will follow if we show that (6.21) is true for all $B \in \mathcal{B}$. We argue as follows.

Taking $B = \Omega$ in (6.21), we get $\mathbb{E}X = \mathbb{E}X_\infty$ Let $\mathcal{M}$ be the collection of all sets $B \in \mathcal{B}$ for which (6.21) holds and note that $\mathcal{M} \supset \mathcal{F}$. Trivially, $\mathcal{M}$ is closed under complementation, since $\mathbb{E}X = \mathbb{E}X_\infty$. Let $A_n \in \mathcal{M}$ and let $A_n \uparrow A$. We note $(X\chi_{A_n})$ and $(X_\infty \chi_{A_n})$ are bounded respectively by the integrable functions X and X_∞. Hence $\int_A X \,\mathrm{d}P = \int_\Omega X\chi_A \,\mathrm{d}P = \int_\Omega \lim_{n\to\infty} X\chi_{A_n} \,\mathrm{d}P = \lim_{n\to\infty} \int_\Omega X\chi_{A_n} \,\mathrm{d}P$, by the Bounded Convergence Theorem, $= \lim_{n\to\infty} \int_\Omega X_\infty \chi_{A_n} \,\mathrm{d}P$ (since $A_n \in \mathcal{M}$) $= \int_\Omega \lim_{n\to\infty} X_\infty \chi_{A_n} \,\mathrm{d}P$ (again by the same Theorem) $= \int_\Omega X_\infty \chi_A \, \mathrm{d}P = \int_A X_\infty \, \mathrm{d}P$. This shows that $\mathcal{M}$ is closed under monotone limits. i.e. $\mathcal{M}$ is a monotone class of sets containing $\mathcal{F}$. Since the smallest monotone class containing a field $\mathcal{F}$ of sets is the σ-field $\sigma(\mathcal{F})$ (ref. Theorem B, p27, [H]), it follows that (6.21) holds for every $B \in \mathcal{B}$.

Hence the proof that $\lim_{n\to\infty} E(X|\mathcal{B}_n) = E(X|\mathcal{B})$ $wp1$ is complete when $\mathbb{E}X^2 < \infty$.

Step 6. Let $\mathcal{B}_n,\ \mathcal{B},\ \mathcal{F}$ be as in Step 5. Let $\mathbb{E}|X| < \infty$ (no higher moment assumed).

Proof of (α) and (β). Recall (α): $E(X|\mathcal{B}_n) \overset{wp1}{\longrightarrow} E(X|\mathcal{B})$ and (β) $E(X|\mathcal{B}_n) \to E(X|\mathcal{B})$ in L_1-norm.

Obviously, it is enough if (α) and (β) are proved for non-negative variables X.

Define for $\nu \geq 1$, $X_\nu = X$ if $X \leq \nu$ and $X_\nu = 0$ if $X > \nu$. We note that, as $\nu \to \infty$, $0 \leq X_\nu(\omega) \uparrow X(\omega)$ for all ω and $\mathbb{E}X_\nu \to \mathbb{E}X$.
$|E(X|\mathcal{B}_n) - E(X|\mathcal{B})| \leq |E(X|\mathcal{B}_n) - E(X_\nu|\mathcal{B}_n)| + |E(X_\nu|\mathcal{B}_n) - E(X_\nu|\mathcal{B})| + |E(X_\nu|\mathcal{B}) - E(X|\mathcal{B})| = a_{n,\nu} + b_{n,\nu} + c_\nu$, say.

By property (v) above, $\lim_{\nu\to\infty} c_\nu = 0$.

Since $\mathbb{E}X_\nu^2 < \infty$, the result just proved applies and we conclude that $wp1$, $\lim_{n\to\infty} b_{n,\nu} = 0$.

$0 \leq a_{n,\nu} = E(X - X_\nu|\mathcal{B}_n)$. Write $u_\nu = \sup_{n\geq 1} a_{n,\nu}$. By property (x),

$$P(u_\nu > \varepsilon) \leq \frac{2\mathbb{E}(X - X_\nu)}{\varepsilon}.$$

Hence $u_\nu \overset{pr}{\to} 0$. Since $u_\nu(\omega) \downarrow$ as $\nu \uparrow$, it follows that $u_\nu \overset{wp1}{\longrightarrow} 0$ *iff* $u_\nu \overset{pr}{\to} 0$ which holds. Thus $0 \leq \lim_{\nu\to\infty} \lim_{n\to\infty} a_{n,\nu} \leq \lim_{\nu\to\infty} u_\nu = 0$ $wp1$.

With this, the proof of (α) is complete that $E(X|\mathcal{B}_n) \overset{wp1}{\longrightarrow} E(X|\mathcal{B})$.

Since $X_\nu \leq X$,it follows $|c_\nu| \leq 2E(X|\mathcal{B})$. Since $\mathbb{E}\{E(X|\mathcal{B})\} = \mathbb{E}X < \infty$ and since $c_\nu \overset{wp1}{\longrightarrow} 0$, we can appeal to the bounded convergence theorem and claim c_ν converges to 0 in L_1-norm.

The proof of the L_1-convergence of $b_{n,\nu}$ to 0 is contained in step 5, since $\mathbb{E}X_\nu^2 < \infty$ and hence the steps of the proof of (α) are applicable.

We note $X - X_\nu \geq 0$. Hence $\int_\Omega a_{n,\nu} \, \mathrm{d}P = \mathbb{E}\{E(X - X_\nu|\mathcal{B}_n)\}$ $= \mathbb{E}X - \mathbb{E}X_\nu \to 0$ (by the monotone convergence theorem). The convergence to 0 in L_1 of each of the three sequences $(a_{n,\nu})$, $(b_{n,\nu})$, (c_ν) imply the truth of the claim in (β). □

(xii) Let rv Y have range contained in $\{0, 1, 2, \ldots..\}$. Let $P(Y = n) = p_n$, $\sum_n p_n = 1$. Let $\sum_n np_n = \mathbb{E}Y < \infty$. Let $\mathcal{B} \subset \mathscr{A}$ be a sub σ-field. Then

$wp1,\ E(Y|\mathcal{B}) = \sum_{n=1}^{\infty} nP(Y = n|\mathcal{B})$.

Proof. Let χ_n be the characteristic function of the event $(Y = n)$. We have $Y = \sum_{n=0}^{\infty} n\chi_n$. Hence

$E(Y|\mathcal{B}) = E(\lim\limits_{n\to\infty} \sum_{\nu=1}^{n} \nu\chi_\nu|\mathcal{B})$ = (by **(6.4)**(v)) $\lim\limits_{n\to\infty} E(\sum\limits_{\nu=1}^{n} \nu\chi_\nu|\mathcal{B})$

$= \lim\limits_{n\to\infty} \sum_{\nu=1}^{n} \nu\, E(\chi_\nu|\mathcal{B})$ (by **(6.4)**(i))$= \sum_{n=1}^{\infty} nP(Y = n|\mathcal{B})$. □

For another proof of this result ref.**(6.5.4)**

(xiii) Recall the discussion in Sec. **(6.3.3)**(vi). We establish now the relation at (6.13).

Follow the notation of **(6.3.3)**(vi). ((6.13) is copied below and renumbered as (6.22))

For every $D \in \mathcal{U}$,

$$P((X,\ Y) \in D|\mathscr{A}_1) = P((X,\ Y) \in D|Y)\ \ wp1. \tag{6.22}$$

Proof. If $D = B \times \mathrm{E},\ B \in \mathcal{X}$ and $\mathrm{E} \in \mathcal{Y}$, then the left side of (6.22) is

$$\begin{aligned} P[X \in B,\ Y \in \mathrm{E}|\mathscr{A}_1] &= E[\chi_{\{X\in B\}}\chi\{Y \in \mathrm{E}\}|\mathscr{A}_1] \\ &= \chi_{\{Y\in \mathrm{E}\}}E[\chi_{\{X\in B\}}|\mathscr{A}_1] \text{ (by } \mathbf{(6.4)}\text{(vi))} \\ &= \chi_{\{Y\in\mathrm{E}\}}\mathbb{E}\chi_{\{X\in B\}} = \chi_{\{Y\in\mathrm{E}\}}P(X \in B)\ wp1. \end{aligned}$$

Similar arguments lead to rightside of (6.22) being equal to the same expression. Thus (6.22) holds if D is a rectangle, $D = B\times\mathrm{E},\ B \in \mathcal{X}$ and $\mathrm{E} \in \mathcal{Y}$. By the additivity property of conditional probability (ref. **(6.4)**(i)), it is immediate from this that (6.22) holds if D is the finite disjoint union of rectangles of the above type. We know (ref. Theorem E, p139, [H]) that the collection of all the finite disjoint union of rectangles $B \times \mathrm{E},\ B,\ \mathrm{E}$ as above, forms a field, say, $\mathbb{R}$.

Let $\mathcal{D}$ be the collection of all sets $D \in \mathcal{X} \otimes \mathcal{Y}$ for which (6.22) holds. We note $\mathbb{R} \subset \mathcal{D}$. Let $D_n \in \mathcal{D},\ n \geq 1,\ D_n \uparrow D$. By the monotone convergence theorem for conditional expectation (ref. **(6.4)**(v)), it follows that (6.22) holds for D. Thus $\mathcal{D}$ is closed under monotone limits. Since it contains the field $\mathbb{R}$, it contains (ref. Theorem A, p 27, [H]) the σ-field generated by $\mathbb{R}$. i.e., it contains $\mathcal{U}$. Thus (6.22) holds for every $D \in \mathcal{U}$, as was to be proved. □

6.4.1. The details promised in (**(6.3.3)**(ii)) are presented here.

Let $(M,\ d)$ be a complete separable metric space, $\mathcal{M}$ its Borel σ-field and $(M, \mathcal{M},\ \mu)$ a probability measure space. Let P denote the completion of the measure μ. Let m denote the corresponding enhanced σ-field containing all the subsets of sets of μ-measure 0. We note (ref. **(1.9.4)**) that P is a tight measure. Let X, Y be real rvs defined on M. Let A a Borel subset of the real line R; let $\mathrm{E} = Y^{-1}(A)$ and note $\mathrm{E} \in \mathcal{m}$.

Define $\widetilde{B}_{k,n} = [\frac{k}{2^n}, \frac{k+1}{2^n})$, $B_{k,n} = X^{-1}(\widetilde{B}_{k,n})$, $\widehat{B}_n = \cup_k \widetilde{B}_{k,n}$. Write $\mathcal{B}_n = \sigma(B_{k,n})$ and note that $\mathcal{B}_n \uparrow$. Let $\mathcal{B} = \sigma(\cup_n \mathcal{B}_n)$ and note that $\mathcal{B}$ is the σ-field generated by X.

By **(6.4)**(xi) there exists $C_1 \in \mathcal{m}$ such that $P(C_1) = 1$ and such that $\omega \in C_1$ implies $P(\mathrm{E}|\mathcal{B})(\omega) = \lim_{n\to\infty} P(\mathrm{E}|\mathcal{B}_n)(\omega)$.

Let $D = X(C_1)$, the image of C_1 under X. It is possible D is not a Borel set. But $X^{-1}(D) = X^{-1}(X(C_1)) \supset C_1$ and hence differs from C_1 by a subset of a set of P-measure 0. Since P is complete it follows $X^{-1}(X(C_1)) \in \mathcal{m}$ (with P-measure 1). If D is not a Borel set then, by **(1.10.10)**, there exist Borel sets $D_1 \subset D \subset D_2$ such that $P(X^{-1}(D_2 \sim D_1)) = 0$. This implies $P_1(D_1) = 1$ where P_1 is the pm of X. (If D is a Borel set, take $D_1 = D_2 = D$). Recall definition above of $\widehat{B}_n$ and note that $P_1(\widehat{B}_n) = 1$. This implies $P_1(\widehat{B}_n \sim D_1) = 0$. Hence every point of D_1 is a possible value of X. Write $C = X^{-1}(D_1) \cap C_1$. We note $P(C) = 1$ and $X(C) = D_1$.

For each $\omega \in C$, $P(E|X)(\omega) = P(\mathrm{E}|\mathcal{B})(\omega) = \lim_{n\to\infty} P(\mathrm{E}|\mathcal{B}_n)(\omega)$. Given $x \in D_1$, find $\omega \in C$ such that $X(\omega) = x$. Now for that ω, $P(\mathrm{E}|X)(\omega) = P(\mathrm{E}|X = x)$.

Further, $P(\mathrm{E}|\mathcal{B}_n)(\omega) = \sum_k P(\mathrm{E}|B_{k,n})\chi_{B_{k,n}}(\omega)$. Since, for n fixed, the $B_{k,n}$ are mutually exclusive and exhaustive, there exists a unique $k = k_n^* = k_n^*(\omega)$ such that $\omega \in B_{k,n}$. If x is a possible value, then $P(B_{k_n^*,n}) > 0$. Thus $P(\mathrm{E}|\mathcal{B}_n) = P(\mathrm{E}|B_{k_n^*,n}) = P(\mathrm{E}|X \in \widetilde{B}_{k_n^*,n}) = P(\mathrm{E}|X \in J_{n;x})$ where $J_{n;x} = \widehat{B}_{k_n^*,n} = [\frac{k_n^*}{2^n}, \frac{k_n^*+1}{2^n})$. Thus $P(\mathrm{E}|X = x) = \lim_{n\to\infty} P(\mathrm{E}|X \in J_{n;x})$. This is true for every $x \in D_1$. i.e. this is true for a set of x with P_1- measure 1. □

6.4.2. As promised, the claim in (4.31) is proved now. Follow the notation in (**(4.6.3)**, Remarks). We note $\mathbb{E}e^{X_n-\theta} = 1$. Now, $\mathbb{E}\{e^{T_{n+1}}|e^{T_\nu}, 1 \le \nu \le n\} = \mathbb{E}\{e^{T_n}e^{X_{n+1}-\theta}|e^{T_\nu}, 1 \le \nu \le n\} = e^{T_n}\mathbb{E}e^{X_{n+1}-\theta} = e^{T_n}$. Hence $(e^{T_n}, n \ge 1)$ is a martingale sequence (for definition ref. **(6.3.3)**). This martingale sequence is L_1-bounded since $\mathbb{E}e^{T_n} = \prod_{k=1}^{n} \mathbb{E}e^{X_k-\theta} = 1$. For such a sequence, the martingale convergence theorem (ref. Theorem 4.1, p 319, [D]) applies and guarantees the convergence $wp1$ of e^{T_n} to a finite valued integrable rv, say $\mathcal{Z}$. Trivially $\mathcal{Z} \ge 0$. $P(\mathcal{Z} > 0) = \alpha > 0$ would imply that $P(T_n$ has a finite limit$) > 0$. But, as already noted in (**(4.6.3)**, Remarks) the probability is zero that T_n has a finite limit. Thus necessarily $P(\mathcal{Z} = 0) = 1$. This is equivalent to saying $T_n \overset{wp1}{\longrightarrow} -\infty$.

[Below we prove a more general result. The just proved result is presented only to place in the hands of the reader a tool for possible uses.]

For a particular value of θ, we have shown that $wp1$, $\sum_n (X_n - \theta) = -\infty$, using the sophisticated tool of the convergence property of L_1 bounded martingales. The question remains : what can we say about the series $\sum_n (X_n - \alpha)$, $\alpha > 0$, arbitrary. We have $T_n = S_n - n\alpha$. We recall $P(A) = 1$ where $A = \{\omega : S_n(\omega) = 0 \text{ i.o.}\}$. It follows that if $\omega \in A$, then $\varliminf_{n\to\infty} T_n(\omega) = -\infty$. i.e. $wp1$, $-\infty$ is a limit point of (T_n). Can ∞ be a limit point of T_n? If it is , we must have $P(S_n - n\alpha > \lambda \text{ i.o.}) = 1$ for every $\lambda > 0$. That would imply $P(\frac{S_n}{n} > \alpha \text{ i.o.}) = 1$. But, by the strong law of large numbers, that is not possible. Hence ∞ is not a limit point of (T_n). Can a finite point u be a limit point of (T_n)? Such a point u would exist if for every $\varepsilon > 0$ it is true that $P(|S_n - n\alpha - u| < \varepsilon \text{ i.o.}) = 1$. i.e., if it is true that $P(|\frac{S_n}{n} - \alpha| < \delta \text{ i.o.}) = 1$ for every $\delta > 0$. If this holds, then it would violate the strong law of large numbers for the X_n sequence. Thus we have succeeded in proving that neither ∞ nor any finit number is a limit point of (T_n). Collecting the results, we conclude $T_n \overset{wp1}{\longrightarrow} -\infty$.

By symmetry, $wp1$, $\sum_n (X_n + \alpha) = \infty$. □

6.5 Regular conditional probability measures

Recall notation : $(\Omega, \mathscr{A}, P)$ is the basic probability space and $\mathcal{B} \subset \mathscr{A}$ is a sub σ-field. In **(6.4)** we established many properties of conditional expectation $E(X|\mathcal{B})$. These apply to conditional probabilities too since $P(A|\mathcal{B}) = E(\chi_A|\mathcal{B})$. Hence by properties (i) and (v) in **(6.4)** we have the following result for mutually exclusive events A_n : if $A = \cup_n A_n$, then $wp1$,

$$P(\cup_n A_n|\mathcal{B}) = E\chi_A|\mathcal{B}) = E(\sum_n \chi_{A_n}|\mathcal{B}) = \sum_n E\chi_{A_n}|\mathcal{B}) = \sum_n P(A_n|\mathcal{B}). \qquad (6.23)$$

This countable additivity tempts us to ask if the set function $P^{\mathcal{B}}$ defined on $\mathscr{A}$ through $P^{\mathcal{B}}(A) = P(A|\mathcal{B})$ is a probability measure. In this connection we note that $P(A|\mathcal{B})$ is a function q of ω and A defined on $\Omega \times \mathscr{A}$ and that (6.23) may fail to be true over a ω-set of zero measure. Further the exceptional set may depend on the sequence (A_n). The union of all the exceptional sets may not even be measurable. Thus there is no guarantee that $q(\omega, .)$ is a pm for all $\omega \in \Omega$, not even for every ω in a set of probability 1.

Definition 6.5.1. Let $(M,\ d)$ be a metric space and m be its Borel σ-field. Let X be a measurable mapping of $(\Omega,\ \mathscr{A},\ P)$ into $(M,\ m)$. Let $\mathcal{B} \subset \mathscr{A}$ be a sub σ-field. Then, a regular conditional probability of X given $\mathcal{B}$, is a function q defined on $\Omega \times m$ such that

$$\text{(i) for each } E \in m,\ q(\cdot, E) \text{ is } \mathcal{B}\text{-measurable} \tag{6.24}$$

$$\text{(ii) for each } \omega \in \Omega,\ q(\omega, \cdot) \text{ is a probability measure on } m \tag{6.25}$$

and

$$\text{(iii) for each E} \in m,\ P(X \in E|\mathcal{B})(\omega) = q(\omega\ ,\ \text{E})\ wp1. \tag{6.26}$$

A function $\mathbb{F}(.\ ,\ .)$ defined on $R^p \times \Omega$ is said to be a regular conditional *df* of the rV **X** if

$$\text{(i) for each } \mathbf{x} \in R^p,\ \mathbb{F}(\mathbf{x}\ ,\ .) \text{ is } \mathcal{B}\text{-measurable} \tag{6.27}$$

$$\text{(ii) for each } \omega \in \Omega,\ \mathbb{F}(.\ ,\ \omega) \text{ is a } df \text{ in } R^p \tag{6.28}$$

and

$$\text{(iii) for each } \mathbf{x} \in R^p,\ \ P(\mathbf{X} \le \mathbf{x}|\mathcal{B})(.) = \mathbb{F}(\mathbf{x}\ ,\ .)\ wp1. \tag{6.29}$$

Theorem 6.5.2. *Let* $\mathbf{Z} = (Z_1, \ldots,\ Z_p)$ *be a* R^p*-valued random vector defined on* $(\Omega,\ \mathscr{A},\ P)$*. Let* $\mathcal{B} \subset \mathscr{A}$ *be a* σ*-field. Then a regular conditional df* $\mathbb{F}(.\ ,\ .)$ *of* **Z** *given* $\mathcal{B}$ *exists.*

Proof. (i) For notation regarding inequalities between vectors in R^p, refer **(1.5.3)**. Call a point in R^p a rational vector if each of its p co-ordinates is a rational number. We recall that the set $\Re$ of all such points is countable and dense in R^p.

To avoid notation clutter, we write down detailed steps for the case $p = 2$.

For $\mathbf{r} = (r_1,\ r_2)$, let $A_{\mathbf{r}} = \mathbf{Z}^{-1}\left((-\infty,\ r_1] \times (-\infty,\ r_2]\right)$. We note that $A_{\mathbf{r}} \subset A_{\mathbf{s}}$ if $\mathbf{r} \le \mathbf{s}$. For each $\mathbf{r} \in \Re$ choose and fix a version of the conditional probability $\mathbb{F}(\mathbf{r}\ ,\ \omega) = P(\mathbf{Z} \le \mathbf{r}|\mathcal{B})(\omega) = \mathbb{E}(\ \chi_{A_{\mathbf{r}}}|\mathcal{B})(\omega)$.

Since, for $\mathbf{r} \le \mathbf{s}$, $\chi_{A_{\mathbf{r}}} \le \chi_{A_{\mathbf{s}}}$, it follows, by **(6.4)**(ii), that $\mathbb{F}(\mathbf{r}\ ,\ .) \le \mathbb{F}(\mathbf{s}\ ,\ .)\ wp1$ if $\mathbf{r} \le \mathbf{s}$. This implies, since pairs $(\mathbf{r},\ \mathbf{s})$ are countable in number, that there exists $B \in \mathscr{A}$ with $P(B) = 1$ such that $(\omega \in B) \Rightarrow (\mathbb{F}(\mathbf{r},\ \omega) \le \mathbb{F}(\mathbf{s},\ \omega))$ for every $\mathbf{r},\ \mathbf{s} \in \Re,\ \mathbf{r} \le \mathbf{s}$. Thus for every $\omega \in B$, the function $\mathbb{F}(.\ ,\ \omega)$ is non-decreasing.

We note that $A_{\mathbf{r}} = \lim_{n\to\infty} A_{\mathbf{s}_n}$ where $\mathbf{s}_n = \mathbf{r} + \frac{1}{n}(1,\ 1)$. This implies, by the monotone convergence theorem applied to $1 - \chi_{A_{\mathbf{s}_n}}$, that $\mathbb{F}(s_n\ ,\ .) \to$

$\mathbb{F}(\mathbf{r}\,,\,.)$ $wp1$. i.e. there exists $B_{\mathbf{r}}$ with $P(B_{\mathbf{r}}) = 1$ such that $(\omega \in B_{\mathbf{r}}) \Rightarrow (\mathbb{F}(\mathbf{s}_n\,,\,\omega) \to \mathbb{F}(\mathbf{r}\,,\,\omega))$. Let $\omega \in B \cap B_{\mathbf{r}}$. Since $\mathbb{F}(.\,,\,\omega)$ is non-decreasing, $\mathbb{F}(\mathbf{t}_n\,,\,\omega) \to \mathbb{F}(\mathbf{r}\,,\,\omega)$, whatever be the sequence $(\mathbf{t}_n)$ with $\mathbf{t}_n \geq \mathbf{r}$ and $\mathbf{t}_n \to \mathbf{r}$. i.e., on $B \cap B_{\mathbf{r}}$, $\mathbb{F}$ is right continuous at $\mathbf{r}$. Let $D = B \cap (\cap_{\mathbf{r}\in\Re} B_{\mathbf{r}})$. Then $P(D) = 1$. We note that for each $\omega \in D$, $\mathbb{F}(.\,,\,\omega)$ is a non-decreasing and right continuous function.

Since $Z_1,\ Z_2$ are finite valued, $A_{\mathbf{r}} \to \Omega$ as $\min(r_1,\ r_2) \to \infty$ and that $A_r \to \phi$ as $\min(r_1,\ r_2) \to -\infty$. Thus
$\lim_{\min(r_1,r_2)\to\infty} \mathbb{F}(\mathbf{r}\,,\,\omega) = \lim_{\min(r_1,r_2)\to\infty} E(\chi_{A_{\mathbf{r}}}|\mathcal{B})(\omega) =$ (by **(6.4)**(v))
$E(\chi_\Omega|\mathcal{B})(\omega) = 1$ for all ω except for a ω-set of probability 0. Similarly $\lim_{\min(r_1,r_2)\to-\infty} \mathbb{F}(\mathbf{r},\,.) = 0\ wp1$.

For $\mathbf{r} \leq \mathbf{s}$, define $A_{2,2} = \mathbf{Z}^{-1}\left((-\infty,\ s_1] \times (-\infty,\ s_2]\right)$;
$A_{1,2} = \mathbf{Z}^{-1}\left((-\infty,\ r_1] \times (-\infty,\ s_2]\right)$;
$A_{2,1} = \mathbf{Z}^{-1}\left((-\infty,\ r_2] \times (-\infty,\ s_1]\right)$;
$A_{1,1} = \mathbf{Z}^{-1}\left((-\infty,\ r_1] \times (-\infty,\ r_2]\right)$;
$Q = (A_{2,2} \cup A_{1,1}) \sim (A_{1,2} \cup A_{2,1})$ and
$X(\omega) = \mathbb{F}((s_1,\ s_2)\,,\,\omega) - \mathbb{F}((r_1,\ s_2)\,,\,\omega) - \mathbb{F}((r_2,\ s_1)\,,\,\omega) + \mathbb{F}((r_1,\ r_2)\,,\,\omega)$.
We note X is $\mathcal{B}$-measurable. Let $B = \{\omega : X(\omega) < 0\}$. We have

$$
\begin{aligned}
0 \geq \int_B X\,\mathrm{d}P &= \int_B \chi_{A_{2,2}}\,\mathrm{d}P - \int_B \chi_{A_{1,2}}\,\mathrm{d}P - \int_B \chi_{A_{2,1}}\,\mathrm{d}P \\
&\qquad + \int_B \chi_{A_{1,1}}\,\mathrm{d}P \\
&= \int_B \{\chi_{A_{2,2}} - \chi_{A_{1,2}} - \chi_{A_{2,1}} + \chi_{A_{1,1}}\}\,\mathrm{d}P \\
&= \int_B \chi_{\mathbf{Z}^{-1}(Q)}\,\mathrm{d}P \geq 0.
\end{aligned}
$$

i.e., $\int_B \chi_{\mathbf{Z}^{-1}(Q)}\,\mathrm{d}P = 0$, which implies $P(B) = 0$. Hence $wp1\ X \geq 0$.

To summarise : As **r, s** run through $\Re$, the exceptional sets of zero probability that we encountered in establishing the above results are at most countably infinite. Hence their union B is a set of probability 0. i.e. there exists a set A of probability 1 such that $\omega \in A$ implies $\mathbb{F}(\mathbf{x}\,,\,\omega)$, as a function of **x**, is a *df* on $\Re$. For $\omega \in A'$ take $\mathbb{F}$ to be a fixed but arbitrary *df* G. Thus for all ω, $\mathbb{F}(.\,,\,\omega)$ is a *df* in $\Re$. By **(1.7.3)**, the *df* on $\Re$ can be extended uniquely to a *df* on R^2, which we denote by the same symbol $\mathbb{F}$.

We claim that $\mathbb{F}$, so defined, is a regular conditional *df* of **Z** given $\mathcal{B}$.

For $\mathbf{z} \in \Re$, $\mathbb{F}(\mathbf{z}\,,\,\omega)$ is a $\mathcal{B}$-measurable rv, since it is, by definition, a version of $P(\mathbf{Z} \leq \mathbf{z}|\mathcal{B})$. If $\mathbf{z} \in R^2 \sim \Re$, a sequence $(\mathbf{r}) \subset \Re$ can be found such that $r \geq \mathbf{z}$ and $\mathbf{r} \to \mathbf{z}$. This has two consequences. Since $\mathbb{F}(.\,,\,\omega)$ is a *df*, $\mathbb{F}(\mathbf{r}\,,\,\omega) \to \mathbb{F}(\mathbf{z}\,,\,\omega)$ for all ω. Hence $\mathbb{F}(\mathbf{z}\,,\,\omega)$ is a $\mathcal{B}$-measurable rv, so (6.27) is satisfied. The second consequence is $\chi_{A_{\mathbf{r}}} \downarrow \chi_{A_{\mathbf{z}}}$. There exists a set $\hat{D}$ with $P(\hat{D}) = 1$ such that $\omega \in \hat{D}$ implies $E(\chi_{A_{\mathbf{r}}}|\mathcal{B})(\omega) \to$

$E(\chi_{A_\mathbf{z}}|\mathcal{B})(\omega)$. Let $E = A \cap \hat{D}$. We note $P(E) = 1$. Now for $\omega \in E$, $\mathbb{F}(\mathbf{z}, \omega) = \lim_{\mathbf{r}\to\mathbf{z}} \mathbb{F}(\mathbf{r}, \omega) = \lim_{\mathbf{r}\to\mathbf{z}} P(\mathbf{Z} \leq \mathbf{r}|\mathcal{B})(\omega) = \lim_{\mathbf{r}\to\mathbf{z}} \mathbb{E}(\chi_{A_\mathbf{r}}|\mathcal{B})(\omega) = \mathbb{E}(\lim_{\mathbf{r}\to\mathbf{z}} \chi_{A_\mathbf{r}}|\mathcal{B})(\omega) = \mathbb{E}(\chi_{A_\mathbf{z}}|\mathcal{B})(\omega) = P(\mathbf{Z} \leq \mathbf{z}|\mathcal{B})(\omega)$, so (6.28) and (6.29) are satisfied. With this, the proof is complete. □

Theorem 6.5.3. *Let $\mathcal{B}$ and $\mathbf{Z}$ be as in* **(6.5.2)**. *Then a regular conditional probability measure of $\mathbf{Z}$ given $\mathcal{B}$ exists.*

Proof. Let $\mathbb{F}$ be as in **(6.5.2)**. Define, for $A \in \mathscr{R}^p$, $q(\omega\,, A) = \int_A \mathrm{d}\,\mathbb{F}(\mathbf{z}\,, \omega)$. We claim q is a regular conditional pm of $\mathbf{Z}$ given $\mathcal{B}$. For each $\omega \in \Omega$, $q(\omega\,, .)$ is a pm on $\mathscr{R}^p$, since for each ω, $\mathbb{F}(.\,, \omega)$ is a df. Hence in order to prove the claim, we must show that for every $A \in \mathscr{R}^p$

(α) $q(.\,, A)$ is $\mathcal{B}$-measurable and

(β) $P(\mathbf{Z} \in A|\mathcal{B})(.) = q(.\,, A)$ $wp1$.

Let $\mathcal{D}$ be the collection of all sets $A \in \mathscr{R}^p$ for which (α) and (β) hold. The problem before us is to show that $\mathscr{R}^p \subset \mathcal{D}$. The crucial step in the proof is to show that $\mathcal{D}$ contains all the open sets.

For $I(\mathbf{t})$ an 'interval', $I(\mathbf{t}) = \times_1^p(-\infty, t_i]$, $q(.\,, I(\mathbf{t})) = \mathbb{F}(\mathbf{t}\,, .)$ which is $\mathcal{B}$-measurable since $\mathbb{F}$ is a regular conditional df. Further $\mathbb{F}(\mathbf{t}\,, .) = P(\mathbf{Z} \in I(\mathbf{t})|\mathcal{B})(.)$ $wp1$ (by **(6.5.2)**). Hence $I(\mathbf{t}) \in \mathcal{D}$ for every $\mathbf{t} \in R^p$.

We note that all finite 'intervals' in R^p are of 4^p types depending on whether a boundary line of a component linear interval is included or not.

Consider a type-1 'interval' : $J((\mathbf{u}, \mathbf{t}]) = \times_1^p(u_i, t_i]$. For such an 'interval', $q(.\,, J((\mathbf{u}, \mathbf{t}]))$ can be written as a finite linear combination of terms of the type $\mathbb{F}(\mathbf{v}\,, .)$ each of which is $\mathcal{B}$-measurable. Hence $q(.\,, J((\mathbf{u}, \mathbf{t}]))$ is $\mathcal{B}$-measurable. Further, since $\mathbb{F}(\mathbf{v}\,, .) = E(\chi_{I(\mathbf{v})}|\mathcal{B})(.)$ $wp1$, it follows, on simplification, that $q(.\,, J((\mathbf{u}, \mathbf{t}])) = P(\mathbf{Z} \in J((\mathbf{u}, \mathbf{t}])|\mathcal{B})(.)$ $wp1$. Thus all type-1 'intervals' belong to $\mathcal{D}$. Consider now a type-2 'interval': $J((\mathbf{u}, \mathbf{t})) = \times_1^p(u_i, t_i)$. Such an 'interval' can be obtained as the limit of a monotonic increasing sequence of typ-1 'intervals'. Hence by **(1.4.9)** and **(6.4)**(v), these intervals satisfy (α) and (β) and hence belong to $\mathcal{D}$. Any of the remaining $4^p - 2$ types of intervals can be obtained as the limit of a monotonic decreasing sequence of type-1 or type-2 'intervals'. Hence by **(1.4.9)** and (**(6.4)** (v) *Note*), these 'intervals too belong to $\mathcal{D}$. Thus all the 4^p types of 'intervals' lie in $\mathcal{D}$.

If G is an open set, then it is the union of a countable number of type-2 'intervals', which can be expressed as the union of a countable number of disjoint 'intervals', not necessarily of type-2. Write $G = \cup_n I_n$, the I_ns being disjoint intervals. If $G_n = \cup_{k=1}^n I_k$, then $G_n \uparrow G$. Since $q(.\,, G_n) = \sum_{k=1}^n q(.\,, I_n)$ and since each term of this sum is $\mathcal{B}$-measurable, it follows that $q(.\,, G_n)$ is

$\mathcal{B}$-measurable. Hence G_n satisfies (α). Further the kth term of this sum is $P(\mathbf{Z} \in I_k|\mathcal{B})\ wp1$, since $I_k \in \mathcal{D}$. If χ_k is the indicator function of the set $\{\mathbf{Z} \in I_k\}$, then

$q(.\,,\,G_n) = \sum_{k=1}^{n} E\,(\chi_k|\mathcal{B})\ wp1\ =$ by **(6.4)**(i) $E(\sum_{k=1}^{n}\chi_k|\mathcal{B})\ wp1 = P(\mathbf{Z} \in G_n|\mathcal{B})\ wp1$. i.e. G_n satisfies (β). Hence $G_n \in \mathcal{D}$.

Since $q(.\,,\,G) = \lim_{n\to\infty} q(.\,,\,G_n)$, it follows that $q(.\,,\,G)$ is $\mathcal{B}$ -measurable. Since $\{\mathbf{Z} \in G_n\} \uparrow \{\mathbf{Z} \in G\}$, we have, by **(6.4)**(v), $q(.\,,\,G) = \lim_{n\to\infty} q(.\,,\,G_n)$ $= \lim_{n\to\infty} P(Z \in G_n|\mathcal{B})\ wp1 = P(\mathbf{Z} \in G|\mathcal{B})\ wp1$. Thus G satisfies both (α) and (β) and hence lies in $\mathcal{D}$.

We have shown that all open sets G are in $\mathcal{D}$.

Arguments similar to the above help us conclude that $\mathcal{D}$ contains all the g_δ sets, since a g_δ set is the limit of a monotonic decreasing sequence of open sets.

Let μ be the pm induced by $\mathbf{Z}$ on $\mathscr{R}^p$. By **(1.9.2)**, μ is outer regular. Hence to every $A \in \mathscr{R}^p$ there can be found a g_δ set $A^* \supseteq A$ with $\mu(A^*) = \mu(A)$. In particular if $\mu(A) = 0$ then $\mu(A^*) = 0$. Since $A^* \in \mathcal{D}$, $q(.\,,\,A^*) = P(\mathbf{Z} \in A^*|\mathcal{B})\ wp1 = 0\ wp1$. Hence if A is arbitrary Borel set with $\mu(A) = 0$, then $q(.\,,\,A) \leq q(.\,,\,A^*) = 0\ wp1$. Hence all μ-null sets belong to $\mathcal{D}$.

Finally, let A be an arbitrary Borel set. We have $q(.\,,\,A) = q(.\,,\,A^*)$ $-q(.\,,\,A_1)$, where $A_1 = A^* \sim A$. We note $\mu(A_1) = 0$. Hence $q(.\,,\,A) = q(.\,,\,A^*) = P(\mathbf{Z} \in A^*|\mathcal{B})\ wp1 = P(\mathbf{Z} \in A|\mathcal{B}) + P(\mathbf{Z} \in A_1|\mathcal{B})\ wp1 = P(\mathbf{Z} \in A|\mathcal{B})\ wp1$. Thus $\mathscr{R}^p \subset \mathcal{D}$, which completes the proof. □

6.5.4. Let $q(.\,,\,.)$ defined on $\Omega \times \mathscr{A}$ be a regular conditional probability measure on $\mathscr{A}$ given $\mathcal{B}$ (ref. **(6.5.1)**). i.e., $(i) q(.\,,\,A)$ is $\mathcal{B}$-measurable for every $A \in \mathscr{A}$. (ii) $q(\omega\,,\,.)$ is a probability measure on $\mathscr{A}$ for every $\omega \in \Omega$. (iii) For each $A \in \mathscr{A}$, $q(\omega\,,\,A) = P(A|\mathcal{B})(\omega)$ for an ω-set of probability 1.

Theorem 6.5.5. *If X is a rv with $E|X| < \infty$, then*

$$E(X|\mathcal{B})(.) = \int_{\Omega} X(u)\,dq(.\,,\,u) \quad wp1. \tag{6.30}$$

Proof. It is obvious that we need only to establish claim for $X \geq 0$.

If $X = \chi_A$, $A \in \mathscr{A}$, then $E(X|\mathcal{B}) = P(A|\mathcal{B})$. For this X, rightside of (6.30) is $q(\omega\,,\,A)$, which by definition of regular conditional probability equals $P(A|\mathcal{B})\ wp1$. Hence in this case (6.30) is verified.

If X is a simple random variable, $X = \sum_{k=1}^{n} a_k\chi_k$, where χ_k is the indicator function of the event A_k, $E(X|\mathcal{B}) = \sum_{k=1}^{n} a_k P(A_k|\mathcal{B})\ wp1$ (by **(6.4)**(i)), while

$\int_\omega X(u)\,\mathrm{d}\,q(.\,,\,u) = \sum_{k=1}^{n} a_k q(.\,,\,A_k) = \sum_{k=1}^{n} a_k P(A_k|\mathcal{B})$.
Thus (6.30) holds true in this case too. To extend this result to a non-negative rv X we invoke the monotone convegence theorem which is available also for conditional expectations (ref. **(6.4)**(v)). □

Remarks 6.5.6. (i) If our interest is only in a single real rv Y with $\mathbb{E}|Y| < \infty$ (and not in all possible real rvs with finite expected values), then it is enough if $q(.\,,\,.)$ is a regular conditional probability measure on $\Omega \times Y^{-1}(\mathscr{R})$ for the stated theorem to hold.

(ii) Let M be the set of all non-negative integers, $M = \{0,\ 1.\ 2, \ldots.\}$ and $\mathcal{M}$ the σ-field consisting of all the subsets of M. Let Y be a rv taking values in M. Let **E** be a countable collection $\{0,\ 1,\ 2,\ ..\}$ of labels and $\mathcal{E}$ the σ-field of all of its subsets. Let X be a rE taking values in **E**. Let $\mathcal{B} = X^{-1}\mathcal{E}$. By **(6.5.3)**, a regular conditional probability $q(.\,,\,.)$ of Y given $\mathcal{B}$, defined on $\Omega \times \mathcal{M}$ exists. Hence by **(6.5.4)**,
$\mathbb{E}(Y|\mathcal{B})(\omega) = \int_M Y(u)\,\mathrm{d}_u q(\omega,\ u) = \sum_{n=1}^{\infty} nq(\omega,\ (Y = n))$
$= \sum_{n=1}^{\infty} nP(Y = n|\mathcal{B})(\omega)$,
leading to $\mathbb{E}(Y|X = i) = \sum_{n=1}^{\infty} nP(Y = n|X = i)$.

For another proof of this result, ref. ((**6.4**)(xii))

Let **X**, **Y**, **U**, $\mathcal{X}$, $\mathcal{Y}$, $\mathcal{U}$, Ω, $\mathscr{A}$, P, X, Y, g, μ, ν be as in **(6.3.3)**(vi). Let $\mathscr{A}_1$ be a sub σ-field of $\mathscr{A}$ such that Y is measurable with respect to it and X is independent of it (and hence independent of Y).

Theorem 6.5.7. *If* $\mathbb{E}|g(X,\ Y)| < \infty$ *then*

$$\psi(Y) = E(g(X,\ Y)|\mathscr{A}_1) = \int_{\mathbf{X}} g(x,\ Y)\,d\,\mu(x)\ \ wp1. \tag{6.31}$$

Proof. By **(6.3.3)**(vi), $P\left((X,\ Y) \in \hat{D}|\mathscr{A}_1\right) = P\left((X,\ Y) \in \hat{D}|Y\right)$ $wp1$ for each $\hat{D} \in \mathcal{U}$. Hence, writing $Z = g(X,\ Y)$, and denoting by $\mathcal{B}$ the σ-field generated by Y, we have

$$P(Z \in D|\mathscr{A}_1) = P(Z \in D|\mathcal{B}) \quad wp1. \tag{6.32}$$

for every $D \in \mathscr{R}$. We note that both sides of (6.32) are $\mathscr{A}_1$-measurable.

The exceptional set may depend on D. If $\mathcal{I}$ is the collection of all linear intervals whose finite end points are rational numbers, then $\mathcal{I}$ is countable. Hence there exists $A \in \mathscr{A}_1$ such that $P(A) = 1$ and

$$\omega \in A \text{ implies } P(Z \in D|\mathscr{A}_1)(\omega) = P(Z \in D|\mathcal{B})(\omega) \tag{6.33}$$

Let $\mathcal{E}$ be the collection of all $D \in \mathscr{R}$ for which (6.33) is true. Since any interval can be obtained as the monotonic limit of members of $\mathcal{I}$ and since the monotone convergence theorem is available for both sides of (6.32), it follows that all intervals belong to $\mathcal{E}$. By the additivity property of conditional expectation, (6.33) holds also when D is the finite union of disjoint intervals. i.e., if $\mathbb{R}$ is the collection of all finite unions of disjoint intervals, then $\mathbb{R} \subset \mathcal{E}$. We note $\mathbb{R}$ is a field. Again by monotone convergence theorem for conditional expectations,we note that (6.33) holds for monotonic limits of members of $\mathbb{R}$. Thus $\mathcal{E}$ contains the field $\mathbb{R}$ and is closed under monotonic limits. Hence $\mathcal{E} \supset \mathscr{R}$. Thus (6.33) holds for every linear Borel set D. i.e., $\mathcal{E} = \mathscr{R}$. By **(6.5.3)**, a regular conditional pm $q(. , .)$ for Z on $\Omega \times \mathscr{R}$ exists. Hence $P(Z \in .|\mathscr{A}_1)(.)$ and $P(Z \in .|\mathcal{B})(.)$ are versions of $q(. , .)$. It then follows by **(6.5.4)** that $E(Z|\mathscr{A}_1) = E(Z|\mathcal{B})$.

That ψ, defined on **Y** is Borel measurable is established by proving it to be true when g is the characteristic function of a set in $\mathcal{U}$, then proving it for simple functions and finally for limits of simple functions. Further by Fubini's theorem, $\mathbb{E}|\psi(Y)| < \infty$.

Let $B \in \mathcal{B}$. This implies there exists a $\hat{B} \in \mathcal{Y}$ such that $B = Y^{-1}(\hat{B})$. Now,

$$\int_B \psi(Y)\,\mathrm{d}P = \int_{\hat{B}} \psi(y)\,\mathrm{d}\nu(y) \text{ (by \textbf{(2.3.6)})} \tag{6.34}$$

$$= \int_{\hat{B}} \Big(\int_{\mathbf{X}} g(x,\, y)\,\mathrm{d}\mu(x)\Big)\,\mathrm{d}\nu(y) \tag{6.35}$$

$$= \int_{\mathbf{X}\times\hat{B}} g(x,\, y)\,\mathrm{d}\mu(x)\,\mathrm{d}\nu(y) \text{ (by Fubini's theorem)} \tag{6.36}$$

$$= \int_B g(X,\, Y)\,\mathrm{d}P \text{ (again by \textbf{(2.3.6)})} \tag{6.37}$$

$$= \int_B \mathbb{E}(Z|\mathcal{B})\,\mathrm{d}P, \tag{6.38}$$

so that $\psi(Y) = \mathbb{E}(Z|\mathcal{B})\ wp1 = \mathbb{E}(Z|\mathscr{A}_1)\ wp1$, proving the claim. □

In the next theorem we give an application of the above description. The proof of the theorem depends on an integral whose evaluation may be of independent interest.

Theorem 6.5.8. *Let $Z = \frac{X}{Y}$ where $X,\ Y$ are iid standard normal variables. (When $Y = 0$, we take $Z = 0$). Then $\varphi(t) = e^{-|t|}$ where φ is the cf of Z.*

The proof depends on the following

Lemma 6.5.9. *Let* $a > 0,\ b \geq 0$. *Then* $I(a,\ b) = \int_0^\infty e^{-(a^2x^2+\frac{b^2}{x^2})}\,\mathrm{d}\,x = \frac{\sqrt{\pi}}{2a}e^{-2ab}$.

Proof of Lemma.

We note differentiation under the integral sign is justified. Hence

$$\begin{aligned}\frac{\mathrm{d}}{\mathrm{d}\,b}I(a,\ b) &= \int_0^\infty e^{-(a^2x^2+\frac{b^2}{x^2})}\frac{-2b}{x^2}\,\mathrm{d}\,x\\ &= -2\int_0^\infty e^{-(a^2x^2+\frac{b^2}{x^2})}(a+\frac{b}{x^2})\,\mathrm{d}\,x\\ &\quad + 2a\int_0^\infty e^{-(a^2x^2+\frac{b^2}{x^2})}\,\mathrm{d}\,x\end{aligned}$$

or

$$\begin{aligned}\frac{d}{\mathrm{d}\,b}I(a,\ b) - 2aI(a,\ b) &= -2\int_0^\infty e^{-(ax-\frac{b}{x})^2-2ab}(a+\frac{b}{x^2})\,\mathrm{d}\,x\\ &= -2\int_{-\infty}^\infty e^{-(u^2+2ab)}\,\mathrm{d}\,u\end{aligned}$$

(putting $ax - \frac{b}{x} = u$ and noting that u is a strictly increasing continuous function of x)

Or, writing $y = I(a,\ x)$, $\frac{\mathrm{d}\,y}{\mathrm{d}\,x} - 2ay = -2e^{-2ax}\int_{-\infty}^\infty e^{-u^2}\,\mathrm{d}\,u = -2\sqrt{\pi}e^{-2ax}$.

i.e. $\frac{\mathrm{d}\,y}{\mathrm{d}\,x} - 2ay = -2\sqrt{\pi}e^{-2ax}$.

(When $P,\ Q$ are functions of x alone, the general solution of the differential equation $\frac{\mathrm{d}\,y}{\mathrm{d}\,x} + Py = Q$, we recall, is given by : $ye^{\int P\,\mathrm{d}\,x} = C + \int Qe^{\int P\,\mathrm{d}\,x}\,\mathrm{d}\,x$). Hence the general solution for the case on hand is given by $ye^{-2ax} = C + \frac{\sqrt{\pi}}{2a}e^{-4ax}$. When $b = x = 0,\ y = \frac{\sqrt{\pi}}{2a}$. Hence $C = 0$ and the desired result follows. The proof of the Lemma is completed.

Proof of the Theorem.

$$\begin{aligned}\mathbb{E}e^{itZ} &= \mathbb{E}\left(\mathbb{E}(e^{\frac{itX}{Y}}|Y)\right) = \mathbb{E}e^{-\frac{t^2}{2Y^2}} = \int_R e^{\frac{-t^2}{2y^2}}\frac{1}{\sqrt{2\pi}}e^{-\frac{y^2}{2}}\,\mathrm{d}\,y\\ &= \frac{\sqrt{2}}{\sqrt{\pi}}\int_0^\infty e^{-(\frac{y^2}{2}+\frac{|t|^2}{2y^2})}\,\mathrm{d}\,y\\ &= e^{-|t|} \text{ (taking } a^2 = \tfrac{1}{2} \text{ and } b^2 = \tfrac{|t|^2}{2} \text{ in the Lemma).}\end{aligned}$$

□

6.5.10. We now present an example to show that a regular conditional probability measure may not always exist.

Let $\Omega = [0,\ 1]$ with the usual metric, $\mathscr{A}$ its Borel σ-field and let P denote the Lebesgue measure. For any $A \subseteq \Omega$, define $P^*(A) = \inf_{B \supseteq A,\ B \in \mathscr{A}} P(B)$ and $P_*(A) = \sup_{B \subseteq A,\ B \in \mathscr{A}} P(B)$. Clearly if $A \in \mathscr{A}$, then $P^*(A) = P_*(A) = P(A)$. Let $C \subset \Omega$ be such that $P^*(C) = 1$ and $P_*(C) = 0$. Refer Theorem E, page 70, [H] to accept that such a set C exists. Of course $C \notin \mathscr{A}$. We observe $P^*(C') = 1$ and $P_*(C') = 0$.

Let $\mathcal{A}$ be the collection of all sets A of the type $(C \cap B_1) \cup (C' \cap B_2)$ for some $B_1,\ B_2 \in \mathscr{A}$. We note $A \in \mathscr{A}$ implies $A \in \mathcal{A}$ since $A = (C \cap A) \cup (C' \cap A)$. Further, $C \in \mathcal{A}$ since $C = (C \cap \Omega) \cup (C' \cap \phi)$. Hence $\mathscr{A} \subset \mathcal{A}$, the inclusion being strict. Let $A_n \in \mathcal{A},\ n \geq 1,\ A_n = (C \cap A_{n,1}) \cup (C' \cap A_{n,2}),\ A_{n,1},\ A_{n,2} \in \mathscr{A}$. Let $B_1 = \cup_n A_{n,1}$ and $B_2 = \cup_n A_{n,2}$. We note that $B_i \in \mathscr{A},\ i = 1,\ 2$ and that $\cup_n A_n \in \mathcal{A}$ since $\cup_n A_n = (C \cap B_1) \cup (C' \cap B_2)$. Thus $\mathcal{A}$ is closed under countable unions. Again, if $A = (C \cap B_1) \cup (C' \cap B_2)$, then $A' = (C' \cap B_2') \cup (C \cap B_1') \cup (B_1' \cap B_2') = \{(C' \cap B_2') \cup (C \cap B_1') \cup\} \cup B$ where $B = B_1' \cap B_2' \in \mathscr{A}$. Rewriting, $A' = \{(C' \cap B_2') \cup (C \cap B_1')\} \cup \{(C \cap B) \cup (C' \cap B)\} = (C \cap D_1) \cup (C' \cap D_2)$ where $D_1 = B \cup B_1' \in \mathscr{A}$ and $D_2 = B \cup B_2' \in \mathscr{A}$. It now follows that $\mathcal{A}$ is a σ-field.

On $\mathcal{A}$, define set function μ : if $A = (C \cap B_1) \cup (C' \cap B_2)$, let $\mu(A) = \frac{1}{2}\{P(B_1) + P(B_2)\}$. We need to make sure that μ is well defined. To do this, let $A = (C \cap B_1) \cup (C' \cap B_2) = (C \cap E_1) \cup (C' \cap E_2)$. Since $C,\ C'$ are disjoint, $(C \cap B_1) = (C \cap E_1)$. This implies that $(B_1 \Delta E_1) \cap C = \phi$. Now, $B_1 \Delta E_1 \in \mathscr{A}$ and $B_1 \Delta E_1 \subset C'$. Since $P_*(C') = 0$, we conclude $P(B_1 \Delta E_1) = 0$ or, what is same, $P(B_1) = P(E_1)$. Similarly $P(B_2) = P(E_2)$. That μ is well defined follows now.

That $\mu \geq 0$ and that $\mu(\Omega) = 1$ are obvious.

Let $A_n,\ A_{n,i},\ i = 1,\ 2,\ B_1,\ B_2$ be as above. Assume $A_n \cap A_m = \phi$ if $n \neq m$. This implies $(C \cap A_{n,1}) \cap (C \cap A_{m,1}) = \phi$ and $(C' \cap A_{n,2}) \cap (C' \cap A_{m,2}) = \phi$. i.e., $A_{n,1} \cap A_{m,1} \subset C'$ and $A_{n,2} \cap A_{m,2} \subset C$. Since $P_*(C) = 0 = P_*(C')$, it follows that $P(A_{n,i} \cap A_{m,i}) = 0$, for $i = 1,\ 2$. Recall $\cup_n A_n \in \mathcal{A}$. Now, $\mu(\cup_n A_n) = \mu((C \cap B_1) \cup (C' \cap B_2)) = \frac{1}{2}\{P(B_1) + P(B_2)\} = \frac{1}{2}\{\sum_1^\infty P(A_{n,1}) + \sum_1^\infty P(A_{n,2})\} = \sum_1^\infty \mu(A_n)$, proving that μ is countably additive. Thus $(\Omega,\ \mathcal{A},\ \mu)$ is a probability space, $\mathscr{A} \subset \mathcal{A}$ and μ restricted to $\mathscr{A}$ is P.

We claim that a regular conditional probabilty measure $q(.\ ,\ .)$ defined on $\Omega \times \mathcal{A}$ given $\mathscr{A}$ does not exist. i.e., $q(.\ ,\ .)$ can not be determined so that $q(.\ ,\ .)$ satisfies the conditions (6.24)–(6.26). Proof is by showing that assumption that it exists leads to a contradiction.

Suppose such a q exists.

Let $D_0 \in \mathscr{A}$, $P(D_0) = 1$ such that $q(\omega\ ,\ .)$ is a pm on $\mathcal{A}$ for each $\omega \in D_0$.

Now $E(\chi_C|\mathscr{A}) = \frac{1}{2}\ wp1$ since for $B \in \mathscr{A}$, $\int_B \chi_C \, \mathrm{d}\,\mu = \mu(C \cap B) = \frac{1}{2}P(B) = \int_B \frac{1}{2} \mathrm{d}\,\mu$. There exists therefore $D_1 \in \mathscr{A}$ such that $P(D_1) = 1$ and such that $\omega \in D_1$ implies $q(\omega\ ,\ C) = \mathbb{E}(\chi_C|\mathscr{A})(\omega) = \frac{1}{2}$.

Let $\mathcal{J}$ be the collection of all the subintervals of Ω with rational end points and their finite unions. It is not difficult to see that $\mathcal{J}$ is a countable collection, that it is a field and that it generates the Borel σ-field $\mathscr{A}$. Hence there exists $D_2 \in \mathscr{A}$ such that $P(D_2) = 1$ and such that $q(\omega\ ,\ F) = \chi_F(\omega)$ for every $F \in \mathcal{J}$ and $\omega \in D_2$.

Let $D = D_0 \cap D_1 \cap D_2$ and note $P(D) = 1$ and that $\omega \in D$ implies $q(\omega\ ,\ .)$ is a pm on the field $\mathcal{J}$. Hence it can be extended uniquely to a pm, say, $q^*(\omega\ ,\ .)$ on $\mathscr{A}$. But for $\omega \in D$, $q^*(\omega\ ,\ .) = q(\omega\ ,\ .)$ on the field $\mathcal{J}$. Hence for $\omega \in D$, $q^*(\omega\ ,\ .) = q(\omega\ ,\ .)$ on $\mathscr{A}$.

Fix $\omega \in D$. Since $q(\omega\ ,\ .)$ is a pm, $q(\omega\ ,\ \{\omega\}) = \lim_{n\to\infty} q(\omega\ , I_n)$ (where $I_n = (\omega - \frac{1}{n}\ ,\ \omega + \frac{1}{n})) = \lim_{n\to\infty} \chi_{I_n}(\omega) = 1$.

Since $q(\omega\ ,\ \{\omega\}) = 1$, since $q(\omega\ ,\ C) = \frac{1}{2}$ and since $q(\omega\ ,\ .)$ is a pm, it follows $q(\omega\ ,\ C \cap \{\omega\}) > 0$. This implies $C \cap \{\omega\} \neq \phi$. i.e., $\omega \in C$. Since $\omega \in D$ is arbitrary, it follows $D \subset C$. This is not possible since $P(D) = 1$ and $P_*(C) = 0$. This contradiction establishes the claim. □

In **(6.5.3)** we proved the existence of a regular conditional probability measure $q(.\ ,\ )$ of $\mathbf{Z}$, given $\mathcal{B}$, for $\mathscr{R}^\nu$, exists . In **(6.5.10)** we proved by an example that a regular conditional probability measure does not always exist. In this context, the following result, stated without proof, is of interest. Follow the notation of **(6.5.1)**.

Theorem 6.5.11. *Let $(M,\ d)$ be a complete, separable metric space. Then a regular conditional probability measure of X given $\mathcal{B}$ exists.*

However we discuss below a particular case where 'X' is a BMP on $[0,\ \infty)$ and where $(M,\ d) = (C[0,\ \infty),\ d)$.

Let $C = C[0,\ \infty)$ denote the set of all continuous functions defined on $[0,\ \infty)$. Define $g_n(t) = 1$ if $0 \le t \le n$; $g_n(t) = n + 1 - t$ if $n \le t \le n + 1$; $g_n(t) = 0$ if $t \ge n+1$. For $x,\ y \in C$, let $d(x,\ y) = \sum_{n=1}^{\infty} \frac{1}{2^n} \frac{\|xg_n - yg_n\|}{1+\|xg_n - yg_n\|}$. So defined, d can be shown to be a metric and $(C,\ d)$ shown to be a complete, separable metric space. Denote by $\mathscr{C}$ its Borel σ-field.

For x a real number and $A \in \mathscr{C}$ we understand by $(x + A)$, the set of all functions $x + f,\ f \in A$ where $x + f$ is the function with value $x + f(t)$ at t.

For any pm ν on $\mathscr{C}$, $\nu(A - x)$ is a Borel measurable function of x defined over R. The proof for this is as follows. Consider the product space $(R \times C,\ \mathscr{R} \otimes \mathscr{C})$ and the measurable transformation T from $R \times C$ into $R \times C$:

$x \in R$, $f \in C$, $T(x, f) = (x, x + f)$. Under this transformation $R \times A$ is mapped onto $T(R \times A)$ whose x-section is $\{f : x + f \in A\} = A - x$. The desired result for $\nu(A - x)$ follows now by Theorem A, p 143, [H].

Let $B(t)$, $t \geq 0$ be a standard BMP defined on a probability space $(\Omega, \mathscr{A}, P)$. For $0 < s < \infty$ fixed, define $Y_s(t, \omega) = B(s + t, \omega)$, $t \geq 0$ and note that $Y_0 = B$. If Y_s is the mapping of Ω into C given by $Y_s(\omega)(t) = B(s + t, \omega)$, then we know (ref. **(5.6)**) Y_s is measurable. Let us also write B_s for $B(s)$. B as well as $Y_s - B_s$ would each induce the pm μ, the Wiener measure, on $\mathscr{C}$. Denote by $\mathscr{A}_r$ the σ-field in Ω generated by the family $\{B(t) : 0 \leq t \leq r\}$, $r > 0$. Let $\mathscr{A}_s^*$ be the σ-field generated by B_s. We note $\mathscr{A}_s^* \subset \mathscr{A}$. Define, for s fixed, $s \geq 0$, $q(. , .)$ on $\Omega \times \mathscr{C} : q(\omega , A) = \mu(A - B(s, \omega))$.

Theorem 6.5.12. *q is a regular conditional pm of Y_s given $\mathscr{A}_s$.*

Proof. We have already noted that $g(x) = \mu(A - x)$ is a Borel measurable function. Hence $\mu(A - B(s, \omega)) = g(B(s, \omega))$ is an $\mathscr{A}_s$-measurable function of ω.

That for each ω, $\mu(. - B(s, \omega))$ is a pm is obvious since μ is one.

To complete the proof, it remains to show that for every $A \in \mathscr{C}$, $\mu(A - B(s, \omega)) = P(Y_s \in A|\mathscr{A}_s)(\omega)$ for all ω in a set of probability 1.

Write $Z_s = Y_s - B_s$. We note Z_s is independent of $\mathscr{A}_s$ and B_s is $\mathscr{A}_s$-measurable. In **(6.3.3)**(vi) set $\mathbf{X} = C$, $\mathbf{Y} = R$, $\mathcal{X} = \mathscr{C}$, $\mathcal{Y} = \mathscr{R}$, $X = Z_s$, $Y = B_s$, $\mathbf{U} = C \times R$ and $\mathcal{U} = \mathscr{C} \otimes \mathscr{R}$ and conclude for $D \in \mathcal{U}$,

$$P[(Z_s, B_s) \in D|\mathscr{A}_s] = P[(Z_s, B_s) \in D|B_s] \quad wp1. \tag{6.39}$$

Define g on $\mathbf{U} : g(f, x) = f + x$ and note that it is a measurable map. Hence for $D \in \mathscr{C}$, $P\{Z_s + B_s \in D|\mathscr{A}_s\} = P\{Z_s + B_s \in D|B_s\}$ $wp1$.

Hence it remains to show that

$$\int_D \chi_E(\omega)\, \mathrm{d}P(\omega) = \int_D \mu(A - B(s,\omega))\, \mathrm{d}P(\omega) \text{ for } D \in \mathscr{A}_s^*,$$

where $E = \{\omega : Y_s(\omega) \in A\}$. i.e., to show

$$P(D \cap E) = \int_D \mu\,(A - B(s, \omega))\, \mathrm{d}P(\omega). \tag{6.40}$$

Now, $P(D \cap E) = \mathbb{E}\chi_{D\cap E} = \mathbb{E}\chi_D\chi_E = \mathbb{E}(\mathbb{E}(\chi_D\chi_E|\mathscr{A}_s)) =$ (since $D \in \mathscr{A}_s^*$, ref. **(6.4)**(vi)) $\mathbb{E}(\chi_D\mathbb{E}(\chi_E|\mathscr{A}_s^*)) = \mathbb{E}(\chi_D\mathbb{E}(\chi_{\{Z_s+B_s\in A\}}\,|\mathscr{A}_s^*)) =$ (by **(6.3.3)**(vi)) $\mathbb{E}(\chi_D \int_{\{x+B(s,\,\omega)\in A\}} \mathrm{d}\mu(x))) = \mathbb{E}(\chi_D(\mu\,(A - B(s, \omega))) = \int_D \mu\,(A - B(s, \omega))\, \mathrm{d}P(\omega)$. □

6.6 Skorohod embedding

Let $(X,\ X_n,\ n \geq 1)$ be $iid\ rvs$ with common *df* F. Let $\mathbb{E}X = 0,\ \mathbb{E}X^2 = 1$. Let (S_n) be the partial sum sequence : $S_n = \sum_{k=1}^{n} X_k,\ S_0 = 0$. Define the process $\mathbf{S}_n(t),\ 0 \leq t \leq 1$ (called the partial sum process) with polygonal sample functions : $\mathbf{S}_n(\frac{i}{n},\ \omega) = \frac{1}{\sqrt{n\varphi(n)}} S_i(\omega),\ i = 0,\ 1, \ldots,\ n$ and $\mathbf{S}_n(.,\ \omega)$ linear on $[\frac{i-1}{n},\ \frac{i}{n}],\ 1 \leq i \leq n$.

We wish to find the limit set of $\mathbf{S}_n(t)$. To be able to talk of the limit set of this sequence, we must satisfy ourselves that each of its convergent subsequence converges $wp1$ to a non-random point in C. For this purpose define, for $k \geq 1,\ y_n(0) = 0,\ y_{n,k}(\frac{i}{n}) = \frac{S_i - S_k}{\sqrt{n\varphi(n)}},\ i \geq k+1$ and linear elsewhere. We note $\|\mathbf{S}_n - y_{n,k}\| \leq \frac{1}{\sqrt{n\varphi(n)}} \max_{1 \leq r \leq k} |S_k| \overset{pr}{\to} 0$. Hence the limit x of a convergent subsequence of $(\mathbf{S}_n)$ is the same as that of a corresponding subsequence of $(y_{n,k})$. For each n, the $rE\ y_{n,k}$ is measurable with respect to the σ-field $\mathfrak{F}_k$ generated by the $rvs\ X_\nu,\ \nu \geq k+1$. Thus x is $\mathfrak{F}_k$- measurable. This being true for all k, it follows that x is measurable with respect to the tail σ-field of the independent sequence (X_n) of rvs, which we know to be the trivial σ-field. Hence x is a non-random element in $C\ wp1$.

The tool needed to determine the limit set is Skorohod embedding or Skorohod representation. Given a *df* F on the line, this consists in finding a non-negative $rv\ \tau$ such that *df* of $B(\tau)$ is F. We refer to τ as the Skorohod time for X or F. As before B is a standard BMP. A special property of the BMP, called the strong Markov property, allows us to find independent $rvs\ \tau_n,\ n \geq 1$ and to examine the sequence $(B(\tau_1 + \tau_2 + \ldots + \tau_n))$ instead of the sequence of sums (S_n).

B is a standard $BMP : \{B(t),\ t \geq 0\};\ C_1 = C[0,\ \infty)$ is the space of all continuous functions x on $[0,\ \infty)$ such that $\overline{\lim}_{t \to \infty} x(t) = \infty$ and $\underline{\lim}_{t \to \infty} x(t) = -\infty$. Endow C_1 with the smallest σ-field Σ with respect to which the projection mappings π_t are measurable. Given $-u < 0 < v$ and $x \in C[0,\ \infty)$, let $\tau = \tau(u,\ v;\ x) = \inf\{t : x(t) = -u \text{ or } v\}$. We establish now a few basic properties of τ and $B(\tau)$.

Theorem 6.6.1. *(i) $\tau(u,\ v;\ B)$ is a finite valued rv, for $u,\ v$ fixed. (ii) $\mathbb{E}\tau^k < \infty$ for every $k \geq 0$. (iii) $B(\tau)$ is a proper rv (iv) $\mathbb{E}B(\tau) = 0$. (v) $\mathbb{E}B^2(\tau) = uv$ (vi) $\mathbb{E}\tau = uv$.*

Proof. (i) It follows immediately from **(5.9.1)** that τ is finite valued. Proving that, for $u,\ v$ fixed, it is a rv is equivalent to showing $\tau(u,\ v;\ x)$, defined on C_1, is measurable with respect to the σ-field Σ. i.e. we must show that to this

σ-field belongs the set $\{\tau(u,\ v;\ x) > t\}$ for every $t > 0$. Define $M_t,\ m_t$ on C_1 as follows. $M_t(x) = \sup_{0\le r\le t} x(r)$ and $m_t(x) = \inf_{0\le r\le t} x(r)$ and note that these are Σ- measurable, since each projection π_r is Σ-measurable and since x is a continuous function. Required follows now since $\{\tau(u,\ v;\ x) > t\} = \{x :\ x \in C_1 \text{ and } M_t(x) < v\} \cap \{x :\ x \in C_1 \text{ and } m_t(x) > -u\}$.

(ii) For $\alpha > 0$, $\mathbb{E}\tau^\alpha < \infty$ if (ref. **(2.6.4)**) $\int_0^\infty t^{\alpha-1} P(\tau > t)\,\mathrm{d}t < \infty$. Let $q = \max\{u,\ v\}$. From the definition of τ, $P(\tau > t) = P(-u < \inf_{0\le r\le t} B(r) < \sup_{0\le r\le t} B(r) < v) \le P(\sup_{0\le r\le t} |B(r)| < q) = P(\sup_{0\le s\le 1} |B(s)| < \frac{q}{\sqrt{t}})$ (ref. **(5.9)**(iii)). From **(5.5.2)** read with **(5.6)**, we see that the df of $\sup_{0\le s\le 1} |B(s)|$ is $\Lambda(x) = \frac{4}{\pi}\sum_{m=0}^\infty \frac{(-1)^m}{2m+1} e^{-\frac{(2m+1)^2\pi^2}{8x^2}}$, $x \ge 0$. Since

$$\Lambda(x) \le \frac{4}{\pi} e^{-\frac{\pi^2}{8x^2}}, \quad \int_0^\infty t^{\alpha-1} P(\tau > t)\,\mathrm{d}t \le \int_0^\infty t^{\alpha-1} \Lambda(\frac{q}{\sqrt{t}})\,\mathrm{d}t$$
$$\le \int_0^\infty t^{\alpha-1} \frac{4}{\pi} e^{-\lambda t}\,\mathrm{d}t < \infty,$$

where $\lambda = \frac{\pi^2}{8q^2}$.

(iii) By $B(\tau)$ we understand the variable which, at ω, takes the value $B(\tau(\omega),\ \omega)$. Since τ is a finite valued rv, there exist a sequence (τ_n) of elementary functions (ref. **(1.4.3)**(iii)) converging to τ uniformly. Now, suppose τ_n assumes the values $a_{k,n}$, $k = 1,\ 2, \ldots$. Then $\{\omega :\ B(\tau_n(\omega),\ \omega) \le x\} = \bigcup_{k=1}^{\infty} \{B(a_{k,n}) \le x)\} \cap \{\tau_n = a_{k,n}\}$, a measurable set. i.e. $B(\tau_n)$ is a rv. Since the sample functions of B are continuous, we have : $\lim_{n\to\infty} B(\tau_n(\omega),\ \omega) = B(\tau(\omega),\ \omega)$. This shows that $B(\tau)$ is a rv.

[Note 1 : *The above steps and the conclusion are valid in the more general context where* $\{B(t),\ t \ge 0\}$ *is any stochastic process with continuous sample paths and* $\tau \ge 0$ *is any* rv].

[Note 2 : *From the definition of a* BMP*, it is immediate that for* $a \ge 0$*, and* $b > 0$*,* $\frac{B(a)}{b}$ *and* $B(\frac{a}{b^2})$ *are identically distributed, namely, normal with zero mean and variance* $\frac{a}{b^2}$*. But if* Y *is a non negative variable, then* $\frac{B(Y)}{b}$ *and* $B(\frac{Y}{b^2})$ *may not be identically distributed, as is clear from the following example*].

Let $\mathrm{E} = \{\omega\ :\ B(1) \le 1\}$ and $Y = \chi_{\mathrm{E}}$. Let $b = 2$. Then $P(\frac{B(Y)}{2} > 1) = P(\frac{B(Y)}{2} > 1,\ Y = 0) + P(\frac{B(Y)}{2} > 1,\ Y = 1) = P(B(0) > 2, B(1) > 1) + P(B(1) > 2,\ B(1) \le 1) = 0$, while $P(B(\frac{Y}{4}) > 1) = P(B(\frac{Y}{4}) > 1,\ Y = 0) + P(B(\frac{Y}{4}) > 1,\ Y = 1) = 0 + P(B(\frac{1}{4}) > 1,\ B(1) \le 1) \ge P(2 >$

$B(\frac{1}{4}) > 1,\ B(1) - B(\frac{1}{4}) + B(\frac{1}{4}) \le 1) \ge P(1 < B(\frac{1}{4}) < 2,\ B(1) - B(\frac{1}{4}) < -1) = P(1 < B(\frac{1}{4}) < 2)P(B(1) - B(\frac{1}{4}) < -1) > 0$.

(iv) Since $B(\tau)$ takes only two finite values, $\mathbb{E}|B(\tau)| < \infty$.

Define $\tau_1 = \inf\{t : B(t) = -v \text{ or } u\}$. Then $P(\tau_1 > t) = P(-v < B(s) < u,\ 0 \le s \le t) = P\big(-u < B(s) < v,\ 0 \le s \le t\big) = P(\tau > t)$. Thus $\tau,\ \tau_1$ are identically distributed. Hence $B(\tau),\ B(\tau_1)$ are identically distributed. But $B(\tau_1) = -B(\tau)$. It follows from this that $\mathbb{E}B(\tau) = 0$. □

(v) We know $B(\tau)$ takes only the values $-u$ and v and we have proved that $\mathbb{E}B(\tau) = 0$. Hence $P\,(B(\tau) = -u) = \frac{v}{u+v}$ and $P\big(B(\tau) = v\big) = \frac{u}{u+v}$.

It follows from this that $\mathbb{E}B^2(\tau) = uv$. □

(vi) The proof that $\mathbb{E}\tau = uv$ is much involved and lengthy. The method of proof is by establishing $\mathbb{E}\{B^2(\tau) - \tau\} = 0$. □

The details are presented in the following subsection.

Theorem 6.6.2. *Let $u,\ v,\ B,\ \tau$ be as in* **(6.6.1)**. *Then $\mathbb{E}\tau = uv$.*

Proof. Step 1.
Write $X(t) = B^2(t) - t,\ t \ge 0$. Denote by $\mathcal{F}(t)$ the σ-field generated by the variables $X(s),\ 0 \le s \le t$. Let $0 < s < t$ and let $A \in \mathcal{F}(s)$. We note $X(t)$ is a martingale (ref. **(6.3.3)**(viii)). Hence $wp1$,

$$E(X(t)\chi_{A}|\mathcal{F}(s)) = \chi_{A}E(X(t)|\mathcal{F}(s)) \text{ (by \textbf{(6.4)}(vi))} = \chi_{A}X(s).$$

Now appealing to the definition of conditional expectation, we have

$$\int_{\Omega} X(t)\chi_{A}\, \mathrm{d}P = \int_{\Omega} \chi_{A}X(s)\, \mathrm{d}P.$$

$$\text{i.e.,} \quad \int_{A} X(t)\, \mathrm{d}P = \int_{A} X(s)\, \mathrm{d}P \tag{6.41}$$

Step 2.
Fix t and define events $E_{j,n} = (\frac{j-1}{n}t < \tau \le \frac{j}{n}t),\ j = 1,\ 2, \ldots,\ n;\ n = 1,\ 2, \ldots$. We note that, for n fixed the $E_{j,n}$s are mutually exclusive and $\cup_{j=1}^{n} E_{j,n} = E = (\tau \le t)$. Define $\tau_n = \frac{j}{n}t$ if $\tau \in E_{j,n},\ 1 \le j \le n$ and $\tau_n = \tau$ if $\tau > t$. We note that if $\omega \in (\tau \le t)$, then $|\tau_n(\omega) - \tau(\omega)| \le \frac{1}{n}$. Thus for all $\omega \in \Omega,\ \tau_n(\omega) \to \tau(\omega)$. Writing $s = \frac{j}{n}t$, we note $E_{j,n} \in \mathcal{F}(s)$. Hence by (6.41), $\int_{E_{j,n}} X(\frac{j}{n}t)\, \mathrm{d}P = \int_{E_{j,n}} X(t)\, \mathrm{d}P$. Hence $\int_{(\tau \le t)} X(\tau_n)\, \mathrm{d}P = \sum_{j=1}^{n} \int_{E_{j,n}} X(\tau_n)\, \mathrm{d}P = \sum_{j=1}^{n} \int_{E_{j,n}} X(\frac{j}{n}t)\, \mathrm{d}P = \sum_{j=1}^{n} \int_{E_{j,n}} X(t)\, \mathrm{d}P = \int_{(\tau \le t)} X(t)\, \mathrm{d}P$.

Step 3.
Now take limit as $n \to \infty$. Since $\tau_n(\omega) \to \tau(\omega)$, $\omega \in \Omega$ and since X has continuous sample functions, it follows that $X(\tau_n(\omega)) \to X(\tau(\omega))$. Further, over $(\tau \leq t)$, $|X(\tau_n)| \leq \sup_{0\leq s\leq t} |X(s)|$. Since
$\mathbb{E} \sup_{0\leq s\leq t} |X(s)| \leq \mathbb{E}\{ \sup_{0\leq s\leq t} |B(s)|\}^2 + t < \infty$ (ref. **(5.9)**(vi)), **(2.6.4)**, **(4.4.5)**), the bounded convergence theorem applies and we get, from step 2,

$$\int_{(\tau\leq t)} X(\tau)\,\mathrm{d}\,P = \int_{(\tau\leq t)} X(t)\,\mathrm{d}\,P.$$

Step 4.
$\int_{(\tau>t)} |X(t)|\,\mathrm{d}\,P \leq \int_{(\tau>t)} \{B^2(t) + t\}\,\mathrm{d}\,P$. Since by **(6.6.1)**(ii), $\mathbb{E}\tau < \infty$, it follows that $\lim_{t\to\infty} tP(\tau > t) = 0$; since over $(\tau > t)$, $|B(t)| \leq \lambda = \max(u,\ v)$, we have $\lim_{t\to\infty} \int_{(\tau>t)} B^2(t)\,\mathrm{d}\,P \leq \lambda^2 \lim_{t\to\infty} P(\tau > t) = 0$. These imply $\lim_{t\to\infty} \int_{(\tau>t)} |X(t)|\,\mathrm{d}\,P = 0$.
Step 5.
We have
$0 = \int_\Omega X(t)\,\mathrm{d}\,P = \int_{(\tau\leq t)} X(t)\,\mathrm{d}\,P + \int_{(\tau>t)} X(t)\,\mathrm{d}\,P$
$= \int_{(\tau\leq t)} X(\tau)\,\mathrm{d}\,P + \int_{(\tau>t)} X(t)\,\mathrm{d}\,P$ (by Step 3). Take limit as $t \to \infty$. The second term on the right side tends to 0 by Step 4. The limit of the first term is $\mathbb{E}X(\tau)$. Thus $\mathbb{E}\{B^2(\tau) - \tau\} = \mathbb{E}X(\tau) = 0$. Since (ref. **(6.6.1)**(v)) $\mathbb{E}B^2(\tau) = uv$, it follows that $\mathbb{E}\tau = uv$. □

Let X be a *rv* independent of the BMP B, all defined on a common probability space $(\Omega,\ \mathscr{A},\ P)$. Let F be the *df* of X and let $\mathbb{E}X = 0$.

In our next theorem we show that there exists a *rv* τ such that $B(\tau)$ has *df* F. Our main interest in this result is to develop it into a tool to establish what has come to be called the functional law of the iterated logarithm ($FLIL$) for *iid rvs* with finite variances (ref. **(6.1)**). It is adequate for our purpose to find the τ under the assumption that

$$F \text{ is a continuous } df \text{ and } F(b) - F(a) > 0 \text{ for every } a < b. \tag{6.42}$$

This implies that F is strictly increasing.

We now introduce the function G determined by F and study its properties. This function G plays a vital role in the construction of τ such that $B(\tau)$ has *df* F.

Given $x < 0$ we claim we can find $G(x) > 0$ such that $\int_x^{G(x)} y\,dF(y) = 0$. For, suppose no such $G(x)$ exists. We note that the continuity of F implies that of $u(.)$ defined on $[x, \infty)$ given by $u(y) = \int_x^y t\,dF(t)$. Now $u(0) \leq 0$ since the integrand is negative in this range. $u(0) \neq 0$ since that would imply $F(0) - F(x) = 0$. Thus $u(0) < 0$.

This and the continuity of u together with the assumption that for no y, $\int_x^y t\,d\,F(t) = 0$ imply that $\int_x^\infty t\,dF(t) \leq 0$. Hence $\int_{-\infty}^\infty t\,dF(t) < 0$, a contradiction. Thus given $x < 0$ there exists, as claimed, $G(x) > 0$ such that $\int_x^{G(x)} t\,dF(t) = 0$. Is $G(x)$ unique? Suppose $y > G(x)$ and $\int_x^y t\,d\,F(t) = 0$. That would imply $\int_{G(x)}^y t\,d\,F(t) = 0$ leading to $F(y) - F(G(x)) = 0$, violating the hypothesis. Similarly $0 < y < G(x)$ is not possible. Thus $G(x)$ is unique. Similarly, corresponding to $x > 0$, there exists a unique number $G(x) < 0$ such that $\int_{G(x)}^x t\,d\,F(t) = 0$. Since, given x, $G(x)$ is unique, it follows that $G(G(x)) = x$. Hence $G^{-1} = G$. Also $G(0) = 0$.

Further G is a continuous function. For, if possible, let $x_0 > 0$ be a discontinuity point of G. Let $x \downarrow x_0$. By the definition of G, $G(x) \uparrow$ as $x \downarrow$ and $G(x) < G(x_0)$. Let $\lim_{x \downarrow x_0} G(x) = \alpha$. If possible, let $\alpha < G(x_0)$ (< 0). From $\int_{G(x)}^x y\,d\,F(y) = 0$, we get $\int_\alpha^{G(x_0)} y\,d\,F(y) = 0$. Hence $I = \int_\alpha^{G(x_0)} (-y)\,d\,F(y) = 0$. But $I \geq (-G(x_0))\{F(G(x_0)) - F(\alpha)\} > 0$, a contradiction. Case where $x_0 < 0$ and other variations in the approach of x to x_0 can be similarly handled.

It follows that if A is a Borel set, then so is $G(A)$ $\big(= G^{-1}(A)\big)$. Denote by μ the *pm* generated by F.

Theorem 6.6.3. *For a Borel set $A \subset R$, define $\nu_1(A) = \int_A y\,d\,\mu(y)$ and $\nu_2(A) = -\int_A G(y)\,d\,\mu G(y)$. Then $\nu_1 \equiv \nu_2$.*

Proof. Remember $\mathbb{E}|X| < \infty$. We note $(\mathfrak{a})$ ν_1 is finite valued $(\mathfrak{b})$ ν_1 is countably additive and $(\mathfrak{c})$ ν_1 assumes the value 0 at ϕ. Since $G = G^{-1}$, **(2.3.6)** lets us conclude $\nu_2(A) = -\int_{G(A)} y\,d\,F(y)$. Since G is a one-to-one map, it follows that ν_2, as also $\nu = \nu_1 - \nu_2$, possess the properties $(\mathfrak{a})$, $(\mathfrak{b})$ and $(\mathfrak{c})$.

Let $\mathcal{E}$ denote the collection of Borel sets E such that $\nu(E) = 0$. Since $G(R) = R$, it follows $R \in \mathcal{E}$, since $\nu_1(R) = \mathbb{E}X = 0$. The additivity property of ν now implies that $\mathcal{E}$ is closed under complementation.

$I = [a,\ b],\ a < 0 < b$. Remembering that F is a continuous df, we have

$$\begin{aligned}\nu(I) &= \int_a^0 y\,\mathrm{d}\,F(y) + \int_0^b y\,\mathrm{d}\,F(y) + \int_{G(b)}^0 y\,\mathrm{d}\,F(y) + \int_0^{G(a)} y\,\mathrm{d}\,F(y)\\ &= \int_a^{G(a)} y\,\mathrm{d}\,F(y) + \int_{G(b)}^b y\,\mathrm{d}\,F(y) = 0.\end{aligned}$$

These arguments and the result holds for any arbitrary interval $(a,\ b)$, closed, semi-closed or open, wholly or partly lying in the right or left half line. Thus $\mathcal{E}$ contains all the intervals and hence, by the countable additivity of ν, the open sets too. Now let E be an arbitrary Borel subset of $(0,\ \infty)$. Restricted to this half line, ν_1 is a finite measure and is outer regular (ref. **(1.9.2)**). Hence, given $\varepsilon > 0$, an open subset $U \supset E$ can be found such that $\nu_1(U) < \nu_1(E) + \varepsilon$. We note that $G(U)$ is an open set, that $G(U) \supset G(E)$ and that $\nu_2(E) \leq \nu_2(U)$. Since U is an open set, $\nu_1(U) = \nu_2(U)$. Hence $\nu_2(E) \leq \nu_1(E) + \varepsilon$. Since $\varepsilon > 0$ is arbitrary, $\nu_2(E) \leq \nu_1(E)$. Similarly it can be shown that $\nu_1(E) \leq \nu_2(E)$. Thus for all Borel sets E, $\nu(E) = 0$, as was to be proved. □

Remark 6.6.4. For future reference we record the following consequence of **(6.6.3)**: For any measurable function f,

$$-\int_R f(y)G(y)\,\mathrm{d}\,\mu(G(y)) = \int_R f(y)y\,\mathrm{d}\,\mu(y) \tag{6.43}$$

in the sense that if the integral on either side exists, so will the other and then the two sides will be equal.

The construction of τ is done in the following theorem.

Theorem 6.6.5. *Let $B,\ X,\ F,\ \mu,\ G$ be as earlier defined. If τ is the smallest root of the equation $(B(t) - X)(B(t) - G(X)) = 0$, then the variable $B(\tau)$ has df F.*

Proof. Since X is independent of B, we have by **(6.6.1)**(v). $P\,(B(\tau) = X|X) = \frac{|G(X)|}{|X|+|G(X)|}$ and $P\,(B(\tau) = G(X)|X) = \frac{|X|}{|X|+|G(X)|}$. Hence

$$\begin{aligned}P(0 < B(\tau) < x) &= E\{P(0 < B(\tau) < x|X)\}\\ &\quad\text{(directly from the definition of conditional expectation)}\\ &= \int_{-\infty}^{\infty} P(0 < B(\tau) < x|X = t)\,\mathrm{d}\,F(t)\end{aligned}$$

$$= \int_{\{0\}} P(0 < B(\tau) < x | X = t)\, \mathrm{d}\, F(t)$$
$$+ \int_{(-\infty,\, 0)} P(0 < B(\tau) < x | X = t)\, \mathrm{d}\, F(t)$$
$$+ \int_{(0,\, \infty)} P(0 < B(\tau) < x | X = t)\, \mathrm{d}\, F(t) = I_0 + I_1 + I_2,$$

say. $I_0 = 0$, since $X = 0$ implies $B(\tau) \equiv 0$. Write $I_2 = I_{2,1} + I_{2,2}$ where $I_{2,2} = \int_{[x,\, \infty)} P(0 < B(\tau) < x | X = t)\, \mathrm{d}\, F(t) = 0$ since $X = t \geq x$ implies $B(\tau) \notin (0,\, x)$. Similarly $\int_{(-\infty,\, G(x)]} P(0 < B(\tau) < x | X = t)\, \mathrm{d}\, F(t) = 0$. Now $0 < t < x$ implies $B(\tau) = t$. Hence,
$I_{2,1} = \int_{t \in (0,\, x)} P(B(\tau) = t | X = t)\, \mathrm{d}F(t) = \int_{(0,\, x)} \frac{|G(t)|}{|t| + |G(t)|}\, \mathrm{d}F(t)$
$= \int_{(0,\, x)} \frac{-G(t)}{t - G(t)}\, \mathrm{d}F(t)$.
Similarly,

$$\int_{(G(x),\, 0)} P(0 < B(\tau) < x | X = t)\, \mathrm{d}\, F(t)$$
$$= \int_{(G(x),\, 0)} \frac{|t|}{|t| + |G(t)|}\, \mathrm{d}\, F(t)$$
$$= \int_{(G(x),\, 0)} \frac{-t}{-t + G(t)}\, \mathrm{d}\, F(t).$$

Thus

$$P\,(0 < B(\tau) < x)$$
$$= \int_{(0,\, x)} \frac{-G(t)}{t - G(t)}\, \mathrm{d}\, \mu(t) + \int_{(G(x),\, 0)} \frac{-t}{-t + G(t)}\, \mathrm{d}\, \mu(t)$$
$$= \int_{(0,\, x)} \frac{-G(t)}{t - G(t)}\, \mathrm{d}\, \mu(t) + \int_{(G(x),\, 0)} \frac{G(G(t))}{G(G(t)) - G(t)}\, \mathrm{d}\, \mu(t)$$
$$= \int_{(0,\, x)} \frac{-G(t)}{t - G(t)}\, \mathrm{d}\, \mu(t) - \int_0^x \frac{G(t)}{G(t) - t} d\mu G^{-1}(t) \text{ (by } \mathbf{(2.3.6)})$$
$$= \int_{(0,\, x)} \frac{-G(t)}{t - G(t)}\, \mathrm{d}\, \mu(t) - \int_0^x \frac{G(t)}{G(t) - t}\, \mathrm{d}\, \mu G(t)$$
$$= \int_{(0,\, x)} \frac{-G(t)}{t - G(t)}\, \mathrm{d}\, \mu(t) - \int_0^x \frac{1}{G(t) - t} t\, \mathrm{d}\, \mu(t), \text{ by (6.43)}$$

$$= \int_0^x \mathrm{d}\, F(t).$$

Similarly it can be shown that $P(x < B(\tau) < 0) = \int_x^0 \mathrm{d}\, F(t)$. The desired result follows from this. □

Recall notation : rv X has mean zero and has a continuous and strictly increasing *df* F with $F(b) - F(a) > 0$ for all $a < b$. The pm determined by F is denoted by μ. Function G is such that $\int_{G(x)}^{x} y \,\mathrm{d}\, F(y) = 0$ for all x. It is instructive to have a numerical example before us. Let $F(x) = \frac{5}{18x^2}$ if $x \leq -1$; $F(x) = \frac{5}{12} + \frac{5}{36}x$ if $-1 \leq x \leq 1$ and $F(x) = 1 - \frac{4}{9x^5}$ if $x \geq 1$. It is easy to verify that for $x \geq 1,\ G(x) = -x^4$; that for $x \leq -1,\ G(x) = |x|^{\frac{1}{4}}$ and that for $-1 \leq x \leq 1,\ G(x) = -x$.

6.7 The strong Markov property of a BMP

We start this section with the definition (ref. (6.44)) of a Markov process. The property described in (6.44) is called the Markov property. U is called the state space. When t takes only a countable number of values, say, $t = 0,\ 1,\ 2, \ldots$ and if U is a countable set, the Markov process is called a discrete time Markov chain (M.C.) with a countable number of states. Such M.C.s are studied in detail in Chapter 7. After showing that a BMP is a Markov process according to this definition, we study filtration, stopping time and strong Markov property. This property stipulates a certain type of dependence among the random variables of the process.

Let $X_t,\ t \geq 0$ be a stochastic process defined on a probability space $(\Omega,\ \mathscr{A},\ P)$ taking values in a measurable space $(\mathbf{U},\ \mathscr{U})$. Let $\mathscr{A}(t) \subset \mathscr{A}$ be the σ-field generated by the family $X_s,\ 0 \leq s \leq t$. It is customary to descrbe X_t as the position at time t of a random system varying in time.

Definition 6.7.1. Let T be a linearly ordered index set. The process $(X_t,\ t \in T)$ is said to be a *Markov Process* if it possesses the Markov property:

$$(s,\ t \in T,\ s < t \text{ and } \Gamma \in \mathscr{U}) \Rightarrow \\ P(X_t \in \Gamma | \mathscr{A}(s)) = P(X_t \in \Gamma | X_s)\ wp1. \quad (6.44)$$

The Markov property is described by saying that the process is memoryless. This is said in the sense that the distribution of the future position of the system, i.e., the distribution of X_t, depends only on the last known past, namely, the state X_s, and not on how the system reached this state. In other words, information about the past of the process *up to time* s is not any more useful than information about the process *at time* s, in describing the position of the process at time t.

The following theorem is an immediate consequence of this definition.

Theorem 6.7.2. *Let $a > 0$ be fixed. Then,*

$$\text{the process } (X_{a+t},\ t \geq 0) \text{ is a Markov process} \tag{6.45}$$

if $(X_t,\ t \geq 0)$ is a Markov process.

Proof. Let $0 < s < t$ and let (X_t) be a Markov process.

$$\begin{aligned}
&P(X_{a+t} \in \Gamma | X_{a+r},\ 0 \leq r \leq s) \\
&\quad = E(\{P(X_{a+t} \in \Gamma | X_{a+r},\ 0 \leq r \leq s)\} | X_u,\ 0 \leq u \leq a+s) \\
&\qquad \text{since } \sigma(X_{a+r},\ 0 \leq r \leq s) \subseteq \sigma(X_u,\ 0 \leq r \leq a+s) \\
&\quad = E(\{P(X_{a+t} \in \Gamma | X_r,\ 0 \leq r \leq a+s)\} | X_{a+r},\ 0 \leq r \leq s) \\
&\qquad\qquad\qquad \text{(ref. } \mathbf{(6.4)} \text{ (vii))} \\
&\quad = E(\{P(X_{a+t} \in \Gamma | X_{a+s})\} | X_{a+r},\ 0 \leq r \leq s), \\
&\qquad\qquad\qquad \text{(since } (X_t) \text{ is a Markov process)} \\
&\quad = P(X_{a+t} \in \Gamma | X_{a+s}), \\
&\qquad\qquad\qquad \text{(since } X_{a+s} \text{ is } \sigma(X_{a+r},\ 0 \leq r \leq s) \text{ -measurable)}
\end{aligned}$$

as was to be proved. □

Theorem 6.7.3. *With ($\boldsymbol{U}$, $\mathscr{U}$) $= (R,\ \mathscr{R})$, a standard BMP B is a Markov process.*

Proof. Write $X_{t+s} = X + Y$ where $X = X_{t+s} - X_s$ and $Y = X_s$ and note that X is independent of $\mathscr{A}(s)$ and that Y is $\mathscr{A}(s)$-measurable. Hence by **(6.3.3)**(vi), $P(X + Y \in \Gamma | \mathscr{A}(s)) = P(X + Y \in \Gamma | Y)$ $wp1$ which is the desired result. □

We now equip the basic probability space (Ω, $\mathscr{A}$, P) with a filtration. i.e. with a non-decreasing family of sub σ-fields. In applications, it is common to interpret filtration $\mathcal{A}_t$ as the collection of all events that are verifiable at *t*. The increase in the size of $\mathcal{A}_t$ as *t* increases is interpreted as the accumulation of information over time. It is through filtration that stopping time variables are defined and stopping time variables are the foundation of Skorohod embedding.

Definitions 6.7.4. (i) A non-decreasing family $\mathcal{A}(t),\ t \geq 0$ of sub σ-fields of $\mathscr{A}$ is called a filtration. Given a filtration $(\mathcal{A}(t))$ and a random element X, denote by $(\mathcal{A}_X(t))$ the filtration where $\mathcal{A}_X(t)$ is the minimal σ-field containing $\mathcal{A}(t)$ and the σ-field generated by X. The family $\mathscr{A}(t)$ defined in **(6.7)** is called the natural filtration of the process $(X(t))$. A stochastic process $(X(t), t \geq 0)$ is said to be adapted to the filtration $\{\mathcal{A}(t)\}$ if, for each t, $X(t)$ is $\mathcal{A}(t)$-measurable.

We note that every process is adapted to its natural filtration.

(ii) A standard Brownian Motion Process $(SBMP)$ $B = \{B(t),\ \mathscr{A}(t); t \geq 0\}$ adapted to the filtration $\{\mathscr{A}(t)\}$ defined on a probability space $(\Omega,\ \mathscr{A},\ P)$ is a process satisfying the following conditions :

(p_1) $B(0) \equiv 0$

(p_2) $B(t,\ \omega)$ is a continuous function of t for every $\omega \in \Omega$

(p_3) for $0 \leq s < t < \infty,\ B(t) - B(s)$ is independent of $\mathscr{A}(s)$

(p_4) $B(t) - B(s)$ is normally distributed with mean zero and variance $|t - s|$. Equivalently, for every $\{t_1,\ t_2, \dots,\ t_k\}$, the vector $(B(t_1),\ B(t_2), \dots,\ B(t_k))$ has the multivariate normal distribution with zero mean vector and covariances $\mathbb{E}B(t_i)B(t_j) = \min(t_i,\ t_j)$.

The natural filtration of B will be denoted by $(\mathscr{A}^B(t),\ t \geq 0)$.

Remark. That a $SBMP$ (adapted to its natural filtration) exists was established in **(5.7.1)**.

Trivially a $SBMP$ is adapted to the larger filtration $\mathscr{A}_X^B$ if the random element X is independent of B.

The conclusions in **(6.3.3)**(viii) continue to hold under the new definition of B.

A particularly useful enlargement of the filtration $(\mathscr{A}^B(t))$ is by augmentation, which we explain now. Denote by $\mathcal{N}$ the collection of all the subsets of sets of P-measure zero in every $\mathscr{A}^B(t),\ t \geq 0$. Define $\hat{\mathscr{A}}^B(t) = \sigma(\mathscr{A}^B(t) \cup \mathcal{N})$. Clearly $(\hat{\mathscr{A}}^B(t),\ t \geq 0)$ is a filtration, called the augmented filtration and the BMP is adapted to it.

We note that relative to this filtration $(B(t),\ t \geq 0)$ is still a BMP. This is so, since augmentation of the σ-fields $\mathscr{A}^B(t)$ does in no way affect the satisfaction of the conditions (p_1) to (p_4) stated earlier.

From here forward, we dispense with the natural filtration $\mathscr{A}^B(t)$ and write $\mathscr{A}^B(t)$ for $\hat{\mathscr{A}}^B(t)$.

(iii) A non-negative rv τ is said to be Markov relative to the filtration $(\mathcal{A}(s),\ s \geq 0)$ if $\{\omega\ :\ \tau(\omega) \leq t\} \in \mathcal{A}(t),\ t \geq 0$. A Markov time is also called a stopping time.

If τ is a Markov time, then $(\tau < t) \in \mathcal{A}(t)$ since $(\tau < t) = \cup_n(\tau \leq t - \frac{1}{n})$.

Let $s > 0$. Then $\tau \equiv s$ is a Markov time for every filtration. This is so, since $(\tau < t) = \phi$ or Ω depending on whether $t < s$ or $t \geq s$.

Remarks. ($\mathfrak{a}$) The rv τ defined in **(6.6.5)** is Markov relative to the filtration $(\mathscr{A}_X^B(t))$. For, if $Y_1(\omega) = \min\{X(\omega), G(X(\omega))\}$ and $Y_2(\omega) = \max\{X(\omega), G(X(\omega))\}$, then $A = \{\omega : \tau(\omega) > t\} = \{\omega : Y_1(\omega) \leq Z_1(\omega) = \min_{0\leq s\leq t} B(s, \omega) \leq Z_2(\omega) = \max_{0\leq s\leq t} B(s, \omega) \leq Y_2(\omega)\}$. This implies that A (and hence A') lies in $\mathscr{A}_X^B(t)$.

($\mathfrak{b}$) The τ discussed above can alternatively be defined as $\tau = \inf\{t > 0, B(t) = \min(X, G(X))\}$. More generally, let $W = \{W(t), t \geq 0, W(0) = 0\}$ be any process with all its sample functions continuous. Further let

$$\overline{\lim_{t\to\infty}} W(t) = \infty \;\&\; \underline{\lim_{t\to\infty}} W(t) = -\infty \; wp1. \tag{6.46}$$

Let $(\mathscr{A}(t), t \geq 0)$ be its natural filtration. Let Y be a rv, defined on the same space as W. Denote by $\mathcal{A}(t)$ the enlarged filtration $\mathscr{A}_Y^W(t)$.

Define $\tau(Y(\omega), \omega) = \inf\{t > 0, W(t, \omega) = Y(\omega)\}$. To claim τ is a Markov time relative to the filtration $\{\mathcal{A}(t)\}$, the argument is as follows.

That τ is finite valued follows from (6.46). To prove the other defining property of a Markov time, that is to show $(\tau \leq t) \in \mathcal{A}(t)$ we proceed along the following steps.

Step 1

Let $Y \equiv c$, a constant. Using the hypothesis that all the sample functions of W are continuous, we have, for $c > 0$, $(\tau(c, .) \leq t) = \bigcap_{s<c} \bigcup_{r<t} \{W(r, .) > s\}$ where r, s run through the rationals.The event $\{W(r) > s\}$ belongs to $\mathscr{A}(r) \subset \mathscr{A}(t)$. Hence the union over $r < t$, r rational, belongs to $\mathscr{A}(t)$. It follows $(\tau(c, .) \leq t) \in \mathscr{A}(t)$.

With $-W$ in place of W and arguing on similar lines we see that $(\tau(c, .) \leq t) \in \mathscr{A}(t)$ holds also when $c < 0$.

If $\tau(\omega) = \inf\{t > 0 : W(t, \omega) = 0\}$, then $\{\omega : \tau(\omega) > t\} = \{\bigcap_{\substack{0<r\leq t \\ r\ rational}} (\omega : W(r) > 0)\} \cup \{\bigcap_{\substack{0<r\leq t \\ r\ rational}} (\omega : W(r) < 0)\}$ and hence lies in $\mathscr{A}(t) \subset \mathcal{A}(t)$.

Thus $\tau(c, .)$ is a Markov time relative to the filtration $\mathcal{A}(t)$ for all values of c.

Step 2

Let $\tau(\omega) = \min\{\tau(Y_1(\omega), \omega); \tau(Y_2(\omega), \omega)\}$ where $Y_1 = Y\, \chi_{(Y\geq 0)}$, $Y_2 = Y\, \chi_{(Y\leq 0)}$. That τ is a Markov time relative to the filtration $(\mathcal{A}(t))$ will follow if we show that each of the two rvs in the double bracket is a Markov time relative to the same filtration.

The other case being similar, it is enough to show that $\tau(Y_1(.),\ .)$ is a Markov time relative to $(\mathcal{A}(t))$. Hence it remains to show, changing to a more convenient notation, that τ is a Markov time relative to the filtration $\mathcal{A}(t) = \mathscr{A}_Z^W(t)$ where $\tau(\omega) = \inf\{t > 0 :\ W(t,\ \omega) = Z(\omega)\}$ and Z is a non-negative rv.

Step 3

Define $Z_n(\omega) = \frac{i-1}{2^n}$ if $\frac{i-1}{2^n} < Z(\omega) \leq \frac{i}{2^n},\ i = 1,\ 2, \ldots\ldots$ Clearly $Z_n(\omega) \uparrow Z(\omega)$ for every $\omega \in \Omega$. In fact the convergence is uniform in ω since $|Z_n(\omega) - Z(\omega)| \leq \frac{1}{2^n}$.

We note : from its definition, it is obvious that $\tau(Z_n(\omega),\ \omega) \leq \tau(Z(\omega),\ \omega)$ and that $\tau(Z_n(\omega),\ \omega) \uparrow$ for every $\omega \in \Omega$. Since $\tau(Z_n(\omega),\ \omega) \leq \tau(Z(\omega) + \frac{1}{2^n},\ \omega)$ and since $W(t,\ \omega)$ is a continuous function of t, for each ω fixed, it follows that $\tau(Z_n(\omega),\ \omega) \uparrow \tau(Z(\omega),\ \omega)$. This implies that $(\tau(Z(.),\ .) \leq t) = \cap_n(\tau(Z_n(.),\ .) \leq t)$. Hence we can conclude that $(\tau(Z(.),\ .))$ is a Markov time relative to the filtration $(\mathcal{A}(t))$ if we show that each $(\tau_1(Z_n(.),\ .))$ is a Markov time relative to that filtration. Again changing notation, what remains to be shown is that $\tau(Z(.),\ .)$ is a Markov time relative to $(\mathcal{A}(t))$ where Z is a discrete variable taking the values $0 = a_0 < a_1 < a_2 < \ldots.$This we proceed to establish.

Step 4

$$(\tau(Z(.),\ .) \leq t) = \cup_j\{(\tau(Z(.),\ .) \leq t) \cap (Z = a_j)\} = \bigcup_j\{(\tau(a_j,\ .)$$

$\leq t) \cap (Z = a_j)\}$. By step 1, $(\tau(a_j,\ .) \leq t) \in \mathcal{A}(t)$ while $(Z = a_j) \in \mathcal{A}(a_j) \subset \mathcal{A}(t)$. With this the proof is complete.

(c) **Theorem**

If $\tau,\ \eta$ are Markov times relative to a filtration $(\mathcal{A}(t),\ t \geq 0)$, then so is $\tau + \eta$.

Proof.

$(\tau+\eta > t) = (\tau = 0,\ \eta > t) \cup (\tau > t,\ \eta = 0) \cup (0 < \tau < t,\ \tau+\eta > t) \cup (\tau \geq t,\ \eta > 0) = A \cup B \cup C \cup D$, say. $A \in \mathcal{A}(t)$ since $(\tau = 0) \in \mathcal{A}(0) \subset \mathcal{A}(t)$ and $(\eta > t) \in \mathcal{A}(t)$. Similrly $B \in \mathcal{A}(t)$. Now $C = \bigcup_{\substack{r\ rational \\ r<t}} \{r < \tau < t,\ \eta > t - r\}$. The event $\{r < \tau < t\} = (\tau < t) \cap (\tau \leq r)'$ and hence belongs to $\mathcal{A}(t)$. The event $(\eta > t - r) \in \mathcal{A}(t - r) \subset \mathcal{A}(t)$. That each term of the union defining C belongs to $\mathcal{A}(t)$. Since C is a countable union of these events, $C \in \mathcal{A}(t)$. Finally $D \in \mathcal{A}(t)$ since $(\tau \geq t) \in \mathcal{A}(t)$ and $(\eta > 0) \in \mathcal{A}(0) \subset \mathcal{A}(t)$. □

Here we define and study the σ-field $\mathcal{A}(\tau)$ where τ is a Markov time relative to the filtration $\mathcal{A}(t)$. Given τ, define $\tau_n : \tau_n(\omega) = \frac{j}{2^n}$ if $\frac{j-1}{2^n} \leq \tau(\omega) < \frac{j}{2^n}$, $j = 1, 2, \ldots$. Since τ is finite valued, $\tau_n(\omega)$ is defined for all $\omega \in \Omega$. Clearly $\tau_n(\omega) \downarrow \tau(\omega)$. We will presently show in (iv) below that each τ_n is a Markov time relative to the same filtration. Since the τ_ns are discrete valued (in fact, they are lattice distributed) it is easier to make use of the Markov property for them. For this reason, approximating τ by the τ_ns often helps.

Let $T = T_1 \geq T_2 \geq \ldots$ be Markov times for the filtration $(\mathcal{A}(t))$ and let $T_n(\omega) \downarrow \tau(\omega)$ for all $\omega \in \Omega$.

Definitions 6.7.5. *(a) Let $\mathcal{A}(t)$, $t \geq 0$ be a filtration. Then $\mathcal{A}(t+)$ is defined as the σ-field*
$\bigcap_{\varepsilon>0} \mathcal{A}(t+\varepsilon)$. Clearly, if $0 \leq u < t$, then, $\mathcal{A}(u) \subset \mathcal{A}(u+) \subset \mathcal{A}(t) \subset \mathcal{A}(t+)$.

A filtration $\mathcal{A}(t)$, $t \geq 0$ is said to be right continuous if $\mathcal{A}(t) = \mathcal{A}(t+)$, $t \geq 0$.

Theorem. Let $\{\mathscr{A}(t), t \geq 0\}$ be an arbitrary filtration Then the filtration $\{\mathscr{A}(t+), t \geq 0\}$ is right continuous.

Proof.

$$\bigcap_{s>t} \mathscr{A}(s+) = \bigcap_{s>t} \bigcap_{u>s} \mathscr{A}(u)$$
$$= \bigcup_{u>t} \bigcup_{t<s<u} \mathscr{A}(u) = \bigcap_{u>t} \mathscr{A}(u) = \mathscr{A}(t+). \quad \square$$

(b) Let τ be a Markov time relative to the filtration $\mathcal{A}(t)$, $t \geq 0$. We define $\mathcal{A}(\tau)$ and $\mathcal{A}(\tau+)$ as follows.

$$\begin{cases} \mathcal{A}(\tau) = \{A : A \in \mathscr{A} \text{ and } A \cap (\tau \leq t) \in \mathcal{A}(t) \text{ for every } t \geq 0\}. \\ \mathcal{A}(\tau+) = \{A : A \in \mathscr{A} \text{ and } A \cap (\tau \leq t) \in \mathcal{A}(t+) \text{ for every } t \geq 0\}. \end{cases} \tag{6.47}$$

Clearly, $\mathcal{A}(\tau) \subset \mathcal{A}(\tau+)$.

We note that the collection $\mathcal{A}(\tau+)$ is closed under complementation, since $A' \cap (\tau \leq t) = (\tau \leq t) \sim A \cap (\tau \leq t)$. It is closed under finite union, since $(A \cup B) \cap (\tau \leq t) = \{A \cap (\tau \leq t)\} \cup \{B \cap (\tau \leq t)\}$. Further $A_n \in \mathcal{A}(\tau+)$ and $A_n \uparrow A$ imply $A \in \mathcal{A}(\tau+)$ since $\{A_n \cap (\tau \leq t)\} \uparrow A \cap (\tau \leq t)$. Thus $\mathcal{A}(\tau+)$ is a σ-field.

On the same lines one can show that $\mathcal{A}(\tau)$ is a σ-field.

$\mathcal{A}(\tau)$ is known as the σ-field of events up to the instant τ or the σ-field of the τ-past. $\mathcal{A}(\tau+)$ is known as the σ-field of events up to immediately after the instant τ.

(c) Lemma.

(i) $\mathcal{A}_1(\tau+) = \mathcal{A}(\tau+)$ where

$$\mathcal{A}_1(\tau+) = \{A : A \in \mathscr{A} \text{ and } A \cap (\tau < t) \in \mathcal{A}(t)\ t > 0\}. \tag{6.48}$$

(ii) If $\eta \geq \tau$ is another Markov time for the same filtration, then $\mathcal{A}(\tau+) \subset \mathcal{A}(\eta+)$ and $\mathcal{A}(\tau+) \subset \mathcal{A}(\eta+)$.

(iii) $\cap_n \mathcal{A}(T_n+) = \mathcal{A}(\tau+)$.

(iv) τ_n is Markov relative to the filtration $\mathcal{A}(t),\ t \geq 0$. Also, $\mathcal{A}(\tau) \subseteq \mathcal{A}(\tau_n)$.

(v) If a filtration $\mathcal{A}(t), t \geq 0$ is right continuous i.e., if $\mathcal{A}(t+) = \mathcal{A}(t),\ t \geq 0$ then so is the filtration $\mathcal{A}_Y$, for every random element Y.

(vi) If τ is a Markov time relative to a right continuous filtration $(\mathcal{A}(t), t \geq 0)$, then $\mathcal{A}(\tau+) = \mathcal{A}(\tau)$.

(vii) If $\tau \equiv s, s > 0$, then $\mathcal{A}(\tau) = \mathcal{A}(s)$.

Proof.

(i) $A \in \mathcal{A}(\tau+)$ implies that $A \cap (\tau \leq t) \in \mathcal{A}(t+),\ t \geq 0$. Since $A \cap (\tau < t) = A \cap \{\cup_n(\tau \leq t - \frac{1}{n})\} = \cup_n\{A \cap (\tau \leq t - \frac{1}{n})\}$ and since $A \cap (\tau \leq t - \frac{1}{n}) \in \mathcal{A}(\overline{t - \frac{1}{n}}+) \subset \mathcal{A}(t - \frac{1}{2n}) \subset \mathcal{A}(t)$, it follows that $\mathcal{A}(\tau+) \subseteq \mathcal{A}_1(\tau+)$.

Suppose now $A \in \mathcal{A}_1(\tau+)$. Hence for every $n = 1, 2, \ldots,\ A \cap (\tau < t + \frac{1}{n}) \in \mathcal{A}(t + \frac{1}{n})$. Since $A \cap (\tau \leq t) = \lim_{n\to\infty} \{A \cap (\tau < t + \frac{1}{n})\}$, it follows $A \cap (\tau \leq t) \in \mathcal{A}(t+)$. Thus $\mathcal{A}_1(\tau+) \subseteq \mathcal{A}(\tau+)$, leading to the claim.

(ii) Since $\eta \geq \tau$, this implies $A \cap (\eta \leq t) = A \cap (\eta \leq t) \cap (\tau \leq t) = \{A \cap (\tau \leq t)\} \cap (\eta \leq t)$. If $A \in \mathcal{A}(\tau)$, then $A \cap (\tau \leq t) \in \mathcal{A}(t)$, since τ is Markov relative to the filtration $(\mathcal{A}(t),\ t \geq 0)$. $(\eta \leq t) \in \mathcal{A}(t)$, since η is Markov relative to the same filtration. Thus $A \in \mathcal{A}(\tau)$ implies $A \in \mathcal{A}(\eta)$. i.e., $\mathcal{A}(\tau) \subset \mathcal{A}(\eta)$.

The proof that $\mathcal{A}(\tau+) \subset \mathcal{A}(\eta+)$ is on similar lines.

(iii) By (ii), $\mathcal{A}(\tau+) \subseteq \cap_1^\infty \mathcal{A}(T_n+)$. To prove inclusion in the other direction, let event $A \in \cap_1^\infty \mathcal{A}(T_n+)$. This implies that for each $n \geq 1$, and $t \geq 0$, $A \cap (T_n \leq t) \in \mathcal{A}(t+)$. Since $T_n \downarrow \tau$, we have $\tau = \inf_n T_n$. Hence $(\tau < t+\varepsilon) = \cup_1^\infty (T_n < t+\varepsilon)$. This leads to $A \cap (\tau < t+\varepsilon) = \cup_1^\infty\{A \cap (T_n < t + \varepsilon)\}$. Thus by (i), $A \cap (\tau < t + \frac{1}{n}) \in \mathcal{A}(t + \frac{1}{n}),\ n = 1, 2, \ldots$. For N fixed and for all $n > N$, we have $A \cap (\tau < t + \frac{1}{n}) \in \mathcal{A}(t + \frac{1}{N})$. Take limit as $n \to \infty$ to get $A \cap (\tau \leq t) \in \mathcal{A}(t + \frac{1}{N})$. This is true for all $N \geq 1$. Hence $A \cap (\tau \leq t) \in \cap_N \mathcal{A}(t + \frac{1}{N}) = \mathcal{A}(t+)$. This is equivalent to the claim $A \in \mathcal{A}(\tau+)$

(iv) Fix $t,\ 0 \leq t < \infty$ and $n \geq 1$. There can be found an integer $j,\ 1 \leq j < \infty$ such that $\frac{j-1}{2^n} \leq t < \frac{j}{2^n}$. If $0 \leq t < \frac{1}{2^n}$, $(\tau_n \leq t) = (\tau_n = 0) = (\tau = 0) \in \mathcal{A}(0) \subset \mathcal{A}(t)$. Let $\frac{j-1}{2^n} \leq t < \frac{j}{2^n}$, $j \geq 2$, then $(\tau_n \leq t) = (\tau < \frac{j-1}{2^n}) \in \mathcal{A}(\frac{j-1}{2^n}) \subset \mathcal{A}(t)$. This proves the claim for τ_n.

The second assertion follows from part (ii).

(v) Let $\mathcal{B}$ be the σ-field generated by Y. We have,
$\mathcal{A}_Y(t+) = \bigcap_{\varepsilon>0} \sigma(\mathcal{A}_Y(t+\varepsilon)) = \bigcap_{\varepsilon>0} \sigma(\mathcal{A}(t+\varepsilon), \mathcal{B}) \supseteq \sigma(\mathcal{A}(t+),\ \mathcal{B})$. Suppose there exists $A \notin \sigma(\mathcal{A}(t+),\ \mathcal{B})$. Hence $A \notin \mathcal{A}(t+)$ and $A \notin \mathcal{B}$. This implies that for some $\varepsilon > 0,\ A \notin \mathcal{A}(t+\varepsilon)$. This together with the fact that $A \notin \mathcal{B}$ imply $A \notin \sigma(\mathcal{A}(t+\varepsilon),\ \mathcal{B})$. Hence $A \notin \bigcap_{\varepsilon>0} \sigma(\mathcal{A}(t+\varepsilon), \mathcal{B})$. Thus, $\mathcal{A}_Y(t+) = \sigma(\mathcal{A}(t+),\ \mathcal{B})$. Now use hypothesis $\mathcal{A}(t+) = \mathcal{A}(t)$ and conclude the proof.

(vi) The assertion follows since, for $t \geq 0$,
$(A \in \mathscr{A}^B(\tau+)) \Rightarrow (A \cap (\tau \leq t) \in \mathscr{A}^B(t+)) = (A \cap (\tau \leq t) \in \mathscr{A}^B(t)) \Rightarrow (A \in \mathscr{A}^B(\tau))$.

(vii) We note $(\tau \leq t) = \phi$ or Ω depending on whether $t < s$ or $t \geq s$. Hence $A \cap (\tau \leq t) = \phi$ or A and will therefore belong to $\mathcal{A}(t)$ if $t \leq s$. i.e., $\mathcal{A}(\tau) \supseteq (s)$. Let the non-null set $A \notin \mathcal{A}(s)$. If $A \in (\tau)$, we must have $A \cap (\tau \leq s) \in \mathcal{A}(s)$. Since $(\tau \leq s) = \Omega$, this implies $A \in \mathcal{A}(s)$. This contradictin establishes the claim. □

(ð) **Theorem.**

The natural filtration of a process, all of whose sample functions are continuous, can fail to be right continuous.

Proof. Denote by ω the elements of $C = C[0,\ 1],\ \omega(0) = 0$. Endow C with its Borel σ-field $\mathscr{C}$ under the uniform metric. Let μ be an arbitrary probability measure on $\mathscr{C}$. Define stochastic process $X_t(\omega) = \omega(t),\ t \geq 0$. Obviously, all the sample functions of the process are continuous. We will show that the natural filtration $(\mathscr{A}(t),\ t \geq 0)$ of the process $(X_t,\ t \geq 0)$ is not right continuous.

Step 1

Let f be an arbitrary real valued function defined on $I = [a,\ b] \subset T$. f is said to have a **local maximum** at $t \in I$ if there exists $\delta > 0$ such that $f(t) \geq f(s)$ for all s with $|t-s| \leq \delta$. If $t = a$ or $t = b$, the range of s is to be suitably modified.

Step 2.

Define, for $t > 0$ fixed, $F = F_t = \{\omega\ :\ \omega(u)$ has local maximum at $t\}$. We note F is non-empty. We further note that for each $m \geq 0$,

$$F = \bigcup_{n=m}^{\infty} \bigcap_{\substack{|t-r|<\frac{1}{n} \\ r \text{ rational}}}^{\infty} \{\omega\ :\ \omega(t,\ \omega) \geq \omega(r,\ \omega)\} = \bigcup_{n=m}^{\infty} G_n,$$

say. Clearly $G_n \in \mathscr{A}(t+\frac{1}{n})$. Hence, since $\mathscr{A}(r)$ is a filtration, $F = \bigcup\limits_{n=m}^{\infty} G_n \in \mathscr{A}(t+\frac{1}{m})$. Since this relation holds for all $m \geq 0$, we conclude $F \in \mathscr{A}(t+)$.

Step 3.

We show now $F \notin \mathscr{A}(t)$. Suppose $F \in \mathscr{A}(t)$. This implies that there exists (ref. **(1.2)**) $\{t_1,\ t_2, \ldots..\} \subset [0,\ 1]$ and $\mathbf{A} \in \mathscr{R}^\infty$ such that $F = \{\omega\ :\ (\omega(t_1),\ \omega(t_2), \ldots) \in \mathbf{A}\}$. It follows that if $\omega \in F$ then every function which coincides with ω on the t_js will also lie in F. In particular, $\omega \in F$ implies $\bar{\omega} \in F$ where $\bar{\omega}(s) = \omega(s)$ for $0 \leq s \leq t$ and $\bar{\omega}(s) = \omega(t) + s - t$ for $s \geq t$. But $\bar{\omega}$ does not have a local maximum at t, since however small $|t - r|$ may be, $\omega(t) - \omega(r) = t - r < 0$ for $r > t$. This proves $F \notin \mathscr{A}(t)$. Hence the filtration $(\mathscr{A}(t), t \geq 0)$ is not right continuous. □

Remark. It may be noted that if μ is the Wiener measure, then the X_t process would be a Brownian Motion Process, called the co-ordinate Brownian Motion Process. We have shown above that for the co-ordinate Brownian Motion Process, the natural filtration is not right continuous.

($\mathfrak{e}$) Theorem.

The augmented filtration $(\mathscr{A}^B(t),\ t \geq 0)$ is right continuous.

Proof.

Fix $s \geq 0$. Let $\mathscr{E}$ be the collection of all sets of the form $\{B(t_k) \in A_k,\ 1 \leq k \leq n\}$ for every choice of $n \geq 1$, $\{t_\nu,\ 1 \leq \nu \leq n\} \subset [0,\ \infty)$ and linear Borel sets A_k. We note that $\mathscr{E}$ is a field of sets and that $\sigma(\mathscr{E}) = \sigma(B(t),\ t \geq 0)(= \mathscr{A}^B_\infty$, say). Let $\mathscr{D} \subset \mathscr{A}^B_\infty$ be the collection of sets F such that $P(F|\mathscr{A}^B(s+)) = P(F|B(s))\ wp1$.

If $F_n \in \mathscr{D}$, $n \geq 1$, if the F_ns are disjoint and if $F = \cup_n F_n$, then

$$\begin{aligned}
P(F|\mathscr{A}^B(s+)) &= P(\lim_{n\to\infty} \cup_{\nu=1}^n F_\nu|\mathscr{A}^B(s+)) \\
&= \lim_{n\to\infty} P(\cup_{\nu=1}^n F_\nu|\mathscr{A}^B(s+)) \text{ (by } \mathbf{(6.4)}\text{(v))} \\
&= \lim_{n\to} \sum_{\nu=1}^n P(F_\nu|\mathscr{A}^B(s+)) \quad \text{by } \mathbf{(6.4)}\text{(i)} \\
&= \lim_{n\to} \sum_{\nu=1}^n P(F_\nu|B(s)) \\
&= P(\cup_n F_n|B(s)) \text{ (reversing the steps).}
\end{aligned}$$

Thus the collection $\mathscr{D}$ of all sets F for which

$$P(F|\mathscr{A}^B(s+)) = P\ (F|B(s))\ wp1$$

is closed under countable disjoint unions. If $A,\ B \in \mathscr{D}$ and $A \subset B$, then

$$\begin{aligned}P(B \sim A|\mathscr{A}^B(s+)) &= E(\chi_{B\sim A}|\mathscr{A}^B(s+)) \\ &= E(\chi_B|\mathscr{A}^B(s+)) - E(\chi_A|\mathscr{A}^B(s+)), \\ &\qquad\qquad \text{by } \mathbf{(6.4)}\text{(i)} \\ &= E(\chi_B|B(s)) - E(\chi_A|B(s)) \\ &= P(B \sim A|B(s)) \text{ (reversing the steps).}\end{aligned}$$

These two results show that $\mathscr{D}$ is a monotone class.

We now show that for every $A \in \mathscr{E},\ P(A|\mathscr{A}^B(s+)) = P(A|B(s))\ wp1$.

Let $0 = t_0 < t_1 < \dots t_\nu \leq s < t_{\nu+1} < \dots < t_n$ and let $A_k,\ 0 \leq k \leq n$ be linear Borel sets. Then

$$\begin{aligned}&P(B(t_k) \in A_k,\ 0 \leq k \leq n|\mathscr{A}^B(s+)) \\ &= E\{\chi_{\{B(t_k)\in A_k,\ 0\leq k\leq n\}}|\mathscr{A}^B(s+)\} \\ &= E\{\chi_{\{B(t_k)\in A_k,\ 0\leq k\leq \nu\}}\ \chi_{\{B(t_k)\in A_k,\ \nu+1\leq k\leq n\}}|\mathscr{A}^B(s+)\} \\ &= \chi_{\{B(t_k)\in A_k,\ 0\leq k\leq \nu\}} E\{\chi_{\{B(t_k)\in A_k,\ \nu+1\leq k\leq n\}}|\mathscr{A}^B(s+)\} \\ &= \chi_{\{B(t_k)\in A_k,\ 0\leq k\leq \nu\}} P\{B(t_k) \in A_k,\ \nu+1 \leq k \leq n|\mathscr{A}^B(s+)\} \\ &= \chi_{\{B(t_k)\in A_k,\ 0\leq k\leq \nu\}} P\{B(t_k)-B(s)+B(s) \in A_k,\ \nu+1 \leq k \leq n|\mathscr{A}^B(s+)\} \\ &= \chi_{\{B(t_k)\in A_k,\ 0\leq k\leq \nu\}} P\{B(t_k) - B(s) + B(s) \in A_k,\ \nu+1 \leq k \leq n|B(s)\},\end{aligned}$$

by **(6.3.3)**(vi), a $B(s)$-measurable function.

Thus $\mathscr{E} \subset \mathscr{D}$. We recall (ref. Theorem B, p27 [H]) that if a monotone class contains a field then it contains the σ-field generated by the field. Hence $\mathscr{A}^B_\infty = \sigma(\mathscr{E}) \subset \mathscr{D} \subset \mathscr{A}^B_\infty$. Hence $\mathscr{D} = \mathscr{A}^B_\infty$.

If $F \in \mathscr{A}^B(s+)$, $P(F|\mathscr{A}^B(s+)) = \chi_F\ wp1$. Hence there exists a $\mathscr{A}^B(s)$-measurable $rv\ Y$ such that $Y = \chi_F\ wp1$. Since $(\mathscr{A}^B(t),\ t \geq 0)$ is the augmented filtration, it follows $F \in \mathscr{A}^B(s)$. Thus $\mathscr{A}^B(s+) \subseteq \mathscr{A}^B(s)$. This implies $\mathscr{A}^B(s+) = \mathscr{A}^B(s)$, as was to be proved. □

Let B be a standard BMP and let X be a rv independent of B. Let $B_1(t) = B(\tau + t) - B(\tau),\ t \geq 0$. We state the following theorem without proof.

Theorem 6.7.6. *Let B be a BMP, let X be a rv defined on a common probability space and let τ be a Markov time of $(\mathscr{A}^B_X(t))$. Then the process*

$$(B(\tau + t) - B(\tau),\ t \geq 0) \text{ is independent of } \mathscr{A}^B_X(\tau). \tag{6.49}$$

Proof of this assertion depends very heavily on deep theorems in Martingale theory and hence is beyond the scope of this book. (For proof ref. Theorem 50.10, p 464, Probability Theory by Heinz Bauer, Narosa Publishing House, 1999).

In case τ has lattice distribution, here is a proof.

Let τ take the values $a,\ 2a,\ 3a,\dots$. Let $\mathrm{E} \in \mathscr{A}_X^B(\tau)$. Given $t > 0$, find $n \geq 1$ such that $(n-1)a \leq t < na$. Let $B_1(t) = B(\tau + t) - B(\tau)$. For arbitrary linear Borel sets A_r and arbitrary positive numbers $t_r,\ 1 \leq r \leq k$,

$$
\begin{aligned}
&P(B_1(t_r) \in A_r,\ 1 \leq r \leq k,\ \mathrm{E}) \\
&\quad = \lim_{t\to\infty} P(B_1(t_r) \in A_r,\ 1 \leq r \leq k,\ \mathrm{E} \cap (\tau \leq t)) \\
&\quad = \lim_{t\to\infty} P(B_1(t_r) \in A_r,\ 1 \leq r \leq k,\ \mathrm{E} \cap \cup_{j=0}^{n-1}(\tau = ja)) \\
&\quad = \lim_{t\to\infty} \sum_{j=0}^{n-1} P(B_1(t_r) \in A_r,\ 1 \leq r \leq k,\ \mathrm{E} \cap (\tau = ja)) \\
&\quad = \lim_{t\to\infty} \sum_{j=0}^{n-1} P(\big(B(ja + t_r) - B(ja\big) \in A_r, \\
&\qquad 1 \leq r \leq k,\ \mathrm{E} \cap (\tau = ja)) \\
&\quad = \lim_{t\to\infty} \sum_{j=0}^{n-1} P(\big(B(ja + t_r) - B(ja\big) \in A_r, \\
&\qquad 1 \leq r \leq k) P(\mathrm{E} \cap (\tau = ja))
\end{aligned}
$$

(since $\mathrm{E} \cap (\tau = ja) \in \mathscr{A}_X^B(ja)$ and since $B(ja + t_r) - B(ja)$ is independent of $\mathscr{A}_X^B(ja)$)

$$
\begin{aligned}
&= \lim_{t\to\infty} \sum_{j=0}^{n-1} P(B_1(t_r) \in A_r,\ 1 \leq r \leq k) P(\mathrm{E} \cap (\tau = ja)) \\
&= \lim_{t\to\infty} P(B_1(t_r) \in A_r,\ 1 \leq r \leq k) \sum_{j=0}^{n-1} P(\mathrm{E} \cap (\tau = ja)) \\
&= \lim_{t\to\infty} P(B_1(t_r) \in A_r,\ 1 \leq r \leq k) P(\mathrm{E} \cap (\tau \leq t)) \\
&= P(B_1(t_r) \in A_r,\ 1 \leq r \leq k) P(\mathrm{E}) \\
&= P(B_1(t_r) \in A_r,\ 1 \leq r \leq k) P(\mathrm{E}).
\end{aligned}
$$

Let $\mathcal{E}$ be the collection of all sets of the type $(B_1(t_r) \in A_r,\ 1 \leq r \leq k)$ for all choices of k and all choices of the t values. We note $\mathcal{E}$ is a field and that $\sigma(\mathcal{E}) = \mathcal{F} = \sigma(B_1(t),\ t \geq 0)$. Let $\mathcal{D} \subseteq \mathcal{F}$ be the collection of all sets A such that, for every $\mathrm{E} \in \mathscr{A}_X^B(\tau)$, the equality $P(A \cap \mathrm{E}) = P(A)P(\mathrm{E})$. Clearly $\mathcal{E} \subseteq \mathcal{D}$ and $\mathcal{D}$ is a monotone class. Hence (ref. Theorem B, p27, [H]) $\mathcal{F} = \mathcal{D}$, thus completing the proof that B_1 is independent of $\mathscr{A}_X^B(\tau)$. □

Theorem 6.7.7. (i) *B_1 is a standard BMP.*
(ii) *$B(\tau)$ is $\mathscr{A}_X^B(\tau)$-measurable*
(iii) *A BMP B is a strong Markov process. i.e., for every rv X independent of B, for every Markov time τ of $(\mathscr{A}_X^B(t))$ and for every linear Borel set Γ,*

$$P(B(\tau+t)\in\Gamma|\mathscr{A}_X^B(\tau)) = P(B(\tau+t)\in\Gamma|B(\tau)) \qquad (6.50)$$

(Compare with (6.44))

Proof.
(i) This claim is immediate from Theorem **(6.7.6)** : Let F be the df of τ. Let $\mathcal{T}$ be the sigma field generated by τ. Since process B_1 is independent of $\mathscr{A}_X^B(\tau)$ and since $\mathcal{T}\subset\mathscr{A}_X^B(\tau)$, it follows that B_1 is independent of τ.

$$\begin{aligned}
P(B(\tau+t_\nu) &- B(\tau)\in A_\nu,\ 1\le\nu\le k)\\
&= E\{P(B(\tau+t_\nu)-B(\tau)\in A_\nu, 1\le\nu\le k)|\tau\}\\
&= \int_0^\infty P(B(\tau+t_\nu)-B(\tau)\in A_\nu,\\
&\qquad 1\le\nu\le k|\tau=x)\,\mathrm{d}F(x)\\
&= P(B(t_\nu)\in A_\nu,\ 1\le\nu\le k),
\end{aligned}$$

proving the claim.

(ii) Given the Markov time τ, define τ_n as in the last section and recall that, so defined, τ_n is a Markov time for the filtration $(\mathscr{A}_X^B(t))$. Given $t>0$, there exists a unique integer $r=r(n;\,t)\ge 1$ such that $\frac{r-1}{2^n}\le t<\frac{r}{2^n}$. Then $(\tau_n\le t)=\cup_{j=1}^{r-1}\{\tau_n=\frac{j}{2^n}\}$. If $0\le t<\frac{1}{2^n}$, then $\tau_n=0$ and $\tau=0$. In this case $(B(\tau_n)\le u,\ \tau_n\le t)=\Omega$ or ϕ and hence belongs to $\mathscr{A}_X^B(t)$. Assume now $r\ge 2$. Since $(\tau_n=\frac{j}{2^n})=(\frac{j-1}{2^n}\le\tau<\frac{j}{2^n})$, $(B(\tau_n)\le u,\ \tau_n\le t)=\bigcup_{j=1}^{r-1}(B(\frac{j}{2^n})\le u,\ \frac{j-1}{2^n}\le\tau<\frac{j}{2^n})$. We note $(B(\frac{j}{2^n})\le u)\in\mathscr{A}_X^B(\frac{j}{2^n})\subset\mathscr{A}_X^B(\frac{r-1}{2^n})\subset\mathscr{A}_X^B(t)$. By **(6.7.5)**(i), $(\tau<\frac{j}{2^n})\in\mathscr{A}_X^B(\frac{j}{2^n})\subset\mathscr{A}_X^B(t)$. That $(\tau\le\frac{j-1}{2^n})$ and its complement lie in $\mathscr{A}_X^B(\frac{j-1}{2^n})$ and hence in $\mathscr{A}_X^B(t)$ derives from τ being a Markov time for the filtration $(\mathscr{A}_X^B(t))$. It now follows that $B(\tau_n)$ is $\mathscr{A}_X^B(\tau_n)$-measurable. Fix N. By **(6.7.5)**(ii), $B(\tau_n)$ is $\mathscr{A}_X^B(\tau_N)$-measurable, for $n\ge N$. It follows $B(\tau)$ is $\mathscr{A}_X^B(\tau_N)$-measurable for every $N\ge 1$ and hence is $\mathscr{A}_X^B(\tau)$-measurable (ref. **(6.7.5)**(iii)).

Note.
If $\tau,\ X$ are as in **(6.6.2)**, if $\mathscr{A}_X^B(\tau)$ is as in **(6.7.4)**, then to show that $B(\tau)$ is $\mathscr{A}_X^B(\tau)$-measurable a simpler and a more direct proof is as follows. This is possible since a BMP has independent increments.

Let Y_1, Y_2, Z_1, Z_2 be as in **(6.7.4)**. We must show $(B(\tau) \leq u) \in \mathscr{A}_X(\tau+)$. We must show that for $t \geq 0$, $(B(\tau) \leq u) \cap (\tau \leq t) \in \mathscr{A}_X^B(t)$. Now, $(\tau \leq t) = (Z_1 \leq Y_1) \cup (Z_2 \geq Y_2)$. Hence $(B(\tau) \leq u) \cap (\tau \leq t) = [(B(\tau) \leq u) \cap (Z_1 \leq Y_1)] \cup [(B(\tau) \leq u) \cap (Z_2 \geq Y_2)] = [(Y_1 \leq u) \cap (Z_1 \leq Y_1)] \cup [(Y_2 \leq u) \cap (Z_2 \geq Y_2)] \in \mathscr{A}_X^B(t)$.

(iii) By Theorem **(6.7.6)**, $B(\tau + t) - B(\tau)$ is independent of $\mathscr{A}_X^B(\tau)$. By part (ii) above $B(\tau)$ is $\mathscr{A}_X^B(\tau)$ measurable. Hence, by sec. **(6.4)**(xiii), $P(B(\tau+t) - B(\tau) + B(\tau) \in \Gamma | \mathscr{A}_X^B(\tau)) = P(B(\tau+t) \in \Gamma | B(\tau))$ for every linear Borel set Γ, as was to be proved. □

Notation 6.7.8. Let $(X, X_n, n \geq 1)$ be a sequence of zero mean independent rvs and independent of B, all defined on the same probability space, with strictly increasing continuous dfs F, F_n. If finite, variance of X, X_n will be denoted by $\sigma^2, \sigma_n^2, n \geq 1$. Let G, G_n be defined using F, F_n like G earlier was defined using F. Let B be a standard BMP adapted to the filtration $\mathscr{A}_{X_1}^B(t), t \geq 0$, the natural filtration of B enlarged by the independent variable X_1.

Let τ, τ_1 be the smallest root, respectively, of the equations $(B(t) - X)(B(t) - G(X)) = 0$ and $(B(t) - X_1)(B(t) - G_1(X_1)) = 0$. We know τ_1 is a Markov time relative to the filtration $\mathscr{A}_{X_1}^B(t)$. We assume as known at this stage that $\mathbb{E}\tau_1 = \mathbb{E}X_1^2$. (proof will be given in **(6.8.3)**). In **(6.6.5)**, we proved that $B(\tau_1)$ has distribution function F_1. In **(6.7.6)** we proved that the process $B_1 = \{B_1(t) = B(\tau_1 + t) - B(\tau_1)\}$ is a $SBMP$ independent of $\mathscr{A}_{X_1}^B(\tau_1)$.

Let the smallest root of the equation $(B_1(t) - X_2)(B_1(t) - G_2(X_2)) = 0$ be τ_2.

As in the case of B, τ_2 is a Markov time relative to the filtration $\mathscr{A}_{X_2}^{B_1}(t)$. Also $\mathbb{E}\tau_2 = \mathbb{E}X_2^2$; $B_1(\tau_2)$ (i.e., $B(\tau_2 + \tau_1) - B(\tau_1)$) has distribution function F_2; the process $B_2 = \{B_2(t) = B_1(\tau_2 + t) - B_1(\tau_2) = B(\tau_2 + \tau_1 + t) - B(\tau_2 + \tau_1)\}$ is a $SBMP$ independent of $\mathscr{A}_{X_2}^{B_1}(\tau_2+)$.

Successively, define $\tau_n, n \geq 3$ to be the smallest root of the equation $(B_{n-1}(t) - X_n)(B_{n-1}(t) - G_n(X_n)) = 0$ where $B_n(t) = B(\sum_{k=1}^{n-1} \tau_k + t) - B(\sum_{k=1}^{n-1} \tau_k)$, $t \geq 0$. In this way we arrive at a sequence (τ_n) of independent rvs and another sequence
$Y_n = \{B(\sum_{k=1}^{n} \tau_k) - B(\sum_{k=1}^{n-1} \tau_k)\}$, $n \geq 1$, $\tau_0 = 0$, of independent rvs such that F_n is the df of Y_n.

Let $\eta_n = \sum_1^n \tau_k$, $n \geq 1$. We talk of X_n being embedded in B at the time η_n, $n \geq 1$ or that $\eta_1, \eta_2, \ldots$ are the Skorohod times for $X_1, X_1 + X_2, \ldots$. Write $S_n = \sum_1^n X_k$ and note that the distribution of the sequence

(S_n) coincides with that of $(B(\eta_n))$.

An examination of the equations defining τ_i, $i \geq 1$ shows that they are identically distributed, if the X_is are identically distributed.

In the Lemma at **(6.8.1)**, we strengthen the above observation.

6.8 Studying the tail behaviour of the Skorohod time

We relate in this section the moments of an embedded variable X and those of its Skorohod time τ.

Lemma 6.8.1. *Let the non-negative rvs α, β be independent of the $SBMP$ B. Then $B(\alpha)$, $B(\beta)$ are identically distributed iff α and β are so.*

Proof. That $B(\alpha)$, $B(\beta)$ are rvs follows from **(6.6.1)**(iii). We have $\mathbb{E}e^{iuB(\alpha)} = \mathbb{E}\left\{\mathbb{E}(e^{iuB(\alpha)}|\alpha)\right\} = \mathbb{E}e^{-\frac{u^2}{2}\alpha^2}$. Thus ($B(\alpha)$, $B(\beta)$ are identically distributed) $\Leftrightarrow$ (α^2 and β^2 have the same Laplace transform) $\Leftrightarrow$ (α^2 and β^2 are identically distributed) $\Leftrightarrow$ (α, β are identically distributed), (since α, β are non-negative). □

Remark 6.8.2. Let X_1, X_2, B, τ_1, τ_2 be as in **(6.7.7)**. Let $X = X_1 + X_2$ and let τ be the Skorohod time for X. Then, since $B(\tau_1 + \tau_2)$ and $B(\tau)$ have each the distribution of $X_1 + X_2$, it follows that $\tau_1 + \tau_2$ and τ are identically distributed.

Let X_1, X_2, B, τ_1, τ_2, F_1, G_1 be as in **(6.7.7)**. Let X_2 and $-X_1$ be identically distributed. Let X, τ be as in **(6.8.2)**. Let μ_1, μ be the distribution measures of X_1 and X.

Theorem 6.8.3. $\mathbb{E}\tau_1^\delta < \infty$ *iff* $\mathbb{E}|X_1|^{2\delta} < \infty$, $\delta > 0$. *Further,* $\mathbb{E}\tau_1 = \mathbb{E}X_1^2$.

Proof. From the definition of τ, we have

$$\begin{aligned}
P(\tau > t) &= P\left(\max_{0\leq s\leq t} |B(s)| \leq |X|\right) \\
&= \int_{-\infty}^{\infty} P\left(\max_{0\leq s\leq t} |B(s)| \leq |x|\right) \mathrm{d}\,\mu(x) \\
&= \int_{-\infty}^{\infty} \Lambda\left(\frac{|x|}{\sqrt{t}}\right) \mathrm{d}\,\mu(x)
\end{aligned}$$

where Λ is the df of $\max\limits_{0\leq s\leq 1} |B(s)|$ (ref. **(6.6.1)**(ii)).

We record for use the inequalities:
The form for Λ (ref. (5.25)),

$$\Lambda(x) = \frac{4}{\pi}\sum_{n=0}^{\infty}\frac{(-1)^n}{2n+1}e^{-\frac{\pi^2(2n+1)^2}{8x^2}}, \ x > 0, \tag{6.51}$$

leadsto
$$\frac{8}{3\pi}e^{-\frac{\pi^2}{8x^2}} \leq \Lambda(x) \leq \frac{4}{\pi}e^{-\frac{\pi^2}{8x^2}}, \ x > 0. \tag{6.52}$$

We have

$$\begin{aligned}\int_0^{\infty} t^{\delta-1}P(\tau > t)\,\mathrm{d}\,t &= \int_0^{\infty} t^{\delta-1}\Big[\int_{-\infty}^{\infty}\Lambda(\frac{|x|}{\sqrt{t}})\,\mathrm{d}\,\mu(x)\Big]\,\mathrm{d}\,t \\ &= 2\int_0^{\infty}\Big[\int_{-\infty}^{\infty}\Lambda(\frac{|x|}{\sqrt{t}})\,\mathrm{d}\,\mu(x)\Big]\,\mathrm{d}\,t\end{aligned}$$

Hence

$$\begin{aligned}2\int_0^{\infty}\Big[\int_0^{\infty}\frac{8}{3\pi}e^{-\frac{\pi^2 t}{8x^2}}\,\mathrm{d}\,\mu(x)\Big]\,\mathrm{d}\,t &\leq \int_0^{\infty} t^{\delta-1}P(\tau > t)\,\mathrm{d}t \\ &\leq 2\int_0^{\infty} t^{\delta-1}\Big[\int_0^{\infty}\frac{4}{\pi}e^{-\frac{\pi^2 t}{8x^2}}\,\mathrm{d}\,\mu(x)\Big]\,\mathrm{d}\,t\end{aligned}$$

Or, by Fubini's theorem (applicable since the terms are non-negative),

$$\begin{aligned}\frac{16}{3\pi}\int_0^{\infty}\Big[\int_0^{\infty} t^{\delta-1}e^{-\frac{\pi^2 t}{8x^2}}\,\mathrm{d}\,t\Big]\,\mathrm{d}\,\mu(x) &\leq \int_0^{\infty} t^{\delta-1}P(\tau > t)\,\mathrm{d}t \\ &\leq \frac{8}{\pi}\int_0^{\infty}\Big[\int_0^{\infty} t^{\delta-1}e^{-\frac{\pi^2 t}{8x^2}}\,\mathrm{d}\,t\Big]\,\mathrm{d}\,\mu(x)\end{aligned}$$

i.e., $\frac{16}{3\pi}(\frac{8}{\pi^2})^{\delta}\int_0^{\infty}\big[\int_0^{\infty} u^{\delta-1}e^{-u}\,\mathrm{d}\,u\big]x^{2\delta-1}\,\mathrm{d}\,\mu(x)$
$\leq \int_0^{\infty} t^{\delta-1}P(\tau > t)\,\mathrm{d}t$
$\leq \frac{8}{\pi}(\frac{8}{\pi^2})^{\delta}\int_0^{\infty}\big[\int_0^{\infty} u^{\delta-1}e^{-u}\,\mathrm{d}\,u\big]x^{2\delta-1}\mu(x)$
From these inequalities it follows (ref. **(2.6.3)**) that $\mathbb{E}\tau^{\delta} < \infty$ *iff* $\mathbb{E}|X|^{2\delta} < \infty$. Since $\tau_1 + \tau_2$ and τ have the same distribution, $(\mathbb{E}\tau^{\delta} < \infty) \Leftrightarrow (\mathbb{E}(\tau_1+\tau_2)^{\delta}) < \infty$. This is possible *iff* $\mathbb{E}\tau_j^{\delta} < \infty$, $j = 1, 2$, since τ_1 and τ_2 are non-negative variables. Since the distribution of τ_1 is the same as that of τ_2, namely, Λ, we conclude $\mathbb{E}(\tau_1+\tau_2)^{\delta} < \infty$ *iff* $\mathbb{E}|X|^{\delta} < \infty$. Use Minkowski inequality **(2.6.5)**(v) and then the symmetrization inequality to conclude that $\mathbb{E}|X_1|^{2\delta} < \infty$ *iff* $\mathbb{E}|X|^{2\delta} < \infty$. These several results imply $\mathbb{E}\tau_1^{\delta} < \infty$ *iff* $\mathbb{E}|X_1|^{2\delta} < \infty$.

Now we prove the second assertion.

$$\begin{aligned}\mathbb{E}\tau_1 = \mathbb{E}\{\mathbb{E}(\tau_1|X_1)\} &= -\mathbb{E}X_1G_1(X_1) \text{ (by } \mathbf{(6.6.2)}) \\ &= -\int_{-\infty}^{\infty} xG_1(x)\,\mathrm{d}\,\mu_1(x) \\ &= -\int_{-\infty}^{\infty} G_1(G_1(x))G_1(x)\,\mathrm{d}\,\mu(x) \\ &= -\int_{-\infty}^{\infty} \lambda(G_1(x))\,\mathrm{d}\,\mu_1(x),\end{aligned}$$

where $\lambda(x) = xG_1(x)$. Thus

$$\begin{aligned}\mathbb{E}\tau_1 &= -\int_{-\infty}^{\infty} xG_1(x)d\mu_1(G_1^{-1}(x)) = -\int_{-\infty}^{\infty} xG_1(x)\,\mathrm{d}\,\mu_1(G_1(x)) \\ &= \text{(by } \mathbf{(6.6.4)}) \int_{-\infty}^{\infty} x\,x\,\mathrm{d}\,\mu_1(x) = \int_{-\infty}^{\infty} x^2\,\mathrm{d}\,\mu_1(x) = \mathbb{E}X_1^2\end{aligned}$$

in the sense that both are finite and equal or both are infinite and equal. □

Lemma 6.8.4. *Let X, X_n, F, B, τ, τ_n be as in* **(6.7.7)** *and Λ as in* **(6.8.3)**. *Assume F is a symmetric df. Define b_n as in* **(5.4.1)** *and $\eta_n = \frac{1}{b_n^2}\sum_1^n \tau_k$. Let $\Psi(\lambda) = \mathbb{E}e^{-\lambda\tau}$, $\lambda \geq 0$.*

($\mathfrak{a}$) $I_\theta = \int_0^\infty y^{-\theta}\Lambda(y)\,d\,y < \infty$ for every $\theta > 1$.

($\mathfrak{b}$) $I_3 = \int_0^\infty y^{-3}\Lambda(y)\,d\,y = \frac{1}{2}$.

($\mathfrak{c}$) Let $\mathbb{E}X^2 = \infty$ and let $H(x) = \int_{-x}^{x} y^2\,d\,F(y)$ be slowly varying. Then $\int_0^x P(\tau > t)\,d\,t \sim H(\sqrt{x})$, as $x \to \infty$.

($\mathfrak{d}$) $1 - \Psi(\lambda) \sim \lambda H(\sqrt{\frac{1}{\lambda}}\,)$ as $\lambda \downarrow 0$.

($\mathfrak{e}$) $\eta_n \overset{pr}{\to} 1$.

Proof. ($\mathfrak{a}$) As noted in **(6.8.3)**, $\Lambda(y) \leq 2e^{-\frac{\pi^2}{8y^2}}$. Hence $\lim\limits_{y\to 0} y^{-3}\Lambda(y) = 0$. We note $\lim\limits_{y\to 0} y^{-\theta}\Lambda(y) = 0$ for every $\theta > 0$. The integrand in I_θ would be a continuous function (and hence integrable) over $[0,\ 1]$ if it is defined to be 0 at 0. Over $[1,\ \infty)$, the integrand is dominated by the function $\frac{1}{y^\theta}$ which is integrable since $\theta > 1$. The finiteness of I_θ, for $\theta > 1$, follows.

($\mathfrak{b}$) The finiteness of I_3 follows from part ($\mathfrak{a}$). Let $\mathbb{E}X^2 = 1$. Then

$$\begin{aligned}1 = \mathbb{E}X^2 &= \mathbb{E}\tau \text{ (by } \mathbf{(6.8.3)}) \\ &= \int_0^\infty P(\tau > t)\,\mathrm{d}\,t \\ &= \int_0^\infty \{2\int_0^\infty \Lambda(\tfrac{u}{\sqrt{t}})\ \mathrm{d}\,F(u)\}\,\mathrm{d}\,t\end{aligned}$$

$= 2\int_0^\infty\{\int_0^\infty \Lambda(\frac{u}{\sqrt{t}})\,\mathrm{d}\,t\}\,\mathrm{d}\,F(u)$
$= 2\int_0^\infty\{\int_0^\infty \Lambda(y)\frac{2u^2}{y^3}\,\mathrm{d}\,y\}\ \mathrm{d}\,F(u)$
$= 2I_3\int_{-\infty}^\infty u^2\,\mathrm{d}\,F(u) = 2I_3.$

(c)

$$\begin{aligned}\int_0^x \{P(\tau > t)\}\mathrm{d}t &= \int_0^x \{\int_{-\infty}^\infty \Lambda(\frac{|u|}{\sqrt{t}})\mathrm{d}F(u)\}\mathrm{d}t \\ &= \int_{-\infty}^\infty \left\{\int_0^x \Lambda(\frac{|u|}{\sqrt{t}})\mathrm{d}t\right\}\mathrm{d}F(u) \\ &= \int_0^\infty \left\{\int_0^x \Lambda(\frac{|u|}{\sqrt{t}})\mathrm{d}t\right\}u^{-2}\mathrm{d}H(u) \\ &= \int_0^\infty \{\int_{\frac{u}{\sqrt{x}}}^\infty \Lambda(y)2y^{-3}\mathrm{d}y\}\mathrm{d}H(u) \\ &= 2\int_0^\infty \{\int_0^{y\sqrt{x}} \mathrm{d}H(u)\}y^{-3}\Lambda(y)\mathrm{d}y \\ &= 2\int_0^\infty H(y\sqrt{x})y^{-3}\Lambda(y)\mathrm{d}y \\ &= 2H(\sqrt{x})J(x),\ \text{say.}\end{aligned}$$

Here $J(x) = \int_0^\infty \frac{H(y\sqrt{x})}{H(\sqrt{x})}y^{-3}\Lambda(y)\mathrm{d}y$. For $0 \le y \le 1$, $\frac{H(y\sqrt{x})}{H(\sqrt{x})} \le 1$. Fix $\varepsilon > 0$ small. For $y > 1$ and for all x large $\frac{H(y\sqrt{x})}{H(\sqrt{x})} \le y^\varepsilon$, for every $\varepsilon > 0$. Take $\varepsilon < 1$. Since the function $y^{-2}\Lambda(y)$ is an integrable function, bounded convergence theorem applies and we get

$$\lim_{x\to\infty} J(x) = \int_0^\infty \lim_{x\to\infty}\frac{H(y\sqrt{x})}{H(\sqrt{x})}y^{-3}\Lambda(y)\mathrm{d}y = \int_0^\infty y^{-3}\Lambda(y)\mathrm{d}y = \tfrac{1}{2}.$$

The desired result follows now.

(d) We have, by integration by parts (ref. **(2.6.3)**),

$\int_0^x e^{-\lambda t}\,\mathrm{d}\,P(\tau \le t) = e^{-\lambda t}P(\tau \le t)\Big|_0^x + \lambda\int_0^x e^{-\lambda t}P(\tau \le t)\,\mathrm{d}\,t$
$= e^{-\lambda x}P(\tau \le x) + \lambda\int_0^x e^{-\lambda t}\{1 - P(\tau > t)\}\,\mathrm{d}\,t$
$= e^{-\lambda x}P(\tau \le x) + 1 - e^{-\lambda x} - \lambda\int_0^x P(\tau > t)e^{-\lambda t}\,\mathrm{d}\,t.$

Let $x \to \infty$ to get $\lambda\int_0^\infty e^{-\lambda t}P(\tau > t)\,\mathrm{d}\,t = 1 - \Psi(\lambda)$. In **Q**12$(iii)$, taking $U(x) = \int_0^x P(\tau > t)\,\mathrm{d}\,t$ and $\omega(\lambda) = \frac{1-\Psi(\lambda)}{\lambda}$, we get

$$\frac{1-\Psi(\lambda)}{\lambda} \sim \int_0^{\frac{1}{\lambda}} P(\tau > t)\,\mathrm{d}\,t \sim H(\sqrt{\tfrac{1}{\lambda}}\,) \text{ as } \lambda \downarrow 0.$$

(e) Since $\lim_{n\to\infty} n\{1 - \Psi(\frac{\lambda}{b_n^2})\} = \lim_{n\to\infty} n\frac{\lambda}{b_n^2}H(\frac{b_n}{\sqrt{\lambda}}) = \lim_{n\to\infty}\lambda n\frac{H(b_n)}{b_n^2} = \lambda$, it follows that $\lim_{n\to\infty}\mathbb{E}e^{-\lambda\eta_n} = \lim_{n\to\infty}\Psi^n(\frac{\lambda}{b_n^2}) = e^{-\lambda}$. This is equivalent to the claim. □

[Note. When $\mathbb{E}X^2 < \infty$ and is equal to 1, we will have $b_n = \sqrt{n}$ and $\eta_n \xrightarrow{wp1} 1$].

6.9 Using the tool of Skorohod representation

We follow the notation in **(6.7.7)** but assume that the X_ns are identically distributed with F for their common *df*. In this section we prove Hartman-Wintner LIL, its converse **(6.9.5)** and a result **(6.9.6)** that highlights the precision of the norming constant. In **(6.10)** we establish Strassen's functional LIL for the partial sum process x_n defined in **(5.6)**. The main tool used in both the cases is Skorohod embedding.

For definition of φ, R refer to **(6.1.1)** and **(6.1.3)**. When in **(6.10.1)** the FLIL is established, it will be seen that its proof does not depend on the Lemma in **(6.9.1)** and that **(6.9.1)** can be obtained as a corollary of the FLIL. Even so a direct proof of the Lemma in **(6.9.1)** was considered desirable for the same reason a direct proof of the Corollary in **(6.2.11)** was considered desirable. Additional reason is that the direct proof introduces us early to the tool of Skorohod embedding.

Lemma 6.9.1. *Recall B is a $SBMP$. Fix $\varepsilon > 0$. Let $2 > c > 1$, $\lambda = \frac{1}{2-c}$, $\alpha(n) = \sqrt{n}\varphi(n)$, $a \geq 1$ an integer and*

$$\hat{E}_n(c) = \hat{E}_n = \left(\max_{0 \leq s \leq n} \sup_{s \leq t \leq c^a s} |B(t) - B(\lambda s)| > \varepsilon\alpha(n) \right).$$

Then a $c_0 > 1$ can be found such that for every c, $1 < c < c_0$, $P(\overline{\lim_{n\to\infty}} \hat{E}_n) = 0$.

Proof. Write $E_n = \left(\max_{1 \leq s \leq n} \sup_{s \leq t \leq c^a s} |B(t) - B(\lambda s)| > \varepsilon\alpha(n) \right)$. Required is equivalent to showing $P(\overline{\lim_{n\to\infty}} E_n) = 0$ since $\max_{0 \leq s \leq 1} \sup_{s \leq t \leq c^a s} | B(t) - B(\lambda s)|$ is a rv free from n. For an arbitrary $c \in (1, 2)$, define $F_k = \cup\{E_\nu : c^k \leq \nu \leq c^{k+1}\}$. Since $\overline{\lim_{n\to\infty}} E_n = \overline{\lim_{k\to\infty}} F_k$, we will show $P(\overline{\lim_{k\to\infty}} F_k) = 0$ for a suitable choice of c. By Borel-Cantelli lemma, this will follow if we show that $\sum_k P(F_k) < \infty$. Towards this, we attempt to find a helpful bound for the kth term $P(F_k)$ of the series. On F_k, s ranges from 1 to c^{k+1}. Hence it has to belong to one of the intervals $[c^j, c^{j+1}]$, $0 \leq j \leq k$. It follows F_k can be written $F_k = \cup\{F_{k,j} : 0 \leq j \leq k\}$ where

$$F_{k,j} \subset \left(\max_{c^j \leq s \leq c^{j+1}} \sup_{s \leq t \leq c^a s} |B(t) - B(\lambda s)| > \varepsilon\alpha(c^k) \right)$$

$\subset A_{k,j} \cup B_{k,j} \cup C_{k,j}$

where $A_{k,j} = \left(\sup_{c^j \le t \le c^{j+a+1}} |B(t) - B(c^j)| > \frac{1}{3}\varepsilon\alpha(c^k) \right)$;

$B_{k,j} = \left(\max_{c^j \le s \le c^{j+a+1}} |B(\lambda s) - B(\lambda c^j)| > \frac{1}{3}\varepsilon\alpha(c^k) \right)$ and

$C_{k,j} = \left(|B(\lambda c^j) - B(c^j)| > \frac{1}{3}\varepsilon\alpha(c^k) \right)$, leading to

$P(F_{k,j}) \le P(A_{k,j}) + P(B_{k,j}) + P(C_{k,j})$. We have,

$$\begin{aligned}
\sum_{j=0}^{k} P(A_{k,j}) &= \sum_{j=0}^{k} P(\max_{0 \le t \le c^{j+a+1} - c^j} |B(t)| > \frac{1}{3}\varepsilon\alpha(c^k)) \\
&\le 4\sum_{j=0}^{k} P(\xi > \frac{1}{3}\frac{\varepsilon\alpha(c^k)}{\sqrt{c^j(c^{a+1}-1)}}) \\
&\le \sum_{j=0}^{k} q\frac{\sqrt{c^j(c^{a+1}-1)}}{\alpha(c^k)} e^{-\frac{1}{18}\frac{\varepsilon^2\alpha^2(c^k)}{c^j(c^{a+1}-1)}} \\
&\le qe^{-\frac{1}{18}\frac{\varepsilon^2}{c^{a+1}-1}\varphi^2(c^k)} \frac{\sqrt{c^{a+1}-1}}{\alpha(c^k)} \sum_{j=0}^{k} \sqrt{c^j} \\
&= qe^{-\frac{1}{18}\frac{\varepsilon^2}{c^{a+1}-1}\varphi^2(c^k)} \frac{c^{\frac{k+1}{2}}}{c^{\frac{k}{2}}\varphi(c^k)} \\
&\le qe^{-\frac{1}{18}\frac{\varepsilon^2}{c^{a+1}-1}\varphi^2(c^k)} \le qe^{-\frac{1}{18\times 2^{a+1}}\frac{\varepsilon^2}{c-1}\varphi^2(c^k)}.
\end{aligned}$$

$$\begin{aligned}
\sum_{j=0}^{k} P(B_{k,j}) &= \sum_{j=0}^{k} P(\sup_{c^j \le s \le c^{j+a+1}} |B(\lambda s) - B(\lambda c^j)| > \frac{1}{3}\varepsilon\alpha(c^k)) \\
&\le \sum_{j=0}^{k} P(\sup_{c^j \le s \le c^{j+1}} |B(\lambda s) - B(\lambda c^j)| > \frac{1}{3}\varepsilon\alpha(c^k)) \\
&= \sum_{j=0}^{k} P(\sup_{\lambda c^j \le t \le \lambda c^{j+1}} |B(t) - B(\lambda c^j)| > \frac{1}{3}\varepsilon\alpha(c^k)) \\
&= \sum_{j=0}^{k} P(\sup_{0 \le t \le \lambda c^j(c-1)} |B(t)| > \frac{1}{3}\varepsilon\alpha(c^k)) \\
&\le 4\sum_{j=0}^{k} P(\xi > \frac{1}{3}\frac{\varepsilon^2\alpha(c^k)}{\sqrt{\lambda(c-1)c^j}})
\end{aligned}$$

$$\leq q\sum_{j=0}^{k}\frac{\sqrt{\lambda(c-1)c^j}}{\alpha(c^k)}e^{-\frac{1}{18}\frac{\varepsilon^2\alpha(c^k)}{\lambda(c-1)c^j}}$$

$$\leq \frac{q_1}{\alpha(c^k)}e^{-\frac{1}{18}\frac{\varepsilon^2\alpha(c^k)}{\lambda(c-1)c^k}}\sum_{j=0}^{k}c^{\frac{j}{2}}\ \text{(assuming } c<\frac{3}{2})$$

$$< \frac{q_1}{\alpha(c^k)}e^{-\frac{1}{18}\frac{\varepsilon^2\varphi^2(c^k)}{\lambda(c-1)}}\frac{c^{\frac{k}{2}}}{c-1}$$

$$= \frac{q_2}{\varphi(c^k)}e^{-\frac{1}{18}\frac{\varepsilon^2\varphi^2(c^k)}{\lambda(c-1)}} < q_2e^{-\frac{1}{36}\frac{\varepsilon^2\varphi^2(c^k)}{c-1}}$$

$$\sum_{j=0}^{k}P(C_{k,j}) = 2\sum_{j=0}^{k}P(\xi>\frac{1}{3}\frac{\varepsilon\alpha(c^k)}{\sqrt{(\lambda-1)c^j}}) \leq q\sqrt{\lambda-1}e^{-\frac{1}{18}\frac{\varepsilon^2}{\lambda-1}\varphi^2(c^k)}$$

$$= q\sqrt{\frac{c-1}{2-c}}e^{-\frac{1}{18}\frac{\varepsilon^2(2-c)}{c-1}\varphi^2(c^k)} \leq qe^{-\frac{1}{36}\frac{\varepsilon^2\varphi^2(c^k)}{c-1}},\ \text{if } c<\frac{3}{2}.$$

Choose c such that $1<c<c_0=\min\{1+\frac{\varepsilon^2}{18\times 2^{a+1}R^2},\frac{3}{2}\}$. Hence $\sum_k P(F_k)\leq q\sum_k e^{-\frac{1}{2}\frac{\varepsilon^2}{18\times 2^{a+1}(c-1)}\varphi^2(c^k)}<\infty$, for every $0<c<c_0$ since for that range of c, $\frac{\varepsilon^2}{18\times 2^{a+1}(c-1)}>\frac{\varepsilon^2}{18\times 2^{a+1}(c_0-1)}>R^2$. □

Theorem 6.9.2 ((Hartman-Wintner LIL) (Bulinskii version)). *Let* $\mathbb{E}X_1^2=1$ *and* $S_n=\sum_{k=1}^n X_k$. *Then* $wp1$

$$\textit{the limit set of } (\frac{S_n}{\sqrt{n}\varphi(n)}) \textit{ is } [-R,\ R]. \tag{6.53}$$

Proof. Let (ξ_n) be a sequence of iid zero mean normal variables with variance σ^2, independent of the X_ns. All the variables are assumed to be defined on a common probability space, which can always be arranged. Let $Y_n=\frac{X_n+\xi_n}{\sqrt{1+\sigma^2}}$. We note (Y_n) is a sequence of iid variables with zero means and unit variances. Their common df G is a strictly increasing and continuous function. Further $G(b)-G(a)>0$ for all $a<b$. Write $T_n=\sum_{k=0}^n\xi_k$ and $\widehat{S}_n=\sum_{k=0}^n Y_k$. Let τ_n be the Skorohod time of Y_n and let $\eta_n=\sum_1^n\tau_j$. As noted in **(6.7)**, the τ_is are independent and identically distributed. By **(6.8.3)**, $\mathbb{E}\tau_1=\mathbb{E}Y_1^2=1$. Hence by the $SLLN$, $\frac{\eta_n}{n}\overset{wp1}{\longrightarrow}1$. Equivalently, given $\delta>0$ and c with $1<c<2$, we can find $M=M(c,\ \delta)$ such that $P(D')<\delta$ where

$$D=\left(\omega\ :\ |\frac{\eta_m}{m}-1|\leq c-1 \text{ for all } m\geq M\right). \tag{6.54}$$

We will first show that

$$wp1 \text{ the limit set of } \frac{\widehat{S}_n}{\sqrt{n}\varphi(n)} \text{ is } [-R,\ R]. \tag{6.55}$$

Since the sequences $(\widehat{S}_n)$ and $(B(\eta_n))$ have the same distribution, establishing (6.55) is equivalent to proving

$$wp1, \text{ the limit set of } \frac{B(\eta_n)}{\sqrt{n}\varphi(n)} \text{ is } [-R,\ R]. \tag{6.56}$$

We had established in **(6.2.11)** that

$$wp1, \text{ the limit set of } \frac{B(n)}{\sqrt{n}\varphi(n)} \text{ is } [-R,\ R]. \tag{6.57}$$

It is clear that (6.56) will follow if we show that $\frac{B(\eta_n)-B(n)}{\sqrt{n}\varphi(n)} \xrightarrow{wp1} 0$ which, in turn will follow if we show that

$$\frac{\max\limits_{1\leq m\leq n} |B(\eta_m) - B(m)|}{\sqrt{n}\varphi(n)} \xrightarrow{wp1} 0.$$

Equivalently, we must show that for every $\varepsilon > 0$,

$$P(\overline{\lim}\, A_n) = 0$$
$$\text{where} \quad A_n = \left(\frac{\max\limits_{1\leq m\leq n} |B(\eta_m) - B(m)|}{\sqrt{n}\varphi(n)} > \varepsilon \right). \tag{6.58}$$

For N fixed, $\frac{\max_{1\leq m\leq N} |B(\eta_m)-B(m)|}{\sqrt{n}\varphi(n)} \xrightarrow{wp1} 0$ since the numerator is a rv free from n. Hence it is enough to show that, for a suitably chosen N, $P(\overline{\lim}\, \tilde{A}_n) = 0$ where now

$$\tilde{A}_n = \left(\frac{\max\limits_{N\leq m\leq n} |B(\eta_m) - B(m)|}{\sqrt{n}\varphi(n)} > \varepsilon \right). \tag{6.59}$$

Choose $N = M$ where M is as in (6.54). Let $\omega \in D$. For $m \geq N$ fixed, we note $m(2-c) \le \eta_m(\omega) \le mc$. Hence

$$\begin{aligned}
\tilde{A}_n \cap D &\subset \left(\max_{M \le m \le n} \sup_{m(2-c) \le j \le mc} |B(j) - B(m)| > \varepsilon\sqrt{n}\varphi(n) \right) \\
&= \left(\max_{M \le \frac{s}{2-c} \le n} \sup_{s \le j \le s\frac{c}{2-c}} |B(j) - B(\frac{s}{2-c})| > \varepsilon\sqrt{n}\varphi(n) \right) \\
&\subset \left(\max_{0 \le s \le n} \sup_{s \le t \le s\frac{c}{2-c}} |B(t) - B(\frac{s}{2-c})| > \varepsilon\sqrt{n}\varphi(n) \right) \\
&\subset \left(\max_{0 \le s \le n} \sup_{s \le t \le sc^3} |B(t) - B(\frac{s}{2-c})| > \varepsilon\sqrt{n}\varphi(n) \right) \\
&\qquad\qquad\qquad\qquad\qquad\qquad (\text{with } c < \frac{4}{3})
\end{aligned}$$

By **(6.9.1)**, this implies $P(\tilde{A}_n \cap D \text{ i.o.}) = 0$. Since $\tilde{A}_n = (\tilde{A}_n \cap D) \cup (\tilde{A}_n \cap D')$ and since $(\tilde{A}_n \cap D' \text{ i.o.}) \subset D'$, it follows $P(\tilde{A}_n \text{ i.o.}) \le \delta$. That $P(\tilde{A}_n \text{ i.o.}) = 0$ follows now since $\delta > 0$ is arbitrary.

Thus we have shown that $wp1$ the limit set of the sequence $(U_n + V_n^\sigma)$ is $[-R\sqrt{1+\sigma^2},\ R\sqrt{1+\sigma^2}]$ where $U_n = \frac{S_n}{\sqrt{n}\varphi(n)}$ and $V_n^\sigma = \frac{T_n}{\sqrt{n}\varphi(n)}$ i.e., there exists a set A_σ with $P(A_\sigma) = 1$ and such that $\omega \in A_\sigma$ implies that $[-R\sqrt{1+\sigma^2},\ R\sqrt{1+\sigma^2}]$ is the limit set of $(U_n(\omega) + V_n^\sigma(\omega))$. This result is applicable to the sequence (ξ_k) since the normal *df* Φ satisfies (6.42). Applying, we get : there exists a set B_σ with $P(B_\sigma) = 1$ and such that $\omega \in B_\sigma$ implies the limit set of $(V_n^\sigma(\omega))$ is $[-R\sigma,\ R\sigma]$.

Let $C_\sigma = A_\sigma \cap B_\sigma$ and note $P(C_\sigma) = 1$. If $\omega \in C_\sigma$, then $\overline{\lim_{n\to\infty}}\ U_n(\omega)$ $= \overline{\lim_{n\to\infty}} \{(U_n(\omega) + V_n^\sigma(\omega)) - V_n^\sigma(\omega)\} \le R\sqrt{1+\sigma^2} + R\sigma$ and, similarly, $\underline{\lim_{n\to\infty}}\ U_n(\omega) \ge -R\sqrt{1+\sigma^2} - R\sigma$. Thus $wp1$, $I_\sigma = [-R\sqrt{1+\sigma^2} - R\sigma,\ R\sqrt{1+\sigma^2} + R\sigma]$ contains the limit set of (U_n). Let $C = \cap_k C_{\frac{1}{k}}$ and note $P(C) = 1$. Let $\omega \in C$. This implies $\omega \in C_{\frac{1}{k}}$ for every $k \ge 1$. Hence the limit set of the sequence $(U_n(\omega))$ is contained in $\cap_k I_{\frac{1}{k}} = [-R,\ R]$. Thus $wp1$ the limit set of (U_n) is contained in $[-R,\ R]$, the exceptional set being C'.

Let $-R < a < R$. Again let $\omega \in C$. Hence there exist subsequences, which we will denote as, $(U_m(\omega) + V_m^{(k)}(\omega))$ and $(V_m^{(k)}(\omega))$ converging respectively to a and some number α_k. (It may be noted that the sequence (m) may depend on ω, k and a and that $|\alpha_k| < \frac{R}{k}$). Hence an m_k can be found such that $|U_{m_k}(\omega) - a| < |\alpha_k| + \frac{1}{k} < \frac{R+1}{k}$ for all k large. Thus for each

$\omega \in C$, a is a limit point of the sequence $(U_n(\omega))$. This completes the proof that $wp1$, the limit set of $\left(\frac{S_n}{\sqrt{n\varphi(n)}}\right)$ is the interval $[-R,\ R]$. □

Remark 6.9.3. Let $0 < r < R$. Determine θ according to (6.7) and define $n_k = [c^{k^\theta}]$. Arguing on the same lines, we conclude that $wp1$,

$$\overline{\lim_{k\to\infty}} \frac{S_{n_k}}{\sqrt{n_k\varphi(n_k)}} = r, \text{ and } \underline{\lim_{k\to\infty}} \frac{S_{n_k}}{\sqrt{n_k\varphi(n_k)}} = -r.$$

Corollary 6.9.4. *Let (X_n) be a sequence of iid variables with zero means and unit variances. Let $(X_{j,n})$, $1 \le j \le \kappa$ be κ independent copies of the sequence (X_n). Let $S_{j,n} = \sum_{\nu=1}^{n} X_{j,\nu}$. Then $wp1$ the limit set A of the vector sequence $\mathbf{U}_n = (U_{1,n},\ U_{2,n}, \ldots,\ U_{\kappa,n}) = \frac{1}{\sqrt{n\varphi(n)}}(S_{1,n},\ S_{2,n}, \ldots,\ S_{\kappa,n})$ is the following subset $\mathbf{A}_R$ of R^κ : $\mathbf{A}_R = \{\mathbf{a} = (a_1,\ a_2, \ldots,\ a_\kappa) : \sum_{\nu=1}^{\kappa} a_\nu^2 \le R^2\}$.*

Proof. All convergent subsequences of each component $\left(\frac{S_{j,n}}{\sqrt{n\varphi(n)}}\right)$ converge $wp1$ only to constants. Hence the same is true of the vector sequence $(\mathbf{U}_n)$. The existence of the limit set $\mathbf{A}_R$ and its description will be established simultaneously by the following arguments.

If $\mathbf{a}$ is a limit point of $(\mathbf{U}_n)$, then given $\varepsilon > 0$, $P(\|\mathbf{U}_n - \mathbf{a}\| < \varepsilon \text{ i.o.}) = 1$. This implies $P(|U_{j,n} - a_j| < \varepsilon\ i.o.) = 1,\ 1 \le j \le \kappa$. Since the limit set of each of the component sequences $(U_{j,n})$ is $[-R,\ R]$, it follows that there exists a set F with $P(F) = 1$ such that $\omega \in F$ implies $\mathbf{U}_n(\omega) \in [-R,\ R]^\kappa$ for all n large and hence that the limit set $\mathfrak{L}$ of $(\mathbf{U}_n)$, if it exists, is contained in $[-R,\ R]^\kappa$.

We will now show that, if it exists, $\mathfrak{L} \subset \mathbf{A}_R$. Let $\mathbf{x} = R(x_1,\ x_2, \ldots,\ x_\kappa) \in \mathfrak{L}$. Since $\mathfrak{L} \subset [-R,\ R]^\kappa$, $\max\limits_{1\le j\le\kappa} |x_j| \le 1$. Since $\mathbf{x}$ is a limit point of $\mathbf{U}_n$ and since for arbitrary $\mathbf{a}' = (a_1,\ a_2, \ldots,\ a_\kappa)$ the mapping carrying $(y_1,\ y_2, \ldots,\ y_\kappa)$ to $a_1y_1 + a_2y_2 + \ldots + a_\kappa y_\kappa$ is continuous, it follows that $\sum_{j=1}^{\kappa} a_j x_j$ is a limit point of $(\sum_{j=1}^{\kappa} a_j\ U_{j,n})$. But this is the partial sum sequence of iid variables : $\sum_{j=1}^{\kappa} a_j\ U_{j,n} = \frac{1}{\sqrt{n\varphi(n)}} \sum_{m=1}^{n} Y_m$ where $Y_m = \sum_{j=1}^{\kappa} a_j X_{j,m}$; $\mathbb{E}Y_m = 0$, $\mathbb{E}Y_m^2 = \sum_{j=1}^{\kappa} a_j^2$. Hence, by the result for one-dimension, the limit set of $\left(\frac{1}{\sqrt{n\varphi(n)}} \sum_{m=1}^{n} Y_m\right)$ is $[-R(\sum_{j=1}^{\kappa} a_j^2)^{\frac{1}{2}},\ R(\sum_{j=1}^{\kappa} a_j^2)^{\frac{1}{2}}]$. It follows $\sum_{j=1}^{\kappa} a_j x_j \in [-R(\sum_{j=1}^{\kappa} a_j^2)^{\frac{1}{2}},\ R(\sum_{j=1}^{\kappa} a_j^2)^{\frac{1}{2}}]$.
i.e., $R^2 \sum_{j=1}^{\kappa} a_j^2 - (\sum_{j=1}^{\kappa} a_j x_j)^2 \ge 0$ for all $\mathbf{a} \in R^\kappa$. The left side in the above inequality is a quadratic expression in $a_1,\ a_2, \ldots,\ a_\kappa$. If this is to be

non-negative always, then the corresponding matrix.

$$\begin{pmatrix} R^2 - x_1^2 & -x_1x_2 & \dots & -x_1x_\kappa \\ -x_2x_1 & R^2 - x_2^2 & \dots & -x_2x_\kappa \\ \dots & \dots & \dots & \dots \\ -x_\kappa x_1 & -x_\kappa x_2 & \dots & R^2 - x_\kappa^2 \end{pmatrix}$$

must be non-negative definite. Consequently, the determinant $R^2 - x_1^2 - x_2^2 - \dots - x_\kappa^2$ of the matrix is non-negative. Thus $x_1^2 + x_2^2 + \dots + x_\kappa^2 \le R^2$, showing that the limit set of $(\mathbf{U}_n)$, if it exists, is contained in $\mathbf{A}_R$.

Given $0 < r \le R$, let n_k be as defined in **(6.9.3)**. Then $wp1$, the limit set of $(\frac{S_{n_k}}{\sqrt{n_k\varphi(n_k)}})$ is $[-r,\ r]$. Starting with $(\frac{S_{n_k}}{\sqrt{n_k\varphi(n_k)}})$ in place of $(\frac{S_n}{\sqrt{n\varphi(n)}})$ and arguing as above we see that there exists a set F_r with $P(F_r) = 1$ such that $\omega \in F_r$ implies $\mathbf{U}_{n_k} \in [-r,\ r]^\kappa$ for all k large.

We will now show that $wp1$ every point of $\tilde{\mathbf{A}}_r = \{r\mathbf{a} : \|\mathbf{a}\| = 1\}$ is a limit point of $(\mathbf{U}_{n_k})$. We note $\tilde{\mathbf{A}}_r = \{r\mathbf{a} : \|\mathbf{a}\| = 1\}$, being a closed subset of R^κ, is separable. Let $(r\mathbf{a}_\nu,\ \|\mathbf{a}_\nu\| = 1,\ \nu = 1,\ 2, \dots)$ be a separability set for $\tilde{\mathbf{A}}_r$. As argued earlier, the sequence $(\sum_{j=1}^{\kappa} a_{j,\nu}U_{j,n_k})$ has limit set $[-r,\ r]$ where $\mathbf{a}'_\nu = (a_{1,\nu}, \dots,\ a_{\kappa,\nu})$. There exists then a set E_ν such that $P(E_\nu) = 1$ and such that $\omega \in E_\nu$ implies that the limit set of $\sum_{j=1}^{\kappa} a_{j,\nu}U_{j,n_k}(\omega)$ is $[-r,\ r]$. Let $E = \cap_\nu E_\nu$ and note $P(E) = 1$. Since the $r\mathbf{a}_\nu$s constitute a dense subset of $\tilde{\mathbf{A}}_r$, it follows that $\omega \in E$ implies $(\sum_{j=1}^{\kappa} a_{j,\nu}U_{j,n_k}(\omega))$ has limit set $[-r,\ r]$ for *every* $r\mathbf{a} \in \tilde{\mathbf{A}}_r$.

If $\omega \in E$, then $\mathbf{U}_n(\omega) \in \mathbf{A}_r$ for all large n. For, if possible, let $\omega_0 \in E$ and, for a subsequence (m) let $\mathbf{U}_m(\omega_0) \to (a_1, \dots,\ a_\kappa)$ where $(a_1, \dots,\ a_\kappa)$ may depend on ω_0 and lies in $[-r,\ r]^\kappa \sim \mathbf{A}_r$. Hence $\sqrt{\sum_{j=1}^{\kappa} a_j^2} > r$. Let $\lambda_j = \frac{a_j}{\sqrt{\sum_{q=1}^{\kappa} a_q^2}},\ 1 \le j \le \kappa$. Hence $\sum_{j=1}^{\kappa} \lambda_j U_{j,m} \to \sum_{j=1}^{\kappa} \lambda_j a_j = \sqrt{\sum_{j=1}^{\kappa} a_j^2} > r$, a contradiction since, as noted above, this limit point must lie in $[-r,\ r]$.

Let $r\mathbf{a} \in \tilde{\mathbf{A}}_r$ be fixed. Let $\omega_0 \in E \cap F_r$. Hence, by **(6.9.3)**, r is a limit point of $(\sum_{j=1}^{\kappa} a_jU_{j,m}(\omega_0))$. There is then a subsequence $(n' = n'(\omega_0))$ such that $\sum_{j=1}^{\kappa} a_jU_{j,n'}(\omega_0) \to r$ and $\mathbf{U}_{n'}(\omega_0)$ is a convergent sequence, converging to, say, $\mathbf{b} = \mathbf{b}(\omega_0) = (b_1,\ b_2, \dots,\ b_\kappa)'$ where the b_js depend on ω_0. As shown in the last para, necessarily $\sum_{j=1}^{\kappa} b_j^2 \le r^2$.

We thus have

$$\sum_{j=1}^{\kappa} a_jb_j = r \tag{6.60}$$

and

$$\sum_{j=1}^{\kappa} a_j^2 = 1. \tag{6.61}$$

These two relations yield

$$\sum_{j=1}^{\kappa} b_j^2 = r^2 \tag{6.62}$$

since $r = \sum_{j=1}^{\kappa} a_j b_j = |\sum_{j=1}^{\kappa} a_j b_j| \le \left(\sum_{j=1}^{\kappa} a_j^2\right)^{\frac{1}{2}} \left(\sum_{j=1}^{\kappa} b_j^2\right)^{\frac{1}{2}}$
$= \left(\sum_{j=1}^{\kappa} b_j^2\right)^{\frac{1}{2}} \le r.$

Let $\mathbb{M}$ be a $\kappa \times \kappa$ orthonormal matrix with first row ($\mathbf{a}'$). $\mathbb{M}$ maps $\mathbf{A}_r$ onto itself in a one-to-one manner. If $\mathbb{M}\mathbf{b} = \Lambda$, then $\sum_{j=1}^{\kappa} \lambda_j^2 = r^2$. Further $\lambda_1 = \sum_{j=1}^{\kappa} a_j b_j = r$. These two relations imply $\Lambda = (r,\ 0,\ 0, \ldots,\ 0)'$. And then $\mathbf{b} = \mathbb{M}^{-1}(r,\ 0, \ldots,\ 0)' = \mathbb{M}'(r,\ 0, \ldots,\ 0)' = r\mathbf{a}$, proving that $wp1$, $r\mathbf{a}$ is a limit point of $(\mathbf{U}_{n_k})$. Since $\tilde{\mathbf{A}}_r$ is separable, it follows that $wp1$ every point in $\tilde{\mathbf{A}}_r$ is a limit point of $(\mathbf{U}_{n_k})$ and hence of $(\mathbf{U}_n)$. Since the collection of these shells $\tilde{\mathbf{A}}_r$, r rational, $r \le R$ is countable and dense in $\mathbf{A}_R$, the proof of the corollary is complete. □

Theorem 6.9.5 (Converse to the LIL). *Let φ, R be as in **(6.1.1)** and **(6.1.3)**. Let (X_n) be a sequence of iid variables and (S_n) their partial sums. If with positive probability*

$$\overline{\lim_{n\to\infty}} \frac{|S_n|}{\sqrt{n\varphi(n)}} < \infty \tag{6.63}$$

then $EX_1^2 < \infty$, and $EX_1 = 0$.

Proof. Since the event involved is in the tail σ-field of the X_n-sequence, the hypothesis implies that the event holds $wp1$. Let (Y_n) be an independent copy of the X_n sequence, let $Z_n = X_n - Y_n$ and $T_n = \sum_{k=1}^{n} Z_k$. Again the hypothesis implies that $wp1$ $\overline{\lim}_{n\to\infty} \frac{|T_n|}{\sqrt{n\varphi(n)}} < \infty$. Proving $\mathbb{E}X_1^2 < \infty$ is equivalent to proving $EZ_1^2 < \infty$. Thus there is no loss of generality in assuming that the *df* of X_1 is symmetric around 0 to prove the part that $\mathbb{E}X_1^2 < \infty$. We will first prove this under the additional assumption the *df* F of X_1 is continuous and $F(b) - F(a) > 0$ for all $a < b$. Let $\eta_n = \sum_{j=1}^{n} \tau_j$ be the Skorohod time of embedding in a $SBMP$ B for X_n, $n = 1,\ 2, \ldots$. If $\mathbb{E}\tau_1 < \infty$, then by **(6.8.3)** it would follow that $\mathbb{E}X_1^2 < \infty$. Suppose $\mathbb{E}\tau_1 = \infty$. This implies (ref. **(4.8.4)**) $\lim_{n\to\infty} \frac{\eta_n}{n} = \infty$ $wp1$. Let A be the ω-set with $P(A) = 1$ such that $\omega \in A$ implies $\lim_{n\to\infty} \frac{\eta_n(\omega)}{n} = \infty$.

Under our assumptions on F, τ_n is the smallest root of the equation $\big(B(\eta_{n-1}+t)-B(\eta_{n-1})-X_n)\big)\big(B(\eta_{n-1}+t)-B(\eta_{n-1})+X_n\big)=0$ (ref. **(6.7.7)**). Hence for $0\le t\le \tau_n$, $|B(\eta_{n-1}+t)-B(\eta_{n-1})|\le |X_n|$ or $|B(t)-B(\eta_{n-1})|\le |B(\eta_n)-B(\eta_{n-1})|$ $wp1$ for all n and for all $t\in[\eta_{n-1},\ \eta_n]$. Hence for $\eta_{n-1}\le t\le \eta_n$, $\quad B(t)\le B(\eta_{n-1})+|B(\eta_n)-B(\eta_{n-1})|\le 2|B(\eta_{n-1})|+|B(\eta_n)|$ $wp1$ so that $wp1$,

$$
\begin{aligned}
\overline{\lim_{t\to\infty}}\ \frac{B(t)}{\sqrt{t\varphi(t)}} &\le \overline{\lim_{n\to\infty}}\ \frac{2|B(\eta_{n-1})|+|B(\eta_n)|}{\sqrt{\eta_{n-1}\varphi(\eta_{n-1})}}\\
&= \overline{\lim_{n\to\infty}}\left(\frac{2|S_{n-1}|+|S_n|}{\sqrt{n\varphi(n)}}\ \frac{\sqrt{n\varphi(n)}}{\sqrt{\eta_{n-1}\varphi(\eta_{n-1})}}\right).
\end{aligned}
$$

By (6.63), $\overline{\lim\limits_{n\to\infty}}\left(\frac{2|S_{n-1}|+|S_n|}{\sqrt{n\varphi(n)}}\right)<\infty\ wp1$.

Let $\omega\in A$. Then $\frac{\alpha_{n-1}(\omega)}{n-1}>2$ for all n large. Hence, for all n large, $\eta_{n-1}>n$ and $\frac{\varphi(n)}{\varphi(\eta_{n-1}(\omega))}\le 1$ for all n large. This and the fact that $\frac{n}{\eta_{n-1}}\to 0$ lead to $\overline{\lim\limits_{n\to\infty}}\ \frac{\sqrt{n\varphi(n)}}{\sqrt{\eta_{n-1}(\omega)\varphi(\eta_{n-1}(\omega))}}=0$. This proves that $wp1$ $\overline{\lim\limits_{t\to\infty}}\ \frac{B(t)}{\sqrt{t\varphi(t)}}=0$. But (ref. A direct proof **(6.2.11)**), $wp1$, $\overline{\lim\limits_{t\to\infty}}\ \frac{B(t)}{\sqrt{t\varphi(t)}}\ge R$. This contradiction establishes the claim that $\mathbb{E}X_1^2<\infty$.

To extend result to general dfs F, we argue as follows. Let (ξ_n) be a sequence of iid standard normal variables, independent of the X_ns. Let $V_n=\sum_{j=1}^n\xi_j$. By **(6.9.2)**, $\overline{\lim\limits_{n\to\infty}}\ \frac{|V_n|}{\sqrt{n\varphi(n)}}<\infty\ wp1$. Hence $\overline{\lim\limits_{n\to\infty}}\ \frac{|U_n|}{\sqrt{n\varphi(n)}}<\infty\ wp1$ where $U_n=S_n+V_n$. We note that G the df of $X_1+\xi_1$ is continuous and $G(b)-G(a)>0$ for all $a<b$. Hence by what has just been proved, $\mathbb{E}(X_1+\xi_1)^2<\infty$. Since X_1 and ξ_1 are independent and sinc $\mathbb{E}\xi_1^2<\infty$, it follows that $\mathbb{E}X_1^2<\infty$.

Let $\mathbb{E}X_1=\mu$ and $varX_1=\sigma^2$. Then by **(6.9.2)**,

$$
\overline{\lim_{n\to\infty}}\ \frac{|S_n-n\mu|}{\sqrt{n\varphi(n)}}<\infty\quad wp1. \tag{6.64}
$$

Now the relation $\frac{\mu\sqrt{n}}{\varphi(n)}=\frac{S_n-(S_n-n\mu)}{\sqrt{n\varphi(n)}}$, (6.63) and (6.64) yield $|\mu|\ \overline{\lim\limits_{n\to\infty}}\ \frac{\sqrt{n}}{\varphi(n)}<\infty$. If $\mu\ne 0$, this implies that there exists $\beta>0$ such that $\frac{\sqrt{n}}{\varphi(n)}\le\frac{1}{\sqrt{2\beta}}$ for all n. This, in turn, implies, for $c=2$, $\sum_k e^{-\frac{1}{2}r\varphi^2(2^k)}\le\sum_k e^{-r\beta k}<\infty$ for all $r>0$. This is not possble since $R>0$. Hence $\mu=0$. □

Proof of the theorem at **(6.9.6)** rests on Egorov's theorem which is quoted below.

Q13 Theorem (Egorov, p88, [H])

Let μ be a finite measure on the measurable space $(\Omega,\ \mathscr{A})$. Let (f_n) be a sequence of finite valued measurable functions converging to a finite valued measurable function f. Then, for every $\varepsilon > 0$, there exists $F \in \mathscr{A}$ such that $\mu(F) < \varepsilon$ and such that the sequence (f_n) converges to f uniformly on F'.

The theorem promised at the start of this section will be presented now, showing the norming constants in the LIL to be precise.

Let (X_n) be a sequence of iid variables with zero means and unit variances. Let $(x_{j,n}),\ 1 \le j \le \kappa$, be independent copies of (X_n); $S_n = \sum\limits_{r=1}^{n} X_r;\ S_{j,n} = \sum\limits_{r=1}^{n} X_{j,r};\ Z_n^2 = \sum\limits_{j=1}^{\kappa} S_{j,n}^2;\ b_\delta(n) = (1+\delta)\sqrt{n}\varphi(n)$

$\mathbf{Y}_n(\delta) = 1$ if $Z_n^2 = \sum\limits_{j=1}^{\kappa} S_{j,n}^2 \ge b_\delta^2(n)$ and zero otherwise.

$\mathbf{N}_\infty(\delta) = \sum\limits_{n=1}^{\infty} Y_n(\delta)$; $Y_n(\delta) = 1$ if $|S_n| \ge b_\delta(n)$ and zero otherwise; $N_\infty(\delta) = \sum\limits_n Y_n(\delta)$; $\mathbf{Y}_n(\delta) = 1$ if $|Z_n| \ge b_\delta(n)$ and zero otherwise; $\mathbf{N}_\infty(\delta) = \sum\limits_n \mathbf{Y}_n(\delta)$.

By cor. at **(6.9.4)**, we know that $\mathbf{N}_\infty(\delta)$ is, for δ fixed, a finite valued rv. The preciseness of the norming constants in the LIL comes through the property that $\mathbb{E}\mathbf{N}_\infty^\lambda(\delta) = \infty$, for every $\lambda > 0$. Define $N_{1,\infty}(\delta)$ as $N_\infty(\delta)$ using the variables $(X_{1,n})$ in place of the variables (X_n) and note that $\mathbf{N}_\infty(\delta) \ge N_{1,\infty}(\delta)$. Hence, it is enough to establish the following

Theorem 6.9.6. $\mathbb{E}N_\infty^\lambda(\delta) = \infty$ *for every* $\lambda > 0$.

Proof.

We note $N_\infty(\delta)(\omega)$ is the number of n for which the inequality $|S_n|(\omega) \ge (1+\delta)\sqrt{n}\varphi(n)$ holds.

Proof depends on the following two lemmas and the Egorov theorem (ref. **Q**13).

Lemma 1.

If $\lambda > 0$, then $n^\lambda - (n-1)^\lambda \sim \lambda n^{\lambda-1}$.

Proof of Lemma 1.

$$\frac{n^\lambda-(n-1)^\lambda}{\lambda n^{\lambda-1}} = \frac{1}{\lambda}\{n-(n-1)(1-\tfrac{1}{n})^{\lambda-1}\}$$
$$= \frac{1}{\lambda}\{n-(n-1)+(n-1)[1-(1-\tfrac{1}{n})^{-(1-\lambda)}]\}$$

$$= \tfrac{1}{\lambda}\{1 + (n-1)[1 - (1 + \tfrac{1-\lambda}{n} + O(\tfrac{1}{n^2}))]\}$$
$$= \tfrac{1}{\lambda}\{1 - (1-\lambda) + O(\tfrac{1}{n})\} \to 1.$$

Lemma 2.

If $0 < \lambda < 1$ and if $\sum_n n^{\lambda-1} P(|S_n| \geq b_\delta(n)) = \infty$, then $\mathbb{E}N_\infty^\lambda(\delta) = \infty$.

Proof of Lemma 2.

Let $N_0(\delta) = 0$. Let $N_m(\delta) = \sum_{n=1}^{m} Y_n(\delta)$. We note $N_m(\delta) \leq N_{m+1}(\delta)$, $m \geq 1$. Since $N_\infty(\delta) \geq N_m(\delta)$ for all m, it follows that

$$\begin{aligned}
\mathbb{E}N_\infty^\lambda &\geq \lim_{m\to\infty} \mathbb{E}N_m^\lambda(\delta) \\
&= \lim_{m\to\infty} \sum_{n=1}^{m} \{\mathbb{E}(N_n^\lambda(\delta)) - \mathbb{E}(N_{n-1}^\lambda(\delta))\} \\
&= \lim_{m\to\infty} \sum_{n=1}^{m} \mathbb{E}\{(N_{n-1}(\delta) + Y_n(\delta))^\lambda - N_{n-1}^\lambda(\delta)\} \\
&= \lim_{m\to\infty} \sum_{n=1}^{m} \mathbb{E}\big(\mathbb{E}\{(N_{n-1}(\delta) + Y_n(\delta))^\lambda - N_{n-1}^\lambda(\delta)|Y_n(\delta)\}\big) \\
&= \lim_{m\to\infty} \sum_{n=1}^{m} \mathbb{E}\{(N_{n-1}(\delta) + 1)^\lambda - N_{n-1}^\lambda(\delta)|Y_n(\delta) = 1\}\times \\
&\qquad\qquad P(Y_n(\delta) = 1).
\end{aligned}$$

Over the region $x > 0$, the function $(x+1)^\lambda - x^\lambda$ is monotonic decreasing when $0 < \lambda < 1$. Hence $\min_{0 \leq x \leq n-1} \{(x+1)^\lambda - x^\lambda\} = n^\lambda - (n-1)^\lambda$. Since $N_{n-1}(\delta)$ can never exceed $n-1$, it follows $\mathbb{E}N_\infty^\lambda(\delta) \geq \lim_{m\to\infty} \sum_{n=1}^{m} \{n^\lambda - (n-1)^\lambda\} P(|S_n| \geq b_\delta(n)) \geq q \sum_{n=1}^{\infty} n^{\lambda-1} P(|S_n| \geq b_\delta(n))$, by Lemma 1. With this the proof of Lemma 2 is complete.

Proof of assertion (ii).

We need consider only small values of λ since $\mathbb{E}N_\infty^\lambda = \infty$ implies $\mathbb{E}N_\infty^\theta = \infty$ if $0 < \lambda < \theta$. Let us then fix a λ, $0 < \lambda < 1$. Part (ii) of the theorem will follow from Lemma 2 if we show that

$$\sum_n n^{\lambda-1} P(|S_n| \geq b_\delta(n)) = \infty. \tag{6.65}$$

We first prove (6.65) holds when the common *df* F of the X_ns satisfies (6.42). In that case, let $\eta_n = \sum_{j=1}^{n} \tau_j$ be the Skorohod time of embedding in a $SBMP$ B for X_n, $n = 1, 2, \ldots$. Then (6.65) is equivalent to

$$\sum_n n^{\lambda-1} P(|B(\eta_n)| \geq b_\delta(n)) = \infty. \tag{6.66}$$

Now

$$\begin{aligned} P(|B(\eta_n)| \geq b_\delta(n)) &= P(|B(\eta_n) - B(n) + B(n)| \geq b_\delta(n)) \\ &\geq P\big(\big||B(\eta_n) - B(n)| - |B(n)|\big| \geq b_\delta(n)\big) \\ &\geq P(|B(n)| \geq b_\delta(n) + |B(\eta_n) - B(n)|) \\ &= P(|B(n)| \geq b_\delta(n)\{1 + \frac{1}{b_\delta(n)}|B(\eta_n) - B(n)|\}). \end{aligned}$$

We know (ref. steps in **(6.9.2)**) $Z_n = \frac{B(\alpha_n)-B(n)}{b_\delta(n)} \overset{wp1}{\longrightarrow} 0$. Hence, given $\varepsilon > 0$, there exists (by Egorov theorem (ref.**Q**13)) a ω-set A_ε and an $n_0 \geq 1$ (possibly depending on A_ε) such that $P(A_\varepsilon) > 1 - \varepsilon$ and $|Z_n(\omega)| < \delta$ for all $\omega \in A_\varepsilon$ and for all $n \geq n_0$. This leads to

$$\begin{aligned} &\sum_n n^{\lambda-1} P\big(|B(\eta_n)| \geq b_\delta(n)\big) \\ &\qquad \geq \sum_n n^{\lambda-1} P\big(|B(n)| \geq b_\delta(n)(1 + |Z_n|)\big) \\ &\qquad \geq \sum_n n^{\lambda-1} P\big(\{|B(n)| \geq b_\delta(n)(1 + |Z_n|)\} \cap A_\varepsilon\big) \\ &\qquad \geq \sum_n n^{\lambda-1} P\big(\{|B(n)| \geq b_\delta(n)(1 + \delta)\} \cap A_\varepsilon\big) \\ &\qquad \geq \sum_n n^{\lambda-1} P\big(\{|B(n)| \geq (1 + 3\delta)\sqrt{n}\varphi(n)\} \cap A_\varepsilon\big). \end{aligned}$$

If this series is divergent for any $\varepsilon > 0$ then (6.66) is true and there is nothing more to show. If this series is convergent for every $\varepsilon > 0$, then

$$\begin{aligned} \sum_n n^{\lambda-1} P\big(|B(\eta_n)| \geq b_\delta(n)\big) &\geq \lim_{k\to\infty} \sum_n n^{\lambda-1} P\big(\{|B(n)| \\ &\qquad \geq (1 + 3\delta)\sqrt{n}\varphi(n)\} \cap A_{\frac{1}{k}}\big) \\ &= \lim_{k\to\infty} \int_Q a_k(n) \, \mathrm{d}\,\mu(n) \end{aligned}$$

where Q is the set of positive integers, μ is the σ-finite measure which puts unit mass at each member of Q and $a_k(n) = n^{\lambda-1} P\big(\{|B(n)| \geq (1+3\delta)\sqrt{n}\varphi(n)\} \cap A_{\frac{1}{k}}\big)$ Since for each n fixed, $a_k(n) \leq a_{k+1}(n)$, monotone convergence theorem (ref. **Q2**) applies and we get

$$\begin{aligned}\sum_n n^{\lambda-1}P(|B(\eta_n)| \geq b_\delta(n)) &\geq \int_Q \lim_{k\to\infty} a_k(n)\,\mathrm{d}\,\mu(n)\\ &= \sum_n n^{\lambda-1}P(|B(n)| > (1+3\delta)\sqrt{2n\log\log n})\\ &= \sum_n n^{\lambda-1}P(|\xi| > (1+3\delta)\sqrt{2\log\log n}) = \infty,\end{aligned}$$

since
$P(|\xi| > (1+3\delta)\sqrt{2\log\log n}) \sim \frac{q}{\sqrt{\log\log n}} \times \frac{1}{(\log n)^\theta}$, $\theta > 1$. Thus (6.66) is satisfied and the proof is complete when F satisfies (6.42).

Let now F be any *df* with mean 0 and variance 1 but not necessarily satisfying (6.42). Let (ξ_n) be a sequence *iid* standard normal variables. Write $T_n = \sum_1^n \xi_i$. Since the *df* of $X_1 + \xi$ has 0 mean, variance 2 and satisfies (6.42), we have $\sum_n \frac{1}{n^{1-\lambda}} P(|S_n + T_n| \geq \sqrt{2}b_\delta(n)) = \infty$. Trivially, this implies $\sum\limits_n \frac{1}{n^{1-\lambda}} P(|S_n + T_n| \geq b_\delta(n)) = \infty$. Since T_n has median 0, we have (ref. **(4.2.3)**) $P(|S_n| \geq b_\delta(n)) = P(|S_n + T_n - T_n| \geq b_\delta(n)) \geq \frac{1}{2}P(|S_n + T_n| \geq b_\delta(n))$ and hence $\sum_n \frac{1}{n^{1-\lambda}} P(|S_n| \geq b_\delta(n)) = \infty$. This completes the proof. □

6.10 Functional LIL

Let (X_n) be a sequence of *iid* variables with zero means, unit variances and with common *df* F. Let $S_n = \sum_{k=1}^n X_k$. Define polygonal functions x_n in C:

$$x_n(\frac{i}{n}) = \frac{S_i}{\sqrt{n\varphi(n)}} \text{ and linear elsewhere on } [0,\ 1]. \tag{6.67}$$

Let B be a standard BMP independent of the X_ns, all defined on the same probability space $(\Omega,\ \mathscr{A},\ P)$. Let K_R be as in **(6.2)**.

In the rest of this section we discuss a $FLIL$ and two applications.

Theorem 6.10.1.

$$wp1, \textit{ the limit set of the sequence } (x_n) \textit{ is } K_R. \tag{6.68}$$

Proof. We establish the claim first when the *df* F satisfies the condition (6.42).

Let τ_n be the Skorohod time of embedding X_n in B and let $\alpha_n = \sum_{k=1}^n \tau_k$. Define polygonal functions y_n in C:

$$y_n(\frac{i}{n}) = \frac{B(\alpha_i)}{\sqrt{n\varphi(n)}} \text{ and linear elsewhere on } [0,\ 1]. \tag{6.69}$$

The required limit set, we note, is the same as that of the sequence (y_n).

We define two more processes, z_n and u_n:

$z_n(t) = \frac{B(nt)}{\sqrt{n\phi(n)}}$ and $u_n(\frac{i}{n}) = z_n(\frac{i}{n})$ and u_n is linear elsewhere on $[0,\ 1]$.

We note $\|u_n - z_n\| \leq \max\limits_{1 \leq i \leq n} \max\limits_{\frac{i-1}{n} \leq t \leq \frac{i}{n}} |u_n(t) - z_n(t)|$. Hence

$P(\|u_n - z_n\| > \varepsilon) \ \leq \sum_{i=1}^{n} P(A_i)$ where

$$\begin{aligned}
P(A_i) &= P(\max_{\substack{j \\ j=i-1,\, i}} \max_{\frac{i-1}{n} \leq t \leq \frac{i}{n}} |B(j) - B(nt)| > \varepsilon\sqrt{n\varphi(n)}) \\
&\leq P(\max_{i-1 \leq t \leq i} |B(i-1) - B(t)| > \varepsilon\sqrt{n\varphi(n)}) \\
&\quad + P(\max_{i-1 \leq t \leq i} |B(i) - B(t)| > \varepsilon\sqrt{n\varphi(n)}) \\
&\leq 2P(\max_{i-1 \leq t \leq i} |B(t) - B(i-1)| > \frac{1}{2}\varepsilon\sqrt{n\varphi(n)}) \\
&\quad + P(|B(i) - B(i-1)| > \frac{1}{2}\varepsilon\sqrt{n\varphi(n)}) \\
&\leq qP(\xi > \frac{1}{2}\sqrt{n\varphi(n)}) \\
&\leq P(\xi > \sqrt{2n}), \text{ for all } n \text{ large.}
\end{aligned}$$

Thus $P(\|u_n - z_n\| > \varepsilon) \ \leq ne^{-n}$, leading to $\sum_n P(\|u_n - z_n\| > \varepsilon) < \infty$. This implies that the limit set of the sequence (u_n) is the same as that of (z_n), which, by **(6.2.10)** is K_R.

$$\|y_n - u_n\| \leq \frac{1}{\sqrt{n\varphi(n)}} \max_{1 \leq m \leq n} |B(\alpha_m) - B(m)|,$$

which $\xrightarrow{wp1}$ 0 (ref. **(6.9.2)**).

The above two results imply $\|y_n - z_n\| \xrightarrow{wp1} 0$. It follows from this that the limit set of the sequence (y_n) (and hence that of (x_n)) is K_R.

Before we proceed to extend the result to dfs F which may not satisfy condition (6.42), we wish to record, for later use, the following property.

Define right continuous step function $\hat{x}_n(t) = \frac{1}{\sqrt{n\varphi(n)}} S_{[nt]},\ S_0 = 0$. We note $\|x_n - \hat{x}_n\| \leq \frac{1}{\sqrt{n\varphi(n)}} \max\limits_{1 \leq k \leq n} |X_k|$. Now, $\sum_n P(|X_n| > \varepsilon\sqrt{n}) = \sum_n P(X_1^2 > \varepsilon^2 n) < \infty$, since $\mathbb{E}X_1^2 < \infty$. Hence by the Borel-Cantelli lemma, $\frac{X_n}{\sqrt{n}} \xrightarrow{wp1} 0$. This implies $\frac{1}{\sqrt{n\varphi(n)}} \max\limits_{1 \leq k \leq n} |X_k| \xrightarrow{wp1} 0$. It follows now $\|x_n - \hat{x}_n\| \xrightarrow{wp1} 0$. Hence the limit set of the sequence $(\hat{x}_n)$ in the metric space $(D,\ \rho)$ is the same as that of the sequence (x_n).

Let (ξ_n) be a sequence of iid zero mean normal variables with common variance σ^2 and independent of the sequence (X_n). All the variables are assumed defined on a common probability space, which can be arranged. Define $Y_n = \frac{X_n+\xi_n}{\sqrt{1+\sigma^2}}$ and note that (Y_n) is a sequence of iid variables with zero means, with unit variances and with their common df satisfying condition (6.42). Hence $wp1$, the limit set of $\left(\frac{\widehat{S}_{[nt]}}{\sqrt{n\varphi(n)}}\right)$ is $K_{R\sqrt{1+\sigma^2}}$ where $\widehat{S}_n = \sum_{\nu=1}^{n} Y_\nu = S_n + T_n$, $S_n = \sum_{\nu=1}^{n} X_\nu$ and $T_n = \sum_{\nu=1}^{n} \xi_\nu$. Since a normal distribution satisfies condition (6.42), the limit set of $(V_{n;\sigma}(t))$, that is of, $\left(\frac{T_{[nt]}}{\sqrt{n\varphi(n)}}\right)$ is $K_{R\sigma}$. If then $U_n(t) = \frac{S_{[nt]}}{\sqrt{n\varphi(n)}}$, its limit set, $wp1$, is contained in $K_{R\sqrt{1+\sigma^2}+R\sigma} \subset K_{R(1+2\sigma)}$. The rest of the proof that it is K_R can be constructed on the lines of tne proof of **(6.9.2)**. □

Corollary 6.10.2. *Let H be a separable metric space and let ψ be a continuous map of C into H. Then the sequence $(\psi(x_n))$ has limit set $\psi(K_R)$.*

Proof. That $\psi(K_R)$ is a compact set (being the continuous image of the compact set K_R) is well known. It is obvious that $wp1$ the limit points of $(\xi(x_n(\omega)))$ are contained in $\psi(K_R)$. Let $u \in \psi(K_R)$ and let G be a neighbourhood of u. There exists $y \in K_R$ such that $\psi(y) = u$ and $\psi^{-1}(G)$ is an open set containing y. Since y is a limit point of (x_n) there exists a ω-set A such that $P(A) = 1$ and $\omega \in A$ implies that a subsequence of $(x_n(\omega))$ converges to y. This implies $\psi(x_n(\omega))$ converges to u. Since $\psi(K_R)$ is a separable set, it follows that it is the limit set of $(\psi(x_n))$. □

We now discuss two applications of the above FLIL, the second of which necessitates the evaluation of an integral of independent interest.

Let x_n be as in (6.67)

(i) $\psi(x) = x(1)$. That, in this case, the limit set of $(\psi(x_n))$ is $[-R,\ R]$ has already been shown in **(6.2.11)**.

(ii) Define ψ over C : $\psi(x) = \int_0^1 x(t)f(t)\,\mathrm{d}t$ where $f \in C$ is fixed. Let $F(t) = \int_t^1 f(s)\,\mathrm{d}s$, $0 \le t \le 1$. We note that, so defined, ψ is a continuous functional. Hence by the Corollary, $P(\overline{\lim_{n\to\infty}}\ \psi(x_n) = \sup_{x\in K_R} \psi(x)) = 1$. We have

$$\sup_{x\in K_R} \psi(x) = \sup_{x\in K_R} \int_0^1 x(t)f(t)\,\mathrm{d}t = \sup_{x\in K_R} \int_0^1 x(t)\{-\tfrac{\mathrm{d}F(t)}{\mathrm{d}t}\}\,\mathrm{d}t$$

$$= \sup_{x\in K_R} \{-F(t)x(t)\big|_0^1 + \int_0^1 F(t)\tfrac{\mathrm{d}x(t)}{\mathrm{d}t}\,\mathrm{d}t\} \text{(using integration by parts and the fact that } x \in K_R)$$

$$= \sup_{x\in K_R} \int_0^1 F(t)\dot{x}\,\mathrm{d}t \ \left(\text{since } x(0) = 0 \text{ and } F(1) = 0\right).$$

Let L_2 denote the Hilbert space $\mathfrak{H}$ of real functions defined on $[0,\ 1]$ and square integrable with respect to the Lebesgue measure. Denote by $\langle .\ ,\ .\rangle$ and $\|\ .\ \|$ respectively the inner product and the norm in $\mathfrak{H}$. We note $F \in L_2$ since it is a continuous function and $\dot{x} \in L_2$ since $x \in K_R$. Hence $\int_0^1 F(t)\dot{x}(t)\,\mathrm{d}t = \langle F,\ \dot{x}\rangle$. Now, $\sup\limits_{x\in K_R} \langle F,\ \dot{x}\rangle = R \sup\limits_{x\in K_R} \langle F,\ \frac{1}{R}\dot{x}\rangle = R \sup\limits_{y\in L_2,\ \|y\|\leq 1} \langle F,\ y\rangle = R\|F\| = R\big(\int_0^1 (F(t))^2\,\mathrm{d}t\big)^{\frac{1}{2}}$. (The claim in the last step is a familiar result in the area of linear functionals in a Hilbert space)

We have $\overline{\lim\limits_{n\to\infty}}\ \psi(x_n) = \overline{\lim\limits_{n\to\infty}} \int_0^1 x_n(t)f(t)\,\mathrm{d}t =$ (approximating the integral by a Riemann sum)

$$\overline{\lim_{n\to\infty}} \sum_{i=1}^{n} x_n(\frac{i}{n})f(\frac{i}{n})\frac{1}{n} = \overline{\lim_{n\to\infty}}\ \frac{1}{n^{\frac{3}{2}}\varphi(n)} \sum_{i=1}^{n} f(\frac{i}{n})S_i.$$

We conclude $P\big(\overline{\lim\limits_{n\to\infty}}\ \frac{1}{n^{\frac{3}{2}}\varphi(n)} \sum_{i=1}^n f(\frac{i}{n})S_i = R\{\int_0^1 (F(t))^2\,\mathrm{d}t\}^{\frac{1}{2}}\big) = 1.$

Let $f(t) = t^\alpha,\ \alpha > -1$. When $-1 < \alpha < 0$, $\int_0^1 f(t)\,\mathrm{d}t$ exists as an improper Riemann integral. Also $\int_0^1 x(t)f(t)\,\mathrm{d}t,\ x \in C$ will exist at least as an improper Riemann integral.

When $-1 < \alpha < 0$, define f_n thus : $f_n(0) = 0,\ f_n(t) = f(t)$ for $t \geq \frac{1}{n}$ and f_n linear over $[0,\ \frac{1}{n}]$. Since $\int_0^1 x_n(t)f(t)\,\mathrm{d}t = \int_0^1 x_n(t)f_n(t)\,\mathrm{d}t + u_n$, where $|u_n| = |\int_0^{\frac{1}{n}} x_n(t)\big(f(t) - f_n(t)\big)\,\mathrm{d}t| \leq \frac{q}{n^{1+\alpha}}$. Hence (arguing as before),

$\overline{\lim\limits_{n\to\infty}}\ \psi(x_n) = \overline{\lim\limits_{n\to\infty}} \int_0^1 x_n(t)f_n(t)\,\mathrm{d}t = \overline{\lim\limits_{n\to\infty}} \sum_{i=1}^n x_n(\frac{i}{n})f(\frac{i}{n})$.

Thus for all $\alpha > -1$,

$$wp1,\ \overline{\lim_{n\to\infty}}\ \frac{1}{n^{\alpha+\frac{3}{2}}\varphi(n)} \sum_{i=1}^{n} i^\alpha S_i = R\{\int_0^1 \left(F(t)\right)^2 \mathrm{d}t\}^{\frac{1}{2}}$$

$$= \frac{R\sqrt{2}}{\sqrt{(\alpha+2)(2\alpha+3)}}.$$

Chapter 7

Discrete Time Markov Chains

Independence of rEs was defined in **(4.1.1)**. A collection of rEs will be called dependent if it fails to satisfy the criterion for independence as set out in **(4.1.1)**. It is possible that a collection of dependent rEs contains an independent sub collection. For an example ref. **(4.1.1)**(α).

7.1 From independence to Markov dependence

Study of dependent variables naturally demands knowledge of the nature of dependence amongst the members of the collection. In **(6.3.3)**, the type of dependence which makes the collection a *martingale* was explained. Markov processes were defined in **(6.7.1)**. In the present chapter we study in some detail sequences $(X_n,\ n \in T,\ T = \{0, 1, \ldots\})$ of rEs taking values in an abstract countable set **E**, equipped with the σ-field of all its subsets, called the state space, which is finite or denumerably infinite, satisfying (7.1) below, or what is same, (6.44). Such a sequence is called a Markov Chain (MC). We label the members of **E** as $\mathbf{E} = \{0,\ 1,\ 2, \ldots\}$ or$\{1,\ 2, \ldots\}$. Thus a MC can be thought of as a system which can be in various states, and which changes randomly in discrete steps. It can be helpful to think of the system as evolving through discrete steps in time, although strictly speaking, the "step" may have nothing to do with time as such.

A MC is referred to as a finite Markov Chain if **E** is a finite set.

Notation 7.1.1. *The symbols* $0,\ 1,\ 2, \ldots$ *making up the state space* **E** *can be labels or abstract points and not numbers. For example,* X_n *may represent the colour of the ball drawn at the* n^{th} *draw from an urn containing balls of different colours. To emphasise they are labels, we write* $\mathbf{E} = \{s_0,\ s_1,\ \ldots.\}$*, choosing and fixing a listing. Consider* $\sum_{j \in A_k} f(s_j)$ *where* $A_k = \{s_k,\ s_{k+1},\ \ldots\}$

and f is a functional on **E**. *It would be so much less cumbersome to write the sum as* $\sum_{j=k}^{\infty} f(j)$. *We will use this simpler notation, with no room for confusion. Again consider the expression* $\lim_{k\to\infty} \sum_{j=k}^{\infty} f(j)$. *This naturally stands for limit of* t_{A_k} *as* $A_k \downarrow \phi$ *where* $t_{A_k} = \sum_{j\in A_k} f(j)$. *Or, the limit can be thought of as the limit of the net* $(t_{A_k},\ A_k \in \mathcal{N},\ \supset)$ *in R, where* $\mathcal{N}$ *is the collection of the sets* $(A_k,\ k = 0,\ 1,\ 2,\ ....)$ *which is a directed set under the set inclusion sign* $\supset$ *(ref. page 65, [K]).*

The following example compels us to adopt an appropriate convention so that certain irksome trivialities are avoided. Let $\Omega = [0,\ 1] \times [0,\ 1]$, let $\mathscr{A}$ be the Borel σ-field of Ω and let P be the Lebesgue measure on $\mathscr{A}$. Denote a generic point in Ω by $(x,\ y)$. Define rvs $X_0,\ X_1,\ X_2$ on Ω : $X_0(x,\ y) = y$; $X_1(x,\ y) = 0$ if $0 \leq x < \frac{1}{2}$, $X_1(x,\ y) = \frac{1}{2}$ if $x = \frac{1}{2}$, $X_1(x,\ y) = 1$ if $\frac{1}{2} < x \leq 1$; $X_2(x,\ y) = 1$ if $\frac{1}{8} \leq x \leq \frac{3}{4}$ and $X_2(x,\ y) = 0$ otherwise. Clearly the vector $(X_1,\ X_2)$ is independent of X_0. Hence trivially $P(X_2 = 1|X_1,\ X_0) = P(X_2 = 1|X_1)$. Thus $X_0,\ X_1,\ X_2$ is a Markov chain and its state space E, according to the above definition, would be the set of numbers $\{0,\ \frac{1}{2},\ 1\}$. A common sense approach would expect that removal of the state $\frac{1}{2}$, which value is taken by X_1 with zero probability, and making the state space $\mathbf{E} = \{0,\ 1\}$, should not alter the nature of dependence among the three variables. This expectation is correct. Formally, it is secured as follows. Define $Y_1 = X_1$ if $X_1 = 0$ or 1 and $Y_1(\frac{1}{2}) = 1$ and note that the joint distribution of $(X_0,\ X_1,\ X_2)$ is the same as that of $(X_0,\ Y_1,\ X_2)$. Since such alteration can be made in a general MC, we assume that for every $a \in \mathbf{E}$, there exists an n such that $P(X_n = a) > 0$.

We start by rephrasing definition **(6.7.1)** so as to make it easier to apply/verify in the case of MCs.

Definition 7.1.2. A sequence $(X_n,\ n = 0,\ 1,\ 2,\ldots)$ of rEs taking values in a finite or a denumerably infinite set **E** is a Markov chain if it possesses the Markov property:

$$\begin{aligned} P(X_n = j_n|X_r = j_r,\ 0 \leq r \leq n-1) \\ = P(X_n = j_n|X_{n-1} = j_{n-1})\ wp1 \quad (7.1) \end{aligned}$$

for every choice of $j_r \in \mathbf{E}$ with $P(X_r = j_r,\ 0 \leq r \leq n-1) > 0$ and for every choice of $n \geq 1$. We describe this condition by saying 'whenever the left side is defined'. When this happens, the right side expression is automatically defined.

It is not assumed that each one of the rEs X_n takes each one of the states as values. To appreciate this point consider the MC $(S_n,\ n \geq 0)$, $S_0 = 0$, $S_n = \sum_0^n X_k$ where the X_ks are *iid* standard Bernoulli variables. i.e., X_1 takes only the values 1 and -1. Here $\mathbf{E} = \{0,\ \pm 1,\ \pm 2, \ldots\}$. For example, S_2 takes only the values $-2,\ 0,\ 2$.

Endow $\mathbf{E}$ with the σ-field $\mathfrak{E}$ of all its subsets. Then all the X_ns are measurable mappings of Ω into $\mathbf{E}$. Let $\mathcal{A}_n,\ n \geq 0$ be the σ-field in Ω generated by $X_\nu, 0 \leq \nu \leq n$.

It follows immediately from the definition of conditional probability, that (7.1) is equivalent to

$$P(X_n = j|\mathcal{A}_{n-1}) = P(X_n = j|X_{n-1})\ wp1,\ n \geq 1. \tag{7.2}$$

Remarks 7.1.3. Let $\mathcal{A}_{n,k}$ denote the σ field generated by the rEs $X_\nu,\ n \leq \nu \leq n+k$ and $\mathcal{A}_{n,\infty}$ that generated by the elements $X_\nu,\ \nu \geq n$.
(i) Let $0 \leq r < s \leq n-1$. Then,

$$P(X_n = j|X_\nu = i_\nu,\ r \leq \nu \leq s) = P(X_n = j|X_s = i_s). \tag{7.3}$$

Proof. Let $f = \chi_{\{X_n = j\}}$. We have,

$$P(X_n = j|\mathcal{A}_{r,s}) = E(f|\mathcal{A}_{r,s}) = E(\{E(f|\mathcal{A}_{r,s})\}|\mathcal{A}_{n-1}).$$

By **(6.4)**(iii), this step is valid since the inner conditional expectation is $\mathcal{A}_{r,s}$-measurable and hence is measurable with respect to the larger σ-field $\mathcal{A}_s$. Appeal now to **(6.3)**(vii) to conclude

$$\begin{aligned} E(f|\mathcal{A}_{r,s}) &= E(\{E(f|\mathcal{A}_s)\}|\mathcal{A}_{r,s}) = E(\{E(f|X_s)\}|\mathcal{A}_{r,s}) \text{ (by (7.2))} \\ &= E(f|X_s), \text{ as was to be proved.} \end{aligned}$$

(ii) *When the left side is defined,*

$$\begin{aligned} P(X_{n+1} = k|X_{n-1} = j,\ X_\nu \in A_\nu,\ 0 \leq \nu \leq n-2) \\ = P(X_{n+1} = k|X_{n-1} = j). \end{aligned} \tag{7.4}$$

Proof. Let $\hat{\mathbf{E}}$ consist of those states i for which $P(X_n = i,\ X_{n-1} = j,\ X_\nu \in A_\nu,\ 0 \leq \nu \leq n-2) > 0$ and let $\tilde{\mathbf{E}}$ consist of those states i for which $P(X_n = i,\ X_{n-1} = j) > 0$.

Left side of (7.4) is equal to

$$\frac{1}{P(X_{n-1}=j,\ X_\nu \in A_\nu,\ 0 \le \nu \le n-2)} \times$$
$$\sum_{i \in \mathbf{E}} P(X_{n+1}=k,\ X_n=i,\ X_{n-1}=j, X_\nu \in A_\nu,\ 0 \le \nu \le n-2)$$
$$= \sum_{i \in \hat{\mathbf{E}}} \{P(X_{n+1}=k | X_n=i,\ X_{n-1}=j,\ X_\nu \in A_\nu,\ 0 \le \nu \le n-2)$$
$$\times P(X_n=i | X_{n-1}=j,\ X_\nu \in A_\nu,\ 0 \le \nu \le n-2)\}$$
$$= \sum_{i \in \hat{\mathbf{E}}} P(X_{n+1}=k | X_n=i) P(X_n=i | X_{n-1}=j).$$

We note that the hypothesis implies $P(X_{n-1}=j) > 0$. We have

$$P(X_{n+1}=k | X_{n-1}=j) = \frac{P(X_{n+1}=k,\ X_{n-1}=j)}{P(X_{n-1}=j)}$$
$$= \frac{1}{P(X_{n-1}=j)} \sum_{i \in \mathbf{E}} P(X_{n+1}=k,\ X_n=i,\ X_{n-1}=j)$$
$$= \frac{1}{P(X_{n-1}=j)} \sum_{i \in \tilde{\mathbf{E}}} P(X_{n+1}=k,\ X_n=i,\ X_{n-1}=j)$$
$$= \sum_{i \in \tilde{\mathbf{E}}} P(X_{n+1}=k | X_n=i,\ X_{n-1}=j) P(X_n=i | X_{n-1}=j)$$
$$= \sum_{i \in \tilde{\mathbf{E}}} P(X_{n+1}=k | X_n=i) P(X_n=i | X_{n-1}=j) \text{ (by (i))}.$$

Thus (7.4) will stand proved if we show $\hat{\mathbf{E}} = \tilde{\mathbf{E}}$. Trivially, $\hat{\mathbf{E}} \subseteq \tilde{\mathbf{E}}$. Let $i \in \tilde{\mathbf{E}}$. Hence $P(X_n=i,\ X_{n-1}=j) > 0$. This implies $P(X_{n-1}=j) > 0$ and $P(X_n=i | X_{n-1}=j) > 0$. Additionally, we have from the hypothesis $P(X_{n-1}=j,\ X_\nu \in A_\nu,\ 0 \le \nu \le n-2) > 0$. We can now conclude that $i \in \hat{\mathbf{E}}$, since

$$P(X_n=i,\ X_{n-1}=j,\ X_\nu \in A_\nu,\ 0 \le \nu \le n-2)$$
$$= P(X_n=i | X_{n-1}=j,\ X_\nu \in A_\nu,\ 0 \le \nu \le n-2) \times$$
$$P(X_{n-1}=j,\ X_\nu \in A_\nu,\ 0 \le \nu \le n-2)$$
$$= P(X_n=i | X_{n-1}=j) \times$$
$$P(X_{n-1}=j,\ X_\nu \in A_\nu,\ 0 \le \nu \le n-2) > 0.$$

Note.

In proofs hereafter, the procedure of separating the terms with zero probabilities and adding them later will not be followed. We will behave as if the

conditioning events have positive probabilities. The reader is advised to write for himself steps on the lines of the proof given above.
(iii) *For subsets* $A_k \subset \boldsymbol{E}$, $0 \le k \le n-1$,

$$P(X_{n+1} = j | X_n = i,\ X_k \in A_k,\ 0 \le k \le n-1) \\ = P(X_{n+1} = j | X_n = i) \quad (7.5)$$

whenever the left side is defined.

Proof. Left side of (7.5) is equal to

$$\frac{P(X_{n+1} = j,\ X_n = i,\ X_k \in A_k,\ 0 \le k \le n-1)}{P(X_k \in A_k,\ 0 \le k \le n-1,\ X_n = i)}$$
$$= \frac{1}{P(X_k \in A_k,\ 0 \le k \le n-1,\ X_n = i)} \times$$
$$\sum_{\substack{i_k \in A_k \\ 0 \le k \le n-1}} \{P(X_{n+1} = j | X_n = i,\ X_k = i_k, 0 \le k \le n-1)$$
$$\times P(X_n = i,\ X_k = i_k, 0 \le k \le n-1)\}$$
$$= P(X_{n+1} = j | X_n = i) \times$$
$$\sum_{\substack{i_k \in A_k \\ 0 \le k \le n-1}} \frac{P(X_k = i_k,\ 0 \le k \le n-1,\ X_n = i)}{P(X_k \in A_k,\ 0 \le k \le n-1,\ X_n = i)}$$
$$= P(X_{n+1} = j | X_n = i), \text{ as was to be proved.}$$ □

(iv) *Let* C, D, A_ν*s denote subsets of the state space* E. *Let* $n \ge 1$ *be arbitrary. Let* $B \in \mathcal{A}_{n+1,\,\infty}$. *Then*

$$P(B | X_n = i,\ X_\nu \in A_\nu,\ 0 \le \nu \le n-1) = P(B | X_n = i). \quad (7.6)$$

We first show, for arbitrary subsets $B \in \bigotimes_{\nu=n+1}^{n+r} \mathcal{A}_\nu$,

$$P(B | X_n = i,\ X_\nu \in A_\nu,\ 0 \le \nu \le n-1) = P(B | X_n = i). \quad (7.7)$$

We establish (7.7) by induction on r. This is done through steps 1, 2 and 3. (7.6) is proved, using (7.7).
Step 1. Let $r = 1$.

$$P(X_{n+1} \in A_{n+1} | X_n = i,\ X_\nu \in A_\nu,\ 0 \le \nu \le n-1)$$
$$= \sum_{j \in A_{n+1}} P(X_{n+1} = j | X_n = i,\ X_\nu \in A_\nu,\ 0 \le \nu \le n-1) \text{ (by } \mathbf{(6.4)}\text{(i))}$$

$$= \sum_{j \in A_{n+1}} P(X_{n+1} = j | X_n = i) \text{ (Markov property)}$$

$$= P(X_{n+1} \in A_{n+1} | X_n = i_n) \text{ (by } \mathbf{(6.4)}\text{(i))}.$$

Thus for every $B \in \mathcal{A}_{n+1}$,

$$P(B | X_n = i,\ X_\nu \in A_\nu,\ 0 \leq \nu \leq n-1) = P(B | X_n = i). \tag{7.8}$$

Step 2. $r = 2$.

$$\begin{aligned}
&P(X_{n+1} \in C,\ X_{n+2} \in D | X_n = i,\ X_\nu \in A_\nu,\ 0 \leq \nu \leq n-1) \\
&= \sum_{j \in C} P(X_{n+2} \in D,\ X_{n+1} = j | X_n = i,\ X_\nu \in A_\nu,\ 0 \leq \nu \leq n-1) \\
&= \sum_{j \in C} \{P(X_{n+2} \in D | X_{n+1} = j,\ X_n = i,\ X_\nu \in A_\nu,\ 0 \leq \nu \leq n-1) \\
&\qquad \times P(X_{n+1} = j | X_n = i,\ X_\nu \in A_\nu,\ 0 \leq \nu \leq n-1)\} \\
&= \sum_{j \in C} P(D | X_{n+1} = j) P(X_{n+1} = j | X_n = i)
\end{aligned} \tag{7.9}$$

(appealing to (7.8) and to Markov property). Again,

$$\begin{aligned}
&P(X_{n+2} \in D,\ X_{n+1} \in C | X_n = i) \\
&= \sum_{j \in C} P(X_{n+2} \in D,\ X_{n+1} = j | X_n = i) \\
&= \sum_{j \in C} \{\sum_{k \in D} P(X_{n+2} = k,\ X_{n+1} = j | X_n = i)\} \\
&= \sum_{j \in C} \{\sum_{k \in D} P(X_{n+2} = k |\ X_{n+1} = j,\ X_n = i) P(X_{n+1} = j | X_n = i)\} \\
&= \sum_{j \in C} \sum_{k \in D} P(X_{n+2} = k | X_{n+1} = j) P(X_{n+1} = j | X_n = i) \\
&= \sum_{j \in C} P(X_{n+2} \in D | X_{n+1} = j) P(X_{n+1} = j | X_n = i).
\end{aligned} \tag{7.10}$$

We conclude from (7.9) and (7.10) that

$$P(E \times F | X_n = i,\ X_\nu \in A_\nu,\ 0 \leq \nu \leq n-1) = P(E \times F | X_n = i)$$

for every pair of sets $E,\ F$ where $E \in \mathcal{A}_{n+1}$ and $F \in \mathcal{A}_{n+2}$.
Step 3. Since $P(.|X_n = i,\ X_\nu \in A_\nu, 0 \leq \nu \leq n-1)$, is an additive set function, it follows that $P(G | X_n = i,\ X_\nu \in A_\nu, 0 \leq \nu \leq n-1) =$

$P(G|X_n = i)$ for every G where G is the finite union of disjoint intervals $E \times F$. The collection of all such unions form a field (ref. Theorem E, p139, [H]). We recall that $\mathcal{A}_{n+1} \bigotimes \mathcal{A}_{n+2}$ is the σ-field generated by this field. That the collection of all sets B in $\mathcal{A}_{n+1} \bigotimes \mathcal{A}_{n+2}$ which satisfy (7.7) form a monotone class follows from $P(.|X_n = i,\ X_\nu \in A_\nu, 0 \leq \nu \leq n-1)$ being a probability measure. Since a monotone class containing a field of sets contains the σ-field generated by the field (ref. Theorem B, p27, [H]), it follows that (7.7) holds for every $B \in \mathcal{A}_{n+1} \bigotimes \mathcal{A}_{n+2}$.

Step 4. $r \geq 3$.

Let claim hold for every $B \in \bigotimes_{\nu=n+1}^{n+m} \mathcal{A}_\nu$ and for every $n \geq 1$. We will show that (7.8) holds for every $B \in \bigotimes_{\nu=n+1}^{n+m+1} \mathcal{A}_\nu$.

Let $F \times E \in \bigotimes_{\nu=n+2}^{n+m+1} \mathcal{A}_\nu \bigotimes \mathcal{A}_{n+1}$. Then for some $C \subset \mathbf{E}$,

$$\begin{aligned}
&P(F \times E|X_n = i,\ X_\nu \in A_\nu,\ 0 \leq \nu \leq n-1)\\
&\quad = \sum_{j \in C} P(F,\ X_{n+1} = j|X_n = i,\ X_\nu \in A_\nu, 0 \leq \nu \leq n-1)\\
&\quad = \sum_{j \in C} P(F|X_{n+1} = j,\ X_n = i,\ X_\nu \in A_\nu,\ 0 \leq \nu \leq n-1) \times\\
&\qquad\qquad\qquad\qquad P(X_{n+1} = j|X_n = i)\\
&\quad = \sum_{j \in C} P(F|X_{n+1} = j) P(X_{n+1} = j|X_n = i)
\end{aligned}$$

(by induction hypothesis). Similarly,
$P(E \times F|X_n = i) = \sum_{j \in C} P(F|X_{n+1} = j) P(X_{n+1} = j|X_n = i)$
(arguing as in proving (7.10)). The arguments set out in Step 3 can be repeated to conclude that (7.8) holds for every
$B \in (\bigotimes_{\nu=n+2}^{n+m+1} \mathcal{A}_\nu) \bigotimes \mathcal{A}_{n+1}$. This completes the induction procedure since there is a natural one-to-one correspondence between the measurable spaces $(\bigotimes_{\nu=n+2}^{n+m+1} \mathcal{A}_\nu) \bigotimes \mathcal{A}_{n+1}$ and $\bigotimes_{\nu=n+1}^{n+m+1} \mathcal{A}_\nu$ which preserves the Markov property.

Step 5. Consider the product measurable space $\mathcal{B}^{(m)} = \bigotimes_{\nu=n+1}^{\infty} \mathcal{B}_\nu^{(m)},\ m \geq 1$, where $\mathcal{B}_\nu^{(m)} = \mathcal{A}_\nu$ if $n+1 \leq \nu \leq n+m$ and $\mathcal{B}_\nu = \{\Omega,\ \phi\}$ if $\nu > n+m$. It is trivial to verify that $\bigcup_{m=1}^{\infty} \mathcal{B}^{(m)}$ is a field, that $\sigma(\bigcup_{m=1}^{\infty} \mathcal{B}^{(m)}) = \mathcal{A}_{n+1,\infty}$ and that for every $E \in \bigcup_{m=1}^{\infty} \mathcal{B}^{(m)}$,

$$P(E|X_n = i,\ X_\nu \in A_\nu,\ 0 \le \nu \le n-1) = P(E|X_n = i)$$

Arguments parallel to earlier ones establish (7.6).

These results are often summarised in the words: the Markov property states that the probability distribution for the system at the next step (and in fact at all future steps) depends only on the current state of the system, and not additionally on the state of the system at previous steps.

However if $A \subset \mathbf{E}$ contains more than one member then it is generally not true that $P(X_{n+1} = j|X_n \in A,\ X_k = j_k,\ 1 \le k \le n-1) = P(X_{n+1} = j|X_n \in A)$, as is clear from the following example.

Let $P(X_0 = 0) = \frac{1}{2}$, $P(X_0 = \pm n) = \frac{c}{n^2}$, $n \ge 1$, $\sum_1^\infty \frac{c}{n^2} = \frac{1}{4}$; $P(X_1 = j|X_0 = i) = \frac{1}{5}$ if $j = i-2,\ i-1,\ i,\ i+1,\ i+2$ and 0 otherwise; $P(X_2 = j|X_1 = i) = P(X_1 = j|X_0 = i)$. A reference to **(1.11.3)** makes it clear that a probability space exists supporting three variables $X_0,\ X_1,\ X_2$ with the above properties. By their very definition, $X_0,\ X_1,\ X_2$ is a MC with $\mathbf{E} = \{0,\ \pm 1,\ \pm 2, \ldots\}$. Now

$$P(X_2 = 1|X_1 \in \{3,\ 6\},\ X_0 = 5)$$
$$= \frac{P(X_2 = 1,\ X_1 = 3,\ X_0 = 5) + P(X_2 = 1,\ X_1 = 6,\ X_0 = 5)}{P(X_1 = 3,\ X_0 = 5) + P(X_1 = 6,\ X_0 = 5)}$$

$$= \frac{1}{10}.$$

We note $P(X_1 = 3) = \frac{1}{5}P(X_0 \in \{1,\ 2,\ 3,\ 4,\ 5\}) = \frac{c}{5}(\sum_1^5 \frac{1}{k^2}) = \frac{c}{5}1.4636$; $P(X_1 = 6) = \frac{c}{5}\sum_4^8 \frac{1}{k^2} = \frac{c}{5}0.1663$. Hence $P(X_2 = 1|X_1 \in \{3,\ 6\}) = 0.1796$.

Since the system changes randomly, it is generally not possible to predict the exact state of the system in the future. However, the statistical properties of the system after a great many steps in the future can often be described. In many applications it is these statistical properties that are important.

The changes of state of the system are called transitions, and the probabilities associated with various state-changes are called transition probabilities.

$P(X_n = j|X_{n-1} = i)$, when defined, is generally a function of $i,\ j$, and n, called the one-step transition probability. If this is independent of the 'time' n i.e., if it is the same for all values of n for which it is defined, then we denote it by $p_{i,j}$ and say that the chain has stationary (one-step) transition probabilities. And in that case we agree to write $P(X_n = j|X_{n-1} = i) = p_{i,j}$ even when $P(X_{n-1} = i) = 0$.

We will be studying only chains with stationary transition probabilities.

By **(1.11.3)**, *there always exists a probability space supporting* MC (X_n) *with a specified state space* **E**, *with a specified distribution for* X_0 *and a specified stochastic matrix for its one-step transition probabilities.*

Since $P(X_{n+1} \in \mathbf{E}|X_n = i) = 1$, we have $\sum_{j\in \mathbf{E}} p_{i,j} = 1$. i.e., the matrix $P = (p_{i,j}),\ i,\ j \in \mathbf{E}$ is a stochastic matrix. (We recall a matrix with non-negative entries is called a stochasic matrix if the entries in each row add up to unity).

Example 7.1.4. Let $(Y_n,\ n \geq 1)$ be a sequence of iid rvs such that $P(Y_1 = k) = p_k$, $k = 0,\ 1, \ldots$. Put $X_0 = 0$ and for $n \geq 1$, $X_n = \sum_{\nu=1}^{n} Y_\nu$. The state space **E** of the X_n process is the set $\{0,\ 1,\ 2, \ldots\}$ of integers. Then $P(X_{n+1} = i_{n+1}|X_0 = 0,\ X_1 = i_1,\ \ldots,\ X_n = i_n) = P(Y_{n+1} = i_{n+1} - i_n|X_0 = 0, \ldots,\ X_n = i_n) = P(Y_{n+1} = i_{n+1} - i_n) = p_{i_{n+1}-i_n}$. Further $P(X_{n+1} = i_{n+1}|X_n = i_n) = P(Y_{n+1} = i_{n+1} - i_n) = p_{i_{n+1}-i_n}$. Thus $(X_n,\ n \geq 0)$ is a MC with transition probabilities $p_{i,j} = p_j$ if $i \leq j$ and $p_{i,j} = 0$ if $i > j$. In matrix form

$$P = (p_{i,j}) = \begin{pmatrix} p_0 & p_1 & p_2 & p_3 & \cdots \\ 0 & p_0 & p_1 & p_2 & \cdots \\ 0 & 0 & p_0 & p_1 & \cdots \\ 0 & 0 & 0 & p_0 & \cdots \\ & & & & \\ 0 & \cdots & & & \end{pmatrix}$$

Example 7.1.5. Let N_n be the number of successes in n independent Bernoulli trials, where the probability of a success in any one is p. This is a particular case of Example 1 when the rvs Y_n have Bernoulli distribution. The transition matrix in this case is

$$\begin{pmatrix} 1-p & p & 0 & 0 & \cdots \\ 0 & 1-p & p & 0 & \cdots \\ 0 & 0 & 1-p & p & \cdots \\ 0 & 0 & 0 & 1-p & \cdots \\ & & & & \\ 0 & \cdots & & & \end{pmatrix}$$

Example 7.1.6. Consider some piece of equipment which is now in use. When it fails, it is replaced immediately by an identical one. When that one fails it is again replaced by an identical one and so on. Let p_k be the probability that a new item lasts for k units of time, $k = 1,\ 2, \ldots$. Let X_n be the remaining lifetime of the item in use at time n and let Z_{n+1} be the lifetime of the item installed at time n. Then $X_{n+1} = Z_{n+1} - 1$ if $X_n = 0$; and $X_{n+1} = X_n - 1$ if $X_n \geq 1$. Clearly, for the sequence (X_n), the state space $\mathbf{E} = \{0, 1, 2, \ldots\}$. The scheme of replacement implies that the distribution of X_{n+1} depends only on the value of X_n. Hence $(X_n, n \geq 0)$ is a MC. The transition probabilities are as follows.

$$p_{0,j} = P(X_{n+1} = j | X_n = 0) = P(Z_{n+1} - 1 = j | X_n = 0) = p_{j+1};$$

for $i \geq 1$,

$$p_{i,j} = P(X_{n+1} = j | X_n = i) = P(X_n - 1 = j | X_n = i) = \begin{cases} 1 & \text{if } j = i - 1 \\ 0 & \text{if } j \neq i - 1 \end{cases}$$

The transition matrix is

$$P = \begin{pmatrix} p_1 & p_2 & p_3 & p_4 & \ldots \\ 1 & 0 & 0 & 0 & \ldots \\ 0 & 1 & 0 & 0 & \ldots \\ 0 & 0 & 1 & 0 & \ldots \\ & & & & \\ 0 & \ldots & & & \end{pmatrix}$$

7.2 Chapman-Kolmogorov equation

The probabilities $P(X_{n+m} = j | X_n = i)$, if defined, will naturally be called the m-step transition probabilities of the chain. It is natural to ask: for a MC with stationary one-step transition probabilities, will the m-step transition probabilities ($m \geq 2$) be stationary? The answer is *yes* and the proof is by induction on m. Suppose the value of $P(X_{n+m} = j | X_n = i)$ is the same for all values of n for which it is defined. i.e., the m-step transition probabilities are stationary, so we can denote them by $p_{i,j}^{(m)}$ with $p_{i,j}^{(1)} = p_{i,j}$. Then

$$P(X_{n+m+1} = j | X_n = i) = \frac{P(X_{n+m+1} = j, X_n = i)}{P(X_n = i)}$$

$$= \sum_{k \in \mathbf{E}} \frac{P(X_{n+m+1} = j, X_{n+m} = k, X_n = i)}{P(X_n = i)}$$

$$= \sum_{k \in \mathbf{E}}{}' \frac{P(X_{n+m+1} = j, X_{n+m} = k, X_n = i)}{P(X_{n+m} = k, X_n = i)} \times \frac{P(X_{n+m} = k, X_n = i)}{P(X_n = i)}$$

where the summation is over those states k for which $P(X_{n+m} = k,\ X_n = i) > 0$. The general term in the summation can be written

$$P(X_{n+m+1} = j|X_{n+m} = k,\ X_n = i)P(X_{n+m} = k|X_n = i) = \\ P(X_{n+m+1} = j|X_{n+m} = k)P(X_{n+m} = k|X_n = i) \text{ (by } \textbf{(7.1)}\text{b(iii)).}$$

We note the first factor is an one-step transition probability and hence is free from n. The second factor is a m-step transition probability and is free from n by the induction hypothesis. The product then is $p_{k,j}p_{i,k}^{(m)}$. Each term in $\sum_{k\in\mathbf{E}}$ but not in $\sum'_{k\in\mathbf{E}}$ is 0 since $P(X_{n+m} = k,\ X_n = i) = 0$. This implies that for such states k, $P(X_{n+m} = k|X_n = i) = 0$. By the induction hypothesis, this is $p_{i,k}^{(m)}$. Hence this term of value 0 can be written $p_{k,j}p_{i,k}^{(m)}$. Thus $P(X_{n+m+1} = j|X_n = i) = \sum_{k\in\mathbf{E}} p_{k,j}p_{i,k}^{(m)}$, free from n, as was to be proved.

Starting now with $p_{i,j}^{(m+1)} = \sum_{k\in\mathbf{E}} p_{i,k}^{(m)}p_{k,j}$ we arrive, by induction on n, at what is called the Chapman-Kolmogorov equation:

$$p_{i,j}^{(m+n)} = \sum_{k\in\mathbf{E}} p_{i,k}^{(m)}p_{k,j}^{(n)}. \tag{7.11}$$

Writing, for $A \subset \mathbf{E}$, $p_{i,A}^{(n)} = p^{(n)}(i,\ A) = \sum_{j\in A} p_{i,j}^{(n)}$, the following relation is immediate from this:

$$p^{(m+n)}(i,\ A) = \sum_{k\in\mathbf{E}} p_{i,k}^{(m)}p^{(n)}(k,\ A). \tag{7.12}$$

Remark 7.2.1. Let us arrange the m-step transition probabilities in a matrix: $P^{(m)} = (p_{i,j}^{(m)})$. That this is a stochastic matrix when $m = 1$ was already noted. That this is again a stochastic matrix for all m is proved by induction on m : Let $\sum_{k\in\mathbf{E}} p_{i,k}^{(m)} = 1$ for some $m \geq 1$. Then, by (7.11),

$$\sum_{j\in\mathbf{E}} p_{i,j}^{(m+1)} = \sum_{j\in\mathbf{E}}\sum_{k\in\mathbf{E}} p_{i,k}^{(m)}p_{k,j} = \sum_{k\in\mathbf{E}} p_{i,k}^{(m)}\{\sum_{j\in\mathbf{E}} p_{k,j}\} = \sum_{k\in\mathbf{E}} p_{i,k}^{(m)} = 1.$$

Thus $P^{(m)}$ is a stochastic matrix for all $m \geq 1$. Further, since the ith row j^{th} column element in $P^{(m+1)}$ is $p_{i,j}^{(m+1)} = \sum_{k\in\mathbf{E}} p_{i,k}^{(m)}p_{k,j}$, we conclude $P^{(m+1)} = P^{(m)}P$. Successive reduction yields : $P^{(m)} = P^m$.

Remark 7.2.2. To find $p_{i,j}^{(n)}$ we have to raise the matrix P to power n and read off the relevant entry. This is not always easy. In the case of example 2 in **(7.1.3)** the computation is easy and we find, for $j = i,\ i+1,\ ...,\ i+n$, $p_{i,j}^{(n)} = \binom{n}{j-i}p^{j-i}(1-p)^{n-j+i}$.

Remark 7.2.3. For any 2 states $i,\ j$, distinct or not, and any positive integers $m,\ n$,

$$p_{i,j}^{(m+n)} \geq p_{i,j}^{(m)} p_{j,j}^{(n)};\ p_{i,i}^{(m+n)} \geq p_{i,j}^{(m)} p_{j,i}^{(n)} \tag{7.13}$$

Classification of states

Definitions 7.2.4. (i) State i is said to lead to state j ($i \to j$) if there exists $n \geq 1$ such that $p_{i,j}^{(n)} > 0$. If $i \to j$ and $j \to i$ we say i and j communicate with each other ($i \rightleftarrows j$).

We note a state may not communicate with itself, as the state 0 in the MC with state space $\{0,\ 1\}$ and transition probabilities $p_{0,0} = 0,\ p_{0,1} = 1,\ p_{1,0} = 0,\ p_{1,1} = 1$. The claim follows since $P^n = P$.

We note that if $i \neq j$,and $i \rightleftarrows j$ then $i \rightleftarrows i$, since $p_{i,j}^{(m)} > 0$ and $p_{j,i}^{(n)} > 0$ imply $p_{i,i}^{(m+n)} \geq p_{i,j}^{(m)} p_{j,i}^{(n)} > 0$.

Similar arguments lead to : $i \to j,\ j \to k$ imply $i \to k$.

(ii) We say that a class of states $C \subseteq \mathbf{E}$ is closed if $i \in C,\ i \to j$ imply $j \in C$. A state i is called an absorbing state if $\{i\}$ is a closed class. (iii) A collection $A \subset \mathbf{E}$ is a communicating class if all the states in A communicate with one another.

A chain is said to be irreducible if its state space $\mathbf{E}$ is a communicating class.

Consider a MC with state space $\{1,\ 2,\ 3,\ 4,\ 5,\ 6\}$ and transition matrix

$$P = \begin{pmatrix} \frac{1}{2} & \frac{1}{2} & 0 & 0 & 0 & 0 \\ 0 & 0 & 1 & 0 & 0 & 0 \\ \frac{1}{3} & 0 & 0 & \frac{1}{3} & \frac{1}{3} & 0 \\ 0 & 0 & 0 & \frac{1}{2} & \frac{1}{2} & 0 \\ 0 & 0 & 0 & 0 & 0 & 1 \\ 0 & 0 & 0 & 0 & 1 & 0 \end{pmatrix}$$

We notice $p_{1,2} > 0,\ p_{2,3} > 0,\ p_{3,1} > 0,$. Hence $A = \{1,\ 2,\ 3\}$ is a communicating class. $p_{4,4} > 0$. Hence $B = \{4\}$ is a communicating class. $p_{5,6} > 0$ and $p_{6,5} > 0$. Hence $C = \{5, 6\}$ is a communicating class. Neither A nor B nor C is a subset of any bigger communicating class. We note C is the only closed collection which is a communicating class. Trivially, the chain is not irreducible.

(iv) We say a return (of the system) to state i is possible at the nth step if $p_{i,i}^{(n)} > 0$; we say a visit from state i to state j is possible at the nth step if $p_{i,j}^{(n)} > 0$. Return to state i is said to be possible in n steps starting from step

k if $P(X_{k+j} = i,$ for some $j,\ 1 \le j \le n|X_k = i) > 0$. Visit to state j from state i in n steps is similarly defined.

State i is said to be periodic with period d if a return to it is possible only in number of steps that is a multiple of d and is not a multiple of any number larger than d. In other words, $d =$ g.c.f. $\{n : p_{i,i}^{(n)} > 0\}$. If $d = 1$, then we say the state is aperiodic.

Consider the MC in (iii) above. $p_{1,1} > 0$; $p_{4,4} > 0$. Hence states 1 and 4 are aperiodic. $p_{2,2}^{(1)} = 0,\ p_{2,2}^{(2)} = 0,\ p_{2,2}^{(3)} \ge p_{2,3}^{(1)} p_{3,2}^{(2)} \ge p_{2,3}^{(1)} p_{3,1}^{(1)} p_{1,2}^{(1)} > 0$; also $p_{2,2}^{(4)} \ge p_{2,3} p_{3,1} p_{1,1} p_{1,2} > 0$. Hence state 2 is aperiodic. Since $p_{5,5}^{(2n+1)} = 0$ and $p_{5,5}^{(2n)} = 1 > 0$, it follows that state 5 is periodic with period 2.

(v) First return / visit : We recall $p_{i,j}^{(n)}$ is the probability that the system starting from state i reaches or visits state j at the nth step, not necessarily for the first time. We define $f_{i,j}^{(n)}$ to be the probability that the system starting from state i reaches state j at the nth step for the first time. i.e., $f_{i,j} = f_{i,j}^{(1)} = P(X_1 = j|X_0 = i) = p_{i,j}$. For $n \ge 2$, $f_{i,j}^{(n)} = P(X_n = j,\ X_\nu \ne j,\ 1 \le \nu \le n-1|X_0 = i)$. We define $f_{i,j}^{*} = \sum_{j \in \mathbf{E}} f_{i,j}^{(n)} = P(X_\nu = j$ for some $\nu,\ 1 \le \nu < \infty|X_0 = i)$. Thus $f_{i,j}^{*}$ is the probability that the system, starting from state i, ever reaches state j as it evolves through time and is hence less than or equal to 1. The convergence of the series defining $f_{i,j}^{*}$ follows from this.

(vi) If there exists a state i in a $MC(X_n,\ n \ge 0)$ such that $f_{j,i}^{*} = 1$ for all $j \ne i$, then $f_{i,i}^{*} = 1$.

Proof.

$$
\begin{aligned}
f_{i,i}^{*} &= P(X_n = i \text{ for some } n \ge 1|X_0 = i) \\
&= P(X_1 = i|X_0 = i) \\
&\qquad + P(X_n = i,\ \text{for some } n \ge 2,\ X_1 \ne i|X_0 = i) \\
&= P(X_1 = i|X_0 = i) \\
&\qquad + \sum_{j \ne i} P(X_n = i,\ \text{for some } n \ge 2,\ X_1 = j|X_0 = i) \\
&= p_{i,i} + \sum_{j \ne i} p_{i,j} P(X_n = i,\ \text{for some } n \ge 1|X_0 = j) \\
&= p_{i,i} + \sum_{j \ne i} p_{i,j} f_{j,i}^{*} = \sum_{j \in \mathbf{E}} p_{i,j} = 1.
\end{aligned}
$$

□

Let $\tau_{i,j}$ be the time taken by the system to reach state j from state i. Is $\tau_{i,j}$ a rv? It is clear that $\tau_{i,j}$ takes only the values $1,\ 2,\ 3, \ldots$. The set

$\{\omega : \ \tau_{i,j}(\omega) = n\}$ is not well defined. However, conceptually, this is an eminently natural function defined on Ω worthy of investigation. In order to deal with these mappings $\tau_{i,j}$, state i fixed, state j arbitrary, we consider the measure space $(\Omega,\ \mathscr{A}_N,\ P_i)$. Here N is the least integer n for which $P(X_n = i) > 0$; $\mathscr{A}_N$ is the σ-field generated by the rEs $\{X_\nu,\ \nu \geq N\}$; for $A \in \mathscr{A}_N$, $P_i(A) = P(A|X_N = i)$. Define $(\tau_{i,j} = 1) = (X_N = i,\ X_{N+1} = j)$; $(\tau_{i,j} = n) = (X_{N+n} = j,\ X_{N+r} \neq j,\ 1 \leq r \leq N-1,\ X_N = i)$, $n \geq 2$. It is clear that the $\tau_{i,j}$ are rvs defined on the the measure space $(\Omega,\ \mathscr{A}_N,\ P_i)$ and $P_i(\tau_{i,j} = n) = P(\tau_{i,j} = n|X_N = i) = f_{i,j}^{(n)}$.

Let us write τ_i for $\tau_{i,i}$.

Definition 7.2.5. If τ_i is a rv, i.e., if $f_{i,i}^* = \sum_{n=1}^{\infty} f_{i,i}^{(n)} = 1$, i.e., if the probability is 1 that the system, starting from state i, will return to it then i is called a persistent or recurrent state. If $f_{i,i}^* < 1$, then i is called a transient state.

Consider a two state MC with $\mathbf{E} = \{0,\ 1\}$ and transition probabilities $p_{0,0} = 0,\ p_{0,1} = 1,\ p_{1,0} = 0$ and $p_{1,1} = 1$. If P is the transition matrix, as noted earlier (ref. **(7.2.4)**(i)), $P^n = P,\ n \geq 1$. This shows $p_{0,0}^{(n)} = 0, n \geq 1$. Hence $f_{0,0}^{(n)} = 0$, leading to $f_{0,0}^* = 0$. State 0 is thus a transient state. Since $f_{1,1}^{(1)} = p_{1,1}^{(1)} = 1$, it follows that 1 is a persistent state.

Definition 7.2.6. Suppose i is a persistent state. It will be called a (persistent) positive state or an ergodic state if $\mathbb{E}\tau_i < \infty$. It will be called a (persistent) null state if $\mathbb{E}\tau_i = \infty$. i.e., a persistent state i would be a positive state if $\sum_{n=1}^{\infty} n f_{i,i}^{(n)} < \infty$ and a null state if this series diverges.

In **(7.4)**, conditions for a state to be transient, to be persistent, to be persistent positive or null will be given in terms of the probabilities $(p_{i,i}^{(n)})$.

We present an example to show that the return time variable of a persistent null state may possess no finite moment of positive orders.

Let $\mathbf{E} = \{0,\ 1,\ 2, \ldots\}$; $p_{0,0} = 0,\ p_{0,n} = a_n,\ n \geq 1$; $\sum_{n=1}^{\infty} a_n = 1$; $p_{n,n-1} = 1,\ n \geq 1$. Set $a_n = \frac{c}{n(\log(n+1))^2}$. We note that all the states form a communicating class and that $f_{0,0}^{(n+1)} = a_n,\ n \geq 1$. $\mathbb{E}\tau_0^\delta = \sum_{n=2}^{\infty} n^\delta a_{n-1} = \sum_{n=1}^{\infty} (n+1)^\delta \frac{c}{n(\log(n+1))^2} = \infty$. Since $\sum_n f_{0,0}^{(n)} = \sum_n a_n = 1,\ 0$ is a persistent state. Since $\sum_{n=1}^{\infty} n f_{0,0}^{(n)} = \infty$, it is a null state.

For states $i,\ j$ fixed, define $A_1 = (X_1 = j),\ A_t = (X_t = j,\ X_\nu \neq j,\ 1 \leq \nu \leq t-1), t \geq 2$ and note $\sum_t P(A_t|X_0 = i) = f_{i,j}^*$. Let B_ν be the event that the sequence $(X_{\nu+n},\ n \geq 1)$ contains infinitely many js and $B_\nu^{(q)}$ the event that the sequence $(X_{\nu+n}, n \geq 1)$ contains at least q js. We note $B_\nu^{(q)} \downarrow B_\nu$ as

$q \uparrow \infty$. Write $g_{i,j} = P(B_0|X_0 = i)$, $g_{i,j}^{(q)} = P(B_0^{(q)}|X_0 = i)$ and note that $g_{i,j}^{(q)} \downarrow g_{i,j}$ as $q \uparrow \infty$. We have

$$\begin{aligned} g_{i,j}^{(q)} &= P(B_0^{(q)}|X_0 = i) = P(B_0^{(q)} \cap \cup_t A_t|X_0 = i) \\ &= \sum_{t=1}^{\infty} P(B_0^{(q)} \cap A_t|X_0 = i) \\ &= \sum_{t=1}^{\infty} P(B_t^{(q-1)}|A_t,\ X_0 = i)P(A_t|X_0 = i) \\ &= \sum_{t=1}^{\infty} P(B_0^{(q-1)}|X_0 = j)P(A_t|X_0 = i) \\ &= f_{i,j}^* P(B_0^{(q-1)}|X_0 = j). \end{aligned}$$

This argument can be continued and, after q steps, we arrive at : $g_{i,j}^{(q)} = f_{i,j}^*(f_{j,j}^*)^{q-1}$. Hence $g_{i,j} = 0$ if j is a transient state and $g_{i,j} = f_{i,j}^*$ if j is a persistent state. Thus $g_{i,i} = 0$ *iff* i is a transient state state. As the following theorem demonstrates, this derived property of a transient state i sometimes proves more useful than the defining property that $f_{i,i}^* < 1$.

Theorem 7.2.7. *In a finite MC, not all states can be transient.*

Proof. Let $\mathbf{E}$ be a finite set and let all the states be transient. Let $D_j,\ j \in \mathbf{E}$ be the event that the sequence $(X_n,\ n \geq 1)$ contains infinitely many js. We have $1 = P(X_n \in \mathbf{E},\ n \geq 1) \leq P(\cup_j D_j) \leq \sum_{j \in \mathbf{E}} P(D_j)$. Now, $P(D_j) = \sum_{i \in \mathbf{E}} P(D_j \cap (X_0 = i)) = \sum_{i \in \mathbf{E}_1} P(D_j|X_0 = i)P(X_0 = i)$ where $\mathbf{E}_1$ consists of those states i with $P(X_0 = i) > 0$. Thus $P(D_j) = \sum_{i \in \mathbf{E}_1} g_{i,j} P(X_0 = i) = 0$. Since this is true for each $j \in \mathbf{E}$, we reach the contradiction $1 \leq 0$. □

For a proof of this theorem based on the limiting property of the sequence $(p_{i,i}^{(n)})$, ref. **(7.4.2)**(iii). *Also refer* **(7.5.2)**(vi).

Remark 7.2.8. Consider a MC in which all the states are transient. Necessarily $\mathbf{E}$ is an (countably) infinite set. (*For an example of such a MC, refer to* **(7.5.2)**(iv)). We want to record here that even though the system will be in $\mathbf{E}$ forever, the probability that the system will visit any particular state infinitely many times is nil.

7.2.1 Taboo states

We denote by ${}_kp_{i,j}^{(n)}$ $[{}_kf_{i,j}^{(n)}]$ the probability that the system starting from state i and reaching state j at the nth step [for the first time] without visiting state k in the intervening $n-1$ steps. Thus ${}_kp_{i,j}^{(1)} = p_{i,j}^{(1)} = f_{i,j}^{(1)} = {}_kf_{i,j}^{(1)}$. For $n \geq 2$, ${}_kp_{i,j}^{(n)} = P(X_n = j,\ X_\nu \neq k,\ 1 \leq \nu \leq n-1 | X_0 = i)$ and ${}_kf_{i,j}^{(n)} = P(X_n = j,\ X_\nu \neq k,\ X_\nu \neq j, 1 \leq \nu \leq n-1 | X_0 = i)$. We note ${}_jp_{i,j}^{(n)} = f_{i,j}^{(n)}$.

We adopt the convention: for all $i,\ j \in \mathbf{E}$, $i \neq j$, $p_{i,i}^{(0)} = 1$; $p_{i,j}^{(0)} = 0$; $f_{i,i}^{(0)} = 0$; $f_{i,j}^{(0)} = 0$; ${}_kp_{i,i}^{(0)} = 0 = {}_kp_{i,j}^{(0)}$.

7.2.2 A series of relations

$$\text{(i)} \qquad p_{i,j}^{(n)} = \sum_{\nu=1}^{n} f_{i,j}^{(\nu)} p_{j,j}^{(n-\nu)}, \qquad i,\ j \text{ arbitrary.} \tag{7.14}$$

Proof. For $n = 1$, the claimed relation is obvious. For $n = 2$,

$$\begin{aligned} p_{i,j}^{(2)} &= P(X_2 = j | X_0 = i) \\ &= P(X_2 = j,\ X_1 \neq j | X_0 = i) + P(X_2 = j,\ X_1 = j | X_0 = i) \\ &= f_{i,j}^{(2)} + P(X_2 = j | X_1 = j, X_0 = i) P(X_1 = j | X_0 = i) \\ &= f_{i,j}^{(2)} + P(X_2 = j | X_1 = j) f_{i,j}. \end{aligned}$$

This verifies (7.14) for $n = 2$.

Here and in all what follows, if events like $\{X_1 = j,\ X_0 = i\}$ *have zero probability, they must be handled as in the derivation of* (7.11). *Explicit simplification steps will not be presented. We will behave as if the conditioning events have positive probabilities.*

Let $n \geq 3$. The event

$$\begin{aligned} \{X_n = j\} = \{X_n = j,\ X_\nu \neq j,\ 1 \leq \nu \leq n-1\} \\ \cup \{X_n = j,\ X_\nu = j \text{ for some } \nu,\ 1 \leq \nu \leq n-1\}. \end{aligned}$$

Hence

$$\begin{aligned} p_{i,j}^{(n)} &= f_{i,j}^{(n)} + P(X_n = j,\ X_1 = j | X_0 = i) \\ &\quad + \sum_{r=2}^{n-1} P(X_n = j,\ X_r = j,\ X_\nu \neq j,\ 1 \leq \nu \leq r-1 | X_0 = i) \end{aligned}$$

$$= f_{i,j}^{(n)} + f_{i,j}^{(1)} p_{j,j}^{(n-1)}$$

$$+ \sum_{r=2}^{n-1} P(X_n = j | X_r = j,\ X_\nu \neq j,\ 1 \leq \nu \leq r-1,\ X_0 = i)$$

$$\times P(X_r = j,\ X_\nu \neq j,\ 1 \leq \nu \leq r-1 | X_0 = i)$$

$$= f_{i,j}^{(n)} + \sum_{r=1}^{n-1} f_{i,j}^{(r)} p_{j,j}^{(n-r)}$$

$$= \sum_{r=1}^{n} f_{i,j}^{(r)} p_{j,j}^{(n-r)}. \qquad \square$$

Note. For $j = i$, (7.14) would read

$$p_{i,i}^{(n)} = \sum_{\nu=1}^{n} f_{i,i}^{(\nu)} p_{i,i}^{(n-\nu)},\ i \text{ arbitrary.} \tag{7.15}$$

Let d be the greatest common factor ($g.c.f.$) of the numbers in $A = \{n : p_{i,i}^{(n)} > 0\}$ and $\mathfrak{d}$ be that of the numbers in $B = \{n : f_{i,i}^{(n)} > 0\}$. We show now that the above relation (7.15) helps prove

Theorem $d = \mathfrak{d}$.

Proof. Since $B \subset A$, it follows that $\mathfrak{d} \geq d$. By the property of integers, there exist an integer $k \geq 2$ and $n_1 < n_2 < \ldots < n_k$ in A such that n_1 is the smallest member of A and n_2 is the next larger member and so on and such that d is the $g.c.f.$ of these k numbers. If $n_1 = 1$, then $f_{i,i}^{(1)} = p_{i,i}^{(1)} > 0$ and then $d = 1 = \mathfrak{d}$. Let $n_1 \geq 2$. We note that if $n_0 = 0$ and if $n_{r-1} < n < n_r$, $r = 1, 2, \ldots, k$, then $p_{i,i}^{(n)} = 0$ and hence $f_{i,i}^{(n)} = 0$. By (7.15) $p_{i,i}^{(n_1)} = f_{i,i}^{(n_1)}$. Hence $f_{i,i}^{(n_1)} > 0$. i.e. $n_1 \in B$. If $n_1 < n < n_2$, then $p_{i,i}^{(n)} = 0$ and, hence $f_{i,i}^{(n)} = 0$. And then, again by (7.15), $p_{i,i}^{(n_2)} = f_{i,i}^{(n_2)}$. Thus $n_2 \in B$. This argument can be kept up and we conclude $n_r \in B$, $1 \leq r \leq k$. Hence $\mathfrak{d}$ divides each of these k numbers. Since this implies $\mathfrak{d}$ divides d and since, as already noted, $\mathfrak{d} \geq d$, it follows $\mathfrak{d} = d$. $\square$

(ii) $$f_{i,j}^{(1)} = {}_i f_{i,j}^{(1)} \text{ and } f_{i,j}^{(n)} = {}_i f_{i,j}^{(n)} + \sum_{\nu=1}^{n-1} {}_j f_{i,i}^{(\nu)} f_{i,j}^{(n-\nu)},$$

$$n \geq 2,\ i \neq j. \tag{7.16}$$

Proof. The result is obvious for $n = 1$.

For $n = 2$,

$$f_{i,j}^{(2)} = P(X_2 = j,\ X_1 \neq j | X_0 = i)$$

$$\begin{aligned}
&= P(X_2 = j,\ X_1 \neq j,\ X_1 \neq i | X_0 = i) \\
&\qquad + P(X_2 = j,\ X_1 = i | X_0 = i) \\
&= {}_i f_{i,j}^{(2)} + P(X_2 = j | X_1 = i,\ X_0 = i) P(X_1 = i | X_0 = i) \\
&= {}_i f_{i,j}^{(2)} + f_{i,j}^{(1)} f_{i,i}^{(1)} = {}_i f_{i,j}^{(2)} + f_{i,j}^{(1)}\, {}_j f_{i,i}^{(1)}.
\end{aligned}$$

Let $n \geq 3$. Then

$$\begin{aligned}
f_{i,j}^{(n)} &= P(X_n = j,\ X_\nu \neq j,\ 1 \leq \nu \leq n-1 | X_0 = i) \\
&= P(X_n = j,\ X_\nu \neq j,\ X_\nu \neq i,\ 1 \leq \nu \leq n-1 | X_0 = i) \\
&\qquad + P(X_n = j,\ X_\nu \neq j,\ 1 \leq \nu \leq n-1, \\
&\qquad\quad X_\nu = i \text{ for some } \nu,\ 1 \leq \nu \leq n-1 | X_0 = i) \\
&= {}_i f_{i,j}^{(n)} + P(X_n = j,\ X_\nu \neq j,\ 1 \leq \nu \leq n-1,\ X_1 = i | X_0 = i) \\
&\qquad + \sum_{r=2}^{n-1} P(X_n = j,\ X_\nu \neq j,\ 1 \leq \nu \leq n-1,\ X_r = i, \\
&\qquad\quad X_t \neq i,\ 1 \leq t \leq r-1 | X_0 = i) \\
&= {}_i f_{i,j}^{(n)} + f_{i,j}^{(n-1)} f_{i,i}^{(1)} \\
&\quad + \sum_{r=2}^{n-1} P(X_n = j,\ X_\nu \neq j,\ r+1 \leq \nu \leq n-1 \\
&\qquad\quad | X_r = i,\ X_t \neq i,\ 1 \leq t \leq r-1, X_0 = i) \times \\
&\qquad\quad P(X_r = i,\ X_\nu \neq j, X_\nu \neq i,\ 1 \leq \nu \leq r-1 | X_0 = i) \\
&= {}_i f_{i,j}^{(n)} + \sum_{r=1}^{n-1} f_{i,j}^{(n-r)}\, {}_j f_{i,i}^{(r)}, \text{ since } f_{i,i}^{(1)} = {}_j f_{i,i}^{(1)}.
\end{aligned}$$

(7.16) can be written as follows, since the added term is zero.

$$f_{i,j}^{(n)} = {}_i f_{i,j}^{(n)} + \sum_{\nu=1}^{n} {}_j f_{i,i}^{(\nu)} f_{i,j}^{(n-\nu)},\ n \geq 1,\ i \neq j. \tag{7.17}$$

(iii) $f_{i,i}^{(1)} = {}_j f_{i,i}^{(1)}$ and

$$f_{i,i}^{(n)} = {}_j f_{i,i}^{(n)} + \sum_{\nu=1}^{n-1} {}_i f_{i,j}^{(\nu)} f_{j,i}^{(n-\nu)},\ n \geq 2,\ i \neq j. \tag{7.18}$$

Proof. Let $i \neq j$. The result is obvious for $n = 1$.

$$\begin{aligned} f_{i,i}^{(2)} &= P(X_2 = i,\ X_1 \neq i | X_0 = i) \\ &= P(X_2 = i, X_1 \neq i,\ X_1 \neq j | X_0 = i) \\ &\qquad + P(X_2 = i,\ X_1 = j | X_0 = i) \\ &= {}_j f_{i,i}^{(2)} + f_{i,j}^{(1)} f_{j,i}^{(1)} = {}_j f_{i,i}^{(2)} + {}_i f_{i,j}^{(1)} f_{j,i}^{(1)}. \end{aligned}$$

Let $n \geq 3$. Then

$$\begin{aligned} f_{i,i}^{(n)} &= P(X_n = i,\ X_\nu \neq i,\ X_\nu \neq j,\ 1 \leq \nu \leq n-1 | X_0 = i) \\ &\quad + P(X_n = i,\ X_\nu \neq i,\ 1 \leq \nu \leq n-1,\ X_1 = j | X_0 = i) \\ &\quad + \sum_{r=2}^{n-1} P(X_n = i,\ X_\nu \neq i,\ 1 \leq \nu \leq n-1,\ X_r = j, \\ &\qquad\qquad X_t \neq j,\ 1 \leq t \leq r-1 | X_0 = i) \\ &= {}_j f_{i,i}^{(n)} + f_{i,j}^{(1)} f_{j,i}^{(n-1)} \\ &\qquad + \sum_{r=2}^{n-1} P(X_n = i,\ X_\nu \neq i, r+1 \leq \nu \leq n-1 \\ &\qquad\quad X_r = j,\ X_t \neq j,\ X_t \neq i,\ 1 \leq t \leq r-1,\ X_0 = i) \\ &\qquad\quad \times P(X_r = j,\ X_t \neq j,\ X_t \neq i,\ 1 \leq t \leq r-1 | X_0 = i) \\ &= {}_j f_{i,i}^{(n)} + \sum_{r=1}^{n-1} f_{j,i}^{(n-r)}\ {}_i f_{i,j}^{(r)}. \end{aligned}$$

□

(7.18) can be written as follows, since the added term is zero.

$$f_{i,i}^{(n)} = {}_j f_{i,i}^{(n)} + \sum_{\nu=1}^{n} {}_i f_{i,j}^{(\nu)} f_{j,i}^{(n-\nu)},\ n \geq 1,\ i \neq j. \tag{7.19}$$

Each of these four relations is valid for denumerably many values of n. In order to effectively use these infinitely many relations, we introduce generating functions : for arbitrary states $i,\ j,\ k$ and z complex,

$$U_{i,j}(z) = \sum_{n=0}^{\infty} p_{i,j}^{(n)} z^n \tag{7.20}$$

$${}_k U_{i,j}(z) = \sum_{n=1}^{\infty} {}_k p_{i,j}^{(n)} z^n \tag{7.21}$$

$$F_{i,j}(z) = \sum_{n=1}^{\infty} f_{i,j}^{(n)} z^n \tag{7.22}$$

$$_kF_{i,j}(z) = \sum_{n=1}^{\infty} {}_kf_{i,j}^{(n)} z^n. \tag{7.23}$$

Clearly $U_{i,j}$ and $_kU_{i,j}$ are defined at least for $|z| < 1$. The functions $F_{i,j}$ and $_kF_{i,j}$ are defined at least for $|z| \leq 1$. This is because the coefficient of the nth term in each of the series is less than or equal to $f_{i,j}^{(n)}$ and $\sum_{n=1}^{\infty} f_{i,j}^{(n)} = f_{i,j}^* \leq 1$.

The relations (7.14) lead to:

$$U_{i,j}(z) = \delta_{i,j} + F_{i,j}(z)U_{j,j}(z). \tag{7.24}$$

The relations (7.17) lead to:

$$F_{i,j}(z) = {}_iF_{i,j}(z) + {}_jF_{i,i}(z)F_{i,j}(z),\ i \neq j. \tag{7.25}$$

The relations (7.19) lead to:

$$F_{i,i}(z) = {}_jF_{i,i}(z) + {}_iF_{i,j}(z)F_{j,i}(z),\ i \neq j. \tag{7.26}$$

(7.24), (7.25) and (7.26) are true for all z in the unit disc. Theorems in section **(7.7)** depend on the functions involved having unique meromorphic extensions (ref. **Definition (7.8.1)**) to larger discs, with center at the origin, and the validity there of these relations.

For future applications we record here two consequent or derived relations: At least for $|z| < 1$, $1 - {}_jF_{i,i}(z) \neq 0$. Eliminating $_iF_{i,j}(z)$ between (7.25) and (7.26) we have

$$1 - F_{i,j}(z)F_{j,i}(z) = \frac{1 - F_{i,i}(z)}{1 - {}_jF_{i,i}(z)},\ i \neq j. \tag{7.27}$$

Interchanging i and j, we get

$$1 - F_{j,i}(z)F_{i,j}(z) = \frac{1 - F_{j,j}(z)}{1 - {}_iF_{j,j}(z)},\ i \neq j. \tag{7.28}$$

Hence if $i \neq j$, then at least for $|z| < 1$,

$$g_{i,j}(z) = \frac{1 - F_{i,i}(z)}{1 - F_{j,j}(z)} = \frac{1 - {}_jF_{i,i}(z)}{1 - {}_iF_{j,j}(z)}. \tag{7.29}$$

Another useful relation is:

$$p_{i,j}^{(n)} = \sum_{\nu=0}^{n-1} p_{i,i}^{(\nu)}\ {}_ip_{i,j}^{(n-\nu)},\ i \neq j. \tag{7.30}$$

Proof of (7.30).

For $n = 1$, right side is $p_{i,i}^{(0)}\, {}_ip_{i,j}^{(1)} = p_{i,j}^{(1)}$. For $n = 2$, $p_{i,j}^{(2)} = P(X_2 = j|X_0 = i) = P(X_2 = j,\ X_1 = i|X_0 = i) + P(X_2 = j,\ X_1 \neq i|X_0 = i) = p_{i,i}^{(1)}p_{i,j}^{(1)} + {}_ip_{i,j}^{(2)}$, thus verifying the formula in this case. For $n \geq 3$,

$$\begin{aligned} p_{i,j}^{(n)} &= P(X_n = j|X_0 = i) \\ &= \sum_{\nu=0}^{n-2} P(X_n = j,\ X_\nu = i,\ X_r \neq i,\ \nu + 1 \leq r \leq n-1|X_0 = i) \\ &\qquad\qquad + P(X_n = j,\ X_{n-1} = i|X_0 = i) \\ &= \sum_{\nu=0}^{n-1} p_{i,i}^{(\nu)}\, {}_ip_{i,j}^{(n-\nu)}. \end{aligned}$$

□

Relation (7.30) can be written equivalently as:

$$p_{i,j}^{(n)} = \sum_{\nu=0}^{n} p_{i,i}^{(\nu)}\, {}_ip_{i,j}^{(n-\nu)}, \quad i \neq j. \tag{7.31}$$

The relation between the corresponding generating functions would be

$$U_{i,j}(z) = U_{i,i}(z)\, {}_iU_{i,j}(z). \tag{7.32}$$

7.3 Solidarity theorems

These theorems investigate whether a property possessed by one state is automatically shared by the other state if the two states are communicating with each other.

Theorem 7.3.1. *If $i \rightleftarrows j$ then their periods are same.*

Proof. Let $p_{i,j}^{(M)} > 0$, and $p_{j,i}^{(N)} > 0$. Hence $p_{i,i}^{(M+N)} > 0$; $p_{j,j}^{(M+N)} > 0$. It follows that both d_i and d_j divide $M + N$ where d_i, d_j are the periods respectively of i and j. If $p_{i,i}^{(n)} > 0$ then $p_{j,j}^{(n+M+N)} \geq p_{j,i}^{(N)}p_{i,i}^{(n)}p_{i,j}^{(M)} > 0$. Hence d_j divides $n + M + N$. But we have already noted that d_j divides $M + N$. Hence d_j divides n. Since this is true for every n for which $p_{i,i}^{(n)} > 0$, it follows that d_j divides d_i. Similarly d_i divides d_j. Hence $d_i = d_j$. □

Theorem 7.3.2. *Let $i \neq j$.* (i) *If $i \to j$ then ${}_iF_{i,j}(1) > 0$.* (ii) *If $i \to j$ and if ${}_jF_{i,i}(1) > 0$ then $F_{i,j}(1) > {}_iF_{i,j}(1)$.*

Proof. (i) Since $i \to j$, there exists an integer $n \geq 1$ with $f_{i,j}^{(n)} > 0$. Let $m = \min\{\nu : f_{i,j}^{(\nu)} > 0\}$. Then clearly ${}_i f_{i,j}^{(m)} = f_{i,j}^{(m)} > 0$. Since ${}_i F_{i,j}(1) \geq_i f_{i,j}^{(m)}$, the claim follows.

(ii) Since ${}_j F_{i,i}(1) > 0$, there exists $m \geq 1$ such that ${}_j f_{i,i}^{(m)} > 0$. Let N denote the least such integer m. Since $i \to j$, there exists $M \geq 1$ such that $f_{i,j}^{(M)} > 0$. We then have $f_{i,j}^{(N+M)} - {}_i f_{i,j}^{(N+M)} = P(X_{M+N} = j,\ X_\nu \neq j,$ and $X_\nu = i$ for some $\nu,\ 1 \leq \nu \leq M+N-1 | X_0 = i) \geq P(X_{M+N} = j,$ $X_\nu \neq j, 1 \leq \nu \leq M+N-1, X_N = i, X_\nu \neq i, 1 \leq \nu \leq N-1 | X_0 = i) =$ ${}_j f_{i,i}^{(N)} f_{i,j}^{(M)} > 0$, implying the desired result. □

Example 7.3.3. Consider the MC with $\mathbf{E} = \{0,\ 1\}$ and $p_{0,1} = 1 = p_{1,0}$. It is easy to verify $F_{0,1}(1) = 1 = {}_0 F_{0,1}(1)$ and this is no contradiction to (ii) of **(7.3.2)** since ${}_1 F_{0,0}(1) = 0$.

Theorem 7.3.4. *(i) If* $i \to j$ *then* ${}_j F_{i,i}(1) < 1$. *In an irreducible infinite state* MC, *it is possible* $\sup_{\substack{j \in \mathbf{E} \\ j \neq i}} {}_j F_{i,i}(1) = 1$. *(ii) If* $i,\ j$ *are communicating persistent positive states with mean times of return* $\mu_i,\ \mu_j$, *then* $g_{i,j}(1) = \frac{\mu_i}{\mu_j}$.

Proof. (i) By hypothesis and **(7.3.2)**, we have ${}_i F_{i,j}(1) > 0$.
By (7.25), $F_{i,j}(1)\{1 - {}_j F_{i,i}(1)\} = {}_i F_{i,j}(1) > 0$. Hence ${}_j F_{i,i}(1) < 1$.

Let the following be the transition probability matrix of a MC with $\mathbf{E} = \{0,\ 1,\ 2,\ ...\}$:

$$P = \begin{pmatrix} 0 & \frac{1}{2} & \frac{1}{2^2} & \frac{1}{2^3} & . & . & . & & & \\ 1 & 0 & 0 & 0 & . & . & . & & & \\ 0 & 1 & 0 & 0 & 0 & . & . & . & & \\ 0 & 0 & 1 & 0 & 0 & 0 & . & . & . & \\ 0 & 0 & 0 & 1 & 0 & 0 & 0 & . & . & . \\ . & . & & & & & & & & \end{pmatrix},$$

We note the chain is irreduciblle. We have, ${}_1 F_{0,0}(1) = 0$ and for $j \geq 2$, ${}_j F_{0,0}(1) = \sum_{\nu=1}^{j-1} \frac{1}{2^\nu}$. Hence $\sup_{j \neq 0} {}_j F_{0,0}(1) = \sum_{\nu=1}^{\infty} \frac{1}{2^\nu} = 1$.

(ii) We have from (7.29), by continuity,

$$g_{i,j}(1) = \lim_{x \to 1} \frac{1 - F_{i,i}(x)}{1 - F_{j,j}(x)} = \lim_{x \to 1} \frac{\frac{1 - F_{i,i}(x)}{1-x}}{\frac{1 - F_{j,j}(x)}{1-x}} = \frac{\mu_i}{\mu_j}. \qquad \square$$

Theorem 7.3.5. *Let* $i \neq j$. *If* $i \to j$ *but* $j \nrightarrow i$, *then* i *is a transient state.*

Proof. Since $j \nrightarrow i$, $F_{j,i}(1) = 0$. Hence by (7.26) followed by **(7.3.4)**, we have $F_{i,i}(1) = {}_j F_{i,i}(1) + {}_i F_{i,j}(1) F_{j,i}(1) = {}_j F_{i,i}(1)$
< 1. i.e. $f_{i,i}^* < 1$. Hence i is a transient state. □

Theorem 7.3.6. (i) *If i is persistent and if $i \to j$, then $F_{i,j}(1) = 1 = F_{j,i}(1)$, $i \rightleftarrows j$ and j is persistent.*
(ii) *If $i \to j$ and if j is a transient state, then i is a transient state.*

Proof. (i) Since i is a persistent state $F_{i,i}(1) = 1$. By **(7.3.4)**, $i \to j$, implies ${}_jF_{i,i}(1) < 1$. Now appeal to (7.27) to get $1 - F_{i,j}(1)F_{j,i}(1) = 0$. Thus $F_{i,j}(1) = 1 = F_{j,i}(1)$. This implies $i \rightleftarrows j$.

Since $j \to i$, it follows by **(7.3.4)** that ${}_iF_{j,j}(1) < 1$. From (7.28), we then get $F_{j,j}(1) = 1$. i.e., j is a persistent state.

(ii) The claim follows since i persistent and $i \to j$ would imply that j is persistent. □

Remarks 7.3.7. (i) Two communicating states are both persistent or both transient.

(ii) Recall $C \subseteq \mathbf{E}$ is called a closed class or closed collection if $i \in C$ and $j \notin C$ implies $p_{i,j} = 0$ (ref. **(7.2.4)**(ii)).

(iii) The collection C of all persistent states communicating with a particular state, say i, is a closed class. For, $j \in C$, $k \notin C$, $j \to k$ imply (by part (i) of theorem) $j \leftrightarrow k$ and hence (refer **(2.7)**) $i \leftrightarrow k$. This means that $k \in C$, a contradiction.

(iv) If a closed communicating class C consists of a single state i, then necessarily $p_{i,i} = 1$ and i is a persistent state. Such a state i is called an absorbing state.

(v) Let C be a closed communicating class of persistent states. Let $j,\ k \in C$. Let $i \notin C$. Then $f^*_{i,j} = f^*_{i,k}$.

For, let $f^*_{i,j,k}$ denote the probability that the system starting from state i reaches state k after visiting state j. $f^*_{i,k,j}$ is similarly defined. Now

$$\begin{aligned}
f^*_{i,j,k} &= P(X_\nu = j \text{ for some } \nu \geq 1 \text{ and } X_{\nu+n} = k \\
&\qquad\qquad \text{for some } n \geq 1 | X_0 = i) \\
&= P(X_1 = j \text{ and } X_{1+n} = k \text{ for some } n \geq 1 | X_0 = i) \\
&\quad + \sum_{\nu=2}^{\infty} P(X_\nu = j,\ X_t \neq j,\ 1 \leq t \leq \nu - 1 \text{ and } X_{\nu+n} = k \\
&\qquad \text{for some } n \geq 1 | X_0 = i) \\
&= f^{(1)}_{i,j} P(X_{1+n} = k \text{ for some } n \geq 1 | X_1 = j) \\
&\quad + \sum_{\nu=2}^{\infty} f^{(\nu)}_{i,j} P(X_{\nu+n} = k \text{ for some } n \geq 1 | X_\nu = j)
\end{aligned}$$

$$= f_{i,j}^* P(X_n = k \text{ for some } n \geq 1 | X_0 = j) = f_{i,j}^* f_{j,k}^*$$

$= f_{i,j}^*$ since $j,\ k$ are persistent communicating states, ref. **(7.3.6)**. Similarly, $f_{i,k,j}^* = f_{i.k}^*$. Since $f_{i,k}^* \geq f_{i,j,k}^* = f_{i,j}^*$ and since $f_{i,j}^* \geq f_{i,k,j}^* = f_{i,k}^*$, it follows $f_{i,j}^* = f_{i,k}^*$.

Note 1. If $i \notin C$ and if i is a persistent state, then $f_{i,j}^* = 0$ for all $j \in C$. For, $f_{i,j}^* > 0$ implies $i \to j$ and then (ref. **(7.3.6)** (i)) i would be a persistent state, communicating with j. That would imply $i \in C$, a contradiction.

Note 2. If $j,\ k$ are two distinct communicating transient states and if i is a third transient state, then it is not always true that $f_{i,j}^* = f_{i,k}^*$. To see this, consider a MC with $\mathbf{E} = \{1,\ 2,\ 3,\ 4\}$ and

$$P = \begin{pmatrix} 0 & 1 & 0 & 0 \\ 0 & 0 & \frac{1}{2} & \frac{1}{2} \\ 0 & \frac{1}{2} & 0 & \frac{1}{2} \\ 0 & 0 & 0 & 1 \end{pmatrix},$$

We note $2 \leftrightarrow 3$. Both are transient states, since $2 \to 4$, an absorbing state. $1 \to 2$, a transient state. Hence 1 is a transient state. Now, $f_{1,2}^* = 1$, since $f_{1,2}^* \geq f_{1,2}^{(1)} = p_{1,2} = 1$. $f_{1,3}^{(1)} = 0$; $f_{1,3}^{(2)} = \frac{1}{2}$; $f_{1,3}^{(n)} = 0,\ n \geq 3$. Hence $f_{1,3}^* = \frac{1}{2}$. Thus $f_{1,2}^* \neq f_{1,3}^*$.

When C is a closed communicating class of persistent states, this common value $f_{i,j}^*$, same for all $j \in C$, will be denoted by $f_{i,C}$. Since the system, once reaching any state in C, can not leave the class C, we call $f_{i,C}$ the probability that the system, starting in state i, is aborbed in C. If $i \notin C$, is a persistent state, then $f_{i,C}^* = 0$. Equations whose solutions will be these absorption probabilities for $i \in T$, the collection of all transient states, will be developed in **(7.5)**.

Any state space $\mathbf{E}$ can be written $\mathbf{E} = T \cup \cup_\nu C_\nu$ where T is the collection of all the transient states and each C_ν is a closed communicating class of persistent states. If the system is in any C_ν at some time, then it will be in that C_ν forever afterwards, since C_ν is closed. If the system is in T and if T is a finite set, then (i) $\mathbf{E} \sim T$ is not empty and (ii) the system can not be in T forever (both by **(7.2.4)**(v)). i.e., the system must leave T and must reach or get absorbed in a C_ν.

$T = \mathbf{E}$, an infinite set, is possible (ref. **(7.4.11)**, case $p \neq q$).

If $T_1 \subseteq T$ is the collction of all transient states i such that i leads to no state $j \in T_1'$, then the matrix $(p_{i,j},\ i,\ j \in T_1)$ would be an infinite stochastic matrix. The system once in T_1 will continue to be in T_1 forever. Though consisting only of transient states, T_1 will be a closed class. It is possible for the system to start from $i \in \mathcal{T} = T \sim T_1$ and get absorbed in T_1. For $i \in \mathcal{T}$, we denote by f^*_{i,T_1}, in analogy with closed communicating class C, the probability that the system starting in i eventually gets absorbed in T_1. But it is generally not true that $f^*_{i,T_1} = f^*_{i,k}$ for any $k \in T_1$.

To enhance understanding, an example is presented now.

Let $\mathbf{E} = \{\mathfrak{a}, 0, 1, 2, \ldots\}$; $a_i > 0$, $i = 1, 2, \ldots$, $\sum_{r=1}^{\infty} a_r = \frac{1}{2}$, $\sum_{r=1}^{\infty} a_{2r-1} = \alpha$, $\beta = \frac{1}{2} - \alpha$; $p_{\mathfrak{a},0} = 1$, $p_{0,\mathfrak{a}} = \frac{1}{2}$, $p_{0,i} = a_i$; $p_{1,3} = 1$; $p_{2i+1,2i+3} = 1 - (\frac{1}{3})^i$, $p_{2i+1,1} = (\frac{1}{3})^i$, $i = 1, 2, 3, \ldots$; $p_{2,4} = 1$, $p_{2i,2i+2} = \frac{1}{2} = p_{2i,2}$, $i = 2, 3, 4, \ldots$. Let $T = \{1, 3, 5, \ldots\}$ and $C = \{2, 4, 6, \ldots\}$. We note that $\mathbf{E} = \{\mathfrak{a}, 0\} \cup T \cup C$, that T and C are disjoint closed sets of communicating states, that $f^{(1)}_{1,1} = 0$, that $f^{(2)}_{1,1} = \frac{1}{3}$, $f^{(n)}_{1,1} = \{\prod_{\nu=1}^{n-2}(1 - (\frac{1}{3})^\nu)\}(\frac{1}{3})^{n-1}$, $n \geq 3$, that $f^*_{1,1} < \frac{1}{3} + \sum_{n=3}^{\infty} \frac{1}{3^{n-1}} < \frac{1}{2}$, that $f^{(1)}_{2,2} = 0$, $f^{(n)}_{2,2} = \frac{1}{2^{n-1}}$, $n \geq 2$ and that $f^*_{2,2} = 1$. Thus T is a closed communicating class of transient states and C is a closed communicating class of persistent states. 0 is a transient state since it leads to the state $2 \in C$ but does not communicate with it. $\mathfrak{a}$ is a transient state since it communicates with state 0. Since $p_{\mathfrak{a},0} = 1$, $f^*_{\mathfrak{a},T} = f^*_{0,T}$ and $f^*_{\mathfrak{a},C} = f^*_{0,C}$. If $f^{(n)}_{0,T}$ is the probability that the system starting from state 0 enters T at the nth step then $f^{(2n)}_{0,T} = 0$ and $f^{(2n-1)}_{0,T} = \frac{1}{2^{n-1}}\alpha$, $n = 1, 2, \ldots$, leading to $f^*_{0,T} = 2\alpha$. Similarly, $f^*_{0,C} = 2\beta$. The probability $\hat{f}^{(n)}_{0,1}$ that the system enters T at the nth step and reach state 1, some time, is 0 if n is even. $\hat{f}^{(2n-1)}_{0,1} = \frac{1}{2^{n-1}}\{a_1 + \sum_{i=1}^{\infty} a_{2i+1} f^*_{2i+1,1}\}$. By **(7.2.4)**(vi) there exists at least one i such that $f^*_{2i+1,1} < 1$. Hence $\hat{f}^{(2n-1)}_{0,1} < \frac{1}{2^{n-1}}\{a_1 + \sum_{i=1}^{\infty} a_{2i+1}\} = \frac{1}{2^{n-1}}\alpha$, leading to $f^*_{0,1} < \alpha \sum_1^{\infty} \frac{1}{2^{n-1}} = 2\alpha$. Parallel steps will show that $f^*_{0,2} = 2\beta$. Thus $f^*_{0,2} = f^*_{0,C}$ while $f^*_{0,1} \neq f^*_{0,T}$.

Sections **(7.3.8)** through **(7.3.11)** study the return time variables τ_i, τ_j of two communicating persistent states i, j. We show that if the moment of a certain order of either of them is finite, then so is that of the other. Then we find the relation between the moments of the two variables. Since the variables are positive valued, use of Laplace transforms is found to be the right tool.

Let ψ_i, ψ_j denote the Laplace transforms of τ_i and τ_j the return time variables of the two communicating persistent states i and j and note that $\psi_i(s) = F_{i,i}(e^{-s})$ and $\psi_j(s) = F_{j,j}(e^{-s})$.

Write

$$g(s) = g_{i,j}(e^{-s}) = \frac{1 - {}_jF_{i,i}(e^{-s})}{1 - {}_iF_{j,j}(e^{-s})}$$

and note that

$$0 < \alpha = 1 - {}_jF_{i,i}(1) \leq g(s) \leq \frac{1}{1 - {}_iF_{j,j}(1)} = \beta < \infty.$$

Relation (7.29) becomes

$$1 - \psi_i(s) = g(s)\{1 - \psi_j(s)\}, \ s \geq 0. \tag{7.33}$$

We note that for any persistent state i, the nth differential coefficient at 0, $\psi_i^{(n)}(0)$ exists finitely *iff* $\mathbb{E}\tau_i^n < \infty$. If the condition holds then $\mathbb{E}\tau_i^n = (-1)^n \psi_i^{(n)}(0)$.

When $\psi_i^{(n)}(0)$ exists finitely (or, equivalently, when $\mathbb{E}\tau_i^n < \infty$), define

$$\Lambda_i^{(n)}(s) = (-1)^{n+1}\{\psi_i(s) - \sum_{j=0}^{n} \frac{s^j}{\underline{|j}} \psi_i^{(j)}(0)\}. \tag{7.34}$$

Let $\mathbb{E}\tau_i^n < \infty$, $n \geq 0$. Then, recall **(3.10.2)**, a necessary and sufficient condition for the finiteness of $\mathbb{E}\tau_i^{n+1}$ is : $\mathbb{E}\tau_i^n < \infty$ and $\lim_{s\to 0} \frac{\Lambda_i^{(n)}(s)}{s^{n+1}}$ exists finitely. And then the limit is equal to $\frac{1}{\underline{|n+1}} \mathbb{E}\tau_i^{n+1}$.

Lemma 7.3.8. *Let τ_i, τ_j be as above. Let $n \geq 1$ be an integer. If $\mathbb{E}\tau_i^n < \infty$ and $\mathbb{E}\tau_j^n < \infty$, then*

(i) the nth differential coefficient $g^{(n)}(0)$ of $g(s)$ at $s = 0$ exists finitely and

(ii) $\mathbb{E}\tau_i^n = \sum_{\nu=0}^{n-1} (-1)^\nu \binom{n}{\nu} g^{(\nu)}(0) \mathbb{E}\tau_j^{n-\nu}$.

Proof. (i) We note

$$(\mathbb{E}\tau_i^k < \infty) \Leftrightarrow (\sum_{n=1}^{\infty} n^k f_{i,i}^{(n)} < \infty) \Rightarrow (\sum_{n=1}^{\infty} n^k \ {}_j f_{i,i}^{(n)} < \infty).$$

Hence ${}_jF_{i,i}(e^{-s})$ has a finite k^{th} order differential coefficient at $s = 0$, namely, $\sum_{n=1}^{\infty} (-1)^k n^k \ {}_j f_{i,i}^{(n)} < \infty$. Since it has been given that $\mathbb{E}\tau_j^k < \infty$, we argue, on similar lines and claim that ${}_iF_{j,j}(e^{-s})$ has a finite k^{th} order differential coefficient at $s = 0$. These two facts and the fact that $1 - {}_iF_{j,j}(e^{-s}) > 0$ for $s \geq 0$ imply that $g(s)$ has a finite differential coefficient of order k at $s = 0$.

(ii) This claim is immediate on differentiating both sides of the relation $1 - \psi_i(s) = g(s)\{1 - \psi_j(s)\}$ n times at $s = 0$. □

Theorem 7.3.9. *Let i, j, τ_i, τ_j, ψ_i, ψ_j be as above. Let integer $n \geq 1$. If $\mathbb{E}\tau_i^n < \infty$, then $\mathbb{E}\tau_j^n < \infty$.*

Proof. We establish the theorem by induction on n. By (7.34),

$$(\mathbb{E}\tau_i < \infty) \Leftrightarrow \left(\lim_{s\to 0} \frac{\Lambda_i^{(0)}(s)}{s} < \infty \right) \Leftrightarrow \left(\lim_{s\to 0} \frac{1 - \psi_i(s)}{s} < \infty \right)$$

$$\Leftrightarrow \left(\lim_{s\to 0} g(s)\frac{1-\psi_j(s)}{s} < \infty\right) \text{ (by (7.33))}$$

$$\Leftrightarrow \left(g(0)\lim_{s\to 0}\frac{1-\psi_j(s)}{s} < \infty\right) \Leftrightarrow (g(0)\mathbb{E}\tau_j < \infty).$$

Further $\mathbb{E}\tau_i = g(0)\mathbb{E}\tau_j$, since $g(0) > 0$, the relation checking with the promised one at (ii). Thus the theorem stands proved for $n = 1$. We note that, by the Lemma above, $g(s)$ has a finite differential coefficient at $s = 0$.

[This result is also available at **(7.3.4)**(ii)]

Suppose the theorem is true for $n = 1, 2, \ldots, m$. We note that, by the above Lemma, $g^{(\nu)}(0)$ exists finitely for $1 \le \nu \le m$.

Further, for $1 \le k \le m$,

$$\mathbb{E}\tau_i^k = \sum_{\nu=0}^{k-1}(-1)^\nu\binom{k}{\nu}g^{(\nu)}(0)\mathbb{E}\tau_j^{k-\nu}.$$

Let now $\mathbb{E}\tau_i^{m+1} < \infty$.

We then have

$$\left(\mathbb{E}\tau_i^{m+1} < \infty\right) \Leftrightarrow \Big(\lim_{s\to 0}\frac{\Lambda_i^{(m)}(s)}{s^{m+1}} \text{ exists finitely}\Big)$$

$$\Leftrightarrow \Big(\lim_{s\to 0}\frac{1}{s^{m+1}}\{\psi_i(s) - 1 - \sum_{r=1}^{m}(-1)^r\frac{s^r}{\underline{|r}}\mathbb{E}\tau_i^r\} \text{ exists finitely}\Big)$$

$$\Leftrightarrow \Big(\lim_{s\to 0}\frac{1}{s^{m+1}}[g(s)\{\psi_j(s) - 1\}$$

$$-\sum_{r=1}^{m}(-1)^r\frac{s^r}{\underline{|r}}\sum_{\nu=0}^{r-1}(-1)^\nu\binom{r}{\nu}g^{(\nu)}(0)\mathbb{E}\tau_j^{r-\nu}] \text{ exists finitely}\Big)$$

$$\Leftrightarrow \Big(\lim_{s\to 0}\frac{1}{s^{m+1}}[g(s)\{\psi_j(s) - 1 - \sum_{r=1}^{m}(-1)^r\frac{s^r}{\underline{|r}}\mathbb{E}\tau_j^r\}$$

$$+ g(s)\sum_{r=1}^{m}(-1)^r\frac{s^r}{\underline{|r}}\mathbb{E}\tau_j^r$$

$$-\sum_{r=1}^{m}\frac{s^r}{\underline{|r}}\sum_{\nu=1}^{r}(-1)^\nu\binom{r}{\nu}g^{(r-\nu)}(0)\mathbb{E}\tau_j^\nu] \text{ exists finitely}\Big)$$

$$\Leftrightarrow \Big(\lim_{s\to 0}\frac{1}{s^{m+1}}[(-1)^{m+1}g(s)\Lambda_j^{(m)}(s) + g(s)\sum_{r=1}^{m}(-1)^r\frac{s^r}{\underline{|r}}\mathbb{E}\tau_j^r$$

$$-\sum_{r=1}^{m}\frac{s^r}{\underline{|r}}\{\sum_{\nu=1}^{r}(-1)^\nu\binom{r}{\nu}g^{(r-\nu)}(0)\mathbb{E}\tau_j^\nu\}] \text{ exists finitely }\Big)$$

$$\Leftrightarrow \left(\lim_{s\to 0} \frac{1}{s^{m+1}}[(-1)^{m+1}g(s)\Lambda_j^{(m)}(s) + A_m(s)] \text{ exists finitely } \right)$$

where

$$A_m(s) = g(s)\sum_{r=1}^{m}(-1)^r \frac{s^r}{\underline{|r}}\mathbb{E}\tau_j^r$$
$$- \sum_{r=1}^{m} \frac{s^r}{\underline{|r}}\{\sum_{\nu=1}^{r}(-1)^\nu \binom{r}{\nu} g^{(r-\nu)}(0)\mathbb{E}\tau_j^\nu\}.$$

In $A_m(s)$, coefficient $a_{\nu;m}(s)$ of $(-1)^\nu \mathbb{E}\tau_j^\nu$ is

$$\frac{s^\nu}{\underline{|\nu}}g(s) - \sum_{r=\nu}^{m}\binom{r}{\nu} g^{(r-\nu)}(0)\frac{s^r}{\underline{|r}} = \frac{s^\nu}{\underline{|\nu}}\{g(s) - \sum_{r=\nu}^{m} g^{(r-\nu)}(0)\frac{s^{r-\nu}}{\underline{|r-\nu}}\}$$
$$= \frac{s^\nu}{\underline{|\nu}}\{g(s) - \sum_{r=0}^{m-\nu} g^{(r)}(0)\frac{s^r}{\underline{|r}}\}.$$

As $s \to 0$, this is asymptotically

$$\frac{s^\nu}{\underline{|\nu}}\frac{s^{m+1-\nu}}{\underline{|m+1-\nu}}g^{(m+1-\nu)}(0) = \frac{s^{m+1}}{\underline{|\nu}\underline{|m+1-\nu}}g^{(m+1-\nu)}(0).$$

This is so since $1 \le m+1-\nu \le m$ for $1 \le \nu \le m$ and since, as already noted, $g^{(m)}(0)$ exists finitely. Thus $\lim_{s\to 0}\frac{1}{s^{m+1}}A_m(s)$ exists finitely and equals $\sum_{\nu=1}^{m}\frac{(-1)^\nu}{\underline{|\nu}\underline{|m+1-\nu}}g^{(m+1-\nu)}(0)\mathbb{E}\tau_j^\nu$. The existence finitely of this limit implies $\lim_{s\to 0}(-1)^{m+1}g(s)\frac{\Lambda_j^{(m)}(s)}{s^{m+1}}$ exists finitely which, in turn, implies the existence finitely of $\mathbb{E}\tau_j^{m+1}$.

To get the relation between $\mathbb{E}\tau_i^{m+1}$ and the $\mathbb{E}\tau_j^\nu$s we can appeal to part (ii) of the above Lemma or we can utilise the preceeding steps as follows.

$$\frac{1}{\underline{|m+1}}\mathbb{E}\tau_i^{m+1} = \lim_{s\to 0}\frac{\Lambda_i^{(m)}(s)}{s^{m+1}}$$
$$= (-1)^{m+1}\lim_{s\to 0}\frac{1}{s^{m+1}}\{(-1)^{m+1}g(s)\Lambda_j^{(m)}(s) + A_m(s)\}$$
$$= (-1)^{m+1}\{(-1)^{m+1}g(0)\frac{1}{\underline{|m+1}}\mathbb{E}\tau_j^{m+1}$$
$$+ \sum_{\nu=1}^{m}\frac{(-1)^\nu}{\underline{|\nu}\underline{|m+1-\nu}}g^{(m+1-\nu)}(0)\mathbb{E}\tau_j^\nu\}.$$

Hence

$$\mathbb{E}\tau_i^{m+1} = g(0)\mathbb{E}\tau_j^{m+1} + \sum_{\nu=1}^{m}(-1)^{m+1-\nu}\binom{m+1}{\nu}g^{(m+1-\nu)}(0)\mathbb{E}\tau_j^{\nu}$$
$$= \sum_{\nu=1}^{m+1}(-1)^{m+1-\nu}\binom{m+1}{\nu}g^{(m+1-\nu)}(0)\mathbb{E}\tau_j^{\nu}. \qquad \square$$

Theorem 7.3.10. *Let i, j be two communicating persistent states with return time variables denoted by τ_i and τ_j. Let m be an arbitrary non-negative integer and let $m < p < m+1$. Then $\mathbb{E}\tau_i^p < \infty \Leftrightarrow \mathbb{E}\tau_j^p < \infty$.*

Proof. Let ψ_i, ψ_j be the Laplace transforms of τ_i and τ_j. We recall (ref. **(3.10.3)**) that $\mathbb{E}\tau_i^p < \infty$ *iff* $\mathbb{E}\tau_i^m < \infty$ and $I_p^{(i)} = \int_0^\infty \frac{\Lambda_i^{(m)}(s)}{s^{1+p}}ds < \infty$. When this is true

$$I_p^{(i)} = c_p E\tau_i^p$$

$$\text{where } c_p = \int_0^\infty \frac{1}{u^{1+p}}(-1)^{m+1}\{e^{-u} - \sum_{\nu=0}^{m}\frac{(-1)^\nu}{\underline{|\nu}}u^\nu\}du.$$

We note $(\mathbb{E}\tau_i^p < \infty) \Rightarrow (\mathbb{E}\tau_i^m < \infty) \Rightarrow$ (by the just proved result) $(\mathbb{E}\tau_j^m < \infty)$. Now,

$$(\mathbb{E}\tau_j^p < \infty) \Leftrightarrow (\int_0^\infty \frac{\Lambda_j^{(m)}(s)}{s^{1+p}}ds < \infty)$$
$$\Leftrightarrow (\int_0^\infty \frac{(-1)^{m+1}[g(s)(-1)^{m+1}\Lambda_j^{(m)}(s) + A_m(s)]}{s^{1+p}}ds < \infty)$$

where $A_m(s) = \sum_{\nu=1}^{m}(-1)^\nu \mathbb{E}\tau_j^\nu a_{\nu;m}(s)$ and $a_{\nu;m}(s)$ is as in the last theorem. We noted there that, as $s \to 0$, $a_{\nu;m}(s) = O(s^{m+1})$. Hence $\frac{a_{\nu;m}(s)}{s^{1+p}}$ is integrable over $[0, 1]$. If $s \geq 1$, $|a_{\nu;m}(s)| \leq qs^m$. This implies, since $p - m > 0$, that $\frac{a_{\nu;m}(s)}{s^{1+p}}$ is integrable over $[1, \infty)$. Thus $\int_0^\infty \frac{A_m(s)}{s^{1+p}}ds$ exists finitely. It now follows $\int_0^\infty \frac{\Lambda_j^{(m)}(s)}{s^{1+p}}ds$ exists finitely. This is equivalent to the existence finitely of $\mathbb{E}\tau_j^p$.

[The above steps further show that
$c_p\mathbb{E}\tau_i^p = \int_0^\infty \frac{1}{s^{1+p}}[g(s)\Lambda_j^{(m)}(s) + (-1)^{m+1}A_m(s)]ds$]. $\square$

Theorem 7.3.11. *Let τ_i, τ_j be the return time variables of the communicating persistent states i and j. Let their Laplace transforms be ψ_i and ψ_j. Let $\int_0^x P(\tau_i > y)\,dy = L_i(x)$; and $\int_0^x P(\tau_j > y)dy = L_j(x)$. As $x \to \infty$, $\lim_{x\to\infty} L_i(x) = \infty$ and L_i is slowly varying at ∞ iff L_j has these properties.*

Proof. We note that both L_i and L_j are monotonic increasing functions. Suppose $\lim_{x\to\infty} L_i(x) = \infty$. Then, $\lim_{x\to\infty} L_j(x) = \infty$, as otherwise $(\lim_{x\to\infty} L_j(x) < \infty) \Leftrightarrow (\mathbb{E}\tau_j < \infty) \Leftrightarrow (\mathbb{E}\tau_i < \infty) \Leftrightarrow \lim_{x\to\infty} L_i(x) < \infty$, a contradiction.

Let L_i be sv at ∞. Refer to **(3.10.8)** to see that $\int_0^\infty e^{-\lambda x}\, dL_i(x)$ exists for all $\lambda > 0$ and is equal to $\frac{1-\psi_i(\lambda)}{\lambda}$. By **Q**12(ii), it now follows that $\frac{1-\psi_i(\lambda)}{\lambda}$ is sv at 0. This implies by (7.33) that $g(\lambda)\frac{1-\psi_j(\lambda)}{\lambda}$ is sv at 0. That $\frac{1-\psi_j(\lambda)}{\lambda}$ is sv at 0 follows from this, since $\lim_{\lambda\to 0} g(\lambda) = g(0)$ exists, finite and positive. We now appeal again to **Q**$12(ii)$ and conclude L_j is sv. □

7.4 Long term behaviour of the transition probabilities

In this section, limits of the sequences $(p_{i,j}^{(n)},\ n \geq 0)$ are studied and the limits are related to the first moments of the return time variables. Parameters connected with the functions $U_{i,j}$ are related to parameters connected with the functions $F_{i,j}$. As promised in **(7.2.4)**(v), conditions in terms of the probabilities $(p_{i,j}^{(n)})$ are found for a state to be transient or persistent or persistent positive or persistent null.

Lemma 7.4.1. (i) *If $a_n \geq 0$, $n \geq 0$ and if $f(x) = \sum_{n=0}^\infty a_n x^n$ exists finitely for $0 \leq x < 1$, then $\lim_{x\to 1} f(x) = s$ exists finitely iff $\sum_{n=0}^\infty a_n < \infty$ and then s would be the sum of this series.*

(ii) *Let the a_ns be arbitrary real numbers, not necessarily non-negative and let $\lim_{n\to\infty} n a_n = 0$. Then $\lim_{x\to 1} f(x) = s$ implies $\sum_n a_n = s$.*

Proof. (i) In the forward part, the result is true also for complex series. Given $\varepsilon > 0$, we can find $N = N(\varepsilon)$ such that $\sum_{n=N+1}^\infty a_n < \varepsilon$. And then, for all z, $|z| \leq 1$, $|\sum_{n=N+1}^\infty a_n z^n| \leq \sum_{n=N+1}^\infty a_n < \varepsilon$. We have,

$$\begin{aligned}|\sum_{n=0}^{\infty} a_n z^n - \sum_{n=0}^{\infty} a_n| &\leq |\sum_{n=0}^{N} a_n z^n - \sum_{n=0}^{N} a_n| + |\sum_{n=N+1}^{\infty} a_n z^n| \\ &\quad + \sum_{n=N+1}^{\infty} a_n \\ &< |\sum_{n=0}^{N} a_n z^n - \sum_{n=0}^{N} a_n| + 2\varepsilon.\end{aligned}$$

This inequality holds for all z, $|z| \leq 1$. The claim that $\lim_{z\to 1} \sum_{n=0}^\infty a_n z^n = \sum_{n=0}^\infty a_n$ follows now by taking (term by term) limit as $z \to 1$ and noting that $\varepsilon > 0$ is arbitrary.

Conversely, suppose now f is defined by the infinite series for $0 \le x < 1$ and suppose $\lim_{x\to 1} f(x)$ is finite. If the claim that $\sum_n a_n < \infty$ is not conceded, assume the series diverges. Now, for every $k \ge 1$, $\lim_{x\to 1} f(x) \ge \lim_{x\to 1}\sum_{n=0}^{k} a_n x^n = \sum_{n=0}^{k} a_n \to \infty$ as $k \to \infty$, a contradiction.

(ii) For a proof of this part, refer pp 10-11, [T]. □

Theorem 7.4.2. (i) *i is a transient state if and only if $\sum_n p_{i,i}^{(n)} < \infty$. Equivalently, i is a persistent state if and only if $\sum_n p_{i,i}^{(n)} = \infty$.*
(ii) *If $j(\neq i)$ is a transient state, then $\sum_n p_{i,j}^{(n)} < \infty$.*
(iii) *All the states of a finite MC can not be transient.*
(iv) *If i is a persistent state and if $i \to j$, then $\sum_n p_{i,j}^{(n)} = \infty$.*

Proof. (i) (7.24) gives

$$U_{i,i}(x) = 1 + F_{i,i}(x)U_{i,i}(x), \quad 0 \le x < 1.$$
$$\text{Or, } U_{i,i}(x)\{1 - F_{i,i}(x)\} = 1.$$
$$\text{Or, } U_{i,i}(x) = \frac{1}{1 - F_{i,i}(x)}, \quad 0 \le x < 1.$$

By definition, i is a transient state if

$$0 < 1 - f_{i,i}^* = 1 - \sum_{n=1}^{\infty} f_{i,i}^{(n)} = \lim_{x\uparrow 1}\{1 - F_{i,i}(x)\}$$
$$= \lim_{x\uparrow 1}\frac{1}{U_{i,i}(x)}.$$

Thus

$$\lim_{x\uparrow 1} U_{i,i}(x) = \frac{1}{1 - f_{i,i}^*},$$

positive and finite. Since the terms are non-negative, an appeal to **(7.4.1)** yields

$$\lim_{x\uparrow 1} U_{i,i}(x) = \sum_n p_{i,i}^{(n)} = \frac{1}{1 - f_{i,i}^*}.$$

The steps are reversible and the claim follows.
(ii) By (7.24), $U_{i,j}(x) = F_{i,j}(x)U_{j,j}(x),\ 0 \le x < 1$. By (i), $U_{j,j}(1) < \infty$. Since $U_{i,j}(x) \le U_{j,j}(x),\ 0 \le x \le 1$, the claim follows.
(iii) Suppose the states are finite in number and suppose all the states are transient. Fix $i \in \mathbf{E}$. By (ii), for every $j \in \mathbf{E}$, $\sum_n p_{i,j}^{(n)} < \infty$. Hence for every

$j \in \mathbf{E}$, $\lim_{n\to\infty} p_{i,j}^{(n)} = 0$. But, if $\mathbf{E} = \{1, 2, ..., K\}$, we have for every $n \geq 1$, $1 = \sum_{j=1}^{K} p_{i,j}^{(n)} = \lim_{n\to\infty} \sum_{j=1}^{K} p_{i,j}^{(n)}$ and hence, as $n \to \infty$, $1 = \sum_{j=1}^{K} \lim_{n\to\infty} p_{i,j}^{(n)} = \sum_{j=1}^{K} 0 = 0$, a contradiction.

For a proof of this result based on the probability of infinitely many returns to a state, refer to theorem **(7.2.4)**(v).

Let T be the set of all the transient states and let Q be the matrix $(p_{i,j})$, i, $j \in T$. *For a proof based on the invertibility of* $I - Q$, *refer to* **(7.5.2)**(v).

(iv) The case $i = j$ is covered by part (i). Assume now $i \neq j$; i persistent and $i \to j$ implies (ref. **(7.3.6)**(i)) $f_{i,j}^* = 1 = f_{j,i}^*$. By (7.24), $U_{i,j}(x) = F_{i,j}(x)U_{j,j}(x)$, $0 \leq x < 1$. As $x \to 1$, $F_{i,j}(x) \to F_{i,j}(1) = 1$ and, by **(7.4.1)** and **(7.4.2)**(i), $\lim_{x\to 1} U_{j,j}(x) = \sum_n p_{j,j}^{(n)} = \infty$. Thus $\lim_{x\to 1} U_{i,j}(x) = \infty$. Again appeal to **(7.4.1)** and conclude $\sum_n p_{i,j}^{(n)} = \infty$. □

The following result from real analysis will be needed to prove **(7.4.3)** *where we present a condition for a persistent state* i *to be positive in terms of the Cesaro limit of the sequence* $(p_{i,i}^{(n)})$.

Q14 [Hardy-Littlewood Theorem](p226, [T]) Let $a_n \geq 0$, $n \geq 0$; $s_n = \sum_{m=0}^{n} a_m$. As $x \to 1-$, $f(x) = \sum_{n=0}^{\infty} a_n x^n \sim \frac{c}{1-x}$, $c > 0$ iff $s_n = \sum_{\nu=0}^{n} a_\nu \sim cn$ as $n \to \infty$.

Theorem 7.4.3. *A persistent state* i *is positive if and only if the sequence* $(\frac{1}{n}\sum_{\nu=0}^{n} p_{i,i}^{(\nu)})$ *of Cesaro sums has a limit c which is positive.*

Proof. We recall that i is a positive state if and only if $\mathbb{E}\tau_i = \mu_i < \infty$. We note $\mu_i \geq 1$.

Let $\mu_i < \infty$. Hence $\frac{1}{\mu_i}$ is a finite and positive number. Now

$$\mu_i = \sum_{n=1}^{\infty} n f_{i,i}^{(n)} = F'_{i,i}(1) = \lim_{x\to 1-} \frac{1 - F_{i,i}(x)}{1 - x}.$$

Or,

$$\frac{1}{\mu_i} = \lim_{x\to 1-} \frac{1 - x}{1 - F_{i,i}(x)} = \lim_{x\to 1-} (1 - x)U_{i,i}(x).$$

i.e., as $x \to 1-$, $U_{i,i}(x) \sim \frac{1}{\mu_i}\frac{1}{1-x}$. By **Q**14, it now follows that

$$\lim_{n\to\infty} \frac{1}{n} \sum_{m=0}^{n} p_{i,i}^{(m)} = \frac{1}{\mu_i}.$$

The steps are reversible and the converse follows. □

Remark 7.4.4. The proof that the Cesaro limit is zero in the null state case will follow from **(7.4.7)** below. The proof on the above lines is not possible. However this much can be asserted: if i is a null state and if the Cesaro limit $\lim_{n\to\infty} \frac{1}{n}\sum_{m=0}^{n} p_{i,i}^{(m)}$ exists then it has to be zero.

Theorem 7.4.5. (i) *If i has period d, then $p_{i,i}^{(nd)} > 0$ for all n large.*
(ii) *Let $j \to i$ and let i be aperiodic. Then $p_{j,i}^{(n)} > 0$ for all large n.*
(iii) *If a finite MC is irreducible and aperiodic, then there exists integer $N \geq 1$ such that $p_{i,j}^{(n)} > 0$ for all $i,\ j \in \mathbf{E}$ and all $n \geq N$.*

Proof. (i) Let $S = \{n\ :\ p_{i,i}^{(n)} > 0\}$. We will show $nd \in S$ for all n large By the definition of period, every member of S is an integral multiple of d. Since $p_{i,i}^{(m+n)} \geq p_{i,i}^{(m)} p_{i,i}^{(n)}$, it follows that S is closed under addition. Hence from the properties of integers, there exist $M,\ N\ \in A$ such that $d = M - N$. i.e, there exists an integer a such that $a,\ a+d\ \in S$.

If $n \geq \frac{a^2}{d^2}$, then nd can be written $nd = ma + t$ for some $m \geq \frac{a}{d}$ and some $t,\ 0 \leq t < a$. If $t = 0$, $nd = ma \in S$. Let $0 < t < a$. Since nd and a are divisible by d, it follows that t can be written $t = rd$, $r < \frac{a}{d} \leq m$. Since $a,\ a+d\ \in S$, it follows $(m-r)a + r(a+d) \in S$. i.e., $ma + rd \in S$. i.e. $nd \in S$ for all $n \geq \frac{a^2}{d^2}$.

(ii) By part (i), i aperiodic implies $p_{i,i}^{(n)} > 0$ for all large n. $j \to i$ implies that there exists m such that $p_{j,i}^{(m)} > 0$. That $p_{j,i}^{(m+n)} > 0$ for all n large follows now from the inequality $p_{j,i}^{(m+n)} \geq p_{j,i}^{(m)} p_{i,i}^{(n)}$.

(iii) Since the chain is irreducible, any two states $i,\ j$ communicate with each other. Since the states are aperiodic, it follows, by part(ii), that to each pair of states $i,\ j$, there corresponds an integer $N_{i,j}$ such that $p_{i,j}^{(n)} > 0$ for all $n \geq N_{i,j}$. For $N = \max_{i,\ j\in\mathbf{E}} N_{i,j}$, it is clear that the claim holds true. □

Remark 7.4.6. If state i has period d, then $p_{i,i}^{(n)} = 0$ if n is not divisible by d. Also $p_{i,i}^{(nd)}$ may not be positive for all n (ref. example in **(7.2.4)**(iv). Our theorem above states that $p_{i,i}^{(nd)} > 0$ for all n large. This result holds whether i is a transient state or a persistent positive state or a persistent null state. We proceed now to examine if $\alpha = \lim_{n\to\infty} p_{i,i}^{(nd)}$ exists and, if it does, what its connection is, if any, to the return time variable. If i is a transient state, we know (ref. **(7.4.2)**) $\sum_n p_{i,i}^{(n)} < \infty$. Hence in this case $\lim_{n\to\infty} p_{i,i}^{(n)} = 0$. Case of persistent state remains. Let i be a persistent positive state and let the limit α exist. Given n, find $k = k(n)$ such that $kd \leq n < (k+$

$1)d$. By **(7.4.3)**, $\frac{1}{\mu_i} = \lim_{n\to\infty} \frac{1}{n}\sum_{\nu=0}^{n} p_{i,i}^{(\nu)} = \lim_{n\to\infty} \frac{1}{n}\sum_{\nu=0}^{k} p_{i,i}^{(\nu d)} = \lim_{n\to\infty} \frac{k}{n}\frac{1}{k}\sum_{\nu=0}^{k} p_{i,i}^{(\nu d)} = \frac{1}{d}\alpha$. Thus in this case $\alpha = \frac{d}{\mu_i}$.

Theorem 7.4.7. *If i is a persistent state with period d then $\lim_{n\to\infty} p_{i,i}^{(nd)}$ exists and is equal to 0 if i is a null state and is equal to $\frac{d}{\mu_i}$ if i is a positive state.*

Proof We prove the theorem under the assumption $f_{i,i}^{(1)} > 0$. This implies (and is more restrictive than demanding) $d = 1$.

Write $u_n = p_{i,i}^{(n)}$ and $f_n = f_{i,i}^{(n)}$ and $\mu_i = \mu$. Then (7.14) can be written

$$u_n = u_0 f_n + u_1 f_{n-1} + \cdots + u_n f_0. \tag{7.35}$$

$$\text{Let} \qquad r_n = \sum_{n+1}^{\infty} f_n,\ n \geq 0. \tag{7.36}$$

We note that $r_0 = 1$ and that $\sum_0^\infty r_n = \mu$, which is equal to ∞ if i is a null state. Now $u_n = \sum_{\nu=1}^{n} f_\nu u_{n-\nu} = \sum_{\nu=1}^{n} u_{n-\nu}\{r_{\nu-1} - r_\nu\}$
or, $u_n + \sum_{\nu=1}^{n} r_\nu u_{n-\nu} = \sum_{\nu=1}^{n} u_{n-\nu} r_{\nu-1}$
or, $w_n = \sum_{\nu=0}^{n} r_\nu u_{n-\nu} = \sum_{\nu=0}^{n-1} r_\nu u_{n-1-\nu} = w_{n-1}$. Continuing this equality we arrive at $w_n = w_0 = r_0 u_0 = 1$. That is, for all $n \geq 0$,

$$\sum_{\nu=0}^{n} r_\nu u_{n-\nu} = 1. \tag{7.37}$$

Let $\lambda = \overline{\lim_{n\to\infty}}\ u_n$. Let $m = m(n)$ be a subsequence such that $\lim_{m\to\infty} u_m = \lambda$. Given $\varepsilon > 0$ find $k = k(\varepsilon)$ such that $\sum_{\nu=k}^{\infty} f_\nu < \varepsilon$ and such that $u_n \leq \lambda + \varepsilon$ for all $n \geq k$. Now, for all $m \geq 2k$,

$$\begin{aligned}
\lambda &= \lim_{m\to\infty} u_m \\
&= \lim_{m\to\infty} \sum_{\nu=1}^{m} f_\nu u_{m-\nu} = \lim_{m\to\infty} \{f_1 u_{m-1} + \sum_{\nu=2}^{m} f_\nu u_{m-\nu}\} \\
&\leq \underline{\lim}_{m\to\infty} \{f_1 u_{m-1} + \sum_{\nu=2}^{k} f_\nu u_{m-\nu} + \sum_{\nu=k+1}^{m} f_\nu u_{m-\nu}\} \\
&\leq \underline{\lim}_{m\to\infty} \{f_1 u_{m-1} + \sum_{\nu=2}^{k} f_\nu (\lambda+\varepsilon) + \sum_{\nu=k+1}^{\infty} f_\nu\} \\
&\leq \underline{\lim}_{m\to\infty}\ f_1 u_{m-1} + (1 - f_1)(\lambda + \varepsilon) + \varepsilon,
\end{aligned}$$

leading to $\lambda f_1 \leq f_1 \lim\limits_{m\to\infty} u_{m-1}$. Now $(f_1 > 0) \Rightarrow (\lim\limits_{m\to\infty} u_{m-1} \geq \lambda)$. Since λ is the limit superior of the sequence (u_n), it follows that $\lim\limits_{m\to\infty} u_{m-1}$ exists and is equal to λ. This argument can be kept up a finite number times to get $\lim\limits_{m\to\infty} u_{m-k} = \lambda$ every $k \geq 1$.

Fix k. By (7.37), $1 \geq \sum_{\nu=0}^{k} r_\nu u_{m-\nu}$. Take limit as $m \to \infty$ and get $1 \geq \lambda \sum_{\nu=0}^{k} r_\nu$.

If i is a null state, $\sum_0^\infty r_\nu = \infty$. Hence $\lambda = 0$. Thus in this case $p_{i,i}^{(n)} \to 0$.

If i is a positive state, the above steps give $\lambda \leq \frac{1}{\mu}$.

Let $\underline{\lim}_{n\to\infty} u_n = \theta$. Let $t = t(n)$ be a subsequence such that $\lim\limits_{t\to\infty} u_t = \theta$. On the same lines as for the limit superior it can be shown that $\lim\limits_{t\to\infty} u_{t-k} = \theta$ for every $k \geq 1$. From (7.37) we have

$$1 \leq \sum_{\nu=0}^{k} r_\nu u_{t-\nu} + \sum_{\nu=k+1}^{\infty} r_\nu.$$

On letting $t \to \infty$, this leads to

$$1 \leq \theta \sum_{\nu=0}^{k} r_\nu + \sum_{\nu=k+1}^{\infty} r_\nu.$$

Now let $k \to \infty$ and obtain $\theta \geq \frac{1}{\mu}$. This with the previous inequality gives

$$\lim_{n\to\infty} p_{i,i}^{(n)} = \frac{1}{\mu}. \qquad \square$$

Theorem 7.4.8. *Let j ($\neq i$) be a persistent aperiodic state and i an arbitrary state leading to j, $(i \to j)$. Then $\lim\limits_{n\to\infty} p_{i,j}^{(n)} = 0$ or $\frac{1}{\mu_j} f_{i,j}^*$ according as j is a null state or a positive state.*

Proof. We have (ref. (7.14)):

$$p_{i,j}^{(n)} = \sum_{\nu=1}^{n} f_{i,j}^{(\nu)} p_{j,j}^{(n-\nu)}.$$

Given $\varepsilon > 0$, find s such that $\sum_{\nu=s+1}^{\infty} f_{i,j}^{(\nu)} < \varepsilon$. Hence for all $n > s$,

$$p_{i,j}^{(n)} \leq \sum_{\nu=1}^{s} f_{i,j}^{(\nu)} p_{j,j}^{(n-\nu)} + \varepsilon.$$

If j is a null state, this gives

$$\overline{\lim_{n\to\infty}}\, p_{i,j}^{(n)} \leq 0 + \epsilon$$

thus completing the proof in this case. If j is a persistent positive state, this gives

$$\overline{\lim_{n\to\infty}}\, p_{i,j}^{(n)} \leq \sum_{\nu=1}^{s} f_{i,j}^{(\nu)} \frac{1}{\mu_j} + \epsilon.$$

Let $s \to \infty$ and conclude that

$$\overline{\lim_{n\to\infty}}\, p_{i,j}^{(n)} \leq \frac{1}{\mu_j} f_{i,j}^{*}.$$

Since for every $n > s \geq 1$,

$$p_{i,j}^{(n)} \geq \sum_{\nu=1}^{s} f_{i,j}^{(\nu)} p_{j,j}^{(n-\nu)},$$

we get

$$\overline{\lim_{n\to\infty}}\, p_{i,j}^{(n)} \geq \frac{1}{\mu_j} \sum_{\nu=1}^{s} f_{i,j}^{(\nu)}.$$

Now let $s \to \infty$ and conclude that

$$\underline{\lim}_{n\to\infty}\, p_{i,j}^{(n)} \geq \frac{1}{\mu_j} f_{i,j}^{*}.$$

The two inequalities prove the claim. □

Remark 7.4.9. (i) When j is a positive state, the Cesaro limit of the sequence $(p_{i,j}^{(n)})$ can be obtained directly from **(7.4.3)** as follows. We recall (7.24) $U_{i,j}(x) = F_{i,j}(x)U_{j,j}(x)$, $0 \leq x < 1$. Hence $\lim_{x\to 1}(1-x)U_{i,j}(x) = F_{i,j}(1)\lim_{x\to 1}(1-x)U_{j,j}(x) = f_{i,j}^{*}\frac{1}{\mu_j}$. The desired result follows now from **Q**14.

(ii) Let the MC be irreducible and aperiodic, with all states persistent positive. By **(7.3.6)**, $f_{i,j}^{*} = 1$ for all choice of states i, j. And then $\lim_{n\to\infty} p_{i,j}^{(n)} = \pi_j = \frac{1}{\mu_j}$. The only way to find the π_js at this stage is to find explicitly the $p_{j,j}^{(n)}$s and take limits. (This is not always easy. Later, in section **(7.6.3)**, the method of finding the π_js as solutions of certain linear equations will be presented). In the case of the following infinite state MC, the $p_{i,j}^{(n)}$s are obtainable explicitly. Let **E** $= \{0,\ 1,\ 2, \ldots\}$ and let

$$P = \begin{pmatrix} \frac{1}{2} & \frac{1}{2^2} & \frac{1}{2^3} & \frac{1}{2^4} & . & . & . \\ 1 & 0 & 0 & 0 & . & . & . \\ 1 & 0 & 0 & 0 & 0 & . & . & . \\ 1 & 0 & 0 & 0 & 0 & 0 & . & . & . \\ 1 & 0 & 0 & 0 & 0 & 0 & 0 & . & . & . \\ . & . \end{pmatrix} \tag{7.38}$$

Let $P^n = (p_{i,j}^{(n)})$, $i,\ j \in \mathbf{E}$. Every $p_{i,j}^{(n)}$ is of the type $\frac{a_{i,j}(n)}{2^{b_{i,j}(n)}}$ where $a_{i,j}(n)$, $b_{i,j}(n)$, $n = 1,\ 2,\ 3, \ldots$ are positive integers. We have, for $n = 1, 2, 3, \ldots$, $a_{0,0}(2n) = 1 + \frac{2}{3}(2^{2n} - 1)$; $a_{0,0}(2n-1) = 1 + \frac{4}{3}(2^{2n-2} - 1)$; $a_{0,j}(n) = a_{0,0}(n-1), j \geq 1, n \geq 2$; $a_{j,0}(n) = a_{0,0}(n-1), j \geq 1, n \geq 2$; $a_{i,j}(1) = 0$, $i \geq 1, j \geq 1$; $a_{i,j}(n) = a_{0,j}(n-1), i \geq 1, j \geq 1, n \geq 2$.

$b_{0,j}(n) = j + n$, $j = 0,\ 1,\ 2, \ldots$; $b_{j,0}(n) = n - 1$, $j \geq 1$; $b_{i,j}(n) = j + n - 1, i \geq 1, j \geq 1$.

$f_{0,0}^{(1)} = \frac{1}{2}$; $f_{0,0}^{(2)} = \frac{1}{2}$. Hence $f_{0,0}^{*} = 1$. Further $\mu_0 = \sum_{n=1}^{\infty} n f_{0,0}^{(n)} = \frac{3}{2}$. Hence 0 is a persistent positive state. Since all the states form a communicating class, it follows by **(7.3.6)** and subsection **(7.3.8)** that all the states are persistent positive. For $n \geq 1$, $p_{0,0}^{(n)} = \frac{2}{3} + (-1)^n \frac{1}{3}\frac{1}{2^n}$ leading to $\lim_{n\to\infty} p_{0,0}^{(n)} = \pi_0 = \frac{2}{3}$.

Let $j \geq 1$. We note $p_{j,j}^{(1)} = 0$; if $n \geq 2$, $p_{j,j}^{(n)} = \frac{1}{3}\frac{1}{2^j} + (-1)^n \frac{1}{3}\frac{1}{2^j}\frac{1}{2^{n-1}}$. This gives $\lim_{n\to\infty} p_{j,j}^{(n)} = \frac{1}{3}\frac{1}{2^j}$.

Collected here are all the transition probabilites for ready reference:

$$\begin{aligned}
p_{0,0}^{(n)} &= \frac{2}{3} + (-1)^n \frac{1}{3}\frac{1}{2^n}, \quad n \geq 1; \\
p_{0,j}^{(n)} &= \frac{1}{3}\frac{1}{2^j} + (-1)^{n-1} \frac{1}{3}\frac{1}{2^{j+n}}, \quad j \geq 1,\ n \geq 1 \\
p_{j,0}^{(n)} &= \frac{2}{3} + (-1)^{n-1} \frac{1}{3}\frac{1}{2^{n-1}}, \quad j \geq 1,\ n \geq 1; \\
p_{i,j}^{(1)} &= 0, \quad i \geq 1,\ j \geq 1; \\
p_{i,j}^{(n)} &= \frac{1}{3}\frac{1}{2^j} + (-1)^n \frac{2}{3}\frac{1}{2^{j+n}}, \quad n \geq 2,\ i \geq 1,\ j \geq 1.
\end{aligned}$$

We will revisit this example in section **(7.7.8)** and in subsection **(7.8.8)**.

Theorem 7.4.10. *In a finite MC no state can be a persistent null state.*

Proof. If possible let i be a persistent null state and let $\mathbf{E}$ be finite. Let $\mathbf{E}_1 \subset \mathbf{E}$ be the collection of all the states that i leads to (and hence, i communicates with (ref. **(7.3.6)**). We note $i \in \mathbf{E}_1$. [Necessarily, $\mathbf{E}_1$ contains states other than

i, since $\mathbf{E}_1$ consisting of only i means that i does not lead to any state other than itself. That would imply $p_{i,i} = 1$, which, in turn, would imply that i is a persistent positive state]. For all $j \in \mathbf{E} \sim \mathbf{E}_1$, $p_{i,j}^{(n)} = 0$ for all n; for $j \in \mathbf{E}_1$, $\lim_{n\to\infty} p_{i,j}^{(n)} = 0$, refer **(7.4.8)**. Since $1 = \sum_{j\in E} p_{i,j}^{(n)}$ for every $n \geq 1$ and since limit term by term can be taken, we arrive at a contradiction, namely $1 = 0$, thereby establishing the claim. □

Note 7.4.11. Examples of infinite irreducible MCs all of whose states are (i) transient and (ii) persistent null.

Let $(X_n,\ n \geq 1)$ be a sequence of independent and identically distributed Bernoulli variables, $P(X_1 = 1) = p$, $P(X_1 = -1) = q$. Let $S_0 \equiv 0$, $S_n = \sum_{\nu=1}^{n} X_n$, $n \geq 1$. (S_n) is a MC with state space the set of all the integers and with stationary transition probabilities. The chain is irreducible and all the states periodic with period 2. Further $p_{0,0}^{(2n+1)} = 0$, $n \geq 0$; $p_{0,0}^{(2n)} = \binom{2n}{n} p^n q^n$, $n \geq 1$. Since $p_{0,0}^{(2n)} \sim$ const. $\frac{1}{\sqrt{n}}(4pq)^n$, it follows that $\sum_n p_{0,0}^{(2n)}$ is convergent or divergent according as $p \neq \frac{1}{2}$ or $p = \frac{1}{2}$. Thus all the states are transient if $p \neq \frac{1}{2}$ and all the states are persistent if $p = \frac{1}{2}$. Also when $p = \frac{1}{2}$, these are null since $\lim_{n\to\infty} p_{i,i}^{(2n)} = \lim_{n\to\infty} \frac{\text{const.}}{\sqrt{n}} = 0$.

For $p = \frac{1}{2}$, $p_{0,0}^{(2n)} = \binom{2n}{n}\frac{1}{2^{2n}} = (-1)^n\binom{-\frac{1}{2}}{n}$. Hence $U_{0,0}(x) = (1-x^2)^{-\frac{1}{2}}$. It follows $1 - F_{0,0}(x) = (1-x^2)^{\frac{1}{2}}$. Let $0 < \delta < 1$. By **(3.10.3)** the return time variable τ_0 will have finite moment of order δ *iff* $\int_0^\infty \frac{1-F_{0,0}(e^{-s})}{s^{1+\delta}} ds < \infty$. i.e. *iff* $\int_0^\infty \frac{(1-e^{-2s})^{\frac{1}{2}}}{s^{1+\delta}} ds < \infty$. Since, as $s \to 0$, $\frac{(1-e^{-2s})^{\frac{1}{2}}}{s^{1+\delta}} \sim \frac{\sqrt{2}}{s^{\frac{1}{2}+\delta}}$, we conclude that the cited integral is finite *iff* $\delta < \frac{1}{2}$. i.e., $\mathbb{E}\tau_0^\delta < \infty$ *iff* $\delta < \frac{1}{2}$.

For another example of an irreducible MC with all states persistent null, refer **(7.2.4)**(v).

7.5 Absorption probabilities

As promised in **(7.3.7)**(v), we now develop a method of finding the absorption probabilities $f_{i,C}$, $i \in T$ (writing $f_{i,C}$ for $f_{i,C}^*$) corresponding to a specified closed communicating class C of persistent states. Write α_i for $f_{i,C}$. Let β_i, $i \in T$, be the probability that the system starting in state i stays in T forever. If T is finite, then that $\beta_i = 0$, $i \in T$ has already been noted. (ref. **(7.2.7)** or **(7.4.2)**(iii))

Let T be infinite. For $i \in T$, define $a_i^{(n)}$ to be the probability that the system, starting from state $i \in T$, enters C for the first time at step n. Then,

$a_i^{(1)} = \sum_{j \in C} p_{i,j}$, $i \in T$.

Theorem 7.5.1. (i) *The* β_i, $i \in T$ *satisfy the following equations:*

$$y_i = \sum_{j \in T} p_{i,j} y_j, \quad i \in T. \tag{7.39}$$

(ii) *The* α_i, $i \in T$ *satisfy the following equations:*

$$x_i = \sum_{j \in T} p_{i,j} x_j + a_i^{(1)}, \quad i \in T. \tag{7.40}$$

Proof. (i) Define $\beta_i^{(n)}$, $n = 1, 2, \ldots$ to be the probability that the system, starting from state $i \in T$ is in T at time n. Clearly

$$\beta_i^{(n+1)} = \sum_{\nu \in T} p_{i,\nu} \beta_\nu^{(n)}.$$

We note

$$\begin{aligned}
\beta_i^{(1)} &= \sum_{\nu \in T} p_{i,\nu} \le 1. \\
\beta_i^{(2)} &= \sum_{\nu \in T} p_{i,\nu} \beta_i^{(1)} \le \sum_{\nu \in T} p_{i,\nu} \\
&= \beta_i^{(1)}.
\end{aligned}$$

Suppose we have successively established $\beta_i^{(t)} \le \beta_i^{(t-1)}$, $1 \le t \le n$, $\beta_i^{(0)} = 1$. Then

$$\beta_i^{(n+1)} = \sum_{\nu \in T} p_{i,\nu} \beta_i^{(n)} \le \sum_{\nu \in T} p_{i,\nu} \beta_i^{(n-1)} = \beta_i^{(n)}.$$

Since $\beta_i^{(n)} \downarrow \beta_i$, the desired result follows.

(ii) Recall that $a_i^{(1)}$ is as already defined.

$$\begin{aligned}
a_i^{(2)} &= P(X_2 \in C,\ X_1 \in T | X_0 = i) \\
&= \sum_{j \in T} P(X_2 \in C,\ X_1 = j | X_0 = i) = \sum_{j \in T} p_{i,j} a_j^{(1)}.
\end{aligned}$$

Similar arguments lead to

$$a_i^{(n+1)} = \sum_{j \in T} p_{i,j} a_j^{(n)}, \quad n \ge 2.$$

For all $n \geq 1$ and for all $i \in T$, the $a_i^{(n)}$s can be calculated using the above recursive relations. We note $\alpha_i = \sum_{n=1}^{\infty} a_i^{(n)}$. Adding the above relations for $n = 1, 2, \ldots$, we get $\alpha_i - a_i^{(1)} = \sum_{j \in S} p_{i,j}\alpha_j$. Thus the α_i, $i \in T$ satisfy the equations (7.40). □

Remarks 7.5.2. (i) The linear equations at (7.39) are homogeneous and hence the null solution is always a solution. If the null solution is the only solution, then necessarily each $\beta_i = 0$. i.e., whatever be the state in T where the system starts, it will, $wp1$, get absorbed in some closed class of states.

The question arises: if (7.39) admits of a non-null bounded solution (and hence a non-null bounded-by-one solution), then which solution represents the β_is? The null solution or the non-null solution? Let $(y_i,\, i \in T)$ be a non-null solution. We have

$$|y_i| \leq \sum_{j \in S} p_{i,j}|y_j| \leq \sum_{j \in S} p_{i,j} = \beta_i^{(1)}.$$

If for some $n \geq 1$, $|y_i| \leq \beta_i^{(n)}$, $i \in T$, then

$$|y_i| \leq \sum_{j \in T} p_{i,j}|y_j| \leq \sum_{j \in T} p_{i,j}\beta_j^{(n)} = \beta_i^{(n+1)}.$$

Taking $n \to \infty$, we get $|y_i| \leq \beta_i$. Thus $(\beta_i,\, i \in T)$ is the maximal solution of the equations at (7.39).

(ii) The equations at (7.40) will have a unique solution *iff* the null solution is the only solution of the equations at (7.39).

If (7.40) has multiple solutions, then the following arguments show that the α_is constitute the minimal solution.

Denote $P(.|X_0 = i)$ by $P_i^*(.)$. We note that for $i \in T$, $\alpha_i = P_i^*(X_n \in C \text{ for some } n \geq 1)$.

Let $(x_i \geq 0,\, i \in T)$ be a solution of (7.40). For $i \in T$,

$$\begin{aligned}
x_i &= \sum_{j \in C} p_{i,j} + \sum_{j \in T} p_{i,j}x_j \\
&= P_i^*(X_1 \in C) + \sum_{j \in T} p_{i,j}x_j \\
&= P_i^*(X_1 \in C) + \sum_{j \in T} p_{i,j}\{\sum_{k \in C} p_{j,k} + \sum_{k \in T} p_{j,k}x_k\} \\
&= P_i^*(X_1 \in C) + P_i^*(X_1 \in T,\, X_2 \in C) + \sum_{j \in T}\Big\{\sum_{k \in T} p_{j,k}x_k\Big\}.
\end{aligned}$$

This argument can be kept up indefinitely. After $n \geq 2$ steps, we arrive at $x_i \geq P_i^*(X_1 \in C) + \sum_{\nu=2}^{n} P_i^*(X_\nu \in C,\ X_t \notin C,\ 1 \leq t \leq \nu - 1) = P_i^*(X_\nu \in C \text{ for some } \nu \leq n)$. Take limit as $n \to \infty$ to get $x_i \geq \lim_{n\to\infty} P_i^*(X_\nu \in C \text{ for some } \nu \leq n) = P_i^*(X_\nu \in C \text{ for some } \nu \geq 1) = \alpha_i$.

An example

Let $\mathbf{E} = \{0,\ 1,\ 2, \ldots\}$. Let $p_{0,0} = 1$, $p_{i,i+1} = p$, $p_{i,i-1} = q$; $p, q \geq 0$, $p + q = 1$, $i \geq 1$. To find α_i, $i \in T = \mathbf{E} \sim \{0\}$. The α_is constitute the minimal non-negative solution to the equations

$$x_1 = px_2 + q, \quad x_i = px_{i+1} + qx_{i-1}. \tag{7.41}$$

The characteristic polynomial corresponding to this difference equation is $c(t) = pt^2 - t + q$. The roots of the equation $c(t) = 0$ are $t = 1$ and $t = \frac{q}{p}$. Hence (7.41) has the general solution $x_i = A + B(\frac{q}{p})^i$ when $q \neq p$ and $x_i = A + Bi$ when $q = p$ where A, B are constants to be determined. If $q \geq p$, the bounded solution is $x_i = A$. The requirement $x_1 = px_2 + q$ leads to $A = 1$. Thus if $q \geq p$, then $x_i = 1$. If $q < p$, the requirement $x_1 = px_2 + q$ leads to $A + B = 1$. Thus in this case $x_i = A + (1 - A)(\frac{q}{p})^i = (\frac{q}{p})^i + A\{1 - (\frac{q}{p})^i\}$. For x_i to be non-negative for all i, it is necessary that $A \geq 0$. The minimal non-negative solution is obtained by taking $A = 0$ and then $x_i = (\frac{q}{p})^i$.

A second example

Let $\mathbf{E}$, T be as in the last example. Let $p_{0,0} = 1$, $p_{i,i+1} = p_i$, $p_{i,i-1} = q_i$, $p_i > 0$, $q_i > 0$, $p_i + q_i = 1$. To find α_i, $i \in T$. The α_is constitute the non-negative minimal solution to the equations

$$x_1 = p_1x_2 + q_1, \quad x_i = p_ix_{i+1} + q_ix_{i-1}. \tag{7.42}$$

The method of solution of the last example fails here. Let $x_0 = 1$; for $i \in T$ write $y_i = x_{i-1} - x_i$. Writing (7.42) in terms of the y_is, we get $p_iy_{i+1} = q_iy_i$. Hence $y_{i+1} = \lambda_i y_1$ where $\lambda_i = \prod_{\nu=1}^{i} \frac{q_\nu}{p_\nu}$. This relation is true also for $i = 0$ if we set $\lambda_0 = 1$. Since

$$1 - x_i = \sum_{\nu=1}^{i} y_\nu = y_1 \sum_{\nu=0}^{i-1} \lambda_\nu,$$

we have

$$x_i = 1 - y_1 \sum_{\nu=0}^{i-1} \lambda_\nu = 1 - y_1\Lambda_i,$$

say. Two cases arise. If $\lim_\nu \Lambda_\nu = \infty$, then to get a non-negative bounded-by-one solution we must set $y_1 = 0$ as otherwise x_i for i large will be negative or larger than 1. So setting, we see that (7.42) has the unique solution : $x_i = 1$. If $\lim_{i\to\infty} \Lambda_i = \Lambda < \infty$, then the solutions x_i of (7.42) which lie between 0 and 1 are $x_i = 1 - a\Lambda_i$ where $a \geq 0$ and $a \leq \frac{1}{\Lambda_i}$. Amongst these, the minimal solution is obtained by choosing for a its maximum permissible value. Now $x_i \geq 0$ *iff* $a \leq \frac{1}{\Lambda_i}$. Since this inequality has to hold for all i, it follows that $a \leq \frac{1}{\Lambda}$ and thus $\alpha_i = 1 - \frac{\Lambda_i}{\Lambda}$.

A numerical illustration

Let

$$p_i = \frac{(i+1)^2}{i^2 + (i+1)^2}.$$

We have

$$\Lambda = \sum_{i=0}^{\infty} \lambda_i, \quad \lambda_i = \prod_{\nu=1}^{i} \frac{\nu^2}{(\nu+1)^2} = \frac{1}{(i+1)^2}$$

$$\text{and } \Lambda = \sum_{i=0}^{\infty} \frac{1}{(i+1)^2} = \frac{\pi^2}{6}.$$

It now follows

$$\alpha_1 = 1 - \frac{\Lambda_1}{\Lambda} = 1 - \frac{3}{2\pi^2}.$$

This can be phrased that the probability is $\frac{3}{2\pi^2}$ that the system starting from state 1 will remain forever in T.

(iii) For the validity of the discussion in the early part of (ii), it is not necessary that T be transient states and that C be closed communicating class of persistent states. T and C as assumed are natural choices. When C is any closed class, we can not talk of $f^*_{i,C}$ but we talk of α_i as the probability of absorption in C. If C is any arbitrary collection of states, we talk of α_i as the probability of hitting C.

(iv) Theorem **(7.5.1)**(i) can be used, as follows, to decide if the states of an irredicble MC are all transient or not:

Let the irreducible MC have state space $\mathbf{E} = \{0, 1, 2, \ldots\}$ and let $T = \{1, 2, 3, \ldots\}$. Then the states are transient *iff* the system of equations

$$y_i = \sum_{j=1}^{\infty} p_{i,j} y_j, \quad i \in T \tag{7.43}$$

possesses a non-null bounded solution.

Proof. Let $MC2$ be a MC with state space **E** and transition probability matrix $(r_{i,j})$, $i,\ j \in \mathbf{E}$ where $r_{0,0} = 1$, $r_{0,j} = 0$, $j \geq 1$; $r_{i,j} = p_{i,j}$ for all other $i,\ j$. Since the original MC is irreducible, it follows that $f^*_{i,0} > 0$, $i \in T$ and hence that all states in T are transient for $MC2$. For the $MC2$, the probabilities $r^*_{i,0}$ of absorption in $\{0\}$ are given by $r^*_{i,0} = f^*_{i,0} > 0,\ i \in T$.

Suppose that all the states of the MC are transient. Then there exists $j \geq 1$ such that $f^*_{j,0} < 1$, as otherwise 0 would be a persistent state by **(7.2.4)**(vi). If β_i, $i \in T$ is the probability that the system $MC2$ starting in state $i \in T$ continues to stay forever in T then the β_is satisfy the equations (7.43). This is true since the arguments of Theorem **(7.5.1)**(i) apply. We note $\beta_i = 1 - f^*_{i,0}$. All the β_is are not zeros since $f^*_{j,0} < 1$. Thus (7.43) admits of a non-null bounded solution, as was to be proved.

Conversely, suppose (7.43) possesses a bounded non-null solution. Let β_i, $i \in T$ be the probability that the system $MC2$, starting from state $i \in T$ stays in T forever. By the arguments in **(7.5.2)**(i), the β_is constitute the maximal solution among all the bounded-by-one solutions of (7.43). Hence $(\beta_i,\ i \in T)$ is a non-null solution of (7.43). i.e., there exists $j \in T$ with $\beta_j > 0$. Since $\beta_j = 1 - f^*_{j,0}$, it follows $f^*_{j,0} < 1$. Now, if the claim is not admitted that all the states of the original MC are transient, then it would follow that all the states would form a communicating class of persistent states. In that case, $f^*_{j,0} = 1$, a contradiction, completing the proof of the converse. □

By way of illustration, Consider the MC with $\mathbf{E} = \{0,\ 1,\ 2,\ ....\}$ and $p_{0,1} = 1$, $p_{i,i-1} = q$, $p_{i,i+1} = p$, $p > 0$, $q > 0$, $p + q = 1$, $q < p$. Clearly the chain is irreducible. The equations (7.43) applied to this MC become $y_1 = py_2$, $y_i = qy_{i-1} + py_{i+1}$, $i \geq 2$. It is easily verified that $y_i = (\frac{q}{p})^i$, $i = 1,\ 2,\ ....$ is a solution of this system of equations. Since this is a non-null bounded solution, we conclude that all the states are transient.

(v) Let the set T of all the transient states of a MC be finite. Let Q be the matrix $(p_{i,j},\ i,\ j \in T)$. Then $(I - Q)$ has an inverse.

*Proof*We know (ref. **(7.4.2)**(ii)) that $\sum_n p^{(n)}_{i,j} < \infty$. $i,\ j \in T$. Denote by $r^{(n)}_{i,j}$ the entry in Q^n corresponding to the states i and j. We claim that if $i,\ j \in T$, then $r^{(n)}_{i,j} = p^{(n)}_{i,j}$, $n \geq 1$. Proof of this claim is by induction on n. The claim is obvious if $n = 1$. Suppose the claim is true for $1 \leq n \leq m$. Now let $i,\ j \in T$. Then

$$p^{(m+1)}_{i,j} = \sum_{k \in \mathbf{E}} p^{(m)}_{i,k} p_{k,j} = \sum_{k \in T} p^{(m)}_{i,k} p_{k,j} = \sum_{k \in T} r^{(m)}_{i,k} r_{k,j} = r^{(m+1)}_{i,j}$$

(since $p_{k,j} = 0$ if k is persistent and j is transient.)

Hence for all $i,\ j \in T$, $\sum_n r_{i,j}^{(n)} < \infty$. This means $\hat{Q} = I + Q + Q^2 + Q^3 + \ldots$ is a well defined finite dimensional matrix where I is the identity matrix. Now the entry in $Q\hat{Q}$ corresponding to $i,\ j \in T$ is

$$\begin{aligned}\textstyle\sum_{k\in T} r_{i,k}\{\sum_{n=0}^{\infty} r_{k,j}^{(n)}\} &= \textstyle\sum_{n=0}^{\infty}\sum_{k\in T} r_{i,k}r_{k,j}^{(n)} \\ &= \textstyle\sum_{n=0}^{\infty} r_{i,j}^{(n+1)} = \sum_{n=0}^{\infty} r_{i,j}^{(n)} - r_{i,j}^{(0)}.\end{aligned}$$

This being true for all $i,\ j \in T$, we have $Q\hat{Q} = \hat{Q} - I$. Hence $(I - Q)\hat{Q} = \hat{Q} - Q\hat{Q} = I$. That $(I - Q)^{-1}$ exists (and is equal to $\hat{Q}$) follows from this. □

(vi) It may be noted that the equations (7.39) can be written $(I - Q)\mathrm{b} = \mathbf{0}$ where b is the vector $(\beta_i,\ i \in T)$. When T is finite, $I - Q$ is non singular (ref. part (v) above) and hence the null solution is the only solution to the equations (7.39). Thus when T is finite, the probability is zero that the system starting anywhere in T will stay forever in T.

For two other proofs of this assertion refer **(7.2.4)** (v) and **(7.4.2)** (iii).

Time to absorption

Let $T \subset \mathbf{E}$ be the collection of all the transient states. Further let the probability be zero that the system starting anywhere in T stays in T forever. We note T' is a closed collection, consisting of all the persistent states. Let $Y = \inf\{n \geq 0\ :\ X_n \in T'\}$. Y is the time to absorption. In this section we show that the expected times to absorption starting from the various transient states can be obtained as the minimal solution of a system of linear equations.

But first we must show that Y is a proper rv. To do this, we argue as follows.

$$\begin{aligned}&\textstyle\sum_{n=0}^{\infty} P(Y = n) = \sum_{n=0}^{\infty}\sum_{i\in\mathbf{E}} P(Y = n|X_0 = i)P(X_0 = i)\\ &= \textstyle\sum_{n=0}^{\infty}\sum_{i\in T} P(Y = n|X_0 = i)P(X_0 = i)\\ &\qquad + \textstyle\sum_{n=0}^{\infty}\sum_{i\in T'} P(Y = n|X_0 = i)P(X_0 = i)\\ &= \textstyle\sum_{n=0}^{\infty}\sum_{i\in T} P(Y = n|X_0 = i)P(X_0 = i)\\ &\qquad + \textstyle\sum_{i\in T'} P(Y = 0|X_0 = i)P(X_0 = i)\\ &= \textstyle\sum_{i\in T} P(X_0 = i)\sum_{n=0}^{\infty} P(Y = n|X_0 = i) + \sum_{i\in T'} P(X_0 = i)\\ &= \textstyle\sum_{i\in T} P(X_0 = i) + \sum_{i\in T'} P(X_0 = i) = 1.\end{aligned}$$

□

Theorem 7.5.3. *Let T be the set of all the transient states. Let $u_i = E(Y|X_0 = i)$, $i \in T$. Then the u_is constitute the minimal non-negative solution to the fol-*

lowing system of linear equations:

$$y_i = 1 + \sum_{j \in T} p_{i,j} y_j, \quad i \in T. \tag{7.44}$$

Proof. That the u_is are non-negative derives from the fact that $Y \geq 0$ (ref. **(6.4)**(ii)). We have

$$\begin{aligned}
u_i &= E(Y|X_0 = i) = \sum_{n=0}^{\infty} nP(Y = n|X_0 = i) \\
&= \sum_{n=1}^{\infty} nP(Y = n,\ X_1 \in T'|X_0 = i) \\
&\qquad\qquad + \sum_{n=1}^{\infty} \sum_{j \in T} P(Y = n,\ X_1 = j|X_0 = i) \\
&= P(Y = 1,\ X_1 \in T'|X_0 = i) \\
&\qquad\qquad + \sum_{n=2}^{\infty} nP(Y = n,\ X_1 \in T'|X_0 = i) \\
&\qquad\qquad\qquad + \sum_{n=1}^{\infty} \sum_{j \in T} nP(Y = n,\ X_1 = j|X_0 = i) \\
&= \sum_{j \notin T} p_{i,j} + 0 + \sum_{j \in T} \sum_{n=1}^{\infty} nP(Y = n,\ X_1 = j|X_0 = i) \\
&= \sum_{j \notin T} p_{i,j} + \sum_{j \in T} \sum_{n=2}^{\infty} nP(Y = n,\ X_1 = j|X_0 = i) \\
&= \sum_{j \notin T} p_{i,j} + \sum_{j \in T} \sum_{n=1}^{\infty} (n+1)P(Y = n+1,\ X_1 = j|X_0 = i) \\
&= \sum_{j \notin T} p_{i,j} + \sum_{j \in T} \sum_{n=1}^{\infty} (n+1)P(Y = n+1|X_1 = j)p_{i,j} \\
&= \sum_{j \notin T} p_{i,j} + \sum_{j \in T} \sum_{n=1}^{\infty} nP(Y = n|X_0 = j)p_{i,j} \\
&\qquad\qquad\qquad + \sum_{j \in T} \sum_{n=1}^{\infty} P(Y = n|X_0 = j)p_{i,j} \\
&= \sum_{j \notin T} p_{i,j} + \sum_{j \in T} p_{i,j} u_j + \sum_{j \in T} p_{i,j} = 1 + \sum_{j \in T} p_{i,j} u_j.
\end{aligned}$$

It follows that the u_is satisfy the equations (7.44).
Let $(v_i,\ i \in T)$ be another non-negative solution to (7.44). Substituting for v_j from (7.44), we get

$$\begin{aligned}
v_i &= 1 + \sum_{j\in T} p_{i,j} v_j \\
&\geq P(Y \geq 1|X_0 = i) + \sum_{j\in T} p_{i,j}\{1 + \sum_{k\in T} p_{j,k} v_k\} \\
&= P(Y \geq 1|X_0 = i) + P(Y \geq 2|X_0 = i) + \sum_{k\in T} v_k \sum_{j\in E} p_{ij} p_{j,k} \\
&= P(Y \geq 1|X_0 = i) + P(Y \geq 2|X_0 = i) + \sum_{k\in T} p_{i,k}^{(2)} v_k.
\end{aligned}$$

Repeating the above substitution process, we arrive at

$$\begin{aligned}
v_i \geq P(Y \geq 1|X_0 = i) &+ P(Y \geq 2|X_0 = i) \\
&+ P(Y \geq 3|X_0 = i) + \sum_{t\in T} p_{i,t}^{(3)} v_t
\end{aligned}$$

and after n steps, we arrive at

$$v_i \geq \sum_{k=1}^{n} P(Y \geq k|X_0 = i) + \sum_{t\in T} p_{i,t}^{(n)} v_t.$$

Since the last term on the right side is non-negative, we have for all n,

$$v_i \geq \sum_{k=1}^{n} P(Y \geq k|X_0 = i)$$

which on letting $n \to \infty$, leads to

$$v_i \geq \sum_{k=1}^{\infty} P(Y \geq k|X_0 = i) = E(Y|X_0 = i) = u_i. \qquad \square$$

Two examples

(i) $\mathbf{E} = \{1,\ 2,\ 3,\ 4\}$; $p_{1,1} = 1 = p_{4,4}$; $p_{2,1} = \frac{1}{2} = p_{2,3}$; $p_{3,2} = \frac{1}{2} = p_{3,4}$. Here $T = \{2,\ 3\}$. Let $C = \{4\}$. The probabilities x_2 and x_3 of starting from states 2 and 3 and getting absorbed in C are the solutions of the equations $x_2 = \frac{1}{2}x_3$ and $x_3 = \frac{1}{2}x_2 + \frac{1}{2}$ (ref. (7.40)), leading to $x_2 = \frac{1}{3}$ and $x_3 = \frac{2}{3}$.

i.e., $f^*_{2,\{4\}} = \frac{1}{3}$ and $f^*_{3,\{4\}}$. Similarly $f^*_{2,\{1\}} = \frac{2}{3}$, and $f^*_{3,\{1\}} = \frac{1}{3}$. Since T is a finite set, the probability is 1 (ref. **(7.5.2)**(vi)) that, wherever in T it starts, the system will leave T and get absorbed in the closed class consisting of the states 1 and 4.

If $u_2,\ u_3$ are the expected times of absorption in the set $\{1,\ 4\}$, if the system starts in the states 2 and 3 respectively, then $u_2,\ u_3$ are the solutions of the equations $u_2 = 1 + \frac{1}{2}u_3$; $u_3 = 1 + \frac{1}{2}u_2$, leading to $u_2 = 2 = u_3$.

(ii) $\mathbf{E} = \{0,\ 1,\ 2, \ldots\}$; $p_{0,0} = 1, p_{1,0} = 1$; $p_{i,0} = \frac{1}{i}, p_{i,i-1} = \frac{i-1}{i}, i \geq 2$. We note 0 is an an absorbing state and that $p_{i,0} > 0,\ i \geq 1$. Hence all the states in $T, T = \{1,\ 2,\ 3, \ldots\}$ are transient. If $\beta_i,\ i \in T$ is the probability that the system starting in state $i \in T$ continues to stay in T for ever, then the β_is satisfy the equations (7.39), which on substitution become

$$y_1 = 0, \quad y_i = \frac{i-1}{i} y_{i-1} \quad i \geq 2. \tag{7.45}$$

Hence $y_i = 0, i \geq 1$. Thus the system, wherever it starts in T will eventually get absorbed in 0.

The expected times $u_i,\ i \in T$ satisfy the equations (7.44). These, on substitution, become

$$y_1 = 1 \text{ and } y_i = 1 + \frac{i-1}{i} y_{i-1}, \quad i \geq 2. \tag{7.46}$$

Or, $iy_i - (i-1)y_{i-1} = i$, Writing these relations for $i = 1,\ 2, \ldots,\ n$ and adding we get $ny_n - 1y_1 = \sum_{r=2}^{n} 1 = \frac{n(n+1)}{2} - 1$. i.e. $ny_n = \frac{n(n+1)}{2} + y_1 - 1$. Even though it is trivial to see that $y_1 = 1$ and hence $y_n = \frac{n+1}{2}$, we would like to evaluate y_1 by invoking the property that the u_is constitute the the minimal solution to the equations (7.44). The expected time y_1 is naturally greater than or equal to 1. The minimal solution is then obtained by taking $y_1 = 1$. And then $u_n = \frac{n+1}{2}, n \geq 1$.

7.6 Stationary distribution

Let the $MC\ (X_n,\ n \geq 0)$ be irreducible with all states aperiodic and persistent positive. Then we know (ref. Theorems **(7.4.7)**, **(7.4.8)**) that $\lim_{n\to\infty} p^{(n)}_{i,j} = \pi_j$ exists and is positive. In this section we show that the π_js add up to 1 (not an unexpected result, since $\sum_j p^{(n)}_{i,j} = 1$) and that the π_js can be obtained as the solution of certain linear equations. We further show that, if X_0 is endowed with the distribution (π_j), then the sequence (X_n) would form a strict sense stationary process (ref. **(7.6.4)**(ii)). For this reason the sequence (π_j) is called the stationary distribution of the chain.

Let the state space be $\mathbf{E} = \{0,\ 1,\ 2, \dots\}$.

Remark 7.6.1. Since all the states be persistent positive, $f^*_{i,j} = 1$ for all i and j and hence $\lim_{n\to\infty} p^{(n)}_{i,j} = \pi_j = \frac{1}{\mu_j}$ (ref. Theorem **(7.4.8)**).

Theorem 7.6.2. (i) $\pi_k = \sum_{j=0}^{\infty} \pi_j p_{j,k},\ k = 0,\ 1,\ 2, \dots$ (ii) $\sum_{k=0}^{\infty} \pi_k = 1$.

Proof.

$$\sum_{k=0}^{m} \pi_k = \sum_{k=0}^{m} \lim_{n\to\infty} p^{(n)}_{0,k}$$
$$= \lim_{n\to\infty} \sum_{k=0}^{m} p^{(n)}_{0,k} \le \lim_{n\to\infty} \sum_{k=0}^{\infty} p^{(n)}_{0,k} = 1.$$

Hence $\sum_{k=0}^{\infty} \pi_k \le 1$. Again

$$\pi_k = \lim_{n\to\infty} p^{(n+1)}_{i,k} = \lim_{n\to\infty} \sum_{j=0}^{\infty} p^{(n)}_{i,j} p_{j,k}$$
$$\ge \sum_{j=0}^{\infty} \lim_{n\to\infty} p^{(n)}_{i,j} p_{j,k} \quad \text{(by Fatou's Lemma)}$$

$= \sum_{j=0}^{\infty} \pi_j p_{j,k}$. If the claim that equality holds for all k is not admitted, then let $\pi_r > \sum_{j=0}^{\infty} \pi_j p_{j,r}$. Adding these inequalities for $r = 0,\ 1,\ 2, \dots,$ we get

$$\sum_{k=0}^{\infty} \pi_k > \sum_{k=0}^{\infty} \sum_{j=0}^{\infty} \pi_j p_{j,k} = \sum_{j=0}^{\infty} \pi_j \sum_{k=0}^{\infty} p_{j,k}$$

(the terms being non-negative, change of order of summation justified) $= \sum_{j=0}^{\infty} \pi_j$. This strict inequality is not possible since $\sum_{k=0}^{\infty} \pi_k$ has been shown to be a finite number (≤ 1).

Hence claim (i) of the theorem follows. From (i),

$$\pi_k = \sum_{j=0}^{\infty} \pi_j p_{j,k} = \sum_{j=0}^{\infty} \{\sum_{r=0}^{\infty} \pi_r p_{r,j}\} p_{j,k}$$
$$= \sum_{r=0}^{\infty} \pi_r \{\sum_{j=0}^{\infty} p_{r,j} p_{j,k}\}$$

$$= \sum_{r=0}^{\infty} \pi_r p_{r,k}^{(2)}.$$

Repeating this proceedure n times we get

$$\pi_k = \sum_{j=0}^{\infty} \pi_j p_{j,k}^{(n)}, \text{ true for all } n \geq 1.$$

Let now $n \to \infty$. Since $\sum_k \pi_k$ is a convergent series the bounded convergence theorem applies and limit termwise can be taken. Doing it, we have

$$\pi_k = \sum_{j=0}^{\infty} \pi_j \pi_k = \pi_k \sum_{j=0}^{\infty} \pi_j.$$

Since $\pi_k \neq 0$, this yields $\sum_{k=0}^{\infty} \pi_k = 1$. □

Theorem 7.6.3. *Let the MC be irreducble and aperiodic. Suppose an absolutely convergent non-null solution (a_k) exists for the following set of equations*

$$x_k = \sum_{j=0}^{\infty} x_j p_{j,k}, \quad k = 0,\ 1,\ 2, \ldots \tag{7.47}$$

Then all the states will be persistent positive and there will exist a constant $c \neq 0$ such that $a_k = c\pi_k$.

Proof. Following the lines of proof of **(7.6.1)**, we arrive at

$$a_k = \sum_{j=0}^{\infty} a_j p_{j,k}^{(n)}.$$

Take limit as $n \to \infty$. The limit can be taken under the summation sign, since $\sum_j |a_j| < \infty$. If the states are transient or null, we get $a_k = 0$ for all k. This is not possible since (a_k) is a non-null solution. Hence the states are all positive and then $a_k = \sum_{j=0}^{\infty} a_j \pi_k = c\pi_k$ where $c = \sum_k a_k$. By hypothesis c exists finitely. It is not possible that $c = 0$ as that would imply that all the a_ks are zeros. □

Remarks 7.6.4. (i) This theorem can be considered as providing a criterion to decide whether the states of an irrreducible aperiodic MC are persistent positive or not.

(ii) Let the MC be irreducible, with all its states aperiodic, persistent and positive. Then define the π_ks as in Theorem **(7.6.1)**. Suppose X_0 has the distribution (π_k). That is $P(X_0 = k) = \pi_k$. Then for $n \geq 1$,

$$P(X_n = k) = \sum_{j=0}^{\infty} P(X_n = k,\ X_0 = j) = \sum_{j=0}^{\infty} \pi_j p_{j,k}^{(n)} = \pi_k.$$

Thus all the X_ns are distributed as X_0. On the same lines it can be shown that for every choice of integers $m,\ r,\ n_1 < n_2 < \cdots < n_r$ the vectors $(X_{n_1},\ X_{n_2}, \ldots,\ X_{n_r})$ and $(X_{n_1+m},\ X_{n_2+m}, \ldots,\ X_{n_r+m})$ have the same distribution. In other words (X_n) is what is called a strict sense stationary process.

(iii) Suppose the X_ns are identically distributed, with $P(X_n = i) = a_i$, $i \in \mathbf{E}$. Then necessarily the states are persistent positive and $a_i = \pi_i$, $i \in \mathbf{E}$. For, let $\mathbf{E}_1 = \{i\ :\ a_i > 0\}$ and note $\sum_{i\in\mathbf{E}_1} a_i = 1$. Since the chain is aperiodic, $\lim_{n\to\infty} p_{i,j}^{(n)} = \theta_j$ exists for all i and j. For all $n \geq 1$ and for all $j \in \mathbf{E}$,

$$a_j = P(X_n = j) = \sum_{i\in\mathbf{E}_1} P(X_n = j,\ X_0 = i) = \sum_{i\in\mathbf{E}_1'} a_i p_{i,j}^{(n)}.$$

If $\mathbf{E}_1' \neq \phi$ and $j \in \mathbf{E}_1'$, then $0 = \theta_j$. Since the chain is irreducible, that would imply $\theta_i = 0$ for all states i. We will then have $a_i = 0$ for all i, which is not possible since the a_is constitute a distribution. We conclude that $\mathbf{E}_1 = \mathbf{E}$ and that all the θ_is are positive. Thus all the states are persistent positive and $a_i = \theta_i = \pi_i$.

Examples 7.6.5. Let $\mathbf{E} = \{0,\ 1,\ 2, \ldots\}$. Let the transition matrix P be:

(i) as given at (7.38). We had noted in **(7.4.8)** that the chain is irreducible and all the states are persistent positive. The equations to determine the stationary distribution are as follows. $\pi_0 = \sum_{j=0}^{\infty} \pi_j p_{j,0} = \frac{1}{2}\pi_0 + \sum_{j=1}^{\infty} \pi_j = \frac{1}{2}\pi_0 + 1 - \pi_0$. Hence $\pi_0 = \frac{2}{3}$. For $k \geq 1$, the equation $\pi_k = \sum_{j=0}^{\infty} \pi_j p_{j,k}$ reduces to $\pi_k = \pi_0 \frac{1}{2^{k+1}} = \frac{1}{3}\frac{1}{2^k}$.

(ii) $p_{0,k} = p_k > 0$, $k \geq 0$, $p_{i,i-1} = 1$, $i \geq 1$. Clearly the MC is irreducible and aperiodic. $f_{0,0}^{(k)} = p_{k-1}$. Hence all the states are persistent. Further the expected return time for state 0 is $\mu = \sum_1^{\infty} n p_{n-1}$. If $\mu = \infty$, then all the states persistent null. When $\mu < \infty$, $\pi_0 = \frac{1}{\mu}$. The π_ks are obtained from the relations $\pi_k = \pi_0 p_k + \pi_{k+1}$, $k \geq 0$. Or $\pi_k - \pi_{k+1} = \pi_0 p_k$, leading to $\pi_0 - \pi_n = \pi_0(p_0 + p_1 + \cdots + p_{n-1})$. This gives $\pi_n = \pi_0(p_n + p_{n+1} + \cdots) = \mu^{-1}\sum_{\nu=n}^{\infty} p_\nu$, $n \geq 0$.

(iii)

$$P = \begin{pmatrix} p_0 & p_1 & p_2 & p_3 & \cdots \\ p_0 & p_1 & p_2 & p_3 & \cdots \\ 0 & p_0 & p_1 & p_2 & \cdots \\ 0 & 0 & p_0 & p_1 & \cdots \\ 0 & \cdots & & & \end{pmatrix}$$

where the p_ks are positive and $\sum_{k=0}^{\infty} p_k = 1$. We note the MC is irreducible *iff* $p_0 > 0$ and $p_0 + p_1 < 1$. When this condition is satisfied, the chain would be aperiodic.

We will show that a necessary and sufficient condition for the states to be persistent and positive is:

$$\sum_{n=1}^{\infty} np_n < 1. \tag{7.48}$$

Let the states be persistent positive. By **(7.6.1)** the stationary distribution exists. Denote it by $(\pi_k,\ k = 0,\ 1,\ 2, \ldots)$. We know each $\pi_k > 0$ and $\sum_{k=0}^{\infty} \pi_k = 1$. This implies, in particular, $\pi_0 < 1$. The relations **(7.6.1)**(i) applied to the present case would read:

$$\pi_k = \pi_0 p_k + \pi_1 p_k + \pi_2 p_{k-1} + \cdots + \pi_{k+1} p_0,\ k \geq 0. \tag{7.49}$$

Form the generating functions: $A(x) = \sum_0^{\infty} \pi_n x^n$; $B(x) = \sum_0^{\infty} p_n x^n$. Both power series are convergent at least for $|x| \leq 1$.

We note that (7.48) is equivalent to $B'(1) < 1$.

Write $C(x) = \sum_{k=1}^{\infty} \pi_k x^{k-1}$ and note that $xC(x) = A(x) - \pi_0$. The relations (7.49) lead to $A(x) = \pi_0 B(x) + \sum_{k=0}^{\infty}(\pi_1 p_k + \pi_2 p_{k-1} + \cdots + \pi_{k+1} p_0)x^k = \pi_0 B(x) + C(x)B(x)$. Hence $xA(x) = \pi_0 x B(x) + \{A(x) - \pi_0\}B(x)$. Transposing, $A(x)\{B(x) - x\} = \pi_0(1 - x)B(x)$. For $0 \leq x < 1$ we then have $A(x)\frac{B(x)-x}{1-x} = \pi_0 B(x)$. Take limit as $x \to 1-$ and appeal to **(7.4.1)** to get

$$\begin{aligned} \pi_0 &= \pi_0(p_0 + p_1 + \cdots) = \pi_0 \lim_{x\to 1-} B(x) = \lim_{x \to 1-} A(x)\frac{B(x) - x}{1 - x} \\ &= \lim_{x\to 1-} A(x) \lim_{x \to 1-} \frac{B(x) - x}{1 - x} = (\pi_0 + \pi_1 + \cdots)\{1 - B'(1)\} \\ &= 1 - B'(1). \end{aligned}$$

Hence necessarily $B'(1) < 1$, as was to be proved.

Conversely, suppose $B'(1) < 1$. For $0 \leq x \leq 1$, the function $B(x) - x \geq 0$ and is 0 only for $x = 1$. This is so since its differential coefficient $B'(x) - 1 \leq$

$B'(1) - 1 < 0,\ B(0) = p_0 > 0$ and $B(1) - 1 = 0$. Define the function

$$A(x) = \frac{1-x}{B(x)-x}B(x) \quad \text{for } 0 \le x < 1.$$

Define

$$A(1) = \frac{1}{1-B'(1)}.$$

Write

$$A(x) = B(x)\{1 - D(x)\}^{-1} \quad \text{where } D(x) = \frac{1-B(x)}{1-x}$$

for $0 \le x < 1$ and $D(1) = B'(1)$. We have $D(x) = \sum_{n=0}^{\infty}(1 - p_0 - p_1 - \cdots - p_n)x^n$, $0 \le x \le 1$. Thus D is a power series with positive coefficients with $D(1) = B'(1) < 1$. Hence $\frac{1}{1-D(x)}$ can be expanded in a power series of positive coefficients, converging for $0 \le x \le 1$. It follows that $A(x) = \frac{B(x)}{1-D(x)}$ is a power series with positive coefficients and converging over $0 \le x \le 1$. Let $A(x) = \sum_{k=0}^{\infty} a_k x^k$. The relation $A(x)\{B(x) - x\} = a_0(1-x)B(x)$ (making the adjustment so the constant terms agree) implies, on equating the coefficients, $a_k = a_0p_k + a_1p_k + a_2p_{k-1} + \cdots + a_{k+1}p_0$, $k \ge 0$. The steps in the necessity part are reversible and we conclude that the positive valued a_ks satisfy the equations (7.47). By **(7.6.3)**, it now follows that all the states are persistent positive.

Note. We will visit this example again in section **(7.8)**.

The discussion and the results in the preceding subsections apply as well to the case of denumerably infinite number of states as to the case of finite number of states. The proof depends on the earlier theorems **(7.4.7)**, **(7.4.8)** that $\lim_{n\to\infty} p_{i,j}^{(n)}$ exists. When the number of states is finite, the following elegant method is available to prove the existence of the stationary distibution without first proving that the limits of $(p_{i,j}^{(n)})$ exist and finding the limits.

Let $(M,\ d)$ be a compact metric space.

Definition 7.6.6. (i) A mapping T taking M into itself is said to be a strict contraction if there exists λ, $0 \le \lambda < 1$, called the contraction constant, such that

$$d(T(\mathbf{x}),\ T(\mathbf{y})) \le \lambda d(\mathbf{x},\ \mathbf{y}) \quad \text{for every } \mathbf{x},\ \mathbf{y} \in M. \tag{7.50}$$

(ii) Let $S \subset R^p$ denote the set of all vectors $\mathbf{x} = (x_1,\ x_2, \ldots,\ x_p)'$ such that $x_i \ge 0,\ 1 \le i \le p$ and such that $\sum_{i=1}^{p} x_i = 1$. We note S is a compact subset of R^p endowed with the metric $d(\mathbf{x},\ \mathbf{y}) = \|\mathbf{x} - \mathbf{y}\| = \sum_{i=1}^{p} |x_i - y_i|$.

It is well known and also easy to prove that the topology induced by this metric is the same as that induced by the Pythagorean metric.

Theorem 7.6.7. *If $(M,\ d)$ is a compact metric space and if T is a strict contraction map, then T has a unique fixed point. i.e., a point $z \in M$ such that $T(z) = z$.*

Proof. Let λ denote the contraction constant. Definition (7.50) implies, it may be noted, that a contraction mapping is automatically continuous.

Define $T^{n+1}(\mathbf{x})$ as $T(T^n(\mathbf{x}))$, $n \geq 0$, $T^0(\mathbf{x}) = \mathbf{x}$. We have

$$d(T^n(\mathbf{x}),\ T^{n+1}(\mathbf{x})) = d(T(T^{n-1}(\mathbf{x}),\ T(T^n(\mathbf{x})))$$
$$\leq \lambda d(T^{n-1}(\mathbf{x}),\ T^n(\mathbf{x}))$$

(by (7.50)). This reduction relation can be used repeatedly to arrive at

$$d(T^n(\mathbf{x}),\ T^{n+1}(\mathbf{x})) \leq \lambda^n d(\mathbf{x},\ T(\mathbf{x})). \tag{7.51}$$

$(T^n(\mathbf{x}))$, being a sequence in a compact metric space, admits of a convergent subsequence, say, $(T^m(\mathbf{x}))$, converging to, say, $\mathbf{z}$. Since T is a continuous mapping, it follows $T(T^m(\mathbf{x})) \to T(\mathbf{z})$. i.e., $T^{m+1}(\mathbf{x}) \to T(\mathbf{z})$. By (7.51), $d(T^m(\mathbf{x}),\ T^{m+1}(\mathbf{x})) \leq \lambda^n d(\mathbf{x},\ T(\mathbf{x})) \to 0$ since $\lambda < 1$. Hence both the sequences $(T^m(\mathbf{x}))$ and $(T^{m+1}(\mathbf{x}))$ have the same limit. Thus $T(\mathbf{z}) = \mathbf{z}$. i.e., $\mathbf{z}$ is a fixed point of T. We have succeeded in showing that T possesses a fixed point $\mathbf{z}$. We claim T can have only one fixed point. i.e., the fixed point $\mathbf{z}$ is unique. For, if possible, let $\mathbf{y}$, $\mathbf{z}$ be two distinct fixed points of T. By (7.50), $d(\mathbf{y},\ \mathbf{z}) = d(T(\mathbf{y}),\ T(\mathbf{z})) \leq \lambda d(\mathbf{y},\ \mathbf{z})$, a contradiction, since $\lambda < 1$. The uniqueness of the fixed point is thus established. □

Theorem 7.6.8. *Let $P = (p_{i,j})$, $1 \leq i,\ j \leq p$ denote the one step transition probability matrix of a MC and let S be as in* **(7.6.6)***(ii). Then, $\mathbf{x} \in S$, and $\mathbf{y}' = \mathbf{x}'P$ imply $\mathbf{y} \in S$.*

Proof. $y_j = \sum_{i=1}^p x_i p_{i,j} \geq 0$. Further,

$$\sum_{j=1}^p y_j = \sum_{j=1}^p \sum_{i=1}^p x_i p_{i,j}$$
$$= \sum_{i=1}^p x_i \sum_{j=1}^p p_{i,j} = \sum_{i=1}^p x_i = 1.$$

□

Theorem 7.6.9. *Let S, P be as above. Then $T\ :\ T(\mathbf{x}') = \mathbf{x}'P$ is a strict contraction mapping of S into S.*

Proof. $\|T(\mathbf{x}') - T(\mathbf{y}')\| = \sum_{j=1}^p\{|\sum_{i=1}^p x_i p_{i,j} - \sum_{i=1}^p y_i p_{i,j}|\}$

$$\leq \sum_{j=1}^p\{\sum_{i=1}^p |x_i - y_i| p_{i,j}\} = \sum_{i=1}^p |x_i - y_i|\{\sum_{j=1}^p p_{i,j}\} = \|\mathbf{x} - \mathbf{y}\|.$$

This inequality does not help prove that T is a strict contraction map but use has not been made of the fact that all the entries in P are positive. Let $2a = \min\limits_{1\le i,\, j\le p} p_{i,j}$. This implies $2pa \le 1$. Hence $0 < 1 - pa < 1$. It is easy to verify that the $p \times p$ matrix $\frac{1}{1-pa}\{P - a\mathcal{I}\}$ is a stochastic matrix where $\mathcal{I}$ is the $p \times p$ matrix with all entries 1. Applying the result just obtained to this matrix, we get

$$\frac{1}{1-pa}\|\mathbf{x}'(P - a\mathcal{I}) - \mathbf{y}'(P - a\mathcal{I})\| \le \|\mathbf{x} - \mathbf{y}\|.$$

Or,

$$\|\mathbf{x}'P - \mathbf{y}'P\| \le (1 - pa)\|\mathbf{x} - \mathbf{y}\|,$$

since the jth entry in

$$(\mathbf{x} - \mathbf{y})'\mathcal{I} \text{ is } \sum_{i=1}^{p}(x_i - y_i) = 0.$$

We conclude the mapping T is a strict contraction, with contraction constant $1 - pa$. □

Note 7.6.10. By **(7.6.7)** and **(7.6.9)**, there exists a unique vector $\mathbf{a} \in S$ such that

$$\mathbf{a}'P = \mathbf{a}' \tag{7.52}$$

Since $\sum_{i=1}^{p} a_i = 1$, $\mathbf{a}$ is not the null vector. In fact, no a_i can be 0. For, suppose $a_k = 0$. There is at least one entry, say, $a_j \ne 0$. Since, by (7.52), $0 = a_k = \sum_{i=1}^{p} a_i p_{i,k} \ge a_j p_{j,k} > 0$, we arrive at a contradiction. Hence all the entries in $\mathbf{a}$ are positive.

Theorem 7.6.11. *A finite irreducible aperiodic MC has a stationary distribution which is unique.*

Proof. Step 1. Since the state space is finite, there exists, by **(7.4.5)**(i), an integer N such that P^N has all entries positive. Hence, by (7.52), there exists a unique vector, say, $\mathbf{a} \in S$ such that $\mathbf{a}'P^N = \mathbf{a}'$.
Step 2. Given $\mathbf{x} \in S$, define $\mathbf{x}_n' = \frac{1}{n}\sum_{\nu=1}^{n} \mathbf{x}'P^\nu$. For each ν, $\mathbf{x}'P^\nu \in S$. (ref. **(7.6.8)**). Hence $\mathbf{x}_n \in S$. Since S is a compact space satisfying the first axiom of countability, sequence $(\mathbf{x}_n)$ admits of a convergent subsequence, say, $(\mathbf{x}_m)$ converging to, say, $\mathbf{z} \in S$. This leads to:

$$\mathbf{z}'P = \lim_{n\to\infty} \frac{1}{n}\sum_{\nu=1}^{m} \mathbf{x}'P^{\nu+1} = \lim_{n\to\infty} \frac{1}{n+1}\{\sum_{\nu=1}^{n+1} \mathbf{x}'P^\nu - \mathbf{x}'P\} = \mathbf{z}'.$$

Step 3. From $\mathbf{z}'P = \mathbf{z}'$, we get: $\mathbf{z}'P^N = \mathbf{z}'$. Since the fixed point of the mapping $T : T(\mathbf{x}') = \mathbf{x}'P^N$ is unique, it follows that $\mathbf{z} = \mathbf{a}$. Thus the equation $\mathbf{x}' = \mathbf{x}'P$ has the unique solution $\mathbf{a}$.

Step 4. By Note **(7.6.9)**, all the entries in $\mathbf{a}$ are positive. The proof that (X_n) is a stationary sequence when X_0 is endowed with the distribution is the same as in **(7.6.4)**(ii). Hence $\mathbf{a}$ is truly the stationary distribution of the MC and the proof is complete. □

7.7 Geometric ergodicity

Definition 7.7.1. The transition probabilities $(p_{i,j}^{(n)})$ of an aperiodic MC are said to be geometrically ergodic if $p_{i,j}^{(n)}$ converges to its limit exponentially fast. i.e., if, to $i,\ j$ there correspond $\rho_{i,j},\ 0 < \rho_{i,j} < 1$ and non-negative numbers $M_{i,j} = M_{i,j}(\rho_{i,j})$ such that for all n,

$$|p_{i,j}^{(n)} - \lim_{n\to\infty} p_{i,j}^{(n)}| \le M_{i,j}\rho_{i,j}^n. \tag{7.53}$$

$\rho_{i,j}$ is called a convergence parameter. If ρ is a convergence parameter, then trivially ρ' is also a convergence parameter if $\rho < \rho' < 1$. Define $\rho_{i,j}^* = \inf\{\rho_{i,j} : \rho_{i,j}$ satisfies (7.53)$\}$. We write $\rho_i,\ M_i$, and ρ_i^* for $\rho_{i,i},\ M_{i,i}$ and $\rho_{i,i}^*$ respectively. We call $\rho_{i,j}^*$ the optimal convergence parameter for the sequence $(p_{i,j}^{(n)})$. We emphasize that $\rho_{i,j}^*$ may not be a convergence parameter but every ρ would be if $\rho_{i,j}^* < \rho < 1$. Refer to **(7.7.8)**(i) where $\rho_0^* = 0$. If sequence $(p_{i,i}^{(n)})$ is geometrically ergodic, we say state i is geometrically ergodic. A MC will be called geometrically ergodic if all the sequences $\{(p_{i,j}^{(n)}),\ i,\ j \in \mathbf{E}\}$ are geometrically ergodic. We call a MC uniformly geometrically ergodic if, in (7.53), the constants $M_{i,j},\ \rho_{i,j}$ are free from i, j. i.e., same for all i and j. Refer to **(7.4.8)**(ii) where $|p_{i,j}^{(n)} - \pi_j| \le \frac{1}{2^n}$ for all i and j. For a general criterion ensuring uniform geometric ergodicity, refer to Theorem **(7.9.2)**.

Clearly if i is a null state, it can not have the geometric ergodicity property as that would imply $\sum_n p_{i,i}^{(n)} < \infty$. Hence geometric ergodicity property is possible for a state only if it is a transient state or a persistent positive state.

Here are trivial examples of a persistent positive state and of a transient state possessing the geometric ergodic property : $\mathbf{E} = \{0,\ 1\}$; $p_{0,0} = 1$, $p_{1,0} = 1$. We note $p_{0,0}^{(n)} = 1,\ p_{1,1}^{(n)} = 0$. For an interesting and non-trivial example, refer to **(7.8)**.

For examples of transient states and of persistent positive states not possessing the geometric ergodicity property, refer **(7.8)**.

Theorem 7.7.2. *Let $i,\ j$ be transient communicating states. If one of them is geometrically ergodic, then so is the other.*

Proof. Let i be geometrically ergodic: $p_{i,i}^{(n)} \leq M_i\rho_i^n,\ 0 < \rho_i < 1$. Since $i \rightleftarrows j$, there exist $\nu \geq 1,\ t \geq 1$ such that $q_1 = p_{i,j}^{(\nu)} > 0$ and $q_2 = p_{j,i}^{(t)} > 0$. From $M_i\rho_i^{\nu+n+t} \geq p_{i,i}^{(\nu+n+t)} \geq q_1 p_{j,j}^{(n)} q_2$, we have $p_{j,j}^{(n)} \leq M_j\rho_i^n$ where $M_j = M_i\rho_i^{\nu+t}q_1^{-1}q_2^{-1}$. □

Remarks 7.7.3. (i) If one state in an irreducible MC is transient and ergodic, then so will be all the other states.
(ii) Let $i,\ j$ be any two communicating states. Then the four power series

$$\sum_n p_{i,i}^{(n)}x^n,\ \sum_n p_{i,j}^{(n)}x^n,\ \sum_n p_{j,i}^{(n)}x^n \text{ and } \sum_n p_{j,j}^{(n)}x^n$$

have the same radius R of convergence. R is independent of $i,\ j$. This claim follows directly from the inequalities (7.13) where $p_{i,j}^{(m)} > 0$ and $p_{j,i}^{(\nu)} > 0$ for some positive integers $m,\ \nu$.

Thus for an irreducible MC this number R is uniquely determined. If the states are persistent then $R = 1$.

(iii) Let i be an arbitray state and let R be the radius of convergence of the series $\sum_n p_{i,i}^{(n)}x^n$. Then $\sum_n f_{i,i}^{(n)}R^n < \infty$ with sum not greater than unity.

Proof If i is a persistent state, then $R = 1$ and the claim that $\sum_n f_{i,i}^{(n)} = 1$ is trivially true. Again if i is a transient state with $R = 1$ (for an example of such a state i, ref. **(7.8)**(i)), the claim is obvious. Suppose i is a transient state with $R > 1$.

Since the relation (ref. (7.24)) $\quad U_{i,i}(x) = \frac{1}{1-F_{i,i}(x)}$

holds for all x for which either side is finite (and, hence, both sides are finite) and since $U_{i,i}(r) < \infty$ for all $r,\ 0 \leq r < R$, it follows $F_{i,i}(r) < 1,\ 0 \leq r < R$. We note $F_{i,i}(x)$ is monotonic increasing function of x for $x \geq 0$ and bounded above by 1 for $0 \leq x < R$. Hence $\lim_{x\uparrow R} F_{i,i}(x)$ exists finitely and is ≤ 1. Now,

$$\begin{aligned}\sum_n f_{i,i}^{(n)}R^n &= \sum_n \lim_{r\to R} f_{i,i}^{(n)}r^n \\ &= \lim_{r\to R}\sum_n f_{i,i}^{(n)}r^n \text{ (byFatou'sLemma)} \\ &= \lim_{r\uparrow R} F_{i,i}(r) \leq 1.\end{aligned}$$ □

In an irreducible MC, the inequalities (7.13) can be used to show that if there exist states $i,\ j$, distinct or not, and constants $M_{i,j},\ \rho_{i,j},\ 0 < \rho_{i,j} < 1$

such that $p_{i,j}^{(n)} \leq M_{i,j}\rho_{i,j}^{n}$ then a similar inequality holds for every pair of states $k,\ r$ with the same convergence constant.

(iv) Let the chain be irreducible and let the states be transient and geometrically ergodic. Hence $\sum_n p_{i,j}^{(n)}|x|^n$ converges if $|x| < \frac{1}{\rho_{i,j}^*}$ and diverges if $|x| > \frac{1}{\rho_{i,j}^*}$. We note that $\frac{1}{\rho_{i,j}^*}$ is the same as the R in (ii). We note too that $\frac{1}{\rho_{i,j}^*} = R > 1$. (If $\rho_{i,j}^* = 0$, then $R = \infty$). Let $r < R$ be arbitrary. This implies $\sum_n p_{i,j}^{(n)} r^n < \infty$. Hence there exists a constant M such that $p_{i,j}^{(n)} \leq M(\frac{1}{r})^n$ for all n. To summarise : In an irreducible MC with all states transient and geometrically ergodic, there is a *common-to-all* i, j best rate of convergence of $p_{i,j}^{(n)}$ to 0.

Notation 7.7.4. The open disc with center at the origin and radius r will be denoted by C_r. Its closure, the closed disc, will be denoted by $\bar{C}_r$.

The following result from the theory of functions is needed for the proof of the next two theorems in **(7.7.5), (7.7.7)** *where we study the analytic properties of the complex valued function* $F_{i,i}(z)$.

Q15. **Theorem.** [Theorem 15.3, p125, [E]] Let the function g be analytic in the disc C_R. If the set of the zeros of g has limit point in C_R, then $g(z) = 0$ for all $z \in C_R$.
[Consequently, if a function g, not identically zero, is analytic in the region $\{C_R, R > 1\}$ and if g has no zeros in the region $\bar{C}_1$, then there exists $r, 1 < r < R$, such that g is zero free in the region $\bar{C}_r$]

Theorem 7.7.5. *Let i denote an aperiodic persistent state. For complex z with $|z| \leq 1$, set $F_{i,i}(z) = \sum_1^{\infty} f_{i,i}^{(n)} z^n$. Then*

(i) $1 - F_{i,i}(z)$ is zero for no z with $|z| \leq 1$, except $z = 1$. Further, $z = 1$ would be a simple zero if i is a positive state.

[Consequently, if i is a positive state, the function

$$\Lambda_i(z) = \pi_i \frac{1 - F_{i,i}(z)}{1 - z}, \ |z| \leq 1, \ z \neq 1$$

and $\Lambda_i(1) = 1$ has no zeros in $\bar{C}_1$.]

(ii) Let i be a positive state. Let function $\Lambda_i(z)$ be as above. It is analytic in C_1. Further $1 - \Lambda_i(z)$ has no zero in $\bar{C}_1$ except at $z = 1$ and $z = 1$ would be a simple zero if

$$\mu_i^{(2)} = \sum_{n=1}^{\infty} n^2 f_{i,i}^{(n)} < \infty.$$

[Thus, if i is a positive state with $\mu_i^{(2)} < \infty$ then $\frac{1-\Lambda_i(z)}{1-z}$, defined at $z = 1$ as $\frac{1}{2\mu_i}\{\mu_i^{(2)} - \mu_i\}$, is zero-free in $\bar{C}_1$]

(iii) The function

$$\mathcal{A}(z) = \frac{1 - \Lambda_i(z)}{1 - F_{i,i}(z)}, \ z \neq 1, \ |z| \leq 1.$$

$$\mathcal{A}(1) = \frac{1}{2\mu_i^2}\{\mu_i^{(2)} - \mu_i\}$$

has no zeros in $\bar{C}_1$ if i a positive state and $\mu_i^{(2)} < \infty$.

(iv)Let $j \neq i$, $j \rightleftarrows i$. Then $1 - F_{j,i}(z)$, also $1 - F_{i,j}(z)$, vanish nowhere in $\bar{C}_1$ except at $z = 1$. If i is a positive state, then

$$\sum_{n=1}^{\infty} n f_{i,j}^{(n)} < \infty, \ \sum_{n=1}^{\infty} n f_{j,i}^{(n)} < \infty$$

and further $z = 1$ is a simple zero of each of the functions $1 - F_{i,j}(z)$ and $1 - F_{j,i}(z)$.

Proof. (i) If $z = re^{i\theta}$, $r < 1$ then

$$|F_{i,i}(z)| \leq \sum_{n=1}^{\infty} f_{i,i}^{(n)} r^n \leq r \sum_{n=1}^{\infty} f_{i,i}^{(n)} = r < 1.$$

Hence the claim is true for all $z \in C_1$. If possible let $F_{i,i}(z) = 1$ for some $z = e^{i\alpha}$, $0 \leq \alpha < 2\pi$. Hence

$$\sum_{n=1}^{\infty} f_{i,i}^{(n)}\{1 - \cos n\alpha\} = 0.$$

Since all the terms of this series are non-negative, each term must be zero. Hence $f_{i,i}^{(n)}$ can be positive at most when $1 - \cos n\alpha = 0$. Define $A = \{n : f_{i,i}^{(n)} > 0\}$. Since i is aperiodic, the greatest common divisor of all the members of A is unity (ref. **(7.2.2)**(i)). Hence there exist a finite number of members of A, say, $n_1, n_2, \ldots, n_p$ such that the greatest common divisor of these p integers is unity. Necessarily $\cos n_j\alpha = 1$, $1 \leq j \leq p$. Let B be the set consisting of all integers of the type $\sum_{\nu=1}^{p} m_\nu n_\nu$ where the m_νs are non-negative integers with at least one $m_\nu > 0$. We note that $n_j \in B$, $1 \leq j \leq p$. Hence the greatest common divisor of the members of B is 1. That B is closed under addition is obvious. These two properties of B, namely the property that B is closed under addition and the property that the greatest common factor of all the members of B is unity, ensure that there exists an integer m such that $m,\ m+1\ \in B$. (In this connection a reference to **(7.4.5)** would be in order).

Now recall $\cos n_j\alpha = 1$. Hence for each n_j, $1 \le j \le p$, $\cos kn_j\alpha = 1$ for every non-negative integer k. This claim is established by induction on k. Next we note $\cos(k_1n_r + k_2n_s)\alpha = \cos k_1n_r\alpha \cos k_2n_s\alpha - \sin k_1n_r\alpha \sin k_2n_s\alpha = 1$. In this way it can be shown that $\cos\nu\alpha = 1$ for every member ν of B.

From $\cos m\alpha = 1$ and $\cos(m+1)\alpha = 1$ we get $\cos\alpha = \cos(m+1)\alpha\ \cos m\alpha + \sin(m+1)\alpha\ \sin m\alpha = 1$. Since $0 \le \alpha < 2\pi$, it follows that $\alpha = 0$ and hence that, in the closed unit disc, $F_{i,i}(z) = 1$ only for $z = 1$.

We note

$$\frac{1 - F_{i,i}(z)}{1 - z} = \sum_{n=0}^{\infty} t_n z^n,$$

valid for $|z| \le 1$ where $t_n = \sum\limits_{\nu=n+1}^{\infty} f_{i,i}^{(\nu)}$.

Since i is a positive state, $\sum\limits_{n} t_n < \infty$. Hence, by **(7.4.1)**, $\lim\limits_{z\to 1} \sum\limits_{n=0}^{\infty} t_n z^n$ exists and the limit is $\sum\limits_{n=0}^{\infty} t_n = \mu_i$. Thus $\lim\limits_{z\to 1} \frac{1-F_{i,i}(z)}{1-z}$ exists finitely and is non-zero. Hence $z = 1$ is a simple zero of $1 - F_{i,i}(z)$.

(ii) Since $\sum\limits_{n=0}^{\infty} t_n = \mu_i < \infty$, it follows that $\Lambda_i(z)$ is analytic in C_1.

For z with $|z| = r < 1$,

$$|\Lambda_i(z)| \le \pi_i \sum_{n=0}^{\infty} t_n r^n < \pi_i \sum_{n=0}^{\infty} t_n = 1.$$

Hence $1 - \Lambda_i(z)$ does not vanish for any z with $|z| < 1$. Suppose now that for some $z, z \ne 1$, $|z| = 1$, $1 - \Lambda_i(z) = 0$. i.e., suppose that for some α, $0 \le \alpha < 2\pi$, $1 - \Lambda_i(e^{i\alpha}) = 0$. This implies, noting $t_0 = 1$, that

$$\sum_{n=1}^{\infty} t_n\{1 - \cos n\alpha\} = 0.$$

Let $\mathcal{E} = \{n\ :\ t_n > 0\}$. To claim that the greatest common divisor of all the members of $\mathcal{E}$ is 1, we argue as follows. Since i can not be an absorbing state there exists $\nu \ge 2$ such that $f_{i,i}^{(\nu)} > 0$. This implies $t_1 > 0$. Since $1 \in \mathcal{E}$, the claim follows.

To conclude that necessarily $\alpha = 0$, argue as in (i).

Writing $s_n = \sum_{\nu=n}^{\infty} t_\nu$,

$$\lim_{z\to 1} \frac{1 - \pi_i \frac{1-F_{i,i}(z)}{1-z}}{1-z} = \lim_{z\to 1} \frac{1-\Lambda_i(z)}{1-z}$$
$$= \lim_{z\to 1} \pi_i \sum_{n=1}^{\infty} s_n z^{n-1} = \pi_i \sum_{n=1}^{\infty} s_n \text{ (by **(7.4.1)**)},$$

since this sum is finite and positive, being equal to $\frac{1}{2}\pi_i \sum_{n=2}^{\infty} (n^2 - n) f_{i,i}^{(n)}$ which is finite if $\mu_i^{(2)} < \infty$. This proves that $z = 1$ is a simple zero of $1 - \Lambda_i(z)$.

(iii) The desired result follows immediately from (i) and (ii) since, under the stated conditions, both $\frac{1-F_{i,i}(z)}{1-z}$ and $\frac{1-\Lambda_i(z)}{1-z}$ are zero-free in $\bar{C}_1$.

(iv) The other case being similar, we will discuss only the function $1 - F_{j,i}(z)$. We note j is aperiodic, since i is aperiodic and $i \leftrightarrows j$. Hence there exists integer n such that $p_{i,i}^{(n)} > 0$ and $p_{i,i}^{(n+1)} > 0$. Since $i \to j$, there exists N such that $p_{i,j}^{(N)} > 0$. We then have $p_{i,j}^{(N+n)} > 0$ and $p_{i,j}^{(N+n+1)} > 0$. Hence the greatest common factor of the members of the set $\{n : p_{i,j}^{(n)} > 0\}$ is unity. The rest of the argument is as in (i) to prove the first assertion.

Let now i be a positive state. For $0 \le x < 1$, we have, by (7.28),

$$\frac{1 - F_{i,j}(x)F_{j,i}(x)}{1-x} = \frac{1}{1 - {}_jF_{i,i}(x)} \frac{1 - F_{i,i}(x)}{1-x}. \tag{7.54}$$

By **(7.3.4)**, $1 - {}_jF_{i,i}(x) > 0$ for all x, $0 \le x \le 1$. Since i is a positive state,

$$\mu_i = \lim_{x\to 1-} \frac{1 - F_{i,i}(x)}{1-x} \text{ exists finitely.}$$

Hence

$$\lim_{x\to 1-} \frac{1}{1 - {}_jF_{i,i}(x)} \frac{1 - F_{i,i}(x)}{1-x} \text{ exists finitely.}$$

Thus

$$\lim_{x\to 1-} \{\frac{1 - F_{i,j}(x)}{1-x} F_{j,i}(x) + \frac{1 - F_{j,i}(x)}{1-x}\} \text{ exists finitely.} \tag{7.55}$$

By **(7.4.1)**(i),

$$\lim_{x\to 1-} \frac{1 - F_{j,i}(x)}{1-x} = \lim_{x\to 1-} \sum_{n=0}^{\infty} u_n x^n \tag{7.56}$$

exists, finitely or infinitely, where $u_n = \sum_{\nu=n+1}^{\infty} f_{j,i}^{(n)}$. Similar statement applies to the first term on the right side of (7.55). Since the left side limit of (7.54) is finite it follows that $F'_{j,i}(1)$ exists finitely and is equal to $\mu_{j,i} = \sum_{n=0}^{\infty} u_n$. Since this series is convergent, again appealing to **(7.4.1)**, we have, from (7.56),

$$\lim_{z\to 1} \frac{1 - F_{j,i}(z)}{1 - z} = \mu_{j,i}, \text{ positive and finite.}$$

Hence $z = 1$ is a simple zero of $1 - F_{j,i}(z)$. □

Remark 7.7.6. Let j be an aperiodic persistent state. For $|z| \leq 1$ and $z \neq 1$, define

$$U_{j,j}(z) = \frac{1}{1 - F_{j,j}(z)}$$

and note that, by **(7.7.5)**, $U_{j,j}$ is well defined. Then for all z, $|z| \leq 1$, $U_{j,j}(z)$ vanishes for no z.

Proof. Since, being ∞, $U_{j,j}(1) \neq 0$, we need consider only those $z \neq 1$ and $|z| \leq 1$. For all such z, we have by **(7.7.5)**, $F_{j,j}(z) \neq 1$. Hence for all these values of z the relation

$$U_{j,j}(z) = \frac{1}{1 - F_{j,j}(z)} \quad \text{holds.}$$

Remembering that $|F_{j,j}(z)| \leq 1$ if $|z| \leq 1$, we conclude

$$|U_{j,j}(z)| \geq \frac{1}{1 + |F_{j,j}(z)|} \geq \frac{1}{2}.$$ □

Theorem 7.7.7. *(i) Let the a_ns be real, let $\lim_{n\to\infty} a_n = a$ exist finitely and let the power series*

$$A(z) = \sum_{n=0}^{\infty} a_n z^n \quad \textit{be analytic in } C_1.$$

Then $|a_n - a| \to 0$ exponentially fast (i.e., sequence (a_n) is geometrically ergodic) iff $(1 - z)A(z)$ has an analytic extension to C_r for some $r > 1$. When the condition holds, $\sum_n (a_n - a)z^n$ will be analytic in C_r.

(ii) An aperiodic state i is geometrically ergodic iff, for some $r > 1$,

$$\sum_{n=1}^{\infty} f_{i,i}^{(n)} r^n < \infty.$$

Proof. (i) Write $b_n = a_n - a$ and note that $b_n \to 0$.

Suppose sequence (a_n) is geometrically ergodic. There exist then numbers $M > 0,\ 0 < \rho < 1$ such that $|a_n - a| \leq M\rho^n$. This implies $\sum_n (a_n - a)z^n$ is analytic in C_r where $r = \rho^{-1}$. For $|z| < 1$, this series sums to $A(z) - \frac{a}{1-z}$. This function and hence the function $(1-z)A(z)$, have analytic extensions to C_r.

Conversely suppose $(1-z)A(z)$ has an analytic extension to C_r, for some $r > 1$.

$$\text{i.e. } a_0 + \sum_{n=1}^{\infty}(a_n - a_{n-1})z^n \text{ is analytic in } C_r.$$

What is same, $a_0 + \sum_{n=1}^{\infty}(b_n - b_{n-1})z^n$ is analytic in C_r. This implies that if $1 < t < r$, then $|b_n - b_{n-1}|t^n \to 0$. We can find N such that for all $n \geq N$, $|b_n - b_{n+1}| \leq \frac{1}{t^{n+1}}$. This yields

$$|b_n - b_{n+m}| \leq \sum_{\nu=n+1}^{n+m} \frac{1}{t^\nu} \leq \frac{1}{t^n}\frac{1}{t-1}.$$

Now allow $m \to \infty$, remember $b_{n+m} \to 0$ and conclude that $|b_n|t^n = O(1)$, which is the desired result.

(ii) Let $r > 1$ and let the series converge. That implies $1 - F_{i,i}(z)$ is analytic in C_r. This will entail $\mu_i < \infty$ if $F_{i,i}(1) = 1$. Hence i can not be a persistent null state.

Consider the case of i being a transient state. Since, in this case, $1 - F_{i,i}(z)$ does not vanish in $\bar{C}_1$ and since it is analytic in C_r, we can find $s,\ 1 < s < r$ such that it does not vanish in C_s. It follows $\frac{1}{1-F_{i,i}(z)}$ and hence $\frac{1-z}{1-F_{i,i}(z)}$ are analytic in C_s. i.e., $(1-z)U_{i,i}(z)$ has an analytic extension to C_s. By part (i), the desired result follows.

Let now i be a persistent positive state. Since $F_{i,i}(z)$ is analytic in C_r, $f_{i,i}^{(n)}\tau^n = o(1)$ for every $\tau,\ 1 < \tau < r$. Hence

$$t_n\tau^n = O(1) \text{ where } t_n = \sum_{\nu=n+1}^{\infty} f_{i,i}^{(n)}.$$

Defining the function $\frac{1-F_{i,i}(z)}{1-z}$ to be μ_i at $z = 1$ and observing it can be written $\sum_{n=0}^{\infty} t_n z^n$, we conclude it is analytic in C_r. Since $\frac{1-F_{i,i}(z)}{1-z}$ is zero-free in $\bar{C}_1$

(ref. **(7.7.5)**(i)), we can find s, $1 < s < r$ such that $\frac{1-F_{i,i}(z)}{1-z}$ is zero-free in C_s. It follows $\frac{1-z}{1-F_{i,i}(z)}$ is analytic in C_s. i.e., $(1-z)U_{i,i}(z)$ has an analytic extension to C_s. Part(i) result applies and the proof is complete.

Conversely, suppose i is geometrically ergodic. Then necessarily i is transient or persistent positive.

If i is transient and geometrically ergodic, then $U_{i,i}(z)$ is analytic in some C_r, $r > 1$. What is same $\frac{1}{1-F_{i,i}(z)}$ is analytic in C_r. Since $|1 - F_{i,i}(z)| \leq 2$ for $|z| \leq 1$, it follows that $\frac{1}{1-F_{i,i}(z)}$ does not vanish in $\bar{C}_1$. Since it is analytic in C_r and zero-free in $\bar{C}_1$, we can find $1 < s \leq r$ such that it is zero-free in $\bar{C}_s$. Hence $1 - F_{i,i}(z)$ is analytic in C_s. This implies

$$\sum_n f_{i,i}^{(n)} t^n < \infty, \quad 1 < t < s.$$

If i is persistent positive and geometrically ergodic, then $(1-z)U_{i,i}(z)$ has an analytic continuation from C_1 to some C_r, $r > 1$. i.e. $\frac{1-z}{1-F_{i,i}(z)}$ has an analytic continuation from C_1 to C_r. Now, for $|z| \leq 1$,

$$|\frac{1 - F_{i,i}(z)}{1-z}| = |\sum_{n=0}^{\infty} t_n z^n| \leq \sum_{n=0}^{\infty} t_n = \mu_i < \infty,$$

it follows that $\frac{1-z}{1-F_{i,i}(z)}$ does not vanish in $\bar{C}_1$. This together with fact that the function has an analytic continuation to C_r helps us conclude that there exists $1 < s < r$ such that $\frac{1-z}{1-F_{i,i}(z)}$ does not vanish in $\bar{C}_s$. This implies that $\frac{1-F_{i,i}(z)}{1-z}$ and, hence, $1 - F_{i,i}(z)$ has an analytic continuation to C_s. That

$$\sum_n f_{i,i}^{(n)} t^n < \infty, \ 1 < t < s \text{ follows from this.} \qquad \square$$

Remarks 7.7.8. (i) If $\Lambda_i(z)$ is an entire function, vanishing nowhere in the finite z-plane, then r in the proof can be any finite positive number and then we will have $|p_{i,i}^{(n)} - \pi_i| \leq M(\rho)\rho^n$ for *every* $\rho > 0$. An example illustrating this fact is as follows. Let the transition matrix be given by: $p_{0,0} = 0$, $p_{0,k} = \frac{1}{\underline{|k}} - \frac{1}{\underline{|k+1}}$; $p_{k,k-1} = 1$, $k \geq 1$. We note $f_{0,0}^{(1)} = 0$; $f_{0,0}^{(n)} = \frac{1}{\underline{|n-1}} - \frac{1}{\underline{|n}}$, $n \geq 2$. Then for $0 \leq x \leq 1$,

$$F_{0,0}(x) = \sum_{n=2}^{\infty} (\frac{1}{\underline{|n-1}} - \frac{1}{\underline{|n}}) x^n = x(e^x - 1) - (e^x - 1 - x).$$

Hence

$$\frac{1 - F_{0,0}(x)}{1-x} = e^x.$$

This implies $\pi_0 = e^{-1}$. We have

$$\sum_{n=0}^{\infty}\{p_{0,0}^{(n)} - e^{-1}\}x^n = U_{0,0}(x) - \frac{e^{-1}}{1-x}$$
$$= \frac{1}{1-F_{0,0}(x)} - \frac{e^{-1}}{1-x} = \frac{e^{-x}}{1-x} - \frac{e^{-1}}{1-x}$$
$$= (\sum_{n=0}^{\infty}\frac{(-x)^n}{\underline{|n}} - e^{-1})\sum_{n=0}^{\infty}x^n.$$

Equating the coefficients of x^n on both sides,

$$p_{0,0}^{(n)} - \pi_0 = -e^{-1} + \sum_{k=0}^{n}\frac{(-1)^k}{\underline{|k}} = -\sum_{k=n+1}^{\infty}\frac{(-1)^k}{\underline{|k}}.$$

This gives

$$|p_{0,0}^{(n)} - \pi_0| \leq \sum_{k=n+1}^{\infty}\frac{1}{\underline{|k}} \leq \frac{1}{\underline{|n}}$$

which, for a given ρ, $0 < \rho < 1$ is $< \rho^n$ for all n sufficiently large, the largeness depending on ρ. In this case $\rho_0^* = 0$ (ref. **(7.7.1)**).

(ii) Let i be aperiodic, persistent positive and geometrically ergodic. Define

$$U_{i,i}^*(z) = \sum_{n=0}^{\infty}\{p_{i,i}^{(n)} - \pi_i\}z^n$$

and note that $U_{i,i}^*(z)$ is analytic in $C_{\mathcal{R}}$ where $\mathcal{R} = \frac{1}{\rho_i^*}$. Further

$$U_{i,i}^*(z) = \widehat{U}_{i,i}(z) = U_{i,i}(z) - \frac{\pi_i}{1-z} \text{ for } |z| < 1.$$

We know by Theorem **(7.7.7)**(ii) that if R is the radius of convergence of the series $\sum_{n=1}^{\infty} f_{i,i}^{(n)}z^n$, then $1 < R \leq \infty$. Denote the sum of this series by $F_{i,i}^*(z)$ and note that it is analytic in C_R.

$$1 - F_{i,i}^*(z) = 1 - F_{i,i}(z) = \frac{1}{U_{i,i}(z)} \text{ for } |z| < 1.$$

Define

$$\Lambda_i(z) = \pi_i\frac{1-F_{i,i}}{1-z} \text{ for } |z| \leq 1,\ z \neq 1$$

and $\Lambda_i(1) = 1$; also

$$\Lambda_i^*(z) = \sum_{n=0}^{\infty} t_n z^n \text{ where } t_n = \sum_{\nu=n+1}^{\infty} f_{i,i}^{(\nu)}, \ n \geq 0.$$

Define

$$\Delta_i(z) = \frac{1-\Lambda_i(z)}{1-z} \text{ for } |z| \leq 1, \ z \neq 1$$

and

$$\Delta_i(1) = \frac{1}{2\mu_i}[\mu_i^{(2)} - \mu_i], \text{ when } \mu_i^{(2)} < \infty;$$

also

$$\Delta_i^*(z) = \pi_i \sum_{n=1}^{\infty} s_n z^n \text{ where } s_n = \sum_{\nu=n}^{\infty} t_\nu, \ n \geq 1.$$

We note that $F_{i,i}^*(z), \Lambda_i^*(z)$ and $\Delta_i^*(z)$ are analytic in C_R and are the analytic continuations respectively of $F_{i,i}(z),\ \Lambda_i(z)$ and $\Delta_i(z)$. We recall (ref. **(7.7.5)**) $\Lambda_i(z)$ does not vanish in $(|z| \leq 1)$. Since it has an analytic continuation to C_R, it follows that there exists $r,\ 1 < r \leq R$ such that $\Lambda_i^*(z)$ does not vanish in $\bar{C}_r$. Let $\bar{C}_r$ be the largest closed disc with center at the origin with the property that $\Lambda_i^*(z) \neq 0,\ z \in \bar{C}_r$. This implies that $\frac{\Delta_i^*(z)}{\Lambda_i^*(z)}$ is analytic in C_r. Let $\bar{C}_\mathfrak{r},\ \mathfrak{r} \geq r$ be the largest disc with centre at origin such that $\frac{\Delta_i^*(z)}{\Lambda_i^*(z)}$ is analytic in $C_\mathfrak{r}$. $\frac{\Delta_i^*(z)}{\Lambda_i^*(z)}$ is the analytic continuation of $\frac{\Delta_i(z)}{\Lambda_i(z)}$ from C_1 to $C_\mathfrak{r}$.

We now claim $\rho_i^* = \frac{1}{\mathfrak{r}}$. We note

$$\widehat{U}_{i,i}(z) = \frac{1}{\mu_i}\frac{\Delta_i(z)}{\Lambda_i((z)}, \ |z| < 1$$

can be analytically continued to $C_\mathcal{R}$ and to no larger disc; also it can be analytically continued to $C_\mathfrak{r}$ and to no larger disc. Hence $\mathcal{R} = \mathfrak{r}$ and the claim is established.

The numerical example we present below is to stress that ρ_i^* determines the largest closed disc, with centre at origin, inside of which $\frac{\Delta_i^*(z)}{\Lambda_i^*(z)}$ is zero-free : it is $\{z \ : \ |z| \leq \frac{1}{\rho_i^*}\}$, if $\rho_i^* > 0$ and it is the entire z-plane if $\rho_i^* = 0$.

Let the transition matrix P of a MC be as given at (7.38). Make use of the calculations made in that section. Then $F_{00}(x) = \frac{1}{2}(x + x^2)$. Hence R_0 can be taken as any number in $(0,\ \infty)$. Since $\pi_0 = \frac{2}{3}$ and

$$|p_{0,0}^{(n)} - \frac{2}{3}| = \frac{1}{3}\frac{1}{2^n},$$

we see that $\rho_0^* = \frac{1}{2}$. We note that for $|z| < 1$,

$$\frac{\Delta_0(z)}{\Lambda_0(z)} = \frac{1}{2+z} = \frac{1}{2}\sum_{n=0}^{\infty}(-\frac{1}{2})^n z^n.$$

This shows, in the notation of (ii), that $\mathfrak{r}_0 = 2$. That $\rho_0^* = \frac{1}{\mathfrak{r}_0}$ checks.

We note that for $j = 1,\ 2, \ldots,\ p_{j,j}^{(1)} = 0$ and

$$p_{j,j}^{(n)} = \frac{1}{3}\frac{1}{2^j}(1 + \frac{(-1)^n}{2^{n-1}}),\ n \geq 2,$$

leading to

$$\lim_{n\to\infty} p_{j,j}^{(n)} = \pi_j = \frac{1}{3}\frac{1}{2^j}.$$

Further from

$$|p_{j,j}^{(n)} - \pi_j| = \frac{1}{3}\frac{1}{2^j}\frac{1}{2^{n-1}},$$

we see that $\rho_j^* = \frac{1}{2}$. To find the radius of convergence of the series defining $F_{j,j}^*(z)$, we observe that for $|z| < 1$,

$$\begin{aligned}
U_{j,j}(z) &= 1 + \sum_{n=2}^{\infty}\frac{1}{3}\frac{1}{2^j}(1 + (-1)^n\frac{1}{2^{n-1}})z^n \\
&= 1 + \frac{1}{3}\frac{1}{2^j}\{\sum_{n=2}^{\infty} z^n + 2\sum_{n=2}^{\infty}(\frac{-z}{2})^n\} \\
&= 1 + \frac{1}{3}\frac{1}{2^j}\{\frac{z^2}{1-z} + \frac{z^2}{2+z}\} \\
&= 1 + \frac{z^2}{2^j(1-z)(2+z)}.
\end{aligned}$$

In particular,

$$U_{1,1}(z) = \frac{4 - 2z - z^2}{2(1-z)(2+z)}.$$

Then, for the stated range,

$$1 - F_{1,1}(z) = \frac{1}{U_{1,1}(z)} = \frac{2(1-z)(2+z)}{4 - 2z - z^2}. \tag{7.57}$$

We note that for $|z| < 1$,

$$\frac{\Delta_1(z)}{\Lambda_1(z)} = \frac{10 + 3z}{2 + z}.$$

Hence $\mathfrak{r}_1 = 2$, again verifying the relation $\rho_1^* = \frac{1}{\mathfrak{r}_1}$.

[Note. It is good to remember the range of z for which the given relations are valid. Thus in (7.57), it would be wrong to substitute $z = -2$ and claim $F_{1,1}(-2) = 1$ or even $F^*_{1,1}(-2) = 1$. In fact, the radius R_1 of convergence of the series $\sum_{n=1}^{\infty} f_{1,1}^{(n)} z^n$ is $\sqrt{5} - 1$. However if $F^{**}_{i,i}$ is the meromorphic extension (ref. **(7.8.1)**) of $F_{i,i}(z)$ to, say, C_3, then it is correct to claim $F^{**}_{i,i}(-2) = 0$. We also note that the radius of convergence of the series $F^*_{i,i}(z)$ is not $\frac{1}{\rho^*_i}$.]

7.8 Examples

Here we present examples of transient states and of persistent states which do not possess the geometric ergodicity property.

(i) Let $\mathbf{E} = \{0, 1, 2, \dots\}$; let $\sum_{2}^{\infty} \frac{c}{n^2} = \frac{1}{2}$; $p_{0,0} = 1$; $p_{1,0} = \frac{1}{2}$, $p_{1,1} = 0$, $p_{1,n} = \frac{c}{n^2}$, $n \geq 2$; $p_{n,n-1} = 1$, $n \geq 2$. We note that 0 is an absorbing state, that $1 \to 0$, that therefore 1 is a transient state, that 1 is an aperiodic state and that all the states $2, 3, 4, \dots$ are transient states since they communicate with state 1.

$$\text{Now, } f_{1,1}^{(1)} = 0, \ f_{1,1}^{(n)} = \frac{c}{n^2}, \quad n \geq 2.$$

Since the series

$$\sum_{1}^{\infty} f_{1,1}^{(n)} x^n = c \sum_{2}^{\infty} \frac{x^n}{n^2}$$

does not converge for any $x > 1$, it follows by **(7.7.7)** that 1 is not geometrically ergodic.

(ii) Let $\mathbf{E} = \{0, 1, 2, \dots\}$; let

$$\sum_{0}^{\infty} \frac{c}{(n+1)^3} = 1 \quad p_{0,n} = \frac{c}{(n+1)^3},$$

$n = 0, 1, 2, \dots$; $p_{n,n-1} = 1$, $n \geq 1$. We note the chain is irreducible and aperiodic. We note $f_{0,0}^{(n)} = \frac{c}{n^3}$, $n \geq 1$. Since

$$\sum_{n} f_{0,0}^{(n)} = \sum_{n=1}^{\infty} \frac{c}{n^3} = 1,$$

it follows that 0 is a persistent state. Since

$$\sum_{n} n f_{0,0}^{(n)} \leq \sum_{n=1}^{\infty} \frac{c}{n^2} < \infty,$$

it follows that 0, and hence every i, is a persistent positive state. Since

$$\sum_n f_{0,0}^{(n)} x^n = \sum_{n=1}^{\infty} \frac{cx^n}{n^3},$$

does not converge for any $x > 1$, it follows that 0 is not geometrically ergodic.

We record the following for future reference.

Definition 7.8.1. A function is said to be meromorphic in a region if it is analytic in the region except at a finite number of points which are poles.

Note.

Let i be an aperiodic persistent state and let $F_{i,i}(z)$ be analytic in C_q for some $q > 1$. (It may be noted that this hypothesis implies that i is a positive state). Then an R, $1 < R < q$ can be found with the property: $F_{i,i}(R) < \infty$ and $1 - F_{i,i}(z)$ is non zero for $|z| \leq R$ except at $z = 1$. This assertion follows from the fact that $1 - F_{i,i}(z)$ vanishes only at $z = 1$ in the region $\bar{C}_1$ (ref. **(7.7.5)**) and the fact that an analytic function can have only a finite number of zeros in a closed bounded region (ref. (**Q**15)).

Theorem 7.8.2. *Let i, j be any two distinct communicating aperiodic and persistent states. If $F_{i,i}(z)$ is analytic in C_q for some $q > 1$, then (i) $F_{j,j}(z)$ has a unique meromorphic extension from C_1 to C_q and (ii) Any zero of $1 - F_{j,j}(z)$ in C_q is a zero of $1 - F_{i,i}(z)$.*

Proof. By our earlier remark (ref. **(7.8.1)**), there exists R, $1 < R < q$ such that $F_{i,i}(R) < \infty$ and such that $1 - F_{i,i}(z)$ has a unique zero at $z = 1$.

The analyticity in C_q of $1 - F_{i,i}(z)$ implies that of $1 - {}_jF_{i,i}(z)$. Being an analytic function, the zeros of $1 - {}_jF_{i,i}(z)$ are atmost finite in number in the bounded region $\bar{C}_q$ (ref. **Q**15). Hence

$$\frac{1 - F_{i,i}(z)}{1 - {}_jF_{i,i}(z)} \text{ is meromorphic in } C_q. \tag{7.58}$$

Recall the relation (ref. (7.26)):

$$F_{i,i}(z) = {}_jF_{i,i}(z) + {}_iF_{i,j}(z)F_{j,i}(z). \tag{7.59}$$

Let $1 < t < q$. The analyticity of $F_{i,i}(z)$ in C_q implies $F_{i,i}(t)$ is finite. Hence ${}_jF_{i,i}(t) < \infty$ and ${}_iF_{i,j}(t)F_{j,i}(t) < \infty$. Since $i \to j$, ${}_iF_{i,j}(t) > {}_iF_{i,j}(1) > 0$ (ref. **(7.3.2)**); since $j \to i$, $F_{j,i}(t) > F_{j,i}(1) > 0$. Hence we conclude that ${}_iF_{i,j}(t) < \infty$ and $F_{j,i}(t) < \infty$. Thus

$${}_jF_{i,i}(z),\ {}_iF_{i,j}(z) \text{ and } F_{j,i}(z) \text{ are all analytic in } C_q. \tag{7.60}$$

Starting now with the fact that $F_{j,i}(z)$ is analytic in C_q and using the relation (ref. (7.25))

$$F_{j,i}(z) = {}_jF_{j,i}(z) + {}_iF_{j,j}(z)F_{j,i}(z) \tag{7.61}$$

and arguing similarly, we conclude

$${}_jF_{j,i}(z) \text{ and } {}_iF_{j,j}(z)F_{j,i}(z) \text{ are analytic in } C_q. \tag{7.62}$$

For $1 < t < q$, use facts $0 < F_{j,i}(t) < \infty$ and ${}_iF_{j,j}(t)F_{j,i}(t) < \infty$ to conclude that ${}_jF_{j,i}(t) < \infty$. This implies ${}_jF_{j,i}(z)$ is analytic in C_q. We then have $F_{j,i}(z)\{1 - {}_iF_{j,j}(z)\} = {}_jF_{j,i}(z)$. If $1 < r < q$, we have $F_{j,i}(r)\{1 - {}_iF_{j,j}(r)\} = {}_jF_{j,i}(r) > {}_jF_{j,i}(1) > 0$ (ref. **(7.3.2)**) and $F_{j,i}(r) > F_{j,i}(1) > 0$. Hence $1 - {}_iF_{j,j}(r) > 0$. This implies

$$1 - {}_iF_{j,j}(z) \text{ can not vanish in } C_q \tag{7.63}$$

since ${}_iF_{j,j}(z) = 1,\ |z| = r < q$ would imply $1 = |{}_iF_{j,j}(z)| \leq {}_iF_{j,j}(r) < 1$, a contradiction. Thus in C_q,

$$F_{j,i}(z) = \frac{{}_jF_{j,i}(z)}{1 - {}_iF_{j,j}(z)}. \tag{7.64}$$

Recall (7.27), namely,

$$1 - F_{i,j}(z)F_{j,i}(z) = \frac{1 - F_{i,i}(z)}{1 - {}_jF_{i,i}(z)}.$$

Right side is defined (ref. (7.58)) for $z \in C_q$ as a meromorphic function while $F_{j,i}(z)$ (ref. (7.60)), defined on C_q, is an analytic function. Also $F_{i,j}(z)$ is analytic in C_1. Hence

$$\frac{1 - F_{i,i}(z)}{1 - {}_jF_{i,i}(z)}$$

is the meromorphic extension of $1 - F_{i,j}(z)F_{j,i}(z)$ from C_1 to C_q.

Parallel to the relation above, we have

$$\{1 - F_{i,j}(z)F_{j,i}(z)\}\{1 - {}_iF_{j,j}(z)\} = 1 - F_{j,j}(z). \tag{7.65}$$

The first factor on the left side has a meromorphic extension from C_1 to C_q and the second factor is (ref. (7.62)) analytic in C_q. Hence the right side function i.e., $1 - F_{j,j}(z)$, has a mermorphic extension to C_q. This concludes the the proof of the first assertion in the theorem.

(ii) In the following, treat the functions as the extended version. For $z \in C_q$,

$$\begin{aligned}(1 - F_{j,j}(z) = 0) &\Leftrightarrow (\{1 - F_{i,j}(z)F_{j,i}(z)\}(1 - {}_iF_{j,j}(z)) = 0)\\ &\Leftrightarrow (1 - F_{i,j}(z)F_{j,i}(z) = 0) \text{ since } 1 - {}_iF_{j,j}(z) \neq 0,\\ &\qquad \text{(ref.(7.63))}\\ &\Rightarrow (\{1 - F_{i,j}(z)F_{j,i}(z)\}(1 - {}_jF_{i,i}(z)) = 0)\text{(since}\\ &\qquad 1 - {}_jF_{i,i}(z) \text{ is analytic in } C_q)\\ &\Rightarrow (1 - F_{i,i}(z) = 0),\end{aligned}$$

thus establishing the second assertion in the theorem. □

Using **(7.7.5)**(i) and **(7.7.7)** we have the following

Corollary 7.8.3. *In an irreducible aperiodic persistent positive and geometrically ergodic MC, there always exists $r > 1$ such that, for all $i \in \mathbf{E}$, $\frac{1-F_{i,i}(z)}{1-z}$ is zero-free in C_r and is meromorphic there.*

Theorem 7.8.4. *Let $i \rightleftarrows j$. Then i is geometrically ergodic if and only if j is.*

Proof. By **(7.7.7)**, (i geometrically ergodic) $\Longleftrightarrow$ ($F_{i,i}(z)$ analytic in C_q for some $q > 1$). This implies that $F_{j,j}(z)$ is meromorphic in C_q (ref. **(7.8.2)**). Since $F_{j,j}(z)$ is bounded in $\bar{C}_1$, the poles, if any, of $F_{j,j}(z)$ must be in $(1 < |z| < q)$. Since $F_{j,j}(z)$ is meromorphic, its poles are finite in number. Hence we can find r, $1 < r < q$ such that $F_{j,j}(z)$ is analytic in C_r. This is equivalent to the claim that j is geometrically ergodic. □

For a simpler proof applicable to the case where the states are transient, refer **(7.7.2)**.

Theorem 7.8.5 (Common convergence rate)**.** *Let the states of an irreducible MC be aperiodic and persistent positive. If any one state i is geometrically ergodic, then all the other states and all the transitions are geometrically ergodic with a common convergence parameter.*

Proof. The first assertion that all states will be geometrically ergodic follows from **(7.8.4)**.

We first consider the diagonal sequences $(p_{i,i}^{(n)})$, $i \in \mathbf{E}$. If every $\rho_i^* = 0$ then there is nothing to prove since any $\rho \in (0,\ 1)$ would serve as a common convergence rate. Assume therefore that there exists a state, label it 0 with $\rho_0^* > 0$. The geometric ergodicity of 0 implies $(1 - z)U_{0,0}(z)$ has an analytic continuation from C_1 to C_R where $R = \frac{1}{\rho_0^*}$ (ref. **(7.7.7)**). Equivalently,

$\frac{1-z}{1-F_{0,0}(z)}$ has this property. Since $\frac{1-z}{1-F_{0,0}(z)}$ does not vanish in $\bar{C}_1$ and since it can be analytically continued to C_R, $R > 1$, we can find r, $1 < r \leq R$ such that it does not vanish in $\bar{C}_r$. Hence $\frac{1-F_{0,0}(z)}{1-z}$ is analytic and does not vanish in $\bar{C}_r$. This implies, by (ref. **(7.8.2)**), that $\frac{1-F_{i,i}(z)}{1-z}$, $i > 0$ is meromorphic and does not vanish in $\bar{C}_r$. Hence $\frac{1-z}{1-F_{i,i}(z)}$ is analytic in $\bar{C}_r$. That $\rho_i^* \leq \frac{1}{r}$ is implied by this.

Any ρ, $\frac{1}{r} < \rho < 1$ will serve as a common convergence rate for the sequences $(p_{i,i}^{(n)})$, $i \in \mathbf{E}$.

Of the off-diagonal sequences $(p_{i,j}^{(n)})$, it is sufficient to consider the sequences $(p_{i,0}^{(n)})$, $i \in \mathbf{E}$, since any state j can take the place of state 0.

We recall (ref. **(7.4.8)**) : $\lim_{n\to\infty} p_{i,j}^{(n)} = \pi_j$.

Since $F_{0,0}(z)$ is analytic in C_r, it follows by (7.60) that $F_{i,0}(z)$ is analytic in C_r. By (7.24), we have the relation

$$(1-z)U_{i,0}(z) = F_{i,0}(z)\{(1-z)U_{0,0}(z)\}, \quad |z| < 1.$$

Since $(1-z)U_{0,0}(z)$ has an analytic extension to C_r (in fact, to C_R) and since $F_{i,0}(z)$ is analytic in C_r, it follows that $(1-z)U_{i,0}(z)$ has an analytic extension to C_r. Now we appeal to **(7.7.7**(i)) and conclude that $\rho_{i,0}^* \leq \frac{1}{r}$. □

[*Note 1. In* **(7.8.5)**, *it must be noted, that the existence of a common rate of convergence does not imply uniform geometric ergodicity since the constants* $M_{i,j}$ *may not have an upper bound.*

Note 2. Refer to **(7.9.2)**. *There, under a condition, uniform geometric ergodicity property is established for irreducible aperiodic* MC*s with countably infinitely many states.*]

The geometric ergodicity of some special Markov chains

The usual method of establishing the geometric ergodicity property of the states of an irreducible MC is to establish it for a conveniently chosen state and then appeal to theorem **(7.8.4)**. Even though a necessary and sufficient condition is known (ref. **(7.7.7)**) for a particular state to have the geometric ergodicity property it is not always easy to verify the condition. It is therefore of interest to examine certain special MCs for the possession of this property.

Theorem 7.8.6. *An irreducible aperiodic finite* MC *has the uniform geometric ergodicity property.*

Proof. Let $\mathbf{E} = \{1, 2, \ldots, s\}$. Refer to **(7.4.2)** and **(7.4.10)** and conclude that all the states are necessarily positive.

Define
$$M_k(n) = \max_{1\le j\le s} p_{j,k}^{(n)}; \quad m_k(n) = \min_{1\le j\le s} p_{j,k}^{(n)}.$$

Now
$$p_{j,k}^{(n+1)} = \sum_{i=1}^{s} p_{j,i}\, p_{i,k}^{(n)} \le M_k(n).$$

Hence $M_k(n+1) \le M_k(n)$. Similarly $m_k(n+1) \ge m_k(n)$. Let $M_k = \lim_{n\to\infty} M_k(n)$; $m_k = \lim_{n\to\infty} m_k(n)$; $d_k(n) = M_k(n) - m_k(n)$.

We note that $0 \le d_k(n+1) \le d_k(n)$. i.e. $d_k(n) \downarrow$.

Since the chain is finite, aperiodic and irreducible, we can find $N \ge 1$ such that
$$c = c_N = \min_{j,k\in\mathbf{E}} p_{j,k}^{(N)} > 0$$

(ref. **(7.4.5)**(ii)). Notice that $0 < c < \frac{1}{2}$ (unless $s = 2$ and the transition probability matrix is
$$\begin{pmatrix} \frac{1}{2} & \frac{1}{2} \\ \frac{1}{2} & \frac{1}{2} \end{pmatrix}$$

and in that case the uniform geometric ergodicity property trivially holds, since $p_{j,k}^{(n)} = \frac{1}{2} = \pi_k$, $k = 1, 2$). Now
$$p_{j,k}^{(\overline{n+1}N)} = \sum_{i=1}^{s} p_{j,i}^{(N)} p_{i,k}^{(nN)}.$$

Choose j such that $p_{j,k}^{(\overline{n+1}N)} = M_k(\overline{n+1}N)$. Choose r such that $p_{r,k}^{(nN)} = m_k(nN)$. Hence
$$\begin{aligned} M_k(\overline{n+1}N) &= p_{j,r}^{(N)} p_{r,k}^{(nN)} + \sum_{i\ne r} p_{j,i}^{(N)} p_{i,k}^{(nN)} \\ &\le p_{j,r}^{(N)} m_k(nN) + M_k(nN)\{1 - p_{j,r}^{(N)}\} \\ &= M_k(nN) - p_{j,r}^{(N)}\{M_k(nN) - m_k(nN)\} \\ &\le M_k(nN) - c d_k(nN). \end{aligned}$$

Similarly it can be shown that
$$m_k(\overline{n+1}N) \ge m_k(nN) + c d_k(nN).$$

These two inequalities yield

$$d_k(\overline{n+1}N) \leq (1-2c)d_k(nN).$$

Repeatedly using this inequality for $n,\ n-1,\ n-2,\dots,\ 1$, we get

$$d_k(\overline{n+1}N) \leq (1-2c)^n d_k(N).$$

Since $0 < 1-2c < 1$, it follows that

$$\lim_{n\to\infty} d_k(nN) = 0.$$

Hence $\lim_{n\to\infty} d_k(n) = 0$, since $(d_k(n))$ is a monotonic decreasing sequence. This implies

$$M_k = m_k = \lim_{n\to\infty} p_{j,k}^{(n)} = \pi_k.$$

It is clear from this that, for all $j,\ k\ \in \mathbf{E}$ and for all $n \geq 1$,

$$|p_{j,k}^{(n)} - \pi_k| \leq d_k(n).$$

For m an arbitrary positive integer, write, $m = nN + r,\ 0 \leq r \leq N-1$ and $n \geq 0$. We recall $d_k(n)\downarrow$. There is no loss in generality in assuming $N \geq 2$.

Case 1. We have,

$$\begin{aligned} d_k(m) &\leq d_k(nN) \leq (1-2c)^{n-1} d_k(N) \\ &= d_k(N)\{(1-2c)^{\frac{1}{N}}\}^{(n-1)N} \\ &= d_k(N)\{(1-2c)^{\frac{1}{N}}\}^{m-r-N} \\ &\leq d_k(N)\{(1-2c)^{\frac{1}{N}}\}^{m}\frac{1}{(1-2c)^2}. \end{aligned} \tag{7.66}$$

Case 2. $N \geq 2,\ n = 1$.We note $m = N + r \leq 2N$. We have

$$\begin{aligned} d_k(m) &\leq d_k(N) \\ &= d_k(N)\{(1-2c)^{\frac{1}{N}}\}^{m}\frac{1}{(1-2c)^{\frac{m}{N}}} \\ &\leq d_k(N)\{(1-2c)^{\frac{1}{N}}\}^{m}\frac{1}{(1-2c)^2}. \end{aligned} \tag{7.67}$$

Case 3. $1 \le m \le N-1$.

$$|p_{j,k}^{(m)} - \pi_k| \le 1 - \pi_k \le 1 - \min_{1\le k\le s} \pi_k = 1 - \alpha, \text{ say.}$$

We note that $\lim_{N\to\infty} c_N = \alpha$. Let N be large such that $c_N \ge \frac{\alpha}{2}$ and such that $16 \max_{1\le k\le s} d_k(N) < 1$, which is possible since $\lim_{n\to\infty} d_k(n) = 0$ for each k, and hence

$$\lim_{n\to\infty} \max_{1\le k\le s} d_k(n) = 0.$$

Find $a \in (0,\ 1)$, call it a_m, such that

$$1 - \alpha = \{1 - (\frac{3}{2} a_m c_N)^{\frac{1}{N}}\}^m.$$

Such an a_m would exist and would exist uniquely if

$$0 < \frac{2}{3} \frac{[1 - (1-\alpha)^{\frac{1}{m}}]^N}{c_N} < 1.$$

That this quantity is positive is obvious. Since $1 - (1-\alpha)^{\frac{1}{m}} \le \alpha$, the cited quantity would be less than unity if $\alpha^N \le \frac{3}{2} c_N$. But this inequality holds atleast for N large since $\lim_{n\to\infty} \alpha^n = 0$ while $\lim_{n\to\infty} c_n = \alpha > 0$.

The proof is now complete that for $i,\ j \in \mathbf{E}$ and for all $n \ge 1$,

$$|p_{i,j}^{(n)} - \pi_j| \le \rho^n$$

where $\rho = 1 - (\frac{3}{2} b c_N)^{\frac{1}{N}}$ and $b = b_N = \min_{1\le m\le N-1} a_m$. □

[*Refer to* **(7.9.2)**. *There, this uniform ergodicity result is extended, under a condition, to the case of countably infinite state space.*]

We consider now the MC with transition probability matrix P given in **(7.6.5)**(iii), which is reproduced below for convenience. This MC arises in queueing theory.

$$P = \begin{pmatrix} a_0 & a_1 & a_2 & a_3 & \cdot & \cdot & \cdot & & & \\ a_0 & a_1 & a_2 & a_3 & \cdot & \cdot & \cdot & & & \\ 0 & a_0 & a_1 & a_2 & a_3 & \cdot & \cdot & \cdot & & \\ 0 & 0 & a_0 & a_1 & a_2 & a_3 & \cdot & \cdot & \cdot & \\ 0 & 0 & 0 & a_0 & a_1 & a_2 & a_3 & \cdot & \cdot & \cdot \\ \cdot & \cdot & & & & & & & & \end{pmatrix}, \quad a_0 > 0,\ a_0 + a_1 < 1$$

As noted in example **(7.6.5)**(iii), all the states form a communicating class. Define $A(x) = \sum_{n=0}^{\infty} a_n x^n$ which is an absolutely convergent series at least for $|x| \le 1$. Define $\alpha = \sum_{n=0}^{\infty} n a_n$ and note that it is equal to $A'(1)$. Write

$$F(x) = F_{00}(x) = \sum_{n=1}^{\infty} f_{00}^{(n)} x^n.$$

We note $f_{00}^{(1)} = a_0$. For $n \ge 2$,

$$f_{00}^{(n)} = \sum_{t=1}^{\infty} a_t f_{t0}^{(n-1)}.$$

Now

$$f_{10}^{(1)} = a_0 = f_{00}^{(1)}; \; f_{10}^{(n)} = \sum_{t=1}^{\infty} a_t f_{t0}^{(n-1)} = f_{00}^{(n)}.$$

Hence $F_{1,0}(x) = F_{00}(x) = F(x)$. Noting that a passage from state 2 to state 0 is possible only through state 1 and noting that by reason of symmetry $f_{21}^{(\nu)} = f_{10}^{(\nu)}$, we have

$$f_{20}^{(n)} = \sum_{\nu=0}^{n} f_{21}^{(\nu)} f_{10}^{(n-\nu)} = \sum_{\nu=0}^{n} f_{10}^{(\nu)} f_{10}^{(n-\nu)}.$$

These relations are equivalent to the relation $F_{20}(x) = (F(x))^2$, $0 \le x \le 1$. Similarly we can show that $F_{t0}(x) = (F(x))^t$. Hence

$$\begin{aligned} F(x) &= a_0 x + \sum_{n=2}^{\infty} f_{00}^{(n)} x^n = a_0 x + \sum_{n=2}^{\infty}\sum_{t=1}^{\infty} a_t f_{t0}^{(n-1)} x^n \\ &= a_0 x + \sum_{t=1}^{\infty} a_t \sum_{n=2}^{\infty} f_{t0}^{(n-1)} x^n = x\{a_0 + \sum_{t=1}^{\infty} a_t \sum_{m=1}^{\infty} f_{t0}^{(m)} x^m\} \\ &= x\{a_0 + \sum_{t=1}^{\infty} a_t F_{t0}(x)\} = x\{\sum_{t=0}^{\infty} a_t (F(x))^t\} \\ &= xA(F(x)); \text{ true for } 0 \le x \le 1. \end{aligned}$$

Thus we have the relation

$$F(x) = xA(F(x)). \tag{7.68}$$

The derivation of this relation shows that whenever either side is finite for a value of x, the other side will be finite too and the two sides will be equal. It is possible that, for some $x > 1$, $F(x) < \infty$ but $A(x) = \infty$ for every $x > 1$. For values of y for which $A(y) < \infty$, consider the equation $y = cA(y)$. If it admits of a finite solution, it would imply $F(c) < \infty$.

F, A are power series convergent in the interval $[0,\ 1]$. Hence at least on $(0,\ 1)$ we get from (7.68), on differentiation,

$$F'(x) = A(F(x)) + xA'(F(x))F'(x). \tag{7.69}$$

For $0 \leq x < 1$, all the terms on both sides are finite. Hence for these values of x,

$$F'(x)\{1 - xA'(F(x))\} = A(F(x)). \tag{7.70}$$

From (7.68), we conclude that $\theta = F(1) = A(\theta)$. In other words

$$x = F(1) \text{ is a solution of } A(x) = x. \tag{7.71}$$

Theorem 7.8.7. *For the MC with transition probability matrix P given above, state 0 is transient or persistent null or persistent positive according as $\alpha > 1$ or $\alpha = 1$ or $\alpha < 1$ where $\alpha = \sum_n na_n$. If $\alpha < 1$ then*

$$\mathbb{E}\tau_0 = \frac{1}{1-\alpha}$$

where τ_0 is the return time variable corresponding to state 0.

[Note. For an alternative proof that $\alpha < 1$ is a necessary and sufficient condition for the MC to be persistent positive, refer **(7.6.5)**(iii)]

Proof. Case (a) $\alpha > 1$. We claim $F(1) < 1$ and hence that 0 is a transient state. If the claim is not conceded, then necessarily $F(1) = 1$. Refer **(7.4.1)** and take limit as $x \to 1-$ in (7.70) :

$$F'(x) \to \sum_n nf_{00}^{(n)} = F'(1),$$

which is positive, finite or infinite. $1 - xA'(F(x)) \to 1 - \alpha$ which is negative. On the right side $A(F(x)) \to 1$. This contradiction shows that $F(1) < 1$.

Case (b). $\alpha \leq 1$. We note that $A(0) = a_0 > 0$ and that over $0 \leq x \leq 1$, $A(.)$ is a convex function, since $A''(x) \geq 0$. Hence in this region the line $y = x$ can cut the curve $y = A(x)$ at most at two points, both with positive x co-ordinates.

The point $(1,\ 1)$ lies on both the line and the curve. They can therefore have atmost one more common point. i.e. the equation $A(x) = x$ may have atmost two solutions, one of which is 1 and the other positive. If $\alpha \leq 1$, i.e. when $A'(1) \leq 1$, the line and the curve have only one common point or, what is same, the equation $A(x) = x$ has only one root namely, $x = 1$. In this case it follows that $\theta = 1$. i.e. $F(1) = 1$. This means that 0 is a persistent state.

Case (b1). $\alpha = 1$. By case (b), $F(1) = 1$. Taking limit in (7.70) as $x \to 1-$ as before, the limit of the right side is 1 while the limit of the left side takes the form $F'(1) \times 0$. Hence $F'(1) = \infty$. i.e. 0 is a null state.

Case (b2) $\alpha < 1$. The above arguments now lead to $F'(1)(1 - \alpha) = 1$. This implies $F'(1) < \infty$. Hence 0 is a positive state. Further $\mu_0 = \frac{1}{1-\alpha}$.

We examine next if state 0 is geometrically ergodic. By **(7.7.7)**, 0 will be so if the series $\sum_n f_{00}^{(n)} x^n$, defining F converges for a value of $x > 1$. We recall that a null state can not be geometrically ergodic. We therefore need consider only the cases $\alpha > 1$ and $\alpha < 1$.

Case $\alpha > 1$. We argue as follows. Since $\alpha > 1,\ F(1) < 1$. Hence an η can be found such that $F(1) < 1-\eta$. Now, $A'(1-0) = \alpha > 1$ or $\lim_{\varepsilon \to 0+} \frac{A(1)-A(1-\varepsilon)}{\varepsilon} = \alpha > 1$. Hence an $\varepsilon < \eta$ be chosen so that $c = \frac{1-\varepsilon}{A(1-\varepsilon)} > 1$. Then $y = 1 - \varepsilon$ is a root of $y = cA(y)$. This implies the equation $F(x) = cA(F(x))$ has $F(x) = 1 - \varepsilon$ for a root. Since $F(x) = 1 - \varepsilon > 1 - \eta > F(1)$, it follows that $x > 1$ and $F(x) = 1 - \varepsilon$. Hence (the transient) state 0 is geometrically ergodic.

Case $\alpha < 1$. The equation $y = A(y)$ is defined only for $y \in [0,\ 1]$. Clearly $y = 1$ is a solution. $y = 0$ is not a solution since $cA(0) = a_0 > 0$. That no solution exists in the interval $(0,\ 1)$ will follow if we show that in this interval $y < A(y)$. Let $g(y) = A(y) - y$. $g'(y) = A'(y) - 1 \leq A'(1) - 1 = \alpha - 1 < 0$, it follows that g is a monotonic non-increasing function. Since $g(0) = a_0 > 0$ and $g(1) = 0$, we conclude that $A(y) \geq y,\ 0 \leq y \leq 1$. Suppose for a $\lambda,\ 0 < \lambda < 1,\ A(\lambda) = \lambda$. The right differential coefficient of A at λ is

$$\lim_{\varepsilon \to 0+} \frac{A(\lambda + \varepsilon) - A(\lambda)}{\varepsilon} \geq \lim_{\varepsilon \to 0+} \frac{\lambda + \varepsilon - \lambda}{\varepsilon} = 1.$$

Similarly its left differential coefficient is ≤ 1. But A is differentiable at λ. Hence $A'(\lambda) = 1$. Since A' is a non-decreasing function, this implies $1 \leq A'(1) = \alpha < 1$. This contradiction shows $A(y) > y,\ 0 \leq y < 1$. Hence the equation $A(x) = x$ has $x = 1$ as the only solution. Since $x = F(1)$ is a solution (ref. (7.71)), it follows that $F(1) = 1$. Thus 0 is a persistent state.

Now let $x \to 1-$ in (7.70) to get $F'(1)(1-\alpha) = 1$. Hence 0 is a persistent positive state with $\mathbb{E}\tau_0 = \frac{1}{1-\alpha}$.

To decide if 0 is geometrically ergodic, we consider two subcases.

(i) $A(y) = \infty$ for all $y > 1$.

If state 0 is geometrically ergodic, then we will have, for some $\lambda > 1$, $1 < \beta = F(\lambda) < \infty$ (ref. **(7.7.7)**(ii)). But that would, by (7.68), imply $A(\beta) < \infty$ contrary to the case assumption. Hence 0 is not geometrically ergodic.

(ii) Let $A(\lambda) < \infty$ for some $\lambda > 1$. Since

$$1 > \alpha = \lim_{\varepsilon \to 0+} \frac{A(1+\varepsilon) - A(1)}{\varepsilon},$$

we can choose $\varepsilon > 0$ sufficiently small so that $A(1+\varepsilon) - A(1) < \varepsilon$. Or, $A(1+\varepsilon) < 1+\varepsilon$. Set now $c = \frac{1+\varepsilon}{A(1+\varepsilon)}$ and note that $y = 1+\varepsilon$ is a solution of $cA(y) = y$. Recall $F(1) = 1$. The equation $F(x) = cA(F(x))$ has a solution $F(x) = 1+\varepsilon$. Since $F(1) = 1$, it follows that $x > 1$. Hence (the persistent positive) state 0 is geometrically ergodic (ref. **(7.7.7)**(ii)). □

Examples 7.8.8. The following numerical examples will help see things better. (i) Suppose $a_n = \frac{1}{2^{n+1}}$, so $A(x) = \frac{1}{2-x}$, $0 \le x < 2$. The equation $y = xA(y)$ yields $y = 1 \pm \sqrt{1-x}$, valid for $0 \le x \le 1$. Thus $F(x) = 1 \pm \sqrt{1-x}$. Since $F(x) \le 1$ for $0 \le x \le 1$, it follows that $F(x) = 1 - \sqrt{1-x}$. This implies 0 is a persistent state. Since $\mu_0 = F'(1) = \infty$, 0 is a persistent null state.

Similarly, if $a_n = \frac{2}{3^{n+1}}$, we get $F(x) = \frac{3}{2}\{1 - \sqrt{1 - \frac{8}{9}x}\}$, $0 \le x \le \frac{9}{8}$, $F(1) = 1$, and the radius of convergence of the series defining F exceeds 1. Hence 0 is a persistent positive and geometrically ergodic state (ref. **(7.7.7**(ii))).

If $a_n = \frac{3^n}{4^{n+1}}$, we would get $F(x) = \frac{1}{3}\{2 - \sqrt{4-3x}\}$. Note $F(1) < 1$ and F is defined for $0 \le x \le \frac{4}{3}$. Hence i is a transient and geometrically ergodic state.

Choose q such that

$$\sum_{n=0}^{\infty} \frac{q}{(n+1)(\log(n+1))^2} = 1.$$

If

$$a_n = \frac{q}{(n+1)(\log(n+1))^2} \text{ then } A'(1) = \infty.$$

Hence in this case all the states are transient and geometrically ergodic.

(ii) Consider the MC with transition matrix P given at (7.38).

$$
\begin{aligned}
p_{0,0}^{(n)} &= \frac{2}{3} + (-1)^n \frac{1}{3}\frac{1}{2^n}, \; n \geq 1; \\
p_{0,j}^{(n)} &= \frac{1}{3}\frac{1}{2^j} + (-1)^{n-1}\frac{1}{3}\frac{1}{2^{j+n}}, \; j \geq 1, \; n \geq 1; \\
p_{j,0}^{(n)} &= \frac{2}{3} + (-1)^{n-1}\frac{1}{3}\frac{1}{2^{n-1}}, \; j \geq 1, \; n \geq 1; \\
p_{i,j}^{(1)} &= 0, \; i \geq 1, \; j \geq 1; \\
p_{i,j}^{(n)} &= \frac{1}{3}\frac{1}{2^j} + (-1)^n \frac{2}{3}\frac{1}{2^{j+n}}, \; n \geq 2, \; i \geq 1, \; j \geq 1.
\end{aligned}
$$

From these relations, it is easy to establish that

$$|p_{i,j}^{(n)} - \pi_j| \leq \frac{2}{3}\frac{1}{2^n}, \; i, \; j \in \mathbf{E}; \; n \geq 1.$$

Various special techniques have been developed for special MCs to study if the geometric ergodicity property is enjoyed or not. A general theorem in this connection is as follows.

Theorem 7.8.9. *A sufficient condition for the state k of a MC to be persistent positive and geometrically ergodic is that* $0 < \inf\limits_{i \in \boldsymbol{E}, \, i \neq k} p_{i,k} \leq 1$.

Proof. Let $\mathbf{E} = \{0, \, 1, \, 2, \ldots\}$. Without loss of generality we assume $k = 0$. By hypothesis $1 - \rho = \inf\limits_{k \geq 1} p_{k,0} > 0, \; 0 \leq \rho < 1$. The probability that the system starting from state 0 and returning to it in n steps is $\alpha_n = \sum\limits_{\nu=1}^{n} f_{0,0}^{(\nu)}$. The probability that the system starting from state 0 and not returning to it in the first n steps is

$$\beta_n = \sum_{k=1}^{\infty} P(X_n = k, \; X_\nu \neq 0, \; 1 \leq \nu \leq n | X_0 = 0) = \sum_{k=1}^{\infty} {}_0p_{0,k}^{(n)}.$$

We note $\alpha_n + \beta_n = 1$. For $n \geq 2$, β_n is equal to the probability that the system starts in state 0 and does not return to it in the first n steps and does not return to it at the nth step

$$= \sum_{k=1}^{\infty} {}_0p_{0,k}^{(n-1)}(1 - p_{k,0}) \leq \rho \sum_{k=1}^{\infty} {}_0p_{0,k}^{(n-1)}.$$

Repeatedly applying this inequality, $\beta_n \leq \rho^n$. Hence $\alpha_n \geq 1 - \rho^n$. Take $n \to \infty$ to get

$$\sum_{\nu=1}^{\infty} f_{0,0}^{(n)} = 1.$$

This proves that 0 is a persistent state. Since

$$\sum_{\nu=n+1}^{\infty} f_{0,0}^{(\nu)} = 1 - \alpha_n \leq \rho^n,$$

it follows that

$$\sum_{\nu=1}^{\infty} \nu f_{0,0}^{(\nu)} = \sum_{n=0}^{\infty} \sum_{\nu=n+1}^{\infty} f_{0,0}^{(\nu)} \leq \sum_{n=0}^{\infty} \rho^n < \infty,$$

proving 0 is a positive state. Further for $x \geq 0$,

$$\sum_{n=1}^{\infty} f_{0,0}^{(n)} x^n \leq \sum_{n=1}^{\infty} x^n \sum_{\nu=n}^{\infty} f_{0,0}^{(\nu)} \leq \sum_{n=1}^{\infty} \rho^{n-1} x^n.$$

This series is convergent for $1 \leq x < \frac{1}{\rho}$. Hence state 0 is geometrically ergodic. □

Remark 7.8.10. If $\rho = 0$, (so, for $i \neq k$, $p_{i,k} = 1$ and consequently $f_{i,k}^* = 1$) then state k will be a persistent state (as proved in **(7.2.4)**(vi)) but may be a null state (and hence can not be geometrically ergodic as is clear from the example in **(7.2.4)**(v)) or may even be a positive state failing to be geometrically ergodic (as is clear from **(7.8)**(ii)) or may be a persistent positive geometrically ergodic state (as is clear from (7.38)).

That the condition stated in the theorem is not a necessary one is brought out by the following example.

Let $p_{0,j} = \frac{1}{2^{j+1}}$, $j = 0, 1, 2, \ldots$; $p_{1,0} = 1$; $p_{j,1} = 1$, $j = 2, 3, \ldots$. The condition of the theorem does not hold in respect of state 0. Yet the conclusion of the theorem applies to it since $0 \rightleftarrows 1$ and the condition of the theorem holds for state 1.

7.9 Hypothesis D and uniform geometric ergodicity

In sec. **(7.10)** we will prove CLT for MCs which possess the uniform geometric ergodicity property. In the present section, we study Doeblin's condition **D** which ensures this property.

We define for $A \subseteq \mathbf{E}$,

$$p^{(\nu)}(i,\ A) = \sum_{j \in A} p^{(\nu)}(i,\ j)$$

where $p^{(\nu)}(i,\ \{j\}) = p^{(\nu)}(i,\ j) = p_{i,j}^{(\nu)}$. We assume that the MC is irreducible and aperiodic. We assume too the following condition, labelled Hypothesis **D** or, simply, **D**:

There exists a probability measure φ on $\mathcal{E}$, the σ-field of all subsets of **E**, an integer $\nu \geq 1$ and an $\varepsilon,\ 0 < \varepsilon < 1$ such that if $A \subseteq \mathbf{E}$ and if $\varphi(A) \leq \varepsilon$, then

$$p^{(\nu)}(i,\ A) = \sum_{j \in A} p_{i,j}^{(\nu)} \leq 1 - \varepsilon$$

for every $i \in \mathbf{E}$.

It may be noted that if **D** holds with $\varphi,\ \nu,\ \varepsilon$, then it holds with $\varphi,\ \nu,\ \eta(< \varepsilon)$. For, $\varphi(A) < \eta$ implies $\varphi(A) < \varepsilon$. Hence

$$P^{(\nu)}(i,\ A) < 1 - \varepsilon < 1 - \eta.$$

Thus if **D** holds with some ε we may and we do take that ε to be less than $\frac{1}{2}$.

If D holds with $\varphi,\ \nu,\ \varepsilon$, then D holds with $\varphi,\ \nu',\ \varepsilon$ for every $\nu' \geq \nu$. For, if $\varphi(A) \leq \varepsilon$, then

$$\begin{aligned} p^{(\nu+n)}(i,\ A) &= \sum_{j \in A} p^{(n+\nu)}(i,\ j) = \sum_{j \in A} \{\sum_{k \in \mathbf{E}} p^{(n)}(i,\ k) \\ p^{(\nu)}(k,\ j)\} &= \sum_{k \in \mathbf{E}} p^{(n)}(i,\ k) p^{(\nu)}(k,\ A) \leq \sum_{k \in \mathbf{E}} p^{(n)}(i,\ k)(1 - \varepsilon) \\ &= 1 - \varepsilon. \end{aligned}$$

Let **D** hold with $\varphi,\ \nu,\ \varepsilon(< \frac{1}{2})$. Let $F = \{i\ :\ \varphi(\{i\}) = 0\}$. Let ψ be the measure on $\mathcal{E}$ with $\psi(\{i\}) = \frac{1}{2^i}$ if $i \in F$; $\psi(\{i\}) = \varphi(\{i\})$ if $i \notin F$. Since $\varphi(\cdot) \leq \psi(\cdot)$, it follows that **D** holds with $\psi,\ \nu,\ \varepsilon$.

Without stating it explicitly each time, it will be assumed that φ is positive valued.

Two examples of Markov chains satisfying Hypothesis **D**.

(i) The MC with transition matrix P given at (7.38) is an example of an irreducible aperiodic MC with all states persistent positive and satisfying condition **D**. That the chain is irreducible and aperiodic is obvious. That all the states are persistent positive follows from state 0 being persistent positive, which is true since for $f_{0,0}^{(1)} = \frac{1}{2}$ and $f_{0,0}^{(2)} = \frac{1}{2}$. To show that **D** holds, take $\nu = 1$, $\varepsilon = \frac{1}{4}$ and $\varphi(\{j\}) = \frac{1}{2^{j+1}}$, $j = 0, 1, 2, \ldots$. Let $A \subset \mathbf{E}$ with $\varphi(A) \leq \varepsilon$. Clearly $0 \notin A$ and $p(i, A) = 0$ if $i \neq 0$. Further $p(0, A) = \sum_{j \in A} \frac{1}{2^{j+1}} \leq \frac{1}{2} < 1 - \varepsilon$.

(ii) Every irreducible aperiodic finite MC saisfies Hypothesis **D**.

To see this, recall that there exists $N \geq 1$ such that all entries in P^N are positive. Let $a = \min_{1 \leq i, j \leq s} p_{i,j}^{(n)}$ where s is the number of states in **E**. We note $sa \leq 1$. Define $\varphi(\{i\}) = \frac{1}{s}$; set $\nu = N$; take $\varepsilon < a$. Hypothesis **D** is trivially satisfied for above choice of φ, ν, ε.

The irreducibility of a MC is equivalent to $f_{i,j}^*$ being positive for every choice of $i, j \in \mathbf{E}$. If, additionally, condition **D** holds for the MC, the stronger result **(7.9.1)** below is available.

Theorem 7.9.1. *If condition* **D** *holds for an irreducible and aperiodic MC, then all its states are persistent and geometrically ergodic.*

Proof. Fix state j and note that $f_{i,j}^* > 0$ for every i. Hence $B_r \downarrow \phi$ as $r \uparrow \infty$ where $B_r = \{i : f_{i,j}^* \leq \frac{1}{r}\}$. Since φ is a finite measure, $\varphi(B_r) \downarrow 0$. Hence given $\varepsilon > 0$, we can find r large such that, writing $\hat{B} = B_r$, $\varphi(\hat{B}) \leq \frac{\varepsilon}{2}$. We note $i \in \hat{B}'$ implies $f_{i,j}^* > \delta = \frac{1}{r}$. Write $\tilde{f}_{i,j}^{(n)} = \sum_{m=1}^{n} f_{i,j}^{(m)}$, $n \geq 1$. The sequence $(\tilde{f}_{\cdot,j}^{(n)})$ of functions defined on the measure space $(\mathbf{E}, \mathcal{E}, \varphi)$ has, for each i, a finite limit (namely, $f_{i,j}^*$). Hence, given $\varepsilon > 0$, there exists , by **Q**13, $\bar{B} \subset \mathbf{E}$ such that $\varphi(\bar{B}) < \frac{\varepsilon}{2}$ and on $\bar{B}'$ the above sequence converges uniformly. Write $B = \hat{B} \cup \bar{B}$ and note that $\varphi(B) \leq \varepsilon$. Since $B' \subset \hat{B}'$, $f_{i,j}^* > \delta$, $i \in B'$. Since $B' \subset \bar{B}'$, it follows the sequence converges uniformly on B'. Hence there exists a positive integer α such that $\tilde{f}_{i,j}^{(\alpha)} \geq \delta$, for all $i \in B'$.

By **D**, $p^{(\nu)}(i, B) \leq 1 - \varepsilon$ or $p^{(\nu)}(i, B') > \varepsilon$, $i \in \mathbf{E}$. For $i \in \mathbf{E}$ arbitrary, we have

$$\begin{aligned}
\tilde{f}_{i,j}^{(\alpha+\nu)} &= P(X_t = j \text{ for some } t,\ 1 \leq t \leq \alpha + \nu | X_0 = i) \\
&\geq P(X_t = j \text{ for some } t,\ \nu + 1 \leq t \leq \nu + \alpha | X_0 = i) \\
&= \sum_{k \in \mathbf{E}} P(X_t = j \text{ for some } t,\ \nu + 1 \leq t \leq \nu + \alpha,\ X_\nu = k | X_0 = i) \\
&\geq \sum_{k \in B'} P(X_t = j \text{ for some } t,\ \nu + 1 \leq t \leq \nu + \alpha,\ X_\nu = k | X_0 = i) \\
&= \sum_{k \in B'} \tilde{f}_{k,j}^{(\alpha)} P(X_\nu = k | X_0 = i)
\end{aligned}$$

$$\begin{aligned}&\geq \delta \sum_{k\in B'} P(X_\nu = k|X_0 = i) = \delta P(X_\nu \in B'|X_0 = i)\\ &= \delta p^{(\nu)}(i,\ B') > \delta\varepsilon.\end{aligned}$$

Write $\alpha + \nu = \beta$ and $\rho = 1 - \delta\varepsilon$. We note that $0 < \rho < 1$ and that ρ depends on j. In terms of the new parameters, what we have shown is:

$$\tilde{f}_{i,j}^{(\beta)} \geq 1 - \rho,\ i \in \mathbf{E}. \tag{7.72}$$

We will show that $\tilde{f}_{i,j}^{(m\beta)} \geq 1 - \rho^m$. This we do by induction on m . Assume then the claim has been established for $m = 1,\ 2, \ldots,\ r$. We note $\tilde{h}_{i,j}^{(n)} = 1 - \tilde{f}_{i,j}^{(n)}$ is the probability that the system starting from state i does not reach state j in the first n steps. Or, what is same, the system stays in $\{j\}'$ during the first n steps. Hence, $\tilde{h}_{i,j}^{(\beta)} \leq \rho$ and

$$\begin{aligned}\tilde{h}_{i,j}^{(\overline{m+1}\beta)} &\leq \sum_{k\neq j} P\big(X_t \neq j,\ 1 \leq t \leq \overline{m+1}\beta,\ X_\beta = k \text{ for some}\\ &\qquad\qquad k \neq j|X_0 = i\big)\\ &= \sum_{k\neq j} P\big(X_t \neq j,\ \beta + 1 \leq t \leq \overline{m+1}\beta|X_\beta = k,\ X_t \neq j,\\ &\qquad\qquad 1 \leq t \leq \beta,\ X_0 = i\big) \times\\ &\qquad\qquad P(X_\beta = k,\ X_t \neq j, 1 \leq t \leq \beta|X_0 = i)\\ &\leq \rho^m \sum_{k\neq j} P(X_\beta = k,\ X_t \neq j,\ 1 \leq t \leq \beta|X_0 = i)\\ &\leq \rho^m P(X_t \neq j,\ 1 \leq t \leq \beta|X_0 = i)\\ &= \rho^m \tilde{h}_{i,j}^{(\beta)}\\ &\leq \rho^m \rho \text{ (by induction hypothesis)}\\ &= \rho^{m+1}.\end{aligned}$$

For n, an arbitrary integer $\geq \beta$,

$$\tilde{f}_{i,j}^{(n)} \geq \tilde{f}_{i,j}^{([\frac{n}{\beta}]\beta)} \geq 1 - \rho^{[\frac{n}{\beta}]} \geq 1 - \rho^{\frac{n}{\beta}-1}.$$

For $n < \beta$, the right side of the inequality is negative and hence the inequality is satisfied trivially.

We note that for all $i \in \mathbf{E}$,

$$f_{i,j}^{*} = \lim_{n\to\infty} \tilde{f}_{i,j}^{(n)} \geq 1 - \lim_{n\to\infty} \rho^{\frac{n}{\beta}-1} = 1.$$

In particular $f_{j,j}^{*} = 1$. i.e., j is a persistent state. This implies, since the chain is irreducible, that all the states are persistent.

Since $\lambda_n = \sum_{t=n+1}^{\infty} f_{j,j}^{(t)} = 1 - \tilde{f}_{j,j}^{(n)}$ tends to zero exponentially fast, it follows that the series $\sum_{n=1}^{\infty} f_{j,j}^{(n)} x^n$ has a radius of convergence greater than 1. Now we appeal to **(7.7.7)** and conclude that j, and hence (ref. **(7.8.4)**) all the states are geometrically ergodic. □

We have shown that if an irreducible and aperiodic MC satisfies condition **D**, then all its states are persistent positive and geometrically ergodic. Let $\mathcal{Q}$ denote the pm generated by the stationary distribution : for $A \subset \mathbf{E}$, $\mathcal{Q}(A) = \sum_{j \in A} \pi_j$. In Theorem **(7.9.2)**, we prove the geometric ergodicity is uniform over subsets of **E**.

The abridged result quoted below from measure theory is needed for the proof of **(7.9.2)**.

Q16 (Theorem E, p121,[H])

Let μ_1, μ_2 be two probability measures on a measurable space $(\mathbf{X}, \mathscr{X})$ and let μ be defined by setting $\mu(A) = \mu_1(A) - \mu_2(A)$, $A \in \mathscr{X}$.

(i) A set E is said to be positive with respect to μ if , for every measurable set F, $E \cap F$ is measurable and $\mu(E \cap F) \geq 0$. Similarly, we call E negative if , for every measurable set F, $E \cap F$ is measurable and $\mu(E \cap F) \leq 0$.

(ii) there exist two disjoint sets A, B such that $A \cup B = \mathbf{X}$ and such that A is positive and B is negative with respect to μ.

Theorem 7.9.2 (Uniform geometric ergodicity). *Let condition* **D** *hold for an irreducible and aperiodic MC. Let $\mathcal{Q}$ denote the stationary measure. Then there exist $\gamma > 0$, $0 < \rho < 1$ such that*

$$\sup_{A \subset \mathbf{E}} \sup_{i \in \mathbf{E}} |p^{(n)}(i,\ A) - \mathcal{Q}(A)| \leq \gamma \rho^n,\ n = 1,\ 2, \ldots \tag{7.73}$$

Proof. An arbitrary state is chosen and fixed and labelled 0.

Step 1. By (7.72), there exists an integer β such that, for all $i \in \mathbf{E}$ and for all $n \geq \beta$, $\tilde{f}_{i,0}^{(n)} \geq \delta_1 (= 1 - \rho) > 0$. The aperiodicity of the states implies that there exists integer τ such that $p_{0,0}^{(n)} > 0$ for all $n \geq \tau\beta$. Since $\lim_{n \to \infty} p_{0,0}^{(n)} = \pi_0 > 0$, it follows that $\mathfrak{q} = \inf\{p_{0,0}^{(n)} : n \geq \tau\beta\} > 0$. Setting $\nu = (\tau + 1)\beta$, we have for all $i \in \mathbf{E}$,

$$p_{i,0}^{(\nu)} = p_{i,0}^{(\tau+1)\beta} = \sum_{t=0}^{(\tau+1)\beta} p_{0,0}^{(t)} f_{i,0}^{(\overline{\tau+1}\beta - t)} \geq \sum_{t=\tau\beta}^{(\tau+1)\beta} \mathfrak{q} f_{i,0}^{(\overline{\tau+1}\beta - t)}$$

$$= \mathfrak{q} \tilde{f}_{i,0}^{(\beta)} \geq \mathfrak{q}\delta_1 = \delta_2, \text{ say.}$$

This leads to

$$\frac{p^{(\nu)}(i,\ 0)}{\varphi(\{0\})} \geq \frac{\delta_2}{\varphi(\{0\})} = \delta_3, \qquad \text{say.}$$

We note that for each $i \in \mathbf{E},\ p^{(\nu)}(i, A)$ is a probability measure on $\mathcal{E}$ with

$$q^{(\nu)}(i,\ k) = \frac{p^{(\nu)}(i,\ k)}{\varphi(\{k\})}$$

for its Radon-Nikodym derivative with respect to φ. In this notation, we have, $q^{(\nu)}(i,\ 0) \geq \delta_3,\ i \in \mathbf{E}$.

Step 2. Define on $\mathcal{E}$ the set functions :

$$\psi_{i,j}(A) = p^{(\nu)}(i,\ A) - p^{(\nu)}(j,\ A),\ i,\ j\ \in \mathbf{E},\ A \in \mathcal{E}.$$

We note that $\psi_{i,j}$ is finite valued and countably additive (and that if the elements of the ith row and of the jth row are the same then $\psi_{i,j}(A) = 0,\ A \in \mathcal{E}$). Let S^+, S^- be respectively the positive set and the negative set with respect to $\psi_{i,j}$ (ref. **Q**16). Since $\psi_{i,j}$ is additive,

$$\begin{aligned}\psi_{i,j}(S^+) + \psi_{i,j}(S^-) &= \psi_{i,j}(S^+ \cup S^-) = \psi_{i,j}(\mathbf{E}) \\ &= p^{(\nu)}(i,\ \mathbf{E}) - p^{(\nu)}(j,\ \mathbf{E}) = 0.\end{aligned}$$

Now,

$$\begin{aligned}\psi_{i,j}(S^+) &= p^{(\nu)}(i,\ S^+) - p^{(\nu)}(j,\ S^+) \\ &= 1 - p^{(\nu)}(i, S^-) - p^{(\nu)}(j,\ S^+) \\ &= 1 - \int_{S^-} q^{(\nu)}(i,\ k)\, d\varphi(k) - \int_{S^+} q^{(\nu)}(j,\ k)\, d\varphi(k).\end{aligned}$$

The right side quantity does not exceed

$$1 - \int_{\{0\}} q^{(\nu)}(i,\ k)\, d\varphi(k) \text{ if } \{0\} \subset S^-$$

and it does not exceed

$$1 - \int_{\{0\}} q^{(\nu)}(j,\ k)\, d\varphi(k) \text{ if } \{0\} \subset S^+.$$

In either case, it does not exceed

$$1-\delta_3\varphi(\{0\}). \text{ Hence } \psi_{i,j}(S^+)\le 1-\delta_3\varphi(\{0\}).$$

Step 3. Define, for $A\in\mathcal{E}$,

$$m^{(t)}(A)=\inf_{i\in\mathbf{E}} p^{(t)}(i,\ A) \text{ and } M^{(t)}(A)=\sup_{i\in\mathbf{E}} p^{(t)}(i,\ A). \text{ By (7.12),}$$

$$\begin{aligned} m^{(t+1)}(A) &= \inf_{i\in\mathbf{E}} p^{(t+1)}(i,\ A)=\inf_{i\in\mathbf{E}}\sum_{j\in\mathbf{E}} p_{i,j}p^{(t)}(j,\ A)\\ &\ge \inf_{i\in\mathbf{E}}\sum_{j\in\mathbf{E}} p_{i,j}\inf_{j\in\mathbf{E}} p^{(t)}(j,\ A)=\inf_{i\in\mathbf{E}}\sum_{j\in\mathbf{E}} p_{i,j}m^{(t)}(A)\\ &= m^{(t)}(A). \end{aligned}$$

Thus $(m^{(t)}(A))$ is a monotonic non-decreasing sequence. It can be shown on similar lines that $(M^{(t)}(A))$ is monotonic non-increasing sequence. We have

$$\begin{aligned} &M^{(\overline{t+1}\nu)}(A)-m^{(\overline{t+1}\nu)}(A)\\ &=\sup_{i\in\mathbf{E}} p^{(\overline{t+1}\nu)}(i,\ A)-\inf_{j\in\mathbf{E}} p^{(\overline{t+1}\nu)}(j,\ A)\\ &=\sup_{i\in\mathbf{E}}\sum_{k\in\mathbf{E}} p^{(\nu)}(i,\ k)p^{(t\nu)}(k,\ A)-\inf_{j\in\mathbf{E}}\sum_{k\in\mathbf{E}} p^{(\nu)}(j,\ k)p^{(t\nu)}(k,\ A)\\ &=\sup_{i,j}\sum_{k\in\mathbf{E}} p^{(t\nu)}(k,A)\big[p^{(\nu)}(i,\ k)-p^{(\nu)}(j,\ k)\big]\\ &=\sup_{i,j}\sum_{k\in S^+} p^{(t\nu)}(k,A)\big[p^{(\nu)}(i,\ k)-p^{(\nu)}(j,\ k)\big]\\ &\quad+\sup_{i,j}\sum_{k\in S^-} p^{(t\nu)}(k,A)\big[p^{(\nu)}(i,\ k)-p^{(\nu)}(j,\ k)\big]\\ &\le\sup_{i,j}\Big\{M^{(t\nu)}(A)\sum_{k\in S^+}\big[p^{(\nu)}(i,\ k)-p^{(\nu)}(j,\ k)\big]\\ &\quad+m^{(t\nu)}(A)\sum_{k\in S^-}\big[p^{(\nu)}(i,\ k)-p^{(\nu)}(j,\ k)\big]\Big\}\\ &=\sup_{i,j}\big\{M^{(t\nu)}(A)\psi_{i,j}(S^+)+m^{(t\nu)}(A)\psi_{i,j}(S^-)\big\}\\ &=\big\{M^{(t\nu)}(A)\psi_{i,j}(S^+)-m^{(t\nu)}(A)\psi_{i,j}(S^+)\big\}\\ &=\psi_{i,j}(S^+)\{M^{(t\nu)}(A)-m^{(t\nu)}(A)\}\\ &\le(1-\delta_3\varphi(\{0\}))\{M^{(t\nu)}(A)-m^{(t\nu)}(A)\}. \end{aligned}$$

Hence,
$$M^{(\overline{t+1}\nu)}(A) - m^{(\overline{t+1}\nu)}(A) \leq (1 - \delta_3\varphi(\{0\}))^{t+1}.$$

For $n \geq 1$ arbitrary, define $t = \left[\frac{n}{\nu}\right]$. Noting that $M^{(n)}(A) - m^{(n)}(A)$ is a monotonic decreasing sequence, we have

$$\begin{aligned} M^{(n)}(A) - m^{(n)}(A) &\leq M^{(t\nu)}(A) - m^{(t\nu)}(A) \\ &\leq (1 - \delta_3\varphi(\{0\}))^t \leq (1 - \delta_3\varphi(\{0\}))^{\frac{n}{\nu}-1} \to 0. \end{aligned}$$

Hence the two sequences $(M^{(n)}(A))$ and $(m^{(n)}(A))$ have a common limit. Denote this limit by $u(A)$. It can be evaluated as the $\lim_{n\to\infty} p^{(n)}(i, A)$ for a conveniently chosen i. We note that u is additive since, for disjoint subsets $A \subset \mathbf{E}$, $B \subset \mathbf{E}$,

$$\begin{aligned} u(A \cup B) &= \lim_{n\to\infty} p^{(n)}(i, A \cup B) \\ &= \lim_{n\to\infty} [p^{(n)}(i, A) + p^{(n)}(i, B)] \\ &= \lim_{n\to\infty} p^{(n)}(i, A) + \lim_{n\to\infty} p^{(n)}(i, B) \\ &= u(A) + u(B). \end{aligned}$$

For every single point set $A = \{j\}$, $u(A) = \pi_j = \mathcal{Q}(A)$. By the additivity of u this implies $u(A) = \mathcal{Q}(A)$ for every finite set A. We claim that even if A has infinite (necessarily countably infinite) number of members, $u(A) = \mathcal{Q}(A)$. If this claim is not admitted, then let A be an infinite set with $u(A) \neq \mathcal{Q}(A)$. Now for any set $E \subset \mathbf{E}$,

$$\begin{aligned} u(E) &= \lim_{n\to\infty} p^{(n)}(i, E) \\ &= \lim_{n\to\infty} \sum_{j\in E} p_{i,j}^{(n)} \\ &\geq \sum_{j\in E} \lim_{n\to\infty} p_{i,j}^{(n)} \text{ (by Fatou's Lemma)} \\ &= \sum_{j\in E} \pi_j = \mathcal{Q}(E). \end{aligned}$$

Hence $u(A) \geq \mathcal{Q}(A)$ while, for the countable set A', we have $u(A') \geq \mathcal{Q}(A')$. If there exists A with $u(A) > \mathcal{Q}(A)$ or $u(A') > \mathcal{Q}(A')$, a contradiction will result since u, $\mathcal{Q}$ being additive set functions we will have $1 = u(\mathbf{E}) = u(A) + u(A') > \mathcal{Q}(A) + \mathcal{Q}(A') = \mathcal{Q}(\mathbf{E}) = 1$. Thus $u(A) = \mathcal{Q}(A)$ for all $A \in \mathcal{E}$. In other words, the limit set function u is the stationary measure $\mathcal{Q}$.

Since for each $i \in \mathbf{E}$, $m^{(n)}(A) \leq p^{(n)}(i, A) \leq M^{(n)}(A)$, it follows that $\sup_{A \subset \mathbf{E}} \sup_{i \in \mathbf{E}} |p^{(n)}(i, A) - \mathcal{Q}(A)| \leq M^{(n)}(A) - m^{(n)}(A) \leq (1 - \delta_3 \varphi(\{0\}))^{\frac{n}{\nu}-1} = (1 - \delta_2)^{\frac{n}{\nu}-1}$.

Now, we chose δ_1 small so that $\delta_2 < 1$. Write $(1 - \delta_2)^{\frac{1}{\nu}} = \rho$ and $(1 - \delta_2)^{-1} = \gamma$ and note that (7.73) is established. □

7.10 The central limit theorem for Markov chains.

Let the irreducible and aperiodic MC, $\{X_n, \ n \geq 0\}$ satisfy condition **D**. Hence (7.73) holds. Let f be a real function defined on **E**. We note the $f(X_n)$s are rvs. As before, $\mathcal{Q}$ denotes the stationary measure of the chain. Write $S_n = \sum_{\nu=0}^{n} f(X_\nu)$. The distribution of S_n is determined by the initial distribution (i.e., the distribution of X_0) and the transition probabilities.

If $\mathfrak{a}(.)$ is the distribution of X_0, expectation operation will be denoted by $\mathbb{E}_{\mathfrak{a}}$. If $\mathfrak{a} = \mathcal{Q}$, the suffix $\mathcal{Q}$ to the operator $\mathbb{E}$ will be dropped.

We recall (ref. **(7.6.4)**) that under $\mathcal{Q}$, the X_ns form a stationary process. Throughout this section, it is assumed $\mathbb{E} f(X_0) = 0$ and $\mathbb{E}\{f(X_0)\}^2 = \sigma^2 < \infty$.

Lemma 7.10.1. *Let g be a real and bounded function, measurable with respect to the σ-field $\mathcal{F}^{(k+1)}$ generated by X_ν, $\nu \geq k+1$. Then*

$$\mathbb{E}(g|X_0) = \sum_{j \in \boldsymbol{E}} \mathbb{E}(g|X_k = j) p^{(k)}(X_0, j).$$

Proof. Let M be a bound for g.

Let $\mathcal{A}$ be the collection of the sets in $\mathcal{F}^{(k+1)}$ such that the lemma is true for their characteristic functions. We note $\mathcal{A}$ is closed under complementation since

$$\begin{aligned}
P(A'|X_0) &= 1 - P(A|X_0) \\
&= 1 - \sum_{j \in \mathbf{E}} P(A|X_k = j) p^{(k)}(X_0, j) \\
&= 1 - \sum_{j \in \mathbf{E}} \{1 - P(A'|X_k = j)\} p^{(k)}(X_0, j) \\
&= 1 - \sum_{j \in \mathbf{E}} p^{(k)}(X_0, j) + \sum_{j \in \mathbf{E}} P(A'|X_k = j) p^{(k)}(X_0, j) \\
&= \sum_{j \in \mathbf{E}} P(A'|X_k = j) p^{(k)}(X_0, j) \text{ and hence } A' \in \mathcal{A}.
\end{aligned}$$

Let $A_n \in \mathcal{A},\ A_n \uparrow A$. We note $A \in \mathcal{F}^{(k+1)}$. Now,

$$\begin{aligned}
P(A|X_0) &= P(\lim_{n\to\infty} A_n|X_0) \\
&= \lim_{n\to\infty} P(A_n|X_0) \text{ (by } \mathbf{(6.4)}\text{(v))} \\
&= \lim_{n\to\infty} \sum_{j\in\mathbf{E}} P(A_n|X_k = j)p^{(k)}(X_0,\ j) \\
&= \sum_{j\in\mathbf{E}} \lim_{n\to\infty} P(A_n|X_k = j)p^{(k)}(X_0,\ j) \\
&\qquad \text{(by the bounded convergence theorem)} \\
&= \sum_{j\in\mathbf{E}} P(A|X_k = j)p^{(k)}(X_0,\ j),
\end{aligned}$$

showing that $\mathcal{A}$ is closed under countable union. That $\mathcal{A}$ is a σ-field follows.

Let A be a cylinder set in $\mathcal{F}^{(k+1)}$. It is necessarily the union of a countable number of sets of the type $\{X_{k+r} = i_r,\ 1 \le r \le t\}$ for some $t \ge 1$.
Let g be the characteristic function of the set $\{X_{k+r} = i_r,\ 1 \le r \le t\}$. Then,

$$\begin{aligned}
\mathbb{E}(g|X_0) &= p^{(k+1)}(X_0,\ i_1)\prod_{r=1}^{t-1} p(i_r,\ i_{r+1}) \\
&= \sum_{j\in\mathbf{E}} p^{(k)}(X_0,\ j)p(j,\ i_1)\prod_{r=1}^{t-1} p(i_r,\ i_{r+1}) \\
&= \sum_{j\in\mathbf{E}} p^{(k)}(X_0,\ j)\mathbb{E}(g|X_k = j)
\end{aligned}$$

proving the Lemma in this case. i.e., all the cylindrical sets belong to $\mathcal{A}$. Hence $\mathcal{A} = \mathcal{F}^{(k+1)}$. In other words, the Lemma is true for the characterstic functions of members of $\mathcal{F}^{(k+1)}$. Extension of the result to simple functions functions follows immediately. Since a bounded non-negative function is the pointwise limit of a uniformly bounded non-negative monotonically increasing sequence of simple functions, the proof is completed by again appealing to (**(6.4)**(v)) and bounded convergence theorem. By the linearity of the conditional expectation operation, the Lemma stands proved for all bounded $\mathcal{F}^{(k+1)}$-measurable functions. □

Lemma 7.10.2. *Let $g,\ M,\ \mathcal{F}^{(k+1)}$ be as in **(7.10.1)**. Then $|\mathbb{E}(g|X_0) - \mathbb{E}g| \le 2\gamma\rho^k$ where $\gamma,\ \rho$ are as in* (7.73).

Proof. Since

$$\begin{aligned}\mathbb{E}g = \mathbb{E}[\mathbb{E}(g|X_k)] &= \sum_{j\in\mathbf{E}} \mathbb{E}(g|X_k = j)P(X_k = j)\\ &= \sum_{j\in\mathbf{E}} \pi_j \mathbb{E}(g|X_k = j),\end{aligned}$$

we have, by **(7.10.1)**,

$$\begin{aligned}|\mathbb{E}(g|X_0 = i) - \mathbb{E}g| &= \Big|\sum_{j\in\mathbf{E}} \mathbb{E}(g|X_k = j)[p^{(k)}(i,\ j) - \pi_j]\Big|\\ &\leq M \sum_{j\in\mathbf{E}} |p^{(k)}(i, j) - \pi_j|.\end{aligned}$$

Let $S^+ = \{j\ :\ p^{(k)}(i,\ j) - \pi_j \geq 0\}$ and $S^- = \mathbf{E} \sim S^+$. Then,

$$\begin{aligned}&|\mathbb{E}(g|X_0 = i) - \mathbb{E}g|\\ &\quad\leq M\{\sum_{j\in S^+} [\,p^{(k)}(i,\ j) - \pi_j] - \sum_{j\in S^-} [\,p^{(k)}(i,\ j) - \pi_j]\}\\ &\quad\leq M\{\,p^{(k)}(i,\ S^+) - \mathcal{Q}(S^+) + |p^{(k)}(i,\ S^-) - \mathcal{Q}(S^-)|\}\\ &\quad\leq 2M\gamma\rho^k \text{ by (7.73).}\end{aligned}$$

□

Lemma 7.10.3. *If integer* $t \geq 1$, *then*

$$|\mathbb{E}f(X_0)f(X_t)| \leq 2\sigma^2\gamma^{\frac{1}{2}}\rho^{\frac{t}{2}}$$

where $\gamma,\ \rho$ *are as in* (7.73).

Proof. Let $S^+,\ S^-$ be as above. We note

$$\mathbb{E}\{f(X_0)\int_{\mathbf{E}} f(y)\mathcal{Q}(\,dy)\} = \{\mathbb{E}f(X_0)\}^2 = 0. \tag{7.74}$$

This result leads to
$|\mathbb{E}f(X_0)f(X_t)| = |\mathbb{E}\{f(X_0)[\mathbb{E}(f(X_t|X_0))]\}|$

$$\begin{aligned}&= |\mathbb{E}\{f(X_0)\int_{\mathbf{E}} f(y)|p^{(t)}(X_0,\ dy)\}|\\ &= |\mathbb{E}\{f(X_0)\int_{\mathbf{E}} f(y)|[p^{(t)}(X_0,\ dy) - \mathcal{Q}(\,dy)]\}|\end{aligned}$$

$$\leq (\mathbb{E}(f(X_0))^2)^{\frac{1}{2}}[\mathbb{E}(\int_{|} \mathbf{E}f(y)\{p^{(t)}(X_0,\ dy) - \mathcal{Q}(\ d|y)\})^2]^{\frac{1}{2}}$$

$$= \sigma[\mathbb{E}\Big(\int_{S^+} f(y)\{p^{(t)}(X_0,\ |dy) - \mathcal{Q}(\ dy)\}$$

$$+ \int_{S^-} f(y)\{p^{(t)}(X_0,\ dy) - |\mathcal{Q}(\ dy)\}\Big)^2]^{\frac{1}{2}}$$

$$\leq \sigma\sqrt{2}[\mathbb{E}(\int_{S^+} |f(y)||\{p^{(t)}(X_0,\ dy) - \mathcal{Q}(\ dy))\})^2$$

$$+ \mathbb{E}(\int_{S^-} |f(y)|\{p^{(t)}(X_0,\ dy)| - \mathcal{Q}(\ dy))\})^2]^{\frac{1}{2}},\ \text{by } \mathbf{(2.6.5)}$$

Write the integrand as $|f(y)| \times 1$ and apply Hölder inequality to get

$$|\mathbb{E}f(X_0)f(X_t)|$$

$$\leq \sigma\sqrt{2}[\mathbb{E}\Big(\{p^{(t)}(X_0,\ S^+) - \mathcal{Q}(S^+)\}$$

$$\int_{S^+} |f(y)|^2\{p^{(t)}(X_0,\ dy) - \mathcal{Q}(\ dy))\})$$

$$+ \mathbb{E}(-\{p^{(t)}(X_0,\ S^-) - \mathcal{Q}(S^-)\}$$

$$\int_{S^-} \{-|f(y)|^2\}\{p^{(t)}(X_0,\ dy) - \mathcal{Q}(\ dy))\}\Big)]^{\frac{1}{2}}$$

$$\leq \sigma\sqrt{2}(\gamma\rho^t)^{\frac{1}{2}}[\mathbb{E}\int_{\mathbf{E}} |f(y)|^2\{p^{(t)}(X_0,\ dy) + \mathcal{Q}(\ dy)\}]^{\frac{1}{2}}$$

$$= \sigma\sqrt{2}(\gamma\rho^t)^{\frac{1}{2}}[\mathbb{E}\{\int_{\mathbf{E}} |f(y)|^2 p^{(t)}(X_0,\ dy)\} + \mathbb{E}(f(X_0))^2]^{\frac{1}{2}}$$

$$= \sigma\sqrt{2}(\gamma\rho^t)^{\frac{1}{2}}[\{\sum_{i\in\mathbf{E}} \pi_i \sum_{j\in\mathbf{E}} |f(j)|^2 p^{(t)}(i,\ j)\} + \sigma^2]^{\frac{1}{2}}$$

$$= \sigma\sqrt{2}(\gamma\rho^t)^{\frac{1}{2}}[\sum_{j\in\mathbf{E}} |f(j)|^2\{\sum_{i\in\mathbf{E}} \pi_i p^{(t)}(i,\ j)\} + \sigma^2]^{\frac{1}{2}}$$

(interchanging the order of summation, which is justified since all the terms are non-negative)

$$= \sigma\sqrt{2}(\gamma\rho^t)^{\frac{1}{2}}[\{\sum_{j\in\mathbf{E}} |f(j)|^2\pi_j\} + \sigma^2]^{\frac{1}{2}}$$

$$= \sigma\sqrt{2}(\gamma\rho^t)^{\frac{1}{2}}(2\sigma^2)^{\frac{1}{2}} \text{ (since } (\pi_j) \text{ is the stationary distribution)}$$
$$= 2\sigma^2\gamma^{\frac{1}{2}}\rho^{\frac{t}{2}}.$$ □

Lemma 7.10.4. *Let*

$$\sigma_1^2 = \sigma^2 + 2\sum_{\nu=1}^{\infty} \mathbb{E}f(X_0)f(X_\nu).$$

Then

$$\lim_{n\to\infty} \mathbb{E}(\frac{S_n}{\sqrt{n}})^2 = \sigma_1^2$$

exists finitely.

Proof. By **(7.10.3)**, $|\mathbb{E}f(X_0)f(X_\nu)| \le q\rho^{\frac{\nu}{2}}$. Hence $\sum_{\nu=1}^{\infty} \nu\mathbb{E}f(X_0)f(X_\nu)$ is absolutely convergent.

$$\mathbb{E}S_n^2 = \sum_{\nu=0}^{n} \mathbb{E}f^2(X_\nu) + 2\sum_{\nu=0}^{n-1}\sum_{t=1}^{n-\nu} \mathbb{E}f(X_\nu)f(X_{\nu+t}) = n\sigma^2 + 2A_n$$

where

$$A_n = \sum_{\nu=0}^{n-1}\sum_{t=1}^{n-\nu} \mathbb{E}f(X_\nu)f(X_{\nu+t}) = \sum_{\nu=0}^{n-1}\sum_{t=1}^{n-\nu} \mathbb{E}f(X_0)f(X_t)$$

since the X_n-process, with X_0 having Q for its distribution, is stationary. Rearranging the terms, $A_n = \sum_{\nu=1}^{n}(n+1-\nu)\mathbb{E}f(X_0)f(X_\nu)$. Hence

$$\mathbb{E}(\frac{S_n^2}{n}) = \sigma^2 + 2\sum_{\nu=1}^{n} \mathbb{E}f(X_0)f(X_\nu) - \frac{2}{n}\sum_{\nu=1}^{n}(\nu-1)\mathbb{E}f(X_0)f(X_\nu)$$
$$\to \sigma^2 + 2\sum_{\nu=1}^{\infty} \mathbb{E}f(X_0)f(X_\nu).$$ □

Example 7.10.5. Consider a 2-state MC with $p_{0,0} = \frac{1}{3}$, $p_{0,1} = \frac{2}{3}$; $p_{1,0} = \frac{3}{4}$, $p_{1,1} = \frac{1}{4}$. The characteristic roots of the transition matrix P are 1 and $\frac{-5}{12}$. The right characteristic vectors corresponding to these roots are respectively $(1,\ 1)'$ and $(8,\ -9)'$; the corresponding left characteristic vectors are $(9,\ 8)'$ and $(1,\ -1)'$. Let

$$A = \frac{1}{\sqrt{17}}\begin{pmatrix} 9 & 8 \\ 1 & -1 \end{pmatrix}; \quad B = \frac{1}{\sqrt{17}}\begin{pmatrix} 1 & 8 \\ 1 & -9 \end{pmatrix}$$

so $AB = I$; and $APB = Q$, where $Q = \begin{pmatrix} 1 & 0 \\ 0 & -\frac{5}{12} \end{pmatrix}$; $P = BQA$; $P^n = BQ^nA$ leading to $p_{0,0}^{(n)} = \pi_0 + \frac{8}{17}(-\rho)^n$; $p_{1,0}^{(n)} = \pi_0 - \frac{-9}{17}(-\rho)^n$; $p_{0,1}^{(n)} = \pi_1 + \frac{-8}{17}(-\rho)^n$; $p_{1,1}^{(n)} = \pi_1 + \frac{9}{17}(-\rho)^n$ where $\rho = \frac{5}{12}$. We note $\pi_0 = \frac{9}{17}$ and $\pi_1 = \frac{8}{17}$.

Define $f(0) = 8$ and $f(1) = -9$. Then $\sigma^2 = \frac{1224}{17}$. To calculate $a_n = \mathbb{E}f(X_0)f(X_n)$ we note that the *rv* $f(X_0)f(X_n)$ takes the value 64 with probability $\pi_0 p_{0,0}^{(n)}$, -72 with probability $\pi_0 p_{0,1}^{(n)} + \pi_1 p_{1,0}^{(n)}$ and 81 with probability $\pi_1 p_{1,1}^{(n)}$. Hence $a_n = 72(-\frac{5}{12})^n$, leading to

$$\lim_{n\to\infty} \mathbb{E}\frac{S_n^2}{n} = \frac{1224}{17} - \frac{720}{17} = \frac{504}{17}.$$

Note 7.10.6. Let $0 \le m_1 < n_1 \le m_2 < n_2 \le \cdots \le m_k < n_k \le n$. Let $Y_r = \sum_{\nu=m_r}^{n_r} f(X_\nu)$. Let $T_k = \sum_{r=1}^{k} Y_r$. If $t_k = \sum_{r=1}^{k}(n_r - m_r) \to \infty$, then arguing on the above lines we get $\lim_{t_k\to\infty} \frac{1}{t_k}\mathbb{E}T_k^2 \to \sigma_1^2$.

Assumption. In what follows, σ_1 is always assumed to be positive.

Our approach to the proof of the CLT i.e., for proving $\frac{S_n}{\sigma_1\sqrt{n}} \xrightarrow{d} \xi$ is as follows. Let $\beta = \beta(n) = [n^{\frac{1}{4}}]$ and $\alpha = \beta^3$. Let $A = \{f(X_\nu),\ \nu = 0, 1, \ldots, n\}$. Let Y_1 denote the sum of the first α of these variables, Z_1 that of the next β variables, Y_2 that of the next α variables, Z_2 that of the next β variables and so on. For our choice of α and β, there will be essentially β blocks, each block the sum of $\alpha + \beta$ variables, namely, the sum $Y_\nu + Z_\nu$ for some ν. We prove that the sum of the Z_νs divided by $\sqrt{n}$ converges to zero in probability. We show that the Y_νs are *nearly* independent. After some adjustments, we appeal to CLT for independent variables and conclude the proof. Now the details.

Define

$$N = N(n) = [\frac{n}{\alpha+\beta}]$$

$$Y_\nu = \sum_{r=(\nu-1)(\alpha+\beta)}^{(\nu-1)(\alpha+\beta)+\alpha-1} f(X_r),\ \nu = 1, 2, \ldots, N;$$

$$Z_\nu = \sum_{r=(\nu-1)(\alpha+\beta)+\alpha}^{\nu(\alpha+\beta)-1} f(X_r),\ \nu = 1, \ldots, N;$$

$$Z_{N+1} = \sum_{r=N(\alpha+\beta)}^{n} f(X_r).$$

By **(7.10.6)**,

$$\mathbb{E}(\sum_{\nu=1}^{N+1} Y_\nu)^2 \sim n\sigma_1^2.$$

This implies

$$\frac{\sum\limits_{\nu=1}^{N+1} Z_\nu}{\sqrt{n}} \overset{pr}{\to} 0,$$

since the expected value of its square, by **(7.10.6)**, being asymptotically $\frac{1}{\sqrt{n}}$, tends to 0 as $n \to \infty$.

Let

$$\Lambda_N = \sum_{\nu=1}^{N} Y_\nu.$$

Thus

$$(\frac{S_n}{\sigma_1\sqrt{n}} \overset{d}{\to} \xi) \text{ is equivalent to } (\frac{\Lambda_N}{\sigma_1\sqrt{n}} \overset{d}{\to} \xi).$$

Again by **(7.10.6)**, $\mathbb{E}\Lambda_N^2 \sim n\sigma_1^2$. We define the expected value of a complex valued random variable in the natural way through the expected values of its real and imaginary parts. By **(6.4)**(vii),

$$\begin{aligned} &E[e^{itY_N}|Y_t,\ 1 \le t \le N-1] \\ &= E\Big\{E\Big(e^{itY_N}|X_\nu,\ 0 \le \nu \le (N-2)(\alpha+\beta)+\alpha-1\Big)\Big|Y_t, \\ &\qquad\qquad 1 \le t \le N-1\Big\} \\ &= (\text{by } \mathbf{(7.10.2)})\ \mathbb{E}e^{itY_N} + u_N \end{aligned}$$

where the error in absolute value is $\le 2\gamma\rho^\beta$.

$$\begin{aligned} &E(e^{it\Lambda_N}|Y_t,\ 1 \le t \le N-1) = e^{it\Lambda_{N-1}}E(e^{itY_N}|Y_t,\ 1 \le t \le N-1) \\ &= e^{it\Lambda_{N-1}}\{E(e^{itY_n}) + u_N\} \end{aligned}$$

(where $|u_N| \le 2\gamma\rho^\beta$ by **(7.10.2)**). Thus

$$\mathbb{E}e^{it\Lambda_N} = \mathbb{E}\Big[e^{it\Lambda_{N-1}}\big\{(\mathbb{E}e^{itY_n}) + u_N\big\}\Big] = (\mathbb{E}e^{itY_n})\mathbb{E}e^{itY_n} + \mathbb{E}v_N$$

where $|v_N| \le 2\gamma\rho^\beta$. This reduction process can be kept up and we get finally, using the fact that Y_νs are identically distributed,

$$\mathbb{E}e^{it\Lambda_N} = (\mathbb{E}e^{itY_1})^N + \sum_{\nu=1}^{N} v_\nu.$$

Since

$$\sum_{\nu=1}^{N} |v_\nu| \le 2\gamma\beta\rho^\beta \to 0 \text{ as } n \to \infty,$$

it follows that the limit of the *cf* of Λ_N is $\lim_{n\to\infty} (\mathbb{E}e^{itY_1})^N$ (if this limit exists). We note $(\mathbb{E}e^{itY_1})^N$ is the *cf* of the sum of N independent rvs identically distributed as Y_1. Let $g_n(t)$ denote the *cf* of $\frac{Y_1}{\sigma_1\sqrt{n}}$. Introduce N fictitious iid variables $Z_{n,\nu}$, $1 \le \nu \le N$ with common *cf* g_n. Denote their sum by $\tilde{Z}_N$. These N variables are infinitesimal, since

$$P(|Z_{n,1}| \ge \varepsilon) \le \frac{1}{\varepsilon^2}\mathbb{E}Z_{n,1}^2 = \frac{1}{\varepsilon^2}\frac{\mathbb{E}Y_1^2}{n\sigma_1^2} \sim q\frac{1}{n^{\frac{1}{4}}} \to 0.$$

Hence *if* the weak limit Z of the sequence $(\tilde{Z}_N)$ (or, for the same reason, the weak limit Z of any of its subsequence) exists, then Z will be infinitely divisible. Further Z will have finite variance, since

$$\lim_{n\to\infty} \mathbb{E}\tilde{Z}_N^2 = \frac{N}{n}\frac{\mathbb{E}Y_1^2}{\sigma_1^2} \sim \frac{N}{n\sigma_1^2}\frac{\alpha\sigma_1^2}{n} \to 1. \tag{7.75}$$

Hence (ref. **(3.4.5)**) the *cf* $g(t)$ of Z will have the Kolmogorov canonical representation

$$\log g(t) = -\frac{1}{2}a^2t^2 + \int_{|x|>0} \{e^{itx} - 1 - itx\}\frac{1}{x^2}\, dK(x) \tag{7.76}$$

for some $a \ge 0$ and some non-decreasing function K with $K(-\infty) = 0$ and $K(\infty) < \infty$.

(7.75) implies that the sequence $(\tilde{Z}_N)$ is a tight sequence. Hence, in order to prove $\tilde{Z}_N \xrightarrow{d} \xi$, the standard normal variable, it is enough to show that every weakly convergent subsequence of the sequence $(\tilde{Z}_N)$ converges weakly to ξ. i.e., it is enough to show that every convergent subsequence of $(g_n(t))^N$ converges to $e^{-\frac{t^2}{2}}$ or, what is same, every convergent subsequence of $(N\{1 - \Re(g_n(t))\})$ converges to $\frac{t^2}{2}$.

We complete the proof assuming

$$x^2 \text{ is uniformly integrable with respect to the sequence } (H_n) \tag{7.77}$$

where H_n is the df of $\frac{S_n}{\sigma_1\sqrt{n}}$. Let the h_n be the cf of H_n. We have the obvious relation

$$g_n(t) = h_{\alpha(n)}\{t\sqrt{\frac{\alpha(n)}{n}}\}.$$

Suppose for a subsequence (n_k), $n_k \uparrow \infty$, the weak limit Z of $(\tilde{Z}_{N(n_k)})$ exists, with *cf* $g(t)$ given at (7.76). This means that

$$-N(n_k)(1 - g_{n_k}(t)) \to \log g(t).$$

Now,

$$N(n_k)\{1 - \Re(g_{n_k}(t))\} = N(n_k)\{1 - \Re(h_{\alpha(n_k)}(t\sqrt{\frac{\alpha(n_k)}{n_k}}))\}$$

$$= N(n_k) \int_R \{1 - \cos tx\sqrt{\frac{\alpha(n_k)}{n_k}}\} \, dH_{\alpha(n_k)}(x).$$

Since

$$\frac{\alpha(n_k)}{n_k} \sim n_k^{-\frac{1}{4}} \sim N(n_k),$$

since $\frac{\sin \theta_n x}{\theta_n x} \to 1$ uniformly in x over bounded intervals as $\theta_n \to 0$ and since x^2 is uniformly integrable with respect to the sequence (H_n), it follows that

$$\lim_{k\to\infty} N(n_k)\{1 - \Re(h_n(t\sqrt{\frac{\alpha(n_k)}{n_k}}))\} = \lim_{k\to\infty} \int_R \frac{1}{2}t^2x^2 \, dH_{n_k}(x)$$

$$= \frac{1}{2}t^2 \lim_{k\to\infty} \operatorname{var} \frac{1}{\sigma_1\sqrt{\alpha(n_k)}} \sum_{\nu=1}^{\alpha(n_k)} f(X_\nu) = \frac{1}{2}t^2.$$

Thus for all t,

$$-\frac{t^2}{2} = -\frac{a^2t^2}{2} - \int_{R\sim\{0\}} \frac{1 - \cos tx}{x^2} \, dK(x).$$

Divide throughout by t^2, allow $|t| \to \infty$ and invoke bounded convergence theorem to arrive at $1 = a^2$. Then necessarily,

$$\int_{R\sim\{0\}} \frac{1 - \cos tx}{x^2} \, dK(x) \equiv 0,$$

leading to $g(t) = e^{-\frac{t^2}{2}}$. With this we have succeeded in proving:

Theorem 7.10.7. *If* (7.77) *holds, then*

$$\frac{S_n}{\sigma_1\sqrt{n}} \xrightarrow{d} \xi. \tag{7.78}$$

Note 7.10.8. The condition (7.77), under which (7.78) is proved, is often difficult to verify in most cases. It is desirable that any required condition is on the component variable $f(X_0)$. We now investigate the possibility of such a condition. Let

$$\mathbb{E}|f(X_0)|^{2+\delta} < \infty \text{ for some } \delta > 0. \tag{7.79}$$

This implies $\mathbb{E}|S_n|^{2+\delta} < \infty$ (ref. **(2.6.5)**). By **(3.6.11)**,

$$g_n(t) = 1 - \frac{t^2}{2}\mathbb{E}|\frac{Y_1}{\sigma_1\sqrt{n}}|^2 + |t|^{2+\delta}\mathbb{E}|\frac{Y_1}{\sigma_1\sqrt{n}}|^{2+\delta}O(1),$$

the constant in the order term being independent of n. If (7.78) holds, then

$$\lim_{n\to\infty}\{1 - \frac{t^2}{2}\mathbb{E}|\frac{Y_1}{\sigma_1\sqrt{n}}|^2 + |t|^{2+\delta}\mathbb{E}|\frac{Y_1}{\sigma_1\sqrt{n}}|^{2+\delta}O(1)\}^{\beta(n)} = e^{-\frac{1}{2}t^2}.$$

This is possible *iff*

$$\lim_{n\to\infty}\beta(n)\mathbb{E}|\frac{Y_1}{\sigma_1\sqrt{n}}|^{2+\delta} = 0$$

(and

$$\lim_{n\to\infty}\beta(n)\mathbb{E}(\frac{Y_1}{\sigma_1\sqrt{n}})^2 = 1,$$

which is true). i.e., *iff*

$$\lim_{n\to\infty}\mathbb{E}|\frac{S_n}{\sqrt{n}}|^{2+\delta}\frac{1}{n^{\frac{\delta}{8}}} = 0.$$

A sufficient condition for this is that there exists a constant a such that

$$\mathbb{E}|\frac{S_n}{\sigma_1\sqrt{n}}|^{2+\delta} \le a,\ n = 1,\ 2,\ldots. \tag{7.80}$$

That (7.80) implies (7.77) follows from (**(2.6.8)**, Remark (i)).

Trivially, (7.80) implies (7.79). For a proof that (7.79) implies (7.80) we refer the reader to Lemma 7.4, p225, [D].

CLT was proved under the assumption that X_0 has the stationary distribution. We now remove this restriction and prove: if (7.77) or (7.79) holds, then CLT holds whatever be the distribution of X_0.

$\mathbb{E}_{\mathfrak{a}}$, $\mathbb{E}$ will respectively denote expectation operation when X_0 has probability measure $\mathfrak{a}$ and when X_0 has the stationary distribution. We note that if h is a measurable function, then $\mathbb{E}_{\mathfrak{a}}h = \mathbb{E}_{\mathfrak{a}}\{E_{\mathfrak{a}}(h|X_0)\} = \sum_{i\in\mathbf{E}} \mathfrak{a}(\{i\})E_{\mathfrak{a}}(h|X_0 = i) = \sum_{i\in\mathbf{E}} \mathfrak{a}(\{i\})E(h|X_0 = i)$. Hence if $m < n$ then

$$\begin{aligned}
\mathbb{E}_{\mathfrak{a}}e^{\frac{it(S_n-S_m)}{\sigma_1\sqrt{n}}} &= \sum_{i\in\mathbf{E}} \mathfrak{a}(\{i\})E(e^{\frac{it(S_n-S_m)}{\sigma_1\sqrt{n}}}|X_0 = i)\\
&= \sum_{i\in\mathbf{E}} \mathfrak{a}(\{i\})\{E(e^{\frac{it(S_n-S_m)}{\sigma_1\sqrt{n}}}|X_0 = i) - \mathbb{E}(e^{\frac{it(S_n-S_m)}{\sigma_1\sqrt{n}}})\}\\
&\qquad + \mathbb{E}(e^{\frac{it(S_n-S_m)}{\sigma_1\sqrt{n}}}).
\end{aligned}$$

This leads to, by **(7.10.2)**,

$$|\mathbb{E}_{\mathfrak{a}}e^{\frac{it(S_n-S_m)}{\sigma_1\sqrt{n}}} - \mathbb{E}(e^{\frac{it(S_n-S_m)}{\sigma_1\sqrt{n}}})| \leq 2\gamma\rho^m.$$

Squaring both sides, the inequality can be written in the form

$$(a_{m,n} - c_{m,n})^2 + (b_{m,n} - d_{m,n})^2 \leq 4\gamma^2\rho^{2m}$$

where

$$\begin{aligned}
a_{m,n} &= \mathbb{E}_{\mathfrak{a}} \cos(\frac{t(S_n - S_m)}{\sigma_1\sqrt{n}}),\\
c_{m,n} &= \mathbb{E} \cos(\frac{t(S_n - S_m)}{\sigma_1\sqrt{n}}),\\
b_{m,n} &= \mathbb{E}_{\mathfrak{a}} \sin(\frac{t(S_n - S_m)}{\sigma_1\sqrt{n}})\\
\text{and } d_{m,n} &= \mathbb{E} \sin(\frac{t(S_n - S_m)}{\sigma_1\sqrt{n}}).
\end{aligned}$$

We are assuming either (7.79) and hence (7.77) or (7.77) directly, so (7.78) holds. This implies that, as $n \to \infty$,

$$c_{m,n} = \Re\mathbb{E}(e^{\frac{it(S_n-S_m)}{\sigma_1\sqrt{n}}}) \to e^{-\frac{1}{2}t^2}$$

and $d_{m,n} \to 0$ since $d_{m,n}$ is the corresponding imaginary part. From these it is easy to show

$$\lim_{m\to\infty}\lim_{n\to\infty} b_{m,n} = 0 \text{ and } \lim_{m\to\infty}\lim_{n\to\infty} a_{m,n} = e^{-\frac{1}{2}t^2}.$$

i.e. the *cf* of $\frac{S_n}{\sigma_1\sqrt{n}} - \frac{S_m}{\sigma_1\sqrt{n}}$ converges to $e^{-\frac{1}{2}t^2}$ as $n \to \infty$ and then as $m \to \infty$ when the *df* of X_0 is $\mathfrak{a}$. But $\frac{S_m}{\sqrt{n}} \xrightarrow{wp1} 0$ as $n \to \infty$, whatever be the distribution of X_0. Hence, under $\mathfrak{a}$ for the distribution of X_0, (ref. **(4.3.2)**(xviii)) $\frac{S_n}{\sigma_1\sqrt{n}} \xrightarrow{d} \xi$ as $n \to \infty$, as was to be proved. □

Index

Texts and Readings in Mathematics

1. R. B. Bapat: Linear Algebra and Linear Models (Third Edition)
2. Rajendra Bhatia: Fourier Series (Second Edition)
3. C. Musili: Representations of Finite Groups
4. H. Helson: Linear Algebra (Second Edition)
5. D. Sarason: Complex Function Theory (Second Edition)
6. M. G. Nadkarni: Basic Ergodic Theory (Third Edition)
7. H. Helson: Harmonic Analysis (Second Edition)
8. K. Chandrasekharan: A Course on Integration Theory
9. K. Chandrasekharan: A Course on Topological Groups
10. R. Bhatia (ed.): Analysis, Geometry and Probability
11. K. R. Davidson: C* – Algebras by Example (Reprint)
12. M. Bhattacharjee et al.: Notes on Infinite Permutation Groups
13. V. S. Sunder: Functional Analysis — Spectral Theory
14. V. S. Varadarajan: Algebra in Ancient and Modern Times
15. M. G. Nadkarni: Spectral Theory of Dynamical Systems
16. A. Borel: Semisimple Groups and Riemannian Symmetric Spaces
17. M. Marcolli: Seiberg – Witten Gauge Theory
18. A. Bottcher and S. M. Grudsky: Toeplitz Matrices, Asymptotic Linear Algebra and Functional Analysis
19. A. R. Rao and P. Bhimasankaram: Linear Algebra (Second Edition)
20. C. Musili: Algebraic Geometry for Beginners
21. A. R. Rajwade: Convex Polyhedra with Regularity Conditions and Hilbert's Third Problem
22. S. Kumaresan: A Course in Differential Geometry and Lie Groups
23. Stef Tijs: Introduction to Game Theory
24. B. Sury: The Congruence Subgroup Problem
25. R. Bhatia (ed.): Connected at Infinity
26. K. Mukherjea: Differential Calculus in Normed Linear Spaces (Second Edition)
27. Satya Deo: Algebraic Topology: A Primer (Corrected Reprint)
28. S. Kesavan: Nonlinear Functional Analysis: A First Course
29. S. Szabó: Topics in Factorization of Abelian Groups
30. S. Kumaresan and G. Santhanam: An Expedition to Geometry
31. D. Mumford: Lectures on Curves on an Algebraic Surface (Reprint)
32. J. W. Milnor and J. D. Stasheff: Characteristic Classes (Reprint)
33. K. R. Parthasarathy: Introduction to Probability and Measure (Corrected Reprint)
34. A. Mukherjee: Topics in Differential Topology
35. K. R. Parthasarathy: Mathematical Foundations of Quantum Mechanics
36. K. B. Athreya and S. N. Lahiri: Measure Theory
37. Terence Tao: Analysis I (Second Edition)
38. Terence Tao: Analysis II (Second Edition)

39. W. Decker and C. Lossen: Computing in Algebraic Geometry
40. A. Goswami and B. V. Rao: A Course in Applied Stochastic Processes
41. K. B. Athreya and S. N. Lahiri: Probability Theory
42. A. R. Rajwade and A. K. Bhandari: Surprises and Counterexamples in Real Function Theory
43. G. H. Golub and C. F. Van Loan: Matrix Computations (Reprint of the Third Edition)
44. Rajendra Bhatia: Positive Definite Matrices
45. K. R. Parthasarathy: Coding Theorems of Classical and Quantum Information Theory (Second Edition)
46. C. S. Seshadri: Introduction to the Theory of Standard Monomials
47. Alain Connes and Matilde Marcolli: Noncommutative Geometry, Quantum Fields and Motives
48. Vivek S. Borkar: Stochastic Approximation: A Dynamical Systems Viewpoint
49. B. J. Venkatachala: Inequalities: An Approach Through Problems
50. Rajendra Bhatia: Notes on Functional Analysis
51. A. Clebsch (ed.): Jacobi's Lectures on Dynamics (Second Revised Edition)
52. S. Kesavan: Functional Analysis
53. V. Lakshmibai and Justin Brown: Flag Varieties: An Interplay of Geometry, Combinatorics, and Representation Theory
54. S. Ramasubramanian: Lectures on Insurance Models
55. Sebastian M. Cioaba and M. Ram Murty: A First Course in Graph Theory and Combinatorics
56. Bamdad R. Yahaghi: Iranian Mathematics Competitions, 1973-2007
57. Aloke Dey: Incomplete Block Designs
58 R. B. Bapat: Graphs and Matrices
59. Hermann Weyl: Algebraic Theory of Numbers (Reprint)
60. Carl Ludwig Siegel: Transcendental Numbers (Reprint)
61. Steven J. Miller and Ramin Takloo-Bighash: An Invitation to Number Theory (Reprint)
62. John Milnor: Dynamics in One Complex Variable (Reprint)